Common Conversion Factors, British Engineering Units

British	×	SI
inch	25.4	millimeters
foot	0.3048	meters
yard	0.9144	meters
mile	1.609	kilometers
square inch	6.452	square centimeters
square foot	0.0929	square meters
acre	0.405	hectares
square mile	2.590	square kilometers
ounce, mass	28.35	grams
pound, mass	0.4536	kilograms
short ton, mass	0.9072	metric tons
pound, force	4.448	newtons
short ton, force	8.896	kilonewtons
psf	47.88	pascals
psi	6.895	kilopascals
psi	6.895×10^{-3}	megapascals
short ton/square foot	95.76	kilopascals
ton/square foot	1.024	kilograms/square meters
atmosphere (STP)	101.3	kilopascals
bar	100	kilopascals
fluid ounce	30	milliliters
quart	0.95	liters
gallon	3.8	liters
pound, force per cubic foot	16.018	kilograms/cubic meters
mile per hour	1.609	kilometers/hour
feet per year	0.966×10^{-6}	centimeters/second
BTU	252	calories
kilowatt-hour	860,421	calories

SI to British conversion factors are the reciprocals of the conversion values shown above.

GEOLOGY APPLIED TO ENGINEERING

Second Edition

Terry R. West
Purdue University

Abdul Shakoor
Kent State University

WAVELAND
PRESS, INC.
Long Grove, Illinois

In all things of nature there is something of the marvelous.

Aristotle 388–322 BC

For information about this book, contact:
Waveland Press, Inc.
4180 IL Route 83, Suite 101
Long Grove, IL 60047-9580
(847)634-0081
info@waveland.com
www.waveland.com

Photo Credits. Chapter 1: Reto Stockli, Alan Nelson, and Fritz Hasler, NASA, Earth Observatory. Chapter 2: Marcel Clemens, Shutterstock. Chapter 3: onime, Shutterstock. Chapter 4: V. J. Matthew, Shutterstock. Chapter 5: corlaffra, Shutterstock. Chapter 6: Les Palenik, Shutterstock. Chapter 7: Skinfaxi, Shutterstock. Chapter 8: farbled, Shutterstock. Chapter 9: Mehmet Cetin, Shutterstock. Chapter 10: Corrado Baratta, Shutterstock. Chapter 11: Matauw, Shutterstock. Chapter 12: Claudio Del Luongo, Shutterstock. Chapter 13: Blur Life 1975, Shutterstock. Chapter 14: tony740607, Shutterstock. Chapter 15: Have a Nice Day Photo, Shutterstock. Chapter 16: Matthew J. Thomas, Shutterstock. Chapter 17: Julien Hautcoeur, Shutterstock. Chapter 18: Andrey VP, Shutterstock. Chapter 19: Labrador Photo Video, Shutterstock. Chapter 20: elvistudio, Shutterstock. Chapter 21: Andrew Zarivny, Shutterstock.

Printed in the United States of America

7 6 5 4 3 2 1

Contents

Preface

Geology Applied to Engineering, Second Edition, represents a significant and insightful update of the first edition. Abdul Shakoor, Emeritus Professor, Kent State University, Ohio, has joined Terry West, Professor, Purdue University, Indiana, in the revision of the first edition. Together, they:

- Effectively and strategically reorganized the arrangement of chapters so that they are cohesive and build on material that is germane to related subjects.

- Added a new chapter that is solely dedicated to the evaluation of construction materials (Chapter 9).

- Expanded the coverage and scope of important topics such as the engineering properties of rocks (Chapter 8), slope stability and ground subsidence (Chapter 15), and significant fields in engineering geology, such as highways, dams, tunnels, and rock blasting (Chapter 21).

The authors have also taken into consideration the use of the text in the classroom. To benefit instructors and students alike, they have expanded the following features of the text to enhance comprehension of the material.

- Throughout the text SI units and their conversion to British engineering units are provided. Solutions to the Example Problems are calculated in both units to facilitate the growing need for the implementation of SI units within the engineering geology and geotechnical fields.

- Equations are numbered for easy reference.

- Numerous high-quality images are included to help better illustrate the physical processes at work below and at the Earth's surface. New examples of minerals, rocks, landforms, engineering structures, and equipment have been supplied.

- In conjunction with the text, the authors have added a new section at the end of each chapter called Engineering Considerations. Insightful and relevant, this material outlines real-world practices, concerns, and issues for today's engineering geologists and geotechnical engineers.

- Throughout the Second Edition current citations and references to pertinent testing procedures are presented to link the material to a professional setting.

Geology Applied to Engineering represents a thorough and up-to-date textbook for courses in Applied Physical Geology, Geology for Engineers, and Engineering Geology at both the undergraduate and graduate levels. It contains appropriate information for geologists and engineers who are involved in designing and constructing engineering structures, as all structures are located either on the Earth or in the Earth, or composed of earth materials. This textbook also provides the fundamentals of subject material included in the Examination for Professional Licensure of Geologists, a growing need for geologists who work in the public sector.

Acknowledgments

Our sincere thanks go to the students, colleagues, teachers, staff, family, and friends who contributed valuable ideas on teaching basic and applied geology at the university level. During the many years that the coauthors have been privileged to teach as university professors, we have taught the basic concepts of geology applied to engineering to hundreds of undergraduate students and dozens of graduate students. This book is a visible product of the lasting feeling of accomplishment that comes with extended university careers.

We wish to dedicate this publication to our wives, Shirley Mueller West and Roohi Shakoor, for their everlasting support and encouragement. Thank you, both.

Terry R. West
Abdul Shakoor
2018

ORIGIN AND DEVELOPMENT OF THE EARTH

1

Chapter Outline

This chapter discusses the Earth's origin and development, beginning with the universe and then focusing on the details of the Earth itself. Information about the solar system and the universe, with some ideas on their origin, is included in the discussion.

The universe (cosmos) extends in all directions as far as we can detect matter. It is approximately 13.7 billion years old and has an estimated diameter of at least 10 billion light-years (8.8×10^{23} km or 5.5×10^{23} mi) (Liddle, 2003; de Bernardis et al., 2008; Lawrence, 2015). For reference, 1 light-year is the distance light travels in a year and is equal to 9.3×10^{12} km (5.8×10^{12} mi). The speed of light is 2.998×10^5 km/sec (1.863×10^5 mi/sec). From this we conclude that the universe is immense beyond human comprehension, with unit distances measured in light-years and with total dimensions measuring in the billions of light-years.

The points of light we see in the sky at night are mostly stars. Only a few are planets moving visibly across the sky and, occasionally, there are hazy patches of light that a telescope is unable to resolve sharply into a single point. These are nebulae, other galaxies that exist beyond the Milky Way, which is the galaxy in which Earth orbits. Galaxies, the building blocks of the universe, vary in size and form, but many are spiral-shaped with curved arms trailing from their centers. When viewed along their spiral diameter, they appear as a disk that bulges outward at the center. The nearest galaxy to the Milky Way is Andromeda, located 2.2 million light-years away. Astronomers estimate that up to 200 billion galaxies exist in the universe. Recent studies from the Hubble telescope suggest the number could be significantly higher than this.

The Milky Way is home to all the stars that we can see clearly without a telescope; they number from 100 to 400 billion stars. In the Milky Way, stars are more or less clustered together to form the constellations, which were named centuries ago with reference to animals, mythical gods, famous people, or common objects. About 90 constellations have been named, based on the forms they seemed to resemble. Today, constellations are used as navigational aids and as a convenience to orient and describe locations in the sky.

Modern Theory of the Origin of the Universe

Light that began its journey hundreds, thousands, millions, and even billions of years ago is only now being received on Earth from different locations in the universe. Because the extent of the universe is so vast, the past is revealed to us on a delayed basis; thus, we can compare the past directly with the present. One fact learned by studying these delayed messages is that energy and matter functioned in the distant past much as they do today. This reinforces the hypothesis proposed by geologists that uniform physical laws have driven the geologic processes throughout the Earth's history.

We know enough about cosmology (the structure of the universe) to construct theories about the origin of the universe. As with most theory development, one must start with a list of known facts and attempt to build a theory consistent with those basic data. Such facts are as follows:

- The universe is vast and extends as far as humans can measure.

- Mass is concentrated into collections of stars known as galaxies.

- Stars undergo stages of development from red giants to white dwarfs.

- The universe is expanding, with stars at the periphery moving away the fastest.

The last item in the above list deserves further consideration. Expansion of the universe is indicated by the Doppler effect. This phenomenon shows that the frequency of light or sound lowers as an object accelerates away from the observer. For example, the pitch (frequency) of an automobile horn decreases as the car accelerates away from a point of reference. Because light emitted from the farthest stars shifts or displaces toward the red end of the visible spectrum, its frequency reduces ($c = \lambda \upsilon$, where c is the speed of light [a constant]; λ is the wavelength [in units of length]; and υ is the frequency of the waves [1/time]).

Therefore, the stars must be accelerating away from the observer.

The Evolving Universe Theory

If matter is accelerating away from a central location and the rate of acceleration and the distance can be estimated, one can estimate how long ago matter expanded from the common source. This forms the basis of the *Evolving Universe* or *Big Bang theory*. Although several theories about the origin of the universe have been proposed in the past, the Evolving Universe theory is currently the widely accepted theory.

According to this prevailing theory, the universe sprang into existence as an extremely dense and intensely hot singularity. It inflated (Big Bang) and expanded, hurling matter and radiant energy into space and bringing about the creation of the present chemical elements (Chaisson and McMillan, 2016). This event happened about 13.7 billion years ago. As expansion occurs, the density of matter becomes less (i.e., volume increases; mass remains constant; mass/volume, or density, decreases).

The initial central mass consisted of subatomic particles not yet organized into elements. The elements formed when protons, neutrons, and electrons united during the expansion process. The elements were built in turn from these building blocks, with neutrons and protons grouping together by fusion to yield elements of greater and greater atomic number. Hence, the abundance (frequency) of the elements in the universe decreases with increasing atomic number. This relationship is illustrated in Figure 1.1.

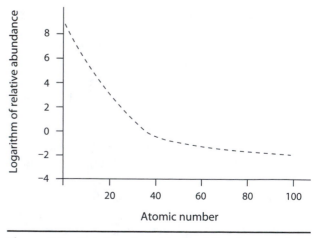

Figure 1.1   Relative abundances of the elements in the universe plotted against atomic number.

Several questions have been raised about the authenticity of this hypothesis, the first having to do with the relative abundance of elements and iron. There are no stable atoms of mass 5 or 8 and the abundance of iron is greater than the smooth curve would suggest. The second is concerned with the effects of the cataclysmic explosion. If such an immense explosion occurred, why is there no mark on the universe as a result? This question may have been answered, at least in part, by the 1978 Nobel Prize winners in astronomy, Dr. Arno Penzias and Dr. R. W. Wilson (1965), who discovered a faint glow of radiation throughout the universe that could be the aftereffects of the Big Bang.

Stages of Star Development

The stages of star development are an observed fact. Gas clouds comprise a large percentage of the matter in the universe. A rotating volume of gas will take on a flattened shape and a spiral will form as a direct consequence of its rotation. As the gas cloud rotates, masses tend to grow by gravitational attraction and, eventually, large masses of about equal size (protostars) will appear. These protostars are not true stars in the sense that they do not give off light. Condensation finally produces dense bodies that begin to glow with light as temperatures reach 1 million °C and their hydrogen undergoes fusion. An initial volume of gas must be compressed (reduced) about a million times to become a star, so that stars in a typical galaxy are separated by great distances. In the Milky Way, the average distance between stars is about 4.3 light-years.

A supernova is the massive explosion of a star, producing the greatest explosion that occurs in space. It happens at the end of a star's lifetime, when the star runs out of nuclear fuel and mass flows into its core. The core collapses, producing a giant explosion. Elements with high atomic numbers are generated by the explosion and are hurled out into space. By this process, a black hole with an extremely high gravitational attraction is created.

Supernovas apparently occur often enough to provide a suitable source of heavy elements for the universe. The abundance and range of heavy elements and related isotopes found on Earth suggests that the matter comprising our solar system has been through at least one supernovation. The Sun, then, is apparently a second- or third-generation star. In fact, several supernovas have been observed in the Milky Way—for example, in 1054, 1572, and 1604—and each, for a time, was much brighter than any planet. No similar events have occurred in the Milky Way since the telescope was invented. However, since 1885, nearly 80 bright supernovas have been observed by telescope beyond the Milky Way.

Origin of the Solar System

The Sun has a family of eight planets and one dwarf planet. The solar system also includes an asteroid belt, which consists of numerous irregularly shaped bodies that lie between the orbits of Mars and Jupiter. The distances from the Sun of the various planets range from 0.4 AU to almost 40 AU and their periods of revolution around the Sun are as short as 88 days (Mercury) and as long as 248.4 years (Pluto). An astronomical unit (AU) is the average distance from the Earth to the Sun, equal to 1.5×10^8 km (9.3×10^7 mi) (Figure 1.2).

A growing body of information, based on recent studies and space probes, now provides a detailed description of the solar system. As in our discussion on the origin of the universe, we begin by listing the facts about the solar system. Any hypothesis must be consistent and explain the known facts about the solar system. These facts are as follows:

1. All the planets revolve around the Sun in elliptical but almost circular orbits that lie nearly in the same plane, and most of the moons revolve in the same direction. Pluto, a dwarf planet only one-fifth the diameter of the Earth, has a very eccentric orbit around the Sun.

2. All planets, except for Uranus, rotate in the same direction as they revolve around the Sun, counterclockwise as one views the Earth from the north. Uranus rotates in a clockwise direction.

3. The planets seem to be spaced at regular intervals from the Sun. The Titius–Bode rule was proposed to explain the geometric progression in the distances of planets from the Sun. Stated verbally, the progression is obtained by writing 0, then 3, and continuing the progression by doubling the previous number to obtain 0, 3, 6, 12, 24, 48, 96, and so forth. Adding 4 to each number in the series yields 4, 7, 10, 16, 28, 52, 100, and so forth.

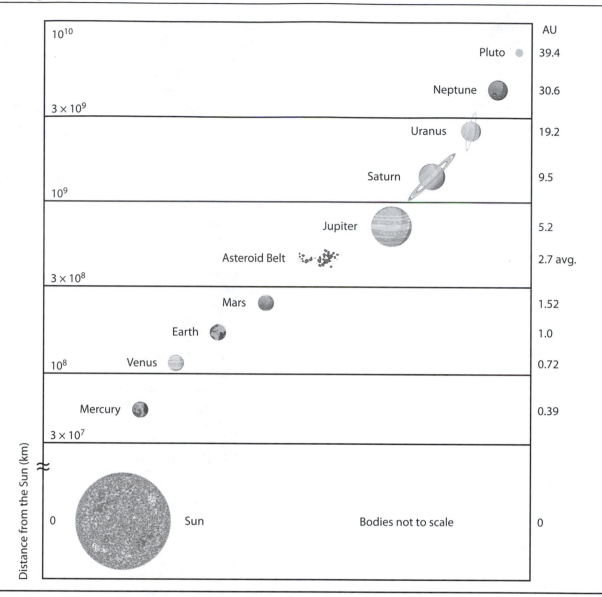

Figure 1.2   The planets, their sizes, and distances from the Sun.

Finally, dividing each number by 10, we get 0.4, 0.7, 1.0, 1.6, 2.8, 5.2, 10, and so forth. This is the approximate distance from the Sun in astronomical units for the planetary orbits. Both Neptune and Pluto deviate from this sequence.

4. Ninety-nine percent of the mass of the solar system is in the Sun but 98% of the angular momentum is in the planets (angular momentum = mass × angular velocity = mass × velocity × distance from center of rotation). Angular velocity is the rate of movement along a path measured as an angle from the center.

5. The terrestrial, or inner, planets (Mercury, Venus, Earth, Mars) are small, dense planets (4 to 5.5 times that of water) and the outer planets (gas or giant planets) have low densities (0.7 to 1.7 times that of water). The outer planets are more similar to the Sun than to the terrestrial planets.

Nebular Hypotheses

In general, the various hypotheses that have been proposed to explain the origin of the solar system fall into two groups. The first requires a catastrophic event, the accidental intervention of another star, to explain the observed facts. These are the so-called "second-body" or "collision" hypotheses. The other group, known collectively as "single-body theories," requires no influences beyond the Sun or solar system to explain the observed facts. Over time, the two-body hypothesis has been discarded because of the low probability of a two-star encounter and the difficulty of a filament of matter to collapse into planets.

After years of discussion and proposals on the origin of the solar system, the single-body hypothesis is preferred. Now termed the nebular or Kant–Laplace hypothesis (Murray and Dermott, 1999; Freedman, 2013), it suggests that a mass of gas cooled and began to contract. As it did, its rotational speed increased, much like skaters can increase their speed when spinning by drawing their arms inward. This is a consequence of the law of conservation of angular momentum. Rings of gases spun off from the center and condensed to form the planets.

Interstellar matter is distributed widely throughout space. In this vast region, rarified matter exists that is about 99% gas and 1% dust. The gases are the ever-present hydrogen and helium, and the dust-size particles have compositions similar to terrestrial material. These include silicon compounds (silicates), iron oxides, ice crystals, and a host of other small molecules and compounds, including some organic compounds.

Light pressure from the stars supposedly causes this matter to consolidate to form a cloud about 9600 billion km (6000 billion mi) in diameter. As the cloud condenses it begins to rotate faster and faster. The rotation increases as it collapses slowly under its own gravity until it reaches a diameter of about 5900 million km (3700 million mi) and then collapses rapidly, perhaps in a few hundred years. The increased pressure of the contracting cloud greatly increases the temperature until it reaches about 1 million °C and nuclear fusion begins. The sun begins to radiate as a star and the planets and satellites are derived from minor dust streams in the original cloud before the last stages of collapse occur.

The nebular hypothesis leaves two important features of the solar system unexplained: the spacing of the planets from the Sun and the angular momentum of the Sun and the planets. From research conducted by Fred Hoyle in 1960 and other related studies, the process of magnetic coupling has been presented to explain the angular momentum problem.

Magnetic Coupling

Magnetic coupling is explained in the following way. Stars are known to have magnetic fields that extend into surrounding space. During formation of the solar system, the more rapidly rotating Sun dragged the less rapidly rotating disk of dust with it. The linkage between the Sun and rotating dust provided a rotating solar magnetic field. Interaction occurred with the gases whose particles served as tiny magnets. The Sun's angular momentum was transferred to the rotating disk of dust. This increased rotation hurled the gaseous portion to the outer reaches of the solar system where it condensed into the great planets. With this outflow of gases, the remaining particles near the Sun consisted mostly of the heavier materials, iron, silicon, and magnesium oxides. These formed into the small terrestrial planets by collision and gravitational attraction (Hoyle, 1960).

There is also some question about whether magnetic coupling completely answers the complicated aspects of how the Sun slowed its rotation and how the planets incorporated the additional angular momentum. These details require further study.

Spacing of the Planets

The spacing of the planets around the Sun still remains to be fully explained. The Titius–Bode rule, discussed earlier, has been proposed as an explanation for planet spacing. It is described as the approximate doubling rule, as each planet is between 1.4 and 2.0 times further away from the Sun than its inside neighbor. Pluto's position does not fit the rule but, as Pluto is no longer considered to be a planet, this can be excluded from consideration. However, the current consensus is that the Titius–Bode rule is not a good enough fit and an accurate explanation has yet to be proposed.

Finally, regarding the two-body hypothesis, it is more likely that a double star, rather than a single star, would approach the Sun as double stars are much more common. About 1 star in 100, such as the Sun, is a single star. Also, it is dynamically unlikely that double stars are capable of holding planets.

The Milky Way

The Milky Way is a spiral-shaped galaxy with two arms extending from its central mass (Marshak, 2013). The diameter of the nebula is about 100,000 light-years and the Sun is located about 3/5 the distance from the center (30,000 light-years) in one of the spiral arms. If viewed along its central plane, the galaxy would appear as a flattened disc with a central bulge (Figure 1.3a). The central bulge is about 30,000 light-years thick. In the spiral arms that dimension reduces to about 10,000 light-years. A photograph of a spiral galaxy, found in Ursa Major (the Big Dipper), and located about 4.2 million light-years from Earth, is shown in Figure 1.3b.

In the Milky Way, our galaxy, there are 100 to 400 billion stars. It is estimated that the number of single stars is 4×10^9. Considering the needs of heat supply, light, size, age, and other requirements, stars with planets similar to the Earth would range from 1 in 1000 to 1 in 1 million but there are about 200 billion galaxies in the universe.

The nearest stars to the Sun are in the Alpha Centauri star system, which can be observed in the Southern Hemisphere. They are about 4.3 light-years away, which is the average distance between stars in the Milky Way. Stars in the galaxy rotate slowly about the central mass, and it takes the Sun and its planets about 200 million years to complete one rotation.

On a clear night, the Milky Way can be seen in a band that stretches across the sky. During the summer, our view from Earth is toward the star Sagittarius and the center of the galaxy. In the winter, the view is away from the center and toward the outer portions, along the spiral arm of the Milky Way.

The Likelihood of Life on Other Planets

In the Milky Way there are from 1000 to 1 million planets on which humanlike forms could exist. Despite the apparent abundance of planets that could conceivably support life, we must remember that distances in space are extremely vast. The minimum distance between such planetary systems is measured in distances of tens of light-years or in units of billions of kilometers or miles. These distances are obviously much beyond the capability of human travel in the foreseeable future.

A number of conditions are necessary before a planet can support life-forms as advanced as humans. The primary requirements include the proper temperature, amount of light, gravity, atmospheric composition and pressure, and water. Other require-

ments, perhaps of lesser significance, involve other life-forms present, wind velocity, dust, and radioactivity. Some conditions that might make a planet uninhabitable are excessive meteorite bombardment, extensive volcanic eruptions, high frequency of earthquakes, and possibly an extreme level of electrical activity (lightning). Consequently, planets must meet some major requirements in order for them to be inhabitable by life-forms similar to humans:

- *Mass of planet*: Must be greater than 0.4 Earth mass to produce and retain a breathable atmosphere and less than 2.35 Earth mass since surface gravity must be less than 1.5 g. The mass of the Earth is approximately 6×10^{24} kg.

- *Period of rotation*: Must be less than about 96 hours (4 Earth days) to prevent excessively high daytime temperatures and excessively low nighttime temperatures.

- *Age of the planet and star about which it orbits*: Must be greater than 3 billion years to allow for the appearance of complex life-forms and the production of a breathable atmosphere.

- *Axial inclination or inclination of equator to the plane of orbit and level of illumination from its sun*: These determine the temperature patterns on the surface. Illumination at low inclinations should lie between 0.65 and 1.35 times the Earth's norm (between 10 and 20 lumens/cm^2). However, certain combinations of illumination up to 1.9 times the Earth's norm and inclinations up to 81° are compatible under marginal conditions.

- *Orbital eccentricity*: Must be less than about 0.2 because greater amounts of eccentricity produce unacceptably extreme temperature effects on the planetary surface.

- *Mass of the star*: Must be less than 1.43 solar mass, because residence time on the main sequence of stars must be more than 3 billion years. Solar mass is the mass of the Sun. It is equal to 3.35×10^5 Earth mass or 2.01×10^{30} kg. Mass also must be more than 0.72 solar mass because smaller stars yield an incompatibility between acceptable illumination levels and tidal retardation of a planet's rotation. For the rare class of planets with extremely large or close satellites, the lower range of the star's mass is extended to 0.35 solar mass.

(a)

(b)

Figure 1.3   (a) The Milky Way from the side showing the location of the solar system. (Produced by the European Southern Observatory.) (b) Photograph of a spiral galaxy in Ursa Major, distance about 4.2 million light-years. (Courtesy of the National Optical Astronomy Observatory/AVRA Image Library.)

• *Binary star system*: If planets orbit this system, the two stars must be quite close together or very far apart to prevent instability of planetary orbits and not produce a level of illumination on the planet that is too variable.

When all of these requirements exist, there is a very good possibility that the planet will be inhabitable. Such planets lie in the circumstellar habitable zone (CHZ), also known as the liquid water belt, HZ, or Goldilocks zone (Petigura et al., 2013). Astronomers report that as many as 11 billion Earth-sized planets within habitable zones of stars could be present in the Milky Way alone. Seven candidates were found between 18 and 385 light-years from Earth during the Keppler observations. However, the vast distance of only 1 light-year indicates that travel to such distant places is far from possible.

The Fate of Life on Earth

The ultimate fate of life on Earth is tied to the evolution of the Sun. It is a fairly old star, about 5.5 to 6 billion years in age. It will continue as a normal or main sequence star for about another few billion years, at which time it will begin to heat up and expand as a consequence of having exhausted much of its hydrogen through the fusion process. It will undergo expansion prior to its demise. The temperature will increase tremendously, boiling the oceans on Earth, killing all life. The Sun may even expand sufficiently to engulf the Earth. Following this red giant stage—in about 7 billion years—the Sun will cool to form a burned-out star.

Major changes in the configuration of the Earth's surface could occur before the Sun begins to heat up and expand. Radioactive decay in the Earth's interior provides the flow of heat to form mountains and move the continents around on the planet. This is discussed later in the section on plate tectonics.

Radioactive decay will continue to diminish with time as the radioactive elements are consumed in the reaction process. Eventually, heat flow will diminish, bringing an end to mountain building and volcanic eruptions. The relentless pounding of the seas against coastal regions over geologic time will flatten the continents, filling the ocean basins with sedimentary debris. The final effect will be a flat continental platform, greatly diminished in size, located at or near sea level. It would cause a major impact to most forms of life on Earth. This is estimated to occur in about 2 billion years.

It was once thought that the Moon had emerged from the Earth well after the onset of planetary development. Moving in an elliptical path away from the Earth, it has been suggested that the Moon would eventually return to Earth, yielding a catastrophic collision and very likely destroying all advanced forms of life.

Studies of lunar rocks have subsequently shown, based on rock and mineral analysis, that the Moon has a slightly different composition than Earth. Within lunar rocks, there is no history of intense oxidation or effects of free water, both of which are so prevalent in rocks on Earth. The currently supported theory on lunar origin is that the Moon separated from the Earth when an extremely large mass impacted the growing Earth mass, during the early planetary accretion stage, about 4.53 billion years ago.

A question sometimes raised is whether life can be rejuvenated on Earth after the Sun cools down from the red giant stage and before it becomes a burned-out star. Based on the history of the Earth it takes about 3 billion years under proper conditions for advanced life-forms to develop. Oceans would have to form anew, along with a proper atmosphere. It would seem unlikely that all of these constraints could be met, but simple forms of life could develop before the Sun cools completely.

Configuration of the Earth's Surface

Early Developments

The planet Earth formed by the accretion of planetesimals, during formation of the solar system about 4.7 billion years ago. Composed of silicon compounds (silicates), iron and magnesium oxides, and lesser amounts of other elements, the planet started out as a cold mass. It began to heat up as three mechanisms contributed heat to the system. The energy of motion of the infalling planetesimals converted to heat and the resulting compression contributed more heat to the planet. These two events accounted for an initial temperature of about 1000°C within the first million years of Earth's existence.

Melting of the Earth

The third mechanism that contributed to heating of the Earth was radioactive decay. Uranium, thorium, potassium, and the other radioactive elements eventually contributed enough heat (reaching temperatures of 2000°C or more) to melt the iron in the Earth, causing the so-called "iron catastrophe" to occur. Iron, being heavier than the other common substances, sank toward the center of the Earth and displaced lighter materials. This marked an event of catastrophic proportions because, as the iron migrated toward the center of the Earth, it released large amounts of gravitational energy that were also converted to heat. This additional heat produced a temperature rise on the order of 2000°C, melting much of the remaining portion of the Earth.

This differentiation process, caused first by the melting of the iron, followed by the additional melting of most other materials, converted a generally homogeneous planet into a zoned body. The resulting configuration consisted of a dense iron core, a surface crust composed of lighter materials that had a lower melting temperature, and an intermediate zone comprising the mantle. The differentiation process also most likely triggered the escape of gases from the interior and eventually led to the formation of the atmosphere and the oceans. Differentiation is thought by many earth scientists to have occurred about 4 billion years ago.

The Earth developed into a stratified planet. The dense core was overlain by a mixture of iron and magnesium silicates, with the lightest minerals forming the crust of the Earth. The overall density of the Earth is 5.5 whereas that of the crust is only 2.7.

Formation of the Oceans and Atmosphere

The water on the Earth's surface today most likely came from chemically bound hydroxyl groups (OH^- groups) attached to minerals in the crust. Such minerals as the micas and amphiboles contain hydrogen and oxygen linked as hydroxyls. As the Earth warmed, water vapor was carried to the surface dissolved in magma. This outgassing is a consequence of differentiation of the Earth that yielded the zones of different composition.

The volcanic gases consisted mainly of water vapor, carbon dioxide, hydrogen, hydrogen chloride, carbon monoxide, and nitrogen. It is likely that much of the outgassing occurred early in the Earth's history, between 4 and 3.8 billion years ago, and coincided with extensive volcanic flows during that time interval. Periodically, since this early time, many volcanic eruptions of short duration have followed this initial, lengthy eruptive phase. The oceans were formed from the early outgassing, as well as the atmosphere, and only small amounts of juvenile or new water were added to the oceans after the initial period. Most of the water delivered to the atmosphere by volcanic eruptions today is recycled, meteoric water.

The early atmosphere did not contain free oxygen in any of the following forms: normal molecular atmospheric oxygen (O_2), ozone (O_3), or rare atomic oxygen (O). There are several reasons why this is indicated: (1) no plausible source of free gaseous oxygen for the early atmosphere has been proposed, (2) the composition of the early atmosphere, generally agreed on by earth scientists, involves gases that would combine with and remove free oxygen, and (3) free oxygen in the early atmosphere would inhibit the origin of life and the fossil record shows that such life existed billions of years ago. The composition of the early atmosphere was probably carbon monoxide, carbon dioxide, nitrogen, hydrogen, hydrogen chloride, and water vapor.

Another hypothesis, no longer seriously considered, suggests that the early atmosphere was primarily methane and ammonia. Geologic evidence does not support the presence of a methane-ammonia-rich atmosphere. The oldest known sedimentary rocks are 3.8 to 4.0 billion years of age, but those that are more than 2.6 billion years old contain relatively small amounts of limestone and are, instead, rich in chemically precipitated silica. An ammonia-rich atmosphere would have favored the deposition of limestone (or dolomite), while greatly limiting the precipitation of silica. Therefore, if an initial methane-ammonia atmosphere existed, it had evolved into an atmosphere dominated by carbon monoxide, carbon dioxide, nitrogen, and water vapor by about 3.8 billion years ago.

These oldest sediments also supply additional information about the past. The small amounts of dolomite found in these rocks contain unoxidized iron compounds, further supporting the position that little or no free oxygen existed in the atmosphere at the time. Abundant Precambrian iron stones suggest a reducing atmosphere, free from oxygen. Also, the presence of dolomite, any at all, suggests a salty ocean, because precipitation of dolomite requires a saline solution. The sea was certainly fully salty by 2 billion years ago, as evidenced by extensive formations of dolomite of that age in southern Africa.

A third item of information shown by the sedimentary column is the continuous existence of liquid water on the Earth for the last 3.8 billion years. This also suggests surface temperatures between the freezing and boiling point of water which, in turn, indicates that either (1) the Sun supplied an extensive amount of heat even at an early time, (2) the Earth was able to retain heat at that level, or (3) loss of heat from the interior kept the temperature in balance.

The production and accumulation of free oxygen apparently came only after life had evolved to at least the level of green algae. It and higher forms of plant life use sunlight to convert carbon dioxide and water into organic matter and oxygen, through photosynthesis. Not until the oxygen produced had satisfied the demands of other atmospheric gases and combined with them, did free oxygen begin to accumulate in the atmosphere. This would also remove any remnants of methane or ammonia that were still present in the atmosphere.

Much of the carbon monoxide and carbon dioxide from the early atmosphere became locked in the carbonate rocks of the geologic column, that is, in the limestones and dolomites. Some hydrogen, carbon dioxide, oxygen, and other elements formed coal and petroleum, which are also stored in the Earth's crust. The small amount of CO_2 still in the atmosphere (a few hundredths of 1%) is quite important because of its role in photosynthesis.

Architecture of the Earth's Surface Today

Many schoolchildren, while studying the globe or maps of the Earth, observe that the continents would fit together if they were moved around. South America could be tucked under Africa, North America against Eurasia, with Antarctica and Australia drawn up from below to fit under the others. This very idea was proposed scientifically by the German meteorologist Alfred Wegener in 1915. Many scientists in the United States scoffed at his proposal of continental drift, where it was generally discounted until intensive studies following World War II provided positive evidence of its occurrence. Today the concepts of seafloor spreading and plate tectonics are firmly established as the likely mechanisms that have continued to move the continents apart during the past 200 million years (Marshak, 2013; Nance and Murphy, 2016).

Prior to the accumulation of the geophysical evidence discussed below, only similarities between rock types and ages of rocks (including fossil evidence) across the continental boundaries, which previously would have been in contact, were advanced in support of continental drift. This evidence was not convincing enough to change the status quo and no consensus favoring Wegener's proposal ever developed.

Seafloor Spreading

Spreading of the seafloor from mid-ocean ridges has been established by oceanographic studies and age determinations of rock. Seafloor geophysics has also provided convincing evidence. Variations in the direction of the magnetic field for the basaltic rocks on either side of the spreading ridges indicate the strong symmetry of the units moving away from the center. During the period of seafloor spreading, a number of reversals in polarity of the Earth's magnetic field occurred; that is, the pole attracting the compass needle has switched directions from north to south a number of times during the 200 million years of spreading. Sequences of magnetic variations on opposite sides of the spreading ridge match each other like opposite fingers on the left and right hands. This symmetry across the spreading ridge is illustrated in Figure 1.4.

The rates of spreading on either side of the ridge range from 1 to 6 cm/yr (0.4 to 2.4 in/yr). This suggests a total spreading rate of from 2 to 12 cm/yr (0.8 to 4.7 in/yr). At this rate it would have taken between 36 and 288 million years for the Atlantic Ocean to attain its present width. This is generally consistent with the 200 million year estimate for the duration of seafloor spreading.

An interesting feature about the mid-ocean spreading ridges is that they are sites for both volcanic and earthquake activity. Iceland, for example, with its well-known volcanic and thermal activities, sits exactly on the crest of the Mid-Atlantic Ridge. Other spreading centers in the oceans that are also locations for earthquakes include the East Pacific Rise, the Mid-

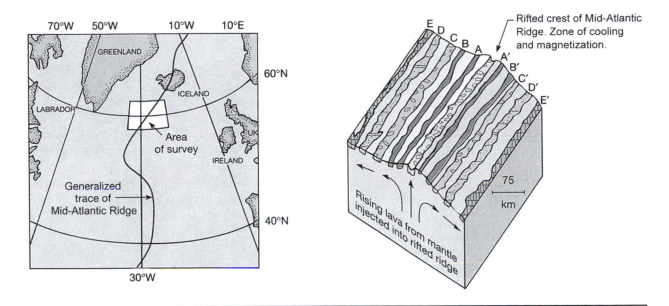

Figure 1.4 Seafloor spreading along the Mid-Atlantic Ridge south of Iceland.

Indian Ocean Ridge, and the ridges or rises that encircle Antarctica. These can be observed in Figure 1.5 on the next page. The earthquakes that occur along these spreading centers are mostly of shallow depth and relatively small magnitude. Because these are locations where heat is welling up from the interior of the Earth, the rocks below are too hot and plastic to accumulate high levels of stress. Therefore, only smaller magnitude earthquakes occur along these ridges.

Plate Tectonics

Plate tectonics is a unifying theory that answers many questions about the Earth. Included are such aspects as varieties and distribution of rocks, history of sedimentary rock sequences, positions and nature of volcanoes, earthquake belts, mountain systems, deep-sea trenches, and ocean basins (Marshak, 2013).

Surfaces along which earthquake foci align are the boundaries of blocks that move independently of each other. These blocks are referred to as *plates* and their movement as *plate tectonics*. Plates are typically thousands of kilometers across but only 100 to 200 km thick (60 to 120 mi). Both oceanic and continental crusts can cap these plates, so they are sometimes referred to individually as *continental plates* and *oceanic plates*. The continent is actually imbedded in the moving plate and is carried passively by it. In fact, many of the plates extend well beyond the continental outline itself; for example,

the African plate (see Figure 1.5) is nearly twice the area of the African continent. Hence, it is about half continental crust and half oceanic crust at the surface. The Pacific plate, by contrast, is nearly all oceanic material except for a sliver of California along the San Andreas Fault. The Eurasian plate is really two ancient plates, now joined or sutured together along the Ural Mountains, and includes both continental areas plus much of the North Atlantic and Arctic Oceans.

The plates are rigid slabs consisting of a continental and/or oceanic crustal cap plus part of the underlying mantle. The combined crust and upper mantle is referred to as the *lithosphere*. These lithospheric plates ride on a weak, plastic zone below known as the *asthenosphere*. A cross section depicting these layers is shown in Figure 1.6 on page 13.

The Earth's surface today is divided into about eight large rigid plates and a dozen smaller plates (see Figure 1.5). All the crustal plates seem to be moving relative to each other except for Africa which, apparently, has remained relatively fixed in position for several tens of millions of years. Well inside the boundaries of some plates are located chains of volcanoes, such as the Hawaiian Islands situated near the center of the Pacific plate. These are thought to mark the locations of fixed hot spots beneath the moving plates and can be used to show rates and directions of movement.

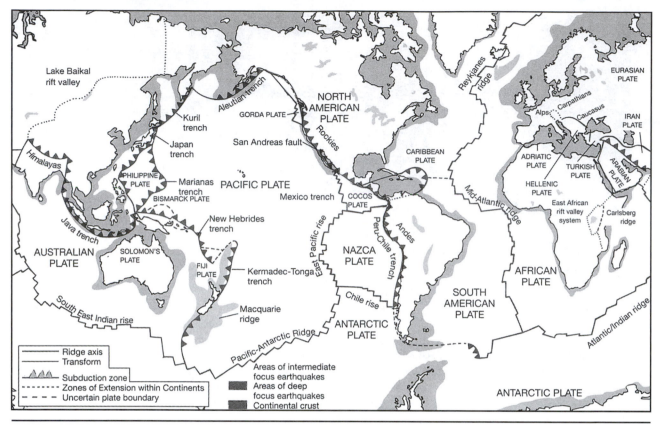

Figure 1.5   Plates and plate boundaries of the Earth.

There are three basic types of plate boundaries or margins: (1) the zones of divergence or spreading, that is, the typical ocean ridges; (2) the transform margins where plates slide sideways past each other; and (3) the zones of convergences where the plates move directly toward each other. The deep-seated earthquakes of the world occur mostly along the zones of convergence and to a lesser extent along the transform margins.

The ocean-spreading ridges are also known as *accretional margins*. New basaltic ocean crust forms where heat and volcanic flows well up from the interior of the Earth. The Mid-Atlantic Ridge, a most prominent spreading center, is also the site for volcanic eruptions and shallow focused earthquakes.

Transform margins or transform fault zones occur where the plates slide sideways past each other. The San Andreas Fault is the boundary where the Pacific plate is moving northward relative to the southward-moving North American plate. Another transform fault lies between the North American plate and the Caribbean plate.

The zones of convergence are essentially of two types, consuming margins and collisional margins. At the consuming margins, the dense oceanic rock dives either beneath the lighter continental crust or beneath a somewhat thinner oceanic crust as in deep-sea trenches. The descending plate is consumed by warming and melting at depth. This melting produces molten rock material, which ascends and provides volcanoes and igneous intrusions above the leading edge of the descending plate. Along the North American coast, from northern California to British Columbia, the Pacific plate is plunging under the North American plate.

Where moving oceanic plates of normal thickness underride the thinner oceanic sequences adjacent to continents, arcuate, deep-sea trenches with paralleling island arcs develop. These island arcs are prominent around the northern and western Pacific and Indonesia. Some of the Caribbean Islands are also examples.

In situations where the moving continent overrides the seafloor, the deep-sea trenches formed have

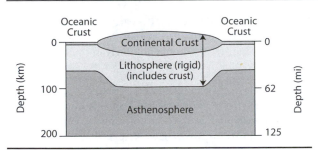

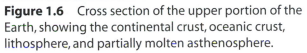

Figure 1.6   Cross section of the upper portion of the Earth, showing the continental crust, oceanic crust, lithosphere, and partially molten asthenosphere.

a straight alignment, and linear chains of volcanoes rise through the continents as in the Andes of South America. Not only is the ocean floor consumed in such cases along with scraped-off islands and seamounts, but also pieces of the adjacent continent are pushed into the trenches and are subsequently carried beneath the advancing continent along with the oceanic rocks. Eventually, these down-thrusted rocks are carried to great depths to mix with molten rocks of continental origin and later to rise as melts of intermediate composition. Ore minerals may be associated with this volcanism in the continental margins because this process concentrates the metals that were finely dispersed throughout the deep-sea sediments and continental crust.

Collisional margins are the final type of plate convergence. Where oceans are drastically narrowed or completely closed by overriding and converging continents, the relatively light, high-standing continental crust is compressed and folded to produce the linear-trending mountain ranges such as the Himalayas, the Alps, and the Appalachians. The latter formed when the ancestral Atlantic Ocean closed about 320 million years ago because the African plate collided with the North American plate. Sediments that had accumulated on the margins of the North American continent were compressed, yielding the folded Appalachians and associated structural features. The Atlantic Ocean reopened when the two plates began moving apart some 200 million years ago.

Plate tectonics provides an explanation for many aspects of global significance. These include volcanoes, earthquakes, folded mountains, island arcs, deep-sea trenches, continents, and ocean basins, in addition to the youthfulness of the oceanic rocks and the great age of their continental counterparts.

The zoned layering of the Earth, formed early in its history, has provided the necessary life-giving oceans and atmosphere. The Earth's internal heat fueled by radioactive decay provides the driving mechanism to propel the rigid plates across the plastic asthenosphere. The Earth is a dynamic body undergoing change. In subsequent chapters, we concentrate on the materials that compose the Earth and on the processes that shape it today.

EXERCISES ON THE ORIGIN AND DEVELOPMENT OF THE EARTH

1. What are the dimensions of the universe in light-years, kilometers, and miles? By what means is this determined?

2. How are galaxies and constellations related to the subdivision of the universe? Approximately how many galaxies are there? Name the galaxy nearest to the Earth. What is its distance?

3. Name our own galaxy. About how many stars does it contain? Based on this, approximately how many stars exist in the universe?

4. What is the currently accepted theory on the origin of the universe?

5. Is the Sun the same size, smaller, or larger than an average star? What will likely happen to it in the future? Explain.

6. How are black holes formed? Is it likely that they are extremely common? Explain.

7. Describe the two groups of proposed hypotheses that explain the origin of the solar system. Which hypothesis is favored today? Why is it so difficult to account for the relationship of angular momentum of the Sun and planets? How is it best explained?

8. What is the Titius–Bode rule? It can be written as $d_p = 0.4 + 0.3(2)^x$ where $x = -\infty$ for $p = 1$ (Mercury) and $x = (p - 2)$ for $p \geq 2$, where p is the number of the planet achieved by counting outward from the Sun and d is the distance of the planet from the Sun in astronomical units. Based on this equation, complete Exercise Table 1.1 on the following page.

Exercise Table 1.1

Planet	Actual Mean Distance from Sun (km × 10⁷)	Actual Mean Distance from Sun (AU)	Distance by Titius–Bode Law (AU)	Percent Error
Mercury	5.85			
Venus	10.80			
Earth	15.00	1.00	1.0	0
Mars	22.80			
Asteroid Belt	33.00–50.00			
Jupiter	78.00			
Saturn	143.10			
Uranus	288.00			
Neptune	459.00			

9. How far away is the Alpha Centauri star system? If a spaceship traveled at twice the escape velocity of the Earth how long would it take to reach this system? The Earth's escape velocity = 11 km/sec or 7 mi/sec.

10. What are the limitations concerning gravitational attraction of a planet if it is to be hospitable to humans?

11. At speeds of one-tenth the speed of light, how long would it take to travel to Alpha Centauri A or B? How far could humans travel at this velocity during a 20-year round-trip mission?

12. It is likely that oceans could be observed on planets at 7 AU whereas forests could be seen at 3 AU from an approaching space rocket. How many years travel away from the surface would this be at one-tenth the speed of light and at 40,000 km/hr (25,000 mi/hr)? Large cities may be visible from 8 million km (5 million mi) away. How many astronomical units does that equal?

13. Why is the melting of the Earth to form a zoned interior important in the development of the Earth as we know it?

14. What is the evidence that the early atmosphere of the Earth contained no free oxygen? Why is the presence of methane and ammonia not likely in the range of 3.8 to 2.6 billion years ago?

15. Why is it likely that less carbon dioxide is present in the Earth's atmosphere today than it was 3 billion years ago? What is the important role of carbon dioxide in the atmosphere today? Explain.

16. Why is it likely that much of the oceans formed early in Earth's history? If this is so, what is the origin for all the water released by volcanoes today? Explain.

17. Why was continental drift discounted by most scientists following Wegener's proposal in 1915? What evidence had been advanced prior to the post-World War II studies of the ocean basins?

18. By what means is seafloor spreading indicated? How does it relate to plate tectonics?

19. Do all plates consist either of all continental or all oceanic crust? Explain this relationship.

20. What is meant by the asthenosphere and the lithosphere? Draw a sketch, and label it properly.

21. Name the different types of plate boundaries. Give examples of each type of plate boundary based on the specific plates and movements on the Earth (see Figure 1.5).

22. What major aspects of global significance have been explained by plate tectonics?

23. Could there have been plate tectonics on the Earth without zoning of the interior? Could it have occurred if the Earth had no oceans? Explain.

24. What evidence is there that plate tectonics on Earth had occurred even prior to the onset of the current movement some 200 million years ago? Explain.

References

Chaisson, E. and McMillan, S. 2016. *Astronomy: A Beginner's Guide to the Universe* (8th ed.). New York: Pearson.

de Bernardis, F., Melchiorri, A., Verde, L., and Jimenez, R. 2008. The cosmic neutrino background and the age of the universe. *Journal of Cosmology and Astroparticle Physics* 20(3).

Freedman, R. 2013. *Universe—The Solar System* (5th ed.). New York: W. H. Freeman.

Hoyle, F. 1960. On the origin of the solar nebula. *Quarterly Journal of the Royal Astronomical Society* 1.

Lawrence, C. R. 2015. *Planck 2015 Results*. Astrophysics Subcommittee, NASA.

Liddle, A. R. 2003. *An Introduction to Modern Cosmology* (2nd ed.). New York: John Wiley and Sons.

Marshak, S. 2013. *Essentials of Geology*. New York: W. W. Norton.

Murray, C. D., and Dermott, S. F. 1999. *Solar System Dynamics*. Cambridge, England: Cambridge University Press.

Nance, D., and Murphy, B. 2016. *Physical Geology Today*. New York: Oxford University Press.

Penzias, A. A., and Wilson, R. W. 1965. A measurement of excess antenna temperature at 4080 Mc/s. *Astrophysical Journal Letters* 142:419–421.

Petigura, E. A., Howard, A. W., and Marcy, G. W. 2013. Prevalence of Earth-size planets orbiting Sun-like stars. *Proceedings of the National Academy of Sciences* 110(48).

MINERALS 2

Chapter Outline

Mineral Definition

Mineral Formation

Occurrence of Minerals

Mineral Identification

Rock-Forming Minerals

Engineering Considerations of Minerals

Geology is the study of the Earth—its surface expression, composition, structure, internal activity, and the processes by which it formed and those that operate on it today. This includes the history of the Earth in both the physical and biological sense. The focus of this text is to describe the physical nature of the Earth and how it impacts the design and construction of engineering structures.

The term *rock* applies to the solid materials that form the outer shell or crust of the Earth. Based on their origin, rocks are classified into three broad groups: igneous, sedimentary, and metamorphic. Igneous rocks are those that have cooled from a molten state; sedimentary rocks are those deposited from a fluid medium (usually water) and, typically, as products of weathering of other rocks; and metamorphic rocks form from preexisting rocks by the action of heat and pressure.

Rocks are assemblages of one or more minerals. Mineral composition and rock texture, or nature and pattern of the assemblage, are useful descriptive features that aid in rock identification. For this reason,

a study of minerals is the proper place to begin our discussion of Earth's materials.

Mineral Definition

A mineral is a naturally occurring solid element or compound, formed by inorganic processes, that has: (1) a definite chemical composition, (2) an ordered internal arrangement of atoms, and (3) a unique set of physical properties (Neuendorf et al., 2005; Marshak, 2013; Nance and Murphy, 2016). According to this definition, oil, coal, volcanic glass (which lacks the ordered internal structure), and manufactured glass do not qualify as minerals. Blast furnace slag, a by-product of steel making and commonly used as road base material and as concrete aggregate, does not qualify because it is glassy and not naturally occurring.

An ordered internal arrangement of atoms in a mineral is known as *crystalline structure*. The crystalline structure refers to a specific repetition of the atomic array, that is, a repeating relationship of constituent atoms comprising the structure. In contrast, a nonordered internal arrangement is termed *amorphous*, or without form, and it occurs in liquids or super-cooled liquids such as glass. Specifically, a mineral is crystalline and glass is amorphous (Figure 2.1). Thus a mineral name such as quartz entails not only the composition SiO_2 but also the specific atomic array of silicon and oxygen that comprises the quartz structure. Calcite, in the same manner, involves the composition $CaCO_3$ but also the calcite structure in which calcium and the carbonate radical are arranged in a specific fashion.

Some minerals have identical compositions but different crystalline structure. Pyrite and marcasite, both FeS_2, are examples of this and both qualify as separate minerals. In pyrite, the iron atoms are equally spaced in all directions, but in marcasite their spacing differs depending on crystallographic direction. Such pairs of minerals are called *polymorphs*. Diamond and graphite (C), calcite and aragonite ($CaCO_3$), and quartz and cristobolite (SiO_2) are other common examples. Thus, crystalline structure is the more diagnostic characteristic of a mineral, not chemical composition.

Mineral Formation

When minerals solidify from the liquid state or form in other ways, they yield an internally ordered, solid material. This process, termed *crystallization*, occurs when many crystals form independently, increase in size, and finally grow together to give a mosaic pattern of interlocking crystals.

○ Oxygen atom   • Silicon atom

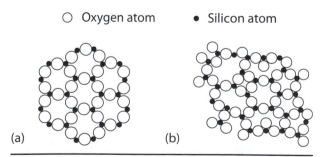

(a)   (b)

Figure 2.1 (a) Crystalline structure of quartz and (b) amorphous structure of glass. Oxygen and silicon atoms have a precise and ordered arrangement in quartz and a random distribution in glass.

Crystallization occurs in nature by solidification of silicate melts (molten rock material) on cooling, by precipitation of crystals from a water solution, by sublimation from the vapor phase, and from other solids (recrystallization) during the alteration and metamorphic processes (Dyar and Gunter, 2007; Haldar and Tišljar, 2013; Marshak, 2013). Silicate melts crystallize when the temperature lowers to the point of fusion (known as freezing in the case of water) and solid crystals begin to form. Ice crystals forming in water under freezing conditions are common examples of crystallization by fusion. Magma, or molten rock within the Earth's crust, begins to crystallize when its temperature falls below the melting point of certain common silicate minerals. This process of silicate crystallization is more complicated than that of ice from water because one phase or mineral will crystallize out of the melt at a certain temperature and subsequently react with the remaining liquid as cooling proceeds.

Precipitation from solution can occur after evaporation of some or all of the liquid. A reduction in the liquid volume yields less of the dissolving agent to hold the ions in solution, and precipitation occurs once saturation is reached. A reduction in temperature of the liquid may yield a reduction in solubility of the constituents, thereby triggering precipitation. A change in pH (hydrogen ion concentration) of the liquid may reduce solubility and cause certain solids to precipitate. A reduction of pressure in the liquid, such as when water flows onto the Earth's surface from within a rock mass, may trigger the precipitation of certain minerals. Biological activity, along with chemical reactions between a mineralizing solution and the host rock (as in ore deposit formation), also causes minerals to precipitate from a fluid medium (Klein and Dutrow, 2007).

Occurrence of Minerals

Minerals comprise the soil and rock materials of the Earth. They are found in all geologic environments, including the alluvial sands along a river bed, soils of a plowed field, and bedrock exposures in a mountainous region. To understand the engineering significance and properties of soils and rocks, it is necessary to have a sound knowledge of minerals and an ability to identify the most commonly occurring minerals.

Although more than 2000 minerals occur in nature, only about 100 are common. These 100 minerals form the basis for study in a typical college course in mineralogy. Included in that collection are metallic and nonmetallic ore minerals and the less common members of the carbonate, sulfate, silicate, and other fundamental mineral groups (Gaines et al., 1997; Nesse, 2011).

Despite the hundreds of known minerals, only 25 or so make up the common rock-forming varieties (Nesse, 2011; Klein and Philpotts, 2012). These commonly occurring minerals primarily influence the engineering properties of soils and rocks. Thus, for many engineering projects, a knowledge of these 25 minerals will provide sufficient mineralogical detail. These minerals can usually be identified, without sophisticated equipment, by careful study of their physical properties.

Mineral Identification

The properties of minerals used for identification are dictated by their composition and crystalline structure. Chemical properties aid in identification of some minerals but mostly physical properties are used for identification of common rock-forming minerals. These properties include color, streak, luster, hardness, specific gravity, cleavage, fracture, crystal form, magnetism, tenacity, diaphaneity, presence of striations, and reaction to acid (Gaines et al., 1997; Dyar and Gunter, 2007; Klein and Dutrow, 2007; Nesse, 2011). Physical properties can be recognized either on sight or after applying several simple tests.

Color, as a distinguishing feature of minerals, can be deceptive, nondefinitive, or, in limited situations, quite diagnostic. Some minerals occur in several different colors, so color alone should be used cautiously for identification of such minerals. For other minerals, such as olivine, color (olive green) is a diagnostic feature. Therefore, color is always noted in mineral identification but, rather than being used alone, is considered with other properties in the identification process.

Streak refers to the color of the finely powdered mineral. By grinding a mineral to powder and then examining its color (easily accomplished by scratching on an unglazed porcelain surface), a common base is obtained for comparison. Surface texture and irregularities on mineral surfaces can produce colors that are considerably different from its streak. Certain minerals have a distinctive streak. For example, hematite (Fe_2O_3) yields a deep-red streak despite the various colors portrayed in sample specimens.

Luster is the overall appearance of the surface of a mineral in reflected light. Luster is divided into two groups for convenience: metallic and nonmetallic. Minerals that show a high degree of reflection from an opaque surface, like that of a metal, are appropriately said to have a metallic luster. Nonmetallic luster includes all other lusters that do not show this extreme degree of opaque-surface reflection. The nonmetallic lusters are further subdivided according to their particular appearance. Common descriptions of nonmetallic lusters are:

- Adamantine—Brilliant luster displayed by gemstones such as diamond.

- Vitreous—Glassy luster, like that of broken glass.

- Pearly—Iridescent luster of a pearl.

- Greasy—Luster such that the surface appears to be coated with a film of oily liquid.

- Silky—Luster like silk caused by a fine fibrous appearance such as that of silky gypsum (satin spar).

- Earthy—Dull luster like that of dry soil.

Hardness is the resistance that a mineral shows to scratching. It is governed by the elements in the mineral, their arrangement, and the strength of the bonds that bind them. Graphite and diamond, two polymorphs of carbon, differ in internal structure and bond strength, thus, having greatly different hardness values.

In geologic studies Mohs hardness scale (Table 2.1) is used to compare mineral hardness. The scale ranges from 1 to 10 with a representative mineral for each integer value. The increments from 1 to 9 are about equal in amount but the difference between 9 and 10 is considerably greater, estimated to be perhaps 30 times larger than the other units.

Several simple tools are used to determine the physical properties of minerals, such as a steel knife with a hardness of about 5.5, a glass plate of a similar

Table 2.1   Mohs Scale of Hardness.

Minerals	Level of Hardness
Talc	1
Gypsum	2
Calcite	3
Fluorite	4
Apatite	5
Orthoclase	6
Quartz	7
Topaz	8
Corundum	9
Diamond	10

hardness, a copper penny with a hardness of 3, fingernails with a hardness of 2.5, and an unglazed porcelain steak plate with a hardness of 7, which is used to streak minerals. Other useful tools are a small bar magnet, a 10× hand lens, which is useful when searching for striations or small crystal forms, and a small bottle of dilute hydrochloric acid. In determining hardness, an attempt is made to scratch the mineral with the tool and, in turn, to scratch the tool with the mineral. A comparison of these effects should tell which is harder.

Specific gravity (SpG) of a mineral is a dimensionless number expressing the ratio between the mass (or weight) of the mineral and that of an equal volume of water. Specific gravity is a function of the chemical composition because elements with high atomic weight increase the specific gravity, as well as differences in mineral structure (packing), because the closer the elements are packed together, the higher the specific gravity.

Specific gravity is measured using the following equation:

$$SpG = \frac{\text{Mass (or weight) of mineral in air}}{\text{Mass (or weight) of equal volume in water}}$$

$$= \frac{\text{Mass (or weight) of mineral in air}}{\text{Mass (or weight) in air} - \text{Mass (or weight) in water}}$$

(Eq. 2-1)

The quantities involved in the above equation are obtained by weighing the mineral in air and then in water. The difference between the two is the buoyancy force of the water, and it equals the mass of water with the same volume as that of the mineral specimen.

The common rock-forming minerals (quartz, feldspar, calcite, etc.), as a group, have a specific gravity near 2.7, which forces the average for common rocks toward this value. Metallic minerals, notably the sulfides, have specific gravities higher than 2.7, more nearly 5.5 or greater. Mercury has a specific gravity of 13.6 and pure gold of 19.3.

Specific gravity is used in a relative sense as a simple physical test. By working with a piece of mineral about the same size each time, for example, about the size of a sugar cube, the heft can be compared to a similar size piece of quartz or feldspar. Low values (around 2.3), average values (nearly 2.7, as in quartz or feldspar), or high values can be indicated by this simple comparison.

Specific gravity of a mineral, or any other material, multiplied by the unit weight of water (or the mass density of water), yields the unit weight of the material (or the mass density of the material). The unit weight or mass density of water is 1 g/cm^3 or 1000 kg/m^3 or 62.4 lb/ft^3. Under the International System of Units (the SI system), mass density is measured in g/cm^3, kg/m^3, or Mg/m^3. As an example, a solid piece of calcite with a SpG of 2.72 would have a mass density of 2.72 (1000) = 2720 kg/m^3 (2.72 Mg/m^3) or 2.72 (1) = 2.72 g/cm^3. Using British engineering units, the mass density would be 2.72 (62.4) = 169.7 lb/ft^3.

Cleavage is the ability of a mineral to break along smooth parallel planes. This property is due to the arrangement of the atoms and the types of bonds between them. Cleavage, in a mineral, indicates a direction of preferred weakness in bond strength between atoms. As the structure or atomic arrangement repeats in space, this weakness repeats as well, so that cleavage occurs in families of planes, all parallel to each other.

Cleavage is described according to its degree of development as good, moderate, or poor, depending on how readily the parallel breaks occur and according to how many planes of cleavage with different orientation occur, or by the shape of the geometric form enclosed by three or more planes. One plane of cleavage or two planes at right angles are common descriptions used for cleavage. Regarding geometric forms, table salt, or halite (NaCl), has three planes of cleavage at right angles, termed *cubic cleavage* because of the resulting cube-shape form. Calcite exhibits cleavage in three directions, but not at right angles, which yields a rhombohedral shape known as *rhombohedral cleavage* (Figure 2.2). Biotite, a type of mica, has a single prominent plane of cleavage, as shown in Figure 2.3.

Fracture is an irregular break that occurs in the absence of cleavage and is the usual consequence when minerals are broken. Some types of fracture can be used for identification purposes. Conchoidal fracture, or fracture that yields curved surfaces, is particularly useful in recognizing quartz. Fibrous fracture and hackly fracture, derived by breaking along irregular, sharp edges, can also be used for identification purposes.

Crystal form involves the display of well-formed crystal faces by a mineral. Crystal faces are an external manifestation of the internal arrangement of atoms. Some minerals exhibit their crystal forms more readily than others. Among the common rock-forming minerals, quartz and garnet most frequently exhibit crystals. Figure 2.4 shows a cluster of quartz crystals.

Crystal forms should be distinguished from cleavage faces. Crystal faces form during the crystallization process, not when the mineral piece breaks, so crystal faces will be destroyed if the specimen is broken. Cleavage faces can be obtained again and

Figure 2.3 Biotite, showing one plane of cleavage. (Karol Kozlowski, Shutterstock.)

again as parallel faces break off when the mineral is struck. Incipient parallel breaks can often be seen within the cleavage fragments, indicating a readiness to form new faces (see Figure 2.2). Crystal forms, however, lack such parallel faces within the specimen that would emerge on breaking. This is why crystals are valued more highly as specimens than are cleavage fragments, because they are a unique form in themselves and will be lost if breakage occurs.

When a specific mineral grows freely in an open space it takes on its own characteristic shape

Figure 2.2 Calcite, exhibiting rhombohedral cleavage. Note the incipient cleavage planes inside the specimens. (Aleksandr Pobedimskiy, Shutterstock.)

Figure 2.4 Cluster of quartz crystals. (Albert Russ, Shutterstock.)

because the angles between adjacent crystal faces are constant for crystals of that mineral. The faces may be different sizes, owing to different growth rates in two directions, thus yielding a deformed crystal, but the interfacial angles will remain the same. Crystal faces in minerals occur along the directions that intersect the maximum number of atoms and, thus, are directly related to the atomic array.

For convenience, the faces are defined in reference to crystallographic axes, three or four in number, which are imaginary lines intersecting at the center of the crystal. Minerals are classified into six crystal systems based on the symmetry demonstrated by their crystal forms. Planes of symmetry and axes of symmetry are used to describe these symmetry relationships. A plane of symmetry divides an object into two halves, each the mirror image of the other. These symmetry planes also contain one or more crystallographic axes.

An axis of symmetry is a line through the crystal about which rotation yields a repeat of the same pattern or crystal face. For example, a square has a fourfold axis of symmetry through its center, thus, it has four planes of symmetry. In Figure 2.5a, rotation about the axis, each 90°, yields the next side of the square. Four turns of 90° are needed to return to the beginning position, and this yields a fourfold axis of symmetry or rotation. A cube is shown in Figure 2.5b. It has three axes of fourfold symmetry, one

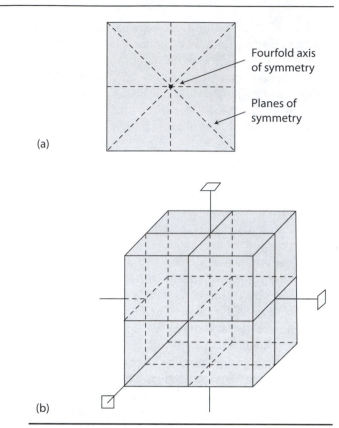

(a)

(b)

Figure 2.5   Planes of symmetry and axes of symmetry for a square and a cube. (a) Square showing basic symmetry elements. (b) Cube showing basic symmetry elements.

Table 2.2   Details on Crystal Systems of Minerals.

System	Axes	Planes of Symmetry (Max.)	Mineral Examples[a]
Cubic (or isometric)	Three equal axes at right angles to one another.	9	Garnet, fluorite, diamond, halite (rock salt), galena, sphalerite, pyrite, magnetite.
Hexagonal	Four axes: three equal and horizontal and spaced at 60° intervals; one vertical axis.	7	Beryl, apatite, tourmaline, calcite, quartz.
Tetragonal	Three axes at right angles: two equal and horizontal, one vertical axis longer or shorter than the others.	5	Zircon, cassiterite, chalcopyrite.
Orthorhombic	Three axes at right angles, all unequal.	3	Olivine, enstatite, topaz, barite.
Monoclinic	Three unequal axes: the vertical axis (c) and one horizontal axis (b) at right angles, the third axis (a) inclined in the plane normal to (b).	1	Orthoclase feldspar, hornblende, augite, biotite, gypsum.
Triclinic	Three unequal axes, no two at right angles.	None	Plagioclase feldspar, turquoise.

[a] Also see mineral descriptions in Appendix A.

through each of the opposing faces. They are each marked with a four-sided form. The cube also has nine planes of symmetry, the three shown plus two diagonal planes each across the three pairs of faces as illustrated in Figure 2.5a.

The cube demonstrates the highest degree of symmetry according to these symmetry elements. In all, 32 different classes of symmetry are possible. Based on the ways these classes can be grouped relative to the crystallographic axes, they are collected into the six crystal systems. Information about the crystal systems is given in Table 2.2.

The crystallographic axes for the six crystal systems are shown in Figure 2.6. A number of crystal forms associated with various minerals are also illustrated. Note that for beryl, an *a* axis can be placed through the center of (perpendicular to) each vertical face with the *c* axis being vertical or through the horizontal faces at the bottom and top, yielding the basic hexagonal system for reference. This method of relating crystal faces to the basic crystallographic axes, that is, recognizing where the axes and planes of symmetry occur, is the way in which crystals are assigned to crystal classes. This overall subject, known as *crystallography*, is typically included as part of a first course in mineralogy (Gaines et al., 1997).

Magnetism is the attraction of a mineral to a magnet. The most notably magnetic mineral is magnetite (Fe_3O_4), which is strongly attracted even to a simple bar magnet. Several other less magnetic minerals exist in nature, for example, pyrrhotite [$Fe_{(1-x)}S$ ($x = 0$ to 0.2)], but they are less common than magnetite.

Tenacity is the resistance a mineral shows to various destructive mechanisms such as crushing, breaking, bending, tearing, and so forth. Minerals can be described according to the following terms.

- Brittle—Breaks or shatters easily.

- Malleable—Can be hammered into thin sheets.

- Sectile—Can be cut into thin shavings with a knife.

- Ductile—Can be drawn into a thin wire.

- Elastic—Bends but resumes its original shape when released.

- Flexible—Bends but does not resume its original shape on release.

Diaphaneity is the ability of a mineral to transmit light. The terms used to express diaphaneity are:

- Transparent—A clear outline of an object can be seen through the mineral.

- Translucent—Light is transmitted through the mineral but objects cannot be discerned.

- Opaque—No light is transmitted through the mineral even on its thinnest edges.

Striations are fine parallel lines that appear to be scribed onto cleavage faces of the plagioclase feldspar. They occur because of twinning of the feldspar crystal; that is, the reversal of direction of component parts

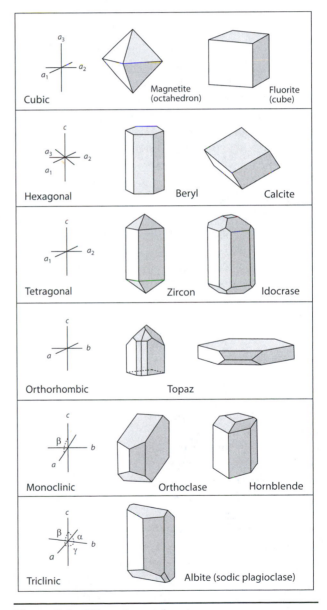

Figure 2.6   Mineral crystal systems.

of the crystal with respect to each other. Striations are useful in telling orthoclase from plagioclase because orthoclase lacks striations whereas plagioclase typically shows them. Striations also occur on crystal faces of quartz.

Reaction to acid is another test. Although not a physical property, effervescence in acid is used in conjunction with the other properties to distinguish minerals. Calcite is easily identified by the application of HCl because it effervesces freely. Dolomite reacts much more slowly and this feature forms a basis for distinguishing between these two carbonates.

Table A.1 in Appendix A can be used for identifying the common rock-forming minerals most likely to be encountered during engineering investigation and construction. Students should become very familiar with the most common minerals because that knowledge forms the basis for rock identification. Table A.3 in Appendix A contains the most common minerals of economic significance. Minerals from this second group are not likely to occur commonly in rocks encountered in most engineering studies but knowledge of them may prove of interest to many engineers. They do illustrate well the range of physical properties discussed previously. If additional knowledge of minerals is desired, a course in mineralogy is recommended for engineers, particularly those specializing in aggregates for construction and in highway materials.

Rock-Forming Minerals

Silicates

Silicates, as the most important group of rock-forming minerals, comprise the majority of the common minerals. In the classification used in mineralogy, subgroups of the silicates are developed on the basis of their silicon to oxygen ratio, which controls the crystal structure (Gaines et al., 1997; Dyar and Gunter, 2007; Klein and Dutrow, 2007; Nesse, 2011). This ratio strongly influences the properties and weathering characteristics of silicates. As an example, olivine has a ratio of 1 to 4 (silicon to oxygen), whereas for quartz the ratio is 1 to 2. These two minerals are vastly different; olivine is a high-temperature ferromagnesian that weathers rapidly, whereas quartz is a low-temperature mineral, highly resistant to weathering. The subgroups of silicates are not detailed here because that task is beyond the scope of a basic applied geology text. It is significant to note that the silicon to oxygen ratio of silicates drastically affects their behavior.

Another means for classifying silicates is to subdivide according to the presence of certain prominent elements (cations). This yields the following grouping: ferromagnesians and nonferromagnesians.

Ferromagnesians

Ferromagnesians are an important group of common silicates that contain iron and magnesium and, generally, are dark in color. Because iron and magnesium are heavy ions, ferromagnesians tend to have a slightly higher specific gravity than 2.7, the average specific gravity of quartz, feldspars, and calcite. Ferromagnesians include olivine, pyroxenes, amphiboles, and biotite.

Olivine is a light-green-colored mineral with a characteristic sugary texture caused by the presence of anhedral or rounded grains. Its hardness of 6.5 to 7 is rather deceptive because the grains break easily, suggesting a lower hardness. The chemical composition is $(MgFe)_2SiO_4$ and the specific gravity can range from 3.2 to 4.2 as the iron content increases. The diagnostic features of olivine are its green color and sugary texture.

Augite is the common member of a group of silicates known as pyroxenes. It is a dark green or black mineral that exhibits two planes of cleavage, nearly at right angles to each other. This right angular cleavage and dark color are diagnostics of augite. In rocks, augite tends to appear as short stubby crystals.

Hornblende is the most common mineral of the amphibole group of silicates. Like augite, it is dark green to black but displays two directions or planes of cleavage at angles of approximately 56° and 124°. This difference in cleavage angles distinguishes hornblende from augite (56° and 124° versus about 90°, respectively). In rocks, hornblende tends to appear more needle shaped as compared to augite. Idocrase, shown in Figure 2.6, is a complex silicate mineral, usually brown in color, that forms in limestones during contact metamorphism.

Biotite is a silicate belonging to the mineral group known as micas. Layers of micas can be peeled off or cleaved easily, yielding sheets of the platy material, depicting its one prominent plane of cleavage (see

Figure 2.3). The plates are flexible and elastic; that is, they will spring back to their original shape when bent and released. Biotite is a dark mica, appearing as either dark green, black, or brown, has a specific gravity of 2.7 to 3.2, and is best identified by its dark color and platy cleavage.

Nonferromagnesians

The nonferromagnesian silicates do not contain iron and magnesium in combination but usually have calcium, sodium, or potassium instead. Muscovite is the nearly colorless mica counterpart of biotite. It has the same properties of platy cleavage and elasticity as biotite but is lighter in color. Its specific gravity is 2.8 to 3.1 and its hardness of 2 to 3 compares well with that of biotite. In thin sheets it is transparent, taking on a yellow color in thicker sheets. Some substitution of aluminum for silicon may occur.

Feldspars

Taken as a whole, the feldspars are the most common rock-forming silicates. Collectively they comprise greater than 50% of all minerals in the Earth's crust and form the basis for igneous rock names and their classification.

Feldspars can be divided into two types: orthoclase, which contains potassium ($KAlSi_3O_8$), and plagioclase, which contains varying amounts of calcium and sodium. Plagioclase forms a continuous series, with sodium and calcium substituting for each other in the silicate structure. The cation composition ranges from all sodium to all calcium with every possible proportion in between. If greater than 50% sodium is present, the mineral is sodic plagioclase; if greater than 50% calcium is present, calcic plagioclase is the term used. In many cases, the sodic plagioclase is assumed to be albite and the calcic plagioclase to be labradorite for hand specimen identification. These are the most common plagioclase minerals.

The feldspars as a group have a vitreous luster, a hardness of 6 to 6.5, and a specific gravity of 2.5 to 2.7. They also have two planes of cleavage at nearly right angles to each other.

After concluding that a mineral is a feldspar, further subdivision is needed to identify the specific feldspar. Commonly, orthoclase ranges from white to pink in color, whereas plagioclase ranges from white to black. As the calcium percentage increases in plagioclase, the mineral becomes darker in color.

Consequently, albite is a white plagioclase and labradorite is a dark-gray plagioclase. The property other than color that separates the two feldspars is the presence or absence of striations. Orthoclase does not exhibit striations. Therefore, striated feldspars are assumed to be plagioclase. A problem arises when white plagioclase does not exhibit striations visible to the naked eye or with a hand lens. These minerals are difficult to identify specifically without the aid of a petrographic microscope.

Quartz Minerals

Quartz (SiO_2) is a common rock-forming mineral that is second only to the feldspars in abundance. Quartz can occur in a variety of colors depending on the presence of minor impurities. This yields such varieties as pink-colored rose quartz, purple-colored amethyst quartz, dark-colored smoky quartz, and cloudy-looking milky quartz. Quartz is best identified by its hardness of 7 and conchoidal fracture. When crystals are present (see Figure 2.4), their six-sided shape is diagnostic of quartz.

Chert is a cryptocrystalline or finely crystalline variety of quartz that occurs primarily in sedimentary rocks. Many geologists refer to the dark variety of chert as flint. Chert is of particular interest to the engineer because it presents certain problems when used as an aggregate material. Therefore, the engineer must be able to identify it. It too has a hardness of 7 (although weathering may suggest a lower value), has a dense appearance, and shows conchoidal fracture.

Chert may be a troublesome material when used in concrete for mechanical and chemical reasons. Lightweight or weathered cherts have a tendency to break or pop out when exposed to freezing and thawing temperatures on a concrete surface. Weak cherts also tend to reduce the strength of concrete if they comprise a high percentage of the aggregate.

Some cherts have been known to react with high alkali cements to cause cracking and expansion of the concrete. This is called the alkali-silica reaction (ASR) problem. A reduction in the Na_2O and K_2O percentage in the cement to below 0.6% usually alleviates this problem.

Other Silicates

Garnet is an accessory mineral found in igneous and metamorphic rocks. There are a number of different types of garnet that form a solid solution group, as do many silicates. The most abundant garnet is red in color, has a hardness of 6.5 to 7.5, and may appear as well-formed multifaced crystal forms belonging to the cubic crystal system. Color and hardness are the most diagnostic features for garnet.

Two common silicates that appear in metamorphic rocks are talc and chlorite. Talc is a soft mineral with the hardness of 1 and is usually white in color with a greasy feel. It has one good cleavage, yielding tiny platy masses with a pearly luster. Rocks with abundant talc are referred to as soapstone by well drillers, construction personnel, and many geologists. Chlorite is a green platy mineral with a hardness value from 2 to 2.5, which is slightly greater than talc. It has a vitreous luster and its plates are flexible but not elastic.

Oxide Minerals

The oxides of importance in regard to common rock-forming minerals are the iron oxides. Hematite (Fe_2O_3) is recognized most easily by its red streak as is limonite ($FeO \cdot H_2O$) by its brown streak. Limonite may occur in gravel deposits in significant quantities, yielding several percent by weight. It makes a poor concrete aggregate because it causes staining and pop outs on the concrete surface after only several cycles of freezing and thawing.

Magnetite (Fe_3O_4) occurs less commonly than the other iron oxides. It is highly magnetic and can be identified readily by this property. It also has a metallic luster and a specific gravity of 5.2.

Sulfide Minerals

The most common sulfide mineral is pyrite (FeS_2). Pyrite, or fool's gold, occurs to some degree in igneous, metamorphic, and sedimentary rocks. It is most easily recognized by its brassy color and metallic luster, hardness of 6 to 6.5, and cubic crystal shape. A similar mineral, chalcopyrite ($CuFeS_2$), is more yellow colored and has a hardness of 3.5 to 4. Chalcopyrite is a valuable source of copper and appears as a primary copper ore in many mining areas.

Pyrite is considered a nuisance mineral if present in gravel pits used for concrete construction. Staining of the surface of concrete may occur following oxidation. Certain varieties of pyrite yield sulfate ions in such quantities that the sulfate is able to attack concrete. If pyrite is present in facing stone for buildings, unsightly stains can occur after weathering. For

these reasons, it is necessary to identify this mineral and to determine its percentage in rock considered for engineering purposes.

Carbonate Minerals

The important carbonates are calcite and dolomite. Calcite ($CaCO_3$) is distinguished by its effervescence in dilute hydrochloric acid, its rhombohedral cleavage, and its relatively low hardness of 3. Dolomite [$CaMg(CO_3)_2$] effervesces only slowly in dilute HCl and is slightly harder than calcite with a hardness of 4. Dolomite will effervesce only when powdered, whereas calcite effervesces actively both in large pieces and in the powdered form.

Sulfate Minerals

The sulfate minerals, though widely represented in nature, have only two members that are reasonably common. These are gypsum ($CaSO_4 \cdot 2H_2O$) and anhydrite ($CaSO_4$). Gypsum is distinguished by its low hardness of 2 and usually white color. It may appear in three forms: the massive alabaster, as satin spar with its silky luster, or in a transparent variety, selenite. It also has a relatively low specific gravity of 2.2 to 2.4. Gypsum occurs as a chemical precipitate from seawater and, when present in thick deposits of several feet or more, is of economic value. Gypsum deposits are quite soluble in groundwater and may be cavernous if located near the Earth's surface. In this respect, the same problems must be considered as with cavernous limestones. These problems are discussed in Chapter 13, Groundwater.

Anhydrite is the water-free form of $CaSO_4$. Its hardness is greater than gypsum, nearly 3 to 3.5, and it usually occurs as a white to blue, massive, vitreous to dull material. It is distinguished from calcite and dolomite by its lack of effervescence in acid. A powdered form should be used for this comparison because dolomite effervesces to a limited extent.

It is important that anhydrite be recognized in nature because it has the property of swelling when wet and converting to gypsum. Accompanying this change is an increase in volume with disastrous effects if present in the foundation of an engineering structure or in a tunnel. Certain cases in which anhydrite was present in the abutment of a dam have resulted in undesirable after effects as the anhydrite absorbed water, expanded, and caused cracks in the structure.

Clay Minerals

The clay minerals are a group of fine-grained minerals (aluminosilicates) that are of major significance to the engineer. They comprise an essential portion of the soil and therefore yield a strong influence on soil behavior.

The term *clay* in the geologic literature has been used to pertain to either a size fraction or to a group of minerals. This is an important distinction because in some soil classifications the clay size can include nonclay minerals. Details concerning clay size are provided in the discussion of soil classifications in Chapter 7. For now, we can assume that clay is that fine portion of the particle size distribution that is smaller than 1/256 mm (0.075 mm) according to the Wentworth scale and of a reasonably similar size for the other important soil classifications.

Types of Clay Minerals

Clay minerals consist of five major groups: kaolinite, halloysite, illite, vermiculite, and smectite (or montmorillonite). They are divided into two categories, those with two layers or sheets in their structural repeat unit and those with three. Kaolinite and

halloysite are two-layer clays; illite, vermiculite, and smectite are three-layer clays (Mitchell, 1993; Holtz et al., 2011).

Clay Mineral Structures

Two kinds of layers comprise the clay minerals, a silica tetrahedral sheet and an octahedral sheet (Mitchell, 1993; Holtz et al., 2011). The silica sheet consists of a series of tetrahedra of four oxygens, with a silicon atom in the center of each tetrahedron, that extends in two directions, yielding the sheet structure. The octahedral sheet consists of octahedrons formed by six oxygens or hydroxls with a magnesium or aluminum atom in the center. It extends in two dimensions to provide the sheet structure. These are illustrated in Figures 2.7 and 2.8, respectively.

These two distinct layers can be joined, yielding a repeat unit of one tetrahedral layer and one octahedral layer, a *t-o structure*, or they can be combined to yield an octahedral layer sandwiched between a tetrahedral layer top and bottom, known as a *t-o-t structure*. The t-o structure is a two-layer clay whereas the t-o-t structure is a three-layer clay. The repeat units are stacked one on top of another, yielding the clay platelets that make up the particles of

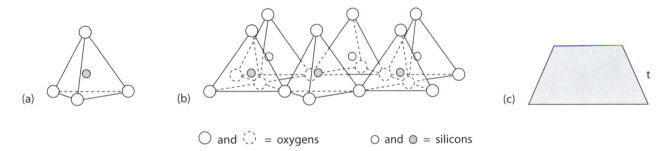

Figure 2.7 Diagrammatic sketch showing: (a) single silica tetrahedron, (b) the sheet structure of tetrahedrons, and (c) the schematic of a tetrahedral sheet.

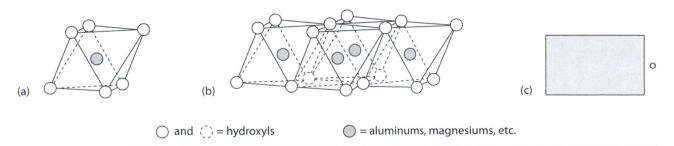

Figure 2.8 Diagrammatic sketch showing: (a) single octahedral unit, (b) the sheet structure of the octahedral units, and (c) the schematic of the octahedral sheet.

clay in nature. The two-layer and three-layer structures are illustrated in Figure 2.9.

Chlorite occurs as a green-colored clay mineral in its megascopic form, as in chlorite schist. It is also relatively common in soils, usually as a minor constituent along with other clay minerals. It is a mixed-layer clay consisting of the three-layer t-o-t structure plus an octahedral layer before the t-o-t portion is repeated. This is illustrated in Figure 2.10.

Two-layer clays are held together by ionic bonds between the sheets with no charge imbalance. They are nonswelling clays; kaolinite and halloysite belong to this category.

Three-layer sheets have some charge imbalance because of substitution of ions in the octahedral and tetrahedral sheets. In illite this is neutralized by potassium ions between the three-layer units, which hold together strongly enough that illite is not a swelling clay. In smectite and vermiculite, the substitution in the tetrahedral and octahedral layers is more extensive and somewhat random so that an orderly balancing by potassium ions at the unit boundaries is not possible. Other cations, typically Ca^{2+}, Mg^{2+}, and Na^{2+}, are present as well. This allows for water to gain access between the clay units, yielding the swelling clays. Smectite

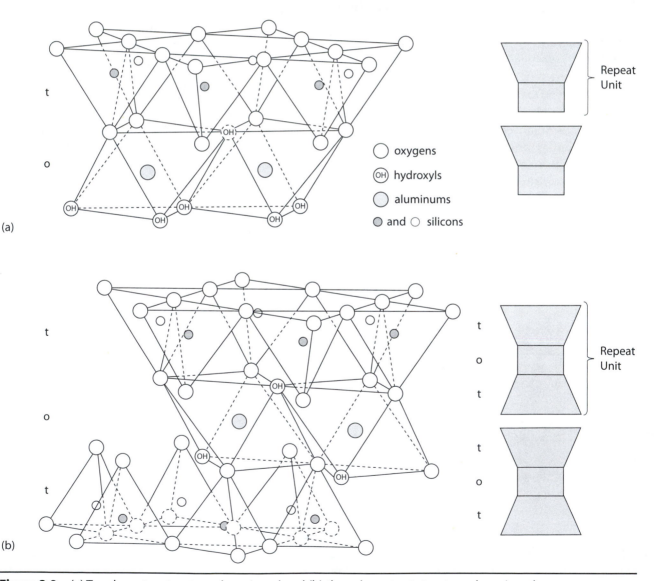

Figure 2.9   (a) Two-layer t-o structure clay mineral and (b) three-layer t-o-t structure clay mineral.

swells more than vermiculite, the specific amount depending on the cations present. Expansive clays are a major geologic hazard in the United States. They are landslide prone and disrupt building foundations, roads, and other structures when they undergo expansion.

Base-Exchange Capacity

Base-exchange capacity is a property ascribed to most clay minerals. It is defined as the ability to attract cations to its surfaces and to exchange them stoichiometrically (valence for valence) for each other. Therefore, a plus-two cation (Ca^{2+} for example) can replace two plus-one cations (Na^+) on the crystal lattice. Negative charges on the clay surfaces provide the charge deficiencies that are satisfied by these cations. The sources of the negative charges on the lattice are crystal terminations and substitutions in the tetrahedral and octahedral sheets of the silicate structure.

Crystal terminations increase as the particle size decreases so that finer grained particles will have a greater base-exchange capacity than coarser grained ones. Clays with a greater degree of substitution in their tetrahedral and octahedral sheets will also have a greater base-exchange capacity. Smectites are both extremely fine grained and prone to major substitution within their silicate structure, which is why these minerals have a high base-exchange capacity.

Smectites have the highest base-exchange capacity and kaolinite has the lowest. The other clays lie somewhere between these extremes. The subject of base-exchange capacity is discussed further in Chapter 7, Elements of Soil Mechanics.

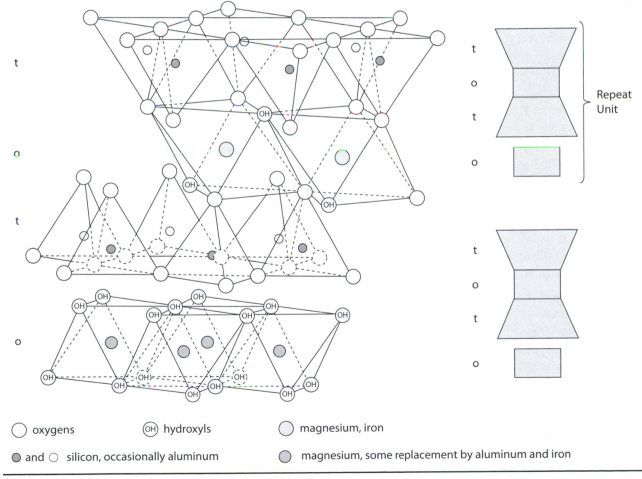

oxygens   (OH) hydroxyls   magnesium, iron

and silicon, occasionally aluminum   magnesium, some replacement by aluminum and iron

Figure 2.10   Mixed-layer structure for chlorite.

Zeolites

Zeolites are silicate minerals related generally to the feldspars, but they have major substitutions in their silicate structures that yield a high density of negative charges on their mineral surfaces. For this reason they are used for softening water, which is removal of Ca^{2+} and Mg^{2+} from a water supply. Na-zeolite is obtained by adding NaCl to the zeolite material. The hard water is allowed to pass over the Na-zeolite so that Ca^{2+} and Mg^{2+} replace the Na+ on the zeolite. Ca^{2+} and Mg^{2+} in water form an insoluble residue when mixed with soap, resulting in bathtub scum, which is a well-known nuisance. Na+ forms no such residue so, by using it to replace Ca^{2+} and Mg^{2+}, the water becomes soft.

Engineering Considerations of Minerals

The problems associated with various minerals may be fully appreciated by the engineer but, first, the problematic mineral must be identified if the problem is to be predicted.

1. Recognizing gypsum in a limestone that lies along a proposed tunnel centerline will provide information on possible problems of swelling in the presence of water and subsequent deterioration of the final concrete lining.

2. The presence of pyrite in a dark shale can suggest ultimate problems of deterioration from swelling and acidic waters.

3. The presence of pyrite and chert in concrete aggregate can create serious problems in concrete.

4. Swelling clays in a shale can provide a warning that slope stability may be a problem or a heaving condition may develop on wetting of the clay.

Exercises on Minerals

1. Name one or more diagnostic properties and the specific value or information for those properties used to differentiate between the following mineral pairs:

   a. calcite and quartz

   b. feldspar and quartz

   c. calcite and feldspar

   d. pyrite and limonite

   e. amphibole (hornblende) and pyroxene (augite)

   f. mica and clay minerals

   g. orthoclase and sodic plagioclase

   h. calcite and dolomite

   i. calcic plagioclase and pyroxene (augite)

   j. biotite and amphibole (hornblende)

2. A mineral with a hardness of 7 appears in two different forms. One is a six-sided barrel-shaped piece with a pyramid at its top, white in color, and transparent. The other is pink, translucent, has no smooth sides but instead has rounded fracture surfaces. What mineral would this be and how do you account for the greatly different appearances?

3. A white mineral has a hardness of 6, two planes of cleavage nearly 90° apart, and shows striations on two of the cleavage faces. What specific mineral would this be?

4. Limestone (which consists mostly of calcite) is being quarried in one location and quartzite (which consists mostly of quartz) is being quarried in another. Which of the two rock materials would cause greater wear on the shovels, trucks, and crushing equipment used to load and size the material for construction use? Explain. Which material would be more resistant to abrasion if used as a road base for a highway or railroad? Explain.

5. Both weathered chert and sound quartz pieces are common constituents in stream gravels. Both have the chemical formula SiO_2. Yet the weathered chert is an unsound material in concrete when subjected to freezing and thawing cycles, whereas quartz does not have this problem. Explain why this is the case.

6. Pyrite (FeS_2) is a common accessory mineral in coal-bearing rocks. What chemical is developed when water comes in contact with pyrite and what is the environmental effect of this?

7. Large fissures and caves develop in rocks composed of calcite. Relate this geologic condition to the properties of calcite studied in this chapter. Would caves also form in dolomite? Explain.

8. Montmorillonite (smectite) is a clay mineral that expands extensively on wetting. What would be the effect of locating a footing for a building in this material if wetting and drying of the clay can occur with seasonal changes?

9. Zeolites are a group of silicates that form from weathering or near-surface processes and appear as secondary fillings and veins in preexisting rocks. They have an expansive quality similar to that of some clay minerals. What adverse effects could occur if zeolite-bearing rocks (a) are used as an aggregate for concrete or (b) form the abutments (or sidewalls) for a concrete dam? Explain.

REFERENCES

Dyar, M., and Gunter, M. E. 2007. *Mineralogy and Optical Mineralogy*. Chantilly, VA: Mineralogical Society of America.

Gaines, R. V., Skinner, C. W., Foord, E. E., Mason, B., and Rosenzweig, A. 1997. *Dana's New Mineralogy: The System of Mineralogy of James Dwight Dana and Edward Salisbury Dana* (8th ed.). Hoboken, NJ: Wiley-Interscience.

Haldar, S. K., and Tišljar, J. 2013. *Introduction to Mineralogy and Petrology*. New York: Elsevier.

Holtz, R. D., Kovacs, W. D., and Sheahan, T. C. 2011. *An Introduction to Geotechnical Engineering* (2nd ed.). New York: Pearson.

Klein, C., and Dutrow, B. 2007. *Manual of Mineral Science* (23rd ed.). New York: Wiley.

Klein, C., and Philpotts, A. R. 2012. *Earth Materials: Introduction to Mineralogy and Petrology*. New York: Cambridge University Press.

Marshak, S. 2013. *Essentials of Geology*. New York: W. W. Norton.

Mitchell, J. K. 1993. *Fundamentals of Soil Behavior*. New York: John Wiley & Sons.

Nance, D., and Murphy, B. 2016. *Physical Geology Today*. New York: Oxford University Press.

Nesse, W. 2011. *Introduction to Mineralogy* (2nd ed.). New York: Oxford University Press.

Neuendorf, K. K. E, Mehl, Jr., J. P., and Jackson, J. A. (eds.). 2005. Glossary of Geology (5th ed.). Alexandria, VA: American Geologic Institute.

IGNEOUS ROCKS

3

Chapter Outline

Definitions of Rock

Geologic Definition

To a geologist, rocks are essential units of the Earth's crust whose origin, classification, history, and spatial aspects are important. Physical properties such as massiveness or durability of the material are not directly involved.

Most physical geology books define rocks as assemblages of one or more minerals (Marshak, 2013; Nance and Murphy, 2016). This is true generally, but there are exceptions if most of the material is nonmineral. For example, coal consists mostly of nonmineral material because it has an organic origin, and volcanic glass (obsidian, etc.) is nonmineral because of its amorphous nature; yet, both are rocks. Therefore, a more acceptable geologic definition for rock is the one already provided above: rock is an essential part of the Earth's crust, and involves details of origin, classification, history, and spatial relationships.

Engineering Definition

An engineering definition of rock differs from that used in geology; engineers consider rock to be a hard, durable material. From an excavation point of view, rock is any material that cannot be excavated without blasting. All other earth materials would be termed *earth* or *soil*. Bedrock is the hard, continuous rock mass lying below the soil, or at the Earth's surface, that must be excavated by blasting. Soil, by contrast, is the loose, unconsolidated material lying above the bedrock that can be excavated by conventional means. In this text, with its emphasis on engineering applications, the engineering definition is generally followed. A discussion of rock versus soil is provided in Chapter 4, Rock Weathering and Soils.

Definition and Occurrence of Igneous Rocks

Igneous rocks form by the cooling of molten rock material (magma or lava). They include the deep-seated, intrusive rocks that cool slowly within the Earth as well as the extrusive lavas and pyroclastics that form at the surface and cool quickly. Pyroclastics are materials ejected from volcanoes that settle out in the form of ash deposits and large rock fragments. Rocks formed at intermediate depth within the Earth, such as dikes and sills, also cool at an intermediate rate. They comprise another important group of igneous rock bodies (Marshak, 2013; Nance and Murphy, 2016).

Each continent of the world has extensive areas in which rocks of predominantly igneous origin are exposed. Igneous rocks make their appearance on the landscape in the centers of mountain belts, in eroded plateaus marking previous mountains, as isolated igneous intrusions, and as extensive outpourings of molten rock. Igneous and metamorphic rocks comprise about 95% of the outer 16 km (10 mi) of the Earth's crust.

A construction engineer, if involved with projects in many portions of the western United States or near the Appalachian region in the east, will encounter igneous rock as building foundation material in excavations, as sources for highway construction materials, and in various other engineering projects. The particular problems associated with igneous rocks and their wide diversity of type and nature are a subject worthy of study for engineers who plan to work with earth materials.

Formation of Igneous Rocks

Igneous rocks can form at the Earth's surface, at shallow depths below the surface, and deep in the Earth's crust. Igneous rocks that cool at the Earth's surface are the product of volcanic activity and are known as volcanic rocks. When eruptions of molten rock, generated within or below the Earth's crust, reach the surface, the opening or crater is enlarged by explosions and slumping of molten and semisolid material. This crater is located at the top of a mound of accumulated lava and pyroclastic debris. The growing mound of material capped by this crater is the well-known volcano.

Lava extruded from a volcano flows over the adjacent land surface, eventually cooling and hardening into solid rock. This is one variety of volcanic or extrusive rock. Intermittently between the lava flows, most volcanoes have periods of explosive eruptions in which pieces of hardened volcanic rock are ejected into the air and settle out as ash deposits and coarser sized materials. These pyroclastics are commonly interbedded with the lava flows.

At depths within the Earth, the molten material or magma intrudes the overlying older rocks, cools slowly, and hardens without reaching the surface. Magma differs from the lava flowing on the surface in its greater content of dissolved fluids, which tend to escape when reaching the Earth's surface. Rocks that crystallize inside the Earth are called intrusive or plutonic igneous rocks. Since intrusive rocks cool very slowly, compared to extrusive or volcanic rocks, their mineral grains are much larger than those comprising volcanic rocks.

Igneous bodies formed at intermediate depth, that is, below the surface but not located as deep as plutons, are called *hypabyssal* rocks. Typically these rocks have textures intermediate in size between coarse-grained intrusive rocks and fine-grained extrusive ones. They may even have a porphyritic texture, one with coarse crystals surrounded by a fine-grained groundmass.

Magmas are hot, viscous, siliceous melts containing dissolved gases. The most abundant elements present are silicon and oxygen and the primary metal ions are potassium, sodium, calcium, magnesium, aluminum, and iron. Dissolved gases typically are H_2O (steam), CO_2, and SO_2 (Gill, 2010). Igneous rocks form as the magma cools and the composition can vary greatly, depending on the original composition of the melt and the history of cooling and emplacement.

The silica (SiO_2) content of igneous rocks ranges from about 40% to more than 80%. Magmas and rocks that contain abundant silica were originally termed *acidic* because of the early notion that these rocks formed from a silicic acid through replacement of the hydrogen atoms by metal cations, as in the formation of salts from acids. Today it is known that minerals form from silicate melts through a complex path of crystallization (Best, 2002; Philpotts and Ague, 2009; Winter, 2009; Frost and Frost, 2013), but the term *acidic* persists to some extent in the literature as a descriptive term. Note, however, that *high-silica igneous rock* is the preferred terminology. High-silica rocks are light in color and are called felsic rocks and low silica rocks are darker in color and are called mafic rocks.

Similarly, low-silica rocks are termed *basic* and those with extremely low silica content (less than 50%) are termed ultrabasic (also ultramafic, meaning extremely dark). These terms describe an important fact about igneous rocks: their extreme range in silica content.

Low-silica magmas are less viscous (more fluid) than high-silica magmas and, in fact, there is considerable question as to whether the high-silica melts could flow as a lava from a volcanic vent. Instead, high-silica extrusive rocks are more likely pyroclastic in origin. For low-silica lavas, temperatures of about 1000°C have been measured in volcanoes. Fluxes, such as dissolved gases, are commonly present, which lowers the melting temperature of the magma.

Types of Volcanoes

Volcanoes and their conduits provide the passageways between bodies of magma in the crust and the Earth's surface (Figure 3.1). Several types of volcanoes can be distinguished as follows (Gill, 2010; Marshak, 2013; Nance and Murphy, 2016):

1. Shield volcanoes with large flat lava cones forming gentle slopes (Hawaiian type).

2. Cinder cones with steep slopes formed by pyroclastic eruptions.

3. Composite volcanoes or stratovolcanoes, also known as composite cones or stratocones, are formed by alternating lava flows and pyroclastic eruptions. These eruptions are accompanied by the emission of large quantities of gas and, sometimes, by violent explosions.

The shield volcanoes, cinder cones, and composite volcanoes are known as central, or vent, eruptions.

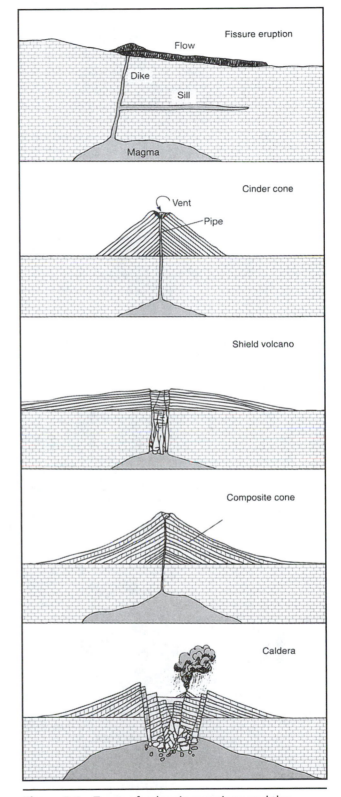

Figure 3.1 Types of volcanic eruptions and the related morphological features.

Shield Volcanoes

Shield volcanoes are large, flat cones consisting almost entirely of lava flows of basaltic composition, which pour out in quiet eruptions from a central vent or closely related fissures. The cones are much broader than their height with slopes seldom steeper than 10° at the summit and 2° at the base.

Mauna Loa, Hawaii, is a shield volcano and is the largest of its kind in the world. It rises 4 km (2.5 mi) above sea level but, measured from the seafloor, it stands 10 km (6 mi) high with a diameter at its base of 100 km (60 mi). This yields an estimated volume of nearly 24,000 km^3 (6000 mi^3) of basalt, all accumulated from individual lava flows only a few meters (tens of feet) thick. The island of Hawaii actually consists of five volcanoes that have built up from the seafloor and have coalesced to form the island.

Cinder Cones

Cinder cones are small mountains, primarily consisting of fragmental (pyroclastic) debris ejected from the central vent. They achieve slopes equal to the angle of repose of the clastic debris, or about 30° to 40°, and they seldom exceed 500 m (1600 ft) in height. Many cinder cones have flows of basalt issuing from their base. Parícutin in Mexico erupted in a farmer's field in 1943 and built a cinder cone 150 m (500 ft) high during its first week of activity. It is currently a dormant volcano about 400 m (1300 ft) high.

Composite Volcanoes

Composite volcanoes or stratovolcanoes are built up by alternating lava flows and beds of pyroclastics. These are the most common form of large continental volcanoes and they are characterized by slopes up to 30° at the summit, tapering to 5° near the base. Composite volcanoes are represented by such famous volcanoes as Fujiyama, Vesuvius, Stromboli, and Mayon on Luzon Island in the Philippines. Mount Shasta in northern California, a stratovolcano in the Cascade Range, is shown in Figure 3.2.

If the central vent of the volcano becomes too high or gets plugged, the lava may find an easier route through the cone and open a subsidiary vent on the volcano's flank. Subsidiary vents commonly occur on major volcanoes.

Figure 3.2 Mount Shasta, northern California, is a composite cone in the Cascade Range. (Jim Feliciano, Shutterstock.)

Morphology of Volcanoes

The features associated with volcanoes and other details of eruptions are illustrated in Figure 3.3. The large mound of lava and pyroclastic debris formed at the Earth's surface is the volcanic cone and the depression at its summit is the crater. Enlarged craters are known as calderas, most of which are formed by collapse, but some occur by explosion. The passageway to the crater is the volcanic pipe, vent, or conduit. Satellite cones, fed by the main conduit, typically form on the flanks.

In some cases, particularly for cinder cones, the volcanic debris can erode, leaving the lava-filled pipe, or volcanic neck or plug, exposed as a resistant column of rock. Dikes feeding away from the center of the neck may be preserved as well. Ship Rock, New Mexico, and Devils Tower, Wyoming, are examples of volcanic necks (Figure 3.4 on the following page). Columnar jointing in Devils Tower is vertical near the top but closer to horizontal near the base. These joints tend to form perpendicular to the cooling surface (or surface of the cone).

Fissure Eruptions

Fissure eruptions occur quietly along linear fractures in the Earth as low-viscosity basalts and spread rapidly over large areas with the emission of only small quantities of gas. In the geologic past, floods of basalt have poured out at locations throughout the world and are known as plateau basalts because of the tendency to form extensive plateaus of layered basalt flows.

The only documented lava flood of historic time occurred in Iceland in 1783 when the Laki fissure, nearly 32 km (20 mi) long, produced lava that spread over an area of about 660 km^2 (260 mi^2). Iceland itself is a remnant of vast lava floods that accumulated over the last 50 million years and blanketed an area of more than 500 million km^2 (200 million mi^2). The hardened lava is believed to be 2800 m (9000 ft) thick.

The plateau basalts of the Columbia Plateau in Washington, Oregon, Idaho, and northern California cover 130,000 km^2 (50,000 mi^2) (Figure 3.5 on page 39). In some areas, more than 1500 m (5000 ft) of extrusive rocks have been built by a series of fissure

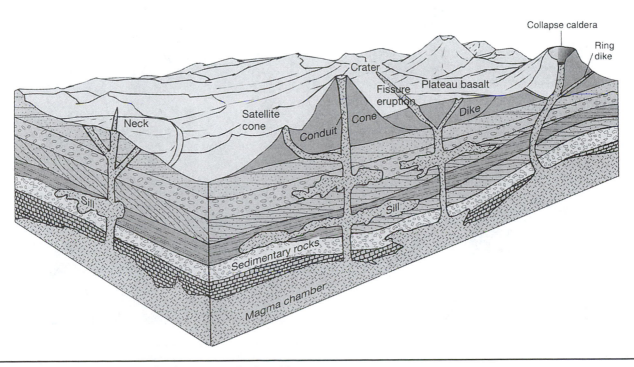

Figure 3.3   Cross section of volcanoes and related features.

(a)

(b)

Figure 3.4   Erosional remnants of volcanic cones at (a) Ship Rock, New Mexico (Zack Frank, Shutterstock) and (b) Devils Tower, Wyoming (Camden Mendel, Shutterstock).

eruptions. Individual eruptions deposited layers from 3 m (10 ft) to 100 m (330 ft) thick. The total volume of rock poured out during this Miocene volcanic episode has been estimated to be more than 100,000 km^3 (25,000 mi^3).

Other extensive areas of the world with major plateau basalt accumulations are the Deccan Plateau of India with an area of 1,000,000 km^2 (400,000 mi^2) and a thickness of more than 1800 m (5500 ft), the Paraná of Brazil and Paraguay with an area of 750,000 km^2 (300,000 mi^2), the Ethiopian Plateau near Victoria Falls on the Zambezi River, north-central Siberia, Greenland, Antarctica, and Northern Ireland. Much older flood basalts (Precambrian in age) are found in northern Michigan and in the Piedmont region of the eastern United States. The basalt flows of the world are estimated to cover more than 2.5 million km^2 (1 million mi^2).

Volcanic Deposits

Lava pouring from volcanoes or fissure eruptions moves down the slope filling low areas in the terrain. Basalt is highly fluid, and can flow at high velocities, 100 km/hr (60 mi/hr), and for distances as long as 50 km (30 mi). Successive flows build individual layers from a few meters (tens of feet) to 100 m (330 ft) thick.

Basaltic flows on Hawaii give rise to two primary surface forms: pahoehoe (pa-hoy-hoy) and aa (ah-ah). Pahoehoe forms ropy features on the surface and, therefore, yields a rough blocky surface. Aa shows a relatively smooth surface. This illustrates that aa is the more fluid of the two lavas and it yields thinner individual flows.

As lava cools and contracts, polygonal shrinkage cracks form perpendicular to the cooling surface.

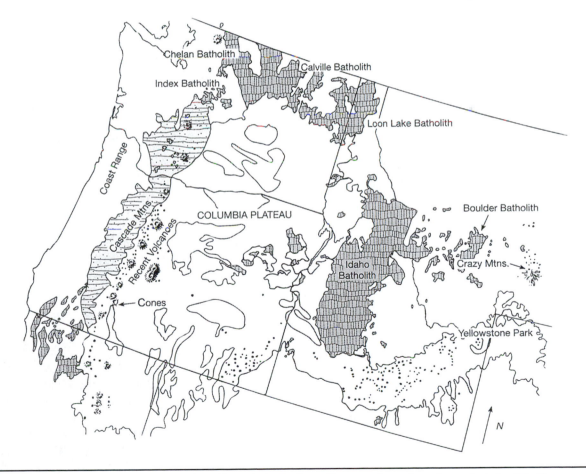

Figure 3.5 Distribution of volcanoes, lava flows (stippled), and igneous intrusions (cross hatched) in the northwestern United States.

Such features are called columnar joints. In thick flows the surface hardens well before the interior and, if the pressure is sufficiently great, lava will break through the sloping crust, extruding molten rock and leaving behind a lava tube or cave. In many cases the pressure pushes up the surface of the flow to yield a pressure ridge, an elongated blister with a central fracture through which the gas escapes. Therefore, lava flows may contain openings that easily transmit water or they can collapse under construction loads.

Pillow lavas are rounded masses a few centimeters (about an inch) to 2 m (6 ft) across that resemble a stack of pillows with their rounded side up. They occur when lava flows into seawater or into fresh-water lakes and ponds. The pillows generally have glassy borders. The rounded shape of the pillows can be used to tell the original top from bottom of a deformed sequence because the pillows are con-cave downward when formed. Pillows are also used to indicate subaqueous flows.

The glassy material in the pillow lavas may alter to a yellowish clay-type mineraloid called palago-nite. Through alteration or weathering, palagonite can change to clay. Palagonite-rich basalts are prone to deterioration in the presence of water when such materials are used as a road aggregate. These materi-als have led to serious pavement problems on high-ways in the northwestern United States.

Lavas can be glassy, finely crystalline, or com-binations of both depending on their cooling rate. When water vapor and other gases are released, gas cavities or vesicles can form. A common rock formed in this way is vesicular basalt.

In addition to lava flows, extrusive igneous rocks are formed as pyroclastic deposits. These consist of fragments of volcanic material blown out of the vent and, collectively, such fragments are called tephra. These ejected particles can consist of rocks, minerals, or glass and are classified according to size as dust (the finest), ash (up to about 100 cm [3 ft]), and larger elliptical-shaped pieces known as volcanic bombs.

Commonly, bombs are ejected in a plastic state and harden before reaching the ground. Large-size fragments of solid rock blown into the air are known as blocks and these can be thrown 10 km (6 mi) or more. Volcanic ash and dust can be carried great distances by volcanic explosions. In the eruption of Krakatoa in 1883, dust reached the upper levels of the atmosphere and was dispersed around the world.

Pyroclastic Deposits

Pyroclastic rocks form when fragments become cemented together or lithified. Volcanic tuffs consist of ash and dust-size pieces; volcanic breccias consist primarily of larger-sized pieces, those above 4 mm in size.

A spectacular and potentially disastrous type of eruption associated with high-silica magmas is known as an ash flow or nuée ardente. It occurs when hot ash, dust, and gases are ejected en masse in a glow-ing cloud that moves downslope at high speed. The solid fragments are buoyed up by the hot gases yield-ing minimal frictional resistance to this incandescent avalanche. Velocities up to 100 km/hr (60 mi/hr) are possible because the expanding gas reduces frictional resistance by forcing the grains apart.

In 1902 an ash flow occurred on Mount Pelée on Martinique in the Caribbean. It engulfed the town of St. Pierre in one minute, killing 28,000 people. Deposits formed by a nuée ardente are poorly sorted and lack bedding.

Typically the fragments remain hot after the ash flow comes to rest. After settlement and com-paction, the fragments are fused by volcanic glass to yield a welded tuff or ignimbrite. Ash flows can be quite vast, forming deposits more than 100 m (330 ft) thick and more than 100,000 km^2 (40,000 mi^2) in extent. Many geologists today consider rock units previously thought to be rhyolite (silica rich, felsic) flows are actually ignimbrite sheets because of the high viscosity of silica-rich melts.

The eruption of Mount Saint Helens, Washing-ton, in the spring of 1980 provides an example of a major pyroclastic explosion in the United States. A detailed account of the eruption can be found at the US Geological Survey (USGS) volcano hazards web-site (https://volcanoes.usgs.gov/index.html).

Fumaroles, Geysers, and Hot Springs

In areas of former or dormant volcanic activity, steam and gases (CO_2, HCl, H_2S, HF) at high temper-ature may be emitted from gas vents (or fumaroles) for a long period of time. Geysers are springs of boil-ing water and steam, supplied from thermal zones below, that erupt on a periodic basis. Geysers differ from hot springs in that hot springs flow constantly and usually have a lower temperature than that of geysers. Yellowstone National Park in northwestern Wyoming contains numerous geysers, fumaroles, and hot springs. Old Faithful, the famous geyser and tourist attraction in Yellowstone, erupts about every hour, shooting water to a height of nearly 60 m (200 ft).

Iceland, a land of numerous hot springs and geysers, is a location where these geothermal sources are used for heating and cooking purposes. Areas in southern California and northern Mexico are currently being exploited for geothermal power generation. Cerro Puerta, southwest of Mexicali, Mexico, is the site of an electrical generating station using the high-temperature source (steam) of the subsurface. Fifty km (30 mi) to the north, near El Centro, California, several American companies have constructed geothermal power plants to produce electrical power.

Intrusive Rock Bodies

Igneous bodies of major areal extent that have solidified inside the Earth are called plutons. Typically they are classified according to their size, shape, and relationship to the rock they have intruded (known as the country rock). Plutons include batholiths, stocks, dikes, sills, and laccoliths (Gill, 2010; Marshak, 2013; Nance and Murphy, 2016). The most common are tabular igneous intrusions (dikes and sills), which feed from other plutons. These various intrusive bodies and topographic expressions of extrusive igneous formations are illustrated in Figure 3.6.

Batholiths are the largest plutons and are discordant in nature, that is, they cut across the bedding in the country rock. By definition they have a cross-sectional area (or map area) of 100 km² (40 mi²) or

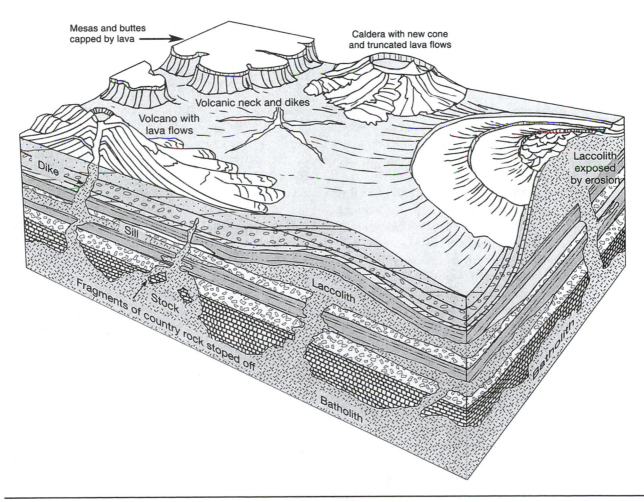

Figure 3.6   Intrusive rock bodies.

more. They have steep boundaries and extend to great depths, estimated to be from 10 to 30 km (6 to 18 mi) into the crust. These enormous igneous bodies, typically composed of granite or granodiorite, crystallized thousands of meters (several miles) below the Earth's surface. Uplift and extensive erosion are required to expose these bodies at the Earth's surface.

Batholiths are found in the centers of mountain belts and in the shield areas of continents such as in the Canadian Shield of North America. They are considered to be the roots of mountains, and the axis of the batholith typically follows that of the mountain range. The Sierra Nevada of eastern California are such an example. The Idaho Batholith is another extensive exposure of these immense plutons; it covers an area of nearly 41,000 km^2 (16,000 mi^2).

Stocks are discordant plutons, similar to batholiths, but smaller. They must have a map exposure of less than 100 km^2 (40 mi^2), but usually occur more in the range of a few square kilometers to tens of square kilometers (1 mi^2 to 10 mi^2). Stocks may grade upward into volcanic necks and many have porphyritic textures; that is, visible crystals in a fine-grained groundmass.

Dikes and sills are tabular bodies that extend outward from other igneous intrusions. Dikes, perhaps the most common of the intrusive bodies, cut across the bedding planes or the foliation of the country rock (and thus are discordant). Typically they are from 0.3 to a few meters (1 to 10 ft) thick, but their thickness can range all the way from a few centimeters to hundreds of meters (1 in to 1000 ft). The largest known dike in the world is the Great Dike of Zimbabwe, which runs cross country for 600 km (375 mi) with an average width of 10 km (6 mi).

Dike emplacement is related to the fracture pattern in the country rock. Magma is forced out through these fractures from the central conduit, stock, or other body. Dikes radiating from volcanic necks are common, such as Ship Rock, New Mexico (see Figure 3.4a). In some locations dikes can radiate outward to form circular features called ring dikes. Large ring dikes yield circular features up to 25 km (15 mi) in diameter. This is illustrated in Figure 3.7. Dike swarms are parallel or radiating dikes consisting of numerous clustered dikes. They may form as fissure fillings or along tension fractures that developed during magma emplacement.

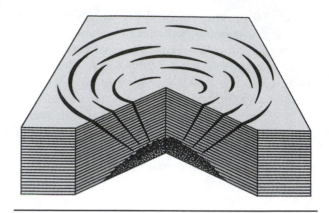

Figure 3.7   Three-dimensional view of ring dikes.

Dikes can be either more resistant or less resistant to weathering than the country rock, which yields either wall-like features cutting across the countryside or elongated narrow trenches.

Sills are tabular bodies formed when magma is injected along bedding planes of layered rocks; hence, they are concordant features. They range in thickness from a few centimeters to hundreds of meters (1 in to 1000 ft) and can extend over broad areas. The Palisades Sill along the Hudson River in New York and New Jersey is a well-known example in the eastern United States. It is 300 m (1000 ft) thick. The Whin Sill in the north of England is about 30 m (100 ft) thick, has a diorite composition, and has an extent of nearly 4000 km^2 (1600 mi^2). It provides an excellent quality aggregate material with high strength and good durability.

Sills are distinguished from flows of extrusive rock in that sills lack the common features of these lava flows, namely, ropy or blocky surfaces or vesicular and pillow structures in the hardened lava. Sills are typically coarser grained than lava flows because they cool more slowly, and the country rock above and below shows thermal effects such as baking. Sills protected by surrounding rock also lack signs of weathering, which may occur in lava flows. Sills can contain pieces of the country rock through which they were injected.

Sills are not necessarily horizontal, but merely concordant to the bedding of the enclosing rock. Tilted beds may also contain sills. Sills are fed from local dikes, volcanic necks, or come directly from stocks and batholiths.

Laccoliths are formed when the injection of magma along bedding planes forces the sediments upward in a dome or mushroom shape. Like sills, the intrusive pressure of the magma is able to overcome the weight of the overlying strata (or the overburden pressure), which causes uplift and makes room for the magma. This suggests a limit in the depth of rock cover during placement, such that the lithostatic pressures can be overcome. Laccoliths have a flat floor and an arched roof. They can be thousands of meters (a mile or so) thick and several kilometers (a mile or so) in diameter. Typically they have porphyritic textures. Laccoliths were first described in the Henry Mountains of southern Utah, a portion of the Colorado Plateau.

Magma rises through the crust in several ways. It can work its way upward by stoping, the process of breaking off blocks of country rock and assimilating them into the melt. A partially digested piece of country rock in a pluton is called a xenolith. By giving off heat, magma can elevate temperatures in the adjacent rocks and make them flow plastically aside. It can also forcibly bow the rock mass upward, causing tension cracks to develop, which are then exploited by upward migration of the intrusion.

Igneous Rock Textures

Texture is a measure of the overall size, shape, and arrangement of mineral grains in a rock mass. It involves a description of the individual grains and the mutual relationship between them. The most common rock textures of igneous rocks are shown in Figure 3.8 on the following page.

Igneous rocks exhibit two types of texture: coarse-grained (or phaneritic) and fine-grained (or aphanitic). In the case of coarse-grained or phaneritic texture, individual grains are sufficiently large to be discerned by the naked eye (see Figure 3.8b). The lower limit of this size is about 0.05 mm in diameter. If individual grains cannot be discerned by the naked eye, the texture is called fine-grained or aphanitic. Plutonic rocks exhibit phaneritic texture because of a slow rate of cooling and volcanic rocks exhibit aphanitic texture because of rapid cooling. For rocks with phaneritic texture, identification of individual mineral constituents and rock classification are easier.

Extremely large grains, those greater than about 10 mm, deserve special attention because of their size (see Figure 3.8a). Such texture is called pegmatitic and the rock type is known as a pegmatite. The mineral composition is used to supply the modifying term. For example, when quartz and orthoclase are present, the rock is termed a granite pegmatite. Extremely coarse-grained pegmatites do occur in nature wherein orthoclase crystals 0.3 m (1 ft) in length can be found. Such deposits are formed from the last magmatic liquid, after most of the intrusive body has solidified. Pegmatites crystallize by slow cooling in the presence of volatiles and rare elements. Crystals are able to grow to considerable size without terminating against a fast-growing neighbor.

Fine-grained rocks with an aphanitic texture have a dull, massive appearance and show subdued conchoidal fracturing. As one would expect, different degrees of fineness occur with regard to these textures.

Aplitic textures, those that occur in the dike rocks known as aplites, are fine sugary textures with crystals just barely discernible by the naked eye. The grains are anhedral, showing a poor degree of crystal outline. Aplites are usually light in color and contain quartz and orthoclase. However, even when using a 10× hand lens, these mineral crystals are difficult to observe.

Basalts represent the typical aphanitic texture with a nearly equal, fine-grain size (see Figure 3.8c). These rocks look massive and have a rather dull appearance because they do not have any large cleavage faces to reflect the light. No individual grains can be discerned so specific mineral identification is not possible without using a petrographic (polarizing) microscope.

A rock with a texture composed of coarse-grained crystals in a fine or aphanitic groundmass is termed a porphyry (see Figure 3.8d). The coarse phaneritic crystals are called phenocrysts and their mineral composition can be determined as easily as can be coarse-grained constituents in a phaneritic rock. Porphyritic textures occur in extrusive lava flows, in dikes and hypabyssal rocks, or even in marginal portions of deep-seated plutons.

A number of other specific textural terms exist that are used by geologists to describe the textural spectrum of igneous rocks. One of particular interest is the diabasic texture. In this case, two minerals of nearly equal size, and usually in the finer range of the phaneritic designation, are intergrown to form a rather unique pattern. The minerals involved are

(a)

(b)

(c)

(d)

(e)

(f)

Figure 3.8 Different textures of igneous rocks: (a) very coarse-grained (Branko Jovanovic, Shutterstock), (b) coarse-grained or phaneritic (Bragin Alexey, Shutterstock), (c) fine-grained or aphanitic (Tyler Boyes, Shutterstock), (d) porphyritic (vvoe, Shutterstock), (e) pumiceous (Suttipon Yakham, Sutterstock), and (f) glassy (vvoe, Shutterstock).

plagioclase feldspar, in lath or rectangular shapes, and the equidimensional pyroxenes. The interlock is particularly strong, therefore the rock is highly durable and massive. Diabase commonly occurs in dikes and sills, as in the Palisades Sill along the Hudson River in New Jersey.

Rocks with glassy textures, for example obsidian, represent the extreme case of rapid cooling (see Figure 3.8f). These rocks are noncrystalline or amorphous. Conchoidal fracture is an obvious feature that accompanies the vitreous or glassy luster. Such rocks form on top of or within lava flows where rapid cooling is possible due to direct contact with the air.

Pyroclastic textures are those composed of clasts or pieces of material that are of volcanic origin and held together by some cohesive or cementing material. The clastic texture can be determined by the ease with which pieces of material are broken from the rock. However, there are also sedimentary rocks with a clastic texture. Mineral content and appearance must be used to distinguish between clastic extrusive rocks and clastic sedimentary rocks.

Some textures of volcanic rocks are due in part to the escape of gasses near the surface of the lava flow. This is why pumice, the frothy top of a lava flow, is full of air bubbles (see Figure 3.8e). Some basalts also contain these holes or vesicles and are called vesicular basalts and occur commonly in volcanic areas.

In some instances, at a time subsequent to the actual lava flow, the vesicles become filled with secondary minerals such as calcite or zeolite. These fillings are called amygdules and a basalt, so filled, is termed an amygdaloidal basalt.

Care should be taken to distinguish between amygdules and phenocrysts, both for academic and practical purposes. As previously noted, zeolites have a tendency to dissolve when in contact with water, which makes them undesirable constituents in an aggregate material. The shape of amygdules is helpful in distinguishing them from phenocrysts, because amygdules are usually rounded or misshapen, much like a compressed air void. Also, a preferred orientation or alignment of the cavity fillings, related to flow directions in the basalt, is commonly exhibited. Phenocrysts typically do not show such preferred alignments.

The relationship between cooling rate and texture should be apparent from the previous discussion. Simply stated, rocks that cool more slowly form larger crystals. Phaneritic rocks cool within the Earth, whereas aphanitic rocks occur as surface flows. Porphyries show a combined history of slow cooling to form the phenocrysts, followed by an accelerated phase that yields the fine groundmass. Glassy textures indicate the extremely high rate of cooling in which crystals are allowed no time to form. Therefore, the texture of an igneous rock is used to infer its origin or mode of formation.

Mineralogy of Igneous Rocks

Igneous rocks are composed primarily of silicate minerals with minor amounts of oxides, sulfides, and other mineral groups. Most silicates are relatively hard, with values on Mohs hardness scale ranging from 5 to 8. This is why igneous rocks as a group are hard and rather resistant to stress in the unweathered state.

It is risky to consider certain minerals as exclusively igneous, metamorphic, or sedimentary because many minerals can originate in different ways or persist from one rock group to another. Yet it is meaningful to recognize that certain minerals commonly occur primarily in igneous, metamorphic, or sedimentary rocks and, therefore, their presence may be indicative of a certain origin. More worthwhile, however, is the combined use of texture, mineralogy, and field relationships to ascertain the rock's origin.

The common silicate minerals in igneous rocks are feldspars (orthoclase and plagioclase), pyroxenes (including augite), amphiboles (including hornblende), quartz, olivine, and the micas (muscovite and biotite) (Best, 2002; Frost and Frost, 2013). Other accessory minerals are pyrite, feldspathoids, zeolites, and other less abundant silicates. These common minerals are often found in igneous rocks but are also present in certain metamorphic rocks. In some cases many will persist as clastic grains in sedimentary rocks, so the student is cautioned not to use constituent minerals exclusively to determine rock names.

Classification of Igneous Rocks

Igneous rock classifications are based on two criteria: texture and mineral composition (mineralogy). Rock mineralogy is sometimes difficult to determine, particularly in fine-grained or aphanitic rocks. Fortunately, a general relationship between color and mineral composition exists. Dark-colored minerals, that is, green, black, brown, and dark gray, are most commonly ferromagnesians. These minerals are low in silica and crystallize early as the magma cools. Calcic plagioclase is also dark in color, crystallizes early from the magma, and is associated with the ferromagnesian minerals. By contrast, the light-colored minerals, white, light gray, pink, and red, are high in silica and crystallize late in the magmatic process. Quartz, orthoclase, and sodic plagioclase are light-colored minerals with a high-silica content.

The classification of igneous rocks appears in chart form in Appendix A, Figure A.1. In this diagram, the igneous rock classification is presented graphically to enable a better understanding and easier identification of igneous rocks. The lower portion indicates the mineral composition of the rock in question. After this is determined, a vertical line is extended upward in the diagram. Based on the textural description, a horizontal line is extended from the right side of the diagram until it intersects the vertical line and a decision is reached on the appropriate rock name.

As discussed previously, texture is also indicative of origin. This relationship can be seen by comparing the left and right sides of the igneous rock classification chart (see Figure A.1). Equal-sized, coarse-grained texture is indicative of the slow cooling at depth for intrusive rock. Fine-grained texture suggests the opposite: rapid cooling at the Earth's surface; and a porphyry typically denotes two stages of cooling rate, first slowly at depth to form the phenocrysts and then rapidly at the surface to form the fine-grained groundmass.

Glassy texture is listed below the fine-grained entry in Figure A.1 and it denotes a rock mass devoid of crystal form. Cooling was so rapid that no nucleation of crystals could occur. In the glassy texture, color can be deceiving because only a minor amount of ferromagnesian minerals will yield a dark appearance. Such is the case for obsidian, which contains only a small amount of dark minerals to provide the black color, yet obsidian is a high-silica rock.

Pyroclastic textures are also listed on the rock classification chart. Pyroclastic rocks comprise an important portion of igneous rocks and are included to provide a complete picture. Pyroclastics are subdivided according to the grain size of their constituent pieces. In a more detailed rock description, the composition of the pyroclastic is added to the fine-grained term in the form of a prefix, yielding such terms as rhyolite tuff or andesite tuff. This degree of description is not typically pursued in a beginning course on applied geology.

Igneous Rock Descriptions

Granite is a coarse-grained igneous rock composed of orthoclase and quartz. Minor amounts of sodic plagioclase, hornblende, biotite, and muscovite may be present. *Syenite* is also a coarse-grained igneous rock similar to granite except that it contains no quartz. Orthoclase is the essential mineral for the rock. *Gabbro* is another coarse-grained igneous rock containing calcic plagioclase and olivine or pyroxene (for example, augite). It is typically dark gray or black in color.

Diorite is a coarse-grained igneous rock intermediate in composition between granite and gabbro. It is usually lighter in color than gabbro because of the presence of light-gray plagioclase (sodic plagioclase). The rock typically has a salt and pepper appearance.

Rhyolite is the fine-grained or aphanitic equivalent of granite; that is, it contains orthoclase and quartz. The aphanitic texture makes it impossible to identify individual mineral grains, so the overall color is used instead.

Trachyte is the fine-grained equivalent of syenite having an aphanitic texture and containing abundant orthoclase but no quartz. The term *felsite* refers to the fine-grained high-silica rocks including both rhyolite and trachyte.

Basalt, the fine-grained equivalent of gabbro, is dark gray to black in color and contains calcic plagioclase. As the most common extrusive igneous rock it is found in many areas of the world. Of slightly coarser texture than basalt, but having a similar mineral composition, is the rock diabase. It normally forms in a near-surface dike or sill. Diabase, basalt, and sometimes other dark fine-grained igneous rocks are designated by the term *trap* or *trap rock*.

Andesite is the fine-grained equivalent of diorite and is composed of sodic plagioclase, pyroxene, and usually amphibole. Andesites are lighter in color than basalts, showing a light gray to green color. Borderline cases between basalt and andesite are difficult to determine without the use of a petrographic microscope.

Ultrabasic rocks (or ultramafic) are those lower in silica than the basaltic composition. These rocks, which contain only minor amounts of calcic plagioclase, are almost entirely composed of olivine, pyroxene, or combinations of the two. *Dunite* is a fairly common ultrabasic rock that consists almost entirely of olivine. *Peridotite* consists of both olivine and pyroxene.

Obsidian is a high-silica glass that is dark in color and exhibits obvious conchoidal fracture. Its distinctive dark color is due to minor amounts of magnetite finely dispersed through the glass. Obsidian forms during extremely rapid cooling on the surface usually above or adjacent to a volcanic vent.

Pumice is the frothy, low-density rock that forms atop a lava flow because of the accumulation of gas bubbles at the surface. The small, numerous holes reduce the bulk specific gravity to a level even below that of water. Pumice is usually white to light gray in color with a fairly high silica content.

Scoria is a coarse, frothy rock of basalt composition and is the low-silica equivalent of pumice. It is usually red to black in color and forms on the surface of basalt flows.

Porphyries are named on the basis of the mineral composition and the relative percentage of phenocrysts versus groundmass. For example, a rock of granite composition is termed a granite porphyry if the phenocrysts make up more than 50% of the rock, and in like manner it is termed a rhyolite porphyry if the groundmass predominates.

Volcanic tuff is a pyroclastic rock composed of ash and dust derived from volcanic explosions. Tuffs include fragments less than 4 mm in size. They may be pieces of pumice, volcanic glass, or fragments of lava that were thrown into the air. Welded tuffs are pyroclastic rocks that become cemented or welded together by the incandescent heat of the deposits. Welded tuffs typically are difficult to tell from rhyolite or andesite.

Volcanic breccia is a pyroclastic rock composed primarily of fragments greater than 4 mm in size. Fragments of lava comprise most of the rock with lesser amounts of glass and pumice. Scoria may be present as well.

Appendix A provides a simplified mineral classification chart (Table A.1) and a mineral tabulation sheet (Table A.2) for igneous rock identification. When both tables are used together, the student can supply information that can indicate the rock name and information on origin.

Engineering Considerations of Igneous Rocks

Some of the engineering problems associated with igneous rocks are listed below:

1. The use of igneous rocks as aggregates in Portland cement concrete can cause problems. In some instances fine-grained siliceous rocks have caused volume expansion because of alkali-silica reaction. The alkali-silica reaction problem can be alleviated by using low-alkali cements or nonreactive aggregates, or by adding pozzolans to the concrete mix. Pozzolans are natural rock materials or fly ash from the smokestacks of coal-burning power plants, which contain finely ground silica. They yield a nonexpansive reaction product when united with the alkalies of the cement.

   The reactive igneous rocks include those that contain volcanic glass with a composition ranging from rhyolite through andesite. Basaltic glass contains too little silica to be reactive. Pyroclastic rocks containing glass with a high-silica composition also can be reactive. This includes most tuff, volcanic breccia, obsidian, and pumice.

   Other siliceous materials in addition to these igneous varieties have proven reactive with high-alkali cement. Opal, or $SiO_2 \cdot nH_2O$, is considered to be the most active mineral constituent in this regard. Chalcedony, a fibrous variety of SiO_2 that occurs as the common agate, may also be reactive.

   Some cherts have proven troublesome but many observers suggest this is due to the presence of opal, and supposedly nonopaline cherts would not be reactive. Some schists, shales, and other rocks have also caused expansion in connection with the alkali-silica problem.

2. Very coarse-grained igneous rocks are undesirable for use as aggregates for construction. With increasing grain size, abrasion resistance is reduced, and the rock is less suitable for use as a base course (road base), concrete aggregate, or source of riprap (large stone used for slope protection along rivers and seacoasts).

3. The presence of certain minerals in igneous rocks makes the rock undesirable for some engineering uses. Zeolite minerals, which are relatively soluble in water, are undesirable in aggregates that will be exposed to the weathering process.

4. In foundations for engineering structures such as dams, bridge piers, and underground installations, weathered igneous rock or any other weathered rock is to be avoided. Excavation must extend through this material into sound rock.

5. Dimension stone includes rock used for tombstones and monuments plus facing stone for buildings. Igneous rocks are commonly used for this purpose because of their resistance to weathering. Rock that undergoes staining, fracturing, and spalling of the surface must be avoided when selecting the proper building stone. Strong, fresh, and unaltered igneous rocks yield the most suitable materials. Common, unweathered, and unaltered phaneritic (coarse-grained) varieties are selected for building stone. These range from granite to gabbro.

EXERCISES ON IGNEOUS ROCKS

1. What is the difference between a rock and a mineral? Are all rocks composed only of minerals? (*Hint:* Consider volcanic glass and coal.)

2. Why are aphanitic igneous rocks more difficult to identify than are phaneritic ones? How is identification accomplished?

3. How does granite differ from diorite? How does basalt differ from obsidian? How does rhyolite differ from tuff?

4. Dikes can be aphanitic, porphyritic, or phaneritic. How is this possible? Explain. Dikes are typically steeply dipping (from the horizontal). How do you think dikes gain access through the existing rock? Explain.

5. Volcanic rocks can vary enormously in strength and in their other physical properties. Relate this to the two different varieties, lava rocks and pyroclastics. Which would be stronger and why?

6. Intrusive rocks are coarse-grained. Why are those of finer crystal size typically stronger than those with larger crystals? Consider rocks of different composition and then of the same composition. Explain.

7. What would be the maximum unit weight for a piece of pumice that floats in water? If the specific gravity of the material that comprises the pumice is 2.65, what is the porosity of the rock? Recall that the unit weight of water = 1000 kg/m^3 (62.4 lb/ft^3) and porosity = volume of the voids/total volume.

8. Volcanic ash deposits sometimes contain expansive clays such as smectite (montmorillonite) and commonly these deposits are water bearing. What engineering problems can be anticipated for such materials?

9. What are the properties of igneous rocks that make them good sources for gravestone monuments and facing stone for buildings? Why would dunite and zeolite-rich basalt not make good facing stone? (*Note:* olivine weathers rapidly in a humid climate.)

10. Refer to the USGS website for discussion on the Mount Saint Helens eruption (https://volcanoes.usgs.gov/index.html). What mass of material was exploded from the mountain on May 18, 1980? How much is this in terms of cubic yards, cubic feet, and cubic meters? What are the individual hazards that developed as a consequence of this eruption?

11. Mount Rainier, Mount Hood, and Mount Lassen are other volcanic cones in the Cascade Range. What cities are they near? What major geologic hazards could they cause? Explain.

12. Based on discussions in the previous chapters, how are flood basalts probably related to plate tectonics?

References

Best, M. G. 2002. *Igneous and Metamorphic Petrology* (2nd ed.). New York: Wiley-Blackwell.

Frost, B. R., and Frost, C. D. 2013. *Essentials of Igneous and Metamorphic Petrology*. New York: Cambridge University Press.

Gill, R. 2010. *Igneous Rocks and Processes: A Practical Guide*. New York: Wiley-Blackwell.

Marshak, S. 2013. *Essentials of Geology*. New York: W. W. Norton.

Nance, D., and Murphy, B. 2016. *Physical Geology Today*. New York: Oxford University Press.

Philpotts, A., and Ague, J. 2009. *Principles of Igneous and Metamorphic Petrology* (2nd ed.). New York: Cambridge University Press.

Winter, J. D. 2009. *Principles of Igneous and Metamorphic Petrology* (2nd ed.). New York: Pearson.

ROCK WEATHERING AND SOILS

Chapter Outline

Rock weathering is the mechanical or chemical breakdown of solid rock (bedrock) in response to the atmosphere, water, and organic matter. Most rocks form under conditions drastically different from those at the Earth's surface. Minerals in igneous and metamorphic rocks reach equilibrium at elevated temperatures and generally must adjust to high confining pressures. Sedimentary rocks are compressed by overburden stresses at temperatures that increase with depth. This compression yields adjustments in the clay constituents, overgrowths on certain crystals, and deposition of cementing minerals between grains. By contrast, in the near-surface environment, low temperatures prevail, there is little or no confining pressure, and water, oxygen, and organic matter are abundant. Consequently, rocks at the surface undergo both physical and chemical changes.

Mechanical weathering involves the physical disintegration or degradation of rock pieces without a change in composition. It is simply the size of the particles that is reduced. This may occur by the disaggregation of separate crystals in a rock or by cross-fracturing of massive, fine-grained rocks.

Chemical weathering occurs by decomposition whereby one mineral species changes into another through various chemical processes. Water plays a major role in chemical weathering.

Factors Controlling Weathering

Role of Water

The amount of water available to the rock is the primary agent controlling the type of weathering. If abundant water, in unfrozen form, is available, chemical weathering will prevail. In the absence of water, mechanical weathering takes control.

Water availability can be related to climate (humid versus arid), elevation (high altitudes yield ice formation), and latitude (high latitudes also yield water in the frozen state). This is why mechanical weathering prevails in the deserts and rapid chemical weathering in the tropics. In mountainous regions, blocks of broken rock line the slopes, and the headward areas of alpine glaciers persistently supply rocks through mechanical breakage. The high latitudes yield colder weather and less liquid water; thus, mechanical weathering prevails.

Water's role in chemical weathering is that of universal solvent. Most chemical reactions on the Earth's surface take place in water because it provides oxygen for reaction and mobility for the ions. With water present, an increased temperature typically yields an increased rate of reaction. This explains the high chemical weathering rate of the tropics. Without water, elevated temperatures do not have this effect. Therefore, mechanical weathering prevails in desert areas.

Topographic Expression

When mechanical weathering prevails, as in arid regions, sharp angular topography develops. In sedimentary terrains, sandstones and limestones (including dolomites) form cliffs. Because water is not abundant, the carbonate rocks are not subject to solution but persist as resistant rocks. The Grand Canyon in Arizona illustrates dramatically the angular forms that develop in an arid region (Figure 4.1).

By contrast, when chemical weathering prevails, rounded topography develops. As in dry areas, sandstones are cliff formers in wet regions, but limestones, dolomites, and shales form slopes. The Appalachian Plateau is a region of horizontal sedimentary rocks that have weathered chemically. The outcrops are more rounded and soil accumulation is deeper than in the arid regions. Thick vegetation, another consequence of humid conditions, blankets the slopes. In these areas the valleys are typically bottomed by shale and limestone, and the ridges are capped with sandstone.

Processes of Mechanical Weathering

Several processes act to reduce the size of massive rock. The primary processes include differential expansion and contraction of rock constituents, wedging action of water freezing in rock fractures, and release of confining stress at the surface.

Differential Expansion and Contraction

Differential expansion and contraction develop between the individual minerals in a coarse-grained rock when heating and cooling occurs. The coefficient of thermal expansion is not the same for all minerals so, when rocks are heated and cooled, the mineral grains are subjected to differential stresses. In some cases this will induce and propagate fractures in rocks whereas in others, individual mineral grains are disaggregated from the mass. Known as granular disintegration, this process is particularly prevalent in coarse-grained, quartz-rich rocks. Quartz has a larger coefficient of expansion than other common minerals and is highly directional, with the greatest expansion occurring along the axis of elongation of mineral crystals (the c axis). Consequently, intrusive igneous bodies in desert environments tend to weather by granular disintegration, yielding a volume of loose, coarse-sand-sized crystals of quartz, feldspar, and hornblende. These deposits are known as grus.

Frost Wedging

The wedging action of water freezing in rock fractures, known as frost wedging, will propagate cracks through rock because of the expansion that occurs when ice crystals form. Each freezing and thawing cycle can extend the fracture until a rock piece is broken from the mass. Most areas of the world experience periods during the year when frost wedging can operate to disintegrate rock.

Frost wedging is caused by a 9% increase in volume that occurs when water freezes to form ice. The force generated by each freezing cycle is about 10.7 MPa (1550 psi), which represents 5 to 50% of the unconfined compressive strength of most rocks. This stress is exerted repeatedly with each freeze-thaw cycle until the rock is broken apart.

The conditions required to cause extensive frost wedging effects are (1) a supply of moisture; (2) fractures, bedding planes, or other openings in the rock for water to enter; and (3) temperature fluctuations across the freezing point of water. Frost wedging is particularly effective in alpine areas where snowmelt forms during the day and refreezes at night after seeping into rock fractures. It is less prevalent where water is permanently frozen or in desert areas where water is scarce.

Figure 4.1   Angular topography of an arid region in the Grand Canyon. (sumikophoto, Shutterstock.)

Piles of angular rock pieces, known as talus, accumulate at the base of steep cliffs as a product of frost wedging. Finer-sized materials downslope from the talus slopes yield half-circle-shaped features called alluvial fans. They form by surface runoff, which carries the finer sediment from the steep cliffs. These deposits are discussed in more detail in Chapter 15 on landslides and related phenomena.

Sheeting and Exfoliation

Release of confining stresses in a rock mass exposed at the surface allows the rock to expand elastically. This occurs in all rock masses and it is the reason why joints in rocks open when the rock is exposed, or conversely why joints tend to tighten at depth. Exfoliation is the term used for the action that takes place when rocks break loose from the surface along a roughly parallel fracture. Massive igneous and metamorphic rocks with high residual stresses are particularly prone to this occurrence. For dimension-stone quarries in granite, this is known as sheeting and special care is taken to prevent its occurrence if possible. In such quarries, large blocks of rock are processed for facing buildings or for gravestones, and sheeting can ruin valuable stone. Quarries may be developed in preferred directions designed to minimize the sudden stress release, or the rock may be kept in compression by long rock anchors that prevent stress release.

Exfoliation is common in massive igneous rock bodies such as stocks and batholiths. It can be observed in Yosemite National Park, California, which is located within the Sierra Nevada Batholith. Half Dome is an exfoliation form that occurs on El Capitan, the prominent wall of rock exposed by alpine glaciation in Yosemite. Exfoliation preserves the vertical face of rock first carved by glacial ice moving down the valley (Figure 4.2).

Spheroidal Weathering

Spheroidally weathered boulders are rounded, large blocks of rock that form in place upon weathering of coarse-grained igneous rocks, especially granite (Figure 4.3). They result from a combination of pressure relief, frost wedging, and expansion of feldspar crystals when weathered to kaolinite. Joints and exfoliation planes form the initial boundaries in the rock mass, then pressure relief and chemical weathering along the surface cause an inward-developing skin of weathered rock to form.

Differential Weathering

Differential weathering is the term used for varying rates of weathering in an area where some rocks are more resistant to weathering than other rock (Plummer et al., 2005). The less resistant rocks weather more rapidly than the resistant rocks, resulting in undercutting of the resistant rock units by the weaker rock units (Figure 4.4).

Other Mechanical Weathering Processes

Several other mechanical weathering processes act, to a lesser extent, to disintegrate rocks. Erosion by wind and running water mechanically

Figure 4.2   Exfoliation of massive rock (Half Dome) in Yosemite National Park. (Kristoff Melier, Shutterstock.)

Figure 4.3   An example of spheroidal weathering. (Photo by Google Maps.)

abrades rock masses but these agents work primarily to move soil or sediments that have already been disaggregated. Root wedging by plants can contribute to mechanical weathering. This activity is observed in some of the most massive rock types such as granite, quartzite, and sandstone where roots of pines and other trees penetrate what seems to be solid rock to obtain support and sustenance.

Forest fires can cause mechanical breakdown of rocks because of the great expansion that occurs at elevated temperatures. From the time of early humans until the sixteenth century, fire and water-quenching were used as a mining procedure to break rocks loose and expose valuable minerals; only after the advent of explosives was this procedure discontinued.

Figure 4.4   An example of differential weathering.

Processes of Chemical Weathering

Mechanical weathering or disintegration, in a sense, aids chemical weathering because it exposes additional surfaces to air and water. This additional exposure is effective, of course, only if liquid water is available to foster the chemical reaction.

Surface Area Effects

Chemical weathering is, basically, a surface phenomenon; therefore, the greater the surface area exposed, the more intense the reaction. Increased surface area results when large blocks are subdivided into smaller ones. The total volume remains the same but the surface area increases logarithmically. A measure of this feature is the specific surface or area to volume ratio, and it also increases logarithmically as the blocks are subdivided.

The relationship between length of sides for cubes and their surface areas is shown in Figure 4.5. From the graph it can be determined that a 1-cm cube, when divided into eight smaller cubes, each in turn subdivided repeatedly for a total of 13 times, will yield cubes 1 μm on a side (10^{-6} m). This provides about 1 trillion (10^{12}) cubes with a total surface area of 5 m^2 or about 10,000 times (10^4) that of the original. It is no surprise, then, based on this example that the subdivision of rock pieces by mechanical means greatly accelerates chemical weathering.

Solution, Oxidation, Hydrolysis, and Effects of Plants

Water is a polar liquid that has the ability, to a certain degree, to dissolve both ionic and organic substances. The water molecule is illustrated in Figure 4.6. Because of its dipolar nature, water is able to dissolve many chemical compounds. In addition to the solution effect, water aids decomposition through acid action, oxidation, and hydrolysis. Acids in the soil are rather weak. Carbonic acid forms when carbon dioxide in the air reacts with rainwater in the following manner:

$$H_2O + CO_2 \rightarrow H_2CO_3 \rightleftarrows H^+ + HCO_3^- \quad \text{(Eq. 4-1)}$$

As water percolates downward through the soil, concentrations of the acid are greatly increased when carbon dioxide is provided by decaying organic matter. Hydrogen ions (H^+) are extremely effective in decomposing minerals. They begin by neutralizing the broken bonds at the edge of the mineral. This is illustrated by the decomposition of potassium feldspar (orthoclase):

$$\underset{\text{(Potassium feldspar)}}{2KAlSi_3O_8} + \underset{\text{(Hydrogen ion)}}{2H^+} + \underset{\text{(Water)}}{H_2O} \rightarrow$$

$$\underset{\text{(Potassium ion)}}{2K^+} + \underset{\text{(Kaolinite)}}{Al_2Si_2O_5(OH)_4} + \underset{\text{(Silica)}}{4SiO_2}$$

$$\text{(Eq. 4-2)}$$

In this chemical equation the reaction between a mineral and water to yield a secondary mineral is known as hydrolysis.

Oxidation of sulfides, native metals, and ferrous oxides occurs when rocks weather. Oxygen is dissolved in the water, which supplies the needed mobility for the reactants.

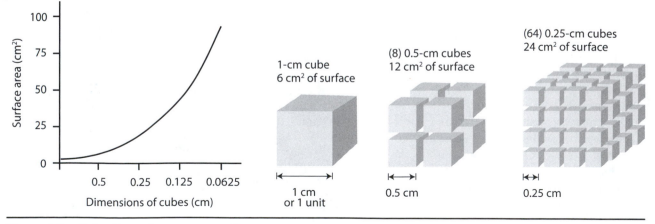

Figure 4.5   Cube dimensions versus surface area.

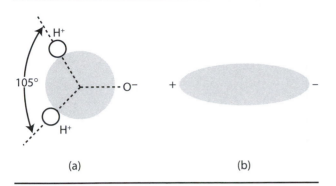

Figure 4.6   (a) Diagram of a water molecule and (b) shown as a dipole.

Soluble materials, including some dissolved silica, are carried away by subsurface water. These include potassium, sodium, magnesium, and calcium cations plus chloride, carbonate, and sulfate anions. Iron and aluminum oxides, clay minerals, and some silica are left behind. This process, known as leaching, involves the removal of soluble materials by water from the bedrock or weathered soil.

The contribution of plants to rock weathering bears further consideration. Roots contribute to mechanical weathering by wedging through fractures in the rock. Chemical weathering is promoted by plants through the presence of hydrogen ions around their roots when they are alive. Following this, on decay of the organic matter, more hydrogen ions are supplied for chemical weathering.

Resistance to Weathering

Some minerals are more susceptible to chemical weathering than others. This feature is related to their temperature of formation, which has a direct influence on the mineral structures that develop on cooling. Silicate minerals, ranging from a high temperature of formation to one of low temperature, portray an increased sharing of oxygens with decreasing temperature of formation. This greater oxygen sharing provides those silicates formed at lower temperatures with a greater resistance to chemical weathering. Figure 4.7 on the following page compares Goldich's mineral stability series with Bowen's reaction series, showing the order of formation of silicate minerals from a silicate melt. The similarities are obvious and illustrate the greater resistance of orthoclase, muscovite, quartz, and clay minerals to chemical attack.

In addition to quartz, other less common minerals are resistant to weathering. Some precious metals (such as gold and platinum) plus ilmenite (a titanium ore), cassiterite (a tin ore), zircon, and diamond persist after rock weathering. The metallic ores also concentrate by differences in particle settling velocity during stream flow, because of their higher density, to yield what are known as placer deposits. Subsequently, these can be mined with great success. In fact, many of the gold strikes of Colorado, Montana, California, and Alaska occurred as placers.

The effects of weathering on different rocks and the rates involved are of particular interest. The Goldich (1938) mineral stability series provides insight into the relative rates of chemical weathering. For all practical purposes, only minor amounts of quartz are decomposed in humid climates, but most other common silicates are ultimately destroyed. In like manner, limestone and dolomite (plus the more soluble rock gypsum and rock salt) are dissolved by the action of groundwater. Only impurities, consisting of clay minerals, some silica, and iron oxide, are left behind. This oxide, in the ferric state, yields the orange to brick red color of terra rosa that is so common in soils derived from limestone.

Effects of Climate

In arid climates, weathering results from mechanical processes that dominate in the absence of water. Rocks simply become smaller in size or disaggregate into their constituent minerals. This occurs in deserts and in high altitudes and high latitudes where liquid water is not generally available.

In humid tropical areas chemical weathering is even more pronounced. One consequence is the increased pH of the soil and of soil water. Under these conditions, silica becomes much more soluble than in other humid areas and, consequently, quartz will be removed from the soil profile during weathering. The residual soil, laterite, is red in color and rich in iron and aluminum oxides. It is soft and earthy when formed but becomes extremely hard on exposure. A crust-like layer, 4.5 m (15 ft) or more thick, may develop in places.

Bauxite is another form of deeply weathered soil. It is common to the tropics and subtropics. Bauxite contains mostly hydrous aluminum oxide with little iron oxide or silica. Apparently bauxite is derived from a parent rock with a low iron content because

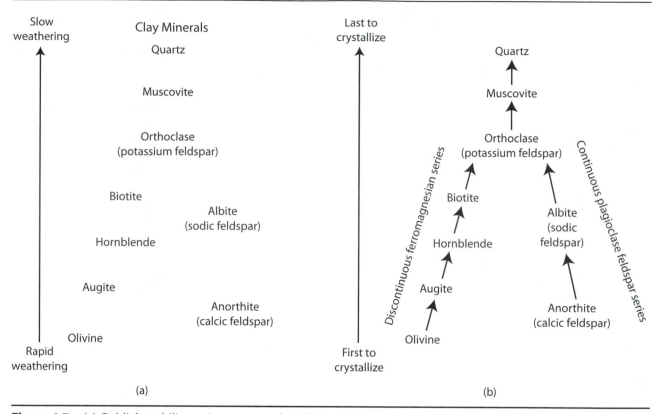

Figure 4.7   (a) Goldich stability series compared to (b) Bowen's reaction series.

iron is absent in the residuum. In view of the low mobility of iron oxide when present, it would persist in the soil if contained in the parent rock. A feldspar-rich granite is probably the original rock from which the silica is removed under tropical weathering conditions to yield bauxite. Bauxite ($Al_2O_3 \cdot nH_2O$) is the primary ore of aluminum.

With this background information on chemical weathering, predictions can be made concerning the end products of weathering for specific rock types under different climatic conditions. The mineral constituents of the rock provide the needed information for this analysis. Examples that depict the various possibilities are shown in Table 4.1.

Classification of Weathering Grade

Weathering leads to a general disintegration of rock through changes in mineral composition, increase in void space, and weakening of interparticle bonds (Ebuk et al., 1993). The degree of disintegration depends on the original composition and texture of the rock, as well as the processes and rates

of weathering, Depending upon the degree of weathering, the engineering properties of weathered rock can be significantly different compared to fresh rock. Therefore, the engineering description of a rock must include a description of its degree or grade of weathering in both qualitative and quantitative terms.

During the past five decades, extensive research has been conducted on the effect of weathering on the engineering properties of different rocks and quantifying the grades of weathering in terms of deterioration of engineering properties (Dearman and Irfan, 1978; Martin, 1986; Martin and Hencher, 1986; Geological Society of London, 1995; Gupta and Rao, 2000; Aydin and Duzgoren-Aydin, 2002; Arikan et al., 2007; Ceryan et al., 2008; Basu et al., 2009; Chiu and Ng, 2014). The literature review indicates that the commonly used properties include water absorption, density, sonic velocity, unconfined compressive strength, Schmidt hammer rebound number (an empirical method for measuring compressive strength), and point load strength index (also an empirical method for measuring compressive strength). Most classifications are based on a

Table 4.1   Weathering Details for Rocks Under Various Climatic Conditions.

Rock Type	Minerals Present	Weathering Conditions	Products
Granite	Quartz, orthoclase	Humid	Quartz pieces, clay, oxides
Gabbro	Calcic plagioclase, olivine	Humid	Iron oxide and clay
Quartz sandstone	Quartz	Humid	Quartz pieces
Granite	Quartz, orthoclase	Arid	Pieces of quartz and orthoclase
Cherty limestone	Calcite, chert, impurities	Humid	Chert, iron oxide, clay
Granite	Quartz, orthoclase	Humid, tropical	Clay, oxides
Muscovite schist	Muscovite, hornblende	Humid	Muscovite pieces, clay, oxides

comparison of the above-stated properties for fresh and weathered rock.

Based on visual observations, the Geological Society of London (1995) proposed six grades of weathering: I—fresh rock; II—slightly weathered; III—moderately weathered; IV—highly weathered; V—completely weathered; and VI—residual soil (Table 4.2). Other researchers propose five grades, not including residual soil. Engineering properties have been related to these grades for different rock types. Because of the variability of these properties for different rocks, it is not possible to give a range of these properties for different grades for all rock types. For important, large-scale projects, site-specific studies should be performed to determine the weathering grades and the corresponding engineering properties.

Table 4.2   Classification of Weathering Grades.

Grade	Term	Typical Characteristics
I	Fresh rock	Unchanged from original state.
II	Slightly weathered	Slight discoloration, slight weakening.
III	Moderately weathered	Considerably weakened, penetrative discoloration, large pieces cannot be broken by hand.
IV	Highly weathered	Large pieces can be broken by hand, does not readily disaggregate (slake) when dry sample is immersed in water.
V	Completely weathered	Considerably weakened, slakes, original texture apparent.
VI	Residual soil	Soil derived by in-situ weathering but retains none of the original texture or fabric.

Source: Geological Society of London, 1995.

Soil Profiles

One consequence of weathering is the formation of the soil profile, the material so vital to human existence. It is certainly one of the primary resources of any country and has much to do with the agricultural base that can be established.

Definitions of Soil

Engineering Definition

The term soil is used in several technical fields, each with a different definition. In civil engineering, soil is the earth material that can be disaggregated in water by gentle agitation, whereas from a construction viewpoint it is material that can be removed by conventional means, that is, without blasting. Both definitions have a practical orientation because the behavior of the material under a specific condition determines how it is classified. These definitions compare favorably to the geologic term regolith, which includes all the loose material lying above bedrock. In the engineering geology specialty, soil and bedrock are commonly considered as construction or foundation materials and the practical definitions for soil, provided above, are used.

Agronomy Definition

In soil science, agronomy, and agriculture, soil consists of the thin upper layers of the Earth's crust, formed by surface weathering, that are able to support plant life. Many geologists adopt this definition as well.

Bedrock

In an engineering sense, material lying below the soil is bedrock. It can be defined more completely as the solid, continuous mass of rock lying below the soil or exposed at the Earth's surface that must be removed by blasting. Other terms for soil and bedrock in the engineering sense are unconsolidated material (or overburden) and consolidated material, respectively. The meanings contrast loose material versus a lithified rock mass because unconsolidated in this sense signifies nonlithified material.

Soil versus Soil Profile

Differences in the meaning of soil between agronomy and engineering can be resolved by referring to the thin upper, weathered layers as the soil profile and the total nonlithified material as soil. The soil profile consists of three major horizons: A, B, and C. These horizons are sometimes referred to as zones, which more accurately indicates their nature. The A horizon, commonly rich in organic matter, is the topsoil or the zone of leaching from which downward percolating water has removed some clays and soluble ions. The B horizon is the subsoil or zone of accumulation. Clays are more prevalent here and organic matter is less abundant. The C horizon marks the transition from the soil profile to the unweathered parent material below.

Figure 4.8 shows an idealized cross section of a soil profile developed on glacial till. Glacial till is the unsorted debris deposited directly by a melting glacier without being reworked by water. The figure further illustrates the distinction between soil profile and soil or the difference between agronomy and engineering definitions. Soil, in the engineering sense, extends for 60 m (200 ft), the entire depth to the sandstone bedrock. By contrast, soil only pertains to the upper 0.6 m (2 ft) or so for the agronomy cross section.

Considering the unconsolidated material as strictly soil material is not without its problems. Take, for example, the glacial till in Figure 4.8. When drilling and sampling the till in a subsurface investigation for a building foundation, the origin or geologic nature of the material (such as "most likely Wisconsinan till") should be noted on the drilling log as well as its soil description ("silty clay with traces of pebbles"). This geologic information should prove useful in making correlations across the site and decisions considering the nature of the soil mass.

Factors Controlling
Soil Profile Development

The soil profile is an end result of surface weathering brought about primarily by downward percolation

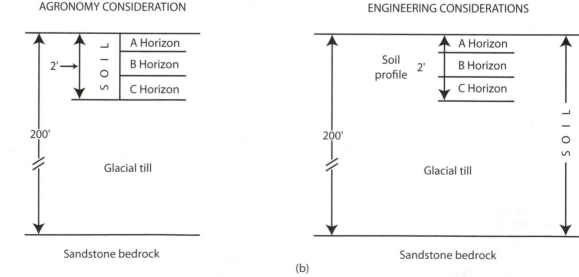

Figure 4.8   Illustrations of the term soil based on (a) agronomy and (b) engineering considerations.

of water. As previously stated, the A horizon is the zone of leaching where organic material may be abundant and the B horizon is the zone of accumulation, mostly of clays left behind after water migration. Figure 4.9 shows typical soil profiles developed on glacial till and on granite bedrock. Saprolite consists of a combination of the granite and other coarse-grained igneous and metamorphic rocks plus residual soil.

The most significant factors controlling soil profile development are parent material, climate, time, topography, and vegetation. Parent material is the original substance from which the soil zones develop. It controls, at least in the early stages of formation, much of the specific details involved with the soil layers such as permeability, acidity, and chemical composition. The two primary subdivisions for parent material are (1) bedrock weathered in place and (2) transported regolith or unconsolidated material, in other words, transported soil in the engineering sense.

Parent Material

Weathered Bedrock

The bedrock parent material can consist of a variety of rocks in sedimentary, igneous, and metamorphic terrains. Shale, limestone, and sandstone dominate in the sedimentary areas, whereas many igneous rock domains are granite to granodiorite in composition, representing intrusive bodies, or basalts and andesites for the extrusive situations. In metamorphic rock areas, gneiss, schist, and phyllite are common parent materials. Each rock type weathers by chemical and/or mechanical means, depending on climatic conditions, to yield the soil. In most cases the overall mineralogy of the rock is sufficient to provide a range of chemical constituents in the soil profile. In some, however, the range is quite limited, for example, in quartz-rich rocks such as quartz sandstone, quartzite, and some siltstones. The residual soils derived from these rocks may be lacking

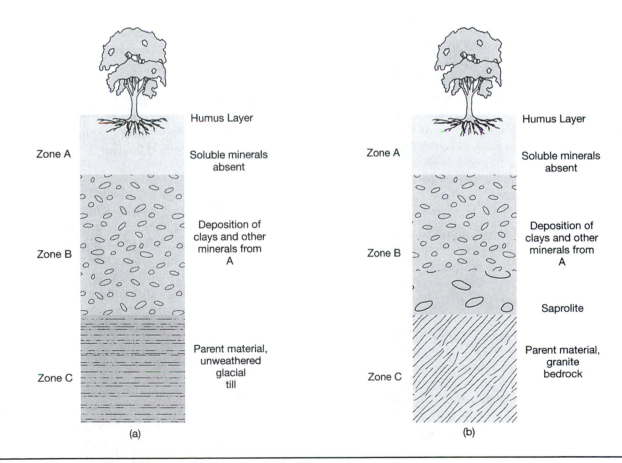

Figure 4.9   Typical soil profiles developed on (a) glacial till and (b) granite bedrock.

in chemical diversity, which reduces their ability to support vigorous plant growth.

Parent material seems to be the primary factor controlling the type of clay prevailing in a soil profile. The clay mineral prevalent in the parent rock is also dominant in the upper soil horizons. Illite appears to dominate in marine shales, whereas kaolinite and illite are prominent in freshwater sediments. Montmorillonite (smectite) is abundant only in the relatively recent sediments, typically Cenozoic in age, and is particularly common in pyroclastic rocks, that is, volcanic ejecta.

Transported Soils

Transported soil is soil that has been carried from one area and deposited in another. It consists of a variety of particle sizes, ranging from clay to gravel size and in some cases even boulders. The transporting agents are those that carve the land surface in the erosive processes. Running water, wind, glacial ice, and gravity are the most important agents. Littoral drift along seacoasts or major lakeshores is another agent. Glacial ice provided the thick blanket of deposits by continental glaciation in the north-central section of the United States as well as the smaller scale deposits in mountainous regions of the western states. Gravity supplies the landslide and colluvial deposits that lie adjacent to most slopes, and the wind yields mostly sand dunes and loess deposits. Water, of course, provides the vast deposits in stream beds and floodplains but also the slope wash that tops most sloping areas.

Climate

Climate is extremely important in soil profile development. The relative abundance of water in liquid form is dictated by the climate and it is this water, moving downward from the surface, that develops the anisotropic or nonisotropic layers of the soil profile. An isotropic substance is one that has the same properties in all directions, which is in direct contrast to conditions in a soil profile with its obviously layered nature. The annual rainfall and annual temperature range determine whether an area is termed hot or cold, humid or arid. The evaporation rate is determined by the climatic conditions, which also determine the local vegetation.

Effect of Climate on Clay Minerals

Climate influences the formation and stability of clay minerals in a soil profile. Although illite is widely distributed, kaolinite seems restricted to areas of high rainfall. To a lesser degree, montmorillonite is associated with soils of a dry climate. Kaolinite development seems to be favored in acid or low pH soils where good drainage and active leaching prevail. The K, Na, Mg, and particularly Ca ions must be removed quickly from the soil as soon as they weather from the parent material if kaolinite is to prevail. These are mobile ions that are easily removed by water, so high rainfall is conducive to kaolinite formation. Formation of illite seems favored by the presence of potassium and to a lesser degree sodium. Montmorillonite develops in the presence of sodium and potassium but magnesium seems to be an essential ingredient.

Effect of Climate on Types of Soil Profiles

The prevailing climate in a region significantly influences the physical, chemical, and biological processes that act on the soil. In warm, wet climates organic activity is high and frost action is absent. In dry regions the dissolving and removal of carbonates, normally quite soluble materials when water is present, are greatly inhibited and instead carbonate precipitation occurs within the B horizon. Different major soil groups form under contrasting climatic conditions. The three most important groups are pedalfers, pedocals, and laterites.

Pedalfers

A pedalfer is a soil in which much clay and iron have been added to the B horizon. It occurs mostly in the eastern half of the United States where the annual precipitation is 63 cm (25 in) or greater. This is a humid region with a moderate to cold climate. In pedalfers the soluble carbonates have been removed and aluminum, iron, and colloidal clay have been carried downward into the B horizon. Podsols are pedalfer soils that have been intensely leached by humic acid. The upper part of the B horizon has an ash gray color in podsols, in contrast to their dark humus-rich A horizon.

In addition to podsols, there are several other varieties of the pedalfer group. Included are the red and yellow soils of the southeastern United States and the gray-brown podsolic soils of the northeastern quarter of the United States and southeastern Canada. This latter group formed in relatively cold regions under heavy forest cover where humus accumulates in the soil. Prairie soils are transitional between the pedalfers in the east and the pedocals in the west.

Pedocals

Pedocals are soils that contain an accumulation of calcium carbonate. These soils are found where the temperature is relatively high, the rainfall is slight, and the vegetation is mostly grass or brush. The pedocals typically occur in the western half of the United States where the annual precipitation is less than 63 cm (25 in). Pedocals contain an accumulation of calcium carbonate in the B horizon. This may range from a few scattered concretions to a solid layer of $CaCO_3$ known as caliche. The depth below the ground surface to the caliche layer is a function of precipitation: Caliche lies deeper in high rainfall areas and nearer the surface in drier ones.

Soils containing abundant calcium carbonate are alkaline in nature, which makes them conducive to the growth of grass instead of trees. Moist, flat areas are typical sites for grassland prairies. A mat of accumulated humus in the A horizon develops as a consequence of the continued grass cover, yielding a dark brown upper zone and a lighter B zone. This feature, typical of prairie soils or chernozems, is in contrast to forest soils, which have a lighter colored A horizon.

Desert areas lack the moisture needed to maintain a grass cover and humus is essentially absent in the A horizon. Therefore, the soil cover is determined by the color of the minerals that persist. Ferric iron oxide, stable where the humic acids of vegetation are absent, yields a red color but the other mineral colors are light in color, including caliche. Typically, desert soils are gray or tan but the red or brown influence of iron oxide is sometimes evident.

Laterites

Laterites, the third group of soil types, were considered briefly in the discussion on the effects of climate. These are the deep red soils of the tropics in which all silicates, including silica, have completely weathered away, leaving mostly aluminum and iron oxides. If little iron was present in the parent rock, for example a high-silica igneous rock such as granite, then bauxite is formed. Tropical soils are not very fertile materials for crop production because the small amounts of humus in these soils are soon used up

by crop growth. Elevated temperatures in the humid climate encourage higher chemical and biological activity, which quickly oxidizes the organic matter. For this reason, many lateritic soils can be used for only a few years before they become barren and must be abandoned. This explains the procedure of clearing by burning and then farming for one or two years before abandoning the land, as is commonly practiced in some of the tropical, developing nations.

Time

Time is a factor in soil development and is most significant for soils developed within the past few 100,000 years. For bedrock terrains exposed throughout the Pleistocene (2 million years or so), the time factor plays a lesser role. In the glacial terrain of the north-central United States, whether the deposit was formed during the Wisconsinan age (ice retreated 8000 to 14,000 years ago) or during the Illinoian or pre-Illinoian age (about 100,000 to 1 million years old) is significant. Soils of central Indiana (Wisconsinan age) can be contrasted with those of central Illinois (Illinoian age). For instance, the Wisconsinan age material has less accumulation of clays in the B horizon and the soil profile in general is not as thick as that in central Illinois. This smaller amount of clay makes the younger soil more permeable and in general helps to promote better soil drainage and more fertility. The nature of the parent material (glacial till) is virtually the same for the two areas so it is the time factor that sets apart the soil profile characteristics. In older glacial deposits, Kansan, for example, which have long been exposed at the surface (northern Missouri area), erosion has prevailed to a major extent and the flat, youthful topography of younger till plains and their highly productive farmland no longer prevail. This erosion yields rolling hills cut by headward eroding tributary streams.

Topography and Vegetation

The concern about erosion provides a link to the next factor in our study of soil profile development, topography. Steep slopes encourage greater runoff and erosion. Consequently, the soil profile thins near the hilltops and thickens in the low lying areas between them. Gravity moves the loose soil to the lower areas, mainly by soil creep, so organic accumulation is greater in these depressions as well. An example of this is provided by the relatively flat, youthful terrain that is common to Wisconsinan age till plains of the north-central United States. They appear lighter in the high areas because of less moisture and organics, and darker in the low portions.

Vegetation controls the extent of runoff from the land surface and, therefore, the amount of erosion. When vegetation is present, soil remains in place, held by the roots of trees and grasses. Farmland, particularly that planted in rows, is subject to surface erosion following plowing and planting and prior to substantial growth of the crop. To a large extent, natural vegetation is dependent on climate and to a lesser degree on soil parent material. In modern agricultural practice, natural vegetation is replaced by crops, pasture, or other vegetation which, in many cases, makes the land more subject to erosion.

Classification of Soil Particle Size and Texture

Particle size and size distribution are important characteristics that affect soil behavior. Various agencies and professional organizations have devised classifications that provide subdivisions that are the most useful for specific applications. Figure 4.10 shows several of the most common classifications based on grain size ranges: the Wentworth scale, the US Department of Agriculture (USDA), the American Society for Testing and Materials (ASTM), the American Association of State Highway and Transportation Officials (AASHTO), and the Unified Soil Classification System (USCS).

The Wentworth scale is used widely by geologists, USDA classification is used by agronomists, AASHTO, and ASTM classifications are mostly used by the highway departments, and USCS is used primarily by geotechnical engineering firms. As can be noticed from Figure 4.10, different classifications use different particle sizes to differentiate between boulders, cobbles, gravel, sand, silt, and clay. The USCS does not differentiate between silt and clay on the basis of particle size. Chapter 7, Elements of Soil Mechanics, describes the AASHTO system and the USCS.

Clay Size versus Clay Mineral

The concept of clay size versus clay mineral has plagued soil workers for many years. Clay minerals are fine-sized platy silicates (phyllosilicates) that have the property of plasticity, that is, they can be rolled into a thin thread that adheres together at low moisture levels. Clay size, by contrast, is a small particle size designation that does not always ensure that its constituents will be plastic. In actual fact, some fine-sized quartz and feldspar grains can occur in the range of 0.005 mm, and it is these nonplastic materials that complicate the problem. However, very few feldspar and quartz grains occur at 0.002 mm and below so, by placing the clay boundary at that lower value, the clay size and clay mineral designations coincide quite well.

Textural Classifications

Soil texture is the appearance and feel of the material as determined primarily by its particle size. The major classes are gravel, sand, silt, and clay. Based on the percentages of sand, silt, and clay, soils are classified into various textural groups. A grain size analysis of the sample is needed prior to classifying soil textures.

Figure 4.11 on page 66 shows the textural classification chart by the USDA. The size fractions are sand (2 to 0.05 mm); silt (0.05 to 0.002 mm); and clay (< 0.002 mm). When gravel is present in a sample in significant amounts, the size fractions of interest must be recalculated, thereby excluding the gravel percentages from consideration. For example, if 25% each of gravel, sand, silt, and clay were present, if we exclude the gravel, there is 25/75 of each constituent in the sample. Readjusting, this yields 33.3% for sand, silt, and clay or by direct calculation it becomes $25/100 \times 100/75 = 33.3\%$.

Agricultural Soils Maps

Agricultural soil maps are large-scale maps (1:4800) included in agricultural soil reports, which depict different soil series. They have been prepared by agronomists based on field study aided by interpretation of aerial photography. A soil series is made up of a location name and a texture name (e.g., Russell silt loam) and it is subdivided according to the slope of the land surface.

Agricultural soils maps are commonly prepared on a county-by-county basis. They are based on the information obtained from the upper 1 m (3 ft) of the soil. Each soil horizon or zone is described on

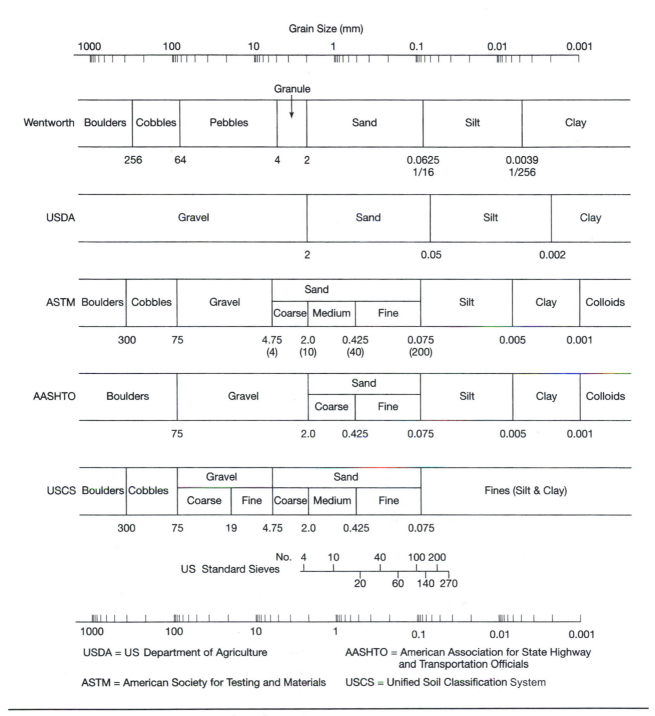

Figure 4.10 Soil classification systems based on grain size ranges.

the basis of color, texture, consistency, depth, porosity, stoniness, moisture aspects, and perhaps other features. The top soil (A horizon), subsoil (B horizon), and parent material (C horizon) comprise the subdivisions described in the report. Soils are subdi-

vided into slope subclasses with the following typical ranges for each: 0 to 2% (level to nearly level); 2 to 6%, gently sloping (undulating); 6 to 12%, sloping (rolling); 12 to 20%, strongly sloping (hilly); 20 to 35%, steep; 35+%, very steep. The maps contain

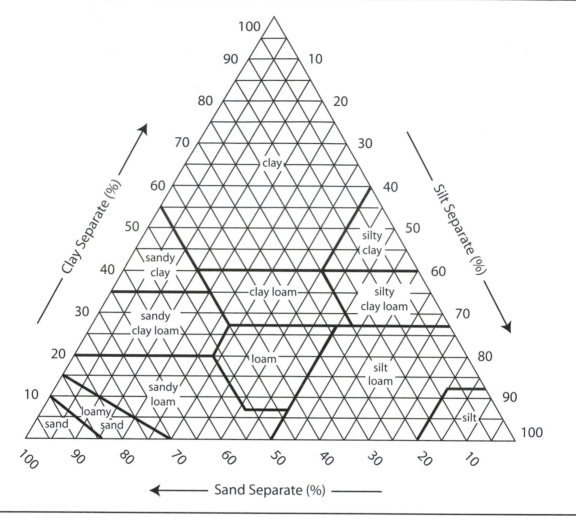

Figure 4.11   USDA textural classification chart.

basic engineering-related information. For example, Russell silt loam, 2 to 6% slope, has an accompanying rating relative to engineering use. Such uses may relate to building foundations, ponds, homesites, roadways, septic tank absorption fields, cemeteries, basements, sanitary landfills, and excavations. The various subclasses are rated good, moderate, and severe relative to various proposed uses.

Agricultural soils maps are used along with other sources of information for site selection for engineering construction. These and other maps are used for reconnaissance purposes prior to detailed subsurface exploration. The contributions of these soils maps are discussed further in Chapters 15 and 19.

Engineering Considerations of Rock Weathering and Soils

Weathering significantly deteriorates the engineering properties of intact rock as well as the engineering behavior of rock mass. The following list illustrates some of the engineering problems associated with rock weathering.

1. Weathered rocks are weaker and more friable than fresh rocks. If weathered rock is to be used for engineering purposes, such as fill material or as riprap, its engineering properties should be determined on samples that are representative of the state of weathering, not on fresh samples.

2. The amount of soil versus rock at a given site depends upon the degree of weathering and can be quite variable because of the geologic conditions of the site. Depth to the soil-rock interface can range from a few centimeters (a few inches) to tens of meters (100 ft or more). This will be an important consideration if the structure is to be located on bedrock or load needs to be transferred to bedrock through deep foundations.

3. Clay minerals can develop in fractures of weathered rock, reducing the shear strength of the rock mass and affecting its hydrologic regime.

4. Weathering by solution of carbonate and sulfate rocks (limestone, dolomite, marble, and gypsum) can result in karst conditions involving formation of underground cavities, caves, and solution channels along fractures that may be quite wide and deep. Rock salt is also subject to solution action. Karst can be very complex and difficult to predict. In karst terrains, the top of bedrock occurs at variable depths below the residual soil. A series of borings is required to determine the range in depth to bedrock and geophysical methods (seismic refraction) are required to locate the solution features. Chert, which is present in many carbonate rocks, remains behind when the calcite or dolomite are dissolved. Chert beds accumulate on the top of bedrock where they complicate subsurface investigation, building support, and soil excavation. Many dams in the United States suffer from serious seepage and piping problems because of karst foundations. Karst conditions also lead to formation of sinkholes.

5. Clay-bearing rocks, such as claystones, mudstones, and shales, can change into a soil-like material upon disintegration due to weathering. Cut slopes in these rocks deteriorate rapidly, resulting in slope stability problems.

6. Differential weathering of sedimentary rocks, consisting of alternating units of harder and softer rocks, results in undercutting which, in turn, results in rock falls, some of which can be quite hazardous.

7. A saprolite is a weak, friable, chemically weathered combination of soil and rock that retains the original structure and fabric of the rock. Saprolite is porous and susceptible to slow leaching of silica, iron, and aluminum. Generally, saprolites are well developed over low-silica rocks that contain abundant quartz and potassium feldspar, whereas ultramafic rocks that lack quartz and potassium feldspar usually do not develop thick saprolite zones. In the Piedmont Plateau of the eastern United States, saprolites are common—determining the soil-bedrock interface is a real challenge. In some situations, hollow stem auger refusal is used to indicate this boundary.

8. Duricrust is a case-hardened crust of a residual accumulation of iron and aluminum (laterite and bauxite) or the precipitation of calcite (calcrete) and silica (silcrete) from groundwater. Laterite is the most common of these, consisting of a red-brown, iron-rich deposit that is generally crumbly and porous. Since duricrust can be deceptive, a detailed subsurface investigation is necessary before locating a structure on it.

9. Corestones are rounded rock mass residuals that are surrounded by completely decomposed rock. They are a common feature of tropically weathered granites but may also occur in volcanic, sedimentary, and metamorphic rocks containing the common discontinuities (bedding, jointing, and foliation). They make it more difficult to locate the soil-bedrock interface where foundation support for structures can be attained.

10. The soil-rock interface, or rockhead, is the idealized surface where the soil ends and hard rock begins. In some locations it is quite sharp but, more commonly, there is a gradational transition between the two or there is a downward increase in rock material. In the last situation, it may involve a corestone or a slab of rock in the soil profile. Location of this interface is critical in foundation design where founding on rock is specified. Building foundations, sewer line construction, and tunnel placement are examples of a critical need to establish this surface. Drilling and sampling, auger refusal, and geophysical exploration are employed, sometimes without success, in an attempt to find the soil-rock interface. Cost of excavation is an issue because excavating rock is much more costly than excavating soil.

EXERCISES ON ROCK WEATHERING AND SOILS

1. Using the information provided in Exercise Table 4.1, fill in the blank spaces.

2. What is a placer deposit? How would it be mined? What is meant by the phrase "mother lode" and how does it relate to placers?

3. What type of weathering would prevail in the following locations? (a) Mojave Desert, California; (b) Houston, Texas; (c) southern Appalachians; (d) Grand Junction, Colorado; (e) Butte, Montana; (f) Portland, Oregon; (g) Honolulu, Hawaii.

4. Examine Figure 4.7 and Table 4.1. Why does muscovite occur as a common mineral in soils of the Piedmont Plateau, a worn-down mountain area? Explain based on rock type and weathering.

5. Rocks such as graywackes contain pieces of basalt, gabbro, and peridotite as grains in the sedimentary rock. How do you account for this relative to weathering? If the source area had a humid climate, how could this occur?

6. The thickness of residual soils in tropical and subtropical areas can exceed 15 m (50 ft). Why would this be the case? If the parent material was a basalt, what would the soil probably consist of?

7. a. In an area in the Ozarks of southern Missouri an accumulation of chert pieces can be observed along the top of the soil-bedrock interface. What is the origin of this assortment of irregular loose chert pieces? (*Hint*: The Ozark Plateau consists of a very thick sequence of cherty, carbonate rocks, i.e., limestones and dolomites.)

   b. If a 0.6-m (2-ft) thick zone of unweathered chert pieces weighed 1.60 Mg/m^3 (100 lb/ft^3) in place and the chert content in the limestone below was 1.3%, what thickness of rock would be indicated by the residual chert layer? Give all assumptions.

8. This question pertains to particle size classifications of soil and is related to Figure 4.10. On the log scale in Exercise Figure 4.1, put in the boundaries for sand, silt, and clay.

Exercise Table 4.1

Rock Type	Minerals Present	Weathering Conditions	Products
Basalt		Humid	
Syenite		Arid	
Gabbro		Arid	
Gneiss		Humid, tropical	
	Muscovite, quartz (with schistosity)	Arid	
Dolomite			Clay minerals and iron oxide
Arkose		Arid	

mm   5   1   .5   .1   .05   .01 .005   .001

Exercise Figure 4.1

Exercise Table 4.2

| Material | Classification | | | |
	1	2	3	4
Gravel	> 2 mm	> 4.76	> 2	> 2
Sand	2 – 0.05	4.76 – 0.074	2 – 1/16	2 – 0.05
Silt	0.05–0.002	< 0.74 nonplastic fines	1/16 – 1/256	0.05 – 0.005
Clay	< 0.002	plastic fines	< 1/256	< 0.005

9. a. Exercise Table 4.2 provides information for four classifications. Identify and name each classification.

    b. Exercise Table 4.3 provides the information obtained from a soil sample that was analyzed for particle size and the nature of the materials found in the sample. What is the percent of sand, silt, and clay for each of the four classifications in the soil sample? Record this information in Exercise Table 4.4.

    c. For classification 1 and 4, locate them on the USDA soil texture classification chart (see Figure 4.11). Give the name for each soil based on this diagram.

10. A soil sample has the following size distribution: 4 to 2 mm = 17%; 2 to 0.05 mm = 12%; 0.05 to 0.005 mm = 22%; 0.005 to 0.002 mm = 10%; < 0.002 mm = 39%. Using the USDA soil texture classification chart (see Figure 4.11), what are the sample's clay, silt, and sand values? What classification name would the sample have?

11. What engineering soils problems are likely to occur in the following landform areas:

    a. Till plain in north-central United States

    b. Gulf Coastal Plain

    c. Alluvial sands of the San Joaquin Valley, California

    d. Floodplain of Mississippi River at St. Louis, Missouri

12. An agricultural soils map is being used to find information concerning the location of a fossil fuel power plant along a large river. What limitations does this type of map have in this application? Explain.

Exercise Table 4.3

Size (mm)	Nature of Material	Percent
2–0.074	Gritty, nonplastic	20
0.074–0.0625	Gritty, nonplastic	30
0.0625–0.05	Gritty, nonplastic	25
0.05–0.005	Nonplastic	15
0.005– 0.0039	Nonplastic	2
0.0039–0.003	Plastic	3
0.003–0.002	Plastic	3
0.002–pan	Plastic	2
		100

Exercise Table 4.4

| Material | Classification | | | |
	1	2	3	4
Sand				
Silt				
Clay				

References

Arikan, F., Ulusay, R., and Aydin, N. 2007. Characterization of weathered acidic volcanic rocks and a weathering classification based on a rating system. *Bulletin of Engineering Geology and the Environment* 66:415–430.

Aydin, A., and Duzgoren-Aydin, N. S. 2002. Indices for scaling and predicting weathering-induced changes in rock properties. *Environmental and Engineering Geoscience* VIII:121–135.

Basu, A., Celestino, T. B., and Bortolucci, A. A. 2009. Evaluation of rock mechanical behaviours under uniaxial compression with reference to assessed weathering grades. *Rock Mechanics and Rock Engineering* 42:73–93.

Ceryan, S., Tudes, S., and Ceryan, N. 2008. A new quantitative weathering classification for igneous rocks. *Environmental Geology* 55:1319–1336.

Chiu, C. F., and Ng, C. W. W. 2014. Relationships between chemical weathering indices and physical and mechanical properties of decomposed granite. *Engineering Geology* 179:76–89.

Dearman, W. R., and Irfan, T. Y. 1978. Classification and Index Properties of Weathered Coarse-Grained Granites from SW England. Proceedings, 3rd International Congress, Madrid, Spain (Vol. 2, pp. 119–130). International Association of Engineering Geology.

Ebuk, E. J., Hencher, S. R., and Lumsden, A. C. 1993. The Influence of Structure on the Shearing Mechanism of Weakly Bonded Soils Derived from Granites. Proceedings, 26th Annual Conference on Engineering Geology of Weak Rocks, Leeds, England, September 9–13, 1990 (Vol. 8, pp. 207–215). Geological Society of London, Engineering Group, Engineering Geology Special Publication.

Geological Society of London. 1995. The description and classification of weathered rock for engineering purposes (Engineering Working Group Party Report). *Quarterly Journal of Engineering Geology* 28(3):207–242.

Goldich, S. S. 1938. A study in rock weathering. *Journal of Geology* 46:17–58.

Gupta, A. S., and Rao, K. S. 2000. Weathering effects on the strength and deformational behavior of crystalline rocks under uniaxial compression. *Engineering Geology* 56:257–274.

Martin, R. P. 1986. Use of Index Tests for Engineering Assessment of Weathered Rocks. Proceedings, 5th Congress, International Association of Engineering Geology, Buenos Aires, Argentina (Vol. 1, pp. 443–450).

Martin, R. P., and Hencher, S. R. 1986. *Principles for Description and Classification of Weathered Rocks for Engineering Purposes* (Special Publication, Vol. 2, pp. 299–308). Geological Society of London, Engineering Geology.

Plummer, C. C., McGeary, D., and Carlson, D. H. 2005. *Physical Geology* (3rd ed.). New York: McGraw Hill.

SEDIMENTARY ROCKS 5

Chapter Outline

Definition and Occurrence of Sedimentary Rocks

In our discussion on igneous rocks (see Chapter 3) we stated that 95% of the outer 16 km (10 mi) of the Earth's crust is composed of igneous and metamorphic rocks. However, 75% of all rocks exposed at the Earth's surface are sedimentary rocks. These rocks form the outer veneer above the more abundant igneous and metamorphic rocks below. Thus, it is likely that many engineers will work more with sedimentary rocks than with the other rock types.

Sedimentary rocks are rocks composed of mineral grains or crystals that have been deposited in a fluid medium and subsequently lithified to form rock. This fluid medium in the broadest sense can be either water or air. However, pyroclastic rocks, which comprise a major portion of those deposited in air, are included in this text in the igneous rock classification. Other authors have chosen to include pyroclastics with sedimentary rocks, because both can display clastic particles and bedding characteristics. In this text, however, pyroclastics are considered to be igneous rocks because both lava flows and pyroclastics have a volcanic origin and both are found in the same geographic location. The three principal types of sedimentary rocks comprise more than 99% of the total: shale, sandstone, and limestone make up 46%, 32%, and 22%, respectively. The great abundance of shale in nature is easily overlooked. Shales form slopes rather than prominent cliffs and do not provide good bedrock exposures, so they are observed less commonly by the casual observer than are the more resistant rocks. Shale slopes weather readily into soil, which supports plant life that soon covers the shale outcrop. Shale is used here as a general term for all clay-bearing rocks, including claystones and mudstones.

Sediments versus Sedimentary Rocks

A distinction must be made between sediments and sedimentary rocks. Sediments are a product of mechanical and chemical weathering. They are pieces of loose debris that have not been lithified, that is, have not been hardened into a rock material. In the engineering sense, such sediments are considered to be soil consisting of a combination of gravel, sand, silt, and clay. Sediments are found in stream bottoms, wide floodplains, deltas, alluvial fans, and bottoms of lakes and oceans where deposition by flowing water has occurred. On the other hand, sedimentary rocks are held together by various types of cementing agents, such as calcite, quartz, and iron oxide, or by the compaction of their mineral grains into an indurated mass.

Lithification

Lithification is the processes by which sediments change into sedimentary rocks. The prominent processes, that may act singly or in combination, are compaction, cementation, and crystallization (Pettijohn, 1983; Tucker, 2001; Nichols, 2009; Boggs, 2011).

In compaction, the pore space between individual grains is gradually reduced by the pressure of the overlying sediments. This occurs by the rearrangement of grains and a reduction in the volume of water filling the pore space. Coarse deposits of sand and gravel experience some compaction, but fine-grained sediments containing silt and clay undergo a much greater reduction in volume. As compaction continues, the particles are forced closer together and the deposit becomes more dense and lithified. Estimates indicate that clay-rich deposits buried to depths of 1000 m (3300 ft) have been compacted some 60% from their original volume whereas sand units are reduced by only 30%. The reader should note that the term *compaction* has a different meaning in engineering geology or geotechnical engineering than it does in a strict geologic sense when applied to lithification. These contrasting definitions are explored in detail in Chapter 15.

Cementation involves filling the spaces between individual particles in a sediment by a binding agent. Calcite, quartz, and iron oxide are the most common cements, but less abundant ones include opal, chalcedony, anhydrite, and pyrite. The cementing material precipitates and crystallizes from the water circulating through the sediment. Coarse-grained deposits with their larger interstitial pore spaces are more easily cemented together than fine-grained deposits. For this reason it is extremely difficult for clay-rich rocks to become well cemented and most are held together by compaction with only a minor contribution from cementation.

Crystallization occurs in sediments after deposition, and in some cases it happens long after lithification. New minerals crystallize or crystals of existing minerals increase in size. Such minerals crystallize from amorphous, colloidal substances in the fine mud. This contributes to lithification in finer-grained sediments.

Formation of Sedimentary Rocks

The constituents comprising sedimentary rocks originate in two ways: (1) breakdown of preexisting rocks and (2) chemical or biochemical precipitation.

Detrital (Latin for "broken down") sedimentary rocks result from weathering of preexisting rocks. Particles of gravel, sand, silt, and clay derived from erosion of the land surface are examples of detrital materials. They are typically carried to the oceans by streams and deposited some distance offshore into a sedimentation basin. Chemical or biochemical sedimentary rocks form by direct precipitation of dissolved materials from the water, by shell deposition, and through the action of plants or animals that extract chemicals from the water (Prothero and Schwab, 2003; Marshak, 2013; Nance and Murphy, 2016).

Many organisms such as corals and clams extract calcium carbonate from seawater and use it to build their shells. Other organisms, for example, a variety of algae known as diatoms, secrete a microscopic skeleton (or test) composed of silica. This is a major source of sedimentary chert. When the animals and plants die, their skeletons settle to the bottom, becoming part of the accumulating sediment. This provides the biochemical portion of the deposit. Precipitation of calcite around detrital and chemically deposited constituents yields lithification of the sediments to form sedimentary rocks. Individual rocks typically contain portions of both constituents, although most are predominantly either detrital or chemical in nature. If organic material, such as shells or coal, is the dominant constituent, the term biochemical sedimentary rock is used.

The sedimentation process controls how sediments are deposited. Factors that influence sedimentation can be enumerated by starting with the basin of deposition and working backward to the landmass. The source of the sediment, how it was transported to the basin, the distance travelled by the sediment, and the environment of deposition all come into play. Variation of these factors contributes to the diverse nature of sedimentary rocks in spite of the fact that only a few minerals comprise the sedimentary rocks (Pettijohn, 1983; Judson and Kauffman, 1990).

The source of detrital material is a preexisting rock mass located on the land surface. The nature of this rock—its composition and texture—influences what form the rock fragments will take. Rock weathering plays an important role in how the constituent mineral grains change into disaggregated particles. Regarding the sedimentary process, we note that both the solid particles (detrital material) and the dissolved constituents are dictated by the original rock and the weathering process.

Transportation, both type and duration, influences how these weathering products will change before arriving at the depositional basin. This is controlled by the geologic agents of transport: stream flow, groundwater, glaciers, ocean currents, wind, and gravity. Each of these agents is discussed in later chapters.

Sedimentation occurs when a geologic agent is no longer able to transport material any further. Suspended particles are deposited at a point of sudden energy loss, such as a decrease in stream velocity. Dissolved solids will precipitate when environmental changes are sufficient to trigger this action, which may be chemical or biochemical in nature. After a subsequent change in conditions, sediment transport may resume but the detention time between travel episodes may consist of weeks, years, or millennia. The final destination can be a stream valley, a lake bottom, or a zone in the ocean. This final sedimentation zone provides additional influence owing to its own chemical and physical makeup. In all, we see that a sedimentary rock is a product of its rock origin plus the geologic developments from weathering to deposition and lithification.

Sedimentary Rock Classification

Based on origin, sedimentary rocks are classified into two broad groups: detrital and chemical or biochemical (Nance and Murphy, 2016). Based on texture, they are subdivided into four major groups (Pettijohn, 1983; Tucker, 2001): (1) clastic, (2) coarsely crystalline or interlocking grains, (3) fine-grained or cryptocrystalline, and (4) whole fossils or their alteration products.

Clastic Sedimentary Rocks

Clastic sedimentary rocks are composed of pre-existing mineral grains or rock fragments (clasts) that settled out of water (or air) and were subsequently cemented or compressed together to form rock. Those held together only by compaction, such as shales, claystones, and mudstones, are some of the weakest sedimentary rocks and tend to slake readily when soaked in water. Clastic sedimentary rocks are further subdivided, based on the size of the clasts that comprise the rock. The rock name is determined by the size of the most prevalent grains. Table 5.1 shows the relationships between particle size and clastic sedimentary rocks. The Wentworth scale of particle size, commonly used by geologists, is followed in the table.

Conglomerates and Breccias

Conglomerates and breccias are composed of gravel-sized particles, the distinction being that the particles comprising a conglomerate are rounded whereas those in a breccia are angular. Rounded particles suggest that rock pieces were transported by stream flow for a longer distance before reaching the

Table 5.1   Sediments and Clastic Sedimentary Rocks.

Sediment	Size (mm)	Sedimentary Rock
Gravel	> 2	Conglomerate (if particles are rounded) Breccia (if particles are angular)
Sand	2 to 1/16	Sandstone
Silt	1/16 to 1/256	Siltstone
Clay	< 1/256	Claystone and mudstone
Silt and clay		Shale (contains laminations referred to as "fissility")

sedimentation basin. Angular pieces denote a short transport distance with quick burial, typical of areas with rapid sedimentation and subsidence of the basin. This denotes an active area of deposition that receives major amounts of sediment as it subsides. Therefore, the degree to which particles are rounded provides information about their mode of transport and deposition.

Sandstones

Sandstone are clastic sedimentary rocks composed of sand-sized grains. Many quartz- and feldspar-rich sandstones are only partially cemented, hence, they contain a large volume of interconnected pores. If the sand grains are loose and can be removed by rubbing the rock surface, the expression *friable sandstone* is applied. The high porosity of sandstones is conducive to high permeability (the ability to transmit fluids) so many sandstones are good reservoir rocks for water, petroleum, and natural gas. For the same reason, such sandstones below the groundwater table can yield great volumes of water, which results in a good aquifer but causes major problems for constructing tunnels. This can be a major problem that significantly reduces both the rate of tunneling and worker safety. Special construction techniques are needed to minimize groundwater inflow.

Based on the composition of clasts, sandstones are further classified into different types. When the grains are essentially all quartz, the rock is called a quartz sandstone. A sandstone composed of feldspar fragments (usually orthoclase), or pieces of preexisting rocks which themselves are rich in feldspar, is called an arkose or arkosic sandstone. Graywackes are sandstones that consist of dark rock particles derived from basalt or other low-silica igneous rocks. They contain clay minerals, including chlorite in the matrix, which surround the rock fragments and hold the rock together. Generally, graywackes are more strongly cemented than are friable quartz sandstones, but not as strongly as limestones or related sedimentary rocks. Graywackes are common in mountainous regions and in plateaus associated with mountains. They are typically interbedded with dark shales and possibly chert beds and subaqueous lava flows, as in the California coastal ranges.

Some limestones consist of sand-sized fragments of broken fossils or of rounded carbonate grains.

These rocks are termed *clastic limestones* because they are composed of individual grains of calcite cemented together. Geologists and engineers should refrain from referring to clastic limestones as a sandstone or calcareous sandstone on both drilling logs and in engineering geology reports. The term clastic limestone is more appropriate and less confusing. In many areas of the United States, sandstone carries with it the connotation of quartz sandstone because most sandstones of the midcontinent region are indeed quartz-rich. Hence, use of the term *calcareous sandstone* on a drilling log suggests a sandstone that is mostly quartz with some calcite as the cementing material, rather than a rock composed primarily of sand-sized pieces of calcite. Clastic limestones can consist of broken fossils, oolites (small rounded grains of calcite), or other calcite fragments.

Shales

Shales are clastic, fine-grained sedimentary rocks composed of silt and clay. An additional feature usually associated with shales is that they show fissility, or the prominent, closely spaced bedding planes along which the rock will break. Fissility indicates a certain strength caused by compaction of the material.

The term shale is frequently used to include all clastic sedimentary rocks finer grained than sandstone, such as siltstone, claystone, and mudstone. This, however, is not a good practice because the engineering behavior of these varieties of silt- and clay-bearing rocks can be highly variable. For example, most siltstones are very resistant to slaking and disintegration, whereas claystones disintegrate very quickly when exposed to moisture (Dick and Shakoor, 1992; Gautam and Shakoor, 2013). Potter et al. (1980) provide a classification of fine-grained clastic rocks based on the percentage of clay-size particles, the presence or absence of laminations (fissility), and the degree of induration. In the original classification by Potter et al. (1980), the percentages of clay-size particles for the three groups were 0 to 32%, 33 to 65%, and 66 to 100%. Dick and Shakoor (1992) modified these percentages to 0 to 32%, 33 to 49%, and 50 to 100 to correspond better with the engineering behavior of these rocks (Table 15.2). This classification is much more representative of the diversity and engineering behavior of fine-grained clastic sedimentary rocks, also referred to as the argillaceous rocks or mudrocks. Table 5.2 shows that mudstones,

Table 5.2   Geologic Classification of Mudrocks.

	Clay-Sized Particles		
	0–32%	33–49%	50–100%
Nonlaminated	Siltstone	Mudstone	Claystone
Laminated	Siltshale	Mudshale	Clayshale
Metamorphosed		Argillite	

Source: Modified after Potter et al., 1980.

like shales, consist of silt and clay, but they lack fissility. Siltstones consist predominantly of silt-sized pieces and, therefore, do not exhibit a plastic nature when cut or rubbed. Instead, the gritty nature of the silt grains can be recognized by grinding a small piece of the rock between one's teeth. Claystones consist predominantly of clay minerals, lacking the grittiness of siltstones and the fissility of shales.

An engineering concern regarding shales and related rocks is whether or not they will require blasting for excavation. Earth moving can be accomplished in some fine-grained clastic sedimentary rocks by wheeled scrapers, dozer blades, or by ripper teeth attached to the back of a dozer and pulled across the rock surface. The rippability of the shales and siltstones is a function of fissility, cemented strength, joint characteristics, and type of interbedding of the units. Thin alternating beds of clayey shale, mudstone, claystone, and sandstone are easily ripped; massive siltstones are not. The seismic velocity of geologic units (sound wave or P wave velocity) has been used to estimate rippability (McCann and Fenning, 1995; Tonnizam et al., 2011; Moustafa, 2015). This technique is considered more fully in Chapter 18, Earthquakes and Geophysics.

Argillites are slightly metamorphosed shales or shale-like rocks. They commonly have greater strength than shales and are used in areas of limited aggregate supply for base course and concrete aggregates. The problems involved with their use are similar to those associated with graywacke sandstones.

Coarsely Crystalline Sedimentary Rocks

Coarsely crystalline sedimentary rocks, with large interlocking crystals, are an important group of nonclastic rocks. The interlocking nature of the crystal grains is similar to that of a phaneritic igneous rock. Instead of consisting of silicates, these rocks

are composed mostly of carbonate, sulfate, and chloride minerals. They form by precipitation from water or by recrystallization of a finer material during and after lithification. Some limestones show a recrystallized mosaic of calcite crystals in which the individual crystals are of a coarse sand size.

Mineral constituents will precipitate from water when the solubility exceeds that of a particular chemical compound. Solutions can become saturated (or oversaturated) with respect to a certain compound either by changes in temperature (usually a decrease), by changes in pH of the solution, or by a reduction of the liquid volume by evaporation. Turbulence in the water may encourage precipitation if the water is nearly saturated relative to a particular chemical.

Mineral crystals become interlocked or intermeshed as they grow larger in size and abut each other. Because of this, the interlocking texture of precipitated and recrystallized sedimentary rocks can be distinguished from clastic rocks with their rounded grains or pieces held together by a cementing agent.

Carbonates

The two most common carbonate rocks are limestone, composed mostly of the mineral calcite ($CaCO_3$), and dolomite or dolostone, composed primarily of the mineral dolomite [$CaMg(CO_3)_2$] plus varying amounts of calcite. Some limestones precipitate directly from seawater. Precipitated limestones commonly are fine grained with a smooth texture. Normally, recrystallization is required for limestones to develop phaneritic crystals. These rocks are called lithographic limestones because limestones of a fine-grained, smooth variety were once used in the early printing industry for lithographic plates. Because of their fine texture, lithographic or dense limestones are not classified with coarsely crystalline limestones even though most other sedimentary rocks formed by precipitation are included in that group. This illustrates a problem of deciding rock origins based on descriptive features. A consistent relationship between origin and texture does not necessarily hold true when extended to a large group of rocks such as chemically deposited sedimentary rocks. Relating the origin and texture for specific rocks still provides the best means for rock study and classification.

A considerable amount of calcium carbonate is known to precipitate directly from solution in the open, warm oceans of the tropics and subtropics.

When it is deposited as a continuous, fine-grained material, lithographic limestone results but, in other cases, the precipitated calcite serves as a cementing agent for other constituents. Dolomite, however, is only rarely deposited directly from the ocean. The solubility of Mg^{2+} is sufficiently great that primary dolomite (precipitated directly from seawater) is fairly rare and is generally formed only when seawater evaporates in an enclosed basin.

Dolomites form, most commonly, by the replacement of Mg^{2+} for some of the Ca^{2+} in limestone long after the rock has lithified. Time is an important factor in this replacement process: older sedimentary sequences have a higher percentage of dolomite in the carbonates than do younger sequences with a generally similar geologic setting. In rocks of the Silurian age or older (400 million years or more) the chances are high that the carbonate rocks will be dolomite rather than limestone. In the replacement process, many dolomites take on an interlocking texture of visible crystals. Those that do this are classified into the coarsely crystalline category. Other dolomites have such small crystals that their outline cannot be seen with the naked eye. These are classified under the fine-grained, dense textural group that also contains lithographic limestone.

Evaporites

Evaporites are sedimentary rocks derived when seawater evaporates in an isolated portion of the ocean. Minerals precipitate sequentially with decreasing water volume and the order of crystallization is dictated by the relative solubilities of the mineral constituents.

Seawater contains about 3.5% dissolved solids in the form of cations, anions, and complex molecules. Nearly 78% of the solids are represented by Na^+ and Cl^-. When seawater is reduced to about one-half its original volume, calcite and some iron oxide precipitate. With additional reduction in water volume to about one-fifth the original, gypsum ($CaSO_4 \cdot 2H_2O$) precipitates. Another sulfate mineral, anhydrite (or $CaSO_4$), also occurs commonly in sedimentary rock sequences.

Since anhydrite is more abundant in the subsurface than gypsum, it is generally assumed, based on field evidence, that gypsum forms from anhydrite by descending groundwater. Anhydrite is likely formed at depth under the heat and pressure of a growing rock column in a sedimentary basin. The apparent sequence is as follows: gypsum precipitates from seawater in an evaporite deposit; during burial gypsum alters to anhydrite over geologic time; gypsum forms again when erosion brings the evaporite deposit close enough to the surface to contact groundwater.

Gypsum has economic importance because it is used as an additive in the manufacture of Portland cement. Anhydrite, which does not hydrate easily in industrial processes, is not as valuable as gypsum but can be used in the manufacture of Portland cement. It is also harder than gypsum (4 versus 2 on Mohs hardness scale) causing greater wear on crushing and handling equipment.

With 10% of the original water volume remaining, NaCl begins to crystallize. Magnesium salts (sulfates and chlorides) form on further evaporation and the last to crystallize are NaBr and KCl. All of these salts, which form after NaCl, precipitate with about 2% of the original water volume remaining.

Commonly the rock name for evaporite deposits is based on the most prevalent mineral, for example, anhydrite rock and gypsum rock. For NaCl-rich rocks, the mineral name halite is not applied; instead, the common name rock salt is used. Bittern salts refer to potassium- and magnesium-rich salts and are so named because of their bitter taste.

Fine-Grained or Cryptocrystalline Rocks

Fine-grained, cryptocrystalline, or amorphous features are sedimentary textures that are too fine to discern with the naked eye. Rocks belonging to this category, previously discussed, are lithographic or dense limestone and dense dolomite. The other prominent member of the group is chert, including flint and opal.

Chert

Chert is a dense, hard, siliceous material (when unweathered) that occurs as rounded to irregular nodules or as persistent layers in sedimentary rocks. It is present, chiefly, in limestones, but also in graywacke sequences. Cherts are widely found in stream gravels that contain a significant portion of sedimentary rock constituents. These may be rounded, polished pieces showing all degrees of weathering from very little (still smooth and dense) to deeply weathered (white, soft, and porous).

Chert, in its unweathered condition, has many of the characteristics of quartz: a hardness of 7, conchoidal fracture, and high resistance to chemical weathering. In fact, the primary constituent of chert is cryptocrystalline or finely crystalline quartz. The dark-gray or black variety of chert is referred to as flint in many areas of the United States; in some localities, however, gray, black, or brown varieties are termed flint. Flint, chert, and volcanic glass (obsidian) were used by Stone Age humans to make arrow points, spear heads, and cutting knives. White chert, which develops a weathered, brown surface coating, is the common variety present in stream gravels.

Chert forms either through primary deposition of silica in the oceans in a colloidal form or as a secondary replacement in carbonate rocks after deposition. Examples of both primary and secondary chert deposition are widespread. Secondary chert may form when minute silica or fragments of siliceous shells distributed throughout a rock mass are first dissolved and then concentrated by precipitation to form a large nodule near the center of the contributing rock.

Whole Fossil Rocks

Whole fossils or their alteration products yield the fourth and final textural group for sedimentary rocks. These textures form from calcite fossils or from materials with a high organic content, including coal.

Fossiliferous Limestone

Fossiliferous limestones consist of abundant calcite fossils of sand size or bigger, many of which are still intact. The fossils are cemented together by minor amounts of calcite where they touch or that fill the interstices between the fossils. In most fossiliferous limestones, the grains are large enough that they can be abraded with ease. Rocks of this type are usually too weak to use as aggregates in concrete, bituminous mixes, or as a base course material for road construction.

If, instead of whole fossils, points of contact between fossil fragments are held by calcite, the rock is known as coquina. Because the shells in much of the coquina are gravel sized the term *shell conglomerate* is sometimes used.

Another calcite-rich rock of marine origin is the familiar material chalk. It is a special variety composed of minute shells or shell fragments cemented lightly together with calcite. Foraminifera, tiny protozoans (small marine animals), have shells, or tests, composed of calcium carbonate that comprise the chalk. These shells are deposited in the deep water beyond the depositional zone of clastic sediments derived from the land.

Rocks Rich in Organic Matter

Organic matter is present in nearly all sedimentary rocks. In marine sediments, the range is about 0.5 to about 10%, the latter occurring in black, pyrite-bearing shales. In general, fine-grained rocks tend to contain more organic debris than do coarse-grained rocks.

Sedimentary deposits also exist that contain up to 100% organic material. These are the freshwater organic accumulations. One of the most important of these is the coal group, which includes peat, brown coal or lignite, bituminous coal, and anthracite coal. A second variety of organic debris comprises the petroleum-rich rocks.

Peat is associated with freshwater swamps. Organic remains are prevented from complete decay by the smothering action of the water, which covers the dead trees and mosses. Organic residues may comprise 70 to 90% of the total accumulation. Coal forms through burial of the peat with an attendant heat and pressure increase. Water is removed through burial and volume reduction, and an increase in carbon content occurs. The progression through the coal series occurs by increased vertical pressure. Anthracite coal forms during intense folding of sedimentary rocks and is classified as a metamorphic rock.

The petroleum series of organics has, as its probable origin, compounds that collect on the bottoms of ocean basins, lagoons, and estuaries. This accumulation is richer in fatty and protein substances than are peats. Normally they are mixed with clay and silt, yielding a lower organic composition than that present in peat beds. With burial and increases in pressure from the overburden, oil shale forms first and then, eventually, petroleum. Oil shale is a black shale containing considerable amounts of hydrocarbons mixed with the silt and clay. Extensive areas of oil shale are present in Colorado and adjacent states, with deposits of lesser extents in many other states of the United States.

Sedimentary Rock Identification

Sedimentary rocks can be identified by using Table A.5 in Appendix A. The sedimentary rock classification is based first on texture and then on additional physical characteristics of the rock.

The first step in the identification process is to determine the characteristics of the texture: (1) clastic, (2) coarsely crystalline, (3) fine grained or dense, or (4) composed of whole fossils or their alteration products. In the case of clastic sedimentary rocks, the grain size is used to further subdivide the groups. Then the shape of the grains (conglomerate versus breccia) or the nature of the grains (siltstone versus shale; quartz sandstone versus arkose) is used to provide the rock name.

For the whole fossil group, the nature of the fossil or organic material is used to determine the rock name. Peat, lignite, and coal are identified in this way. For the fine-grained or dense rocks, the mineralogy is determined in order to name the rock. The amount of chert, calcite, or dolomite present determines what rock name will apply.

In the case of limestones the procedure may be reversed. After the sedimentary rock is observed to effervesce freely in cold hydrochloric acid, the rock is recognized as a limestone. The texture is used to give the rock its complete name. Textures may be clastic (composed of broken pieces of calcite), coarsely crystalline, fine-grained or cryptocrystalline, or consisting of whole fossils, with two extremes for the last case: large gravel-sized shells (coquina) or tiny microscopic ones (chalk). These limestones typically are deposited in marine environments.

Distinguishing Features of Sedimentary Rocks

Bedding

Bedding, stratification, or layering of sedimentary rocks is the most significant rock feature. From an engineering point of view, bedding provides the extremes in directional properties of sedimentary rocks and, to a large extent, the anisotropic nature of these materials. Extreme differences in strength, permeability, electrical conductivity, and seismic velocity are introduced by stratification. Figure 5.1 shows horizontal bedding planes in a limestone outcrop.

Other features of sedimentary rocks that merit discussion are the primary structures of bedding planes (cross bedding, ripple marks, mud cracks, basal conglomerates, graded bedding, and fossil

Figure 5.1   Example of horizontal bedding planes. (Photo by Rhododendrites, Wikimedia Commons [https://creativecommons.org/licenses/by-sa/4.0/deed.en].)

position) and the secondary structures of these rocks (concretions, nodules, geodes, and stylolites).

A valuable characteristic of the primary bedding plane features is that they can be used to tell the original top from the original bottom of the sedimentary sequence. These sedimentary features are examined following a further discussion of bedding.

The bedding of sedimentary rocks marks the interface between the sediment and the water medium at the time of deposition. As additional sediment is added the previous boundary is preserved by minute differences in sedimentation.

Bedding planes manifest by changes in the depositional materials. They are characterized by slight changes in: color, mineral content, or texture. Reduction in water content and eventual lithification reduces the sediment thickness but, under most circumstances, the subtle changes of bedding are preserved in the rock.

Partial weathering of a sedimentary rock enhances these bedding features, making them appear more obvious and sometimes promoting fracture along these planes. Fresh rock cores of shale may show only subtle indications of bedding but, if the cores are exposed to wetting and drying conditions (left lying out in the weather for a period of time), they will form thin wafers a few millimeters thick. Sedimentary rocks used as dimension stone on the exterior facing of buildings or for stone walls should possess only subtle bedding effects because these planes will be exploited by weathering. In most situations, bedding planes are oriented in the horizontal rather than vertical position to minimize the ingress of water into the dimension stone.

Bedding planes for most sedimentary rocks were originally oriented in essentially the horizontal position. This occurs because most sedimentary rocks have a shallow marine origin and these rocks form on the continental shelf where the ocean bottom closely approaches the horizontal. A basic concept of sedimentary rocks known as the *principle of original horizontality* states this horizontal condition (see Figure 5.1).

There are some exceptions to the original horizontality of beds, such as reef deposits of limestone or dolomite. In this situation the ocean bottom slopes away from the reef and different animal and plant assemblages thrive at different water depths, ranging from the surf zone to as deep as 100 m (300 ft).

Figure 5.2a is a sketch of a limestone reef forming in the ocean and 5.2b is a photograph of a stone quarry in a dolomite reef (see next page). In Figure 5.2a the fossil reef continues to accumulate shells as the biological life cycles occur and organisms die. If the reef sinks under its own weight it can continue to grow while utilizing the same water depth and perhaps persist for millions of years. The Silurian period (about 400 million years ago), which can be found at various locations in the United States, is marked by the development of extensive fossil reefs with compositions that have changed from limestone to dolomite over geologic time. Much of the original porosity of these reefs has been preserved despite the dolomitization process.

Bedding Plane Features

Bedding associated with dunes developed in sand is commonly inclined to the base of the dune. This feature, called cross bedding or false bedding, occurs when sand dunes migrate in the direction of current movement. Three types of beds comprise a single dune, the bottom set, the foreset, and the topset beds, as illustrated in Figure 5.3a on page 81. The foreset beds, which represent the sloping face of the dune, may be inclined at angles from 5 to 30°. When a change of current direction occurs, the topset beds erode, truncating the foreset beds. New topset beds are deposited that intersect the old foreset beds at an acute angle, whereas the foreset bed below retains its smooth curve, which becomes tangent to the base. This provides a means of telling the original bottom (tangent portion) from the original top of the bed (truncated), regardless of how much the sequence has been tilted subsequently by rock folding. In Figure 5.3b, the original beds have been overturned and now lie upside down (see page 81). For vertical beds one can also determine the original top from the original bottom in this manner.

Ripple marks are undulations along the bedding plane of a sandstone preserved through the lithification process (Figure 5.4; see page 82). They develop on sand dunes, beaches, or stream bottoms. Ripple marks are of two varieties: symmetrical ones with pointed cusps and asymmetrical ones showing rounded troughs and ridges. Symmetrical ripple marks form by the back and forth motion of water along the seacoast but beyond the surf zone. These are oscillation ripple marks and can be used to tell the original top from original bottom of the sedimentary sequence. The cusps point toward the original top of the unit.

Asymmetrical ripple marks depict current ripple effects or those with a persistent direction of movement. Because of their rounded troughs and ridges, no difference is indicated between the original top direction and that of the original bottom in the sequence. Whether it is the portion below the rippled surface or the imprint of that surface being observed cannot be discerned. If the top of the bed is ascertained by another means, the direction of the current movement during deposition can be found based on the sloping surface.

A basal conglomerate sometimes comprises the first layers formed above an erosion surface or when an abrupt change in sedimentation occurs. It consists of rock pebbles or cobbles surrounded by sand, which grades upward into a typical sandstone. The pebbles mark the residual weathered rock supplied to the basin when the new sedimentation sequence began. A basal conglomerate therefore can be used to tell the original top from the original bottom of the sequence; the conglomerate marks the bottom. This is illustrated in Figure 5.5 on page 83.

Unfortunately, conglomerates do not always mark the beginning of a sandstone unit because intraformational (within the formation) conglomerates also occur. If these are repeated within a

(a)

(b)

Figure 5.2   (a) Formation of a limestone reef and (b) stone quarry in a dolomite reef. (Photo by James St. John [https://creativecommons.org/licenses/by/2.0].)

sandstone unit because of changes in the supply of detrital sediment, the top cannot be discerned from the bottom based on these conglomerates.

Graded bedding is indicated when the particles in a sedimentary bed vary from coarse at the bottom to fine at the top. Therefore, this feature can be used to discern the original top of the bed. Graded bedding forms by the rapid deposition of particles from turbid water carrying a range of different sizes. It is

generally agreed that these are deposited by turbidity or density currents that sweep down steep slopes in an ocean or large lake. They were responsible for breaking the transatlantic cable in 1929 and it is suspected that in human-made reservoirs they occasionally transport large volumes of sediment from the upstream areas toward the dam. Graywackes sometimes show graded bedding and, consequently, it is believed that such rocks are deposited in deep marine environments by turbidity currents.

Mud cracks and position of fossils are the final primary features of bedding planes considered in this discussion. Mud cracks occur when silt or clay dries out and shrinks to yield polygonal cracks. If the cracks are filled by sediment, particularly sand, the mud cracks may be preserved after the rock is lithified. The presence of mud cracks in ancient rocks shows that the sequence was exposed to the air and that the mud cracks pinching out with depth indicate the original top of the sedimentary

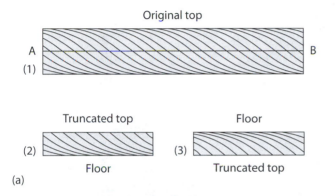

(a)

(b)

Figure 5.3   (a) Cross bedding showing: (1) original complete set of beds, (2) the truncated top portion along A–B, and (3) beds overturned from their original positions. (b) Cross bedding in sandstone, Checkerboard Mesa, Zion National Park, Utah. (richsantaclaus, Shutterstock.)

(a)

(b)

Figure 5.4   (a) Symmetrical (photo by Ron Wolf) and (b) asymetrical ripple marks (Linda Armstrong, Shutterstock).

sequence. These details of mud cracks are illustrated in Figure 5.6 on the following page.

When large fossils, such as the valves (half shell) of a clam, come to rest on the ocean floor, shifting currents will ensure that the shell lies concave downward. Therefore, the original bottom of the sedimentary sequence is in the concave direction of the shells. This can be observed in the cross section of the sedimentary rock sequence of Figure 5.7 on page 85. Raindrop impact impressions and animal tracks can sometimes be observed along sedimentary bedding planes. These can be useful features concerning the environment of deposition.

Secondary Structures

Many sedimentary rocks contain structures that formed after the sediment was deposited and the rock lithified. These secondary structures, unlike the primary features, do not indicate conditions prevalent during deposition. They do document the sedimentary origin of rocks that contain them and many prove helpful during identification and study of geologic history. The features of interest are nodules, geodes, concretions, and stylolites.

A nodule is an irregular, knobby body of material that differs in composition from the sedimentary rock that contains it. Nodules are usually aligned parallel to the bedding of the rock and several may be joined to form a continuous bed. Nodules are typically 2.5 to 10 cm (1 to 4 in) in diameter. Chert or flint comprises most common nodules and they typically occur in limestone or dolomite but also in shale. Nodules probably form when silica, distributed throughout the sediment, is concentrated in a central location during and after lithification. A nodule is shown in Figure 5.8a on page 85.

Geodes, attractive objects commonly available at rock and mineral shops, are roughly spherical in shape with a hollow interior partially or completely filled with mineral crystals. They typically range from about fist size to 30 cm (1 ft) in diameter. A layer of banded chalcedony normally forms the outside layer and crystals of quartz, calcite, or dolomite project inward toward the center. Various sulfide minerals and other less common varieties are observed on an infrequent basis. Figure 5.8b on page 85 shows a quartz-filled geode.

Most geodes are found in limestones and typically are most abundant within a specific zone of a

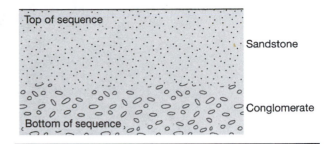

Figure 5.5 Basal conglomerate in sandstone sequence.

geologic unit. Some geodes may be nearly solid and very dense if the crystals inside have grown together. Geodes released from their geologic units by weathering or erosion are sometimes found along stream beds in limestone terrain.

The method of formation of hollow geodes in a solid rock is an intriguing question for geology students and mineral collectors alike. It seems reasonable that a simple opening would close as a result of pressure from the overlying rocks. First a water-filled cavity develops in the sediments, probably when plant remains decay within a buried deposit. As the sediment consolidates during the rock-forming process a wall of silica, colloidal in nature, forms around the water cavity, isolating it from the surrounding material. With time, freshwater washes through the sediments expelling the saltwater, but the protected cavity retains its highly saline water. This sets up an osmotic effect as the higher concentration inside the void tries to equalize with the water outside through the semipermeable membrane, the silica wall. As long as osmosis continues, pressure is exerted outward from the cavity, which expands little by little until the salt concentrations are equalized. At this juncture, osmosis stops, the outward pressure dissipates, and the cavity no longer expands. With time, the silica wall dries, chalcedony crystallizes from the colloidal silica, and the wall contracts, yielding cracks. At some later time groundwater seeps through these cracks and crystals begin to grow inward, toward the center, from the chalcedonic lining. As a consequence, a crystal-lined geode is formed. A significant point is that in geodes crystals grow inward, but in concretions they grow outward.

Concretions are local concentrations of material in a sedimentary rock that have a rounded, smooth surface and range in size from less than 1 cm (1/4 in)

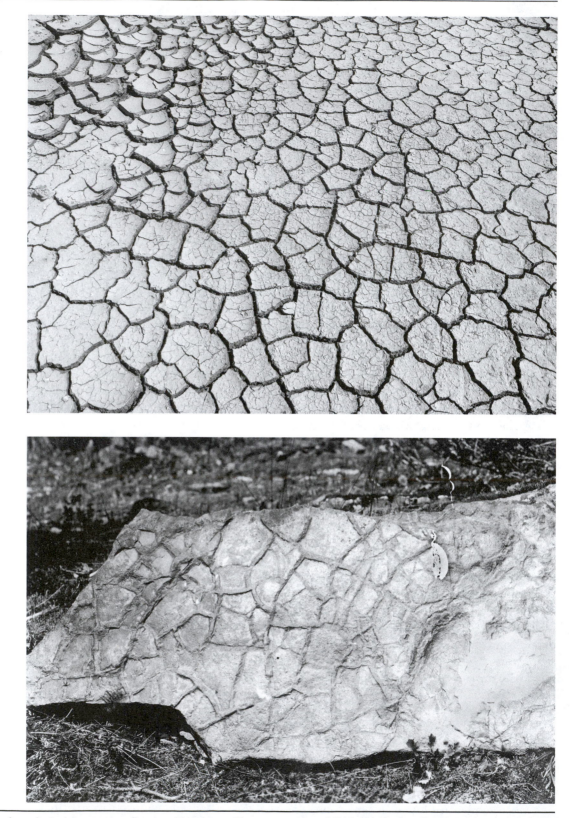

(a)

(b)

Figure 5.6 Mud cracks in (a) recent sediments (Komjomo, Shutterstock) and (b) in lithified rock (US Geological Survey).

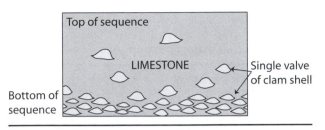

Figure 5.7 Fossil orientation in sedimentary sequence. Shells lie with concave portion downward.

to 1 m (3.3 ft) in diameter. They typically consist of the same cementing material that lithified the rock and, therefore, are probably composed of calcite, iron oxide, or silica, the common cements for sedimentary rocks. Concretions, typically, are more strongly held together than is the host rock; a large concretion is illustrated in Figure 5.8c.

Stylolites are serrated planes or surfaces that occur parallel to bedding in limestones. They form by the dissolving action of water as it moves along the bedding planes followed by deposition of impurities. In a cross section perpendicular to bedding they appear as irregular lines along selected bedding planes and are common features in polished limestone and marble slabs used in building interiors. In some instances people have mistaken stylolites for fossils, worm burrows, or primary sedimentary features. Stylolites are illustrated in Figure 5.9 on the following page.

Figure 5.8 Nodules, geodes, and concretions. (a) Chert nodule in limestone. (Richard Bradford, Shutterstock.) (b) Quartz-filled geode. (Vladimir Dokovski, Shutterstock.) (c) Large septarian concretion showing cracks filled with secondary quartz. (Dirk van der Walt, Shutterstock.)

Environments of Deposition

Sedimentary rocks form in many different locations or sites in the geologic environment. These areas of sediment accumulation and subsequent lithification may be referred to as sedimentation basins, indicating areas of low elevation that receive sediments from streams flowing into them. The environments of deposition can occur at any intermediate point between the source of detrital material and the ocean basins. The oceans, of course, provide the ultimate, low elevation toward which particles move by gravity. The various environments are shown in Table 5.3 (see next page).

Continental Environment

Sedimentary rocks of continental origin form when sediments accumulated in continental environments undergo lithification. Examples are

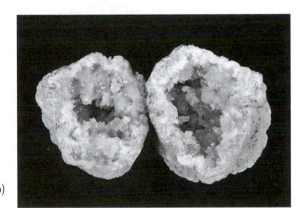

(b)

(c)

Figure 5.9 Stylolites in limestone appear as wavy lines and a dark surface.

lithified sand dunes from desert environments (see Figure 5.3) or sandstones formed from continental stream deposits. Both are common occurrences in the geologic past, as indicated by the rock column.

Fluvial materials are those deposited by flowing water. They can originate in piedmont areas (inter-mountain basins that accumulate debris at the edge of the mountain front) where the sudden change in stream gradient causes coarse particles to be deposited. Valley flat or floodplain deposits, known as alluvium, can also be lithified into sedimentary rocks.

Sediment deposited by glaciers or sediment in lakes, swamps, or caves can undergo lithification as well to yield sedimentary rocks. Swamp environments provide lignite and coal after compaction and burial.

Mixed Environments

The mixed continental and marine environments are locations along the boundary of the landmass and the ocean. Deltaic deposits may provide extensive rock units, for example, in the Catskill area of upstate New York. Littoral deposits, those formed by ocean currents paralleling the coast, are important reservoir rocks for petroleum production. Locations of the coastlines in the geologic past provide valuable information regarding oil migration and accumulation. Rock sequences in an area are studied in great detail by oil geologists prior to subsurface drilling.

Marine Environment

For marine environments, the shallow sea category (continental shelf) provides the most significant

depositional area. In Figure 5.10 the continental shelf is shown in relation to the other marine environments. The major portion of all sedimentary rocks of the continental United States was formed in large inland seas, which lapped up on the continent at various times in the geologic past. The common limestones, shales, and sandstones so prevalent in the central United States formed on the continental shelf. These are referred to as the rocks of the stable shelf or stable interior. The rivers of the world drop most of their sediment on the continental shelves where marine rocks are forming today.

The intermediate sea zone receives some sediment from the continental landmass but only in selected areas of the world. These rocks may occur in great volumes in certain areas. Such basins include the depositional troughs that will eventually give rise to new mountain chains. Sedimentation basins and mountain building are discussed in Chapter 11.

The deep sea receives very small quantities of sediment from the landmass, primarily because of the great distance sediment must travel in the oceans to reach the abyssal depths. For water depths in excess of 1800 m (6000 ft) the pressure is so great that $CaCO_3$ will redissolve before reaching the ocean bottom. Animal life is sparse because of the lack of nutrients, which are supplied by rivers entering the ocean at the coastline. Therefore, the only sedimentation that occurs accumulates very slowly and is provided by fallout from the atmosphere.

Shallow Marine Environment

An important characteristic of shallow marine sedimentary rocks is the definite pattern and sequence of rock types that parallel the ancient

Table 5.3   Environments of Sediment Deposition.

Continental	Mixed Continental and Marine	Marine
Desert	Littoral	Shallow sea
Glacial	Lagoon	Intermediate sea
Fluvial	Estuary	Deep sea
Piedmont	Delta	
Valley flat		
Lake		
Swamp		
Cave		

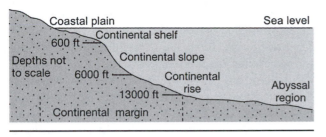

Figure 5.10 General features of continental margins and ocean basins.

coastline. This can be used to establish the geologic history of an area and to explain the details of the rock sequence.

One of the commonly used examples to illustrate shallow marine deposition is a pattern of sandstone, shale, and limestone formed on the gently sloping continental shelf. The basis for this distribution of rock types is that the coarsest detrital material would settle first, causing sandstones to form a strip closest to shore; the finer detrital material would settle next, yielding a band of shale beyond the sandstone; and limestone, absent of detrital material, would form the third band from shore through precipitation of $CaCO_3$ from seawater. This is illustrated in a series of cross sections in Figure 5.11.

To continue this example, we now observe what occurs as the shoreline adjusts to sea level changes. In the second and third diagrams of Figure 5.11 the sea has advanced (transgression or on-lap), yielding deeper water for all locations, and the zones of deposition are displaced to the right (toward the landmass). In the fourth diagram the water level has dropped dramatically, showing retreat (regression or off-lap), and the zones of deposition have been displaced seaward. In cross section this yields wedge-shaped deposits of sandstone, shale, and limestone.

This example provides the detail needed to determine the change in sea level indicated by the occurrence of different sedimentary rocks. A stratigraphic column illustrating sea level changes is presented in Figure 5.12. This is based on what is observed in a borehole obtained at a single location. Beginning at the bottom (which represents the event that occurred first) an increase in water depth is shown, indicating on-lap or an advancing sea. The next three units show off-lap, then on-lap again, and finally a sustained off-lap, yielding continental deposits and coal formation. Then on-lap occurs again.

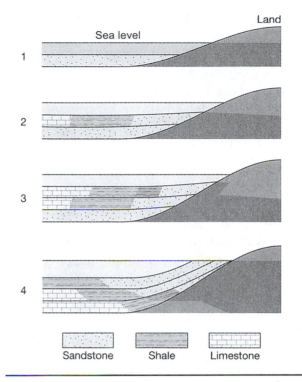

Figure 5.11 Diagrams showing progressive on-lap onto the landmass, followed by off-lap.

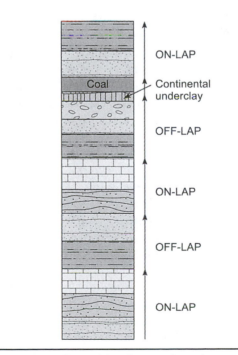

Figure 5.12 Deposition by an advancing and retreating sea.

Additional Details of Carbonate Deposition

Several varieties of calcite-rich materials are deposited in continental environments rather than under marine conditions. Examples are tufa, caliche, marl, and travertine. Tufa is a calcareous material found where hot or cold springs and seeps emerge at the ground surface. $CaCO_3$ precipitates because of evaporation of the water that previously contained the dissolved materials. Tufa is usually porous, spongy, and rather fragile because of its mode of formation. It may contain twigs, soil, or other debris, which is incorporated in the evaporating moisture. By contrast, siliceous sinter or geyserite is a silica-rich deposit formed by evaporation of water from hot springs and geysers. Its origin is somewhat similar to tufa but, of course, differs in composition.

Caliche is a calcareous deposit that forms within the soil profile in semiarid regions. Two schools of thought exist concerning its origin. In one explanation, the calcium carbonate is deposited at the base of the soil profile by water percolating downward from the surface. The second possibility is the deposition, by capillary action, of water moving upward from below. Caliche may consist of a rather indurated deposit up to 15 cm (6 in) thick and found at some depth in the soil profile.

Marl is a term that has been applied to several types of sedimentary deposits that are mixtures of calcium carbonate plus mud and/or sand. Marl also may refer to impure limestones found in freshwater lakes or to marine muds composed primarily of calcium carbonate. These are thought to form near the outer margins of shale deposition. The term has been applied to certain calcareous green sands, for example, those found in New Jersey.

Travertine is a deposit of calcium carbonate commonly found in limestone caves. It is also referred to as dripstone, flowstone, or Mexican onyx. Travertine precipitates from downward percolating waters that seep into the cave system through cracks and joints in the limestone. Stalactites, stalagmites, pillars, and curtains of calcium carbonate are the well-known wonders composed of travertine. Dolomite deserves further explanation because of its significance as a common sedimentary rock. It is composed mostly of dolomite mineral $[CaMg(CO_3)_2]$, plus varying amounts of calcite. The term *dolostone* has been proposed for this rock to remove the confusion caused by the use of the term dolomite for both a rock and a mineral. In many areas of applied geology the term dolostone has not been readily accepted, therefore dolomite still prevails as the commonly used rock term.

Dolomites can contain considerable amounts of noncarbonate minerals, typically quartz or clay minerals—as much as 25%, with 10% being relatively common. Calcite can be present in large percentages in a dolomite, up to one-half of the carbonate fraction. These would be termed calcic dolomites; in like manner, a limestone that contains appreciable amounts of dolomite is called a dolomitic limestone.

Engineering Considerations of Sedimentary Rocks

1. The alkali-silica reaction problem in Portland cement concrete was discussed in Chapter 3. Sedimentary rocks that can be involved in this reaction are chert (including opal and chalcedony) and graywacke. As mentioned in Chapter 2, cherts are subject to two problems when used as concrete aggregates. First, because of the high porosity of weathered chert, it results in surface pop-outs when used in concrete that undergoes freezing. Secondly, certain cherts result in an alkali-silica reaction with high-alkali cements. This reaction leads to cracking and expansion of concrete and, ultimately, to failure of the material. Cherts can consist of three different constituents: opal, chalcedony, and fine or cryptocrystalline quartz. Two of these constituents lead to the alkali-silica reaction: chalcedony and opal. Opal is an amorphous variety of SiO_2 containing some water of hydration, and chalcedony is SiO_2 with a fine, fibrous texture, which can be discerned only by using a petrographic microscope. Opal is highly reactive with alkalis—only 0.25% of the aggregate by weight is needed to cause considerable expansion in alkali-rich concrete. For chalcedony a percentage of about 3% of the coarse aggregate is needed to cause this expansion.

To prevent the alkali-silica reaction problem, three procedures can be followed: (a) use low-alkali cement (< 0.6% alkali as Na_2O); (b) use nonreactive aggregate or add nonreactive aggregate to reduce the percent of reactive material; or

(c) add pozzolans, which include certain volcanic rocks and fly ash or ash collected from smoke-stacks where coal is burned. Adding these finely ground high-silica materials produces a product that has a nonexpansive reaction with the alkali in the cement.

To solve the pop-out problem of weak cherts, the percentage of these low-density materials (specific gravity less than 2.45) is kept to a minimum by selectively choosing gravel sources and by reducing the maximum allowable size of the coarse aggregate. This latter procedure is effective because large, weak pieces are more prone to freeze-thaw damage than are smaller ones. Larger pieces are unable to expel all the water prior to freezing and must absorb the entire volume increase when ice crystals form.

2. The alkali-carbonate reaction problem that develops in some Portland cement concrete is related to a specific carbonate composition and texture. Cracking and an expansion of the concrete occurs, eventually causing failure of the structure. Fine-grained calcic dolomites that are rich in clay comprise the reactive rocks. Nearly equal amounts of dolomite mineral and calcite plus 10% clay or more in the fine-textured rock characterize this problem aggregate. High-alkali cement (greater than 0.6% Na_2O) is needed to trigger the reaction. The procedures for preventing this reaction are to (a) avoid dolomites with this composition and texture, (b) reduce the amount of the troublesome aggregate by adding nonreactive rock materials, or (c) use low-alkali cements.

3. In spite of the above listed chemical reactions, limestone and dolomite provide the best sedimentary aggregates for construction materials. Siltstone, shale, quartz sandstone, and conglomerate are generally not acceptable; graywacke is marginal.

4. Stream and terrace gravels commonly contain weak rock pieces that yield nondurable aggregates in concrete. Weathered chert, shale, siltstone, clayey carbonates, iron oxide nodules, and friable sandstones can cause pop-outs at the concrete surface after a number of freeze-thaw cycles.

5. Graywackes, when used for concrete aggregate and as base course (road base) materials for highways, can result in degradation (reduction in the size of the aggregate pieces). When used as concrete aggregates some graywackes have proven to be alkali-silica reactive, causing expansion and extensive cracking of the concrete.

6. Coarse-grained limestones abrade too severely to be used for aggregates for construction. Such rock particles lose gradation owing to a reduction in particle size.

7. Sedimentary rocks used as dimension stone for the facing of buildings should be nonstaining and resistant to weathering effects. High purity, clastic limestones have proven durable; clayey carbonates and quartz sandstones may be subject to spalling. Quartz sandstones are used for flagstone walls.

8. Shales and siltstones can provide a suitable foundation for buildings, dams, and bridges. A deep weathering profile is one concern, and the slabbing loose of rock due to stress release after excavation is another. Placement of a concrete slab over the shale and possibly the use of rock bolts to restrain the rock are common practices.

9. Limestones, dolomites, and evaporite deposits can exhibit an irregular soil-rock interface in their weathering profiles. Pinnacles (of rock) and pipes (of soil) are common. Care must be taken to ensure that heavy structures are founded completely on solid rock.

10. Sinkholes and underground conduits in limestone and dolomites must be recognized and properly dealt with when founding buildings in these terrains. Changes in the existing ground surface or of subsurface drainage by construction should be evaluated with great care.

11. When water is impounded behind a dam, if limestone lies at the rim, within the reservoir area, or in the foundation, careful consideration is required. The presence of solution channels in the limestone that extend to another surface drainage area will lead to leakage unless the channels are filled by grouting. Many embankment dams built on carbonate rocks in the United States have experienced serious seepage and piping problems, and are currently being remediated.

12. Compaction shales have caused problems when used as rock fills in highway embankments. When such shales become wetted in the embankment, slaking occurs; this material fills the voids between the rock pieces and surface subsidence and slope instability occur. To solve the problem, these shales need to be broken during placement and compacted into a solid mass in the same manner used for soil materials.

13. Conglomerates are basically weak sedimentary rocks because they are poorly cemented and highly porous. Water movement through this rock removes the cement and increases permeability. When encountered in dam abutments and foundations, conglomerates require special treatment to increase their strength and reduce permeability.

14. Sedimentary rocks containing anhydrite are troublesome to engineering structures such as dams, highways, and tunnels because the mineral will alter to gypsum in the presence of water, yielding an increase in volume and considerable stress on the structure adjacent to it. The presence of anhydrite must be recognized and steps taken to reduce its effect.

EXERCISES ON SEDIMENTARY ROCKS

1. Bedding is one of the most prominent features of sedimentary rocks. Would you expect a difference in strength parallel and perpendicular to bedding? In what way? How would permeability be affected? Explain.

2. Limestones can be dark in color. What is the likely cause of this coloration? How would a block of fine-grained limestone be distinguished from basalt? Which would be stronger and why?

3. Clastic limestones are composed of broken fossil fragments; fossiliferous limestones are composed of whole fossils. How do you think the environment of deposition differed between the areas where these rocks formed? (Assume both are marine in origin.) Explain.

4. Quartz sandstones commonly are only partly cemented and contain a high volume of pore space. Would quartz sandstone be a good supplier of groundwater? Shale has a high porosity; would it have a high permeability? How does shale differ from sandstone as a supplier of groundwater?

5. A ledge of rock in a quarry is described as a cherty, argillaceous (clay-rich), fine-grained dolomitic limestone. The chert is deeply weathered. What problems might develop if this rock were used as an aggregate for Portland cement concrete?

6. Large pieces of limestone are being considered for use as slope protection stone for a harbor area. The apparent specific gravity of the rock was found to be 2.52; how much would a large rectangular piece that is 0.91 × 1.12 × 1.51 m (3 × 4 × 5 ft) weigh in kg (lb)? What would be its mass in kg/m^3? If diabase were used instead, with an apparent specific gravity of 2.96, what would be its mass for a piece of the same size as the limestone? What would be the difference in effect of the two pieces relative to resisting the energy of ocean waves? Explain.

7. Explain the differences between volcanic breccia and sedimentary breccia. How can these two rocks be distinguished from each other? Explain in detail and give information on the origins of the two rocks.

8. Fine-grained dolomite might not effervesce in cold, dilute hydrochloric acid and chert will not. What tests should be performed to tell these two rocks apart? What might be done to make the acid test more helpful? Explain.

9. Why are fossils less common in shale than they are in limestone? Consider the origin of the materials that comprise these rocks.

REFERENCES

Boggs, Jr., S. 2011. *Principles of Sedimentology and Stratigraphy* (5th ed.). New York: Pearson.

Dick, J. C., and Shakoor, A. 1992. Lithologic controls of mudrock durability. *Quarterly Journal of Engineering Geology* 25:31–46.

Gautam, T. P., and Shakoor, A. 2013. Slaking behavior of clay-bearing rocks during a one-year exposure to natural climatic conditions. *Engineering Geology* 166:17–25.

Judson, S., and Kauffman, M. E. 1990. *Physical Geology* (8th ed.). Upper Saddle River, NJ: Prentice Hall.

Marshak, S. 2013. *Essentials of Geology*. New York: W. W. Norton.

McCann, D. M., and Fenning, P. J. 1995. *Estimation of Rippability and Excavation Conditions from Seismic Velocity Measurements*. Engineering Geology Special Publications, Vol. 10 (pp. 335–343). Geological Society of London.

Moustafa, S. S. R. 2015. Assessment of excavability in sedimentary rocks using shallow refraction method. *Electronic Journal of Geotechnical Engineering* 20(4):1373–1382.

Nance, D., and Murphy, B. 2016. *Physical Geology Today*. New York: Oxford University Press.

Nichols, G. 2009. *Sedimentology and Stratigraphy* (2nd ed.). New York: Wiley-Blackwell.

Pettijohn, F. J. 1983. *Sedimentary Rocks* (3rd ed.). New York: HarperCollins.

Potter, P. E., Maynard, J. B., and Pryor, W. A. 1980. *Sedimentology of Shale*. New York: Springer-Verlag.

Prothero, D. R., and Schwab, F. 2003. *Sedimentary Geology* (2nd ed.). New York: W. H. Freeman.

Tonnizam, M. E., Rosli, S., Muhazian, M., and Fauzi, M. 2011. Assessment on excavatability in weathered sedimentary rock mass using seismic velocity method. *Journal of Materials Science and Engineering—A: Structural Materials: Properties, Microstructure and Processing* 1(2):258–263.

Tucker, M. E. (ed.). 2001. *Sedimentary Petrology: An Introduction to the Origin of Sedimentary Rocks* (3rd ed.). New York: Wiley-Blackwell.

METAMORPHIC ROCKS 6

Chapter Outline

Definition and Occurrence of Metamorphic Rocks

Metamorphic rocks form within the Earth's crust from preexisting rocks through the action of heat and pressure. Metamorphism changes mineral composition and texture of the preexisting rocks through recrystallization and growth of existing minerals, formation of new mineral species, and rearrangement of mineral grains (Marshak, 2013). The rocks retain their solid form as metamorphism proceeds. If complete melting occurs, the process is igneous rather than metamorphic. In some cases, chemically active hot fluids permeate into the rock mass to cause the change in mineral composition. Together the heat, pressure, and chemically active fluids are known as the agents of metamorphism. The original rocks can be igneous, sedimentary, or metamorphic and are referred to as the protoliths (Marshak, 2013).

Metamorphic rocks occur adjacent to intrusive igneous rocks, in the roots of old, folded mountains and in Precambrian continental shields. As with intrusive igneous rocks, they are exposed at the Earth's surface when the overlying rocks are removed by erosion. For rocks formed by regional metamorphism this involves removal of at least 1000 m (3300 ft) of overlying rock.

Types of Metamorphism

There are two main types of metamorphism: contact metamorphism and regional metamorphism. Contact metamorphism occurs at or near the contact of an intrusive body of magma. The extent of change in the intruded country rock depends on: (1) temperature of the intruding magma; (2) composition of the intruding magma; and (3) composition of the intruded rock. Temperatures of contact metamorphism range from 300 to 800°C and pressures from 100 to 3000 atmospheres or bars (1 bar = 101.3 kPa or 14.7 lb/in^2). Contact metamorphism generally occurs in zones called *aureoles* that may vary in thickness from a few centimeters (an inch) to a few kilometers (1 to 2 miles), depending upon the size of the igneous body (Philpotts and Ague, 2009; Winter, 2009; Marshak, 2013; Nance and Murphy, 2016). These zones surround batholiths, stocks, or other large intrusive igneous bodies.

Examples of rock types formed along igneous contacts are skarns and hornfelses, both fine-grained massive rocks. Mineral assemblages of these rocks include wollastonite, diopside, andalusite, cordierite, garnet, and, occasionally, some ore minerals such as sulfides and oxides. Because these rocks contribute only a minor portion to the total volume of metamorphic rocks, they will not be discussed further in this text.

Regional metamorphism occurs over extremely large areas, involving thousands of square kilometers (thousands of square miles) and thicknesses of 1000 m (3300 ft) or more. When most minerals of preexisting rocks are subjected to the heat and pressure of regional metamorphism, they become unstable under the changed conditions. The increased heat weakens chemical bonds and accelerates the rate of chemical reaction, thus encouraging new minerals to form that are stable under these extreme conditions. The pressure encourages growing crystals to become elongated and align themselves parallel to each other. The final outcome is the formation of new minerals of increased size oriented in a preferred direction. Regional metamorphic minerals include (in order of increasing temperature and pressure) chlorite, muscovite, biotite, staurolite, garnet, andalusite, kyanite, K-feldspar, and sillimanite (Marshak, 2013).

Regional metamorphism occurs within the subduction zones of convergent plate boundaries (Philpotts and Ague, 2009; Winter, 2009; Marshak, 2013) and in sedimentary basins where a thickness of 9000 to 12,000 m (30,000 to 40,000 ft) of sedimentary rock has accumulated. As the depth of the subduction zone increases, the subducting plate is exposed to increasingly higher temperatures and pressures. For metamorphic rocks formed at such great depths to be exposed at the surface, extensive uplift and erosion of thousands of meters (thousands of feet) of overlying rock must have occurred. Typically, surface outcrops

of regionally metamorphosed rocks are extremely old, commonly Precambrian in age, because of the great expanse of time needed for the sedimentation, metamorphism, and erosion processes to occur.

Geologic Rock Cycle

Definite relationships exist among igneous, sedimentary, and metamorphic rocks. Geologic processes change one rock type into another. The way in which rock types and processes are related is shown in Figure 6.1 in a diagram known as the rock cycle.

Beginning with magma, crystallization yields igneous rock, which can be either intrusive or extrusive. These rocks weather, yielding broken rock fragments and other weathering products known as sediments. When sediments become lithified, they yield sedimentary rocks. Metamorphism of these rocks produces metamorphic rocks, which, in turn, on melting provide magma once more. Shortcuts across the rock cycle are possible; igneous rocks can be metamorphosed to metamorphic rocks and sedimentary or metamorphic rocks can weather to sediments.

The intricate history of a rock unit can be illustrated by following it around the rock cycle (see Figure 6.1). For example, quartzite is a metamorphic rock composed of metamorphosed quartz sandstone. The quartz grains in the sandstone existed as sediments before the rock became lithified. At some

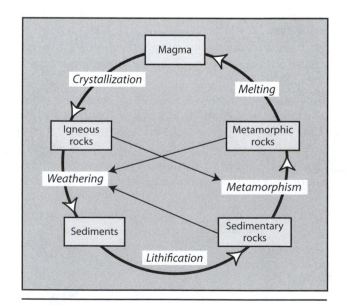

Figure 6.1   Geologic rock cycle.

point in its geologic history, the quartz sand weathered from a coarse-grained igneous rock, most likely a granite. Quartz in the granite probably comprised from 10 to 20% of the rock.

For example, consider a rock unit such as the Baraboo Quartzite, which is about 300 m (1000 ft) thick and likely once covered much of the state of Wisconsin. Five to 10 times this volume of igneous rock had to erode to provide the quartz. All of this took place prior to formation of the Baraboo Quartzite, about 800 million years ago. Wisconsin has an area of about 169,645 km^2 (65,500 mi^2) and if the quartzite averaged 300 m (1000 ft) thick, a rock volume of 51,685 km^3 (12,400 mi^3) is obtained for the Baraboo Quartzite. Five times this volume gives 258,425 km^3 (62,000 mi^3) for the original igneous rock if it contained about 20% quartz and none of the quartz was chemically destroyed during the weathering process.

Textures of Metamorphic Rocks

Most rocks that have undergone the extremes of heat and deforming pressure during regional metamorphism contain minerals arranged in parallel, flat layers or elongated grains in segregated bands of color. This feature of metamorphic rocks is known as foliation. Based on this property, metamorphic rocks can be subdivided into two groups, foliated and nonfoliated. The four distinct types of foliation are slaty, phyllitic, schistose, and gneissic, in order of increasing metamorphic grade or level of metamorphism. Foliated rocks also have a tendency to break along planes of weakness associated with the foliation. This property is known as rock cleavage and should not be confused with the cleavage of individual mineral grains discussed in Chapter 2.

Slaty cleavage consists of fractures in a rock along smooth planes that are separated by distances measuring less than a millimeter. This feature is much more pronounced than breaks along bedding in a sedimentary rock.

Phyllitic cleavage consists of parallel fractures that are closely spaced but generally farther apart than slaty cleavage. It occurs in rocks with a visible sheen of muscovite on the cleavages and these sometimes appear as crenulations or wavy undulations forming the cleavage surfaces.

Schistose cleavage consists of flakes or plates that are visible to the naked eye. The plates are formed by mica sheets or by needle-shaped minerals such as hornblende.

Gneissic structure consists of alternating bands of coarsely crystalline minerals a few millimeters or centimeters thick. Bands commonly consist of light-colored silicate minerals with thinner bands of dark minerals. These bands may or may not exhibit rock cleavage or actual foliation. With increases in platy minerals, the gneissic structure tends more and more toward schistose foliation. Figure 6.2 on the following page shows photographs of slate, phyllite, garnet-mica schist, and banded gneiss, exhibiting different degrees of foliation.

Types of Metamorphic Rocks

Names for foliated metamorphic rocks are typically based on their texture. For this reason, when the texture of these rocks is recognized, the naming of the rock is accomplished as well. In the case of schists, the minerals present provide the modifying term for the rock, such as hornblende schist or garnet-muscovite schist. For nonfoliated metamorphic rocks, mineral composition forms the basis for the rock name. Commonly occurring metamorphic rocks include slate, phyllite, schist, gneiss, marble, and quartzite.

Slate is a rock formed from shale through low-grade metamorphism. It is fine grained and its slaty cleavage is a result of the alignment of microscopic flakes of muscovite, formed from the clay minerals of the shale. It occurs in various colors including black, gray, green, and red. The black color is caused by carbonaceous material or fine pyrite and the green and red by the ferrous and ferric states of iron oxide, respectively.

Phyllite is a metamorphic rock with a composition similar to slate but with larger pieces and greater quantities of muscovite present. This provides a sheen to the cleavage surfaces. Crenulations of the cleavage surfaces are sometimes present, indicating a slightly higher grade of metamorphism than slate.

Schist is the most abundant rock formed by regional metamorphism. Typical of schists are the clearly visible flakes of platy minerals, for example, muscovite, biotite, talc, chlorite, or hematite.

Needle-shaped minerals such as hornblende may also dominate the composition. The prominent minerals present provide modifiers for the rock name, for example, muscovite schist or hornblende schist. Considerable amounts of quartz and feldspar may be present but in lesser amounts than are the platy and needle-shaped minerals. Garnet, magnetite, or other accessory minerals may also occur.

Gneiss is a coarse-grained metamorphic rock with a banded appearance caused by alternating layers of silicate minerals. It forms during high-grade, regional metamorphism. Although gneiss does exhibit rock cleavage to a certain extent, its cleavage is much less pronounced than that of a schist. As the bands of mica increase in a gneissic structure, rock cleavage becomes more developed. Gneisses formed from coarse-grained igneous rocks (e.g., granite and gabbro) consist of alternating bands of quartz,

feldspar, and hornblende or augite, whereas those derived from graywacke or other clayey sedimentary rocks contain muscovite and chlorite in addition to the other minerals. These platy minerals tend to increase rock cleavage characteristics. Figure 6.3 shows a strongly foliated and faulted banded gneiss.

Marble is a common metamorphic rock composed almost entirely of calcite or dolomite. It is coarse grained and is formed by either contact or regional metamorphism, although the latter occurrence is more common. Marble differs from its sedimentary counterparts by its larger grains and absence of sedimentary features, such as fossils and bedding planes, which are destroyed during metamorphism.

During metamorphism, mineral grains enlarge and tend to align themselves in a preferred direction or interlock, but the rock shows no foliation because only a single mineral species is present. Viewed

(a)

(b)

(c)

(d)

Figure 6.2   Hand specimens of (a) slate (nito, Shutterstock), (b) phyllite, (c) garnet-mica schist, and (d) banded gneiss, exhibiting varying degrees of foliation or rock cleavage. (b–d www.sandatlas.org, Shutterstock.)

Figure 6.3 Example of a strongly foliated banded gneiss.

under a petrographic microscope, the preferred orientation or interlocking of the carbonate crystals can be observed.

Marble is an important dimension stone used for interior decoration for walls, panels, and stairways of buildings but also as an exterior facing for buildings. Its various colors and structures result from impurities in the original sedimentary rock and from geologic occurrences such as faulting and brecciation with subsequent void infilling. Common colors are black, green, red, brown, and various shades of gray.

Quartzite, a common, nonfoliated metamorphic rock, forms by the metamorphism of quartz sandstone. The grains of quartz in the sandstone are firmly bonded by silica, which fills the pore spaces during metamorphism. It is distinguished from quartz sandstone by its more massive, dense nature, interlocking texture, lack of clastic particles, which are easily abraded, and the way it breaks through the constituent sand grains rather than around them.

Quartzite is distinguished from marble by its much greater hardness and the effervescence in acid evidenced by marble but not quartzite.

Details of the textural features of quartzite can be discerned only through use of a petrographic microscope. Both the original quartz grains and the overgrowths of quartz yielding an interlocking texture can be seen under the microscope.

Quartzites are white in color when no impurities are present in the rock. Several other colors occur as a consequence of iron or other minor constituents being present. Pink, red, green, and purple are common colors for quartzite.

Metamorphic Grade

In the metamorphic process a sequence of changes occurs as the intensity of heat and pressure increases. This is a measure of metamorphic grade, with low-grade rocks forming first and medium-grade and high-grade rocks thereafter if the degree of metamorphism continues to increase. Based on this concept, a sequence of changes can be noted with increasing extent of metamorphism. The changes that occur as shale and basalt metamorphose are as follows:

- shale → argillite → slate → phyllite → mica schist → garnet-mica schist → feldspar-silliminite gneiss → migmatite → granite

- basalt → greenstone → hornblende schist → feldspar amphibolite → hornblende migmatite → granodiorite

In these two examples, feldspar-silliminite gneiss and feldspar amphibolite are high-grade metamorphic rocks, whereas slate and greenstone are low grade. Hornblende is a type of amphibole, so that feldspar amphibolite is a banded rock of abundant feldspar and hornblende. Regional metamorphic grades are identified by the presence of diagnostic *index minerals* as follows (Philpotts and Ague, 2009; Winter, 2009; Marshak, 2013):

- low grade—first appearance of chlorite

- middle grade—first appearance of garnet

- high grade—first appearance of silliimanite

Metamorphic Rock Identification

Identification of metamorphic rocks is accomplished by use of Table A.7, provided in Appendix A. The first step is to recognize the type of foliation. This foliation determines the rock name and only a modifier based on the mineral content needs to be added. For the nonfoliated rocks, the mineral content is used to distinguish quartzite from marble. This is based on the distinction between quartz and calcite or quartz and dolomite; these differences were described in Chapter 2.

In Appendix A, a tabulation sheet for metamorphic rock identification is also provided (Table A.8). The information indicated by the table headings should be found for each sample. This leads directly to the proper identification of metamorphic rocks.

Rock Identification for All Three Varieties

In the discussions on igneous and sedimentary rocks, and again for metamorphic rocks, each genetic type was considered independently of the other two. After identifying rocks with this approach, one is presented with the challenge of distinguishing individual rocks from a collection comprised of all three rock types. This requires that, when assigning a name to an unknown specimen, a distinction be made among all the common rocks. This clearly is more difficult than when only one genetic group of rocks is considered at a time. To aid in this identification process of the total group of rocks, the following approach should be followed.

Several general rules of thumb are used to isolate one genetic rock type from the other two. These are based on the following considerations: hardness of rock; diagnostic minerals; directional features such as bedding, foliation, and orientation of grains; and other features, including presence of fossils, clastic texture, sedimentary structures, and organics.

Hardness

Igneous and metamorphic rocks are generally considered to be hard, whereas sedimentary rocks are soft. This is a result of the predominance of silicate minerals in igneous and metamorphic rocks whereas sedimentary rocks include carbonates and evaporites, which are inherently soft. The silicates measure 6 or more on Mohs hardness scale compared to carbonate and sulfates, which are in the range of 3 to 4. Also contributing to the soft aspect of sedimentary rocks is the clastic texture that many of these rocks possess. The fragments are only loosely cemented together so pieces are easily abraded from the rocks. Even though the grains themselves may be hard, as in a quartz sandstone, the friable nature of the rock gives it an appearance of being soft. Exceptions exist, as would be expected, such as hard sedimentary rocks (chert), soft igneous rocks (tuff), and soft metamorphic rocks (marble). However, these exceptions are easily remembered simply because they do not conform to the general rule.

Diagnostic Minerals

Certain minerals occur predominantly in only one of the genetic rock types. Garnet, muscovite, graphite, chlorite, and talc are more common in metamorphic rocks. Olivine occurs in igneous rocks whereas calcite, clay minerals, chlorides, sulfates, and chert most commonly occur in sedimentary rocks. Quartz and feldspar can occur in all three types.

Directional Features

Directional features are most prominently displayed as bedding in sedimentary rocks and as foliation in metamorphic rocks. Bedding planes are indicated by color and textural changes, by separation in the rock, and by the overall shape of rock pieces. Most sedimentary rocks show at least some indication of bedding, particularly those with clastic textures. Pyroclastics may show layering and some lava rocks show flow banding. Both are typically less prominent than is bedding in a sedimentary rock. Other details such as mineral composition can be used to tell layered igneous rocks from bedded sedimentary rocks.

Foliation in metamorphic rocks is more prominent than is bedding in most sedimentary rocks. The individual types of foliation—slaty cleavage, schistosity, and gneissic structure—are typically easy to identify, leading directly to rock identification.

Other Features

Other features include the presence of fossils, indicating a sedimentary origin; a clastic texture, suggesting either sedimentary rocks or pyroclastics (with rounded clastic pieces suggesting sedimentary origin); the presence of organics, depicting sedimentary rocks; certain features unique to specific rock types, such as small dikes running through igneous or metamorphic rocks; and the sedimentary features—stylolites, geodes, and nodules, for example. The absence of certain features can also be used for identification purposes. Lack of directional features may suggest an igneous origin, and coarse calcite grains without fossil evidence may suggest a metamorphic rock, such as marble.

By applying these concepts collectively, a major step toward rock identification is accomplished. Hardness and reaction to acid separate quartzite from marble, mineral content suggests the difference between tuff and graywacke, and degree of crystal interlock between granite and sedimentary breccia. Table 6.1 on the next page presents a summary of rock identification aids based on genetic rock type.

Engineering Considerations of Metamorphic Rocks

1. Foliated metamorphic rocks commonly yield rock pieces with elongated shapes when crushed. Gneiss, argillite, and phyllite may be used as concrete aggregates. Flat and elongated pieces cause mixing problems in fresh concrete and directional properties in hardened concrete.

2. Gneiss containing abundant mica can yield problems when used as a concrete aggregate or as building stone. Freeze-thaw and wetting-drying effects cause micas to fail along their prominent cleavage plane. Flaking of aggregates from the concrete surface results. Schists typically are not used as aggregates because of the abundant mica present.

3. Foliated rocks possess prominent directional properties. Strength, permeability, thermal conductivity, and seismic velocity are affected by the direction of foliation. Care should be taken that loads from bridges, dams, and building foundations are not transferred to foliated rock masses in a direction closely parallel to the foliation.

4. Metamorphic rocks may be deeply weathered in the eastern United States (Piedmont Plateau) and the depth to bedrock is quite variable. Saprolite, a partially weathered rock, is extensive. Care must be taken to found heavy structures, or to locate tunnel alignments, in sound rock whenever possible.

5. Coarse-grained gneisses, like granites of a similar size, abrade severely when used as aggregates for construction. These rocks lose gradation by abrasion, resulting in a reduction of particle size.

6. Slate, schist, and phyllite are subject to rock overbreak during blasting of rock cuts or tunnels because of their pronounced rock cleavage. High stress concentrations in tunnels may occur for the same reason.

7. The stability of rock slopes is greatly affected by the attitude of foliation with respect to the rock slope direction. When foliation dips steeply into an opening, rock slides commonly occur. Rock bolts or tendons may be needed to prevent such failures.

8. As mentioned in the sedimentary rock discussion (see Chapter 5), phyllite and argillite can yield alkali-silica reactive aggregates.

9. Marble is subject to the same problems as limestone. Solution cavities and channels may develop, resulting in similar problems of leakage of reservoirs and collapse of newly formed sinkholes.

10. Quartzite is a massive, hard rock and has a major abrasive effect on crushing and sizing equipment. It is more expensive to process than most rocks because of this property. Blasting would be a preferred method for underground excavation in quartzite over a tunnel boring machine.

Table 6.1   Identification Aids for Determining Genetic Rock Types.

Common Minerals	Textures	Structures in the Field	Other Features
Igneous Rocks			
Olivine[a]	Nondirectional, coarse to fine	Columnar jointing	Strong crystal interlock
Augite	Porphyritic	Flow banding	Nonfossiliferous
Hornblende	Vesicular		Usually hard rocks
Plagioclase	Amygdaloidal		Pyroclastics composed of
Orthoclase	Pyroclastic		angular, igneous pieces
Biotite, muscovite, quartz			
Sedimentary Rocks			
Quartz	Layered texture	Bedding	May be fossiliferous
Calcite	Clastic particles, usually	Concretions	Usually soft
Clay minerals[a]	rounded	Cross bedding	Mud cracks
Chemical precipitates[a]	Cement or matrix present	Ripple marks	
Halite	May be friable	Mud cracks	
Gypsum		Stylolites	
Organic material[a]		Geodes	
Coal, peat			
Chert[a]			
Metamorphic Rocks			
Garnet[a]	Commonly foliated:	Rock cleavage	Fossils absent
Muscovite, abundant[a]	slaty, phyllitic,	Minor folds and faults, dikes	Large crystals
Hornblende	schistose, gneissic		Nonclastic
Chlorite, abundant[a]			
Graphite[a]			
Talc[a]			
Quartz			
Orthoclase			

[a]   Good indicator of each individual genetic rock type.

EXERCISES ON METAMORPHIC ROCKS

1. Quartzite can be confused with basalt. What do they have in common? How would you distinguish between the two?

2. Prepare a list of the rocks and minerals listed in the identification tables for common minerals and rocks supplied in Appendix A. Put them in alphabetical order or in some other sequence that mixes the terms so they are no longer grouped into minerals and individual genetic groups of igneous, sedimentary, and metamorphic rocks. Place the correct symbol adjacent to each term, M for mineral, I for igneous, S for sedimentary, and X for metamorphic. Review the answers to ensure that you can accomplish this from memory.

3. Metamorphic rocks are found in the roots of old mountain systems and in continental shields. Describe what is meant by these two locations. Where in eastern North America would such rocks be found? Why are metamorphic rocks present in the stream gravels of the upper Midwest?

4. Coarse crystals of calcite can exist in limestone and they also occur in marble. How would you tell these two rocks apart? Describe how their origins differ. Be specific.

5. In certain geologic terrains, quartzite and granite are found in contact. Assume that both are essentially the same age and both were subjected to equal extremes of heat and pressure during the metamorphic process. Why has the quartz sandstone metamorphosed to quartzite but the granite has not changed to gneiss?

6. If a conglomerate is metamorphosed, what does it become? How would it be distinguished from the original conglomerate?

7. When shale is metamorphosed it changes into different rocks depending on the extent of metamorphism. Show the steps that occur with higher grades of metamorphism. What minerals form in this process?

8. Along a certain steep-sided stream valley, slate dips into the hillside on one side of the stream and into the valley on the other. Draw this in cross section. A road is to be placed along the base of the valley wall by removing rock in a side hill cut. On which side of the stream should the cut be placed? Explain. Why is this so? In addition to the foliation, what other rock weaknesses would be of concern? How would this affect the road cut?

9. A schist sample has an unconfined compressive strength (strength at failure) of 1200 kg/cm^2 (17,070 psi). If the sample tested was 0.8 cm (2 in) in diameter, what total load in kg was applied to the sample to cause failure? What is the value of the compressive strength in kg/cm^2, kg/m^2, kN/m^2, and tons/ft^2?

10. Analyze and then summarize into a short sentence each of the engineering considerations discussed for igneous rocks (Chapter 3), sedimentary rocks (Chapter 5), and metamorphic rocks (Chapter 6).

REFERENCES

Marshak, S. 2013. *Essentials of Geology*. New York: W. W. Norton.

Nance, D., and Murphy, B. 2016. *Physical Geology Today*. New York: Oxford University Press.

Philpotts, A., and Ague, J. 2009. *Principles of Igneous and Metamorphic Petrology* (2nd ed.). New York: Cambridge University Press.

Winter, J. D. 2009. *Principles of Igneous and Metamorphic Petrology* (2nd ed.). New York: Pearson.

ELEMENTS OF SOIL MECHANICS

7

Chapter Outline

Index Properties

Soil Classification

Soil Structure

Compaction

Effective Stress

Shear Strength of Soils

Stress Distribution

Consolidation

Engineering Considerations of Soil Mechanics

Soils are loose, unconsolidated agglomerations of mineral particles that can be easily separated by hand pressure or by immersion in water (Johnson and DeGraff, 1988). They are the byproducts of mechanical and chemical weathering of rocks (Marshak, 2013).

Soil mechanics is the discipline that deals with the action of forces on soil masses and the response of soils to those forces. It involves the scientific approach to understanding and predicting the behavior of soils for engineering purposes, such as when soils are used in building foundations, highway embankments, levees, embankment dams, and as structural fills. Soil engineering is the engineering specialty that involves the principles of soil mechanics.

A term related to this field of study is *geotechnical engineering*. This term refers to the knowledge of soil mechanics and other engineering disciplines combined with a basic knowledge of geology. This combined knowledge is applied with judgment to develop an engineering design. Geotechnical engineering consists primarily of soil mechanics, rock mechanics, foundation engineering, and some aspects of highway engineering.

In this chapter, only the basic principles of soil mechanics are considered. The purpose is to provide an overview of the subject with an emphasis on concepts related to geologic investigation and analysis. Some additional topics related to soil mechanics are presented in other chapters of the text.

Index Properties

Index properties of soils serve as indices of their engineering behavior. They include soil texture, grain size distribution, particle shape, phase relationships, void ratio, degree of saturation, water content, density, relative density, and Atterberg limits.

Soil Texture

The term *texture* involves the appearance or feel of a soil as determined by its particle size, shape, and gradation. The major textural classes are gravel, sand, silt, and clay. Sands and gravels are coarse-textured soils, whereas silts and clays are fine-textured. Fine-textured soils consist primarily of grains too small to discern individually with the naked eye. In addition, for engineering purposes, soils are subdivided into two types: granular and cohesive. Granular soils include sands and gravels and cohesive soils include clays. Cohesion is the property by which clay particles stick together without any confinement, and is attributed to the presence of electrostatic charges on very fine clay particles (Mitchell and Soga, 2005; Holtz et al., 2011). Mud sticking to one's shoes is an example of soil cohesion. Cohesive soils exhibit the property of plasticity, the ability to be rolled into a thin thread before breaking into small pieces. Sands and gravels are nonplastic and cohesionless, clays are plastic and cohesive, and silts have properties intermediate between the extremes of sands and clays; they are fine grained but typically nonplastic and cohesionless.

Cohesionless soils have no shear strength unless confined; that is, they lack the ability to hold together without confinement. Instead they obtain their strength from grain-to-grain contact, which provides frictional resistance between the grains. This requires that an external force be applied to provide the contact and the strength.

Grain Size Distribution

Grain size distribution (range of grain sizes) has an important effect on the engineering behavior of granular soils. Soil particles can range from boulder size (10^3 mm) down to extremely fine colloidal materials (10^{-5} mm), or across a range of 10^8. Particle sizes in the United States are designated by either British engineering or SI units. Millimeters are used for particles smaller than 4.75 mm and inches or fractions of inches are commonly used for larger particles. This system applies to grain size distributions of soil as well as for crushed stone.

Grain sizes between 4.75 mm and 0.074 mm are classified based on US standard sieve numbers. Table 7.1 shows mesh sizes for the most commonly used sieves. In the past, microns were used for designating sizes finer than the No. 200 sieve. For example, the No. 200 size is 0.074 mm or 74 μ. Under the SI designation, the term micrometer (μm) is preferred rather than micron, but both equal 10^{-6} m (10^{-3} mm).

The American Society for Testing and Materials (ASTM) has developed industry accepted standards outling the procedures for the testing of soil properties. For example, ASTM method D6913 is used for determining grain size distribution of coarse-grained soils. The dry sample of soil is shaken mechanically over a stack of successively finer sieves. For sizes smaller than the No. 200 standard sieve, sieve analysis is impractical because of the extremely small openings needed in the wire mesh for these sieves. Consequently, for fine-grained soils a hydrometer analysis (ASTM D7928) is used instead. The test is based on Stokes' law, which pertains to the settling velocity of spheres in a liquid.

The results of sieve or hydrometer analyses are plotted in the form of a grain size distribution curve. Particle size in millimeters is plotted against the cumulative percentage by weight (or mass) finer than that size. The ordinate is an arithmetic scale along which the percent finer is plotted, whereas the abscissa is the logarithmic scale used for plotting grain size. Typically, five-cycle semilog paper is required for grain size distribution curves. Figure 7.1 illustrates several grain size distribution curves.

The curve for a well-graded soil (e.g., glacial till) has an even distribution of particle sizes over a wide range. By contrast the curve for a poorly graded soil (e.g., dune sand), composed mainly of one size of particles, has the grain sizes concentrated in a rather narrow zone, yielding a very steep plot. A gap-graded or skip-graded soil lacks certain sizes in the complete gradation. This yields the flat portions on the grain size distribution curve where gaps in grading occur.

Table 7.1   Common US Standard Sieve Numbers and Openings.

US Standard Sieve No.	Sieve Opening (mm)
4	4.75
10	2.00
20	0.85
40	0.425
60	0.25
100	0.15
200	0.074

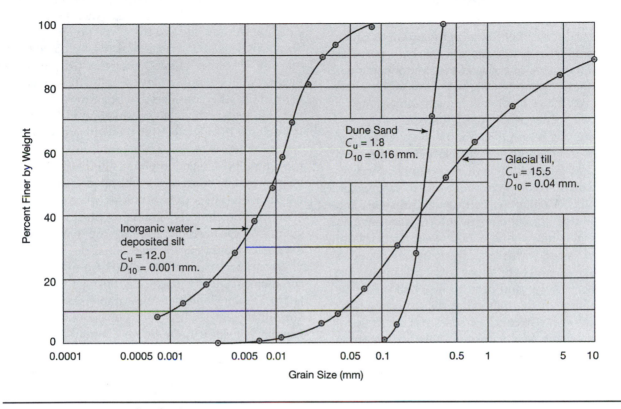

Figure 7.1   Grain size distribution curves.

Several measures are used to describe gradation curves or to quantify the information obtained from them. A common convention is to employ selected particle sizes on the curve, designated by the letter D with a subscript number to indicate the particle diameter for a given percent passing. For example, D_{10} represents a particle diameter that corresponds to 10% passing. More simply, 10% of the sample by weight is smaller than diameter D_{10}. In this example, D_{10} is also known as the effective particle size.

Coefficient of Uniformity

Coefficient of uniformity (C_u) is a measure of the degree of gradation. It is defined by:

$$C_u = \frac{D_{60}}{D_{10}} \qquad \text{(Eq. 7-1)}$$

where D_{60} = particle diameter (mm) corresponding to 60% passing and D_{10} is the diameter (mm) corresponding to 10% passing.

The smaller the value for C_u the more uniform the gradation (or the steeper the curve); $C_u = 1$ is the

minimum possible value and indicates a material of only one size. For example, standard Ottawa sand has a C_u of 1.1; ordinary beach sand, which has undergone some winnowing action by the waves, would have a C_u from 2 to 6. A well-graded soil would have a C_u from 15 to 25 and, typically, glacial tills of the Midwest would have a $C_u = 30$. Values of C_u as high as 1000 have been measured.

Coefficient of Curvature

A second parameter used in conjunction with the uniformity coefficient is the coefficient of curvature (C_c). It is defined by:

$$C_c = \frac{(D_{30})^2}{(D_{10})(D_{60})} \qquad \text{(Eq. 7-2)}$$

C_c is a measure of the second moment of grain distributions and it indicates more than the central tendency of the curve shown by C_u. A soil with a C_c between 1 and 3 is considered to be well graded as long as C_u is greater than 4 for gravels and 6 for sands.

Particle Shape and Roundness

Particle shape is a measure of the extent to which particles are equidimensional or the extent to which they deviate from that shape. On this basis particles can be equant, rod-like, flat, elongated, or combinations of these. Sphericity, the degree to which a particle approximates a sphere's equant dimensions, is another way of describing particle shape.

Roundness is an indication of the angularity of the edges and corners of a particle and is not related to shape. Based on the degree of roundness, particles are described as rounded, subrounded, subangular, and angular. A sphere is both equant in shape and highly rounded, a cube is equant but very angular. In soil mechanics the equant-shaped particles are described as bulky, whereas others are designated as needle-like (rod-like) or flakey (flat). Mica flakes are flakey and Ottawa sand particles are bulky. Particle shape influences the maximum density, compressibility, and shear strength of soils. In fact, in some instances, shape has a greater influence on these properties than does particle size.

Soils containing flakey particles, regardless of the particle size, are typically more compressible than those composed of bulky particles, either rounded or angular. Only a moderate amount of mica in sand will increase the compressibility markedly. Coarse-grained soils with angular particles have both greater strength and bearing capacity than those with rounded particles. In the case of clays, the addition of interlocking rod-shaped particles such as those comprising the mineral halloysite will provide added strength.

Phase Relationships

A mass of soil consists of solid grains of minerals with voids between them. The voids can be filled with air or water or filled in part by both of these constituents. The total volume of the soil mass (V_t) is equal to the volume of the solid (V_s) plus the volume of voids (V_v). In the same fashion, V_v is equal to the volume of the water (V_w) plus the volume of air (V_a). In equation form,

$$V_t = V_s + V_v = V_s + V_w + V_a \qquad \text{(Eq. 7-3)}$$

The three phases of the soil—solid, air, and water—can be represented schematically as though the soil were compartmentalized into these three components. Figure 7.2 is a phase diagram, or solid-water-air diagram, for the soil. Volumes are indicated on the left side and the corresponding masses on the right side. The total mass (M_t) is equal to the mass of the solid mineral particles (M_s) plus the mass of the water (M_w) present in the voids (the mass of the air is considered to be zero). Volumes are in cubic meters or cubic centimeters in the metric system and in cubic feet for the British engineering system, whereas masses are in grams or kilograms in the metric system and in pounds in the British engineering system of units. For saturated soils the phase diagram simplifies to two phases when V_a goes to zero. In geotechnical engineering practice, the total volume V_t, the mass of water M_w, and the mass of the dry solid M_s are typically measured. The other terms or parameters can be calculated from these measurements using the phase relationships. Most of the terms are not dependent on sample size and some are dimensionless. In solving phase-relation problems, it is helpful if the phase diagram is sketched first. The symbols used in the following discussion are shown in Table 7.2.

Volumetric Relationships

The three volumetric ratios commonly used to describe soils can be developed directly from the phase diagram. These are void ratio (e), porosity (n), and degree of saturation (S). Void ratio (e) is defined by:

$$e = \frac{V_v}{V_s} \qquad \text{(Eq. 7-4)}$$

Void ratio is expressed as a decimal. The maximum mathematical value for e is infinity and the minimum is zero. Practically speaking, however, values for granular soils range from about 0.3 to about 1.0, and for cohesive soils (clays) they range from 0.4 to about 1.5 with values as high as 5.0 for organic clays.

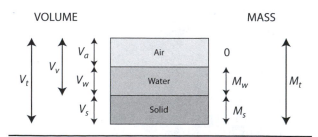

Figure 7.2   Phase diagram, or solid-water-air diagram, for soil.

Table 7.2 Symbols and Terminology for Phase Relations and Related Parameters.

Symbol	Term	Preferred Units
e	Void ratio	Decimal
D_r	Relative density	Decimal
G_s	Specific gravity of solid	Decimal
M_s	Mass of solid	kg
M_t	Total mass	kg
M'_t	Submerged mass	kg
M_w	Mass of water	kg
n	Porosity	%
S	Degree of saturation	%
V_a	Volume of air	m^3
V_s	Volume of solids	m^3
V_t	Total volume	m^3
V_v	Volume of voids	m^3
w	Water content	%
ρ	Total or wet density	kg/m^3
ρ'	Buoyant density	kg/m^3
ρ_d	Dry density	kg/m^3
ρ_s	Density of solids	kg/m^3
ρ_{sat}	Saturated density	kg/m^3
ρ_w	Density of water	kg/m^3
γ	Unit weight (used in British engineering units in same capacity as is in the preferred SI units)	lb/ft^3

Soils with higher void ratios are more compressible and have lower strength (Holtz et al., 2011; Das and Sobhan, 2018).

Porosity (n) is defined by:

$$n = \frac{V_v}{V_t} \times 100(\%) \qquad \text{(Eq. 7-5)}$$

Porosity is expressed as a percentage. The maximum, mathematically, is 100% and the minimum zero, but common values range from about 10 to 90% for soils, with clays and organic clays typically having higher porosities than granular soils. Although n is expressed as a percentage, in equations and calculations in geotechnical engineering, its decimal form is commonly used.

By substituting equivalent terms into the preceding equations for e and n, the following relationships between e and n can be obtained:

$$n = \frac{e}{1+e}$$
$$e = \frac{n}{1-n} \qquad \text{(Eq. 7-6)}$$

The degree of saturation (S) is defined by:

$$S = \frac{V_w}{V_v} \times 100(\%) \qquad \text{(Eq. 7-7)}$$

The degree of saturation indicates the percentage of the total volume of voids that is filled with water. Completely dry soil has $S = 0\%$ and for fully saturated soil, $S = 100\%$. Most cohesive soils, even those exposed to the Earth's surface under drying conditions, contain some moisture and therefore have a value of S greater than zero.

Mass Relationships

Moving to the right side of the phase diagram, the mass relationships are examined next. The first term of interest is the ratio known as the water content (w). It is one of the most significant soil parameters and is defined as:

$$w = \frac{M_w}{M_s} \times 100(\%) \qquad \text{(Eq. 7-8)}$$

Water content is typically expressed as a percentage and values range from zero (completely dry soil) to more than 200%. Because M_s is used in the denominator rather than M_t, values greater than 100% are possible. Values greater than 300% have been measured for organic soils and some marine clays.

The one remaining property of major significance that can be calculated based on the phase diagram relationship is density (ρ), or the mass per unit volume. Density equals mass divided by volume and, therefore, relates the right side to the left of the phase diagram in Figure 7.2. The unit used for density in the SI system is kilograms per cubic meter. For those unfamiliar with the magnitude of these units, the density of water = 1.000 g/cm^3, or 1000 kg/m^3, thus equating the cgs units with the SI units. In like manner, 1000 kg/m^3 = 1.000 Mg/m^3 where Mg stands for a megagram or 1 million grams. Therefore, 1.000 g/cm^3 = 1000 kg/m^3 = 1.000 Mg/m^3.

In the British system, unit weight is used despite a persistent confusion between weight and mass. The equivalence in this respect is 1.000 g/cm^3 = 62.4 lb/ft^3 = 1000 kg/m^3. Therefore, soil with a unit weight of 100 lb/ft^3 = (100/62.4)(1000) = 1603 kg/m^3 = 1.603 g/cm^3 = 1.603 Mg/m^3. Table 7.3 lists the factors used to convert SI units to British engineering units. The inside cover at the end of this textbook also show conversion factors used in geotechnical engineering and engineering geology. It should be noted that in this book we use kN/m^2 and kPa interchangeably (1 kN/m^2 = 1 kPa).

Several density measurements are used in geotechnical engineering. Two of these are total or wet density (ρ) and density of the solids (ρ_s). These densities are defined.

$$\rho = \frac{M_t}{V_t} = \frac{M_s + M_w}{V_t} \quad \text{(Eq. 7-9)}$$

and

$$\rho_s = \frac{M_s}{V_s} \quad \text{(Eq. 7-10)}$$

where ρ can range from just above 1000 kg/m^3 to as high as 2400 kg/m^3 or so. The density of the solids (ρ_s) equals G_s, the specific gravity of the mineral, times 1000 kg/m^3, the density of water. Hence,

$$\rho_s = G_s\rho_w \quad \text{(Eq. 7-11)}$$

where ρ_w = 1000 kg/m^3 (in SI units) and G_s ranges from about 2.0 to 3.0 for common rock-forming minerals. For most soil minerals, it lies between 2.65 (quartz) and 2.80.

Three other densities are used in geotechnical engineering: the dry density (ρ_d), the saturated density (ρ_{sat}), and the submerged density (ρ'). They are defined as follows:

$$\rho_d = \frac{M_s}{V_t} \quad \text{(Eq. 7-12)}$$

$$\rho_{sat} = \frac{M_s + M_w}{V_t} \quad \text{(Eq. 7-13)}$$

where $V_a = 0$ and $S = 100\%$.

$$\rho' = \rho_{sat} - \rho_w \quad \text{(Eq. 7-14)}$$

A useful relationship for soil density involves w, ρ, and ρ_d:

$$\rho_d = \frac{\rho}{1 + w} \quad \text{(Eq. 7-15)}$$

A final consideration before leaving the subject of density is the concept of relative density. It is extremely important for granular soils. Experience indicates that it is not the absolute or numerical value of density, expressed as mass per unit volume, that is important for granular soils. Instead, the *degree* of density relative to its loosest and densest possible conditions controls the behavior of cohesionless soils. This relationship is known as relative density (D_r) and is defined as follows:

$$D_r = \frac{e_{max} - e}{e_{max} - e_{min}} \quad \text{(Eq. 7-16)}$$

where e is the void ratio of the sample, or that for which D_r is being calculated; e_{max} is the maximum value of the void ratio (the loosest state) for a dry or submerged condition; and e_{min} is the minimum value of the void ratio (the densest state). Numerically D_r ranges from 0 for the loosest state to 1 for the densest state. Based on D_r, granular soils are categorized as very loose (D_r = 0–0.15), loose (D_r = 0.15–0.35), medium (D_r = 0.35–0.65), dense (D_r = 0.65–0.85), and very dense (D_r = 0.85–1.0) (Bowles, 1996).

The following examples of phase-relation problems will begin with a two-phase example on page 110.

Table 7.3   Conversion Factors for SI Units to British Engineering Units.

Linear Measurements

1 in = 25.4 mm = 0.0254 m

1 ft = 0.3048 m $\qquad$ 1 m = 3.28 ft

1 yd = 0.9144 m

1 mi (statute) = 1.609×10^3 m = 1.609 km

1 mi (nautical) = 1.852×10^3 m = 1.852 km

1 Å = 1×10^{-10} m

Area

1 in^2 = 6.452 cm^2

1 ft^2 = 0.0929 m^2 $\qquad$ 1 m^2 = 10.764 ft^2

1 acre = 43,560 ft^2 = 4046.8 m^2 = 0.405 hectares

1 hectare = 2.47 acres = 10,000 m^2

1 m^2 = 2.590 km^2 $\qquad$ 1 km^2 = 0.3861 mi^2

Weights and Masses

1 lb = 0.4536 kg

1 short ton = 2000 lb = 907.2 kg

1 g = 10^{-3} kg

1 metric ton = 1 tonne; t = 10^3 kg = 10^6 g = 1 Mg

1 slug (1 lb$_f$/ft/sec^2) = 14.59 kg

Velocity

1 mi/hr = 17.60 in/sec = 1.6093 km/hr = 44.71 cm/sec

10^{-6} cm/sec = 1.035 ft/yr

Force

The SI unit of force is obtained from F = Ma and is known as the newton (N); 1000 N = 1 kilonewton (1 kN).

1 lb$_f$ = 4.448 N

1 short ton = 8.896×10^3 N = 8.896 kN

1 kip = 1000 lb$_f$ = 4.448×10^3 N = 4.448 kN

1 metric ton$_f$ = 1000 kg$_f$ = 9.807×10^3 N = 9.807 kN

1 dyne (g-cm/sec^2) = 10^{-5} N

Pressure

The SI unit for pressure or stress is the pascal (Pa) and is defined as 1 N/m^2.

1 psi (lb$_f$/in^2) = 6.895×10^3 Pa = 6.895 kPa = 6.895×10^{-3} MPa

1 psf (lb$_f$/ft^2) = 1/144 psi = 47.88 Pa

1 kg$_f$/cm^2 = 9.807×10^4 Pa = 98.07 kPa

1 metric ton$_f$/m^2 = 9.807×10^3 Pa = 9.807 kPa

1 short ton$_f$/ft^2 = 9.576×10^4 Pa = 95.76 kPa

1 ksi (kip/in^2) = 1000 psi = 6.895×10^6 Pa = 6.895×10^3 kPa

1 atm at standard temperature and pressure = 1.013×10^5 Pa = 101.3 kPa

For calculating overburden pressure, $\sigma_v = \rho g h$ in SI units (N/m^2).

In terms of British engineering units, $\sigma_v = \gamma h$ (lb$_f$/ft^2), or for previous SI units, $\sigma_v = \gamma h$ (kg$_f$/m^2)

Density and Unit Weight

Density or mass per unit volume in the SI system is in kg/m^3.

The density of water (ρ_w) at 4°C is exactly 1 g/cm^3.

Hence, ρ_w = 1 g/cm^3 = 1000 kg/m^3 = 1.000 Mg/m^3 and is used in density calculations.

Unit weight (γ) is expressed in lb/ft^3. The unit weight of water on Earth at 5°C is 62.4 lb/ft^3.

62.4 lb/ft^3 = 1000 kg/m^3

1 lb/ft^3 = 16.018 kg/m^3

Example Problem 7.1

Given ρ_w, e, and G_s for a saturated soil, find w in terms of the given parameters.

Answers

First draw a two-phase diagram. For purposes of this solution we can assume $V_s = 1$. This is possible because any volume of soil can be assumed for the problem.

Because e is given and $e = V_v/V_s$, assume $V_s = 1$ because 1 in the denominator eliminates the fraction. Hence,

$$e = \frac{V_v}{1} = V_v$$

Because the soil is saturated, $V_v = V_w = e$. Now $M_w = V_w\rho_w$ because $\rho_w = M_w/V_w$:

$$M_w = e\rho_w$$

$$M_s = V_sG_s\rho_w$$

because

$$G_s = \frac{M_s}{V_s\rho_w}$$

$$M_s = 1G_s\rho_w = G_s\rho_w$$

Enter these values on the two-phase diagram you drew (see Figure 7.3). Realizing that $w = M_w/M_s$, we have

$$w = \frac{e\rho_w}{G_s\rho_w} = \frac{e}{G_s}$$

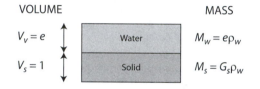

VOLUME		MASS
$V_v = e$	Water	$M_w = e\rho_w$
$V_s = 1$	Solid	$M_s = G_s\rho_w$

Figure 7.3   Phase diagram, two-phase, e given.

Example Problem 7.2

Given ρ_w, n, and w for a saturated soil (hence another two-phase problem), find G_s in terms of the given parameters.

Answers

First draw a two-phase diagram. Because n is given and $n = V_v/V_t$, let $V_t = 1$ to simplify the calculation. Therefore,

$$n = \frac{V_v}{1} = V_v = V_w$$

for the saturated case. By inspection, $V_s = V_t - V_w = 1 - n$. Also, $M_w = V_w\rho_w = n\rho_w$. These can be added to the diagram. Next we note that w is given and by definition

$$w = \frac{M_w}{M_s}$$

or

$$M_s = \frac{M_w}{w}$$

but $M_w = n\rho_w$. Therefore,

$$M_s = \frac{n\rho_w}{w}$$

To find G_s, note that: $M_s = V_s G_s \rho_w = (1 - n)G_s \rho_w$.
But, $M_s = n\rho_w/w$, so

$$\frac{n\rho_w}{w} = (1-n)(G_s)\rho_w$$

$$G_s = \frac{n\rho_w}{w(1-n)\rho_w} = \frac{n}{w(1-n)}$$

Enter these values on the two-phase diagram you drew (see Figure 7.4).

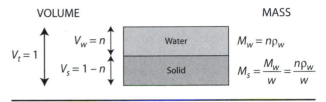

Figure 7.4   Phase diagram, two-phase, n given.

Example Problem 7.3

This is a three-phase problem. Given ρ_w, e, G_s, and S, find w in terms of the given parameters.

Answers

First draw a three-phase diagram. With e given, ($e = V_v/V_s$), let $V_s = 1$, then $e = V_v$. Because $S = V_w/V_v$,

$$V_w = SV_v = eS$$

The mass side of the equation can now be completed:

$$M_s = V_s G_s \rho_w = G_s \rho_w$$

$$M_w = V_w \rho_w = eS\rho_w$$

Then

$$w = \frac{M_w}{M_s} = \frac{eS\rho_w}{G_s\rho_w} = \frac{eS}{G_s}$$

Enter these values on the three-phase diagram you drew (see Figure 7.5).

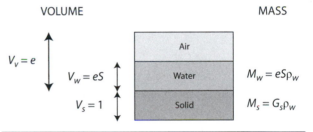

Figure 7.5   Phase diagram, three-phase, e given.

Example Problem 7.4

Given ρ_w, n, w, and S, find G_s in these given terms.

Answers

First draw a three-phase diagram. With n given, we observe $n = V_v/V_t$. Let $V_t = 1$ and $V_v = nV_t = n$.

$$S = \frac{V_w}{V_v}$$

so $V_w = SV_v = Sn$. By subtraction

$$V_s = V_t - V_v = 1 - n$$

$$M_w = \rho_w V_w = \rho_w Sn$$

Observing that $w = M_w / M_s$ and solving for M_s, we have

$$M_s = \frac{M_w}{w} = \frac{\rho_w nS}{w}$$

Also, $M_s = \rho_w G_s V_s$ or $\rho_w nS / w = \rho_w G_s (1 - n)$.

Cancelling ρ_w and solving for G_s,

$$G_s = \frac{nS}{w(1-n)}$$

Enter these values on the three-phase diagram you drew (see Figure 7.6).

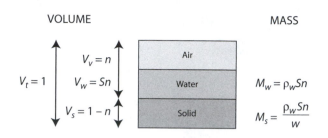

Figure 7.6 Phase diagram, three-phase, n given.

Example Problem 7.5

Here we determine w from soil measurements. A soil sample taken from the field had a mass of 396 g. After drying, the sample had a mass of 327 g. What was its natural moisture content?

Answers

$$w = \frac{M_w}{M_s} \times 100\%$$

$$M_s = \text{mass of the dry soil} = 327 \text{ g}$$

$$M_w = \text{mass of the water} = M_t - M_s = 396 - 327 = 69 \text{ g}$$

$$\therefore w = \frac{69}{327} \times 100 = 21.1\%$$

The above example problems show how void ratio, porosity, water content, and the specific gravity of solids are related. These example problems do not consider various densities of the soil. The following section explains how to determine various types of density.

The densities are best considered relative to void ratio, degree of saturation, and the specific gravity of solids. We begin with a two-phase system, that is, a saturated condition ($S = 100\%$). This is illustrated in Figure 7.7.

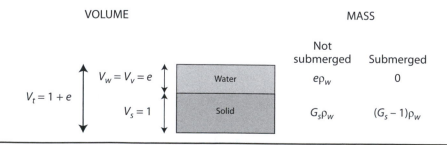

Figure 7.7 Phase diagram, two-phase, submerged versus not submerged.

With e given, as before, we let $V_s = 1$. Hence, $V_w = V_v = e$. Considering the not submerged situation, $M_w = \rho_w V_w = e\rho_w$ and $M_s = V_s G_s \rho_w = G_s \rho_w$ as $V_s = 1$ is assumed. Therefore,

$$\rho_{sat} = \frac{M_w + M_s}{V_t} = \frac{e\rho_w + G_s \rho_w}{1+e} = \frac{\rho_w(e+G_s)}{1+e}$$

$$\rho_d = \frac{M_s}{V_t} = \frac{G_s \rho_w}{1+e}$$

Next we consider the submerged mass. This equals the mass in air minus the buoyancy effect. The buoyancy effect is the volume times ρ_w, the density of water. For the water portion on the mass side we have

$$M_w - \text{buoyancy} = e\rho_w - e\rho_w = 0$$

For the solid portion on the mass side we have

$$M_s - \text{buoyancy} = G_s \rho_w - 1\rho_w = (G_s - 1)\rho_w$$

$$\rho_{submerged} \text{ or } \rho' = \frac{(G_s - 1)\rho_w}{1+e} = \frac{M_t'}{V_t}$$

Another way of expressing this is

$$\rho' = \rho - \rho_w = \frac{e\rho_w + G_s \rho_w}{1+e} - \rho_w$$
$$= \frac{e\rho_w + G_s \rho_w - \rho_w(1+e)}{1+e} = \frac{G_s \rho_w - \rho_w}{1+e}$$
$$= \frac{(G_s - 1)\rho_w}{1+e}$$

The final example involves the three-phase system for submerged densities, as illustrated in Figure 7.8. Working in terms of e, S, and G_s we start with e and set $V_s = 1$, $V_v = e$, $V_w = eS$, and $V_a = e(1 - S)$. For the mass side, not submerged, this yields

$$M_s = V_s G_s \rho_w = G_s \rho_w$$
$$M_w = V_w \rho_w = eS\rho_w$$
$$M_a = 0$$

Hence

$$\rho = \frac{M_w + M_s}{V_t} = \frac{eS\rho_w + G_s \rho_w}{1+e} = \frac{\rho_w(eS + G_s)}{1+e}$$

$$\rho_d = \frac{M_s}{V_t} = \frac{G_s \rho_w}{1+e}$$

The ρ_{sat} situation would occur when the air voids are filled with water so that $V_w = e$ and $M_w = e\rho_w$ and

$$\rho_{sat} = \frac{G_s \rho_w + e\rho_w}{1+e} = \frac{\rho_w(e + G_s)}{1+e}$$

Now we consider the submerged mass. For the water portion,

$$M_w - \text{buoyancy} = eS\rho_w - eS\rho_w = 0$$

For the solid portion,

$$M_s - \text{buoyancy} = G_s \rho_w - 1\rho_w = (G_s - 1)\rho_w$$

For the air portion,

$$M_a - \text{buoyancy} = 0 - e(1 - S)\rho_w = -e(1 - S)\rho_w$$

$$\rho_{submerged} \text{ or } \rho' = \frac{(G_s - 1)\rho_w + 0 - e(1 - S)\rho_w}{1+e}$$
$$= \frac{\rho_w[(G_s - 1) + e(S - 1)]}{1+e}$$

Note that $\rho_s = G_s \rho_w$ and ρ_w is known to be 1000 kg/m^3 or 1 g/cm^3; therefore, ρ_s or G_s can be determined directly when the other is provided.

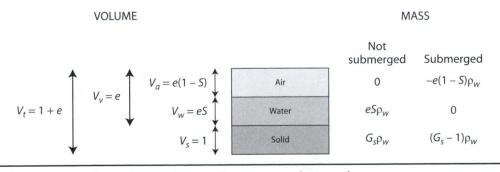

VOLUME					MASS	
					Not submerged	Submerged
		$V_a = e(1-S)$	Air		0	$-e(1-S)\rho_w$
$V_t = 1+e$	$V_v = e$	$V_w = eS$	Water		$eS\rho_w$	0
		$V_s = 1$	Solid		$G_s\rho_w$	$(G_s - 1)\rho_w$

Figure 7.8   Phase diagram, three-phase, submerged versus not submerged.

Two additional useful relationships are:

$$\rho_d = \frac{\rho_s}{1+e}$$

and

$$\rho_d = \frac{\rho}{1+w}$$

and from these, by substitution,

$$\rho = \rho_s \frac{(1+w)}{(1+e)}$$

Another example problem is presented to illustrate density calculations.

Example Problem 7.6

Given $e = 0.75$, $w = 18\%$, and $\rho_s = 2.65$ Mg/m³, find ρ, ρ_d, S, ρ_{sat}, ρ', and w for $S = 100\%$.

Answers

First draw a three-phase diagram. Let

$$V_s = 1$$
$$V_v = e$$
$$V_t = 1.75$$
$$M_s = V_s G_s \rho_w = V_s \rho_s = 1 \times 2.65$$
$$= 2.65$$
$$w = M_w / M_s \text{ or } M_w = wM_s$$
$$= (0.18)(2.65) = 0.477$$
$$M_w = \rho_w V_w$$
$$V_w = M_w / \rho_w = 0.477$$

Then

$$\rho = \frac{M_w + M_s}{V_t} = \frac{0.477 + 2.65}{1.75} = 1.79 \text{ Mg/m}^3$$

$$\rho_d = \frac{M_s}{V_t} = \frac{2.65}{1.75} = 1.51 \text{ Mg/m}^3$$

$$S = \frac{V_w}{V_v} \times 100\% = \frac{0.477}{0.75} \times 100 = 63.6\%$$

$$M_{w(sat)} = \rho_w V_v = 1.0(0.75) = 0.75$$

$$\rho_{sat} = \frac{M_s + M_{w(sat)}}{V_t} = \frac{2.65 + 0.75}{1.75} = 1.94 \text{ Mg/m}^3$$

$$\rho' = \rho_{sat} - \rho_w = 1.94 - 1.00 = 0.94 \text{ Mg/m}^3$$

$$w \text{ for } S \text{ @ } 100\% = \frac{0.75}{2.65} \times 100(\%) = 28.3\%$$

Enter these values on the three-phase diagram you drew (see Figure 7.9).

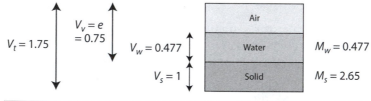

VOLUME

$V_t = 1.75$ $V_v = e$ $= 0.75$ $V_w = 0.477$ $V_s = 1$

MASS

Air

Water $M_w = 0.477$

Solid $M_s = 2.65$

Figure 7.9   Phase diagram, three-phase, density calculation.

Atterberg Limits

Atterberg limits, or consistency limits, are the water contents at which marked changes occur in the engineering behavior of fine-grained soils. They were proposed by Albert Atterberg (1911), a Swedish soil scientist. Consistency denotes the relative ease with which fine-grained soils can be deformed. It depends greatly on the nature of clay minerals present and on the water content. A comparison of natural water content with Atterberg limits can be used to predict the consistency or engineering behavior of fine-grained soils.

The Atterberg limits relevant to engineering practice include liquid limit (LL), plastic limit (PL), and shrinkage limit (SL). The consistency of soils in relation to the Atterberg limits is shown in Figure 7.10.

The liquid limit is the lowest water content at which a soil will behave as a viscous liquid, the plastic limit is the lowest water content at which a soil will behave as a plastic material, and shrinkage limit is the lowest water content beyond which no further change in volume occurs as the soil-water mixture dries. The range in water content over which soils behave plastically is the plasticity index (PI) and is equal to LL – PL. The behavior of a soil depends greatly on its present water content as compared to the LL or PL. If a soil's water content is near its liquid limit, it will be more compressible and probably less permeable than if it is near the plastic limit, where it is stronger and less compressible.

The liquid limit and plastic limit are determined through standardized tests because, otherwise, the definitions become too arbitrary. Liquid limit is defined as the water content at which a groove cut through a soil pat, using a standard grooving tool, will close for a distance of 1 cm when the cup containing the soil pat in the standard liquid limit device is dropped 25 times for a distance of 1 cm onto a hard rubber base. It is virtually impossible to moisten soil to just the right moisture content such that exactly 25 blows will close the groove by the standard distance. Therefore, ASTM method D4318 for liquid limit determination recommends testing five to six samples so that approximately half require fewer than 25 blows to close the groove for 1 cm and half need more than 25 blows, and plotting water contents (determined by oven drying the tested samples for 24 hours at 105°C) versus the logarithm of the corresponding number of blows. A straight-line relationship exists between the water content and the log of the number of blows required to close the groove. Where the straight line crosses the 25 blows defines the liquid limit. If the slope of the line is known in advance, only one point is needed to define the line. The slope of the line depends on the geologic origin of the soil.

The plastic limit test (ASTM method D4318) requires some experience before reproducible results can be obtained. The plastic limit is the water content at which a thread of soil crumbles when rolled into a diameter of 3 mm (1/8 in). It should break into segments 3 to 10 mm (1/8 to 3/8 in) long.

The range of liquid limits can vary from 0 to over 600, but most soils have values below 100. Plastic limit values range from 0 to about 125, but the PL for most soils is 40 or less.

Atterberg limits provide the basis for classification of fine-grained soils, as indicated by Casagrande's (1932, 1948) plasticity chart. Atterberg limits, including PI, also exhibit significant correlations with a number of engineering properties of fine-textured soils (Mitchell and Soga, 2005; Holtz et al., 2011).

Liquidity Index

The Atterberg limits also provide a rough measure of sensitivity of a soil to disturbance or shearing. This is indicated by the liquidity index, which is defined by

$$LI = \frac{w_n - PL}{PI} \qquad \text{(Eq. 7-17)}$$

where w_n is the natural water content for the soil in question. When the liquidity index is 1.0, the soil is at the liquid limit and it possesses little strength. This suggests a high sensitivity. Values greater than 1.0 indicate ultrasensitive or quick clays (values as great as 1.9 have been measured but most lie between 1 and 1.2). These values indicate soils that are extremely sensitive to a sudden collapse in structure when sheared. If disturbed in any way they can flow like a viscous liquid. The St. Lawrence River Valley in eastern Canada and clays in Scandanavia are well-known examples of these ultrasensitive soils.

A liquidity index of zero indicates that the soil is at its plastic limit and is probably not sensitive. If a negative value for LI is obtained, it means the soil is below its plastic limit and will fail as a brittle material when sheared. The advantage of the liquidity index is that a good indication of soil sensitivity can be obtained from the index properties of the soil: w_n, LL, and PL.

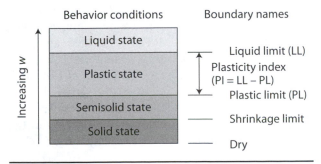

Figure 7.10 Consistency of soils as indicated by Atterberg limits.

Activity Index

The final consistency index for consideration is known as the activity index. This is a measure comparing the clay-like behavior of a soil to the portion of the soil that contributes to that behavior. Therefore, the higher the activity the more clay-like the soil must be.

The activity (A) of a clay is defined as

$$A = \frac{PI}{\text{clay fraction}} = \frac{PI}{\% < 2\,\mu\text{m}} \qquad \text{(Eq. 7-18)}$$

The 2-μm (μm = 10^{-6} m) level is the upper boundary for colloidal clay. Clays with activity index values less than 0.75 are inactive clays; those with activities between 0.75 and 1.25 are active clays. A fair to good correlation exists between activity and clay mineral types. For the common clays, montmorillonite (smectite) generally has the highest activity and kaolinite and halloysite have the lowest (Mitchel and Soga, 2005). Activity index has been also correlated with the swelling potential of clay soils (Seed et al., 1962; van der Merwe, 1964).

Soil Classification

The two widely used soil classifications in engineering practice are: (1) Unified Soil Classification System (USCS) and (2) American Association of State Highway and Transportation Officials (AASHTO) classification. The USCS is used by most engineering firms, US Army Corps of Engineers, and US Bureau of Reclamation (USBR), whereas the AASHTO classification is used primarily by the US Department of Transportation.

Unified Soil Classification System

The USCS, developed by Casagrande (1948), is one of the most commonly used classification systems in engineering. According to this system, coarse-grained soils are classified on the basis of grain size distribution and fine-grained soils on the basis of plasticity characteristics as indicated by Atterberg limits. Soils for which more than 50% by weight is retained on sieve No. 200 (0.074 mm) are considered coarse-grained and those with more than 50% passing the No. 200 sieve are classified as fine-grained. Coarse-grained soils are categorized

as gravels if more than 50% material is retained on No. 4 sieve (4.75 mm) and sands if more than 50% material passes the No. 4 sieve. Gravel is considered coarse if it is between 19 to 75 mm and fine if it is between 4.75 to 19 mm. Sand is further classified into coarse sand (2.00 to 4.75 mm), medium sand (0.425 to 2.00 mm), and fine sand (0.074 to 0.425 mm)

Silts and clays, according to USCS, are distinguished on the basis of plasticity characteristics, not particle size. This is accomplished by plotting LL and PI values on Casagrande's plasticity chart (see Figure 7.11). All points falling above the A-line in Figure 7.11 represent clays and those falling below the A-line indicate silts or organic soils. Further subdivision is based on whether the LL is more or less than 50. According to Casagrande (1948), plasticity characteristics are more predictive of the engineering behavior of fine-grained soils than particle sizes.

In the USCS, the following abbreviations are used for the classification of soils: gravel (G), sand (S), silt (M), clay (C), organic soil (O), and peat (Pt). Descriptive terms can be combined with soil abbreviations to provide more information: well-graded (W), poorly graded (P), high plasticity (H), and low plasticity (L). For example, GW is used for well-graded gravel, ML for silt of low plasticity (LL < 50), and CH for clay of high plasticity (LL > 50). Dual symbols are used for coarse-grained soils with 5 to 12% fines (material finer than 0.074 mm) or for fine-grained soils that have their LL and PI values fall between 12 to 25 and 4 to 7, respectively. Some examples of the use of dual symbols are GP-GM, SP-SM, and CL-ML.

AASHTO Classification System

The AASHTO classification uses grain size distribution and plasticity characteristics of material passing the No. 40 sieve. However, only three sieves (Nos. 10, 40, and 200) are used for grain size distribution analysis. Based on the percentages of the material passing Nos. 10, 40, and 200 sieves and plasticity characteristics of minus 40 material, the soil is divided into eight groups designated as A-1 through A-8. Groups A-1, A-2, and A-7 are further divided into subgroups. Additionally, a parameter referred to as the group index is used to rate the quality of the soil as highway subgrade material. The group index is defined as:

$$(F - 35)\,[0.2 + 0.005\,(LL - 40)]$$
$$+ 0.01\,(F - 15)\,(PI - 10) \qquad \text{(Eq. 7-19)}$$

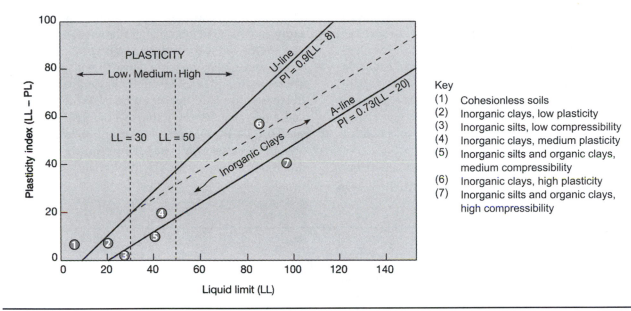

Figure 7.11   Plasticity chart showing classification of fine-grained soils based on LL and PI.

where F = percentage passing No. 200 sieve expressed as a whole number; LL = liquid limit; and PI = plasticity index. Negative values of group index are reported as zero. The group index indicates the percentage and nature of the material passing No. 200 sieve (fines). The amount and nature of fines significantly influence the engineering behavior of soil used as highway subgrade. The higher the value of the group index, the more undesirable are the properties of the soil for highway construction. Groups A-1 and A-2 represent excellent to good rating, groups A-3 to A-7 indicate fair to poor material, and group A-8 indicates unsatisfactory soil material. Table 7.4 on the following page shows the details of AASHTO classification.

Soil Structure

The structure of a soil includes both the geometric arrangement of the particles and the interparticle forces acting between them. Soil fabric, by contrast, pertains only to the geometric arrangement of the particles. For granular or cohesionless soils the forces between particles are so small that soil structure, in fact, is the same as soil fabric. This is particularly true for gravel and sand but to a lesser extent for silt.

In cohesive soils, the interparticle forces are much greater so that both the fabric and the attractive forces must be considered in the analysis of structure.

Since the structure in both cohesive and cohesionless soils has much to do with their engineering behavior, a detailed consideration of the subject is warranted. Because of basic differences in interparticle forces, it is important to discuss soil structure for cohesionless and cohesive soils separately.

Structure of Cohesionless Soils

Individual soil particles that settle, as independent grains, from a fluid medium yield an open packing known as a single-grain structure. These grains are generally coarse silt size or larger than 0.02 mm, and typically consist of gravel, sand, or mixtures of sand and silt. Gravity rather than surface forces controls the settling process of these particles. The medium of deposition can be either water or air. Materials formed in water include stream deposits and alluvial fans, beaches, and deltas; those settling from air include loess, sand dunes, and pyroclastic deposits (mostly volcanic ash).

Single-grain structures of granular soils are described according to their relative density. These soils can be loose (high void ratio, low density, low relative density) or dense (low void ratio, high density, high relative density) or somewhere between these extremes. The relative density of granular soils has a strong influence on their engineering behavior. Loose soils ($D_r < 0.5$) tend to densify under sudden vibrational loading, yielding a sudden decrease in volume and momentary loss of strength. Saturated fine sands and silts are particularly prone to a sudden loss of shear strength during earthquake vibrations. This phenomenon, known as liquefaction, is considered in Chapter 15 in the discussion on slope stability.

A type of open fabric that forms in some cases is the honeycombed structure, which has a very high void ratio. It is metastable and sensitive to collapse under dynamic loading or vibrations.

Note, however, that relative density alone may not fully describe the structure of a granular soil.

This can be illustrated, for example, by two sand samples that have identical gradations and identical void ratios but very different soil fabrics. An open structure versus an evenly dense one is an example. Therefore, their engineering behavior will be quite different although the standard tests of gradation and void ratio determination would suggest closely similar soils. A measure of the grain contacts per unit area in cross section is needed to show a numerical difference between these two materials.

Table 7.4   AASHTO Classification.

Group	Subgroup	% Passing US Sieve No. 10	% Passing US Sieve No. 40	% Passing US Sieve No. 200	Character of Fraction Passing No. 40 Sieve — Liquid Limit	Character of Fraction Passing No. 40 Sieve — Plasticity Index	Group Index No.	Soil Description	Subgrade Rating
A-1			50 max	25 max		6 max	0	Well-graded gravel or sand; may include fines	
	A-1-a	50 max	50 max	15 max		6 max	0	Largely gravel but can include sand and fines	
	A-1-b		50 max	25 max		6 max	0	Gravelly sand or graded sand; may include fines	
A-2[a]				35 max			0–4	Sands and gravels with excessive fines	Excellent to good
	A-2-4			35 max	40 max	10 max	0	Sands, gravels with low plasticity silt fines	
	A-2-5			35 max	41 min	10 max	0	Sands, gravels with clastic silt fines	
	A-2-6			35 max	40 max		4 max	Sands, gravels with clay fines	
	A-2-7			35 max	35 min		4 max	Sands, gravels with highly plastic clay fines	
A-3		51 max	10 max			Nonplastic	0	Fine sands	
A-4				36 min	40 max	10 max	8 max	Low compressibility silts	
A-5				36 min	41 min	10 max	12 max	High compressibility silts, micaceous silts	
A-6				36 min	40 max	11 min	16 max	Low to medium compressibility clays	Fair to poor
A-7				36 min	41 min	11 min	20 max	High compressibility clays	
	A-7-5[b]			36 min	41 min	11 min	20 max	High compressibility silty clays	
	A-7-6[b]			36 min	41 min	11 min	20 max	High compressibility, high volume-change clays	
A-8								Peat, highly organic soils	Unsatisfactory

[a] Group A-2 includes all soils having 35% or less passing a No. 200 sieve that cannot be classed as A-1 or A-3.

[b] Plasticity index of A-7-5 subgroup is equal to or less than LL − 30. Plasticity index of A-7-6 subgroup is greater than LL − 30.

Structure of Cohesive Soils

As mentioned previously, the structure of cohesive soils is dependent on both the interparticle forces between the grains and on the fabric. In this discussion on cohesive soils these two factors are considered.

Interparticle Forces of Clays

An analysis of the interparticle forces of clays begins with a view of clay mineralogy. This pertains to the two-layer and three-layer clays and the properties that develop because of these individual clay structures. In this text, clay minerals are considered in Chapter 2 as part of the discussion on minerals. A review of that section will provide details regarding clay mineralogy. Figure 7.12 shows that clay mineralogy is related to their plasticity characteristics with individual clay-mineral groups plotting in specific locations on the plasticity chart (Mitchell and Soga, 2005). Glacial clays from the Great Lakes region consist predominantly of illite and they plot just above the A-line on this figure.

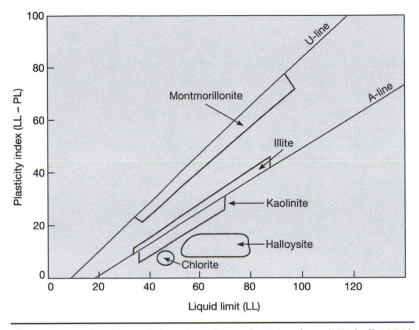

Figure 7.12   Location of clay minerals on plasticity chart. (Mitchell, 1976.)

Specific Surface and Particle Size

Specific surface is the ratio of the surface area of a solid to its volume, that is,

$$\text{Specific surface} = \frac{\text{surface}}{\text{volume}} \qquad \text{(Eq. 7-20)}$$

For a sphere it equals

$$\frac{\pi d^2}{\pi d^3/6} = \frac{6}{d} \qquad \text{(Eq. 7-21)}$$

The units are in reciprocal millimeters or reciprocal inches.

Specific surface is inversely proportional to grain size. Since water molecules are attracted to the surface of soil grains, the amount of moisture per unit volume of soil will increase with a decrease in grain size. This is why cohesive soils tend to have a higher water content than do granular or cohesionless soils with their much larger grain diameters.

Individual particles comprising the different clay minerals are by no means the same size. The more active clays tend to be much smaller in size than those with low activity. This is shown diagrammatically in Figure 7.13 on the next page. Montmorillonite (smectite) is considerably smaller than illite or chlorite (about one-tenth the thickness) and considerably smaller than kaolinite. It is no wonder that the amount of swelling for montmorillonite is much greater than that of kaolinite. The relative sizes of montmorillonite and kaolinite with regard to a layer of water is shown in Figure 7.13. Clay particles are almost always hydrated by a layer of water surrounding each crystal. This is known as adsorbed water, that is, water held as an extremely thin layer on the surface of the clay.

Types of Interparticle Bonds in Clays

Several types of forces or bonds are known to act on small particles such as cohesive clays and considerable differences exist in the strength of these forces. The strongest are primary valence bonds, which bind atoms together. It is not likely that any of these are broken in engineering soils applications but they serve as a point of comparison.

Hydrogen bonds occur because of the dipole nature of water, and they cause the positive end of the water dipole to attach to oxygens or hydroxls on the clay surface. Hydrogen bonds have about one-tenth the strength of primary valence bonds and act over a distance of about 2 to 3Å (Å = 10^{-8} cm = 10^{-10} m).

Another force that attracts a water layer is the cation bond. Cations in the water are attracted by the negative charges on the clay layers. These bonds are somewhat weaker than hydrogen bonds and are less important as attractive forces between the clay surface and the water layer.

Additional forces that act on particles of molecule size and larger are known as van der Waals forces. These are weak forces about 1/100 the strength of primary valence bonds and 1/10 the strength of hydrogen bonds. They act over distances of about 5 Å, and are inversely proportional to r^3 up to r^7 depending on the specific molecules involved (where r = radius of the molecule).

The electrostatic bond is the last and the weakest of the forces involved in the attraction of water layers. They act over distances much greater than 5 Å.

Causes for Negative Charge on Clays

The source of the negative charges on the clay surfaces, which are neutralized by the water layers, is related to two primary causes. The substitution of cations in the silica and the octahedral sheets of the clays yield charge deficiencies, and terminated crystal lattices or broken edges provide the other. In this regard, the smaller the crystal the greater the specific surface and the more crystal terminations.

Cation Exchange Capacity

Various types of clay have different crystal sizes and different charge deficiencies and, thus, different capacities to accept these neutralizing cations. One cation can be exchanged stoichiometrically (valence for valence) for another. Therefore, a plus-two cation (Ca^{2+}, for example) can replace 2 plus-one cations (Na^+, for example) on the lattice. The ability to take on cations by a clay is known as cation exchange capacity. The values for this parameter are given in Table 7.5.

Some exchangeable cations, such as calcium and magnesium, are particularly common in soils whereas potassium and sodium are less abundant. Aluminum and hydrogen are the common cations of acidic soils. Marine clays contain primarily sodium and magnesium, which are the most common cations in the oceans. The presence of organic matter complicates cation exchange in soils.

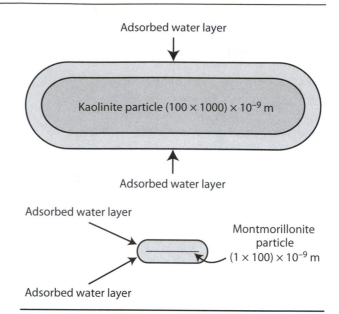

Figure 7.13   Relative sizes of different clay mineral particles.

In attenuation-type sanitary landfills, where soils are used to enclose solid wastes, the clays in the containment soils remove heavy cations from the leachate by their base-exchange ability. Cohesionless soils have a limitation in this regard because their higher permeability and low clay content allow leachate to be more quickly released through the soil and not be cleansed of the heavy cations.

Some cations have a greater ability to replace existing ions on the clay than do others. Higher valence cations can replace those of lower valence. When the ions are the same valence, the larger the ion the greater the replacement capability. The approximate order of replacement ability is displayed here. The exact order depends on the type of clay involved, the ions to be replaced, and the concentration of

Table 7.5   Cation Exchange Capacity of Clay Minerals.

Clay Minerals	Exchange Capacity (milliequivalents per 100 g of dry clay)
Kaolinite	3–15
Halloysite	5–40
Illite	10–40
Montmorillonite (smectite)	80–150

the ions in the water. The order of replacement in increasing order is essentially

$$Li^+ < Na^+ < H^+ < K^+ < NH4^+ < Mg^{2+} < Ca^{2+} < Al^{3+}$$

Interaction of Forces

Individual clay particles interact with each other through their layers of adsorbed water. When two particles approach each other, they are attracted by their van der Waals forces, cation bonds, and electrostatic bonds. When the particles become so closely associated that their adsorbed water layers overlap, repulsion occurs. Hence the particles arrange themselves in response to these attractive and repulsive components.

The presence of various ions and organics and their relative concentrations influence the nature of these forces. The net effect is that individual particles can flocculate together, yielding a flocculated arrangement, or be repelled from each other, providing a dispersed arrangement. A schematic of these structures is shown in Figure 7.14. The flocculation that occurs most commonly in water with a high salt concentration, such as seawater, is shown in Figure 7.14a. Under these conditions the particles stick together in the fashion in which they first make contact. This is an example of face-to-face contact.

The edge-to-edge attraction tends to occur in acidic water where the hydrogen ions attach themselves to the edge of the particles. This is known as low salt flocculation and is illustrated in Figure 7.14b. It is an example of an edge-to-face structure. The dispersed structure occurs when the adsorbed water layer increases, yielding an overlap of the water layers between clay particles, as illustrated in Figure 7.14c. The dispersed structure commonly predominates in clays carried in suspension by freshwater. When this water flows into the ocean, the high salt concentration causes flocculation to occur. This converts the clay structure to that shown in Figure 7.14b.

In summary, the three possible configurations of flocculated particles are edge-to-face, edge-to-edge, and face-to-face. Of these, edge-to-face is the most common and can result in a honeycomb structure. The tendency toward flocculation is enhanced by an increase in (1) concentration of electrolytes, (2) valence of the ion, and (3) temperature; or by a decrease in (1) dielectric constant of the fluid, (2) size of the hydrated ion, (3) pH, and (4) anion adsorption (Lambe, 1958).

Two properties of clay structures are of major interest in this discussion: sensitivity and thixotropy. Sensitive clays are those clays whose flocculent structure becomes unstable because of the leaching of salts and is readily dispersed upon vibrations. Sensitivity is the ratio of the undisturbed strength to the remolded strength for a soil sample tested at the same water content. It occurs because of disruption of the structure when the clay is remolded. Sensitivity for clay soils ranges from 1.5 to more than 8. Ordinary clays have a sensitivity of less than 4, whereas sensitive clays range from 4 to 8. Clays with sensitivities greater than 8 are called extrasensitive or quick clays.

The reduction in strength occurs when edge-to-face bonding is changed to face-to-face bonding on remolding. This pushes the attractive forces farther apart, yielding a lower shear strength in the remolded

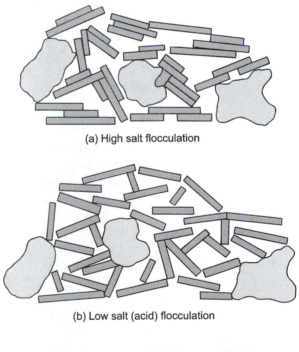

(a) High salt flocculation

(b) Low salt (acid) flocculation

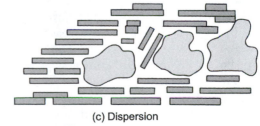

(c) Dispersion

Figure 7.14 Schematic diagram of clay structure.

state. Quick clays are also formed by the leaching of marine clays by freshwater. The addition of rainwater lowers the cation concentration, thus reducing the attractive forces in the clays. Cation bonds are reduced in this way, giving the clay a reduced shear strength.

Thixotropy is the gain in shear strength with time following remolding. It is a recovery of some of the strength lost by sensitive soils on remolding and it develops with time at a constant water content. Remolding of the clay initially causes a change in structure toward the dispersed arrangement. With time the particles begin to reorient toward a more stable flocculent structure to result in a stronger material. Thixotropic behavior is more pronounced for montmorillonite and other expansive clays than it is for kaolinite and similar minerals.

Cohesive Soil Fabrics

Cohesive soil fabrics are not simple collections of grains or particles that have become intergrown. Instead, they are thought to be aggregates of flocculated particles of differing sizes and assemblages with pore spaces between them. The structure is most easily understood by considering this feature on two scales, the macrostructure and the microstructure of the clays.

Macrostructure versus Microstructure

Macrostructure has an important influence on the behavior of soils with regard to their engineering properties. The behavior of a soil mass is dependent on the weakness zones or defects within it and these include, for example, the effects of bedding. This may be manifested as sand or silt layers, varves, zones of increased organic content, or layers in the soil profile. Other defects usually lie at a high angle to the bedding and they include joints, fissures, root holes, and root fillings. Great care must be taken to observe macrostructure effects in soils during site investigation because shear strength, permeability, and compressibility may have critical values at these locations.

Details of clay microstructure have been discerned by scanning electron microscope (SEM) studies. Results show that large collections of aggregated or flocculated particles known as peds can be observed with the naked eye. Peds consist of smaller assemblages of clay particles known as clusters. These clusters are observable with a petrographic microscope using transmitted, visible light when the particles lie in the range of 1 to 5 μm.

The smallest aggregates of clay particles congregate into submicroscopic units known as domains. Domains are aggregates of the clay crystal platelets that collect because of surface forces. Therefore, in order of increasing size the units are domains, clusters, and peds. These are illustrated in Figure 7.15.

The microstructure of clays is more important from a fundamental viewpoint, and the macrostructure is more significant where engineering behavior is of interest. The microstructure records the geologic history of the deposit, including details of deposition, environment, and weathering history.

Compaction

Compaction is densifying a soil through rearranging particles by applying mechanical energy. It

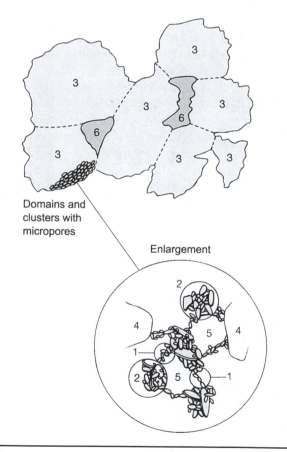

Domains and clusters with micropores

Enlargement

Figure 7.15   Schematic representation of soil microstructure and macrostructure: (1) domain, (2) cluster, (3) ped, (4) silt grain, (5) micropore, and (6) macropore. (Adapted from Holtz et al., 2011.)

may involve modifying the water content but outflow of water is not required. By contrast, consolidation is the process of densification that occurs in clayey soils as water is squeezed from its pores under load. Compaction and consolidation have different meanings in geology where they are applied to the process by which sediments change into sedimentary rocks.

Compaction is required when soils are used as a construction material in applications such as structural fill, highway and railroad embankments, earth dams and levees, cover and liner material for sanitary landfills, foundation material, and reclamation of mine waste embankments.

Compaction is measured in terms of dry density (ρ_d), which is defined as the mass of solids (mineral particles) per unit volume. In the field or laboratory, the bulk or wet density (ρ) and water content (w) are measured first and the dry density (ρ_d) is calculated using the following equations:

$$\rho = \frac{M_t}{V_t} \qquad \text{(Eq. 7-22)}$$

$$\rho_d = \frac{\rho}{1+w} \qquad \text{(Eq. 7-23)}$$

where M_t = total mass of soil and V_t = total volume of soil.

Engineering Significance of Compaction

Compacting soils in the field improves almost all desirable properties of soils. It:

1. Increases soil strength through increased density, which improves, among other characteristics, the slope stability of the fill.

2. Improves bearing capacities of pavement subgrades and base courses.

3. Reduces postconstruction settlement.

4. Reduces the tendency for volume changes such as those caused by frost action or by expansive soils.

Factors Affecting Compaction

Procedures for compacting soils were set forth in the early 1930s by R. R. Proctor. According to Proctor (1933), the degree of compaction a given soil can achieve depends on three factors: (1) water content, (2) compactive effort, and (3) soil type (coarse-grained versus fine-grained; grain size distribution; amount and type of clay minerals). The influence of these factors on density, a measure of the degree of compaction, is discussed in the following section on laboratory tests.

Laboratory Tests

Standard and Modified Proctor Tests

The standard Proctor test, with procedures outlined in ASTM method D698 and AASHTO T991, is a common laboratory test used to evaluate soil compaction. Equipment for the test includes a 2.495 kg (5.5 lb) hammer and a cylinder or mold 944 cm^3 or nearly 1 liter (1/30 ft^3) in volume. Three layers of soil are added to the cylinder with each layer compacted by dropping the hammer a distance of 0.3048 m (1 ft) 25 times on each of the layers. The layers are scarified with a knife to encourage better contact between them. This procedure is repeated on four or five samples of the same soil, each at different water content. A plot of water content versus dry density yields the standard Proctor compaction curve shown in Figure 7.16. Figure 7.16 shows that the dry density first increases and then decreases with increasing water content. The maximum dry density (MDD) is achieved at a water content known as the optimum water content (OWC).

Another laboratory test that involves a higher compactive effort is the modified Proctor test (see

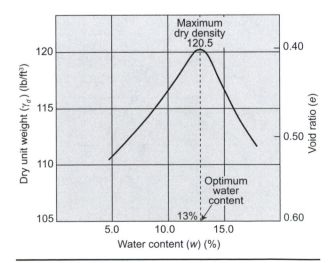

Figure 7.16   Standard Proctor compaction curve.
Note: 62.4 lb/ft^3 = 1 Mg/m^3.

methods ASTM D1557 and AASHTO T180). In the modified Proctor test, the hammer weighs 4.536 kg (10 lb) and is dropped a distance of 0.457 m (1.5 ft). The soil is compacted in five layers with each layer receiving 25 blows. The same mold is used (944 cm³ or 1/30 ft³) as in the standard Proctor test. The surface is scarified before addition of the next soil layer. The modified Proctor test results (Figure 7.17) show that for a given soil, an increase in compactive effort increases MDD and decreases OWC.

Figure 7.17 also shows a plot of the zero air voids line. This represents saturated conditions (or zero unfilled voids) for different water contents. The equation for this zero air voids (ZAV) condition can be derived in terms of ρ_w, G_s, and w using a two-phase soil relationship, as shown in Figure 7.18. Let

$$V_s = 1$$
$$M_s = V_s G_s \rho_w$$
$$w = M_w / M_s$$

or

$$M_w = w M_s = w G_s \rho_w$$
$$M_w = V_w \rho_w$$

or

$$V_w = \frac{M_w}{\rho_w} = \frac{w G_s \rho_w}{\rho_w} = w G_s$$

Finally

$$\rho_{(ZAV)} = \frac{G_s \rho_w}{1 + w G_s} \qquad \text{(Eq. 7-24)}$$

The effect of different soil types on dry density is shown in Figure 7.19, which illustrates two points. The first is that clayey soils have broader curves with a wider range of moisture contents, whereas silty and sandy soils have more steeply peaked compaction curves. This signifies that moisture control is very critical for silty and sandy soils if the optimum dry density is to be obtained. The second feature shown is that densities increase from clayey soils to silty soils to sandy silty soils and the optimum water content decreases. This again yields a line of optimums that slopes to the right.

Calculating Compactive Effort for the Standard and Modified Proctor Tests

In SI units, energy input per unit volume is given in joules per cubic meter (J/m³) and in British engineering units it is in ft-lb/ft³. Since 1 ft-lb/ft³ = 47.88 J/m³, the calculation of compactive effort (CE) for the standard Proctor test is as follows.

- SI units:

$$CE = \frac{2.495 \text{ kg} \times 9.81 \text{ m/sec}^2 \times 0.3048 \text{ m} \times 3 \times 25}{0.944 \times 10^{-3} \text{ m}^3}$$
$$= 592.7 \text{ kJ/m}^3$$

or

$$12,375(0.04788 \text{ kJ/m}^3) \times 1 \text{ ft-lb}_f/\text{ft}^3 = 592.5 \text{ kJ/m}^3$$

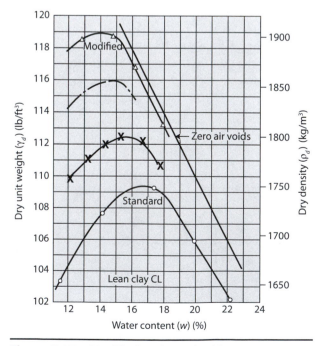

Figure 7.17 Modified and standard Proctor compaction curves. Note: 62.4 lb/ft³ = 1 Mg/m³.

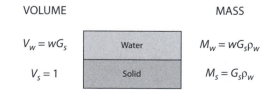

Figure 7.18 Derivation of dry density under zero air voids condition.

- British engineering units:

$$CE = \frac{5.5\ \text{lb}_f \times 1\ \text{ft} \times 3 \times 25}{1/30\ \text{ft}^3}$$

$$= 12{,}375\ \text{ft-lb}_f / \text{ft}^3$$

yielding the same answer when round-off effects are considered.

For the modified Proctor test:

- SI units:

$$CE = \frac{4.536 \times 9.81\ \text{m/sec}^2 \times 0.457\ \text{m} \times 5 \times 25}{0.944 \times 10^{-3}\ \text{m}^3}$$

$$= 2692.8\ \text{kJ/m}^3$$

or

$$56{,}250(0.04788\ \text{kJ/m}^3) \times 1\ \text{ft-lb}_f/\text{ft}^3 = 2693.2\ \text{kJ/m}^3$$

- British engineering units:

$$CE = \frac{10\ \text{lb}_f \times 1.5 \times 5 \times 25}{1/30\ \text{ft}^3}$$

$$= 56{,}250\ \text{ft-lb}_f / \text{ft}^3$$

which is the same answer when round-off effects are considered. We see that the modified Proctor compaction test has 56,250/12,375 = 4.55 times the compactive effort as the standard Proctor test.

Properties of Compacted Cohesive Soils

With respect to the optimum water content, as indicated by the Proctor compaction curve, soils can be compacted *dry of optimum*, *near optimum*, or *wet of optimum*. The engineering behavior of compacted cohesive soils depends upon these three conditions (Holtz et al., 2011).

If compactive effort is held constant, with increasing water content the soil fabric becomes increasingly more oriented. Dry of optimum, soils have a flocculated fabric, but wet of optimum clay minerals are more oriented or dispersed in nature. Permeability at a constant compactive effort decreases as the water content increases, reaching a minimum at about the optimum water content. The coefficient of

permeability is about an order of magnitude greater (10 times more) for soils compacted dry of optimum compared to those compacted wet of optimum. If the compactive effort is increased, permeability decreases because of the reduction in void ratio (Holtz et al., 2011).

Compressibility of a soil depends, along with several other factors, on the level of stress imposed on it. At low stress levels, clays are more compressible if compacted wet of optimum. For high stress levels the opposite is true, because clays compacted wet of optimum are greatly compressed under such loads, making only a small amount of further settlement possible (Holtz et al., 2011).

Clays compacted dry of optimum have a greater tendency to swell than those compacted wet of optimum. On the dry side, water is more deficient and so soils can absorb more water, when available, and swell more. Therefore, soils dry of optimum are more subject to change with changes in environment, water content included. For shrinkage, the opposite situation exists: Soils compacted wet of optimum undergo the greatest shrinkage (Holtz et al., 2011).

With regard to the strengths of compacted clays, those compacted dry of optimum have greater strengths than those compacted wet of optimum. However, if the samples are soaked, because of swelling effects the wet side soils can be somewhat stronger. Relative to compactive effort, the strength increases on the dry side of optimum with increasing compactive effort but on the wet side the strength

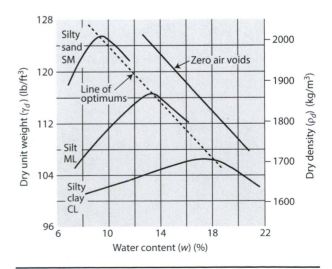

Figure 7.19   Compaction curves for different soils, but same compactive effort. Note: 62.4 lb/ft³ = 1 Mg/m³.

decreases with increasing compactive effort. This fact is most significant in the design and construction of earth fills for engineering works. A summary of the properties of soil relative to the optimum water content is given in Table 7.6.

Compaction and Excavation Equipment

Soil for placement in a compacted fill is obtained from a borrow pit. The equipment needed includes draglines, power shovels, and scrapers to excavate the borrow material. Selfpropelled scrapers, known as pans (Figure 7.20a), remove the soil from the borrow pit and transport it to the fill where it is placed in thin layers or lifts. Dozers (Figure 7.20b) may be used to help the scrapers during loading by pushing them through the borrow area.

If considerable distances are involved, trucks may be used to transport fill material (Figure 7.21a). These are loaded by shovels and draglines, or front-end loaders. The soil is spread over the fill area where drying can lower the water content, if needed, or water trucks can add water to increase the water content, if so desired (Figure 7.21b).

In the fill area, dozers and motor graders, known informally as blades (Figure 7.22a, page 128), are used to spread the soil in lifts. Lift thickness ranges from 0.15 to 0.5 m (6 to 18 in) depending on the compaction equipment, gradation and size of the soil particles, and design of the embankment. Much of the compaction is accomplished by equipment with a roller of some sort. The heavy equipment rollers induce compaction in several ways: by pressure, impact, vibration, or kneading.

Cohesionless soils are most easily compacted by vibration. On construction projects, compacting sands and gravels may be accomplished with hand operated vibration plates and motor-driven, large vibratory rollers. Loose sand can also be compacted using rubber-tired equipment and, more dramatically, by large weights (demolition or "headache" balls) dropped from heights by construction cranes (Leonards, 1980; Lukas, 1995).

Cohesive soils are compacted in the field with sheepsfoot rollers, rubber-tired rollers, and, for jobs with little clearance, by hand-operated tampers. A common practice for compacting soil fills is to route the soil-hauling equipment over the entire area during construction rather than allowing localized traffic in a restricted location.

Table 7.6   Properties of Compacted Soils Relative to Optimum Water Content.

Property	Comparison
Structure	
Particle arrangement	Dry side more random or flocculated.
	Wet side more oriented or dispersed.
Swelling	Dry side imbibes more water, yielding greater swelling.
Effect of environmental change	Dry side more sensitive to change.
Permeability	Dry side more permeable.
Compressibility	Wet side more compressible at low pressure levels, dry side more compressible at high pressure levels.
Strength	
As molded	Dry side higher.
After saturation	Dry side somewhat higher if swelling prevented, wet side likely higher if swelling permitted.
Pore water pressure at failure	Wet side higher.
Stress strain modulus	Dry side much greater.
Sensitivity	Dry side more likely to be sensitive.

Types of Rollers

A smooth wheel or drum roller has 100% contact with the ground (or 100% coverage) and provides contact pressures up to 380 kPa (55 psi). It can be used on all types of soils except those containing large rocks (Figure 7.22b, page 128) (Holtz et al., 2011). These rollers are commonly used in the final rolling of subgrades (proof rolling) and in compacting asphalt pavements. They are also effective in micaceous soils where the mica flakes are forced into a flat-lying position with greater density than would occur with most other rollers. When additional drying of the soil is required, discing equipment is pulled through the fill to expose more surface area for drying (Figure 7.22d).

The pneumatic or rubber-tired roller provides about 80% coverage (from closely spaced tires) and has contact pressures up to 700 kPa (100 psi). As in the

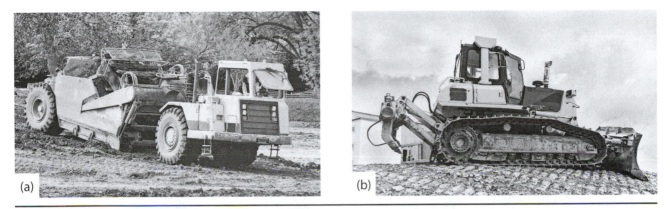

Figure 7.20 Excavation and compaction equipment. (a) Scraper or pan. (TFoxFoto, Shutterstock.) (b) Dozer—used to push scrapers in the borrow area or pull rollers across the fill. (Daniel Jedzura, Shutterstock.)

Figure 7.21   (a) Borrow pit where trucks are loaded by track excavator. (TFoxFoto, Shutterstock.) (b) Water truck adding water to fill area. (TFoxFoto, Shutterstock.)

case of the smooth wheel roller, the rubber-tired roller is used in both cohesive and noncohesive soils for both highways and dam construction (Holtz et al., 2011).

Perhaps the best known piece of compaction equipment is the sheepsfoot roller (Figure 7.22c). This roller has many round, rectangular, or footshaped protrusions, known as feet, that extend from a central steel drum. In the past, other names were used, based on the shape of the protrusion, but sheepsfoot is the overall term commonly applied today.

Sheepsfoot rollers provide about 8 to 12% coverage and very high contact pressures, from 1400 to 7000 kPa (200 to 1000 psi) depending on drum size and whether or not the drum is water filled. The most common rollers on heavy construction projects are the 1.25-m (4-ft) and 1.5-m (5-ft) diameter rollers. On a project where a high degree of compaction is required,

the 1.25-m (4-ft) rollers are preferred because the feet do not shear or knead the soil as much as the 1.5-m (5-ft roller), thereby providing a denser soil structure.

The feet extend about 0.15 to 0.25 m (6 to 10 in) from the drum and they compact the soil immediately below this depth. The roller compacts the soil higher and higher up in the lift with successive passes until it eventually "walks out" as the upper portion is compacted. Sheepsfoot rollers are best suited for cohesive soils (Holtz et al., 2011).

Another roller with protrusions is the tamping foot roller, which can be either towed or self-propelled. These provide about 40% coverage and develop high contact pressures, from 1400 to 8400 kPa (200 to 1200 psi) depending on roller size and whether or not the drum is water filled. These rollers have special hinged feet that rotate in the soil under load to provide a kneading action. They are best suited for fine-grained soils and will "walk out" when proper compaction occurs.

Mesh or grid pattern rollers provide about 50% coverage and pressures from 1400 to 6200 kPa (200 to 900 psi). With their open mesh they are ideally suited for rocky soils, gravels, and sands. A vibrational motion is also provided when the roller is towed at high speed across the fill (Holtz et al., 2011).

Vibratory compaction is particularly effective for granular soils because of the particle rearrangement that vibrations produce. Compaction equipment with vertical vibrators are available on both smooth wheel and tamping foot rollers to provide efficient compaction of granular soils. Small pieces of equipment that compact granular soils in restricted areas are also available. Vibrating plates and rammers,

ranging in size from 58 to 290 cm^2 (9 to 45 in^2), that weigh from 50 to 3000 kg (100 to 6000 lb), are available. They have limited depths of penetration for compaction, extending downward only 1 m (3 ft) or less. This requires that granular soils be placed in lifts of 0.5 m (1.5 ft) or less between vibrational tamping.

Field Compaction Control

The common concern on earth-moving projects is that the soil fill or embankment becomes adequately compacted to yield the density and water content desired. A more complete view of the situation involves other important engineering properties of the soil such as permeability, compressibility, strength, and sensitivity, which should come as a consequence of proper density and water content.

Figure 7.22   (a) Motor grader commonly used to smooth haul roads for earth-moving equipment. (MaZiKab, Shutterstock.) (b) Self-propelled smooth wheel or drum roller. (Juan Enrique del Barrio, Shutterstock.) (c) Sheepsfoot roller. (Roman023_photography, Shutterstock.) (d) Discing equipment used to promote drying in fill area. (Markus Hagenlocher [http://creativecommons.org/licenses/by-sa/3.0/deed.en].)

Despite these details, density and water content are the two parameters usually controlled.

In the past density alone was specified for compaction control without the full realization that very high or very low water contents, although accomplishing the desired dry density, would not provide the desired engineering properties. Current specifications indicate the percent relative density required (relative to either the standard Proctor or the modified Proctor compaction) and a range of water contents between which the compacted soil can vary and still be acceptable. This provides the contractor with the information needed to manipulate the soil in the fill area in order to achieve the acceptable compaction.

Developing Specifications

The procedure for developing compaction control specifications is accomplished in a series of steps. First, samples are obtained from the borrow area for laboratory testing. About 14 kg (31 lb) of sample are needed to perform a Proctor compaction curve. Based on this and other information, the earth structure is designed and the compaction specifications are written in keeping with the design. Field compaction control is specified, such as "95% standard Proctor density with a 3% moisture range on either side of optimum," or a "90% modified Proctor density within 2% water content on the dry side and 3% on the wet side of optimum," or other similar designations. The inspectors for construction control will conduct tests to ensure that these specifications are met.

The compaction control procedure described previously is for end-product specifications. This procedure is followed on most highways and building foundation projects. The equipment or method for achieving the specified density or water content is not dictated; the contractor can use whatever means deemed best to achieve these end-product requirements. Competitive bids and the economics of the marketplace are assumed to provide the incentive for use of the most efficient equipment available.

Another procedure sometimes used involves a different approach, known as method specifications. In this case, the type of roller, number of passes, and lift thickness (the methods) are specified. Material types may also be specified, based on different borrow areas designated for the project. If compaction control testing indicates that the embankment soil does not meet the necessary density or moisture content, then the contractor is paid extra for additional rolling. Method specifications require that the engineer/owner know in advance the detailed behavior of the soil during compaction in order to predict its behavior accurately. Such knowledge requires detailed testing, which typically can be justified only on very large earth-moving projects such as major dam construction.

Testing Procedures

Test procedures for field control of density and moisture content, like many kinds of testing, fall into two groups: destructive and nondestructive.

Destructive Test Methods

For destructive testing, the sample is excavated from the fill area. The mass of the sample in its wet condition is determined and, after oven drying, its dry mass is obtained. These measurements yield the water content. The volume of the hole left by the excavated sample (and of the soil sample itself) is obtained by means of: (1) the sand cone method (ASTM D1556), (2) the balloon method (ASTM D2167), or (3) by pouring water or oil of a known density into the hole, assuming the soil has a low permeability. Another method is by driving a steel tube of known volume into the ground (the drive-cylinder method, ASTM D2937).

In the sand cone method, dry sand of a known density is allowed to flow into the hole. The volume is determined based on the weight of the sand used. For the balloon method, air pressure is used to push water into a balloon against the side of the hole. Changes in water volume within the apparatus are noted as this occurs. In the third method, the soil sample, which extends from a steel tube, is cut level on both ends, yielding a full tube of soil. The volume and weight of the tube are known in advance. The soil, following extraction, is weighed before and after drying. For all methods, the dry density is obtained from the dry mass and volume measured.

Nondestructive Test Methods

Nondestructive testing is accomplished by nuclear density and water content meters (ASTM D6938). These tests have several advantages over the destructive tests: (1) they can be performed quickly with the results available in minutes rather than in hours and (2) because more tests are possible, a greater distribution of tests over the fill area provides a better statistical evaluation.

Nuclear density meters are not without disadvantages. They have a relatively high initial cost and there is a potential danger of exposure to radioactivity. Strict safety standards must be followed when using these devices.

Two different types of radiation are used in non-destructive nuclear testing. Gamma rays are used to determine soil density because the amount of scatter is proportional to total mass. Neutrons are scattered by hydrogen atoms and this provides the means for measuring water content. Thus, two emitters, one for gamma rays and another for neutrons, are included in the nuclear density devices. The accuracy of water content determinations using these devices has been a concern. Periodic checks finding water contents by weighing and drying procedure are recommended.

Effective Stress

Definition

The intergranular or effective stress is defined by the equation:

$$\sigma' = \sigma - u \qquad \text{(Eq. 7-25)}$$

where:

σ' = intergranular stress, grain-to-grain pressure, or effective stress

σ = total normal stress or total stress

u = pore water pressure, pore pressure, or neutral stress

The total stress can be calculated using the densities and thicknesses of the soil layers involved, and the neutral stress is found by calculating the hydrostatic pressure that develops below the groundwater table (u = depth below groundwater table × density of water). Effective stress must be found by subtraction because it cannot be measured directly ($\sigma' = \sigma - u$).

In SI units, pressure is given by $\sigma = \rho g h$. In the case of a 5-m-thick section of soil with a density of 1800 kg/m³,

$$\sigma = 1800 \text{ kg/m}^3 \times 9.81 \text{ m/sec}^2 \times 5 \text{ m}$$
$$= 88,290 \text{ kg/m/sec}^2$$

But 1 N = 1 kg m/sec². Therefore, σ = 88,290 N/m² = 88.29 kN/m² = 88.29 kPa (1 N/m² = 1 Pa). The neutral stress 3 m below the groundwater table is

$$u = \rho_w g h$$
$$= 1000 \text{ kg/m}^3 \times 9.81 \text{ m/sec}^2 \times 3 \text{ m}$$
$$= 29,430 \text{ N/m}^2 = 29.43 \text{ kN/m}^2$$
$$= 29.43 \text{ kPa}$$

For the British engineering system, total pressure is obtained by the equation $\sigma = h\gamma$. For example a 10-ft thick section of soil weighing 120 lb/ft³ has σ = 10 ft (120 lb/ft³) = 1200 lb/ft². Similarly $u = h\gamma_w$ so the neutral stress 5 ft below the groundwater table is u = 5 ft (62.4 lb/ft³) = 312 lb/ft².

Conversions from pressure in British engineering units (in lb_f/ft^2) to SI units (in kN/m²) can be made directly using 1 lb_f/ft^2 = 47.88 N/m². For example,

$$\sigma = 1200 \text{ } lb_f/ft^2 = 1200(0.04788) = 57.46 \text{ kN/m}^2$$

$$u = 312 \text{ } lb_f/ft^2 = 312(0.04788) = 14.93 \text{ kN/m}^2$$

and conversely

$$\sigma = 88.29 \text{ kN/m}^2 = 1/0.04788 = 1844 \text{ } lb_f/ft^2$$

$$u = 29.43 \text{ kN/m}^2 = 1/0.04788 = 614.7 \text{ } lb_f/ft^2$$

Basic Calculations

The calculation of total, neutral, and effective stresses is best illustrated by an example problem. In Figure 7.23, the value of each stress is shown in cross section at soil-layer boundaries.

- Total Stress Calculations

$$\text{At 4 m deep, } \sigma = \rho g h = \frac{1900 \times 9.81 \times 4}{1000}$$
$$= 74.5 \text{ kN/m}^2$$

$$\text{At 10 m deep, } \sigma = 74.5 + \frac{2180 \times 9.81 \times 6}{1000}$$
$$= 74.5 + 128.3 = 202.8 \text{ kN/m}^2$$

$$\text{At 18 m deep, } \sigma = 202.8 + \frac{1925 \times 9.81 \times 8}{1000}$$
$$= 202.8 + 151.1 = 353.9 \text{ kN/m}^2$$

$$\text{At 30 m deep, } \sigma = 353.9 + \frac{1990 \times 9.81 \times 12}{1000}$$
$$= 353.9 + 234.3 = 588.2 \text{ kN/m}^2$$

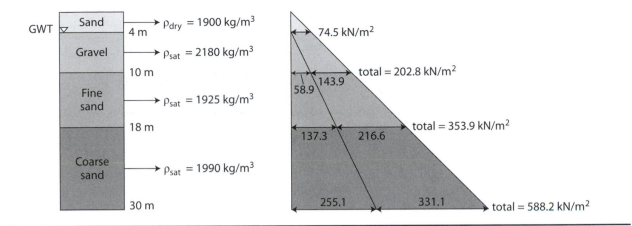

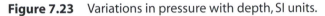

Figure 7.23   Variations in pressure with depth, SI units.

- Neutral Stress Calculations

  At 4 m, $u = 0$

  At 10 m, $u = 0 + \dfrac{6(1000) \times 9.81}{1000}$

  $\quad = 58.9 \text{ kN/m}^2$

  At 18 m, $u = 14 \times 9.81 = 137.3 \text{ kN/m}^2$
  At 30 m, $u = 26 \times 9.81 = 255.1 \text{ kN/m}^2$

  Now the *effective stress* is $\sigma' = \sigma - u$.

- Effective Stress Calculations

  At 4 m, $\sigma' = 74.5 \text{ kN/m}^2$
  At 10 m, $\sigma' = 202.8 - 58.9 = 143.9 \text{ kN/m}^2$
  At 18 m, $\sigma' = 353.9 - 137.3 = 216.6 \text{ kN/m}^2$
  At 30 m, $\sigma' = 588.2 - 255.1 = 331.1 \text{ kN/m}^2$

  The same problem will now be solved using British engineering units, as shown in Figure 7.24.

- Total Stress Calculations

  At 6 ft deep, $\sigma = h\gamma_d = 6(127) = 762 \text{ lb/ft}^2$
  At 10 ft deep, $\sigma = 762 + 4(136) = 1306 \text{ lb/ft}^2$
  At 21 ft deep, $\sigma = 1306 + 11(120) = 2626 \text{ lb/ft}^2$
  At 32 ft deep, $\sigma = 2626 + 11(125) = 4001 \text{ lb/ft}^2$

- Neutral Stress Calculations

  At 6 ft, $u = 0$
  At 10 ft, $u = 4(62.4) = 250 \text{ lb/ft}^2$
  At 21 ft, $u = 15(62.4) = 936 \text{ lb/ft}^2$
  At 32 ft, $u = 26(62.4) = 1622 \text{ lb/ft}^2$

  Now the *effective stress* is $\sigma' = \sigma - u$.

- Effective Stress Calculations

  At 6 ft, $\sigma' = 762 - 0 = 762 \text{ lb/ft}^2$
  At 10 ft, $\sigma' = 1306 - 250 = 1056 \text{ lb/ft}^2$
  At 21 ft, $\sigma' = 2626 - 936 = 1690 \text{ lb/ft}^2$
  At 32 ft, $\sigma' = 4001 - 1622 = 2379 \text{ lb/ft}^2$

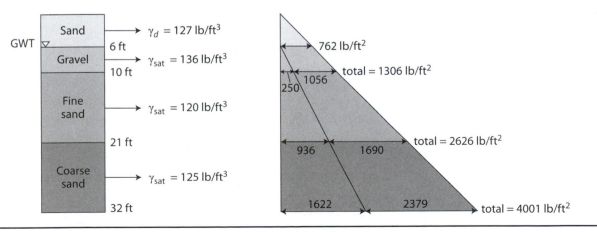

Figure 7.24   Variations in pressure with depth, British engineering units.

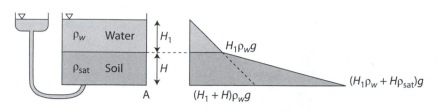

Figure 7.25   Case 1, no flow conditions.

Effects of Water Flow

The effects of water flow can now be considered by looking at three cases: (1) no flow conditions; (2) downward flow through the soil mass; and (3) upward flow through the soil mass. Case 1 is shown in Figure 7.25 (above). At the bottom of the soil (or Point A)

$$\sigma = (H_1\rho_w + H\rho_{sat})g$$
$$= [(H_1 + H)\rho_w + H\rho']g \qquad \text{(Eq. 7-26)}$$

Also at Point A:

$$u = (H_1 + H)\rho_w g \qquad \text{(Eq. 7-27)}$$

$$\sigma' = [(H_1 + H)\rho_w + H\rho' - (H_1 + H)\rho_w]g$$
$$= H\rho'g$$
$$\text{(Eq. 7-28)}$$

Note that the effective stress σ' is independent of the height of water (H_1) above the saturated soil column. Therefore, effective stress in the sediment at the bottom of the ocean basins is related only to the thickness of sediment and not to the water depth. Deep water does not make the sediment compress because the effective stress (intergranular stress) controls the settlement of the sedimentary column and it is independent of water depth.

Case 2, downward flow through the soil mass, is shown in Figure 7.26 (below). At Point A

$$\sigma = (H_1\rho_w + H\rho_{sat})g$$
$$= [(H_1 + H)\rho_w + H\rho']g \qquad \text{(Eq. 7-29)}$$

Also at Point A

$$u = (H_1 + H - h)\rho_w g \qquad \text{(Eq. 7-30)}$$

In this case h is negative because the water would flow downward from the water-soil column toward the standpipe. Downward flow is the opposite of the buoyancy or uplift force (u) and, therefore, it is subtracted from that value. Therefore, at Point A

$$\sigma' = \sigma - u$$
$$= [(H_1 + H)\rho_w + H\rho' - (H_1 + H - h)\rho_w]g$$
$$= (h\rho_w + H\rho')g$$
$$\text{(Eq. 7-31)}$$

The effective stress is increased by a downward flow of water through the soil system.

Case 3, upward flow through the soil mass, is shown in Figure 7.27 on the following page. At Point A

$$\sigma = (H_1\rho_w + H\rho_{sat})g$$
$$= [(H_1 + H)\rho_w + H\rho']g \qquad \text{(Eq. 7-32)}$$

Also at Point A

$$u = (H_1 + H + h)\rho_w g \qquad \text{(Eq. 7-33)}$$

where h is positive in this case because the water flows upward from the standpipe through the water-soil column. This is in the same direction as the

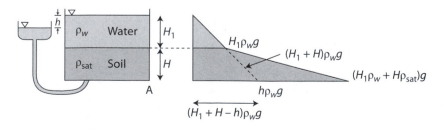

Figure 7.26   Case 2, downward flow through the soil mass.

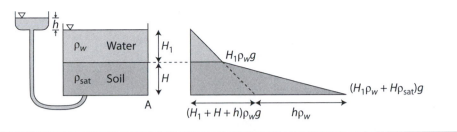

Figure 7.27 Case 3, upward flow through the soil mass.

buoyancy force (upward) and, thus, is added to that value. Therefore, at Point A

$$\sigma' = \sigma - u$$
$$= [(H_1 + H)\rho_w + H\rho']g - (H_1 + H + h)\rho_w g$$
$$= (H\rho' - h\rho_w)g$$

(Eq. 7-34)

This indicates that the effective stress is reduced by an upward flow of water through the soil column.

Critical Hydraulic Gradient

Another significant feature concerning effective stress occurs when the upward flow is sufficient to reduce σ' to zero. These details are shown next.

If $\sigma' = 0$, $\sigma = u$, or $0 = \sigma - u$, then substituting,

$$0 = [(H_1 + H)\rho_w + H\rho']g - (H_1 + H + h)\rho_w g$$
$$= (H\rho' - h\rho_w)g$$

(Eq. 7-35)

Cancelling g and transposing, $H\rho' = h\rho_w$ or $\rho'/\rho_w = h/H = i_{cr}$, where i_{cr} = the critical hydraulic gradient and h/H is a measure of head loss (h) over a distance of flow (H), which is a hydraulic gradient. The i_{cr} value is the hydraulic gradient for a zero effective stress level where the soil has a zero intergranular strength or zero shear strength. This denotes a quick sand condition, which is a critical case for stability.

The value of i_{cr} turns out to be about equal to one ($i_{cr} \sim 1$) for cohesionless soils. This is shown in the next calculation, which is based on Figure 7.28 for a two-phase soil condition. The soil is saturated and considered in terms of e, G_s, and ρ_w.

$$\rho' = \frac{\rho_w G_s - \rho_w(1)}{1+e} = \frac{\rho_w(G_s - 1)}{1+e}$$

(Eq. 7-36)

In this situation, $G_s \sim 2.7$ and e for cohesionless soil is about 0.7. Therefore,

$$\rho' \simeq \rho_w \frac{(1.7)}{(1.7)} \quad \text{or} \quad \rho' \simeq \rho_w \quad \text{and} \quad \frac{\rho'}{\rho_w} \sim 1$$

(Eq. 7-37)

But

$$\frac{\rho'}{\rho_w} = \frac{h}{H} \quad \text{or} \quad \frac{h}{H} = i_{cr} \simeq 1$$

(Eq. 7-38)

An example of a critical hydraulic gradient is shown for a concrete dam in Figure 7.29.

$$i = \frac{h}{H} = \frac{50 \text{ ft}}{50 \text{ ft}} = \frac{15 \text{ m}}{15 \text{ m}} = 1$$

(Eq. 7-39)

thus signaling that a quick condition exists at the toe of the dam. Note: 1 ft = 0.3048 m.

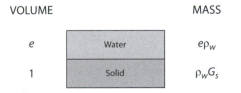

Figure 7.28 Phase diagram for calculating submerged density.

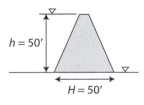

Figure 7.29 Configuration for dam with a critical hydraulic gradient problem.

Shear Strength of Soils

If the load or stress on a soil is increased until deformation is excessive or if the soil mass suddenly experiences a major displacement, the soil is said to have failed. This indicates the strength of the soil, or the maximum stress that the soil can sustain. In soil mechanics, the shear strength is the main concern because most failures result from applications of excessive shear stresses.

Stress at a Point

The usual approach is to consider the state of stress at a point. This is accomplished by viewing an infinitesimally small cube of material acted on by normal stresses and shear stresses, as shown in Figure 7.30. The three stresses acting perpendicular and inward on the cube faces are normal compressive stresses. Those acting parallel to the faces are shear stresses. In addition to the nine stresses shown, another nine are not provided. These nine additional stresses are distributed over the three remaining faces (the left, back, and bottom of the cube) to yield equal and opposite stresses to those shown. Because the cube is infinitesimal (a point), these stresses would all be in equilibrium.

Principal Stresses

A fundamental principle of mechanics states that "through any point there exists three mutually perpendicular planes that have only normal stresses acting on them" (no shear stresses). These comprise the three principal planes with the three principal stresses acting perpendicular to them. They are shown in Figure 7.31.

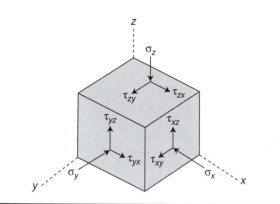

Figure 7.30   Shear and normal stresses acting through a point.

The greatest stress is called the major principal stress or σ_1, the least is the minor principal stress or σ_3, and the third is the intermediate principal stress or σ_2. In many soil mechanics problems the behavior is considered to be independent of σ_2, which yields a two-dimensional problem involving only σ_1 and σ_3. Based on this, we can now show the relationship in two dimensions in Figure 7.32. Planes parallel to σ_1 or σ_3 will have only normal stresses acting on them but all other planes have both a normal stress (σ) and shear stress (τ) that act collectively.

Equations for σ and τ on a plane inclined at the angle θ from the maximum principal plane are:

$$\sigma = \sigma_1 \cos^2\theta + \sigma_3 \sin^2\theta = \sigma_3(\sigma_1 - \sigma_3)\cos^2\theta \quad \text{(Eq. 7-40)}$$

$$\tau = (\sigma_1 - \sigma_3)\sin\theta \cos\theta \quad \text{(Eq. 7-41)}$$

Checking at the boundaries to show that the equations apply there, for $\theta = 0$, $\sin\theta = 0$ and $\cos\theta = 1$:

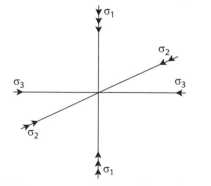

Figure 7.31   Three principal stresses acting through a point.

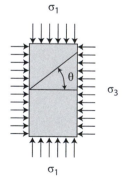

Figure 7.32   Two-dimensional stress field.

$$\sigma = \sigma_1 + 0 = \sigma_1$$

$$\tau = (\sigma_1 - \sigma_3)(1)(0) = 0$$

for $\theta = 90°$, $\sin\theta = 1$ and $\cos\theta = 0$:

$$\sigma = 0 + \sigma_3(1) = \sigma_3$$

$$\tau = (\sigma_1 - \sigma_3)(1)(0) = 0$$

and in another form,

$$\sigma = \frac{\sigma_1 + \sigma_3}{2} + \frac{\sigma_1 - \sigma_3}{2}\cos^2\theta$$

$$\tau = \frac{\sigma_1 + \sigma_3}{2}\sin^2\theta$$

Checking again at the boundaries for $\theta = 0$, $\cos^2\theta = 1$, and $\sin^2\theta = 0$:

$$\sigma = \frac{\sigma_1 + \sigma_3}{2} + \frac{\sigma_1 - \sigma_3}{2}(1) = \sigma_1$$

$$\tau = \frac{\sigma_1 + \sigma_3}{2}(0) = 0$$

for $\theta = 90°$, $\cos^2\theta = 1$ and $\sin^2\theta = 0$:

$$\sigma = \frac{\sigma_1 + \sigma_3}{2} + \frac{\sigma_1 - \sigma_3}{2}(-1) = \sigma_3$$

$$\tau = \frac{\sigma_1 - \sigma_3}{2}\sin^2\theta = \frac{\sigma_1 - \sigma_3}{2}(0) = 0$$

Mohr Circle

If all values of σ and τ are plotted on a σ-τ diagram as they vary with θ from 0 to 360°, a circle is obtained. The circle, known as the Mohr circle, represents all the possible combinations of σ and τ acting in different directions through a point, as is shown in Figure 7.33. As indicated before, σ_1 and σ_3 are the values of the major and minor principal stresses, respectively, and the locations where $\tau = 0$.

The center of the circle is located at $(\sigma_1 + \sigma_3)/2$ and the radius of the circle is $(\sigma_1 - \sigma_3)/2$. The dotted vertical line through σ_3 is the plane of minor principal stress and the σ-axis is the plane of major principal stress. The maximum shear stress is equal to the radius of the circle of $(\sigma_1 - \sigma_3)/2$. The circle can be used to find the state of stress, σ and τ, on any plane inclined at an angle to the maximum principal plane. The resultant stress at any point R is equal to $\sqrt{\sigma^2 + \tau^2}$. In addition, $\sigma_1 - \sigma_3$ is referred to as the stress difference or as the deviator stress.

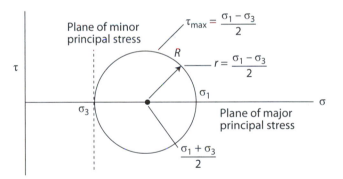

Figure 7.33   Mohr circle diagram for stress at a point.

Example Problem 7.7

Figure 7.34 shows the details for this problem. Given the major and minor principal stresses at a point of 485 and 175 kN/m² (70 and 25 psi), respectively, solve for the following:

a. Find the value of the maximum shearing stress at the point.
b. What is the value of the normal stress on the plane of maximum shearing stress?
c. What are the values of σ, τ, and the resultant stress on a plane at an angle of 22° to the plane of major principal stress?
d. What are the values of σ, τ, and the resultant stress on a plane at an angle 38° to the plane of minor principal stress?

Answers

Using SI units:

a. Maximum shear stress occurs at Point (A) in Figure 7.34a. It equals the value of the radius or

$$\frac{\sigma_1 - \sigma_3}{2} = \frac{485 - 175}{2}$$

$$= 155 \text{ kN/m}^2$$

b. Normal stress at Point (A). It equals the same as the value for the center of the circle, or

$$\frac{\sigma_1 + \sigma_3}{2} = \frac{485 + 175}{2}$$

$$= 330 \text{ kN/m}^2$$

c. A plane 22° from the plane of major principal stress ($\theta = 22°$) is measured from the horizontal,

$$\sigma = \sigma_1 \cos^2\theta + \sigma_3 \sin^2\theta$$
$$= 485(0.9272) + 175(0.3746)$$
$$= 449.7 + 65.6$$
$$= 515.3 \text{ kN/m}^2$$

$$\tau = (\sigma_1 - \sigma_3)\sin\theta \cos\theta$$
$$= (485 - 175)(0.3746)(0.9272)$$
$$= 107.7 \text{ kN/m}^2$$

$$R = \sqrt{\sigma^2 + \tau^2}$$
$$= \left(515.3^2 + 107.7^2\right)^{1/2}$$
$$= 526.4 \text{ kN/m}^2$$

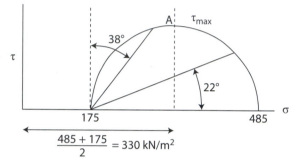

$$\frac{485 + 175}{2} = 330 \text{ kN/m}^2$$

Figure 7.34   (a) Diagram for Mohr circle using SI units.

d. A plane 38° to the plane of minor stress is $90 - 38 = 52°$ from the plane of major stress ($\theta = 52°$).

$$\sigma = \sigma_1 \cos^2\theta + \sigma_3 \sin^2\theta$$
$$= 485(0.6156)^2 + 175(0.7880)^2$$
$$= 183.8 + 108.7$$
$$= 292.5 \text{ kN/m}^2$$

$$\tau = (\sigma_1 - \sigma_3)\sin\theta \cos\theta$$
$$= (485 - 175)(0.6156)(0.7880)$$
$$= 150.4 \text{ kN/m}^2$$

$$R = \sqrt{\sigma^2 + \tau^2}$$
$$= \left(292.4^2 + 150.4^2\right)^{1/2}$$
$$= 328.9 \text{ kN/m}^2$$

Using British engineering units:

a. Maximum shear stress occurs at Point (A) in Figure 7.34b. It equals the value of the radius or

$$\frac{\sigma_1 - \sigma_3}{2} = \frac{70 - 25}{2} = 22.5 \text{ psi}$$

b. Normal stress at Point (A) equals the same as the value for the center of the circle, or

$$\frac{\sigma_1 + \sigma_3}{2} = \frac{70 + 25}{2} = 47.5 \text{ psi}$$

c. A plane 22° from the plane of major principal stress ($\theta = 22°$) is measured from the horizontal,

$$\sigma = \sigma_1 \cos^2\theta + \sigma_3 \sin^2\theta$$
$$= 70(0.9272) + 25(0.3746)$$
$$= 64.90 + 9.37 = 74.3 \text{ psi}$$

$$\tau = (\sigma_1 - \sigma_3)\sin\theta \cos\theta$$
$$= (70 - 25)(0.3746)(0.9272) = 15.6 \text{ psi}$$

$$R = \sqrt{\sigma^2 + \tau^2}$$
$$= \left(74.3^2 + 15.6^2\right)^{1/2}$$
$$= (5520.5 + 243.4)^{1/2} = 75.9 \text{ psi}$$

d. A plane 38° to the plane of minor stress is 90 – 38 = 52° from the plane of major stress ($\theta = 52°$).

$$\sigma = \sigma_1 \cos^2\theta + \sigma_3 \sin^2\theta$$
$$= 70(0.6156)^2 + 25(0.7880)^2$$
$$= 26.53 + 15.52 = 42.0 \text{ psi}$$

$$\tau = (\sigma_1 - \sigma_3)\sin\theta \cos\theta$$
$$= (70 - 25)(0.6156)(0.7880) = 21.8 \text{ psi}$$

$$R = \sqrt{\sigma^2 + \tau^2}$$
$$= \left(42.0^2 + 21.8^2\right)^{1/2}$$
$$= (1764 + 475.2)^{1/2} = 47.3 \text{ psi}$$

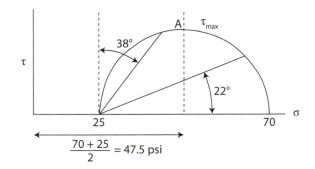

Figure 7.34   (b) Diagram for Mohr circle using British engineering units.

The obliquity angle is the angle between the origin (or the origin of stresses) and the point of interest on the Mohr circle diagram. This is related in the following way:

$$\alpha = \tan^{-1}(\tau/\sigma) \qquad \text{(Eq. 7-42)}$$

and is illustrated in Figure 7.35. The maximum obliquity angle occurs at the point of tangency to the circle. By recalling that the radius equals $(\sigma_1 - \sigma_3)/2$ and the distance to the center of the circle is $(\sigma_1 + \sigma_3)/2$ we have

$$\sin\alpha_m = \frac{\sigma_1 - \sigma_3}{\sigma_1 + \sigma_3} \qquad \text{(Eq. 7-43)}$$

$$\theta_{cr} = 45 + 1/2\alpha_m \qquad \text{(Eq. 7-44)}$$

Note that the point of maximum shear stress, or $(\sigma_1 - \sigma_3)/2$, is not the point of maximum obliquity, and conversely the point of maximum obliquity has a shear stress less than the maximum. This will become more significant later in this discussion.

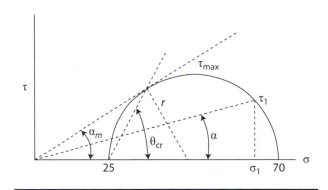

Figure 7.35   Obliquity angle (α) on Mohr circle diagram.

Another example problem is now appropriate for consideration.

Example Problem 7.8

Given the same data as in Example Problem 7.7, find the angle of maximum obliquity and determine the τ and σ values for that orientation.

Answers

Using SI units:

$$\sin\alpha_m = \frac{\sigma_1 - \sigma_3}{\sigma_1 + \sigma_3}$$

$$= \frac{485 - 175}{485 + 175}$$

$$= \frac{310}{660} = 0.4697$$

$$\alpha_m = 28.01$$

$$\theta_{cr} = 45 + 1/2\alpha_m$$
$$= 45 + 1/2(28.01)$$
$$= 59.01°$$

$$\sigma = \sigma_1 \cos^2\theta + \sigma_3 \sin^2\theta$$
$$= 485(0.5149)^2 + 175(0.8573)^2$$
$$= 128.6 + 128.6 = 257.2 \text{ kN/m}^2$$

$$\tau = (\sigma_1 - \sigma_3)\sin\theta \cos\theta$$
$$= (485 - 175)(0.8573)(0.5149)$$
$$= 136.8 \text{ kN/m}^2$$

Check:

$$\tau = \sigma \tan\alpha_m$$
$$= 257.2 + \tan 28.01$$
$$= 136.8 \text{ kN/m}^2$$

Using British engineering units:

$$\sin\alpha_m = \frac{\sigma_1 - \sigma_3}{\sigma_1 + \sigma_3}$$

$$= \frac{70 - 25}{70 + 25}$$

$$= \frac{45}{95} = 0.4736$$

$$\alpha_m = 28.27°$$

$$\theta_{cr} = 45 + 1/2\alpha_m$$
$$= 45 + 1/2(28.27)$$
$$= 59.13°$$

$$\sigma = \sigma_1 \cos^2\theta + \sigma_3 \sin^2\theta$$
$$= 70(0.5131)^2 + 25(0.8583)^2$$
$$= 18.42 + 18.42 = 36.84 \text{ psi}$$

$$\tau = (\sigma_1 - \sigma_3)\sin\theta \cos\theta$$
$$= (70 - 25)(0.8583)(0.5131)$$
$$= 19.82 \text{ psi}$$

Check:

$$\tau = \sigma \tan\alpha_m$$
$$= 36.84 \tan 28.27$$
$$= 19.81 \text{ psi}$$

Mohr–Coulomb Failure Criterion

The Coulomb equation for the strength of soil is written as

$$\tau = c + \sigma \tan\phi \qquad \text{(Eq. 7-45)}$$

and is illustrated in Figure 7.36.

The angle ϕ is known as the angle of internal friction, which provides for an increase in shear strength with an increase in normal stress. The other term, c, is the cohesion of the soil and is the inherent strength of a soil that persists even when unconfined, that is, at zero normal stress. ϕ and c are collectively known as the strength parameters.

For certain types of soil, either the c or ϕ parameter can equal zero. When $\phi = 0$, $\tau = c$ and when $c = 0$, $\tau = \sigma \tan\phi$. These are illustrated in Figure 7.37.

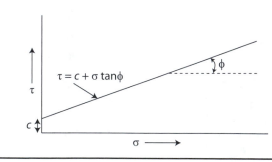

$$\tau = c + \sigma \tan\phi$$

Figure 7.36 Mohr–Coulomb failure criterion.

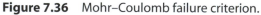

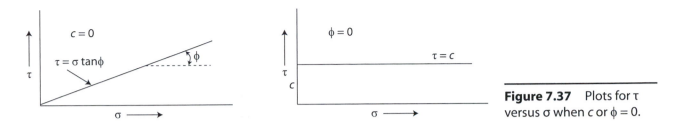

Figure 7.37 Plots for τ versus σ when c or $\phi = 0$.

In 1900 Mohr stated that the shear stress on the failure plane at failure has a value that is a function of the normal stress alone on that plane. This can be written $\tau_{ff} = f(\sigma_{ff})$ where τ and σ are the usual shear stress and normal stress, respectively. The first f in the subscript stands for the failure plane and the second f stands for at failure. So the equation is read shear stress on the failure plane at failure is a function of the normal stress on the failure plane at failure.

If a series of soil tests is run with different σ_1 and σ_3 values and loaded to failure, a line tangent to the failure circles can be drawn. This is known as the Mohr failure envelope. No Mohr circle can exist for soil that extends beyond the envelope because failure would occur before extending beyond that limit. In like manner, the soil cannot fail if its Mohr circle lies within the envelope without touching it. These details are illustrated in Figure 7.38.

When the Coulomb strength equation is combined with the Mohr failure criterion, the Mohr–Coulomb strength criterion is obtained. We do not know who first combined these relationships, but today this is one of the most widely used concepts applied to soil studies. It is given by

$$\tau_{ff} = c + \sigma_{ff} \tan \phi \qquad \text{(Eq. 7-46)}$$

This can now be shown graphically, as in Figure 7.39 on the following page, including an example of a Mohr circle for failure conditions. From Figure 7.40 (also on the next page) we can observe that

$$\sin \phi = \frac{R}{D}$$
$$= \frac{\dfrac{(\sigma_1 - \sigma_3)}{2}}{\dfrac{(\sigma_1 + \sigma_3)}{2} + c \cot \phi} \qquad \text{(Eq. 7-47)}$$

If $c = 0$ then this becomes

$$\sin \phi = \frac{\sigma_1 - \sigma_3}{\sigma_1 + \sigma_3} \qquad \text{(Eq. 7-48)}$$

which is the same as the obliquity relationship developed previously, and ϕ can be substituted for α_m.

Several other relationships can now be developed regarding cohesionless soils ($c = 0$). Figure 7.41 (next page) is used to illustrate this. In Figure 7.41, θ_{cr} is the angle of inclination from the maximum principal plane to the failure plane, β is the opposite angle measured from the maximum principal plane, and 2β is the corresponding central angle.

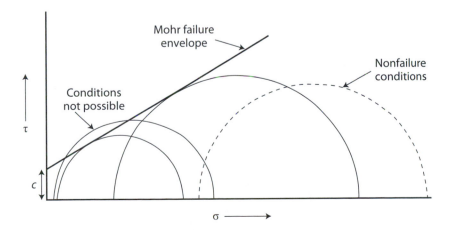

Figure 7.38 Mohr envelope and failure conditions.

$$2\beta + \phi + 90 = 180°$$
$$2\beta = 90 - \phi$$
$$\beta = 45 - \frac{\phi}{2}$$
$$2\beta + 2\theta_{cr} = 180°$$
$$\beta = 90 - \theta_{cr}$$

Equating relationships for β,

$$45 - \frac{\phi}{2} = 90 - \theta_{cr}$$
$$\theta_{cr} = 45 + \frac{\phi}{2}$$

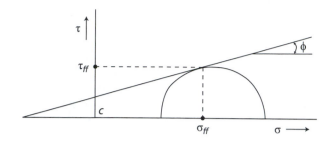

Figure 7.39 Mohr–Coulomb failure, including cohesion.

In addition it can be shown by trigonometry

$$\frac{\sigma_1}{\sigma_3} = \frac{1 + \sin\phi}{1 - \sin\phi}$$
$$= \tan^2\left(45 + \frac{\phi}{2}\right)$$
$$\frac{\sigma_3}{\sigma_1} = \tan^2\left(45 - \frac{\phi}{2}\right)$$
$$\sigma_{ff} = \sigma_3\left(1 + \sin\phi\right)$$
$$= \sigma_1\left(1 - \sin\phi\right)$$
$$= \left(\sigma_1 - \sigma_3\right)\frac{\cos^2\phi}{2\sin\phi}$$

$$\tau_{ff} = \sigma_{ff}\tan\phi$$
$$= \sigma_3\tan\phi\left(1 + \sin\phi\right)$$
$$= \sigma_1\tan\phi\left(1 - \sin\phi\right)$$
$$= \left(\sigma_1 - \sigma_3\right)\frac{1}{2}\cos\phi$$

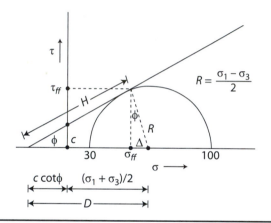

Figure 7.40 Trigonometric relationships, including cohesion.

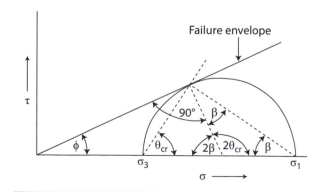

Figure 7.41 Trigonometric relationships, cohesionless case.

Example Problem 7.9

Given a Mohr circle for failure conditions for a cohesionless sand as follows:

$\sigma_1 = 700 \text{ kN/m}^2 \text{ (100 psi)}$

$\sigma_3 = 200 \text{ kN/m}^2 \text{ (30 psi)}$

$c = 70 \text{ kN/m}^2 \text{ (10 psi)}$

find ϕ, θ_{cr}, σ_{ff}, and τ_{ff} (refer to Figure 7.40). If $c = 0$,

$$\sin\phi = \frac{\sigma_1 - \sigma_3}{\sigma_1 + \sigma_3} = \frac{500}{900}$$

$$\phi = 33.75°$$

$$\theta_{cr} = 45 + \frac{\phi}{2} = 45 + 16.87 = 61.9°$$

$$\sigma_{ff} = \sigma_3(1 + \sin\phi) = 200(1 + \sin 33.75°) = 311.1 \text{ kN/m}^2$$

$$\tau_{ff} = \sigma_{ff}\tan\phi = 207.9 \text{ kN/m}^2$$

Answers

Using SI units:

$$R = \frac{(\sigma_1 - \sigma_3)}{2} = \frac{700 - 200}{2} = \frac{500}{2} = 250$$

$$D = \frac{(\sigma_1 + \sigma_3)}{2} + c\cot\phi$$

$$\frac{(\sigma_1 + \sigma_3)}{2} = \frac{700 + 200}{2} = \frac{900}{2} = 450$$

$$\sin\phi = \frac{R}{D} = \frac{(\sigma_1 - \sigma_3)/2}{(\sigma_1 + \sigma_3)/2 + c\cot\phi}$$

$$\sin\phi = \frac{250}{450 + 70\cot\phi}$$

$$\sin\phi(450 + 70\cot\phi) = 250$$

or

$$450\sin\phi + 70\cos\phi = 250$$

Try solving for ϕ by using an interactive method. If $450\sin\phi + 70\cos\phi = 250$

ϕ	Solution
30°	286
25°	254
24°	247
24.5°	250

Therefore,

$$\phi = 24.5°$$

$$\theta_{cr} = 45 + \frac{\phi}{2} = 45 + 12.3 = 57.3°$$

$$\frac{R}{H} = \tan\phi$$

(continued)

$$H = \frac{R}{\tan\phi} = \frac{250}{\tan 24.5} = 548.6$$

$$D = 450 + 70\cot\phi = 494.9$$

$$R = 250$$

Check:

$$D^2 = R^2 + H^2 = 250^2 + 548.6^2$$

$$D = 602.9$$

$$\frac{\tau_{ff}}{H} = \sin\phi$$

$$\tau_{ff} = H\sin\phi = 548.6 \ \sin 24.5 = 227.5 \ \text{kN/m}^2$$

$$\frac{\tau_{ff}}{\sigma_{ff} + c\cot\phi} = \tan\phi$$

$$(\sigma_{ff} + c\cot\phi)\tan\phi = \tau_{ff}$$

$$\sigma_{ff} = \frac{\tau_{ff} - c}{\tan\phi} = \frac{227.5 - 70}{\tan 24.5} = 345.6 \ \text{kN/m}^2$$

$$\frac{\Delta}{R} = \sin\phi$$

$$\Delta = R\sin\phi = 250 \ \sin 24.5 = 103.7 \ \text{kN/m}^2$$

$$\frac{(\sigma_1 + \sigma_3)}{2} - \Delta = \sigma_{ff} = 450 - 103.7 = 346.3 \ \text{kN/m}^2$$

Using British engineering units:

$$\sin\phi = \frac{\sigma_1 - \sigma_3}{\sigma_1 + \sigma_3} = \frac{70}{130}$$

$$\phi = 32.57°$$

$$\theta_{cr} = 45 + \frac{\phi}{2} = 45 + 16.28 = 61.3°$$

$$\sigma_{ff} = \sigma_3(1 + \sin\phi) = 30(1 + \sin 33.75°) = 46.15 \ \text{psi}$$

$$\tau_{ff} = \sigma_{ff}\tan\phi = 23.5 \ \text{psi}$$

If $c = 10$ psi, find ϕ, σ_{ff}, and τ_{ff}. Refer to Figure 7.40.

$$R = \frac{(\sigma_1 - \sigma_3)}{2} = \frac{100 - 30}{2} = \frac{70}{2} = 35$$

$$D = \frac{(\sigma_1 + \sigma_3)}{2} + c\cot\phi$$

$$\frac{(\sigma_1 + \sigma_3)}{2} = \frac{100 + 30}{2} = \frac{130}{2} = 65$$

$$\sin\phi = \frac{R}{D} = \frac{(\sigma_1 - \sigma_3)/2}{(\sigma_1 + \sigma_3)/2 + c\cot\phi}$$

$$\sin\phi = \frac{35}{65 + 10\cot\phi}$$

$$\sin\phi(65 + 10\cot\phi) = 35$$

or

$$65\sin\phi + 10\cos\phi = 35$$

Try solving for ϕ by using an interactive method. If $65\sin\phi + 10\cos\phi = 35$

ϕ	**Solution**
30°	41.1
25°	36.5
23°	34.5
23.5°	35.08
23.4°	34.992

Therefore,

$$\phi = 23.4°$$

$$\theta_{cr} = 45 + \frac{\phi}{2} = 45 + 11.7 = 56.7°$$

$$\frac{R}{H} = \tan\phi$$

$$H = \frac{R}{\tan\phi} = \frac{35}{\tan 23.4} = 80.88$$

$$D = 65 + 10\cot\phi = 88.10$$

$$R = 35$$

Check:

$$D^2 = R^2 + H^2 = 35^2 + 80.88^2$$

$$D = 88.1$$

$$\frac{\tau_{ff}}{H} = \sin\phi$$

$$\tau_{ff} = H\sin\phi = 80.88 \ \sin 23.4 = 32.12 \text{ psi}$$

$$\frac{\tau_{ff}}{\sigma_{ff} + c\cot\phi} = \tan\phi$$

$$(\sigma_{ff} + c\cot\phi)\tan\phi = \tau_{ff}$$

$$\sigma_{ff}\tan\phi + c = \tau_{ff}$$

$$\sigma_{ff} = \frac{\tau_{ff} - c}{\tan\phi} = \frac{32.10 - 10}{\tan 23.4} = 51.12 \text{ psi}$$

$$\frac{\Delta}{R} = \sin\phi$$

$$\Delta = R\sin\phi = 35 \ \sin 23.4 = 13.9$$

$$\frac{(\sigma_1 + \sigma_3)}{2} - \Delta = \sigma_{ff} = 65 - 13.9 = 51.1$$

For clay-rich soils (CH by Unified Soil classification) the ϕ angle is equal to zero. In this case the Coulomb equation for shear strength reduces to $\tau = \sigma$ and the Mohr–Coulomb strength criterion is $\tau_{ff} = c = 1/2q_u$, where q_u is the unconfined compression strength. This is illustrated in Figure 7.42.

Example Problem 7.10

Given that the unconfined compression strength for a highly plastic clay is 145 kN/m² (3000 lb/ft²) and its unit weight is 19.6 kN/m³ (125 lb/ft³), find the following:

a. What is its shear strength, cohesion, or cohesive strength and is it affected by depth?
b. At what depth is the vertical overburden stress σ_v equal to the shear strength?
c. At a depth of 9 m (30 ft), what is the ratio of vertical overburden stress to the shear stress? How would this affect the ability to tunnel through this plastic clay at that depth?

Answers

Using SI units:

a. $\tau_{ff} = \dfrac{q_u}{2} = \dfrac{145}{2} = 72.5 \text{ kN/m}^2$

This is not affected (or increased) with depth as $\phi = 0$.

b. $Z = \dfrac{\tau_{ff}}{\gamma} = \dfrac{72.5 \text{ kN/m}^2}{19.6 \text{ kN/m}^3} = 3.7 \text{ m}$

c. $P_Z = \gamma_Z$
$P = (19.6)(9) = 176.4 \text{ kN/m}^2$

Ratio $= \dfrac{\text{Overburden stress}}{\text{Shear strength}} = \dfrac{176.4}{72.5} = 2.4$

At this high stress ratio, the soil will yield dramatically when excavated. Loss of ground will occur above the tunnel and surface subsidence will result.

Using British engineering units:

a. $\tau_{ff} = \dfrac{q_u}{2} = \dfrac{3000}{2} = 1500 \text{ lb/ft}^2$

This is not affected (or increased) with depth as $\phi = 0$.

b. $Z = \dfrac{\tau_{ff}}{\gamma} = \dfrac{1500 \text{ lb/ft}^2}{125 \text{ lb/ft}^3} = 12 \text{ ft}$

c. $P_Z = \gamma_Z$
$P = (125)(30) = 3750 \text{ lb/ft}^2$

Ratio $= \dfrac{\text{Overburden stress}}{\text{Shear strength}} = \dfrac{3750}{1500} = 2.5$

At this high stress ratio, the soil will yield dramatically when excavated. Loss of ground will occur above the tunnel and surface subsidence will result.

The small difference in the overburden stress to shear strength ratios between SI and British engineering units (2.4 versus 2.5) is due to conversions and rounding off.

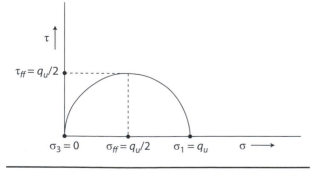

Figure 7.42   Shear strength relationship, $\phi = 0$.

Stress Distribution

When a load or stress is applied at a boundary (the Earth's surface for example), the stress is distributed or transferred to the different levels below. Depending on the size of the loaded area, the added stress will reduce with increasing depth, that is, with increasing distance from the boundary stress. The manner in which this stress dissipates with depth is the stress distribution.

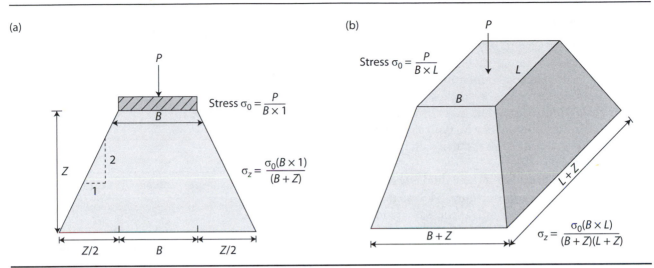

Figure 7.43   The 2:1 approximation for distribution of vertical stress with depth: (a) strip footing and (b) rectangular footing.

Two-to-One Method

A simple way to estimate this distribution of stress with depth is to use the two-to-one (or 2 to 1 or 2:1) method. As the same vertical force is spread over a larger and larger area, the unit stress must decrease with depth, as illustrated in Figure 7.43.

The 2:1 method provides data on the average stress at some depth; however, the maximum stress at that level (vertically below the center of the footing) is somewhat greater. A value of 50% is typically added to yield the maximum value.

Boussinesq Formula

Another method used to estimate the vertical stress at a point below a concentrated applied load at the boundary is by using the Boussinesq formula (Boussinesq, 1885):

$$\sigma_Z = N_B \frac{P}{Z^2} \qquad \text{(Eq. 7-49)}$$

It allows the calculation of stress directly below P and at horizontal distances away from the vertical location. This is illustrated in Figure 7.44a.

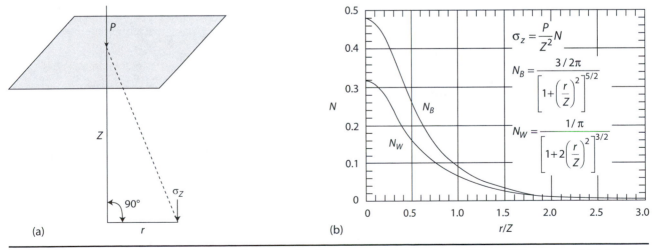

Figure 7.44   (a) Boussinesq relationship for stress distribution and (b) Boussinesq (N_B) and Westergaard (N_W) factors for stress reduction with depth and lateral distance from point load. (After Taylor, 1948.)

Westergaard Formula

The Westergaard (1938) theory was derived assuming thin horizontal sheets and likely is more appropriate for bedded sedimentary rocks and soils. It also typically provides the smallest stress level of the three methods discussed here. This may be of concern to design engineers who typically would not select the lowest value of stress using alternative calculation methods.

Both the Boussinesq (1885) and Westergaard (1938) factors are shown in Figure 7.44b as they vary with r/Z. For values of $r/Z < 1.5$, Boussinesq indicates higher stresses than does Westergaard. When $r/Z \geq 1.5$, both provide about the same answer. In practice the 2:1 ratio method, though fairly crude, is commonly used because it involves a contact area for the load rather than a point load, and it is easy to apply. Example Problem 7.11 provides a comparison of the three methods and is illustrated by Figure 7.45.

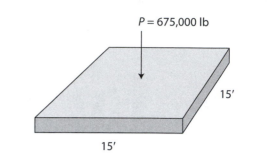

Figure 7.45   Example Problem 7.11 on stress distribution.

Example Problem 7.11

Given a footing 5 m × 5 m (15 ft × 15 ft) with a column load of 300,000 kg (675 kips), find the stress level at 5 m (15 ft) and 10 m (30 ft) directly below the center of the footing using (a) 2:1, (b) Boussinesq, and (c) Westergaard methods.

Answers

Using SI units:

a. 2:1 method:

$$\sigma_Z = \frac{P}{(B+Z)(L+Z)}$$

For a square footing

$$\sigma_Z = \frac{P}{(B+Z)^2}$$

$$P = \sigma_0 = \frac{300,000}{5^2} = 12,000 \text{ kg/m}^2 = 117.7 \text{ kN/m}^2$$

$$\sigma_{5_{avg}} = \frac{300,000}{10^2} = 3000 \text{ kg/m}^2 = 29.4 \text{ kN/m}^2$$

$$\sigma_{10_{avg}} = \frac{300,000}{15^2} = 1333 \text{ kg/m}^2 = 13.1 \text{ kN/m}^2$$

For the maximum stress

$$1.5\sigma_{10_{avg}} = 2000 \text{ kg/m}^2 = 19.6 \text{ kN/m}^2$$

$$1.5\sigma_{15_{avg}} = 4500 \text{ kg/m}^2 = 44.1 \text{ kN/m}^2$$

b. Boussinesq:

$$\sigma_Z = N_B \frac{P}{Z^2}$$

$$\sigma_5 = 0.48 \frac{300,000}{5^2} = 5760 \text{ kg/m}^2 = 56.5 \text{ kN/m}^2$$

$$\sigma_{10} = 0.48 \frac{300,000}{10^2} = 1440 \text{ kg/m}^2 = 14.1 \text{ kN/m}^2$$

c. Westergaard:

$$\sigma_Z = N_W \frac{P}{Z^2}$$

$$\sigma_5 = 0.32 \frac{300,000}{5^2} = 3840 \text{ kg/m}^2 = 37.6 \text{ kN/m}^2$$

$$\sigma_{10} = 0.32 \frac{300,000}{10^2} = 960 \text{ kg/m}^2 = 9.4 \text{ kN/m}^2$$

Using British engineering units:

a. 2:1 method:

$$\sigma_Z = \frac{P}{(B+Z)(L+Z)}$$

For a square footing

$$\sigma_Z = \frac{P}{(B+Z)^2}$$

$$P = \sigma_0 = \frac{675,000}{15^2} = 3000 \text{ psf}$$

$$\sigma_{15_{avg}} = \frac{675,000}{30^2} = 750 \text{ psf}$$

$$\sigma_{30_{avg}} = \frac{675,000}{45^2} = 333 \text{ psf}$$

For the maximum stress

$$1.5\sigma_{15_{avg}} = 1125 \text{ psf} = \sigma_{max}$$

$$1.5\sigma_{30_{avg}} = 499.5 \text{ psf} = \sigma_{max}$$

b. Boussinesq:

$$\sigma_Z = N_B \frac{P}{Z^2}$$

$$\sigma_{15} = 0.48 \frac{675,000}{15^2} = 1440 \text{ psf}$$

$$\sigma_{30} = 0.48 \frac{675,000}{30^2} = 360 \text{ psf}$$

c. Westergaard:

$$\sigma_Z = N_W \frac{P}{Z^2}$$

$$\sigma_{15} = 0.32 \frac{675,000}{15^2} = 960 \text{ psf}$$

$$\sigma_{30} = 0.32 \frac{675,000}{30^2} = 240 \text{ psf}$$

Newmark Influence Chart

A final method for computing vertical pressure σ_Z that can accommodate a uniformly loaded area of any shape is provided by the Newmark influence chart. The point in the subsurface for which the stress level is desired may be inside or outside the loaded area. This is accomplished by placing the point of interest over the center of the chart.

A Newmark influence chart is shown in Figure 7.46. The scale of the chart is set with respect to the depth of interest for which the stress level is being measured. On the chart the yardstick on the scale of distance is shown. This is set equal to the depth Z, the distance below the surface for which the vertical stress σ_Z is desired. This distance is used as a scale to draw the outline of the loaded area. The vertical stress is found according to the following equation:

$$\sigma_Z = q_0 I \times \text{number of blocks covered} \quad \text{(Eq. 7-50)}$$

where q_0 is the contact pressure of the loaded area, I is the influence value specified on the chart, and the number of blocks is the individual small area segments that occur within the boundary of the contact area. As indicated above, the point in the subsurface for which the stress level is desired is set at the center of the chart.

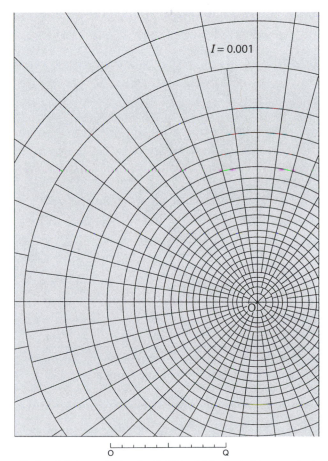

$I = 0.001$

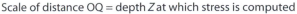

Scale of distance OQ = depth Z at which stress is computed

Figure 7.46 Newmark chart for vertical stress on horizontal planes at depth. (After Newmark, 1942.)

Example Problem 7.12

You are given two different loaded areas. The first is the rectangular-loaded area; it is 15 ft × 15 ft. The pressure on the footings is 3000 psf (see Figure 7.47). The second area is an irregular L-shape. The pressure on the footings is 5000 psf (see Figure 7.48). Find the stress at 15 and 30 ft for both loaded areas at Points A and B. Note: Newmark's chart was developed using the British engineering units system, however, we have also provided the solution converted to SI units.

Answers

For the rectangular-loaded area:

- For a 15-ft (4.57-m) depth, about 14.5 blocks are involved:

$$\sigma_Z = q_0 I \times \text{number of blocks}$$
$$= 3000 \text{ lb/ft}^2 (0.02) \times 14.5$$
$$= 870 \text{ lb/ft}^2 = 177.9 \text{ kg/m}^2 = 674 \text{ kN/m}^2$$

- For a 30-ft (9.14-m) depth, about 5 blocks are involved:

$$\sigma_Z = q_0 I \times \text{number of blocks}$$
$$= 3000 \text{ lb/ft}^2 (0.02) \times 5$$
$$= 300 \text{ lb/ft}^2 = 61.3 \text{ kg/m}^2 = 232.4 \text{ kN/m}^2$$

For the L-shaped loaded area:

- For a 15-ft (4.57-m) depth, about 19 blocks are involved:

$$\sigma_Z = q_0 I \times \text{number of blocks}$$
$$= 5000 \text{ lb/ft}^2 (0.02) \times 19$$
$$= 1900 \text{ lb/ft}^2 = 388.5 \text{ kg/m}^2 = 1472 \text{ kN/m}^2$$

- For a 30-ft (9.14-m) depth, about 9 blocks are involved:

$$\sigma_Z = q_0 I \times \text{number of blocks}$$
$$= 5000 \text{ lb/ft}^2 (0.02) \times 9$$
$$= 900 \text{ lb/ft}^2 = 184 \text{ kg/m}^2 = 697.2 \text{ kN/m}^2$$

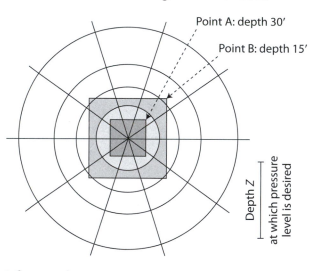

Figure 7.47 Example problem using the Newmark influence chart for a rectangular loaded area.

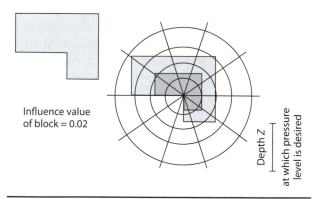

Figure 7.48 Example problem using the Newmark influence chart for an L-shaped loaded area.

Consolidation

Consolidation is the reduction in volume of clays under external loading as water drains from the pores. This reduction of excess pore water pressure or dissipation of pore pressure depends on the permeability of the soil and, therefore, is a time-dependent phenomenon. Settlement takes place as the drainage occurs.

Consolidation can be illustrated using the phase diagram in Figure 7.49. The saturated case is involved. Considered in terms of e, for one-dimensional consolidation, the change in height as compared to the original height can be related as follows.

$$\frac{\Delta H}{H_0} = \frac{\Delta e}{1+e_0} \qquad \text{(Eq. 7-51)}$$

and

$$\Delta H = H_0 \frac{\Delta e}{1+e_0} \qquad \text{(Eq. 7-52)}$$

where:

H = sample thickness or thickness of a soil layer
ΔH = change in sample or soil layer thickness upon consolidation (settlement)

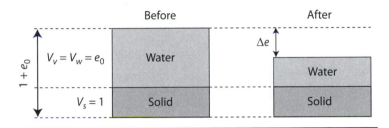

Figure 7.49   Derivation of consolidation equation.

Example Problem 7.13

Given a very compressible soil layer 2.4-m (8-ft) thick with an original void ratio of 1.1, but following consolidation it has a void ratio of 0.95, estimate the settlement of the soil layer.

Answers

Using SI units:

$$\Delta H = H_0 \frac{\Delta e}{1+e_0}$$

$$= (2.4)\frac{1.1-0.95}{1+1.1} = 0.171\,m = 17.1\,cm$$

Using British engineering units:

$$\Delta H = H_0 \frac{\Delta e}{1+e_0}$$

$$= (8)\frac{1.1-0.95}{1+1.1} = 0.57\,ft = 6.8\,in$$

Laboratory Testing

A consolidation test can be performed in the laboratory to determine the compressibility of the clay (ASTM D2435). This is accomplished by compressing the soil under load in a device known as a consolidometer (Figure 7.50 on the following page).

An undisturbed soil specimen is trimmed and placed in the confining ring. Porous stones are placed above and below the sample. A load is added to the loading plate and deformation of the sample is care- fully measured. The stress is determined by dividing the load by the area of the specimen.

Loads are added in increments, but after each increment the sample is allowed to consolidate until little or no further reduction occurs. This signals that very little, if any, excess pore pressure remains and the final stress is an effective stress. Load increments are added until sufficient data points are obtained to define the consolidation curve.

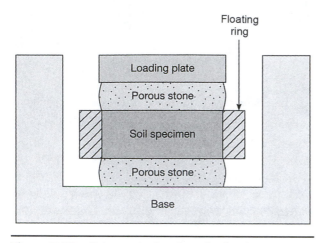

Figure 7.50   Schematic drawing of a floating ring consolidometer.

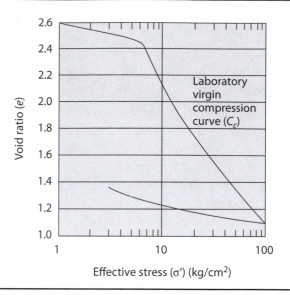

Figure 7.51   Consolidation test data, void ratio versus log of effective stress.

Typically, the void ratio is plotted against the logarithm of the effective consolidation pressure or effective stress (Figure 7.51). The plot is known as the compression curve and has two parts, one with a gentler slope and the other with a steeper slope. The gentler part of the curve is known as the recompression curve, whereas the steeper part is known as the virgin compression curve. The consolidation pressure corresponding to the point where the change in slope occurs is referred to as the preconsolidation pressure (σ'_p). It represents the maximum overburden pressure the soil experienced in its geologic history at the depth from where the sample was obtained for the consolidation test. Based on a comparison of the existing overburden pressure (σ'_v) with the preconsolidation pressure (σ'_p), clays are divided into two types: (1) normally consolidated clays and (2) overconsolidated clays. For normally consolidated clays, $\sigma'_v = \sigma'_p$; for overconsolidated clays, $\sigma'_v < \sigma'_p$. Another distinction between the two types of clays is their natural water content (w_n). For normally consolidated clays, w_n is approximately equal to LL; for overconsolidated clays, it is close to PL. The consolidation or settlement of overconsolidated clays is usually much less than that of normally consolidated clays. Therefore, it is important to differentiate between the two types of clays for settlement calculations.

The slope of the recompression curve in Figure 7.51 is called the recompression index (C_r) and the slope of the virgin compression curve is designated as the compression index (C_c). It can be seen that

$$C_c = \frac{e_1 - e_2}{\log \sigma'_2 - \log \sigma'_1} = \frac{e_1 - e_2}{\log \sigma'_2 / \sigma'_1} \qquad \text{(Eq. 7-53)}$$

The void ratio is dimensionless and the units cancel in σ'_2/σ'_1, so C_c is a dimensionless value.

If you wish to find C_c graphically, note that if you select $\log \sigma'_2/\sigma'_1 = \log 100/10 = \log 10 = 1$, then C_c is equal to Δe for that one log cycle.

Terzaghi and Peck (1967) proposed the following empirical equation for determining C_c for clays of low to medium sensitivity:

$$C_c = 0.009(\text{LL} - 10) \qquad \text{(Eq. 7-54)}$$

where LL = liquid limit.

Settlement of Normally Consolidated Clays

Substituting the value of C_c in Equation 7-51 gives rise to the following equation for settlement of normally consolidated clays:

$$\text{Settlement} = \Delta H = \frac{H}{1 + e_0} C_c \log \left(1 + \frac{\Delta \sigma'}{\sigma'} \right) \text{(Eq. 7-55)}$$

The answer to Example Problem 7.14 provides the derivation for Equation 7-55.

Example Problem 7.14

Find C_c for Figure 7.51.

Answers

For the portion from 10, 2.1 (effective stress and void ratio, respectively) to 43, 1.45

$$C_c = \frac{2.1 - 1.45}{\log 43 - \log 10}$$

$$= \frac{0.65}{\log 43/10} = \frac{0.65}{\log 4.3} = 1.02$$

Alternately, using one log cycle for the portion 10, 2.1 to 100, 1.1

$$\Delta_e = 2.1 - 1.1 = 1.0 = C_c$$

for 1 cycle. If we next examine the two equations presented previously,

$$\Delta H = H_0 \frac{\Delta e}{1 + e_0}$$

and

$$C_c = \frac{\Delta e}{\log \sigma'_2 - \log \sigma'_1}$$

solving the second equation for Δe yields

$$\Delta e = C_c \log(\sigma'_2 - \sigma'_1) = C_c \Delta \log \sigma'$$

but

$$\Delta \log \sigma' = \log(\sigma' + \Delta \sigma') - \log \sigma'$$

$$= \log \frac{\sigma' + \Delta \sigma'}{\sigma'} = \log\left(1 + \frac{\Delta \sigma'}{\sigma'}\right)$$

and

$$\Delta e = C_c \log\left(1 + \frac{\Delta \sigma'}{\sigma'}\right)$$

Finally

$$\Delta H = \frac{H}{1 + e_0} C_c \log\left(1 + \frac{\Delta \sigma'}{\sigma'}\right)$$

Example Problem 7.15

Given the following information, calculate the settlement of the clay. In a 3-m (10-ft) thick clay layer with a compression index of 0.95, the initial void ratio is 2.1 and the effective stress is 75 kN/m² (1550 psf). The pressure increase for the clay is 24 kN/m² (500 psf).

Answers

Using SI units:

$$\Delta H = \frac{H}{1 + e_0} C_c \log\left(1 + \frac{\Delta \sigma'}{\sigma'}\right)$$

$$= \frac{3}{1 + 2.1}(0.95)\log\left(1 + \frac{24}{75}\right) = 0.112 \text{ m} = 11.2 \text{ cm}$$

(continued)

Using British engineering units:

$$\Delta H = \frac{H}{1+e_0} C_c \log\left(1 + \frac{\Delta\sigma'}{\sigma'}\right)$$

$$= \frac{10}{1+2.1}(0.95)\log\left(1 + \frac{500}{1500}\right) = 0.372 \text{ ft} = 4.5 \text{ in}$$

Many details of consolidation have not been included in the above discussion. Settlement of over-consolidated clays is not considered nor is the time rate of settlement discussed. The reader is referred to textbooks on soil mechanics or geotechnical engineering for a detailed study of this subject matter.

Engineering Considerations of Soil Mechanics

A multitude of problems can develop when different soils are used to support engineering structures. They may vary considerably from one area to another in the United States and only a general list is presented below:

1. *Compressible soils:* Compressible soils include organic soils, some glacial deposits, certain floodplain soils, and highly plastic clays. Problems associated with compressible soils include excessive settlement, low bearing capacity, and low shear strength.

2. *Collapsible soils*: Collapsible soils are also known as metastable soils. They are unsaturated soils, primarily sands and silts, that undergo a large volume change upon saturation. The sudden and usually large volume change could cause considerable structural damage. The volume change may or may not occur due to an additional load. These soils occur in sandy coastal plain areas, sandy glacial deposits, and alluvial deposits of intermountain regions in the western United States.

3. *Expansive soils*: Expansive soils contain swelling clays, primarily montmorillonite (smectite). They increase in volume upon wetting and shrink upon drying. Climate is closely related to the severity of the problem. Semiarid to semihumid areas with swelling clays suffer most because the soil moisture active zone has the greatest thickness under such conditions. Foundation supports should be placed below this active soil zone. Expansive soils are most prevalent on the Atlantic and Gulf Coastal Plains and in some areas of the central and western United States.

4. *Corrosive soils*: Corrosive soils have high acidity that causes the corrosion of underground metal pipes. Predictions can be made based on the pH of the soil or on its electrical resistivity. Values of 700 ohm cm or less typically indicate a high corrosion potential for soils, 750 to 1000 is moderate, greater than 1000 is low to moderate, and greater than 1750 is very low corrosion potential.

5. *Fine-textured soils*: These soils can yield problems related to highway performance. Plastic subgrades introduce a problem known as rigid pavement pumping for concrete highways. Rigid pavement pumping is the removal of fines in the subgrade and base course by water when it is forced out from pavement joints with the passing of heavy wheel loads, typically from trucks. This yields a void under the concrete slab and eventually the concrete fails by flexure adjacent to the joint. A rough pavement develops and concrete adjacent to the joint must be replaced.

Silt-sized soils are frost susceptible. Water moves through this material by capillary action to provide a source of water in a freezing soil. Ice lenses form which, on melting, provide insufficient support for the highway. Asphalt highways are particularly prone to spring breakup problems when wheels punch through the flexible pavement causing potholes to form. Gravel secondary roads are subject to intense, spring breakup problems, making some impassable until gravel fill and road grading can remedy the situation. Base courses for flexible pavements must contain 2% or less of minus No. 200 size material (fines) in their gradation to prevent spring breakup from freezing and thawing.

6. *Saturated fine sands and silts*: These are subject to liquefaction failures during earthquake shaking. Alluvial deposits are particularly susceptible to this problem in earthquake-prone areas.

7. *Soils of varying permeability*: This is a problem for many different types of construction. Values for permeability vary over nine orders of magnitude from coarse gravel to colloidal clay—this is an enormous difference. For water supply, high permeability is desirable; in underseepage for dams and containment of contaminated liquids, low permeability is desirable. Problems of dewatering and grouting are related to soil permeability. Slope stability, excavations, road construction, and many other construction specifics are also related to soil permeability.

EXERCISES ON ELEMENTS OF SOIL MECHANICS

1. *Grain size distribution curves*: The following sieves were used to determine grain size distribution: No. 4, No. 10, No. 20, No. 40, No. 60, No. 100, and No. 200. Exercise Table 7.1 provides the weight of material retained on each sieve and the fraction that passed the No. 200 sieve.

   a. Complete the table by calculating the values based on the information supplied.

   b. Using five-cycle semilog paper, plot the grain size distribution for this sieve analysis.

   c. Give the D_{60} and the D_{10} values for this sieve analysis. Calculate the k value for this analysis using Hazen's approximation, which states that k (cm/sec) $= 100D_{10}^2$ (D_{10} in mm).

   d. Calculate the coefficient of uniformity and the coefficient of curvature for this analysis. Is the sample well graded? Explain why or why not.

2. What is the difference between shape and roundness? Describe a cube and a sphere relative to these terms. What is an equant particle called in soil mechanics terms?

3. Given that $n = V_v/V_t$ and $e = V_v/V_s$ show that $n = e/(1 + e)$.

4. The total volume V_t, the total mass M_t of a fully saturated soil, the dry mass M_s, and the specific gravity of the solids G_s are known. Draw the phase diagram for this situation and, using the given parameters, derive the equation $e = wG_s$.

5. The void ratio e, the specific gravity of solids G_s, and the degree of saturation S of a soil are known. In terms of these parameters derive an equation for ρ, the wet density. Begin by drawing the phase diagram.

6. The porosity n and specific gravity of solids G_s of a saturated soil are known. Derive the equation for ρ, the wet density, and for w, the water content, based on this information. Begin by drawing the phase diagram.

7. A moist sand sample has a volume of 40.5 cm^3 in its natural state and a mass of 50.2 g. When dry its mass is 48.3 g; the specific gravity of solids is 2.68. Determine the void ratio, the porosity, water content, and degree of saturation. Begin by drawing the phase diagram.

8. A sample of saturated clay has a mass of 1526 g in its natural state and 1053 g after drying. The value of G_s was found to be 2.70. Calculate w, ρ_d (the dry density), and ρ (the wet density) of the soil.

9. For saturated conditions of a soil, derive the equation for ρ', the submerged density in terms of n, G_s, and ρ_w. Derive the formula in two ways, the first based on the fact that $\rho' = \rho - \rho_w$ and the second using

$$\rho' = \frac{M_t - V_s\rho_w - V_w\rho_w}{V_t}$$

that is, the mass minus the buoyancy effect divided by the volume.

Exercise Table 7.1

US Standard Sieve No.	Sieve Opening (mm)	Weight Retained (g)	% Retained on Sieve	% Passing Sieve
4	4.75	12.7	6.2	93.8
10	2.00	24.8	12.1	81.7
20	0.85	21.5	___	___
40	0.425	38.8	___	___
60	0.25	23.8	___	___
100	0.15	19.5	___	___
200	0.074	46.7	___	___
Minus No. 200		17.3	___	___
Total		205.1		

10. A sample of mica flakes has a dry mass of 145 g. The G_s for the sample is 2.82. In the loose state, the sample had a volume of 1000 cm³. After vibration, the volume of the sample was reduced to 400 cm³, and finally after vibrating the sample with a load placed on the mica flakes, the volume was reduced to 200 cm³.

    a. For the three conditions described (loose, vibrated, vibrated and loaded), calculate the void ratio and porosity for each.

    b. For the three conditions described (loose, vibrated, vibrated and loaded), calculate the dry density, saturated density, and submerged density for each. Record the answers in kg/m³.

11. Given: ρ_s = 2680 kg/m³, ρ = 2050 kg/m³, ρ_d = 1765 kg/m³. Find: e, w, S, n, and ρ'. Assume V_t = 1 m³ and use the appropriate phase diagram to illustrate the problem.

12. For a certain sandy soil, e_{max} = 0.72 and e_{min} = 0.45. In its natural state this soil was found to have a density of 2.01 Mg/m³, a water content of 17%, and a G_s of 2.65. What is the relative density and the degree of saturation? Use the appropriate phase diagram to illustrate the problem.

13. Given: An unsaturated soil has a density of 1850 kg/m³, a water content of 8.6%, and a G_s of 2.65. Find: The relative density D_r if e_{max} = 0.642 and e_{min} = 0.462.

14. The liquid limit of a soil sample is 35 and the plastic limit is 22. What is the plasticity index? What is meant by liquid limit, plastic limit, and plasticity index? Where would this soil plot on the plasticity chart shown in Figure 7.11? If the natural water content of the soil was 30%, what is the liquidity index? What does this value signify about the soil?

15. An inorganic clay with a liquid limit of 80 plots directly on the A-line (see Figure 7.11). What would be the PI value of this soil? Referring to Figure 7.12, what clay mineral would likely make up this sample?

16. A certain soil was found to have an LL of 28 and a PL of 17. What is its PI? Another soil has a PI of 48 and a PL of 32. What is its LL? Is it a low compressibility soil? Explain. Plot both soils on the plasticity chart (see Figure 7.11). In which area do they fall? Supply a descriptive name if possible. Give a typical PI and LL for an ML soil.

17. Distinguish between the terms *soil fabric* and *soil structure*. Why is only soil fabric important for cohesionless soils, whereas both soil fabric and soil structure are of interest to cohesive soils? What is a single-grain structure? Why are loose grain structures subject to sudden densification under vibratory loads?

18. The specific surface of a sphere equals $6/d$ where d is the diameter of the sphere. What would be the specific surface of a fine sand grain 0.05 mm in diameter?

19. List the types of interparticle bonds in clays. What is their strength relative to primary valence bonds? Over what distance do they act? An answer in the form of a table should prove useful.

20. What is meant by the term *cation exchange capacity*? Explain why the mineral kaolinite might have a value of 10 milliequivalents per 100 g of sample whereas smectite could equal 140. In a water softener, Mg^{2+} and Ca^{2+} ions are replaced by the cation from the salt added. What is that cation?

21. What are the three possible configurations for clay structures? Under what circumstances do each of these occur? Which configuration would be most compact? Explain.

22. A clay sample was found to have a maximum strength of 38.3 kPa (800 lb/ft²) and a residual strength of 7.2 kPa (150 lb/ft²). Is this a sensitive clay? Explain.

23. Macrostructure and microstructure of clays are of importance. Bedding is one of the types of macrostructure; name several others. How are the effects of bedding indicated or manifested? Concerning microstructures, what are the different aggregated particles that have been designated for clays? List them from the smallest to the largest.

24. Distinguish between compaction and consolidation in terms of soil mechanics definitions. What are the construction objectives of compaction? What four variables control the extent of compaction?

25. Refer to Figure 7.16. What would 90% of the maximum dry density equal for this curve? If the specifications require 95% of standard Proctor density and within ±2% of the optimum water content, what range of values would be acceptable for this test?

26. Refer to Figure 7.17. Which is a greater density: 100% standard Proctor density or 90% modified Proctor density? What is the optimum moisture content for the standard and the modified Proctor tests? How much of a decrease does this indicate from the lesser compactive effort to the greater one?

27. For a dry density of 1795 kg/m³ and a G_s of 2.65, what water content is required to yield zero air voids? What would be the wet density and void ratio for this situation?

28. Refer to Table 7.6 on the properties of compacted soils. Why are soils compacted dry of optimum more permeable than those compacted wet of optimum, assuming the same compactive effort? Refer also to the text. If a compacted soil dry of optimum had a permeability of about 10^{-6} cm/sec what would it be wet of optimum? On the dry side of optimum the compacted soil is also more sensitive to cracking. Why is this detrimental in earth dam construction? Explain.

29. A specific sheepsfoot roller has a coverage of 11% and a contact pressure of 4.14 MPa (600 psi). How many passes of the roller are needed before 100% coverage of the fill is obtained? Why is it important that the specified lift thickness in the fill not be exceeded? What would be a likely lift thickness allowed for this roller? Explain.

30. Distinguish between end-product and method specifications. Which of these requires more study prior to letting the bids? Explain.

31. What are the various methods used for field control of density and moisture content? Which of these is a nondestructive test? How does it work?

32. Given the cross section in Exercise Figure 7.1, calculate the values related to effective stress. Use SI units exclusively:

   a. Calculate σ, u, and σ' for depths of 5, 10, 18, 25, and 33 m and complete Exercise Table 7.2.

   b. Prepare a drawing of the soil profile. Show σ, u, and σ' values, label it, and draw it to scale. (See Figures 7.23 and 7.24 as examples.)

Exercise Table 7.2

Depth (m)	Total Pressure σ (kPa)	Neutral Pressure u (kPa)	Effective Pressure σ' (kPa)
5	_____	_____	_____
10	_____	_____	_____
18	_____	_____	_____
25	_____	_____	_____
33	_____	_____	_____

33. Given the cross section in Exercise Figure 7.2, calculate values for effective stress. Use both SI and British engineering units. Note: 1 ft = 0.3048 m. In this soil profile, the groundwater table occurs at a depth of 1.2 m (4 ft). Assume that $G_s = 2.65$ for all materials.

   a. Calculate the unit weight for the four different zones in the cross section. Assume the soil above the water table is completely dry and that below the water table it is saturated.

   b. Calculate the total stress, neutral stress, and effective stress for depths of 1.2, 3, 5.5, and 9.1 m (4, 10, 18, and 30 ft). Supply the answers in table form similar to that in Exercise 32.

   c. Draw a pressure diagram of the soil profile showing total stress, neutral stress, and effective stress. Draw to scale.

34. See Exercise Figure 7.3 on the next page. A standpipe consists of 20 cm of water and 20 cm of saturated sand below. A column of water held at a constant head is attached to the base of the standpipe as shown.

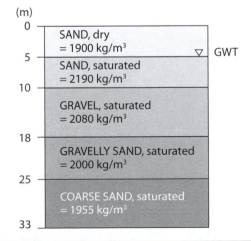

Exercise Figure 7.1

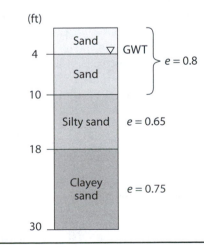

Exercise Figure 7.2

a. Consider that the water level is maintained at location A, 8 cm below the water level in the standpipe. Calculate the total, neutral, and effective stresses at 20 and 40 cm for the standpipe. Draw a pressure diagram to the right of the standpipe showing this relationship.

b. Consider that the water level is maintained at location B, 10 cm above the water level in the standpipe. Calculate the total, neutral, and effective stresses at 20 and 40 cm for the standpipe. Draw a pressure diagram showing this relationship.

c. Calculate what is needed to yield a quick condition at the base of the sand column. What is required of the neutral stress and of the effective stress for this to occur?

35. Give the equation for the critical hydraulic gradient i_{cr} in terms of submerged density of the soil mass and density of water. What would be the critical hydraulic gradient for a 1.5-m (5-ft) thick layer of barite ($BaSO_4$) placed to reduce uplift effects? Barite has a specific gravity of 4.5; assume the barite layer, consisting of a gradation of gravel-sized pieces, has a porosity of 30%.

36. A gravelly sand channel slopes away from a reservoir, through the reservoir rim, to an adjacent drainage area. The channel slopes at a gradient of 23 m (75 ft) vertically to 30.5 m (100 ft) horizontally. In cross section the channel is 1.8 m (6 ft) thick and 4.6 m (15 ft) wide. The channel intersects the adjacent drainage area at a horizontal distance of 152.4 m (500 ft) from the reservoir.

a. At minimum pool elevation in the reservoir the water is 1.5 m (5 ft) above the sandy gravel channel. What is the total pressure head in the gravelly sand seam at the upstream end (or at the reservoir)? What is the total pressure head at the downstream end of the channel where it daylights into the adjacent drainage area 152.4 m (500 ft) away? What is the hydraulic gradient? What is the average velocity of flow if $k = 10^{-2}$ cm/sec? What is the flow loss in m^3/sec (ft^3/sec) and in gal/min for the channel (use Darcy's law, $v = ki$ relationship, see also Chapter 13, Groundwater)? How long does it take for the water to flow from the reservoir to the adjacent drainage area?

b. At maximum pool the water stands at 15 m (50 ft) higher than minimum pool. What is the flow loss in m^3/sec (ft^3/sec) and gal/min for this case? Does a quick condition occur when the pool reaches this maximum level? If not, what is required to cause a quick condition? Explain.

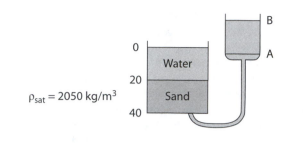

Exercise Figure 7.3

37. Explain why the sediment on the ocean bottom can be just as loosely compacted when it lies at a water depth of 305 m (1000 ft) as it can be at a water depth of only 3 m (10 ft). What is required to compact the sediment? Explain.

38. This question pertains to the Mohr circle procedure for shear strength of soils. The major and minor principal stresses at failure for a point in a cohesionless sand are 448 and 138 kPa (65 and 20 psi), respectively. Begin by drawing the Mohr diagram (see Figure 7.33).

a. What is the value of the maximum shearing stress at that point?

b. What is the value of the normal stress on that plane of maximum shearing stress?

c. What are the normal and shearing stresses on a plane 27° from the plane of major principal stress? What is the value of the resultant stress on this plane?

d. What is the value of σ and τ on a plane 30° from the plane of minor principal stress? What is the resultant stress on this plane?

e. What is the value of the maximum obliquity angle? This angle is α_m or ϕ.

f. Find $\sigma_{failure}$ and $\tau_{failure}$ using the following relationships:

$$\sigma_{failure} = \sigma_3(1 + \sin\phi)$$
$$\tau_{failure} = (\sigma_1 - \sigma_3)\cos 1/2\phi$$

g. If $\theta_{failure} = 45 + 1/2\ \phi$, find $\theta_{failure}$. Show all relationships on the Mohr diagram.

39. Triaxial test results on a clayey silt indicate a c value of 43 kPa (900 lb/ft^2) and a ϕ value of 28°.

a. Draw a Mohr envelope for the soil.

b. If the effective stress σ' on this material was 207 kPa (30 psi), what would the maximum shear stress be under failure conditions?

c. Assume that the water table occurs at the ground surface of the clayey silt deposit under study. The saturated unit weight was found to be 2.08 Mg/m³ (130 lb/ft³). At what depth would a vertical normal effective stress of 207 kPa (30 psi) occur? Show calculations.

40. The unconfined compression strength q_u for a highly plastic clay is 96 kPa (2000 lb/ft²).

    a. What is its q_u in psi, kg/cm², and N/m²?

    b. What is its cohesive strength in these units?

    c. What is its ϕ value? Why?

    d. If the unit weight of the clay is 2 Mg/m³ (125 lb/ft³), at what depth is the overburden stress σ_v (vertical stress) equal to the cohesive strength?

    e. At a depth of 9.8 m (32 ft), what is the ratio of overburden stress to compressive strength?

    f. What effect would this ratio have on the ability to tunnel through this clay at a depth of 9.8 m (32 ft)?

41. A sandy deposit ranges from a clean sand to a slightly clayey sand. For the clean sand, failure conditions were found to be $\sigma_1 = 621$ kPa (90 psi), $\sigma_3 = 172$ kPa (25 psi), and $c = 0$. For the clayey sand, failure conditions were $\sigma_1 = 634$ kPa (92 psi), $\sigma_3 = 172$ kPa (25 psi), and $c = 55$ kPa (8 psi).

    a. Find ϕ, θ_{cr}, σ_{ff} and τ_{ff} for the clean sand. Use $c = 0$ in this calculation.

    b. Find ϕ, θ_{cr}, σ_{ff} and τ_{ff} for the clayey sand. Use $c = 55$ kPa (8 psi) in this calculation.

42. A fine-grained granite was tested in a series of triaxial tests. A q_u value of 207 MPa (30,000 psi) was also determined. For the triaxial tests, a ϕ value of 55° and an S_0 value (τ intercept) of 55 MPa (8000 psi) was obtained. Draw the Mohr circle for the q_u test. Why does the q_u diagram not fall in line with the Mohr envelope for the triaxial test? Explain.

43. A footing 3 m × 3 m (10 ft × 10 ft) is designed to carry a 200-tonne (220-ton) load. Find the stress level at 0, 5, 10, 15, and 20 ft using the 2:1, Boussinesq, and Westergaard methods. Compare the answers obtained. Which would likely be used for checking the adequacy of the soil? Explain.

44. If a footing 2.4 m × 3.7 m (8 ft × 12 ft) is used instead of the footing in Exercise 43, with the same 200-tonne (220-ton) load, find the stress level at a depth of 3 and 6 m (10 and 20 ft) using the Newmark influence chart from Figure 7.46.

45. Convert Exercise 43 to a problem with SI units. Use kilograms, meters, and N/m² as units. Calculate answers for the same depths but use only the Boussinesq method.

46. *Consolidation of Clay*: Given Exercise Figure 7.4 and the accompanying information, what is the ultimate settlement in the clay under the applied load? Note: 1 ft = 0.3048 m; 62.4 lb/ft³ = 1 Mg/m³.

    Assuming the sand will not settle, determine the ultimate settlement in the clay using:

    $$\Delta H = \frac{H}{1+e_0} C_c \log\left(1 + \frac{\Delta\sigma'}{\sigma'}\right)$$

    Proceeding in steps:

    a. Determine the effective stress for the cross section prior to loading. Show the stresses at depth using an appropriate diagram.

    b. What is $\Delta\sigma'$ on the clay layer caused by the imposed load? (*Hint:* Calculate this for the center of the clay layer at 5.8 m [19 ft] deep.)

    c. What is the value of e_0 based on the original porosity of the clay?

    d. How much settlement will occur in the clay?

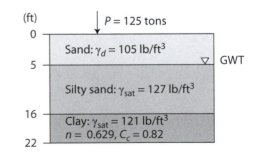

Exercise Figure 7.4

References

Atterberg, A. 1911. Leronas Forhallande till Vatten, deras plasticitetsgranser och plasticitetsgrader (The behavior of clays with water, their limits of plasticity and their degrees of plasticity). *Kungliga Lantbruksakademiens Handlingar och Tidskrift* 50(2):132–158.

Boussinesq, J. 1885. *Application des potentiels a l'estude de l'equilibre et due movement des solides elastiques.* Paris: Gauthier-Villars.

Bowles, J. E. 1996. *Foundation Analysis and Design* (5th ed.). New York: McGraw-Hill.

Casagrande, A. 1932. Research on Atterberg limits of soils. *Public Roads* 13(8):121–136.

Casagrande, A. 1948. Classification and identification of soils. *Transactions of the American Society of Civil Engineers* 113:901–930.

Das, B. M., and Sobhan, K. 2018. *Principles of Geotechnical Engineering* (9th ed.). Boston: Cengage.

Holtz, R. D., Kovacs, W. D., and Sheahan, T. C. 2011. *An Introduction to Geotechnical Engineering* (2nd ed.). New York: Pearson.

Johnson, R. B., and DeGraff, J. V. 1988. *Principles of Engineering Geology.* New York: Wiley.

Lambe, T. W. 1958. The structure of compacted clay. *Journal of the Soil Mechanics and Foundations Division, American Society of Civil Engineers* 84(2).

Leonards, G. A. 1980. Dynamic compaction of granular soils. *Journal of the Geotechnical Engineering Division, American Society of Civil Engineers* 106(1):35–44.

Lukas, R. G. 1995. *Dynamic Compaction.* Geotechnical Engineering Circular No. 1, Federal Highway Administration, Publication 1 (FWHA-SA-95-037). Washington, DC: Office of Technology Applications.

Marshak, S. 2013. *Essentials of Geology.* New York: W. W. Norton.

Mitchell, J. K. 1976. *Fundamentals of Soil Behavior.* New York: Wiley.

Mitchell, J. K., and Soga, K. 2005. *Fundamentals of Soil Behavior* (3rd ed.). New York: Wiley.

Newmark, N. M. 1942. *Influence Charts for Computation of Stresses in Elastic Foundations.* University of Illinois Engineering Experiment Station Bulletin, Series No. 338, 61(92). Urbana, Illinois.

Proctor, R. R. 1933. Fundamental principles of soil compaction. *Engineering News Record* 111(9, 10, 12, and 13).

Seed, H. B., Woodward, R. J., and Lundgren, R. 1962. Prediction of swelling potential for compacted clays. *Journal of the Soil Mechanics and Foundations Divisions, American Society of Civil Engineers* 88(SM4):107–131.

Taylor, D. W. 1948. *Fundamentals of Soil Mechanics.* New York: Wiley.

Terzaghi, K., and Peck, R. B. 1967. *Soil Mechanics in Engineering Practice* (3rd ed.). New York.

van der Merwe, D. H. 1964. The prediction of heave from the plasticity index and the percentage of clay fraction of soils. *Civil Engineering in South Africa* 6(6):103–107.

Westergaard, H. M. 1938. A problem of elasticity suggested by a problem in soil mechanics—a strong material reinforced by numerous strong horizontal sheets. In *Contributions to the Mechanics of Solids,* Stephen Timoshenko 60th Anniversary Volume (pp. 268–277). New York: Macmillan.

ENGINEERING PROPERTIES OF ROCKS

8

Chapter Outline

Properties of Intact Rock

Engineering Classification of Intact Rock

Rock Mass Properties

Engineering Classification of Rock Mass

Rock Strength and Petrography

Engineering Considerations of Rock Properties

From an engineering point of view, rocks are significant for two major reasons: (1) they are an important building material with numerous applications in construction engineering and (2) many engineering structures are founded on rock; their safety depends on the stability of the rock foundation and the adjacent rock mass.

The engineering properties of rocks determine their behavior as construction materials and as a structural foundation. There are two classes of rock properties: properties of intact rock and rock mass properties. An intact rock contains no visible discontinuities or planes of weakness (joints, bedding, foliation, etc.) whereas a rock mass is traversed by many discontinuities. Properties of intact rock are measured on small samples in the laboratory, whereas rock mass properties are evaluated on large outcrops in the field. Typically, properties of intact rock are controlled by petrography (mineral composition and texture) and rock mass properties are controlled by the planes of weakness. In engineering practice, properties of intact rock are needed when rock is used as construction material and rock mass properties are required for design of foundations, underground openings, and cut slopes.

Properties of Intact Rock

Properties used for characterizing intact rock as a building material include: specific gravity, absorption, porosity, degree of saturation, unit weight (density), unconfined compressive strength, tensile strength, shear strength, Young's modulus, Poisson's ratio, and durability (Johnson and De Graff, 1988).

Specific Gravity, Absorption, Porosity, Degree of Saturation, and Unit Weight

Specific gravity is the ratio of the weight in air of a given volume of rock to the weight of an equal volume of water. In order to account for the presence of pores in a rock, the American Society for Testing and Materials (ASTM) recommends using three different types of specific gravity in engineering practice. The laboratory test for determining specific gravity and absorption is ASTM D6473-15 . The method requires that the rock specimen be weighed in air in a dry condition, weighed in air in a saturated condition, and weighed in water in a saturated condition. From these data, specific gravity and absorption values are obtained as follows:

$$\text{Bulk specific gravity } \left(\text{SpG}_d\right) = \frac{A}{B-C} \qquad \text{(Eq. 8-1)}$$

$$\text{Bulk specific gravity } \left(\text{SpG}_s\right) = \frac{B}{B-C} \qquad \text{(Eq. 8-2)}$$
(saturated, surface dried)

$$\text{Apparent specific gravity } \left(\text{SpG}_a\right) = \frac{A}{A-C} \qquad \text{(Eq. 8-3)}$$

$$\text{Absorption} = \frac{B-A}{A}100 \qquad \text{(Eq. 8-4)}$$

where:
- A = mass of rock in air, oven dried for 24 hours
- B = mass of rock in air, saturated for 24 hours, surface dried
- C = mass of rock in water, saturated

Porosity is the ratio of the volume of voids (V_v) to the total volume (V_t) of a rock, expressed as a percentage. It can be determined by using phase relations, as described in most soil mechanics textbooks (Holtz et al., 2011; Das and Sobhan, 2018) and in Chapter 7. Porosity can range from 0.1% for dense rocks like diabase and quartzite to 5 to 25% for sandstones, and even higher for volcanic rocks like tuff (Gonzalez de Vallejo and Ferrer, 2011).

Degree of saturation is the ratio of the volume of water (V_w) to the volume of voids (V_v) in a rock, expressed as a percentage. It can also be determined by using phase relations, and ranges from 0% for completely dry rock to 100% for completely saturated rock.

The unit weight, or density, of rock is defined as the mass per unit volume and can be obtained by multiplying bulk specific gravity by density of water (1 g/cm^3; 1000 kg/m^3; 1 Mg/m^3; 9.81 kN/m^3), or by dividing the mass by volume of a core sample. The general range of unit weight is 20 kN/m^3 (127 lb/ft^3) to 30 kN/m^3 (191 lb/ft^3) (Gonzalez de Vallejo and Ferrer, 2011).

Specific gravity, absorption, porosity, degree of saturation, and density show strong correlations with compressive strength (Shakoor and Bonelli, 1991; Shakoor and Barefield, 2009). Rocks with higher specific gravity and density and lower percent absorption, porosity, and degree of saturation have higher

compressive strength. Table 8.1 lists the porosity and unit weight values for various rock types.

Rock Strength

Strength of a rock is its ability to resist movement along internal surfaces under the application of stresses. In general, a rock can be subjected to three primary types of stress: compressive, tensile, and shear. Compressive stress tends to decrease the volume of the rock by forces acting inward and directly opposite to each other. Tensile stress tends to pull a substance apart by outward-acting, equally opposing forces. Shear stress is caused by two equal forces acting in opposite directions. Compressive, tensile, and shear stresses are illustrated in Figure 8.1. Corresponding to the three types of stresses, there are three types of rock strength: unconfined compressive strength, tensile strength, and shear strength.

Unconfined Compressive Strength

The unconfined compressive strength, also known as uniaxial compressive strength, is one of the most commonly used properties of rock (Bieniawski, 1989). The term unconfined pertains to rock specimens unconfined at their sides while the load is applied vertically until failure occurs (Figure 8.2 on page 162). Either ASTM method D7012-14 or International Society for Rock Mechanics (ISRM) (2014) methods are used to determine unconfined compressive strength. The test method involves failing an N_x-size (54 mm) core sample with a length to diameter ratio of 2.0 to 2.5. The strength is obtained by:

$$\sigma_c = \frac{P}{A} \qquad \text{(Eq. 8-5)}$$

where P is the failure load in newtons (or lb), A is the cross-sectional area of the sample in m^2 (or in^2), and

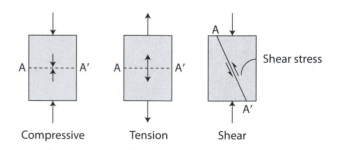

Figure 8.1   Compressive, tensile, and shear stress diagrams.

Table 8.1 Engineering Properties of Different Rocks.

	Compressive Strength σ_c ($\times 10^6$ N/m²)	Shear Strength S_0 ($\times 10^6$ N/m²)	Tensile Strength T ($\times 10^6$ N/m²)	Modulus of Elasticity E ($\times 10^9$ N/m²)	Angle of Shear Resistance ϕ	Poisson's Ratio μ	Unit Weight γ (kN/m³)	Porosity n (%)
Igneous								
Granite, unweathered	138–240	13.4–47.6	6.7–23.9	19.1–57	45–60	0.15–0.24	25.5–27.5	0.5–1.5
Coarse	53–72	10.3–13.8			48–56			
Pegmatitic	47	7.2			58			
Fine	219	18.6			70			
Slightly altered	65	9.8			58			
Syenite	135			57.4–76.5			25.5–26.7	0.5–1.5
Quartz monzonite	210	25.1		64.2–95.8	63	0.17		
Monzonite porphyry	125	16.5			59			
Diorite	172–287	113.8	14.3–28.7	9.31–13.9	54		26.4–27.5	0.1–0.5
Diabase (dolerite)	186–330	23.4–57.4	14.3–33.5	76.5–105	55–60		26.4–29.9	0.1–0.5
Gabbro	172–287	19.9–57.4	14.3–28.7	64.2–105			27.5–30.3	0.1–0.2
Basalt	138–287		9.6–28.7	8.33–13.9	50–55		27.5–28.5	0.1–1.0
Andesite	129–132			57.4–95.8			21.5–22.6	10–15
Tuff	7.0		1.1	3.1		0.11		
Metamorphic								
Gneiss	48–186		4.8–19.3	19.6–59.2	48–73	0.15–0.24	27.5–29.4	0.5–1.5
Massive granite gneiss	221	31			66			
Granite gneiss	53–69	10.3–12.4			48–56			
Schistose	69	12.4			73			
Weathered schistose	53–92	12.4–15.1			59			
Quartzite	143–287	19.1–57.4	9.6–28.7	5.7–8.3	50–60		25.5–25.9	0.1–0.5
Marble	49–239	14.3–28.8	6.7–19.3	4.1–19.3	35–50	0.25–0.38	25.5–25.9	0.5–2.0
Slate	83–239		6.7–19.3				25.5–25.9	0.5–2.0
Schist	8–117					0.08–0.20	24–26	
Biotite	53–83							
Biotite-chlorite	39–117							
Sedimentary								
Sandstone	19–158	7.6–37.9	3.9–23.9	4.8–76.5	35–50	0.17	18.9–25.3	0.5–26
Graywacke	54	11.7		9.6–33.1	47			
Shale, general	90–96	2.9–23.8			15–30		23.5–27.4	
Clayshale	1.2–7	0.3–1.1						
Siltstone	28–50	7.0–30			30–40			
Mudstone				19.3–47.5				
Coal	4.8–48		1.9–4.8	9.6–19.3			11–13.7	
Limestone	29–239	7.6–47.8	4.8–23.98	9.6–75.8	35–50	0.16–0.23	21.5–25.5	5–20
Chalk	5.3	0.4			23–30			30–40
Dolomite	76–239		14.3–28.9	37.9–80.0	35–50		24.5–25.5	1–5

Note: Compressive strength, English units, 1 psi = 6.895 kN/m² or 6.895 kPa in SI units; 1 MPa = 10^6 Pa = 10^3 kPa ≈ 145 psi. Unit weight, γ = lb/ft³ × 1/62.4 × 9.81 = kN/m³.

Figure 8.2   Example of a core sample failed in unconfined compression.

σ_c is the unconfined compressive strength in N/m^2, kPa, MPa, or psi (1 psi = 6.895 kN/m^2 = 6.895 kPa = 0.006895 MPa).

The unconfined compressive strength of rocks ranges from about 1 MPa (145 psi) for weak shales, claystones, and mudstones to more than 280 MPa (40,600 psi) for diabases and some basalts and quartzites. Representative values for compressive strengths of individual rock types are given in Table 8.1. An example problem is provided next.

Example Problem 8.1

A rock core of limestone is 7.6 cm (3 in) in diameter and 15.2 cm (6 in) long. It is loaded to failure in an unconfined compression testing machine. If the failure load was 28,250 kg (62,150 lb), what is the unconfined compressive strength of the limestone sample?

Answers

$$\sigma_c = \frac{P}{A}$$

$$= \frac{28,250}{(3.14)(3.80)^2}$$

$$= 619.8 \text{ kg/cm}^2 = 680 \text{ MPa} = 8816 \text{ psi}$$

The ASTM method D7012-14 for measuring compressive strength is time consuming and core samples required for the test are not always available. For this reason, several empirical tests for estimating compressive strength have been developed. Point load and Schmidt hammer tests are the most frequently used of the empirical tests. The point load test consists of placing an unprepared core sample or an irregular lump of rock between two conical platens and applying compressive load until the sample fails in tension (Broch and Franklin, 1972; ASTM method D5731-16; ISRM, 2014). The

Point load index $I_s = \dfrac{P}{D^2}$

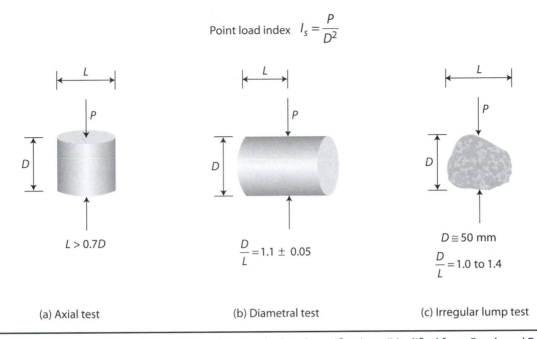

(a) Axial test

$L > 0.7D$

(b) Diametral test

$\dfrac{D}{L} = 1.1 \pm 0.05$

(c) Irregular lump test

$D \cong 50$ mm

$\dfrac{D}{L} = 1.0$ to 1.4

Figure 8.3   Various versions of the point load test and related specifications. (Modified from Broch and Franklin, 1972.)

three variations of the point load test are shown in Figure 8.3 (Broch and Franklin, 1972). From the failure load P, and platen separation D (Figure 8.3), the point load index, I_s, is determined as follows:

$$I_s = \frac{P}{D^2} \qquad \text{(Eq. 8-6)}$$

Unconfined compressive strength of a rock is related linearly to point load index by the following equation:

$$\sigma_c = k(I_s) \qquad \text{(Eq. 8-7)}$$

The value of k depends on core diameter. For N_x-size (54 mm) samples of most hard rocks, k is approximately 24 (Broch and Franklin, 1972; Bieniawski, 1989; Cargill and Shakoor, 1990). For weaker rocks (shales, claystones, mudstones) the k values are significantly less, in the range of 11 to 16 (Greene, 2001). For irregular samples, Broch and Franklin, (1972) have developed correction charts that can be used to normalize I_s values to 50 mm standard size.

The Schmidt hammer (Type L) is a portable device that can be used to estimate compressive strength in both the laboratory and the field. The hammer is pressed against the rock and a rebound number (N) is noted from the scale provided on the hammer sleeve. The rebound number has been correlated previously with unconfined compressive strength as shown in

Figure 8.4 on the next page. The Schmidt hammer is considered to be a less reliable estimator of compressive strength than the point load test (Johnson and De Graff, 1988; Cargill and Shakoor, 1990).

Other indices of compressive strength include shore hardness, indentation hardness, and block punch strength index. Test procedures for determining these indices can be found in ISRM's (2014) suggested methods.

Tensile Strength

The tensile strength of rocks is important in the design of roof spans for underground excavations or in situations where rocks are subjected to bending stresses. On average, tensile strength of rocks is approximately 10% of their compressive strength, the range being 5 to 15% (Jaeger et al., 2007; Gonzalez de Vallejo and Ferrer, 2011). Table 8.1 shows the ranges of tensile strength for various rocks. The tensile strength can be determined either directly by applying a tensile load on a core sample, referred to as the direct pull test (ASTM D2936-08; ISRM, 2014), or indirectly by applying a compressive stress on a disc-shaped sample and failing it in tension (Figure 8.5 on page 165), called the Brazilian test (ASTM D3967-16; ISRM, 2014). In the case of the direct pull test, the tensile strength is obtained by dividing the failure load by the cross-sectional area of the sample. For the more frequently performed Brazilian test, the tensile strength is obtained by the following equation:

$$\sigma_t = \frac{2P}{\pi Dt}$$

$$= \frac{0.636P}{Dt}$$

(Eq. 8-8)

where:

P = failure load
D = disc diameter
t = disc thickness

ASTM recommends a D to t ratio of close to 2:1.

Tensile strength is influenced by the same geologic parameters as compressive strength.

Shear Strength

Shear strength of rocks is evaluated by determining the shear strength parameters of cohesion and friction angle (c and ϕ). This is accomplished by establishing the Mohr envelope by either performing a direct shear test (ASTM D5607-16) or a triaxial test (ASTM D7012-14; ISRM, 2014). The triaxial compression test is performed on rock specimens to determine how the rock will behave under different confining pressures. In this test, the sample, usually a rock core with a length twice its diameter, is placed in a heavy steel cylinder. A rubber jacket is placed around the rock core, with fluid filling the space between the rock and the cylinder wall. The ends of the sample are subjected to increments of increasing load, while confining pressure is applied to the fluid and held constant. Failure occurs when the rock fractures, yielding displacement. The confining pressure and maximum axial load are recorded. The axial load required to cause failure of a rock increases with increasing confining pressure. By performing two or three tests at different confining pressures, the relationship between axial compressive strength and confining pressure can be determined for the rock material. Data are plotted on a shear stress versus com-

pressive stress diagram known as a Mohr diagram. A tangent drawn to the circles is called the Mohr envelope (Figure 8.6). Details on the Mohr circle procedure for representing stresses are provided in Chapter 7, which discusses soil mechanics.

The parameters shown in Figure 8.6 are σ, τ, S_0, and ϕ. The points of tangency (A, B, and C) mark

Figure 8.4   Relationship between Schmidt hammer rebound number and unconfined compressive strength. (After Deere and Miller, 1966.)

Figure 8.5 Example of indirect tensile strength test (Brazilian test).

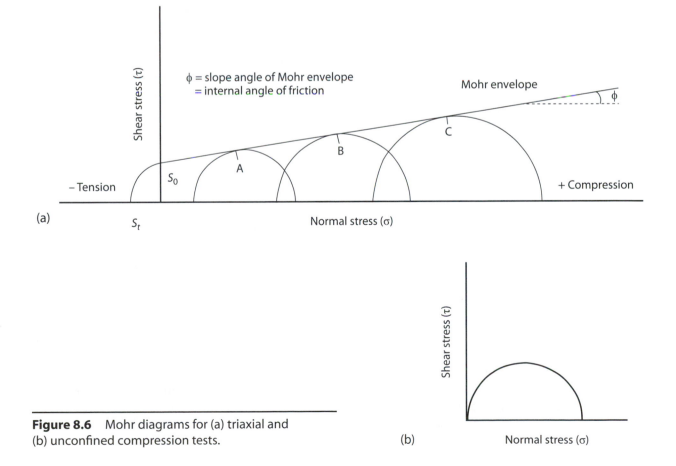

Figure 8.6 Mohr diagrams for (a) triaxial and (b) unconfined compression tests.

the points of failure, indicating the shear stresses at failure for the corresponding values of confining and axial pressure. The slope of the Mohr envelope is a measure of the increase in τ needed to cause failure as the confining pressure increases, and represents the angle of internal friction or shearing resistance. The shear strength at zero normal load for the intact rock is S_0, also referred to as cohesion (c). On the basis of this relationship, the shear strength τ for any confining pressure σ can be determined from the following equation:

$$\tau = S_0 + \sigma \tan\phi \qquad \text{(Eq. 8-9)}$$

Table 8.1 lists the values of friction angle and cohesion for various rock types. The following example problem explains the concept of triaxial testing and interprets the test results.

Example Problem 8.2

Two triaxial compression tests were performed on two samples prepared from the same sandstone. The samples were 5.1 cm (2 in) in diameter and 10.2 cm (4 in) long. For the first sample, a confining pressure of 6.9 MPa (1000 psi) was used. The sample failed at an axial pressure of 75.9 MPa (11,000 psi). For the second sample, a confining pressure of 20.7 MPa (3000 psi) was used. The sample failed at an axial pressure of 120.7 MPa (17,500 psi). Plot the Mohr circles for each test and determine S_0 and ϕ.

Answers

For the first test, $\sigma_1 = 75.9$ MPa (11,000 psi) and $\sigma_3 = 6.9$ MPa (1000 psi). These points are located on the σ axis. Using a diameter of $75.9 - 6.9 = 69$ MPa (11,000 $- 1000 = 10,000$ psi), the first Mohr circle is drawn. For the second test, $\sigma_1 = 120.7$ MPa (17,500 psi) and $\sigma_3 = 20.7$ MPa (3,000 psi). These points are located on the σ axis. Using a diameter of $120.7 - 20.7 = 100$ MPa (17,500 $- 3000 = 14,500$ psi), the second Mohr circle is drawn. See Figure 8.7 for the solution.

A common tangent to the two circles is drawn and extended downward to intersect the τ axis. The intercept value is S_0 or cohesion, which equals 14.5 MPa (2100 psi). The slope angle of the straight line (Mohr failure envelope) is 36°. Based on this information the equation is

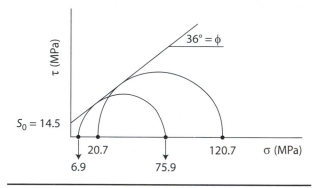

Figure 8.7   Mohr diagram for the triaxial test example.

$$\tau = S_0 + \sigma \tan\phi$$
$$= 14.5 + \sigma \tan 36 \text{ in MPa units or}$$
$$= 2100 + \sigma \tan 36 \text{ in psi units}$$

The actual value of τ will depend on the value of σ. The shear stress on the failure plane, inclined from σ_3 to the point of tangency, is indicated by the point of tangency for each test. For the first test $\tau_{ff} = 17.8$ MPa (2575 psi); for the second test, $\tau_{ff} = 38.4$ MPa (5575 psi).

Elastic Properties of Rocks

Elastic properties indicate deformational behavior of rocks. A cylindrical sample subjected to axial compression will decrease in length and increase in diameter. Upon removal of the compressive force, some, but not all, of the deformation may be recovered. The recoverable deformation is the elastic deformation and the nonrecoverable deformation is the plastic deformation. In engineering, deformation is expressed as percent strain, the ratio of the change in dimension or volume to the original dimension or volume, expressed as a percentage. Figure 8.8 shows a typical stress-strain curve for rocks, as well as elastic and plastic deformations. The two elastic properties that are used most frequently to evaluate the deformational behavior of rocks are Young's modulus and Poisson's ratio. Methods for determining these two properties have been standardized by ASTM (D7012-14) and ISRM (2014).

Young's Modulus

Commonly, the elastic deformation of rock is directly proportional to the applied stress. When this is the case, a proportionality coefficient E can be determined from the following equation:

$$E = \frac{\sigma}{\varepsilon}$$
$$= \frac{P/A}{\Delta L/L} \quad \text{(Eq. 8-10)}$$

where:

σ = compressive stress
ε = axial strain
P = applied load (lb or N)
A = the loaded cross section (in^2 or m^2)
ΔL = the decrease in length (in or m) = axial deflection
L = the original length (in or m)

The proportionality coefficient E is known as the Young's modulus or modulus of elasticity. In a practical sense it is simply the slope of the stress-strain curve.

The modulus of elasticity is used to determine the deformation that will occur in a foundation when the load of the structure is placed on it or when rock pillars or columns are used to support structures. Using the known load and measured or estimated E value, the strain or deformation can be determined. As an example, deformation of the concrete lining in a water tunnel under internal water pressure, such as in penstocks for dams or in some irrigation projects, can be a problem. If the rock deforms much more than the concrete, the lining will lose contact with the rock and become unsupported, thus causing the lining to rupture under the load. In the case of foundations, dams, and tunnels, the in situ modulus of elasticity will

apply. Returning to the statement that E is the slope of the stress-strain curve, one can see from Figure 8.8 that this is not a unique value for a loading condition. Figure 8.8 shows three possible E values: the initial tangent modulus (E_i), the secant modulus at any point selected on the curve (E_s), and the tangent modulus at any point selected on the curve (E_t). The proper modulus value to use is that which best relates to the behavior of the rock under load. It is apparent that a secant modulus or a tangent modulus can be determined at any point and, hence, the possibilities become infinite. The accepted practice, however, is to use the tangent modulus at 50% maximum load $\left(E_{t_{50}}\right)$ as the representative value for the curve. This is appropriate because rock is loaded to much less than 50% of the failure load in a rock foundation so that one-half the failure stress represents a high upper limit of loading. That stress level would provide only a safety factor of 2 relative to failure in compression; a factor of 10 or more is typical for rock foundations.

Rock is not an isotropic material; that is, its properties are not the same in all directions. Instead it is anisotropic, particularly those rocks with prominent bedding or foliation. Therefore, the elastic

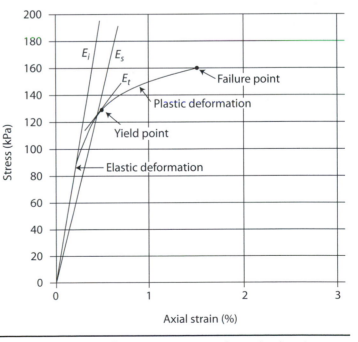

Figure 8.8   A typical stress-strain curve for rocks showing elastic versus plastic deformation and the three types of moduli of elasticity.

properties of rocks will vary depending on the direction of measurement. A lower E value is obtained for rocks when measured perpendicular to bedding or foliation than when measured parallel to these directional features. This results in a larger amount of strain or deflection when loading is perpendicular to bedding than when parallel. For this reason, during testing, rocks should be loaded in the same direction as will occur during loading of the engineering structure. For horizontally bedded rocks, samples should be loaded perpendicular to bedding to depict gravity effects. If vertical rock cores are taken, the loading effects of the engineering structure on the rock foundation are determined by testing these cores along the vertical axis.

In general, rocks with higher compressive strength also exhibit higher E values (Shakoor and Bonelli, 1991) because both properties are controlled by the same petrographic characteristics. Average E values can range from 13.7 GPa for shales and claystones to 79.9 GPa for quartzites (Johnson and De

Graff, 1988; Jaeger et al., 2007; Gonzalez de Vallejo and Ferrer, 2011).

Poisson's Ratio

Also of interest regarding compression testing of rock is the extension that occurs perpendicular to the direction of loading. This lateral extension is typically expressed as a fraction of the vertical strain. The term is known as *Poisson's ratio* (μ) and is expressed by the following equation:

$$\mu = \frac{(\Delta B / B)}{(\Delta L / L)} \qquad \text{(Eq. 8-11)}$$

where B is the lateral dimension and L is the length or axial dimension. Poisson's ratio, for a perfectly elastic material, equals 1/3. Rocks, however, are not perfectly elastic so Poisson's ratio for different rocks ranges from 0.10 to 0.50. As with the modulus of elasticity, Poisson's ratio varies depending on the direction of loading.

Values of Young's modulus and Poisson's ratio for different rocks are given in Table 8.1.

Example Problem 8.3

An N_x core of fine-grained syenite, 5.4 cm (2.125 in) in diameter and 10.8 cm (4.25 in) long, is tested in unconfined compression. The modulus of elasticity for the rock is 63.45 GPa (9.2×10^6 psi), Poisson's ratio is 0.21, and its unconfined compressive strength is 169 MPa (24,500 psi). If the core is loaded to one-quarter its unconfined compressive strength, give the answer to the following questions:
  a. What is the load on the sample?
  b. What axial strain occurs at this load; what deflection?
  c. What lateral strain occurs at this load; what lateral deflection?

Answers

Note: 1 lb = 4.448 N

Using SI units:

a. $\sigma = \dfrac{P}{A}$

$P = \sigma A$

$\quad = \dfrac{169,000 \text{ kPa}}{4} \times \dfrac{\pi \left(0.54 \text{ m}^2\right)}{4}$

$\quad = \dfrac{169,000 \text{ kPa}}{4} \times \dfrac{0.00916 \text{ m}^2}{4} = 96.76 \text{ kN}$

$\quad = 96,760 \text{ N} = 21,754 \text{ lb} = 9888 \text{ kg}$

b. $\varepsilon = \dfrac{\sigma}{E}$ as $E = \dfrac{\sigma}{\varepsilon}$

$\varepsilon = \dfrac{169 \times 10^3 \text{ Pa}}{63.45 \times 10^6 \text{ Pa}} \times \dfrac{1}{4} = 0.6659 \times 10^{-3}$

$\varepsilon = \dfrac{\Delta L}{L}$ so $\Delta L = \varepsilon L$

$\quad = 0.6659 \times 10^{-3} \left(10.8 \text{ cm}\right) = 7.19 \times 10^{-3} \text{ cm}$

c. $\mu = \dfrac{\Delta B / B}{\Delta L / L}$

$\Delta B / B =$ lateral strain $= \mu \dfrac{\Delta L}{L}$

$= 0.21\left(0.6659 \times 10^{-3}\right)$

$= 1.398 \times 10^{-4}$

$\dfrac{\Delta B}{B} = 1.398 \times 10^{-4}$

$\Delta B =$ lateral deflection

$= B\left(1.398 \times 10^{-4}\right)$

$= 5.4\left(1.398 \times 10^{-4}\right)$

$= 7.54 \times 10^{-3}$ cm

Using British engineering units:

a. $\sigma = \dfrac{P}{A}$

$P = \sigma A = \dfrac{24{,}500 \text{ psi}}{4} \times \dfrac{\pi (2.125)^2}{4} = 21{,}723 \text{ lb}$

b. $\varepsilon = \dfrac{\sigma}{E}$ as $E = \dfrac{\sigma}{\varepsilon}$

$\varepsilon = \dfrac{24{,}500 \text{ psi}}{9.2 \times 10^6 \text{ psi}} \times \dfrac{1}{4} = 0.6657 \times 10^{-3}$

$\varepsilon = \dfrac{\Delta L}{L}$ so $\Delta L = \varepsilon L$

$= 0.6657 \times 10^{-3}(4.25 \text{ in}) = 2.83 \times 10^{-3}$ in

c. $\mu = \dfrac{\Delta B / B}{\Delta L / L}$

$\Delta B / B =$ lateral strain $= \mu \dfrac{\Delta L}{L}$

$= 0.21\left(0.6657 \times 10^{-3}\right)$

$= 1.398 \times 10^{-4}$

$\dfrac{\Delta B}{B} = 1.398 \times 10^{-4}$

$\Delta B =$ lateral deflection

$= B\left(1.398 \times 10^{-4}\right)$

$= 2.125\left(1.398 \times 10^{-4}\right)$

$= 2.97 \times 10^{-4}$ in

Rock Durability

Durability is the resistance of a rock to climatic changes such as heating and cooling, wetting and drying, and freezing and thawing, i.e., to weathering and disintegration. Shales, especially clay shales, claystones, and mudstones frequently exhibit nondurable behavior upon wetting and drying. On Interstate 74 in Dearborn County, southeast Indiana, the highway underwent considerable settlement when the soil-like shale in the embankment slaked upon water entering the fill (Lovell, 1979).

Several durability evaluation tests have been developed since the early 1970s, the most important of these being the slake durability index test developed by Franklin and Chandra (1972). Both ASTM (D4644-16) and ISRM (2014) have standardized the procedure for slake durability testing. The test consists of placing an oven-dried sample consisting of 10 to 12 pieces, each weighing 40 to 60 g with a total weight of 450 to 500 g, in a 2 mm meshed drum and rotating the drum through water for 10 minutes at a fixed speed.

The sample that remains in the drum is oven-dried and weighed. The slake durability index (Id) is calculated as the ratio of the weight of the remaining sample to the initial weight, expressed as a percentage. Repeating the test on the remaining sample provides the second-cycle slake durability index (Id_2). Id_2 can range from 0% for some claystones to nearly 100% for some silty shales or siltstones (Gautam and Shakoor, 2016). Id_2 is frequently used as the standard for classification purposes.

Engineering Classification of Intact Rock

Deere and Miller (1966) proposed an engineering classification of intact rock based on the uniaxial compressive strength and the modulus of elasticity of the rock. The compressive strength used in the classification is determined in accordance with ASTM method D7012-14, and the modulus of elasticity is the tangent modulus at one-half the ultimate strength of the rock.

Rocks are subdivided into five strength categories: A—very high, B—high, C—medium, D—low, and E—very low (Table 8.2). These categories follow a geometric progression. In the A category, 220 MPa (32,000 psi) marks the upper limit of most common rocks with only quartzite, diabase, and dense basalt falling above it. The B category (between 110 to 220 MPa or 16,000 and 32,000 psi) includes most igneous rocks, the stronger metamorphic rocks, most limestones and dolomites, and the well-cemented sandstones and shales. The C category (between 55 to 110 MPa or 8000 and 16,000 psi) includes most shales, the porous sandstones and limestones, and the prominent schistose varieties of metamorphic

rocks such as chlorite, mica, and talc schists. The D and E categories (< 55 MPa or 8000 psi) include friable sandstone, porous tuff, clay-shale, rock salt, and weathered or altered rocks of various lithologies.

The second property involved in Deere and Miller's (1966) classification is the modulus of elasticity. However, rather than using the modulus alone for the comparison, the modulus ratio is used, that is, the ratio of the modulus of elasticity to the unconfined compressive strength. Table 8.3 shows the three subdivisions based on modulus ratio.

Table 8.3   Engineering Classification of Intact Rock Based on Modulus Ratio.

Class	Level of Modulus Ratio	Modulus Ratio Value
H	High	500
M	Average or medium	200–500
L	Low	200

Source: Adapted from Deere and Miller, 1966.

The summary plot for igneous rocks is shown in Figure 8.9. The middle, diagonal zone in this figure is designated as that of M, medium or average modulus ratio. Rocks with interlocking textures and little anisotropy fall into this category and it contains the majority of the igneous rocks. Rocks are classified by both strength and modulus ratio with such examples as BH, CM, and BM.

Figure 8.10 on page 172 is the summary plot for sedimentary rocks. Limestones and dolomites fall primarily within strength categories B and C, although some plot as high as A and as low as D; they have from high to medium modulus ratios. Both sandstones and shales extend into the zones of low modulus ratios.

Figure 8.11 on page 173 presents the summary plot for metamorphic rocks. The scatter in results is due to the great range in mineralogy and fabric (and anisotropy) for metamorphic rocks as a whole. Many quartzites plot in similar fashion with other dense, equigranular, interlocking fabrics, i.e., diabase and dense basalt. Gneisses plot mostly in the BM location in a similar fashion to granite except that gneisses show a somewhat lower average strength and greater scatter in the modulus ratio. This is in keeping with the considerable variation in foliation (gneissic structure and some schistosity) that gneisses exhibit as a group.

Table 8.2   Engineering Classification of Intact Rock Based on Unconfined Compressive Strength.

Class	Level of Strength	Uniaxial Compressive Strength	
		MPa	psi
A	Very high	Over 220	32,000
B	High	110–220	16,000–32,000
C	Medium	55–110	8000–16,000
D	Low	27.5–55	4000–8000
E	Very low	Less than 27.5	4000

Source: Adapted from Deere and Miller, 1966.

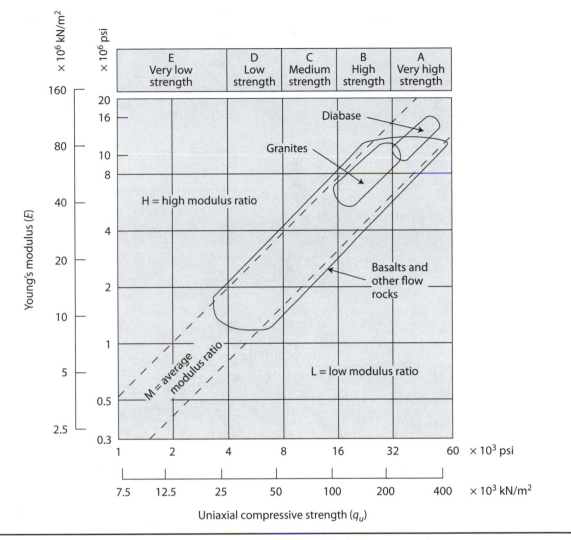

Figure 8.9   Engineering classification for intact rock—summary plot for igneous rocks.

Schists are represented by two zones in Figure 8.11. The first zone, 4A, is for schists with steeply dipping foliation or a high angle (45° or more) between a horizontal plane and the direction of foliation. Because specimens were tested along a vertical load axis, foliation was nearly parallel to the load axis and, thus, strengths were reduced. This also yields a high modulus ratio by virtue of the low strength, σ_c, in the denominator (modulus ratio = E/σ_c).

By contrast, for the 4B zone, the foliation is at a low angle from the horizontal and, therefore, at a high angle to the vertical load axis. The strength is increased somewhat but the modulus of elasticity is reduced because of the closure of microcracks that parallel the foliation. This yields a lower modulus ratio for the 4B group. For marble (zone 3), a high modulus ratio is consistent along with a medium strength.

Deere and Miller (1966) suggest that a lithologic description of rock should be provided with engineering classification. The engineering classification of rock has been related to the stability of rock slopes (West, 1979).

Numerous durability classifications for clay-bearing rocks have been proposed by various researchers. Table 8.4 shows the ISRM (1979) classification based on Id_2.

Rock Mass Properties

The design and stability of large engineering structures such as dams, tunnels, highway cuts, and surface and underground mines depend on the properties of rock masses that are controlled by the presence of discontinuities such as bedding planes, joints, foliation, faults, and shear zones. Also, rock masses are significantly more anisotropic than intact rock.

There are seven aspects of discontinuities that are significant with respect to the stability of rock masses. These include geometry, continuity, spacing, surface irregularities, physical properties of adjacent rock, nature of infilling material, and groundwater (Wyllie and Mah, 2004).

Table 8.4   Durability Classification Based on Second-Cycle Slake Durability Index.

Second-Cycle Slake Durability Index (Id_2)	Classification
0–30	Very low
30–60	Low
60–85	Medium
85–95	Medium high
95–98	High
98–100	Very high

Source: After ISRM, 1979.

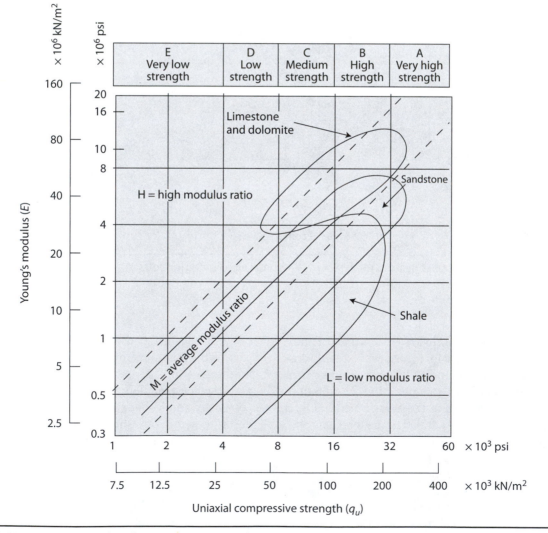

Figure 8.10   Engineering classification for intact rock—summary plot for sedimentary rocks.

1. Geometry deals with the orientation of the discontinuities and plays a fundamental role in the stability of rock slopes (Wyllie and Mah, 2004) and roofs of underground openings (Hoek and Brown, 1980).

2. Continuity indicates the persistence of the discontinuities. The more continuous the discontinuities, the weaker the rock mass.

3. Spacing represents the frequency of discontinuities, with spacing and continuity being interrelated. Table 8.5 shows a classification of discontinuities based on spacing by Deere (1964). Closely spaced discontinuities represent a weaker rock mass with greater potential for slope failure and deformation.

Table 8.5   Descriptive Classification of Discontinuity Spacing.

Bedding	Spacing	Joints
Very thin	< 5 cm	Very close
Thin	5–30 cm	Close
Medium	30 cm–1 m	Moderately close
Thick	1 –3 m	Wide
Very thick	> 3 m	Very wide

Source: After Deere, 1964.

4. Surface irregularities contribute to increased resistance against failure by either overriding the irregularities or shearing through them (Patton,

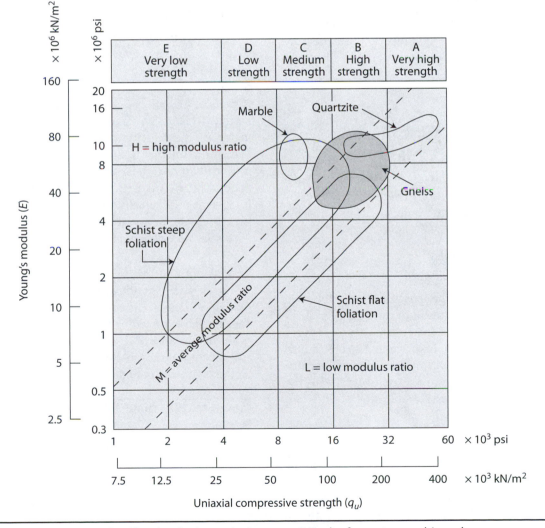

Figure 8.11   Engineering classification for intact rock—summary plot for metamorphic rocks.

1966; Wyllie and Mah, 2004). When a discontinuity separates two different rock types, such as a bedding plane between sandstone and shale units, the properties of the weaker rock unit will control the shear strength along the discontinuity.

5. Physical properties of adjacent rocks jointly contribute to the shearing resistance along a discontinuity. For example, if a sandstone is in contact with a shale, both sandstone and shale will contribute to shearing resistance or the engineer may use the properties of the weaker rock unit to evaluate shearing resistance for design purposes.

6. Infilling includes all soil-like material filling the discontinuities. The properties and thickness of the infilling material significantly influence the resistance against shearing (Wyllie and Mah, 2004).

7. Groundwater decreases the shear strength of a rock mass through buildup of pore pressure (Wyllie and Mah, 2004; Gonzalez de Vallejo and Ferrer, 2011).

Engineering Classification of Rock Mass

The following sections discuss briefly the various indices and classification schemes that describe the quality of rock mass and quantify its engineering behavior.

Percent Core Recovery

Percent core recovery is the ratio of the length of the core obtained to the length drilled, expressed as a percentage. It indicates both the quality of drilling and the soundness of the rock. A core recovery of 90% indicates a sound, homogeneous rock, a 50% recovery suggests rock with seams of weak, weathered material, and very low or no recovery means the rock is highly decomposed (Deere, 1964).

Rock Quality Designation

Rock quality designation (RQD), developed by Deere and Miller (1966), is one of the most important and universally used indices of rock mass quality. It is defined as the ratio of the sum of N_x-size core pieces that are equal to or greater than 10 cm to the total length drilled, expressed as a percentage. Table

8.6 shows the rock mass quality bands based on RQD. The RQD has been used to estimate Young's modulus (Coon and Merritt, 1970), loads on tunnel support systems (Cording et al., 1975), and bearing capacity of foundation rock (Peck et al., 1974). However, while using RQD, one should keep in mind that: (1) RQD depends on the driller's experience; (2) schistose rocks may have a high RQD value but still contain many planes of failure; and (3) joints filled with clay seams may be widely spaced but can still result in failure.

Fracture Index

Fracture index or fracture frequency is the number of fractures per meter length of core (Farmer, 1983). The higher the fracture index, the poorer is the quality of the rock mass.

Velocity Index

Velocity index or velocity ratio is the ratio of the square of seismic wave velocity through a rock mass in the field $(V_F)^2$ to the square of seismic wave velocity through an intact rock sample in the laboratory $(V_L)^2$ (Onedera, 1963; Farmer, 1983; Gonzalez de Vallejo and Ferrer, 2011). As the fracture frequency in rock mass increases, the velocity index decreases. Conversely, a decrease in fracture frequency will result in an increase in velocity index. Table 8.7 (Farmer, 1983) shows the relationship between rock mass quality, RQD, fracture frequency, and velocity index. For a given direction, the correlation between velocity index and RQD is 1:1 (Gonzalez de Vallejo and Ferrer, 2011).

Rock Mass Classification Systems

Rock mass classification systems provide a quantitative ranking of the rock mass in terms of its engineering behavior. Instead of using a single parameter, as discussed above, a series of relevant parameters are selected based on their condition and influence on the engineering behavior.

Table 8.6   RQD Quality Bands.

RQD (%)	Description
0–25	Very poor
25–50	Poor
50–75	Fair
75–90	Good
90–100	Very good

Source: After Deere and Miller, 1966.

Table 8.7   Relationship between RQD, Fracture Frequency, and Velocity Index.

Quality Classification	RQD (%)	Fracture Frequency (per m)	Velocity Index $[(V_F)^2/(V_L)^2]$
Very poor	0–25	> 15	0–0.2
Poor	25–50	15–8	0.2–0.4
Fair	50–75	8–5	0.4–0.6
Good	75–90	5–1	0.6–0.8
Excellent	90–100	1	0.8–1.0

Sources: After Farmer, 1983. Quality classification and RQD data after Deere and Miller, 1966.

One of the earliest rock mass classification systems for estimating tunnel supports was developed by Terzaghi (1946), who divided rock mass, based on discontinuity spacing and degree of weathering, into categories such as intact rock, stratified rock, moderately jointed rock, blocky and seamy rock, squeezing rock, and swelling rock. However, the more frequently used quantitative classification systems, which take into account a number of parameters, are the rock structure rating (RSR) developed by Wickham et al. (1972), the geomechanics classification or rock mass rating (RMR) system developed by Bieniawski (1973), and the rock mass quality or Q-system developed by Barton et al. (1974). The parameters considered in developing these classification systems include discontinuity spacing, discontinuity orientation, discontinuity surface properties, intact rock strength, and groundwater conditions. These parameters are assigned varying scores, based on the conditions they represent, which are then added or multiplied to obtain the final rating index.

The RSR system is based on three parameters (designated A, B, and C) that represent the general geology (rock type and structure), joint pattern (joint spacing and orientation), and groundwater and joint condition, respectively. The equation for determining RSR is as follows:

$$RSR = (A + B) + C \qquad \text{(Eq. 8-12)}$$

The details of this system, including tables for determining the values of A, B, and C parameters, and applications of the system can be found in Wickham et al. (1972), Farmer (1983), and Bieniawski (1989). The system is used specifically for designing support systems for mines and tunnels.

The RMR classification divides rock mass into five classes: very good, good, fair, poor, and very poor. The rock mass is placed into a class based on its scores on the following parameters: (1) intact rock strength, (2) RQD, (3) joint spacing, (4) joint condition (joint separation, joint continuity, degree of hardness of joint wall rock), and (5) groundwater inflow. The following equation is used to estimate the RMR value:

$$RMR = [(1) + (2) + (3) + (4) + (5)] - (b) \qquad \text{(Eq. 8-13)}$$

In Eq. 8-13, (1), (2), (3), (4), and (5) represent the scores of each of the five parameters. Parameter (b) represents the adjustment of the rating based on favorable or unfavorable orientation of joints.

High RMR scores indicate very good to good quality rock mass and low RMR scores represent poor to very poor quality rock mass. The RMR has been related to modulus of deformation (Bieniawski, 1989) as well as cohesion and friction parameters (Hoek and Brown, 1980). Complete details of the RMR system are provided in Hoek and Brown (1980), Farmer (1983), and Bieniawski (1989).

The Q-system of the Norwegian Geotechnical Institute (NGI), developed specifically to evaluate tunnel roof stability and the design of a support system, uses six parameters to obtain the Q-value as follows:

$$Q = \left(\frac{RQD}{J_n}\right)\left(\frac{J_r}{J_a}\right)\left(\frac{J_w}{SRF}\right) \qquad \text{(Eq. 8-14)}$$

where:

RQD = rock quality designation
J_n = number of joint sets
J_r = joint roughness
J_a = joint alteration
J_w = water inflow in joints
SRF = stress reduction factor

In the equation for Q-value, RQD/J_n represents the block size, J_r/J_a the interblock shear strength, and J_w/SRF the active state of stress (loosening load during excavation, squeezing load in incompetent rock, residual stress relief in competent rock). The higher the Q-value, the better the quality of rock mass with respect to tunneling. Tables for assigning scores to various parameters comprising the Q-system can be found in Hoek and Brown (1980), Farmer (1983), and Bieniawski (1989).

Rock Strength and Petrography

Petrography involves detailed description of rocks in terms of their mineralogy and texture. Rock strength, measured by various means, depends on these petrographic details (Hajdarwish et al., 2013).

Rock texture, without considering origin directly, is limited to a few basic varieties: (1) interlocking crystals, (2) clastic pieces cemented by various amounts of crystalline material, (3) clastic pieces with infilling of matrix, and (4) glassy materials or noncrystalline varieties (West et al., 1970).

The interlocking crystals can range in size from microscopic (as small as 5 µm or 5×10^{-4} cm) to those more than 3 cm (about 1 in) in size, and can occur in igneous, sedimentary, or metamorphic rocks. In some respects glassy rocks act like those with extremely small crystals because they are essentially homogeneous and form conchoidal fractures on breaking.

For both clastic grains and interlocking crystals, fractures or failure planes occur either through or around the grains and crystals. Rocks composed of clastic grains surrounded by finer, noncemented matrix typically fail through the weaker matrix material. These rocks are held together by compaction forces originating from overburden stresses developed during rock burial, so they are also easily weakened by wetting, drying, or freezing and, hence, have poor weathering resistance. Such rocks are seldom, if ever, used as building materials.

Clastic rocks with minor amounts of cementing agent, such as a friable sandstone, tend to fail through the cement, sometimes after only minor agitation. Many quartz sandstones have a high pore volume, much of which is interconnected and provides the high permeability typical of these rocks. Sandstone outcrops, contrary to their friable nature, are extremely resistant to weathering simply because water seeping from the porous rocks deposits minerals, calcite primarily, on evaporation. This increases rock strength at the surface. The effect, known as case hardening, provides quartz sandstone outcrops with a high level of resistance to both mechanical and chemical weathering.

Clastic rocks with a major portion of their interstices filled with crystallized cement can develop fractures either through the grains or through the cement. Commonly the cement is calcite, which is weaker than most grains and, thus, fracturing occurs mostly through the cement. These fractures occur both through the center of the cement filling and at the intersection between grain and cement. As explained later, grain size (or crystal size) and mineralogy determine whether failure will occur at the boundaries or through the crystallized cement.

For interlocking crystals or for clastic rocks with strong well-crystallized cement, the fractures can occur either around or through the crystals. Rocks composed of small particles have a much greater total boundary surface between particles than do coarse-grained rocks. This is a fundamental feature indicated by increased specific surface values (the surface-to-volume ratio) that occur with decreasing grain size.

Coarse-grained rocks composed of minerals with good cleavage are much weaker than their fine-grained equivalents, which consist of much smaller crystals, because cleavage yields a planar break directly across the entire crystal, whether large or small. A group of smaller, randomly oriented grains will yield a more irregular surface composed of many small breaks with changing orientation. By contrast the larger grains provide a single straight-line fracture yielding a shorter total distance with less resistance to failure.

For both clastic rocks with crystallized cement and rocks with interlocking crystals, coarse-grained rocks yield lower strengths than fine-grained rocks because the surface of fracture is smaller for coarser rocks. This develops for two reasons: (1) fracture around grain boundaries (coarse-grained rocks are weaker because the boundary surface area is smaller) and (2) fractures through the grains (coarse-grained rocks are weaker because the fracture takes a shorter, less circuitous path through the larger crystals).

The mineralogy of the rock has an effect on strength because some minerals are harder and stronger than others and some minerals cleave more readily than others. This explains why marble, with an average grain size of 1 mm, would be weaker than a granite of the same size, assuming both are unweathered.

The final feature of texture, related to both origin and mineralogy, is the interlocking of the crystals. Two considerations are involved: the shape or configuration of grain interlock and the mutual orientation between elongated grains. Granites contain rounded crystals of quartz and orthoclase that do not interlock as well as the angular grains of plagioclase and augite that occur in diorites. Therefore, a diorite with the same grain size as a granite would have a greater strength. The other feature is the interlocking texture caused by randomly oriented, elongated rectangular grains yielding the so-called "tepee structure" of a diabase. These features have the general repeat pattern of an A-frame building. Fractures around the grains must take a circuitous route following the random orientation of the elongated plagioclase. This yields increased strength because of the increased distance, making diabase one of the strongest rocks with an unconfined compressive strength of 275 MPa (40,000 psi) or more.

Engineering Considerations of Rock Properties

1. Data on rock properties are needed in the design of engineering structures built in the Earth, on the Earth, or composed of earth materials.

2. A distinct difference exists between rock properties and rock mass properties. Discontinuities in a rock mass tend to reduce the strength of the rock mass compared to intact rock and, therefore, a detailed investigation of the nature of discontinuities is required for stability analysis of such projects as dams, tunnels, mine openings, and slopes.

3. Specific gravity, absorption, porosity, degree of saturation, and unit weight can be used as indicators of integrity and rock strength. Higher specific gravity and lower absorption values indicate a more competent rock.

4. Unconfined compressive strength and modulus ratio (Young's modulus/compressive strength) can be used to classify intact rock, in terms of strength and deformability, for situations where intact rock is used as building blocks or as columns to support superstructure.

5. Several empirical tests, which are easier to conduct than unconfined compression testing, can be used economically to determine the unconfined compressive strength of rock. Point load and Schmidt hammer tests are the most frequently used, with varying results.

6. Tensile and shear strength tests provide data regarding bending and shearing displacement. Rocks are considerably weaker in tension and shear than in compression.

7. Rock durability is an important concern for rocks used in highway construction, exposed in rock cuts, and in landslide analysis. Slake durability testing provides this information.

8. The nature and distribution of discontinuities strongly influence rock mass properties. Joints, bedding planes, foliation, faults, shear zones, and any other fractures weaken a rock mass. It is important to characterize the seven aspects of discontinuities, which include: (a) geometry, (b) continuity, (c) spacing, (d) surface irregularities, (e) physical properties of adjacent rocks, (f) nature of infilling material, and (g) groundwater.

9. Rock mass classifications incorporate percent core recovery, rock quality designation, fracture index, and velocity index to rate the overall quality of the rock mass.

10. Rock mass classification systems, including Terzaghi's classification, rock structure rating (RSR), rock mass rating (RMR), and rock mass quality (or Q-system), can be used to estimate rock loads and standup time for underground excavations, and to design the required support system.

11. Rock strength can be related to its petrography (composition and texture). Independent of origin, textures can be divided into four types: (a) interlocking crystals, (b) clastic pieces cemented with crystalline materials, (c) clastic pieces with infilling of matrix, and (d) glassy material or noncrystalline material. Interlocking crystals range in size from 5 μm to more than 3 cm (1 in) and may be igneous, sedimentary, or metamorphic in origin. Fractures in both clastic grains and interlocking crystals occur either across these larger grains or around them. Rocks with clastic grains surrounded by finer, uncemented matrix typically fail through this weaker material and such rocks are seldom used as building materials. Interlocking crystals or clastic rocks with minor amounts of cementing agent, such as friable sandstones, tend to fail through the weaker cement. Clastic rocks filled with crystallized cement can develop

factures either around or through the crystals. If the cement is calcite, the fractures typically occur through it because of lower strength and a greater tendency for cleavage.

12. Crystal size and mineralogy determine whether failure occurs at the boundaries or through the crystallized cement.

13. Coarse-grained rocks containing minerals with strong cleavage are much weaker than their fine-grained equivalents. Crystalline limestone with sand-sized crystals or fossils will not qualify as a concrete aggregate because of low strength.

14. Mineralogy has a significant effect on strength, as some minerals are harder and stronger than others, and also the degree of cleavage varies. Marble with a grain size of 1 mm would be considerably weaker than a granite of the same grain size.

15. Interlocking of grains is the final feature of significance. Both the shape and mutual orientation of grains are involved. Rounded grains have less interlock than angular grains. A random orientation of elongated grains gives rise to a "tepee" structure, typical of the plagioclase crystals in a diabase. This is why diabase is one of the strongest rocks in nature (> 275 MPa or 40,000 psi).

EXERCISES ON ENGINEERING PROPERTIES OF ROCKS

1. Specific gravity information on an argillaceous dolomite is desired. It weighs 250 g in the dry state, 259 g when saturated (surface dried), and 144 g when saturated and weighed in water.

   a. What is the bulk specific gravity; bulk specific gravity, saturated surface dried; apparent specific gravity; and absorption of the rock?

   b. What is the volume of the rock sample?

   c. What would be its weight per ft^3 in the dry state?

   d. Would the rock make a suitable concrete aggregate? Explain.

2. Two triaxial compression tests were performed on two prepared samples of the same limestone. At a confining pressure of 13.8 MPa (2000 psi), the first sample failed at 62.1 MPa (9000 psi). Then at a confining pressure of 34.5 MPa (5000 psi), the other sample failed at 144.8 MPa (21,000 psi).

   a. Plot the two test results on a shear stress-normal stress diagram and draw in the Mohr envelope.

   b. Determine the ϕ value for the limestone.

   c. What is the S_0 value?

   d. What was the shear stress on the failure plane at failure for both tests?

3. Limestone similar to that tested in Exercise 2 exists in a mountainous region at a depth of 182 m (600 ft).

   a. If the vertical stress is the confining stress of the rock at that depth, what is this value in MPa (psf and psi)? Assume a unit weight of 2.56 Mg/m^3 (160 lb/ft^3) for the rock.

   b. At this confining pressure, what horizontal stress would be required to cause a shear failure in the rock? Which is σ_1 and which σ_3? Use the plot constructed in Exercise 2 to obtain the answer. What is the shear stress on the failure plane?

4. a. Under what conditions would the tensile strength of rock be significant? (Consider bending conditions for a rock mass.)

   b. A sample of clastic limestone yielded a compressive strength of 41.4 MPa (6000 psi) and a shear strength of 7.6 MPa (1100 psi). What tensile strength value would you select for this rock if testing was not possible? Explain how you would arrive at this value.

5. A core sample of basalt, 10.2 cm (4 in) long and 5.1 cm (2 in) in diameter, was tested in unconfined compression. The basalt is known to have a modulus of elasticity = 6.28×10^4 MPa (9.1×10^6 psi).

   a. At a pressure of 69 MPa (10,000 psi), what would be the reduction in length of the sample?

   b. At this pressure what would be the vertical load on the sample?

   c. At 69 MPa (10,000 psi) stress the core showed an increase in diameter of 0.014 mm (5.5×10^{-4} in). What is the Poisson's ratio for the sample?

6. A concrete gravity dam with a trapezoidal shape will be founded on sedimentary rock (Exercise Figure

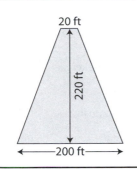

20 ft

220 ft

←—200 ft—→

Exercise Figure 8.1

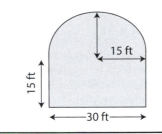

15 ft

15 ft

←—30 ft—→

Exercise Figure 8.2

8.1). The concrete has a unit weight of 2.4 Mg/m³ (150 lb/ft³). The height of the dam is 67.1 m (220 ft), the crest is 6.1 m (20 ft) wide, and the base is 61 m (200 ft) wide. *Note*: 1 ft = 0.3048 m.

a. What is the average pressure exerted on the rock foundation in MPa (lb/ft²)? What is the maximum pressure? Explain.

b. Refer to the different sedimentary rocks listed in Table 8.1. Using average compressive strength and shear strength values for these sedimentary rocks compare them to the average pressure calculated in part (a). Are any so low to be of concern? Explain.

c. Assume that the pressure from the weight of the dam is dissipated within 30.3 m (100 ft) below the rock surface and the full pressure acts over this distance. How much settlement will occur in the center of the dam because of the gravity force for the different sedimentary rocks listed in Table 8.1 (use average values)?

7. Excavation for a gravity dam foundation revealed a unit of poorly cemented sandstone 30.3 m (100 ft) wide (plan view) within a massive dolomite sequence. The dolomite has an E value of 5.5×10^4 MPa (8×10^6 psi) and the sandstone has an E value of 6.9×10^3 MPa (1×10^6 psi).

a. What would be the concern if both rocks were loaded equally by the gravity dam? What would be the effect on the dam?

b. How could this problem be alleviated? Consider both replacement and strengthening possibilities.

8. A 333.3-m (1100-ft) long, horseshoe-shaped tunnel is to be driven through massive granite with a unit weight of 2.72 Mg/m³ (170 lb/ft³). The tunnel is 9.1 m (30 ft) high and 9.1 m (30 ft) wide in cross section (Exercise Figure 8.2). The maximum rock cover above the tunnel in the middle of the mountain is 366.5 m (1200 ft). *Note*: 1 ft = 0.3048 m.

a. What is the total overburden stress at the midway point in the tunnel in MPa (psi and psf)?

b. Because of arching effects that transfer load around the tunnel opening, the roof or crown of the tunnel does not commonly support the full overburden stress, but instead is typically much less. If the roof stress was found to be 0.12 MPa (2500 psf), how many meters (feet) of rock above the tunnel does this represent? How do you suppose this load would be supported in the tunnel?

c. If the tunnel excavation was completed in 80 working days with two shifts per day, how many lineal meters (feet) of tunnel were excavated per day? How many meters (feet) per shift? What was the average volume of rock removed per day, per shift? (*Hint*: Calculate the area of the cross section.)

d. The tunnel project was bid at $1,878,000. What would be the cost per lineal foot of the tunnel?

9. An arch dam transfers much of the water load onto the abutments (the rock mass on the sides of the dam) (see Exercise Figure 8.3 for a plan view). The E value for the concrete is 4.14×10^4 MPa (6×10^6 psi).

a. What would occur if the abutments were shale with an E of 2.07×10^4 (3×10^6 psi)?

b. What if the abutments were diabase with an E of 9.66×10^4 MPa (14×10^6 psi)? Which of the two rocks would be preferred, the shale or the diabase? Explain why.

Upstream

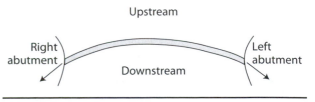

Right abutment

Left abutment

Downstream

Exercise Figure 8.3

c. What would be the effect if the thrust (force direction from the dam) were parallel to the slaty cleavage direction in the abutments? What could be a conclusion of this arrangement?

d. What type of force is preferred within the arch dam: compression or tension? Explain why. Which type of force would develop if the abutments had a high modulus of elasticity?

10. For room and pillar underground mines, the pillars must carry the total weight of the overburden.

a. Why does the arching effect that acts in tunnels not contribute any support for these mines?

b. If a coal mine is 152 m (500 ft) deep, can 40% of the coal be removed and still yield a factor of safety (FS) of 3 relative to compressive strength of the coal in the pillars? Assume an unconfined compressive strength of 2.7 MPa (3000 psi) for the coal and use the equation

$$FS = \frac{\text{rock strength } (1 - \text{fraction removed})}{\text{rock load}}$$

Hint: Use a unit weight of rock = 2.56 Mg/m^3 (160 lb/ft^3). If the answer is no, what is the FS?

c. What percent of the coal can be removed at a depth of 343 m (800 ft) using an FS = 3 and the same compressive strength as in part (b)?

11. A massive hillside of rhyolite contains joint planes that dip at an angle of 40° toward a stream valley. A rectangular-shaped rock bounded by other joint planes rests on the dipping joint plane. The block measures 1.5 m (5 ft) high, 4.5 m (15 ft) wide, and 3 m (10 ft) long. The unit weight of the rhyolite is 2.76 Mg/m^3 (172 lb/ft^3). Assume a dry slope.

a. What is the weight of the rock block?

b. Assuming that only friction holds the block in place (no cohesion along the joint plane), what is the minimum frictional force needed to resist sliding?

c. What coefficient of friction is needed to produce this minimum force?

d. What would be an actual coefficient of friction for this rock block, that is, rhyolite sliding against rhyolite? A friction angle is commonly used, the coefficient of friction being equal to the tangent of that angle. Determine that angle of friction.

12. From the following rocks prepare a list in order of their decreasing unconfined compressive strength. For each rock, give the reason for your selection.

a. fine-grained limestone

b. fine-grained basalt

c. medium-grained diabase

d. porous tuff

e. friable quartz sandstone

f. coarse marble

g. medium-grained granite

13. When shale is excavated from a deep foundation for a heavy structure (such as a power plant or concrete dam), layers of shale may loosen from the bedrock after a few hours of exposure. To prevent this, a concrete slab is placed over the shale soon after excavation and rock bolts may also be placed through the concrete into the shale.

a. Why does shale have this problem whereas limestone, for example, does not?

b. How is this related to the petrography of the rocks, including texture composition and origin?

14. a. Why is tuff an extremely weak rock whereas welded tuff is so massive that it is commonly mistaken for rhyolite or andesite?

b. How does the porosity of a rock generally relate to its strength?

c. The strength of sandstones as a group ranges from about 19.31 to 158.62 MPa (2800 to 23,000 psi). Describe the types of sandstones that would provide the two extremes of this range.

15. Recent volcanic rocks are some of the most hazardous rocks through which to tunnel. List some of the possible problems that could develop when tunneling through such materials.

16. Refer to Figure 8.9. For the granite family, what is the uniaxial compressive strength and modulus of elasticity? What is the modulus ratio? Does this turn out to be average as indicated in Table 8.3? Discuss.

17. Refer to the entry for diabase in Table 8.1. What is the highest compressive strength given for diabase? How does this compare with designations in Table 8.2? What is the highest *E* value given for this rock in Table 8.1? What modulus ratio value is obtained using these two values? What level of modulus ratio does this yield relative to Table 8.3? Check also the highest value for basalt in this respect. What two-letter designation would these examples yield?

18. A certain sandstone rates a CL designation by the Deere classification. What does this mean specifically according to Tables 8.2 and 8.3? Where would it plot on Figure 8.10?

19. Coal is listed in Table 8.1. Using the lowest values for unconfined compressive strength and for the modulus of elasticity, determine where this would plot on Figure 8.10. Calculate the modulus ratio in the process.

References

Barton, N., Lien, R., and Lunde, J. 1974. Engineering classification of rock masses for the design of tunnel support. *Rock Mechanics* 6(4):189–236.

Bieniawski, Z. T. 1973. Engineering classification of jointed rock masses. *Transactions of the South African Institute of Civil Engineering* 15(12):335–343.

Bieniawski, Z. T. 1989. *Engineering Rock Mass Classifications*. New York: John Wiley & Sons.

Broch, E., and Franklin, J. A. 1972. The point load strength test. *International Journal of Rock Mechanics and Mining Sciences* 9:669–697.

Cargill, J. S., and Shakoor, A. 1990. Evaluation of empirical methods of measuring the uniaxial compressive strength. *International Journal of Rock Mechanics and Mining Sciences & Geomechanics Abstracts* 27(6):495–503.

Coon, R. F., and Merritt, A. H. 1970. *Predicting in situ Modulus of Deformation Using Rock Quality Indexes: In situ Testing for Rock* (Special Technical Publication 477, pp. 154–173). West Conshohocken, PA: American Society for Testing and Materials.

Cording, E. J., Hendron, Jr., A. J., MacPherson, H. H., Hansmire, W. H., Jones, R. A., Mahar, J. W., and O'Rourke, T. D. 1975. *Methods of Geotechnical Observations and Instrumentation in Tunneling* (Report No. UILU-ENG 75-2022 1 & 2). Department of Civil Engineering, University of Illinois, Urbana, Illinois.

Das, B. M., and Sobhan, K. 2018. *Principles of Geotechnical Engineering* (9th ed.). Boston: Cengage.

Deere, D. U. 1964. Technical description of cores for engineering purposes. *Rock Mechanics and Engineering Geology* 1:16–22.

Deere, D. U., and Miller, R. P. 1966. *Engineering Classification and Index Properties for Intact Rock* (Technical Report No. AFWL-TR-65-116). University of Illinois, Urbana, Illinois.

Farmer, I. 1983. *Engineering Behavior of Rocks* (2nd ed.). New York: Chapman and Hall.

Franklin, J. A., and Chandra, A. 1972. The slake durability test. *International Journal of Rock Mechanics and Mining Sciences* 9:325–341.

Gautam, T. P., and Shakoor, A. 2016. Comparing the slaking of clay-bearing rocks under laboratory conditions to slaking under natural climatic conditions. *Rock Mechanics and Rock Engineering* 49(1):19–31.

Gonzalez de Vallejo, L. I., and Ferrer, M. 2011. *Geological Engineering*. London: CRC Press.

Greene, B. H. 2001. Predicting the Unconfined Compressive Strength of Mudrocks for Design of Structural Foundations. PhD Dissertation, Department of Geology, Kent State University, Kent, Ohio.

Hajdarwish, A., Shakoor, A., and Wells, N. A. 2013. Investigating statistical relationships among clay mineralogy, index engineering properties, and shear strength parameters of mudrocks. *Engineering Geology* 159:45–58.

Hoek, E., and Brown, E. T. 1980. *Underground Excavation in Rock*. London: The Institution of Mining and Metallurgy.

Holtz, R. D., Kovacs, W. D., and Sheahan, T. C. 2011. *An Introduction to Geotechnical Engineering* (2nd ed.). New York: Pearson.

International Society for Rock Mechanics (ISRM). 1979. Suggested methods for determining water content, porosity, density, absorption, and related properties and swelling and slake durability index properties. *International Journal for Rock Mechanics and Mining Sciences & Geomechanics Abstracts* 16(2), Parts 1 and 2:143–156.

International Society for Rock Mechanics (ISRM). 2014. *The ISRM Suggested Methods for Rock Characterization, Testing and Monitoring: 2007–2014*. R. Ulusay (ed.). Switzerland: Springer International Publishing.

Jaeger, J., Cook, N. G. W., and Zimmerman, R. 2007. *Fundamentals of Rock Mechanics* (4th ed.). Malden, MA: Blackwell.

Johnson, R. B., and De Graff, J. V. 1988. *Principles of Engineering Geology*. New York: John Wiley & Sons.

Lovell, C. W. 1979. Embankments of shale and similar materials. In Selected Geotechnical Design Principles for Practicing Engineering Geologists, Short Course. Association of Engineering Geologists Annual Meeting, Chicago, Illinois.

Onedera, T. F. 1963. Dynamic Investigations of Foundation Rocks in situ. Proceedings, 5th US Rock Mechanics Symposium (pp. 517–533), University of Minnesota.

Patton, F. D. 1966. Multiple Modes of Shear Failure in Rock. Proceedings, 1st International Congress of Rock Mechanics (Vol. 1, pp. 509–513). Lisbon, Portugal.

Peck, R. B., Hanson, W. E., and Thornburn, T. H. 1974. *Foundation Engineering*. New York: John Wiley & Sons.

Shakoor, A., and Barefield, E. H. 2009. Relationship between unconfined compressive strength and degree of saturation for selected sandstones. *Environmental and Engineering Geoscience* XV(1):29–40.

Shakoor, A., and Bonelli, R. E. 1991. Relationship between petrographic characteristics, engineering properties, and mechanical properties of selected sandstones. *Bulletin of the Association of Engineering Geologists* 28(1):55–71.

Terzaghi, K. 1946. Rock defects and loads on tunnel supports. In R.V. Proctor and T. White (eds.), *Rock Tunneling with Steel Supports* (pp. 15–99). Youngstown, OH: Commercial Shearing and Stamping Company.

West, T. R. 1979. Rock properties, rock mass properties and stability of rock slopes. In Selected Geotechnical Design Principles for Practicing Engineering Geologists, Short Course. Association of Engineering Geologists Annual Meeting, Chicago, Illinois.

West, T. R., Johnson, R. B., and Smith, N. M. 1970. *Tests for Evaluating Degrading Base Course Aggregates* (SBN 309-01885-4). National Cooperative Highway Research Program Report 93. Washington, DC: Highway Research Board.

Wickham, G. E., Tiedemann, H., and Skinner, E. H. 1972. Support Determinations Based on Geologic Predictions. Proceedings, 1st Rapid Excavation Tunneling Conference (pp. 43–64), American Institute of Mechanical Engineers.

Wyllie, D. C., and Mah, C. W. 2004. *Rock Slope Engineering—Civil and Mining* (4th ed.). New York: CRC Press.

EVALUATING CONSTRUCTION MATERIALS

9

Chapter Outline

Introduction

Testing Construction Materials

Portland Cement Concrete

Engineering Considerations of Construction Materials

Introduction

Rock is used extensively as a source of construction materials for a variety of purposes, such as building stone, concrete aggregate, base course material, riprap material, rock-fill, and as tombstones. Many of the historic buildings and monuments in the United States, and rest of the world, are made of rock. Two frequent sources of obtaining rock from the Earth are a quarry or a pit. A quarry is a surface excavation where rock is removed from its natural location. Crushed stone is obtained by crushing rock material from a quarry and dimension stone is rock cut into specific sizes for use as facing stone for buildings or as stone monuments. A pit refers to a surface excavation where gravel, sand, or other loose or unconsolidated material is removed from its source. It is the job of an engineer or engineering geologist to evaluate the quality of rocks for various applications.

A common and important use of rock in construction material is as aggregate. *Aggregate* is defined as an assortment of sand, gravel, crushed stone, slag, or other material of mineral composition used alone (as in railroad ballast, filter beds, base courses, and

various manufacturing processes such as fluxing) or in combination with a blending medium to form bituminous and Portland cement concrete, macadam, mastic, mortar, and plaster (West et al., 1970). Coarse aggregate consists of pieces larger than 0.5 cm (3/16 in) in size and fine aggregate is smaller than 0.5 cm (3/16 in) in size (Bobrowski, 1998).

Another function of rock in construction materials is support. As a base course in highway construction, the rock base is placed below the concrete or asphalt pavement to provide support and drainage for the highway. Gravel and crushed stone are used for both concrete aggregates and for highway base courses. Rock-fill is used for the construction of highway embankments and rock-fill dams, which are primarily composed of compacted rock materials (Figure 9.1).

Riprap consists of large pieces of broken rock or boulders that are used as a protective layer on the upstream face of earth dams or on river banks, lake shores, and harbor structures for protection from wave action, currents, and general erosion by water.

There are many more uses for rock in engineering and construction materials. However, the focus of this chapter is on concrete and concrete aggregates, the most common use of rock. Many of the tests used for concrete aggregates are also applicable to other uses of rock.

Testing Construction Materials

Besides economics, the primary concerns regarding using rock as construction material are strength and durability. The index engineering properties of rock, including specific gravity, absorption, porosity, degree of saturation, and unit weight (discussed in Chapter 8), are good indicators of strength and durability. Rocks with higher specific gravity and unit weight, lower absorption, and lower porosity tend to have higher strength and durability (Barksdale, 1991; Shakoor and Bonelli, 1991; Hale and Shakoor, 2003; Shakoor and Barefield, 2009). These properties can be tested and used for preliminary evaluation of rock as construction material (Dolar-Mantuani, 1983; Lienhart, 1998).

Strength Test

When rock is evaluated specifically for use as concrete aggregate, the strength is determined by an abrasion test rather than by compression tests, as for rock foundations.

Los Angeles Abrasion Test

The Los Angeles (LA) abrasion test (ASTM C-131) is used to measure abrasion resistance of aggregate material. The test uses an oven-dried sample of specific gradation weighing 5 kg (11 lb). The sample, along with a specified number of steel spheres, is loaded into a steel drum that contains an interior, projecting shelf. The drum is rotated for 500 rotations at a speed of 25 rotations per minute. At the end of this treatment, the sample is sieved through a No. 12 sieve (0.141 mm). The LA abrasion loss is determined as the ratio of the weight of the

Figure 9.1   A rock-fill dam. (West Virginia Department of Environmental Protection.)

material finer than No. 12 sieve divided by the original weight, expressed as a percentage. The maximum allowable abrasion loss for concrete aggregates and base courses is established by highway construction specifications for each state in the United States and by federal agencies involved in construction. This value is in the range of 35 to 50% for various highway departments for concrete aggregates and, often, a somewhat higher value is allowed for base courses.

Durability Tests

The durability of an aggregate is a measure of its ability to withstand deterioration due to wetting and drying, heating and cooling, and freezing and thawing during the period of performance. Several tests are used to determine durability with each state highway department and federal agency designating the test to be used to determine acceptance or rejection of materials. The two most common aggregate tests with regard to durability are the sulfate soundness test and the freeze-thaw test.

Sulfate Soundness Test

The sulfate soundness test entails soaking a specific, graded sample of the aggregate in a saturated solution of Na_2SO_4 or $MgSO_4$ followed by complete drying of the aggregate in an oven (ASTM C88-13). Weights of specific size gradations, temperatures of sulfate solution and oven drying, and minimum times of soaking and drying are given in the ASTM specifications. In this procedure the sulfate solution forms crystals in aggregate pores on drying. The crystal growth exerts pressure on the internal pores of the rock, tending to disrupt its structure and break pieces from it. Five cycles of soaking and drying are performed followed by sizing of the pieces using a specified sieve. A ratio equal to the material finer than this sieve size divided by the original weight yields the sulfate soundness loss. A loss of 12 to 15% is a typical maximum allowed for concrete aggregates, with a 15 to 18% loss allowed for base course materials.

Several concerns have been raised regarding the sulfate soundness test. Reproducibility of results within a single testing laboratory and between testing laboratories is difficult to achieve if the specified ASTM procedure is not followed carefully. Temperature is important regarding saturation of the solution and the solution should be replaced periodically as prescribed by the specifications. Another concern is

the relationship between specifics of the test and what actually occurs in nature. The question is one of how well the formation of sulfate crystals simulates the effects of ice crystal growth during freezing and thawing, or expansion during wetting and drying and heating and cooling. On this basis, freezing and thawing or wetting and drying tests would seem more applicable for testing purposes but, another consideration is involved. To predict the long-term performance of rock by means of a short-duration procedure, the test must be more severe than the natural weathering effects so that results can be obtained in a reasonable period of time. The sulfate soundness test is severe and certainly qualifies as an accelerated test.

The sulfate soundness test can be used in those states that receive few, if any, freeze-thaw cycles in a winter season. In states such as Florida, it is difficult to convince aggregate producers that the freeze-thaw test should be used to evaluate their materials, but other weathering effects such as heating and cooling and wetting and drying do occur and these are also simulated by the sulfate soundness test. In areas with less severe weather, aggregates with a higher allowable soundness loss can be used. The sulfate soundness test is particularly destructive for argillaceous (clayey) rocks and to those with extremely coarse-grained textures. Experience has shown that clay-rich carbonate rocks yield concrete with low durability and poor long-term quality.

Freeze-Thaw Test

The freeze-thaw test is also run on a specific, graded sample of loose aggregate material (AASHTO T-103). Depending on the specified test, the sample may be frozen in air or in water and thawed in air or in water. The primary concern is that the rock pieces are completely frozen, and completely thawed during each cycle of the test. Typically the number of freeze-thaw cycles is 25. Following completion of the test, the sample is sized using a specified sieve. A ratio equal to the amount finer than the specified sieve size divided by the original weight of the sample yields the percent loss. Typically the same amount of loss is allowed for this test as for the sodium sulfate soundness test.

To determine the performance of concrete in freezing and thawing, concrete beams, rather than unconfined aggregate pieces, can be tested (ASTM C666/C666M-15). At the beginning, and after every 25 cycles of the test, the fundamental transverse frequency N of the beam is determined using appropriate laboratory equipment. The relative dynamic modulus of elasticity E_r of the beam is determined based on the following equation, and when E_r has been reduced by 60%, failure of the beam is indicated. Three hundred fifty cycles of the test are run (unless the beam reaches the reduced value sooner) and, if the relative dynamic modulus remains greater than 40%, the concrete beam passes the test.

$$E_r = \left(\frac{N_c^2}{N_0^2} \right) \times 100 \qquad \text{(Eq. 9-1)}$$

where N_0 is the frequency of wave propagation before testing and N_c is the frequency after c number of freeze-thaw cycles. Freeze-thaw durability of concrete beams can also be determined based on average dilation (percent expansion). Three hundred fifty freeze-thaw cycles are performed and failure is considered to have occurred if greater than 0.06% expansion of the beam's length develops.

Freeze-thaw testing of concrete beams examines not only the performance of the aggregate under these conditions but also that of the hardened paste that binds the aggregate. The concrete is air-entrained at 6% to provide durability for the paste. Beam testing is a more complete procedure than testing aggregates alone but it is quite time consuming and may cause delays in the decision-making process for aggregate selection (Desta et al., 2015).

Other Tests for Evaluating Aggregates

Water Absorption

The amount of water absorption is another specified test used for aggregate selection. A commonly used maximum allowable absorption value is 5% for concrete aggregates and for those used in bituminous pavements. The absorption value is a general indication of the freeze-thaw resistance of aggregates and, although there are exceptions, high absorption values tend to indicate nonresistant materials.

Deleterious Materials

A final criteria for judging the suitability of aggregates for concrete involves the amount of weak-rock constituents present. Weak materials include shale, siltstone, weathered argillaceous carbonates, iron concretions, friable sandstones, deeply weathered rocks, coal, wood, and low density cherts (specific gravity < 2.40). As weak particles fracture when exposed at the concrete surface during freeze-thaw cycles, these materials will pop out from the concrete.

ASTM Standard C33/C33M indicates a maximum allowable percentage of these deleterious materials relative to the intended use of the concrete. For pavements and driveways that experience below freezing temperatures, permissible maximums of 3% clay lumps and friable particles, 5% low density chert, and 0.5% coal are allowed, with the sum of clay lumps, friable particles, and low density chert set at 5%. For exposed architectural concrete under severe weather conditions these values are reduced to 2% clay lumps and friable particles, 3% low density chert, and 0.5% coal, with a total of 3% allowed for the first two categories. Some state highway departments stipulate a 3% maximum for low density chert for aggregates used in concrete pavements.

The tests described above, except the test for deleterious materials, are also used to evaluate the suitability of rock for other purposes such as building stones, riprap, and tombstones.

Portland Cement Concrete

Portland cement concrete, or simply concrete, is an engineering material in wide use throughout the world today. It can be precast into structural members or cast in place during the construction process. It is both strong and durable and its widespread use has made possible many of the construction achievements of the twentieth century. Dams, bridges, highways, buildings, tunnel linings, retaining walls, and sidewalks are constructed from concrete. Therefore, it is appropriate to include a brief discussion of Portland cement concrete in a chapter on construction materials.

Concrete Composition

Concrete is a composite material consisting of a mineral filler, the aggregate (which ranges in size from fine sand to pebbles or larger gravel fragments or crushed stone), and a binding medium, or cement paste, that surrounds the pieces of relatively inert mineral filler. The binder is formed as a reaction product from Portland cement and the addition of water.

The purpose of the aggregate is threefold: (1) to provide an inexpensive filler for the cementing medium, (2) to provide a considerable volume of the material that is resistant to applied loads, abrasion, moisture penetration, and weathering action, and (3) to reduce the volume change effects that occur in the binding medium during hardening and from moisture changes thereafter.

The cement paste has two major functions: (1) to fill the space between aggregate particles, thereby providing (a) lubrication in the fresh, plastic concrete during placement and (b) water tightness after the concrete sets and (2) to provide strength to the hardened concrete. For the hardened paste, the properties depend on the following aspects: (1) the characteristics of the Portland cement itself, (2) the relative proportions of cement and water, usually represented as the water-cement ratio by weight, and (3) the extent of chemical reaction between the cement and water. Cement, water, and aggregate are combined in overall proportions to accomplish the following requirements: (1) when mixed the fresh concrete is workable such that it can be placed in the forms, which will hold it until hardening has occurred, (2) when hardened, the concrete will be sufficiently strong and durable to accomplish the purpose intended, and (3) the cost of the total product is as low as possible in keeping with the necessary quality required.

In hardened concrete the aggregate comprises about 75% of the volume. The space between the aggregate, the remaining 25% or so, is filled with cement paste and air voids. Some air remains trapped in the concrete after placement even though it has been well compacted. This volume is usually between 1 and 2% of the total volume. Since the 1950s, special air-entraining agents have been incorporated when proportioning the concrete that will be exposed to freeze and thawing conditions. Structural concrete is not air-entrained. The air-entraining agents provide small air voids throughout the paste and their primary purpose is to provide freeze-thaw resistance and salt scaling resistance to the concrete during its service life. During freezing of concrete, water in concrete expands and exerts pressure, causing concrete deterioration. For freeze-thaw resistance the desired amount of air content is from 5 to 6% of the total volume of the concrete. This would include both the entrapped air and the entrained air.

The presence of air voids, however, has an effect on the strength of the hardened concrete. This must be considered when proportioning the concrete. Typically more cement must be included in air-entrained concrete to counteract the strength reduction brought about by the addition of the entrained air. If considerably more air is included in the paste, a further reduction in strength occurs. Hence, a detrimental effect of entraining too much air in the paste (above 6%) is a reduced compressive strength of the hardened concrete, perhaps below that which is required.

The hardened cement paste may include some unreacted cement particles and hydrated cement (calcium silicate hydrate or CSH), which is a product of reaction between the cement and water. Proper curing of the concrete is needed to ensure that the cement has full opportunity to react with water to yield the cement hydrate.

Cement hydration requires time, favorable temperatures, and the availability of moisture. Curing takes place as the concrete is subjected to suitable moisture and temperature conditions. The curing period for specimens in the laboratory is 28 days under special moist room conditions. The 28-day compressive strength of standard concrete cylinders is commonly used to determine the acceptability of concrete placement (Table 9.1). In construction work the curing period typically varies from 3 to 14 days.

As a final consideration of concrete composition, the contribution of the aggregate to the properties of concrete can be enumerated: (1) the aggregate particles contribute to strength, elasticity, and durability in a direct way, (2) the nature of the surface of the particles is important (for example, roughness decreases workability of fresh concrete and increases bond with the cement paste after hardening), (3) a dense gradation of the aggregate reduces workability and increases the density of the mix, and (4) the higher the percentage of aggregate, the lower the cost and volume changes that occur on drying.

Table 9.1   Compressive Strength of Concrete for Varying Curing Periods.

Compressive Strength		
(MPa)	(psi)	Days
20.7	3000	7
24.1	3500	14
27.6	4000	21
34.5	5000	28

Petrographic Examination of Concrete

Concrete and concrete-making materials can be studied in much the same way as minerals and rocks are evaluated. This involves petrography, the detailed description of materials based on an evaluation of their texture and mineralogy. Extensive work on concrete petrography has been done by Mielenz (1962) and others (Erlin and Stark, 1990; French, 1991; Oyen et al., 1998; Poole and Sims, 2015). Table 9.2 provides a list of the ways in which petrography can be used to evaluate concrete. Determining the air-void system in concrete can be accomplished according to ASTM C457/C457M, Standard Test Method for Microscopical Determination of Parameters of the Air-Void System in Concrete. In this procedure the percentage of air voids is determined along with the average chord length (diameter) of the voids, the specific surface of the void system, and the spacing factor (the average distance water must travel to reach an air void). The purpose of the analysis is to determine whether or not the concrete is resistant to freeze-thaw attack and to salt attack.

Well-distributed small bubbles in the paste, totaling about 6% of the total concrete volume, will provide resistance to such attack. This determination is made on a cut surface of the concrete that has been polished to a smooth planar surface (Figure 9.2).

Petrographic examination of the polished concrete surface can also provide information on the percentage of constituents: paste, coarse aggregate, fine aggregate, and air content. This is valuable information when working out details of the original mix design. Cement-aggregate reaction products can be viewed using petrographic analysis as well. Alkali-silica and alkali-carbonate reactivity are described in Chapters 3 and 5, respectively.

Petrography of concrete can be accomplished in a number of ways, including examining (1) concrete structures in the field to obtain an overall view of the performance, (2) thin sections of the concrete, and (3) polished sections. These are the same procedures used, traditionally, to examine rocks and minerals in the field of geology. A thin section is a

---

Table 9.2   Uses for Petrographic Examination of Concrete and Concrete-Making Materials.

- **Description of the Concrete**
  1. Mix proportions
  2. Internal structure
  3. Cement-aggregate relationships
  4. Deterioration

- **Description of the Cement**
  1. Composition, especially presence of free CaO or MgO in undesirable amounts
  2. Relative fineness
  3. Identification of certain additives

- **Description of the Aggregate**
  1. Composition, grading, and quality
  2. Identification of the type, kind, and source of the aggregate
  3. Presence of coatings
  4. Detection of contamination

- **Determination of the Cause of Superior Quality and Performance of the Concrete**

- **Identification of Certain Admixtures**
  1. Mineral admixtures
  2. Pozzolans
  3. Siliceous aids to workability

- **Evaluation of the Microscopic and Megascopic Void System**
  1. Air content of the concrete
  2. Size and spacing of the voids in the cement paste

- **Determination of the Cause of Inferior Quality or Failure of Concrete**
  1. Cement-aggregate reaction
  2. Attack by aggressive waters
  3. Freezing and thawing
  4. Unsound aggregate
  5. Unsound cement, especially excessive free CaO and MgO
  6. Inadequate proportioning, mixing, placing, curing, or protection
  7. Structural failure, abrasion, or cavitation

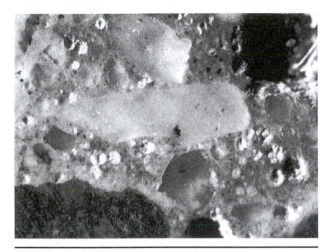

Figure 9.2   Microstructure of concrete exhibiting partial filling of the entrained air-void system. Micrograph field width: 4 mm. (Stutzman, 1999.)

thin slice of rock (or concrete) mounted on a glass slide with optical cement. The section transmits light by virtue of it being ground to a thickness of 0.03 mm or less. Table 9.3 gives a list of features that can be studied in a thin section of concrete. A polished section, by contrast, is a smooth, planar surface that has been highly polished to provide a plane on which the constituents and the voids between them can be observed. These are viewed in reflected light. Both the constituents of cement clinker and concrete can be studied using polished sections.

Engineering Considerations of Construction Materials

1. Rock is used as building stones, concrete aggregate, highway base courses, riprap, rock-fill in embankments and dams, and tombstones.

2. Coarse and fine aggregates comprise about 80% of concrete and bituminous construction materials. Tests to evaluate aggregates can also be applied to evaluate other rock uses such as building foundations and underground construction.

3. Gravel aggregate is extracted from pits with only minor processing whereas crushed stone is obtained from a quarry by drilling, blasting, and crushing of bedrock. Crushed stone is more expensive and usually of a higher quality than gravel.

4. For aggregates, the Los Angeles abrasion test measures strength, whereas durability or resistance to weathering is determined by various durability tests. Testing is performed on aggregates alone or on concrete beams. Freeze-thaw durability of beams is the best measure of the final product, although time consuming.

5. Since 1900, Portland cement concrete has provided a major improvement in the construction industry. Cement, aggregates, water, and some additives react to form a rock-like material held together by calcium silicate hydrate (CSH). Entrained air

Table 9.3   Features Examined in a Thin Section of Concrete.

Fine aggregate
    Composition
    Cement-aggregate reaction
    Alteration in place
Coarse aggregate
    Composition
    Cement-aggregate reaction
    Alteration in place
Cement paste
    Unhydrated clinker particles
    Maximum and average size
    Frequency
    Composition
    Evidence of associated cracking

Hydration products
    Texture and microstructure
    Admixtures and additions
    Calcium hydroxide
    Alteration of the hydration products
Microfractures
    Extent, width, continuity
    Origin and relationships
    Secondary deposits
Voids
    Frequency
    Size
    Special relationships
    Secondary deposits

is added to improve freeze-thaw durability of the paste. Concrete is strong in compression but lacks sufficient tensile strength, which, when needed, is provided by steel reinforcing.

6. Petrographic examination is used to evaluate the quality of concrete. The technique is similar to analysis of sedimentary rocks as the grains and groundmass are evaluated microscopically. Percentages of coarse aggregate, fine aggregate, paste, and voids can be ascertained, and entrapped versus entrained air can be determined. Water-cement ratios can also be estimated making it possible to better diagnose concrete problems.

7. Concrete deterioration can be caused by problems with the aggregate or with those of the paste. For exterior concrete, scaling is due to too much mix water (high water-cement ratio) or lack of air entrainment, pop outs are caused by poor quality coarse aggregates, pattern cracking and expansion are due to alkali-silica and alkali-carbonate reaction, and D-cracking is a long-term deterioration caused by freeze-thaw problems of the coarse aggregate. Helpful information can also be obtained to decide if delamination of reinforcing steel or honeycombing of the concrete has occurred.

Exercises on Evaluating Construction Materials

1. A stream gravel is to be used as a concrete aggregate for a concrete road in central Ohio. What tests should be run on the aggregates to determine if they are suitable materials? What allowable maximum values would be specified for the individual tests?

2. In hardened concrete three major materials are present: coarse aggregate, fine aggregate, and cement paste. Describe the nature of these three constituents.

3. What are the two types of voids that are present in the paste? How did they get into the paste and what is their function, if any?

4. What is meant by hydration of the cement and by the term curing? How long are standard cylinders cured?

5. Why does the aggregate tend to reduce the volume change of concrete during heating-cooling and wetting-drying cycles? Why does a higher water-cement ratio in the paste yield a higher volume change effect?

6. What are the two primary functions of the paste in concrete? On what, primarily, do the properties of the hardened paste depend?

7. Petrography can be used to examine hardened concrete just as it is used to examine various types of rocks. Why does this seem to be a logical extension from rocks to concrete? What similarities are involved?

8. Megascopic examination, thin section, and polished section viewing of concrete are useful petrographic procedures. Describe their differences, the methods involved, and their benefits.

9. Why is the air-void system of hardened concrete of interest in an investigation? Why is an air content that is too high or too low a serious concern? Explain.

10. Petrography can be used to examine both the raw materials for concrete construction and the hardened concrete that results. What materials does this include?

11. Review the discussion on alkali-silica and alkali-carbonate reaction in concrete (Chapters 3 and 5). How can petrography be used to evaluate this problem? Provide a thorough answer.

References

Barksdale, R. (ed.). 1991. *The Aggregate Handbook*. Washington, DC: National Stone Association.

Bobrowski, P. T. (ed.). 1998. *Aggregate Resources: A Global Perspective*. Rotterdam: A. A. Balkema.

Desta, B., West, T., Olek, J., and Whiting, N. 2015. Evaluation of D-Cracking Durability of Indiana Carbonate Aggregates for Use in Pavement Concrete. Proceedings, 66th Highway Geology Symposium, Bainbridge, MA.

Dolar-Mantuani, L. 1983. *Handbook of Concrete Aggregates*. Saddle River, NJ: Noyes.

Erlin, B., and Stark, D. (eds.). 1990. *Petrography Applied to Concrete and Concrete Aggregates* (ASTM Special

Publication 1061). West Conshohocken, PA: American Society for Testing and Materials.

French, W. J. 1991. Concrete petrography: A review. *Quarterly Journal of Engineering Geology and Hydrogeology* 24:17–48.

Hale, P. A., and Shakoor, A. 2003. A laboratory investigation of the effects of cyclic heating and cooling, wetting and drying, and freezing and thawing on the compressive strength of selected sandstones. *Environmental & Engineering Geoscience* IX(2):117–130.

Lienhart, D. A. 1998. Rock engineering ratings system for assessing the suitability of armourstone sources. In J.-P. Latham (ed.), *Advances in Aggregates and Armourstone Evaluation* (Engineering Geology Special Publication 13, pp. 65–85). London: Geological Society.

Mielenz, R. C. 1962. Petrography applied to Portland-cement concrete. *Reviews in Engineering Geology* (Vol. 1). Geological Society of America.

Oyen, C. W., Fountain, K. B., McClellan, G. H., and Eades, J. L. 1998. Thin-section petrography of concrete aggregates: Alternative approach for petrographic number evaluation of carbonate aggregate soundness. *Transportation Research Record* 1619:18–25.

Poole, A. B., and Sims, I. (eds.). 2015. *Concrete Petrography: A Handbook of Investigative Techniques* (2nd ed.). New York: Taylor and Francis.

Shakoor, A., and Barefield, E. H. 2009. Relationship between unconfined compressive strength and degree of saturation for selected sandstones. *Environmental and Engineering Geoscience* XV(1):29–40.

Shakoor, A., and Bonelli, R. E. 1991. Relationship between petrographic characteristics, engineering properties, and mechanical properties of selected sandstones. *Bulletin of the Association of Engineering Geologists* 28(1):55–71.

Stutzman, P. E. 1999. *Deterioration of Iowa Highway Concrete Pavements: A Petrographic Study* (HR-1065). Washington, DC: National Institute of Standards and Technology.

West, T. R., Johnson, R. B., and Smith, N. M. 1970. *Tests for Evaluating Degrading Base Course Aggregates* (SBN 309-01885-4). National Cooperative Highway Research Program Report 93. Washington, DC: Highway Research Board.

STRATIGRAPHY AND GEOLOGIC TIME 10

Chapter Outline

A detailed study of the Earth's history is not typically included in a discussion on physical geology, or in a geology text for engineers. Historical geology, which pertains to the chronological events of Earth's history, including the development of life (Wicander and Monroe, 2015), is a semester-long course. Some knowledge of this history, however, is required for engineers who work with earth materials in order to understand the impact of geologic units on engineering construction. Such information is also needed to understand the details depicted on geologic maps. This chapter provides the basic information needed to understand and appreciate the concepts of geologic time and the geologic past with regard to engineering design and construction.

The chronological sequence of rock units and their field relationships are a product of the geologic history of an area. Many geologic features, including planes of weakness and conditions of the rock mass, are related to the historical development of rocks. Therefore, a full appreciation of the impact of geology on construction is not possible without proper consideration of geologic history.

For example, jointing in a basalt may occur at different times following solidification of the lava. Joint planes formed early in the rock's history may later become filled with secondary minerals such as calcite. This tends to knit the rock mass together, greatly increasing its strength. If it can be established, based on details of geologic history, that the joints are healed in this fashion, a lack of through-going fractures may be confirmed. As a result, a rock excavation in the basalt can be made at a steeper angle or be stabilized with considerably less support.

Groundwater flow is another condition of concern. If an erosion surface is recognized at a certain elevation, increased water flow in rock cuts or tunnels can be anticipated at that level. Establishing the geologic history may lead to the presence of erosion surfaces or other permeable zones, which can then be considered during design and construction.

The shape and orientation of geologic units are also related to geologic history. Whether a body of sand extends as a two-dimensional layer (a sheet) or as an elongated lens is related to both its origin and geologic history. Drainage, groundwater inflow, variation in soil strength, and numerous other factors would differ greatly depending upon the origin of soil and rock units.

Stratigraphy

Stratigraphy refers to the study of rock strata or layers. It also includes the chronological sequence of rock unit origins and the field relationships among them (Harold, 2013; Wicander and Monroe, 2015). All rock types are involved, not only sedimentary rocks, but also igneous and metamorphic rocks. Because the description and origin of rock units are of importance to stratigraphy, a carryover of knowledge from rock identification is required. The texture and composition of rocks are a consequence of their origin and they determine the name assigned to a rock. A foliated metamorphic rock, a clastic sedimentary rock, or a glassy igneous rock illustrates the significance of this relationship. Stratigraphic concepts are also applied to soil or unconsolidated materials.

Rock units are subdivisions of the geologic column, the total sequence of rocks formed since the cooling of the Earth. The basic rock unit is known as a formation. It is defined as a distinct lithologic (rock type) unit that is recognizable in the field based on its physical characteristics. It also must be sufficiently thick (several hundreds of meters or tens to thousands of feet) and have sufficient lateral extent for proper inclusion and easy detection on a geologic map. A formation name consists of two parts, the location name, indicating where the rock unit was first described, and a lithologic name, indicating the type of rock that prevails in the unit. Examples are St. Louis Limestone, Manhattan Schist, Pierre Shale, Monterey Chert, Salem Limestone, and Navajo Sandstone. If no dominant rock type prevails throughout the unit, the word formation is used instead, such as the Chinle Formation, which is found in the Painted Desert of Arizona.

Subdivisions of a formation are called members and a collection of formations is known as a group. The Borden Group in Indiana, a collection of siltstone units plus a few thin limestone beds, is such an example.

Basic Geologic Principles for Relative Age Dating

In discussing the chronology of rock units, several basic principles must be established at the outset. These principles are used to develop the relative ages of geologic units. They include (1) principle of original horizontality and continuity, (2) principle of superposition, (3) principle of faunal assemblage, (4) principle of crosscutting relationships, (5) principal of baked contacts, (6) principle of inclusions, and (7) principle of uniformitarianism (Harold, 2013; Marshak, 2013; Nance and Murphy, 2016).

Principle of Original Horizontality and Continuity

The principle of original horizontality and continuity pertains to most sedimentary rocks because they are usually deposited in parallel, nearly horizontal, continuous layers. Marine sedimentary rocks comprise the major portion of the sedimentary column and are deposited on the continental shelf, which has a gentle slope toward the ocean basins. This principle suggests that sedimentary rocks that deviate significantly from the horizontal position have been tilted or folded by earth stresses to attain the inclined position.

There are exceptions, of course. Some sedimentary sequences are formed in an inclined position. Steeply dipping limestone or dolomite reefs are formed at the sloping edges of islands in the ocean (see Figure 5.2a). These features are fairly common in Silurian aged rocks (about 400 million years old), which are prevalent in areas surrounding the Great Lakes. Other rocks such as lava flows are typically deposited on gently sloping terrain, also yielding a nonhorizontal orientation. All in all, however, the vast majority of sedimentary rocks were originally deposited horizontally.

Principle of Superposition

The principle of superposition states that for a sequence of undisturbed sedimentary rocks (when viewed in cross section), the lowest layer is the oldest and the layers become successively younger toward the top. This assumes that the rocks have not been overturned by folding. When rocks are inclined, this simple concept is a valuable tool for working out the sequential details. Superposition of sedimentary rocks above an old erosion surface, developed on either igneous, sedimentary, or metamorphic rocks, is a natural occurrence.

Principle of Faunal Assemblage

The principle of faunal assemblage is related to age dating of rocks by means of the fossils they contain. Properly stated: like assemblages of fossil organisms indicate like geologic ages for the rocks that contain them. This is related to the evolutionary process of organisms through geologic time.

Principle of Crosscutting Relationships

The principle of crosscutting relationships states that faults, dikes, folds, unconformities, and other crosscutting features are younger than the rock units they cross. Quite simply, a rock must predate a feature that cuts across it. A simple example of this is a tree stump with saw marks on the cut surface: the tree had to exist before the saw was able to cut through it. This is true for microscopic mineral assemblages as well as entire mountain ranges.

Principle of Baked Contacts

The principal of baked contacts states that when a body of magma intrudes the country rock, it bakes (metamorphoses) the rock adjacent to the intrusion. Therefore, the rock that has been baked must be older than the intrusion.

Principle of Inclusions

The principle of inclusions states that a rock containing an inclusion (piece of another rock) must be younger than the inclusion. For example, a conglomerate containing pebbles of sandstone must be younger than the sandstone.

Principle of Uniformitarianism

The principle of uniformitarianism establishes the background for determining occurrences in the geologic past. It proposes that the natural laws operating on Earth today are the same as those that prevailed in the geologic past. The past can be studied by viewing these processes at work on Earth today, allowing for reasonable variations in the magnitude of these effects. Thus, analyzing the details of deposition patterns in a modern beach can aid in answering questions about ancient beach building in a similar environment. Uniformitarianism is sometimes expressed by the phrase "the present is a key to the past."

Unconformities

An unconformity is a surface of nondeposition, or erosion, separating older rocks below from younger rocks above. During the geologic history of an area, rock masses were uplifted from below sea level where they were deposited. In the uplifted position the rocks are subject to erosion followed by (in some cases) lowering below sea level again and a renewal of sedimentation. This gives rise to a buried erosion surface, which separates the two distinct intervals of deposition. Following another cycle of uplift, these surfaces can be observed, in profile view, near the Earth's surface.

The interval of time represented by an unconformity is an important consideration in the geologic history of a region. It indicates not only that lowering and uplift of the Earth's surface occurred, but that a sequence of sedimentary rocks is missing when compared to another contiguous region that received continuous sedimentation. Some unconformities represent breaks in the rock record of a few thousand years, whereas others may designate several hundred million years. Regional studies may be needed to determine how much of the rock record probably eroded or was simply not deposited during the interval of uplift.

Types of Unconformities

The four primary types of unconformities observed in nature are (1) angular unconformity, (2) disconformity, (3) paraconformity, and (4) nonconformity (Harold, 2013; Marshak, 2013; Nance and Murphy, 2016).

Angular Unconformity

In an angular unconformity the older strata (below the unconformity as suggested by superposition) dip or slope at a different angle from that of the younger strata (Figure 10.1a). A significant amount of historical detail is indicated by this unconformity. After initial deposition, the older rocks were folded (creating the slope) and uplifted above sea level. Erosion occurred on the dipping beds to form the erosion surface. Following this episode, the rock sank below the ocean surface and additional sediments were deposited. Later, both units were uplifted to the Earth's surface.

Disconformity

A disconformity is an unconformity in which sedimentary rock beds parallel each other on opposite sides of an irregular unconformity (Figure 10.1b). The minor relief of the erosion surface caused by gullies or slope movement or the missing record of fossils provides evidence of the interruption of sedimentation. These form when layered rocks are uplifted, undergo erosion to yield an irregular surface, and then are lowered below sea level to receive more sedimentation. No folding occurs during this episode.

Paraconformity

A paraconformity is similar to a disconformity except that it lacks any relief on the erosion surface and is not discernable from other bedding plane surfaces (Figure 10.1c). Detailed regional study and considerable experience are required to locate the precise unconformable surface within the rock sequence. Fossil evidence and lithologic detail are used to establish the location of the paraconformity.

Nonconformity

A nonconformity is an unconformity that develops when igneous or metamorphic rocks are exposed to erosion and sedimentary rocks are subsequently deposited above the erosion surface (Figure 10.1d). A typical example consists of granite lying below a sandstone or carbonate rock with an erosion surface between them. The geologic history depicted in this situation is, first, the intrusion of granite into the country rock followed by extensive erosion. Thousands of feet of rock must be eroded to intercept the igneous intrusion. Then the ocean advances over the granite mass and deposits sediments. The sedimentary rock lithifies and the area is uplifted to expose the nonconformity.

Nonconformity versus Intrusive Contact

The discussion on nonconformities brings into focus a challenging interpretation involving geologic

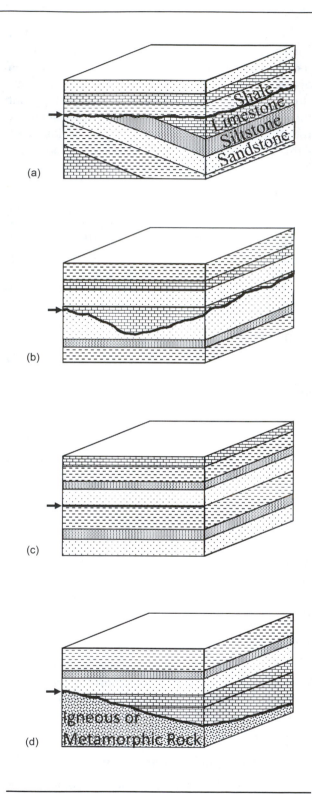

Figure 10.1 Types of unconformities: (a) angular unconformity, (b) disconformity, (c) paraconformity, and (d) nonconformity. (Modified from Boggs, 2001.)

field studies. Another, seemingly similar, exposure consists of a contact between granite and sedimentary rock but it marks the boundary where the igneous intrusion came to rest within the sedimentary rock mass. In this case, the granite is younger than the rock it intruded (crosscutting relationships). In the case of the nonconformity, the granite did not intrude the sedimentary rock above it, and is the older of the two (superposition). The sedimentary rock was deposited on the erosion surface.

To resolve the problem of igneous contact versus nonconformity, the surface between the two rock units must be examined in detail. For the nonconformity, the erosion surface is reasonably smooth without small irregularities. Pieces of granite may occur as conglomerate pebbles in the sedimentary rock and the bedding tends to parallel the unconformity (Figure 10.2a). Joints in the granite may be filled with sand grains from the rock above and dikes may cut through the granite but stop at the unconformity.

The igneous contact would appear differently. The contact may be quite irregular with small protrusions of the granite into the rock above. The overall trend of the contact may not be parallel to the bedding planes of the sedimentary rock above and pieces of that rock may be present as partially digested remnants in the granite mass. A zone of contact metamorphism in the sandstone may surround the granite or an outer zone of the granite may be finer grained than the rest (known as a chilled border), indicating that it cooled more rapidly because of its contact with cooler rock above. Crosscutting dikes may penetrate both the granite and the sandstone (Figure 10.2b).

Correlation

Correlation is the process of tying one rock sequence in one place to another in some other place. Correlation of sedimentary rocks helps extend a chronology of relative time from region to region and continent to continent. Correlation is accomplished by comparing similarities of physical features and faunal assemblages (Harold, 2013).

Because of gaps in the sedimentary sequence, there is no location on Earth where a continuous rock column from the oldest to the youngest rocks exists. We must piece the total column together from one place to another by correlating rock units and their local sequences.

When sedimentary rocks show a fairly constant and distinct lithology over a widespread area, the boundaries between units can be connected or extended from one locality to another. An example is illustrated by Figure 10.3 on the next page. Here, two valley walls at the edge of a wide stream valley are shown. Sandstone, exposed at the surface, is the youngest rock in the sequence. These natural exposures of bedrock at the Earth's surface are called outcrops. Below the sandstone is a siltstone, then a coal seam, and below it, a shale. A sandstone bed is exposed near the base of the valley wall and, on the west, a limestone bed lies below the sandstone. By extending imaginary lines across the valley, we can conclude that the beds correlate with each other.

On the east side of the valley, correlation strongly suggests that a limestone bed exists below the lower sandstone unit in the sequence. Because the rock is not exposed, the presence of the limestone must

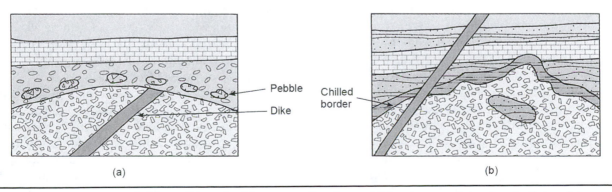

(a)   (b)

Figure 10.2   Contrasting field conditions for (a) a nonconformity and (b) an igneous intrusion, both involving sedimentary rock units.

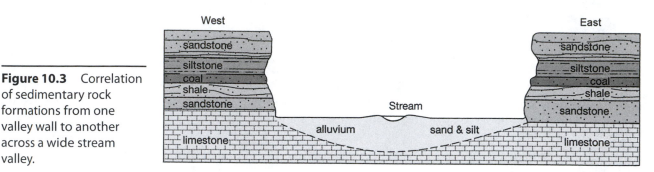

Figure 10.3 Correlation of sedimentary rock formations from one valley wall to another across a wide stream valley.

be verified by another means. If this information is needed for engineering or economic purposes, rock cores can be obtained from exploratory drill holes extending downward from the east valley wall to a depth below the base of the stream valley.

Correlation can be accomplished using rock cores when no rock exposures are available for examination. In engineering studies, correlation from one boring to another is used to determine the nature of the subsurface conditions. Drilling procedures are discussed in detail in Chapter 19 on subsurface investigation and site selection.

Rocks can also be correlated using fossils rather than lithology alone. In many rock sequences, such as the Mississippian System of the midcontinental area of the United States, thick deposits of one major rock type, in this case limestone, occur. Details concerning the differences in color and texture between limestone units are used along with fossil assemblages to determine which formations are present.

Complications in Correlation

Correlation over a large area for an engineering site can be considerably more complicated than the example shown in Figure 10.3. Sedimentary beds are not always continuous over large areas because (1) the rocks may be cut by an unconformity, (2) the rock units may disappear (pinch out) if the basin of deposition did not extend across the entire area, and (3) the rock unit may grade laterally into a different lithology because of changes in the sediments deposited across the basin. Lateral changes that occur across a rock unit are called facies changes. Rock descriptions relate to lithofacies, and the fossil content to biofacies. Changes in lithofacies are the specific concern when rock units grade laterally.

Dating the Earth

By all indications, the human race has been interested in the age of the Earth since earliest recorded time. Before reviewing some early speculations about Earth history, it is worthwhile to examine the reasons why humans seek to know the past.

Two apparent reasons for this interest are (1) natural curiosity and (2) economic considerations. Included in curiosity is the seeking out of past events for religious or theological reasons. Contained also would be pure science with its thirst for knowledge for the sake of knowledge alone. Applied science more likely falls in the realm of economic considerations.

The economic aspects of age relationships become more apparent when we realize that certain types of minerals are found in specific kinds of rocks of an appropriate geologic age. Coal in the United States is found mostly in rocks of Pennsylvanian and Permian age (230 to 320 million years ago). The greatest amounts of petroleum are found in the younger marine rocks, those of Cenozoic age (last 65 million years), although older sedimentary rocks also contain oil. Younger rocks contain more oil because life was more abundant in the oceans during the Cenozoic age and, with more organisms, there was more organic material available to produce oil. Also, less time has passed since the petroleum formed, yielding less chance for loss from erosion.

Metallic ore deposits are commonly associated with certain of the oldest igneous and metamorphic rocks. Gold, silver, and other base metal areas occur generally in isolated zones of the earliest of rocks, Precambrian in age (more than 600 million years old), although younger igneous and metamorphic rocks can also be ore bearing.

Certain construction problems can be anticipated for certain rock types and geologic ages. The expansive shales of the Cretaceous period (65 to 136 million years) are well known, as is the scarcity of carbonate rocks for construction in the Pennsylvanian of the Midwest. Rock salt and prevalent solution cavities are typical in the Permian. Solution cavities and sinkholes in the Mississippian and Ordovician limestones and dolomites are well documented across the United States. The age and history of Earth materials play an important role in their nature and engineering properties.

Early Speculation and Scientific Estimates about the Earth's Age

Every society, religion, or cult seems to develop an explanation of how the Earth began and how long ago it happened. Humans, obviously, have an inherent desire to pursue the details of their beginnings, which is manifested through theological thought and scientific endeavor.

For example, the Brahmins in ancient India thought the Earth was eternal, whereas astrologists of Babylon claimed humans appeared 500,000 years ago. The Persians of 100 BC thought the Earth was 12,000 years old and would last about 3000 years more. The priests of Chaldea said the Earth was 2 million years old.

Careful observations of natural phenomena formed the basis for unraveling the mysteries concerning the Earth's age. Herodotus, the great Greek historian, noted around 450 BC that the Nile River delta had to be the product of many floods because individual floods deposited only thin layers of sediment, yet the river alluvium was quite thick (Gaines, 1994). He reasoned that thousands of years must have occurred to form the Nile delta.

Aristotle and the Greek and Roman naturalists and philosophers who followed, continued the scholarly approach of combining observation with deduction. This method was lost during the Dark Ages (approximately AD 476–1000) and when inquiries about the age of the Earth developed again it was through the biblical interpretations of the seventeenth century. To question such religious pronouncements was heresy, which brought dire consequences to naturalists and scientists of that day.

Seventeenth-Century Biblical Scholars

In the seventeenth century, scholars of Western civilization studied the question of the Earth's age. Interpreters of the Old Testament attempted to determine the time of creation by working back through the lineage of people in the Bible. In 1642, John Lightfoot, a scholar at Cambridge University in England, deduced that the moment of creation was 9:00 AM, September 17, 3928 BC. Later, in 1658, Archbishop Ussher of Ireland claimed that the Earth was created on the evening of October 22, 4004 BC (Barr, 1985). This date appeared in some subsequent editions of the Bible. The basis for these two interpretations was the assumption that humans were created soon after the Earth so, by working back through human history in the Bible, the time of planetary creation could be ascertained. The Earth was thereby designated to be about 6000 years old. It was not until after 1830 or so (about 200 years later) that the scientific community openly professed the belief that the Earth was considerably older than 6000 years.

Eighteenth- and Nineteenth-Century Scientific Studies

James Hutton

In 1785, James Hutton, a Scottish geologist, was one of the early scientists to cast doubt on the accuracy of the biblical scholars. He proposed the principle of superposition for sedimentary rocks but also made some astute observations about the Earth's features. He noted that geologic processes carved the landscape at a relatively slow rate. If the gorges, mountain passes, and plateaus were carved by these geologic processes, it would take much longer than a few thousand years (Hutton, 1988 [1785]).

William Smith

Around 1800, William "Strata" Smith, a civil engineer and surveyor, noted that certain of the flat sedimentary rocks in southern England contained fossils unlike those in any other layer. He could predict the rock strata and the fossils they contained based on the elevation of the rocks due to the horizontal nature of the strata in the area (Smith, 1817). This showed that former animal life and a marine origin were involved, suggesting a need for more time to transpire than only a few thousand years.

Although Hutton and Smith proposed no specific age for the Earth, they pointed out the apparent conflict between the biblically based time scale and that indicated by natural features.

Georges-Louis Buffon

Physicists also began to labor with the problem of such a short time for the age of the Earth. After Buffon, Kant, and Laplace proposed their theories (between 1749 and 1796) on the origin of the solar system there was a basis for estimating the relationship between time and the planetary orbits. The time necessary for the formation and motion of the Sun and planets seemed too great to fit into the brief time span demanded by the biblical studies. Buffon also studied the rates of melting and cooling of iron balls, because he had concluded earlier that the Earth contained an interior not unlike iron because it was so dense. His estimate of 75,000 years (Buffon, 1778) for the Earth to cool seemed low to some geologists at the time but, nevertheless, made the biblical fundamentalists most unhappy.

Lord Kelvin

Different estimates for the age of earth materials came rapidly in the nineteenth century. In 1854, Hermann von Helmholtz, one of the founders of thermodynamics, established that the Sun, based on gravitational contraction, would have burned for 20 to 40 million years. Later, in 1897, Lord Kelvin (William Thomson) indicated that the Earth, by his estimates, had taken 20 to 40 million years to cool. In other papers his estimates had been as high as 75 million years (Burchfield, 1990). We know today that Lord Kelvin's calculations were much too low, but several problems beyond his control placed them in error. He had no accurate measure of heat flow from the Earth; values for the thermal conductivity of rock were virtually unknown at the time and, because radioactivity had not yet been discovered, the radioactivity of elements in the Earth's crust, which supply much of the heat, was not considered.

Darwin and Evolution

In 1859, Charles Darwin put forth his famous theory of evolution (Darwin, 1859). He knew it would have taken a considerable amount of time for the life he observed as fossils to evolve from those simple forms to mammals and finally to humans. Darwin believed it would require at least 100 million years for this evolution and he was concerned that Kelvin's 20 to 40 million-year figure was much too low.

Sediment Accumulation

Estimates on the age of the Earth, based on the accumulation of sediments and sedimentary rocks, were presented during the latter half of the nineteenth century. In 1854, a statue of Ramses II, the renowned Egyptian pharaoh, was found at Memphis, Egypt, under 9 ft of river-laid sediment. The statue was known, from historical data, to be 3200 years old. The 9 ft of sediment cover yielded an accumulation rate of 3.37 in per century. Since the total sediment thickness in the river was 40 ft, a total age of 14,200 years for sediment accumulation in the Nile River was obtained. This alone is more than twice the time suggested by the biblical scholars.

Salt Accumulation in the Sea

Two other estimates of the Earth's age were obtained in a similar way. First, a rate of accumulation is determined, followed by a measurement of the total accumulated amount. Elapsed time is found by dividing the accumulated thickness or amount by the calculated rate. Salinity in the sea was used by John Joly in 1899 for such an estimate. Assumptions included the following: (1) the rate at which Na^+ is added to the sea is constant through time, (2) oceans were fresh (salt free) at the start, (3) only a small amount of Na^+ removal, with time, is considered, and (4) the rate is determined based on a current estimate of runoff to the oceans and the Na^+ concentration in the water delivered to the sea. Using these assumptions, a value of 90 million years was obtained. There are two complications to these assumptions: more Na^+ is removed from the oceans than originally assumed and today's rate of Na^+ input is probably not a good average of that contributed throughout geologic time.

Limestone Deposition

A final estimate for consideration involves the accumulation of limestones around the world. Through correlation, the duplicate sections of rocks deposited simultaneously are excluded and a total thickness of limestone accumulation is obtained. Using the rate of limestone accumulation observed in the oceans (currently about 2.5 cm [1 in] every 200 years), the total elapsed time can be found by dividing the total thickness by the accumulation rate.

$$\text{Time} = \frac{\text{total thickness}}{\text{accumulation rate}} \qquad \text{(Eq. 10-1)}$$

Some 18 determinations were presented between 1860 and 1909 with values ranging from 3 million to 1.5 billion years. Most values were somewhat less than 100 million years.

Common Shortcomings

All of these calculations based on early scientific methods suffer from similar shortcomings. The rates of accumulation are only approximate, the total accumulations cannot be accurately determined, and the assumptions are usually oversimplified. For example, in limestone accumulation it is now known that these rocks did not form during the earliest geologic time so the complete history of the Earth is not represented by the limestone column. By and large, all the age estimates for the Earth turned out to be too low.

Geologic Time Scale

William Smith's use of correlation had shown that the ages of rocks could be compared across distances of tens of miles by means of fossil content and physical appearance of the rock units. By use of fossil content alone, it soon became possible to correlate across hundreds and then thousands of miles.

By the middle of the nineteenth century a general geologic rock column had been fairly well developed. It is a diagram combining in succession, from youngest to oldest, the sequence of all known strata compiled on the basis of fossil or other evidence of relative age. The geologic column provided the basis for geologic time scale (Table 10.1, see following pages). Originally, dates were fixed only in relation to other events with the age increasing downward. Values in years before the present (BP) were added when absolute dating became available.

The geologic column represents not only a succession of layers but a passage of geologic time (Figure 10.4, page 204). Prior to the 1930s and 1940s no absolute values in years could be assigned to the geologic rock column. Although it could be used only to assign relative dates, many of the details of folding, uplift, and erosion had been worked out and placed properly in the sequence. Only the advent of radiometric dating was needed to complete the story.

Absolute or Radiometric Dating

In 1895 and 1896, several events occurred that prepared the way for absolute dating of Earth materials. Within a few months of each other, Henri Becquerel discovered radioactivity in uranium salts, Wilhelm Röntgen discovered X-rays, and Madam Curie isolated radium, a radioactive element. Between 1905 and 1913, the nature of radioactivity and isotopes was clarified and by the 1930s the difficulties concerning various applications had been resolved. Radioactive decay could be used to date rocks that contained these elements.

Emission of Particles

Radioactivity functions in the following way. A few elements, among them uranium and thorium, disintegrate spontaneously into lighter elements when their nuclei give off particles of three different types: alpha, beta, and gamma radiation. Alpha radiation is the emission of helium (He) atoms from the nucleus at the speeds of thousands of kilometers per second. This emission converts the original atom into another having an atomic weight of four less and an atomic number of two less, designated by $_2\text{He}^4$ (the atomic number = 2 and atomic mass = 4). The emitted particles collide with surrounding atoms to generate a considerable amount of heat. This supplies a substantial portion of the heat given off today by the Earth's interior.

Beta particles consist of electrons derived from the disintegration of neutrons in the nucleus, which form protons after the ejection of electrons. Although electrons are expelled at an even higher velocity than are alpha particles, they have so little mass that the heat generated is negligible. Gamma rays are short-wavelength X-rays emitted at the speed of light.

Decay of U^{238}

Emission of either alpha or beta particles from the nucleus of a radioactive atom converts it into a new element (reduces its atomic number). For example, $_{92}\text{U}^{238}$ decays through a series of seven emissions of alpha particles and six emissions of beta particles until it reaches a stable nonradioactive isotope $_{82}\text{Pb}^{206}$. This is shown in Figure 10.5 in a plot of atomic number versus atomic mass for the $_{92}\text{U}^{238} \rightarrow$ $_{82}\text{Pb}^{206}$ reaction (see page 205).

Table 10.1 Geologic Time Scale.

Era	Period	Epoch	Duration	BP	Derivation of Names	Aspects of the Life Record	Aspects of Physical Events
CENOZOIC	Quaternary	Holocene	11,800 Y	11,800 Y	Geologic periods in the Cenozoic era were originally named Primary, Secondary, Tertiary, and Quaternary. The first three are no longer used; Tertiary was the last one replaced.	Recorded history	
		Pleistocene	2.58 MY			Humans	Glaciation
	Neogene	Pliocene	2.7 MY	2.59 MY			Formation of Pacific Coast Range
		Miocene	17.7 MY				
	Paleogene	Oligocene	10.9 MY	23 MY		Grass becomes abundant	Formation of the Alps and many mountain chains
		Eocene	21.9 MY				
		Paleocene	9.7 MY	65.5 MY		Horses appear	Volcanic activity in western United States
MESOZOIC	Cretaceous		80 MY		Derived from Latin word for chalk (creta) and first applied to extensive deposits that form white cliffs along the English Channel.	Extinction of dinosaurs	Early folding of Rocky Mountains
	Jurassic		54.3 MY		Named for the Jura Mountains, located between France and Switzerland, where rocks of this age were first studied.	Birds appear	
	Triassic		51.4 MY	251 MY	Taken from word "trias" in recognition of the threefold character of these rocks in Europe.	Dinosaurs appear	

Era	Period	Epoch	Time in Years		Aspects of the Life Record	Aspects of Physical Events
			Duration	BP		
PALEOZOIC	Permian		48 MY			Folding of Appalachian Mountains
	Carboniferous					Widespread glaciations
	Pennsylvanian		19.1 MY			Coal-forming swamps
	Mississippian		41.1 MY			Paleozoic Alps
	Devonian		56.8 MY			
	Silurian		27.7 MY			
	Ordovician		44.6 MY		Vertebrates appear (fish)	
	Cambrian		53.7 MY	542 MY	First abundant fossil record (marine invertebrates)	
PRECAMBRIAN	Proterozoic Eon		1958 MY		Scanty fossil record	
	Archean Eon		2642 MY	4600 MY	Primitive marine plants and invertebrates, one-celled organisms	Origin of the Earth

Derivation of Names

- **Permian** — Named after the province of Perm, Russia, where these rocks were first studied.
- **Pennsylvanian** — Named for the State of Pennsylvania where these rocks have produced much coal.
- **Mississippian** — Named for the Mississippi River Valley where these rocks are well exposed.
- **Devonian** — Named after Devonshire, England, where these rocks were first studied.
- **Silurian** — Named after Celtic tribes, the Silures and the Ordovices, that lived in Wales during the Roman conquest.
- **Cambrian** — Taken from the Roman name for Wales (Cambria), where rocks containing the earliest evidence of complex forms of life were first studied.
- **Proterozoic Eon** — The time between the birth of the planet and the appearance of complex forms of life. 88% of the Earth's estimated 4.6 billion years falls within the Precambrian era.

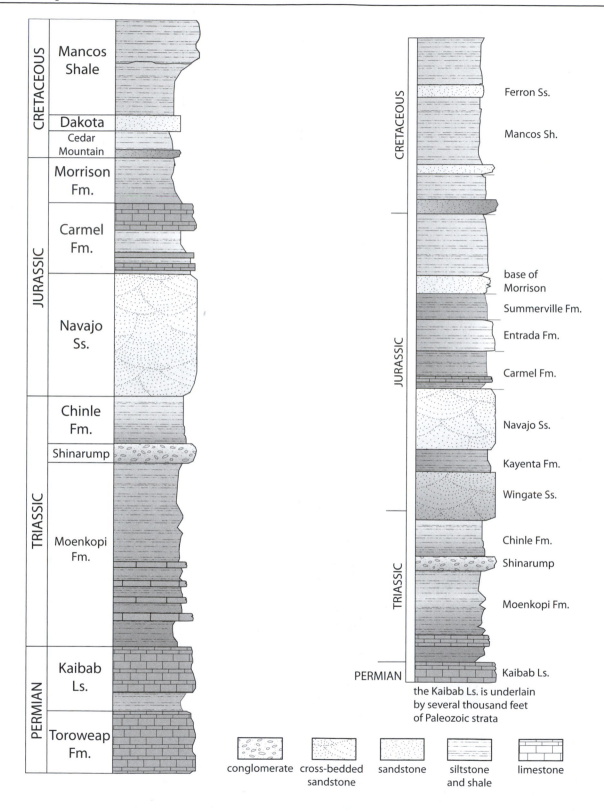

Figure 10.4   Geologic rock column. (Ritter-Petersen, 2015. Courtesy of Scott Ritter.)

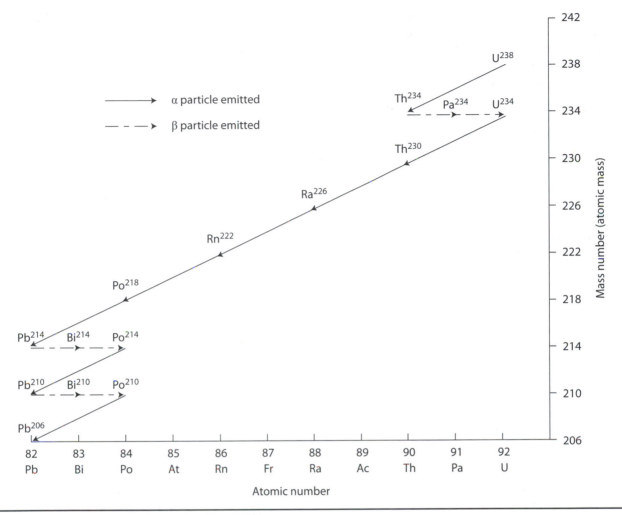

Figure 10.5 Radioactive decay of U^{238} to Pb^{206}.

Uranium has two radioactive isotopes, U^{238} and U^{235}, which provide 99.28 and 0.72% of natural uranium, respectively. U^{235} is used in nuclear power plants to generate electricity. Several other radioactive elements are also used in dating geologic materials and are discussed later in this section.

Decay Constants

The rate of decay is constant for each radioactive isotope, but rates differ considerably from one isotope to another. Disintegration rates are not affected by temperature, pressure, or the chemical environment, which makes it possible to determine the age of geologic materials containing radioactive elements. Prior to 1950, gravimetric techniques (accurate weight determinations) were used to find the mass of the end member elements of the decay

sequence (e.g., $_{92}U^{238}$ and $_{82}Pb^{206}$) and, based on the decay constant, the age of the rock was calculated. This typically required large-sized minerals from pegmatites. Today, using the mass spectrometer, only a few milligrams of the two end members are needed so that small-sized grains can be used to determine age (Rink and Thompson, 2015).

The decay constant "A" is used to express the proportion of atoms of an isotope that decays in a unit of time. The decaying or parent isotope continually decreases in amount while the end member isotope, the daughter, continues to increase. The fraction of the total number of parent atoms that decay during a given interval of time is constant but the actual number that decay will decrease because the parent atoms are constantly being depleted. The abundance of the parent isotope decreases exponentially with time, as shown in Figure 10.6. Disintegration rates are expressed as the "half-life" of the radioactive substance, i.e., the time required for half of the atoms to disintegrate or decay. Half-life units are shown on the abscissa of Figure 10.5.

Calculations for Decay Constant and Half-Life

Disintegration of radioactive isotopes occurs according to a first-order differential equation, which can be written as:

$$-\frac{dc}{dt} = \lambda c \qquad \text{(Eq. 10-2)}$$

where c = the concentration, t = time, and λ = the decay constant. Separating variables and adding the limits of integration,

$$-\int_{c_0}^{c_1} \frac{dc}{c} = \int_0^t \lambda dt$$

or
$$\ln c \Big|_{c_0}^{c_1} = -\lambda t \Big|_0^t$$

$$\ln c_1 - \ln c_0 = -\lambda t$$

$$\frac{c_1}{c_0} = e^{-\lambda t}$$

$$c_1 = c_0 e^{-\lambda t}$$

which is a common form of the equation. Solving for λ,

$$-\lambda t = \ln c_1 - \ln c_0$$

$$\lambda t = \ln c_0 - \ln c_1 = \ln \frac{c_0}{c_1}$$

$$= 2.303 \log \frac{c_0}{c_1}$$

$$\lambda = \frac{2.303}{t} \log \frac{c_0}{c_1}$$

If $c_0 = a$ and $c_1 = a - x$, where a is the initial amount and x is the amount remaining after time t, then

$$\lambda = \frac{2.303}{t} \log \frac{a}{a-x} \qquad \text{(Eq. 10-3)}$$

at the time when one-half of the parent isotope is used up, which is the half-life or $t_{1/2}$, $a - x = 1/2$ if $a = 1$. Then

$$\lambda = \frac{2.303}{t_{1/2}} \log \frac{1}{1/2}$$

$$= \frac{0.693}{t_{1/2}} \text{ or } t_{1/2} = \frac{0.693}{\lambda} \qquad \text{(Eq. 10-4)}$$

Limitations of Isotope Dating

Table 10.2 gives a list of the radioactive isotopes commonly used in age dating. Radioactive decay, as

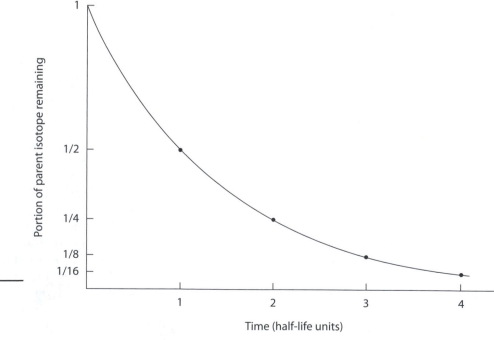

Figure 10.6 Graph of decay of an isotope with time.

Portion of parent isotope remaining

Time (half-life units)

stated previously, finally ends in the formation of a stable end product, although a number of steps is involved in the process. Some additional information about the isotopes is provided in Table 10.2.

Stringent conditions must be met in order to provide an accurate age for a rock. In the case of uranium and thorium, different isotopes of lead result as the end products. The minerals analyzed must be fresh, that is, not altered or weathered, because circulating solutions might leach out the lead and the parent isotopes at different rates. Metamorphism of the mineral would also invalidate the age calculation because lead may be driven out faster than uranium or thorium. Therefore, the rock or mineral must represent a closed system since the time it formed. No loss or gain of either parent or daughter isotope could have occurred. Also, if the radioactive mineral happened to be closely associated with a lead mineral containing mostly common lead, it would be difficult to sort out the various lead isotopes present.

Age Dating Based on Uranium, Thorium, and Lead

Many minerals meet the rigorous conditions just described. For samples containing uranium isotopes, independent dates can be obtained based on U^{235}/Pb^{207} and U^{238}/Pb^{206}. A date can also be determined on Pb^{207}/Pb^{206} and these are usually quite reliable. If any leaching or other loss occurred to the sample, this ratio should be unaffected, although the quantities themselves may change markedly because the two lead isotopes are chemically identical and would be removed in the same fashion. In addition, Th^{232}/Pb^{208} may be used in the same rock for age verification. If the age value obtained by these different ratios agrees within a few percent, the determination is considered to be reliable. Because of the long half-life of these isotopes, rocks must be more than a few million years old to provide accurate dates (Rink and Thompson, 2015).

Age Dating Using Rubidium and Strontium

The $_{37}Rb^{87}$ isotope decays to $_{38}Sr^{87}$ by the emission of a β particle. Since the half-life of $_{37}Rb^{87}$ is about 47 billion years, the method is limited to rocks more than a few million years old. As shown in Table 10.2, it is useful for dating minerals in metamorphic rocks.

Common strontium consists of four isotopes, Sr^{84}, Sr^{86}, Sr^{87}, and Sr^{88}, but not all Sr^{87} is derived from the disintegration of Rb^{87}. Hence, it is necessary to correct for the Rb^{87} in the mineral not generated by radioactive decay. Because of this needed

Table 10.2   Principal Isotopes Used in Radiometric Age Dating.

Isotopes	Half-Life (years)	Effective Dating Range[a] (years)	Earth Materials That Can Be Dated	
Uranium-238/Lead-206	4.50×10^9	10^7 to age of Earth	Zircon Uraninite	Pitchblende
Uranium-235/Lead-207	0.71×10^9	10^7 to age of Earth		
Thorium-232/Lead-208	15×10^9	10^7 to age of Earth	Zircon	
Potassium-40/Argon-40	1.30×10^9	10^4 to age of Earth	Muscovite Biotite Hornblende Whole volcanic rock	Arkose[b] Sandstone[b] Siltstone[b]
Rubidium-87/Strontium-87	4.7×10^{10}	10^7 to age of Earth	Muscovite Biotite	Microcline Whole metamorphic rock
Carbon-14	5730 ± 30	0 to 50,000	Wood Charcoal Peat Grain Tissue Charred bone	Cloth Shells Tufa Groundwater Ocean water

[a] Age of the Earth is about 4.6×10^9 years.

[b] For paleogeographic studies.

correction to a mass that is already small, Rb-Sr age calculations are extremely sensitive to errors in the youngest rocks within their dating span, that is, those rocks only a few million years old. They are, instead, much more reliable for dating rocks Mesozoic in age or older (greater than 65 million years). However, because the Rb-Sr decay path does not contain a gaseous phase, as does the K-Ar disintegration (argon is an inert gas), Rb-Sr determinations are affected much less by the alteration that occurs during slight degrees of metamorphism. Hence, many of the older ages of metamorphic rocks are derived from Rb-Sr for increased accuracy.

Experiments show that biotite heated above 350°C no longer gives stable ages for the Rb-Sr determination, whereas muscovite must be heated to about 550°C before its lattice permits the loss of strontium. The cooling history of the rock may be dated by such comparisons, or metamorphic occurrences may be so established. Such metamorphism is sometimes said to "reset the clock" for age determination.

Age Dating from Potassium and Argon

The potassium isotope $_{19}K^{40}$ decays simultaneously into two daughters, $_{20}Ca^{40}$ by emitting a β particle and $_{18}Ar^{40}$ by capturing an electron from the innermost shell. Of the decaying nuclei, 89% take the Ca^{40} route whereas 11% become Ar^{40}. The conversion to Ca^{40} is of no value in geochronology or age dating because Ca is so abundant in nature and the Ca^{40} isotope is not easily separated from the other Ca isotopes.

The Ar^{40} from the sample deterioration must be discerned from the Ar^{40} of the atmosphere, but this can be accomplished with accuracy. In the atmosphere, a constant ratio of Ar^{36} to Ar^{40} persists so that, by measuring both Ar^{36} and Ar^{40} in the sample, a correction for the Ar^{40} contribution from the atmosphere can be made. The half-life of the K-Ar disintegration process is 1.30 billion years, and rocks approximately 100,000 years or older can be dated. Under extremely favorable conditions, rocks as young as 50,000 years may also be dated.

Argon is an inert gas and therefore does not react to form any chemical compounds. It readily enters and leaves some crystal lattices by diffusion but, in certain lattices, it seems to be retained indefinitely. The minerals biotite, muscovite, hornblende,

and sanidine (the high-temperature form of potassium feldspar, which is common in many volcanic rocks) are particularly good recipients of argon. Any minerals that readily lose argon are not suitable for radiometric dating.

When subjected to high temperatures, all rocks and minerals lose argon; thus, the K-Ar dates are extremely sensitive to the thermal history of the sample. Hornblende seems to retain argon at higher temperatures than does biotite so that K-Ar dates on the same rock using these two minerals may yield quite different dates. This can be used to date an occurrence of metamorphism documented by the lower age value obtained. Commonly a U-Pb analysis would be performed as well to corroborate the time formation of the rock.

Radiometric dates on some sedimentary rocks can be obtained using the K-Ar method. Although most minerals analyzed are from igneous or meta-igneous rocks, the sedimentary mineral glauconite, a greenish clay-like material, contains potassium and has been dated successfully. It forms on the seafloor prior to cementation of the sediment and is fairly common in shale, siltstone, and sandstone. It is widely accepted that argon diffuses from the glauconite lattice more readily than from the micas and, thereby, the accuracy of radiometric dates for sedimentary rocks is less reliable. Typically, sedimentary ages would be bracketed by dates on intrusive rocks, which include those both younger and older than the sedimentary unit.

Radiocarbon Dating

Developing the Method

All of the radiometric methods discussed previously apply to extremely old rocks. In 1947, a method based on C^{14}, or radiocarbon, for determining the ages of younger materials was discovered. Carbon-14 is formed continuously in the upper atmosphere by the bombardment of N^{14} by neutrons from cosmic radiation. Radiocarbon decays by β radiation to N^{14} with a half-life of 5730 years. Previously a half-life value of 5568 years for C^{14} had been used so that earlier calculated carbon dates must be adjusted by a factor of 1.03.

The reaction for the C^{14} relationship is

$$_6C^{14} - \beta = _7N^{14} \qquad \text{(Eq. 10-5)}$$

It is not necessary, and it would actually be impossible, to measure accurately the amount of daughter N^{14} to obtain an age determination. Instead, the amount of C^{14} is determined relative to the other carbon isotopes (mostly C^{12} with about 1% C^{13}) and the age is found by that means. The details involving the carbon system are described in the following section.

C^{14} to C^{12} Ratio

Radioactive carbon mixes with ordinary carbon, diffusing rapidly through the atmosphere, hydrosphere, and biosphere. The proportion of radiocarbon is essentially constant throughout the system because of the rate of mixing. As long as the production rate of C^{14} remains constant, the ratio of C^{14} to C^{12} is also constant, because the production rate and decay rate are in equilibrium. Introducing old carbon from burning coal since the Industrial Revolution, and the increase on Earth in generating C^{14} into the atmosphere, complicate the interpretations but do not invalidate the method.

As long as an organism is alive, it ingests air and water and maintains the equilibrium proportion of C^{14}. After death, the equilibrium is lost as replenishment of air or water ceases and C^{14} is reduced by radioactive decay. Because of the short half-life, C^{14} dates (time elapsed since death) of as low as 100 years can be measured. After 40,000 to 50,000 years (seven to nine half-lives), so little C^{14} remains that dating is no longer possible. For organic materials less than 50,000 years old, radiocarbon is an invaluable measurement tool. It completely revolutionized the study of archaeology and provides great insight into the Holocene and late Pleistocene time spans.

Sources of Error

Several sources of error are associated with radiocarbon dating. Introducing "old carbon" by burning coal and the variation in production of C^{14} with latitude location on Earth are parts of the problem. A variation of C^{14} concentration in the atmosphere with time is also involved. The flux of cosmic rays varies inversely with the strength of the Earth's magnetic field and directly with solar flares. We know that the strength and polarity of the magnetic field have varied significantly in the geologic past, and, as a consequence, we know that the C^{14} concentration has varied as well.

Algae in springs, whose CO_2 is partly obtained from dissolved limestone or "old carbon," yields tufa with a much lower C^{14} content than the air. It is indeed "born old." Mollusk shells (clams, oysters, etc.) in areas where groundwater springs flow upward into the ocean also obtain old carbon from the water, which, typically, is several hundred years old. They too would register older than the time of death of the mollusk. By contrast, porous reefs composed of coral and other animal shells, long dead, would be rejuvenated by the present-day water from the sea spray, providing a younger age for the corals.

There are problems associated with dating parchment (animal carbon), tree rings, and ancient artifacts. Commonly the radiocarbon dates do not agree with the established historical dates. The problems of flux and mixing account for part of this but, for the tree rings, an additional complication occurs. Each tree ring is active for only one year and does not re-equilibrate with the next succeeding ring. Therefore, only the last ring to form before the death of the tree would indicate the correct length of time since its demise. These problems notwithstanding, radiocarbon dating has greatly expanded the knowledge of prehistory and of geologic events in the late Pleistocene and Holocene.

Fission Track Dating

Uranium atoms, in addition to disintegration yielding α and β radiation, also break down by fissioning into two essentially equal-sized nuclei. This involves an extremely small number of atoms, at a rate of 1 in 69×10^{16} atoms per year. These fragments fly apart with a great amount of force, striking the crystal lattices of surrounding minerals and leaving extremely small tracks, about 50 Å wide (5×10^{-7} cm) and 10 to 20 µm (10 to 20×10^{-4} cm) long. The small imperfections are below the resolving power of most microscopes but they are enlarged easily by etching in sodium hydroxide or hydrofluoric acid.

Such etched surfaces are used for geologic dating. Minerals known to concentrate uranium are selected from the rock, embedded in plastic, and etched with sodium hydroxide. For glassy rocks, polished surfaces are etched with hydrofluoric acid. The pits and cones of the lattice imperfections on a measured area are counted. Following this determination, the specimens are bombarded by neutrons in a nuclear reactor to produce additional fission tracks from the U^{235} in the samples. More etching and the new tracks are counted. Since the flux of neutrons is known in the nuclear reactor, the increase in fission tracks is a measure of the uranium atoms present. From this it is possible to compute the time required to cause the first tracks.

About 10 minerals and natural glasses are known to contain fission tracks suitable for dating. It has also been learned that tracks will heal in samples heated to several hundred degrees Celsius and samples exposed at the Earth's surface will give erroneous results because of the effects of cosmic rays. The best results are likely to be obtained on materials less than 100 million years old. To ensure the accuracy of this measurement, radiometric dates are obtained on the same samples.

Absolute Age of the Earth

The tremendous expanse of geologic time is no longer debatable. It is likely that the Earth is nearly 4.6 billion years old. Rocks from eastern Siberia, approaching this age, have been reported. In Greenland rocks dated at 3.75 billion years old are noted and in South Africa very old granitic rocks, 3.50 billion years in age, are found. The oldest rocks in North America are gneisses from Minnesota dated at about 3.35 billion years old.

The radiometric dating methods best suited for these ancient rocks are the U-Pb methods. K-Ar is too susceptible to loss of argon during subsequent heating of the rocks. Rb-Sr and U-Pb methods are not readily susceptible to "resetting of the atomic clock" by reheating but the U-Pb method applied to zircon is a particularly accurate means for measuring the age of very old rocks.

It is of interest that the 4.6 billion year duration of the Earth is equal to about one half-life of U^{238}/Pb^{206} and more than seven times the half-life of U^{235}/Pb^{207}. Hence, the Earth originally contained 40 times as much U^{235} and twice as much U^{238}. The heat flow from the Earth generated by this radioactive decay must have been many times that occurring today and, consequently, many geologic processes were much more active.

Some meteorites have been dated at 4.6 billion years old, which is thought to be the age of the Earth as a planet. These likely indicate the nature of the planetesimals from which the Earth and other inner planets formed.

The Awesome Span of Geologic Time

As mentioned, the age of the Earth is now estimated at 4.6 billion years. The Precambrian era covers the time span from the Earth's beginning until the start of Cambrian time, a duration of 4 billion years. The earliest organic structures in the form of limey secretions, carbonaceous residues, and other indirect evidences of life first appeared about 3 billion years BP. Simple one-celled animals appeared about 2 billion years BP and shelled animals evolved to become abundant 600 million years ago at the start of Cambrian time. The first land plants became established about 500 million years BP and dinosaurs appeared 230 million years BP and became extinct 65 million years BP. The early horse appeared at 60 million years and humans at about 2 million years BP. Recorded history covers only the last 10,000 years.

The 4.6 billion years can be represented by a 24-hour day to emphasize the amount of time between events. The Earth would form at midnight and the Precambrian era would extend until 8:52 PM. The first indication of life would occur at 8:35 AM with simple one-celled life at 1:34 PM and abundant shelled animals at 8:52 PM. The first land plants evolve at 9:23 PM. Dinosaurs make their appearance at 10:48 PM and became extinct at 11:40 PM. Humans appear about 38 sec before midnight and recorded history represents about 0.2 sec. The average life span of an individual human, nearly four score years, is equivalent to 1.5×10^{-3} sec of this 24-hour day.

Human Time versus Geologic Time

The great span of geologic time provides the backdrop for a difficult comparison for the beginning earth scientist and other students of geology to fully appreciate. That is, the considerable difference between the concept of time in the geological sense and in the engineering or worldly sense. The life of an engineering structure may be 50 or 100 years, the 100-year flood is a long interval of concern, and concrete pavements that last for 40 years without major repair are a resounding success. But, erosion continues over millions of years to establish a certain stream gradient, and limestone cave formation may require a similar period of time. Thousands of years are required to develop a fertile soil profile and retreat of continental glaciers involves much the same span of time.

A point of some interest to the engineer involves limestone bedrock and leaking of a surface reservoir impounded behind a dam. Fissured and cavernous limestones provide conduits to transport leakage away from the reservoir and such limestones are a concern if the regional groundwater gradient encourages such movement. However, limestones free of fissures will not develop conduits to carry sizable quantities of water away in the limits of 50 or 200 years, a normal life span for a dam and reservoir. Natural, underground plumbing systems take many times that number of years to develop.

What is a reasonably long time span in human years may be insignificant from a geological standpoint. Geologically speaking, lakes are a temporary interruption in the erosion cycle of streams and, in time, their outlets will be destroyed or their basins filled with sediment. This is not to say that the Great Lakes will soon be lost for, in human time, they should enjoy many centuries of existence; but, measured in millenniums or in large units of time, their demise is to be expected. Of course, humans may be able to stave off the effects of natural erosion and extend this time period considerably.

The downslope movement of soil and rock under the force of gravity is a natural occurrence and part of the erosional process on Earth. Humans attempt to prevent significant slope movements in populated or economically important areas over periods of human years. The purpose, of course, in geology applied to engineering is to minimize the effects of geologic hazards in a variety of ways including restraint, strength improvement, and avoidance. This does not suggest that it is easy to switch from the concept of geologic features in terms of geologic age to a concern for engineering structures designed and built in terms of human years. The two must be considered in turn because they both impact the study of geology applied to engineering.

Geologic Maps

Although engineers working in construction would not be expected to prepare a geologic map of a site, they will need to understand engineering geology reports for construction sites or areas of study. By contrast, engineering geologists and geological engineers do have the training and skills to prepare geologic maps. Engineering geology reports usually contain, among other drawings, a geologic map, which is the most compact manner for supplying detailed geologic information. A preliminary geologic map may be the basis for planning the subsurface exploration program for a project or to investigate possible supplies of construction materials. Following this detailed work, geologic maps and cross sections may summarize the information found in the exploration program.

The two primary types of geologic maps are bedrock and surficial. Bedrock geologic maps show the distribution of rocks on the Earth's surface as it would appear with all soil or unconsolidated material removed. Surficial geologic maps show the unconsolidated materials that occur at the Earth's surface. These maps are available mostly in areas covered by glacial deposits and, hence, have few bedrock outcrops. Consequently, they are sometimes called glacial geology maps. Agricultural soils maps also depict surface materials and are included in this general category. They describe and classify the soils based on agricultural concerns, but some useful relationships to the engineering properties of soil are supplied as well. Surficial maps are useful sources of information in terms of foundation support for engineering structures and locations of construction materials including sand, gravel, and common borrow. They are used for planning detailed subsurface investigations that involve drilling and sampling.

Geologic maps show the position in space (attitude) of rock units and indicate their relative age. Symbols show the attitude of the beds whereby the arrangement of the rocks in the subsurface can be determined. This subject requires knowledge of structural geology, which is presented in detail in Chapter 11. Figure 10.7 presents a geologic map depicting inclined sedimentary beds. As previously mentioned, the formation is the smallest geologic unit commonly depicted on geologic maps, but the scale and dimensions of the map will determine this specifically. For example, a geologic map of an entire state of the United States typically can show only groups of rocks or collections of formations.

The line of intersection between geologic units on a map is called a geologic contact, or simply a

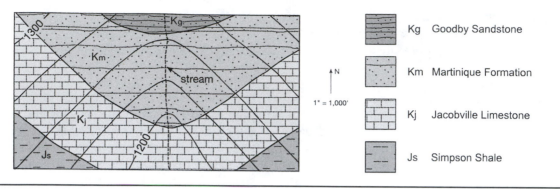

Figure 10.7   Geologic map showing inclined sedimentary rocks.

contact. This represents the surface that separates the two units, appearing as a line on the map or in a plan view. Geologic maps produced by the US Geological Survey are drawn on a topographic map base. This makes it possible to determine the elevations of rock contacts and to see the relationship between elevations and rock units (see Appendix B, Topographic Maps for more information).

Map units are depicted by different colors and abbreviations and are described in the legend. Two letters typically make up the symbol for a formation, such as Sw, which would designate the Wabash Formation of the Silurian period. Patterns are used to denote the lithology of various units (Figure 10.8). The rock units are also listed in vertical stratigraphic rock columns in chronological order in the map legend, beginning with the youngest at the top.

Other maps can be developed from geologic maps by combining or highlighting certain features of the rock units. These may be termed engineering geology maps, environmental geology maps, or other designations. Keep in mind that these are interpretive maps, developed through interpretation of geologic map information, that are one-step removed from the specific field data, and must be viewed with that understanding.

Engineering Considerations of Statigraphy and Geologic Time

1. A stratigraphic description of rock provides the basis for examining the geologic history and setting of a construction site.

2. The basic mapping and descriptive unit for rocks is the formation, which consists of a place name and a lithologic name.

3. The geologic history of a site has a direct influence on its rock properties and engineering behavior.

4. Basic principles of stratigraphy are used to determine geologic occurrences and the sequence of events. They include original horizontality and continuity, superposition, faunal assemblage, crosscutting relationships, baked contacts, inclusions, and uniformitarianism.

5. Unconformities are surfaces of erosion and age gap. Similar to stratigraphy, they help to reconstruct the geologic history of a project site. Nonconformities versus igneous contacts with sedimentary rock provide a challenging field problem, which is resolved based on careful field observations.

6. Correlation involves the lateral extension of stratigraphic units across a site. It provides the basis for comparing subsurface information from one borehole to another, and to extrapolate and interpret the geologic conditions between adjacent boreholes.

7. Two methods are used to date Earth materials: relative dating and absolute dating. Absolute dating involves radioactive decay of certain isotopes whose decay constants are known. Uranium and other heavy isotopes are used to date rocks from 100,000 to 4.6 billion years old, whereas organic substances from 0 to 50,000 years old are dated by the carbon-14 method. For most construction projects, the ages of various rock units can

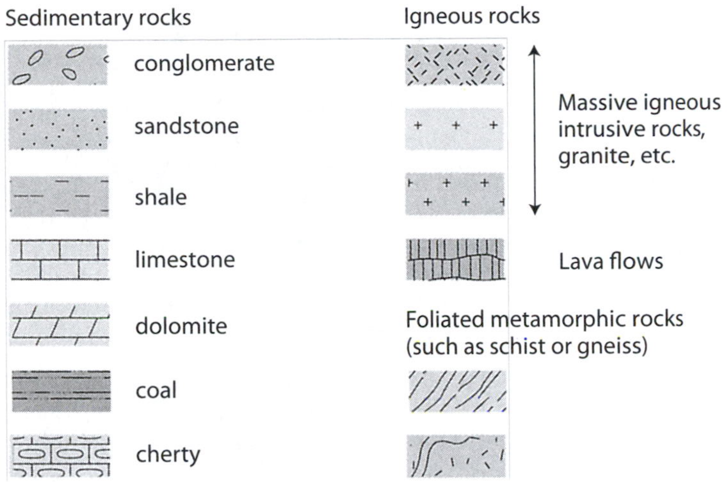

Sedimentary rocks

conglomerate

sandstone

shale

limestone

dolomite

coal

cherty

Igneous rocks

Massive igneous intrusive rocks, granite, etc.

Lava flows

Foliated metamorphic rocks (such as schist or gneiss)

(a)

Cenozoic	Cz	Paleozoic	Pz
Quarternary	Q	Permian	P
Tertiary	T	Pennsylvanian	ℙ
Neogene	N	Mississippian	M
Paleogene	Pz	Devonian	D
Mesozoic	Mz	Silurian	S
Cretaceous	K	Ordovician	O
Jurassic	J	Cambrian	€
Triassic	Tr	Precambrian	p€

(b)

Figure 10.8   (a) Examples of lithology patterns used on geologic cross sections and maps. (b) Examples of geologic time scale abbreviations.

be found from literature review. However, the current practice requires that all artifacts found during excavation for major engineering works, such as dams, be preserved and dated by suitable methods, including the carbon-14 method.

8. The geologic time scale with its divisions of eras, periods, and epochs provides a part of the basic description used in the geologic literature and in engineering reports.

9. In the study of geology and modern humans, we are faced with the contrast of the awesome span of geologic time versus the extent of a human

life, now four score or so in length. These two extremes must be considered when planning engineering constructions.

10. Geologic maps provide the detailed information of a site, city, county, state, or larger region. A map scale is a measure of the detail provided by these maps. Both surficial maps, showing materials at the Earth's surface, and bedrock maps, depicting the solid rock below the unconsolidated materials, are important tools in a site investigation. With a knowledge of Earth processes, stratigraphy, and structural geology, specific geologic details can be discerned.

EXERCISES ON STRATIGRAPHY AND GEOLOGIC TIME

1. Give definitions of the terms *formation*, *member*, *group*, and *stratigraphy*.

2. Discuss the three principles of superposition, original horizontality, and crosscutting relationships.

3. Define unconformity. Identify the names of the four different unconformities. On a separate piece of paper, draw an appropriate cross section for each.

4. See Exercise Figure 10.1. Give the sequence of events that occurred in each of the four cross sections provided. Begin with the oldest unit. Locate and describe any unconformities. Indicate how you determined the relative ages of the units (such as superposition, crosscutting, etc.).

5. *Correlation Studies.* See Exercise Figure 10.2. Diagram (a) depicts two exposures in a plateau region separated by 50 miles; diagram (b) is based on three drill holes spaced a number of miles apart. For (a) and (b), correlate from one section to another. Connect unconformities, igneous intrusions, and formation contacts as appropriate using dotted lines. For beds that do not carry across, indicate what has occurred (lithofacies change, pinch out, etc.).

6. *Estimating the Age of the Earth.*

   a. What are the two primary reasons why humans seek to learn the age of Earth materials? Into which category would applied science and engineering fall? Why?

   b. How did the biblical scholars estimate the age of the Earth in the seventeenth century? What age value did they obtain?

   c. How did James Hutton and William Smith arrive at the conclusion that the Earth's age was greater than that obtained by the biblical scholars? How did the physicists of that time conclude the Earth had to be older? Explain.

   d. Several physical processes have been used to try to determine the age of the Earth: cooling of the Earth; sedimentation in the Nile; sodium in the

(a)

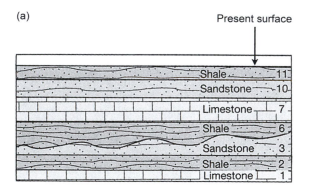

(b)

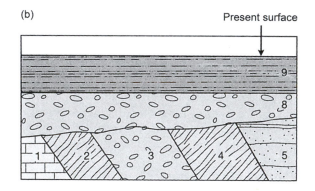

(c)

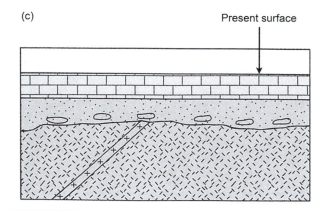

(d)

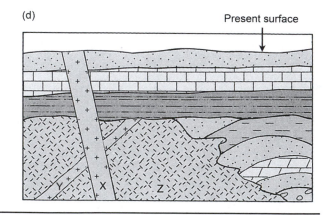

Exercise Figure 10.1

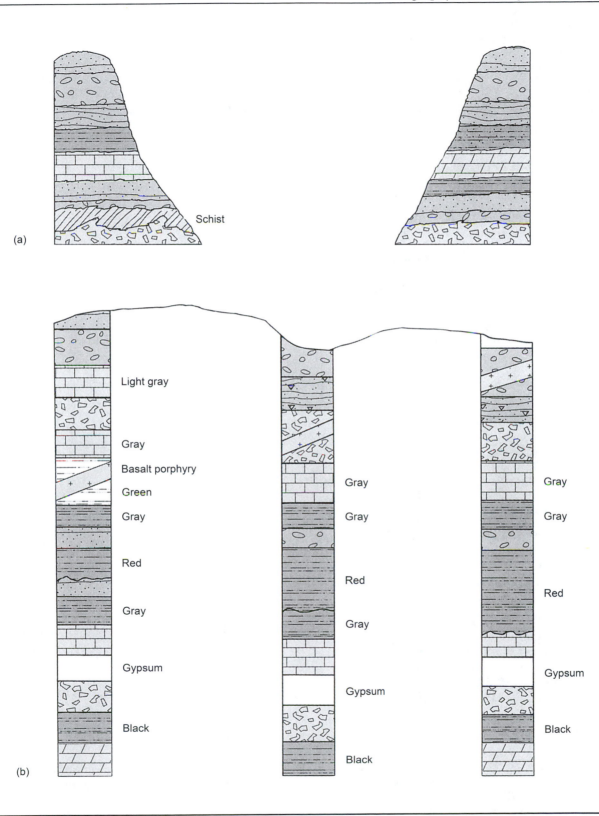

(a)

Schist

(b)

Light gray

Gray

Basalt porphyry

Green

Gray

Red

Gray

Gypsum

Black

Gray

Gray

Red

Gray

Gypsum

Black

Gray

Gray

Red

Gypsum

Black

Exercise Figure 10.2

oceans; and limestone formation throughout the world. How were estimates on the age of the Earth made using these considerations? What values were obtained?

7. *Geologic Time.* Refer to Table 10.1 for assistance.

   a. How long was the Mesozoic era? The Paleozoic era?

   b. How many years elapsed between the end of the Cambrian and the beginning of the Permian? How long was it between the beginning of the Eocene and the end of the Miocene? How long did it take for Mississippian rocks to form?

   c. What percentage of the age of the Earth is represented by Precambrian time? By the Cenozoic era?

   d. Humans first appeared on Earth about 2 million years ago and recorded history consists of the last 10,000 years. What percentage of the age of the Earth do each of these represent?

8. *Radiometric Age Dating.*

   a. What are the three different particles given off by radioactive decay of isotopes? What happens to the isotopes in the process? Explain.

   b. Thorium-232 decays to form lead-208 by a 10-step reaction in this order α, β, β, α, α, α, α, β, β, α. In the same form as shown for the $_{92}U^{238}$ to $_{82}Pb^{206}$ reaction (see Figure 10.5), show this for the $_{90}Th^{232}$ to $_{82}Pb^{208}$ reaction.

   c. Calculate the value of λ, the decay constant, for U^{238}/Pb^{206}, U^{235}/Pb^{207}, and Th^{232}/Pb^{208}. Include the correct units for λ.

   d. What is the half-life for carbon-14? How many half-lives occur in 50,000 years, the limit of age dating by this method? What fraction of the original C^{14} remains after 50,000 years? (Calculate as accurately as possible.)

   e. Why is Rb-Sr dating more accurate in dating some metamorphic rocks if muscovite is used rather than biotite? Explain. What is meant by "resetting the clock" in this respect? If a muscovite sample indicated an age of 138 million years and a biotite sample from the same rock indicated 122 million years, what does this indicate relative to the previous question? Explain.

   f. Why is the K-Ar dating method particularly subject to errors if considerable metamorphism of the rock has occurred? In a sample metamorphic rock the U^{238}/Pb^{206} date made on a zircon crystal was 275 million years, whereas the K-Ar date on a hornblende crystal was 204 million years and the K-Ar date on a biotite crystal was 160 million years. Assuming all of these values are useful, what overall history of this rock can be given?

   g. How is the clay mineral glauconite used to measure the age of sedimentary rocks? What problem exists with this radiometric dating method?

   h. Relative to C^{14} dating, what is meant by "old carbon" and rocks being "born old"? Why do tree rings and carbon dates not always agree in age? Explain.

   i. How are fission tracks measured? What age of materials is best measured by this method? Why?

   j. How do we know that the Earth once contained twice as much U^{238} and 40 times as much U^{235} as it currently does? What effect would that have on volcanic eruption, heat flow, and continental drift, if any? Explain.

9. *Great Span of Geologic Time, Human Time versus Geologic Time.*

   a. In the text, the divisions of geologic time and geologic events were compared to a 24-hour day. Other comparisons have been made relative to tall buildings. How would such a comparison be made to the Willis (Sears) Tower in Chicago or the Empire State Building in New York City? Explain. Other than comparisons to tall buildings, name at least one more way this aspect could be dramatized.

   b. Why is it difficult to keep the comparison between human time and geologic time in perspective? Why must the construction engineer and the engineering geologist deal with this repeatedly? How does the subject of faulting relate to these considerations? Explain.

10. *Geologic Map Reading, Preliminary Aspects.* Refer to Figure 10.7. The four formations shown on the map are the Simpson Shale, the Goodby Sandstone, the Jacobville Limestone, and the Martinique Formation.

   a. What is the area of the map and what is its contour interval?

   b. In what direction does the stream flow? How is this determined? What is its gradient?

   c. Are the sedimentary rocks horizontal? If not, how do you know that? If they are, what tells you so?

   d. Of the following two age sequences (youngest to oldest) for the formations in Figure 10.7, which is correct? Why?
   • Js, Kj, Km, and Kg
   • Kg, Km, Kj, and Js

REFERENCES

Barr, J. 1985. Why the world was created in 4004 BC: Archbishop Ussher and biblical chronology. *Bulletin of the John Rylands Library* 67(2):575–608.

Boggs, Jr., S. 2001. *Principles of Sedimentology and Stratigraphy* (3rd ed.).Upper Saddle River, NJ: Prentice Hall.

Buffon, G. L. 1778. *Époques de la nature*. Paris.

Burchfield, J. D. 1990. *Lord Kelvin and the Age of the Earth*. Chicago: University of Chicago Press.

Darwin, C. 1859. On the Origin of Species by Means of Natural Selection, or the Preservation of Favoured Races in the Struggle for Life (Chapter X). London: John Murray.

Gaines, A. G. 1994. *Herodotus and the Explorers of the Classical Age*. New York: Chelsea House.

Harold, L. L. 2013. *The Earth Through Time* (10th ed.). New York: Wiley.

Hutton, J. 1988 [1785]. Theory of the Earth—an investigation of the laws observable in the composition, dissolution, and restoration of land upon the globe. *Transactions of the Royal Society of Edinburgh* (Vol. I, Part II, pp. 209–304).

Marshak, S. 2013. *Essentials of Geology*. New York: W. W. Norton.

Nance, D., and Murphy, B. 2016. *Physical Geology Today*. New York: Oxford University Press.

Rink, W. J., and Thompson, J. W. (eds.). 2015. *Encyclopedia of Scientific Dating Methods*. Netherlands: Springer.

Ritter, S., and Petersen, M. 2015. *Interpreting Earth History: A Manual in Historical Geology* (8th ed.). Long Grove, IL: Waveland Press.

Smith, W. 1817. Stratigraphical system of organized fossils: With reference to the specimens of the original geological collection in the British Museum: Explaining their state of preservation and their use in identifying the British strata. London.

Wicander, R., and Monroe, J. S. 2015. *Historical Geology* (8th ed.). Boston: Cengage Learning.

STRUCTURAL GEOLOGY 11

Chapter Outline

Rock Deformation

Folds in Rock

Types of Rock Fractures

Folds and Faults Combined

Direction of Stress and Fault Orientation

Engineering Considerations of Structural Geology

Structural geology is the study of rocks that have been deformed by earth stresses and includes a description of their position in space, or attitude, which occurs as a consequence of the deformation. Sedimentary rocks are of particular interest because their layers or bedding express the extent of deformation. Vertical and/or horizontal movements caused by bending and rupture are shown by the relative displacement of the bedding. The spatial relationships of igneous bodies emplaced by magmatic intrusions are also described in terms of structural geology.

Applying the principle of original horizontality and continuity, sedimentary rocks that show tilting or folding indicate deformation caused by compression or tension and by uplift of the strata. In areas such as plateaus, large volumes of rock have been uplifted while preserving the original horizontal attitude. For rocks of marine origin, which comprise much of the sedimentary rock sequence, this involves thousands of meters (feet) of uplift.

Geologic structures range in size from the microscopic features of a fine-grained specimen visible only under high magnification to large geometric forms in a rock mass that may be measured in kilometers (miles). Foliation in metamorphic rocks, flow banding in extrusive igneous rocks, and bedding in sedimentary rocks are examples of structures that can be observed with the unaided eye.

Several related terms are significant in this discussion. *Tectonics* or *geotectonics* is structural geology on a regional scale. Another related subject is rock mechanics, which involves the mechanics of rock deformation with regard to the design and construction of engineering works. Much of the material discussed in Chapter 8 would fall within the scope of rock mechanics. Petrofabrics is concerned with the directional aspects of rock texture (e.g., rock cleavage) and its relationship to the stress field that deformed the rock.

Rock Deformation

Rocks deform under load in the manner discussed in Chapter 8, Engineering Properties of

Rocks. In the Earth's crust, rocks are under high confining pressures and elevated temperatures, thus causing them to be less brittle and allowing plastic deformation to occur (Davis et al., 2012). The general relationship between stress and strain is shown in Figure 11.1. Stress is the applied load per unit area, whereas strain is the deformation per unit length caused by the stress.

For deformation in the elastic range, the rock returns to its original shape when the stress is removed. An example of this type of deformation occurs when earthquake waves pass through rock.

Folding in rocks occurs in the plastic range of the curve (van der Pluijm and Marshak, 2004). Compressive stress slowly deforms the rock but, upon removal of this stress by uplift and erosion, the rock retains the folded shape. Examples of folding can be observed in highway rock cuts and mountainous terrain. Rocks deform into a series of troughs and peaks like a sinusoidal wave trace.

Faulting occurs when the rocks rupture as a brittle material (van der Pluijm and Marshak, 2004; Fossen, 2016). As with folding, on release of the stress, the strain is not recovered.

Folds in Rock

Fold Terminology

Folds are caused by compressional stresses that buckle rock units. The trough or downwarped portion of the fold is called a *syncline* and the crest portion is an *anticline*. If only one direction of dip prevails in a fold system, it is called a *monocline*.

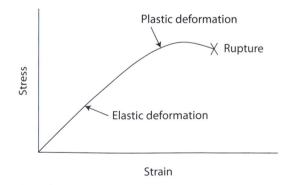

Figure 11.1 Schematic diagram of rock deformation in the Earth's crust.

Figure 11.2 shows anticlines and synclines in cross section view and a photograph of these folds in a mountainous region.

The limb of a fold is the sloping portion that connects the crests and troughs. The limbs dip toward the center of the syncline and away from the center of the anticline unless the folds are overturned.

Another essential part of the fold system is the axial plane. This is an imaginary plane used to divide the fold into two equal, or nearly equal, portions. The intersection of the axial plane with the bedding surface of the rock unit is a line called the axis. As shown in Figure 11.3a, the axis is horizontal, indicating that the fold system trends parallel to a horizontal plane (the Earth's surface). Figure 11.3b shows the axis sloping to the rear of the drawing. This indicates that the fold system is not horizontal but plunges (loses elevation) to the rear. This concept is explained later in the discussion.

The axial plane is used to describe the degree of symmetry of the fold system (Fossen, 2016). In Figure 11.3 the axial plane is vertical, which indicates that the fold is symmetrical. If the axial plane is tilted, with the limbs dipping in opposite directions but not at the same angle, the fold is asymmetrical (Figure 11.4a, see page 222). With further inclination of the axial plane from the vertical, both limbs dip in the same direction to yield an overturned fold (Figure 11.4b), with the bed on the right side overturned. A recumbent fold occurs when the axial plane is essentially horizontal (Figure 11.4c).

Strike and Dip

Before describing the attitude of a fold system in detail, we must first define the terms *strike* and *dip*. Consider Figure 11.5 on page 222 with rock beds inclined in space. Strike is the bearing or compass direction (measured from true north) of the line of intersection between a horizontal plane or bed and an inclined plane or bed. Dip is the maximum slope of an inclined bed, measured with respect to horizontal. When a persistent bed intersects the Earth's surface, it forms a ridge running along the strike of the bed. If the beds are steeply dipping (30° dip or more), they are called hogbacks and can be seen, for example, in the Colorado Rockies west of Denver. Gently dipping resistant beds form cuestas, which dip at angles of less than 10°.

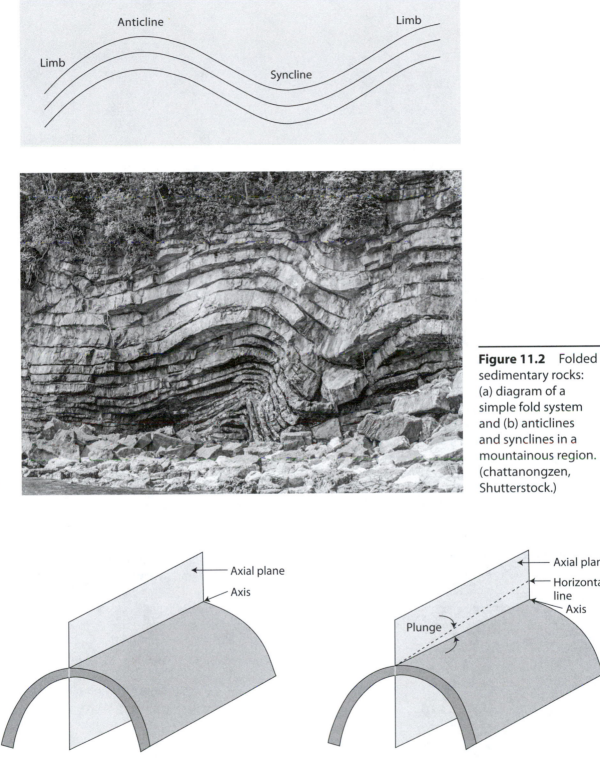

Limb

Anticline

Limb

Syncline

(a)

(b)

Figure 11.2 Folded sedimentary rocks: (a) diagram of a simple fold system and (b) anticlines and synclines in a mountainous region. (chattanongzen, Shutterstock.)

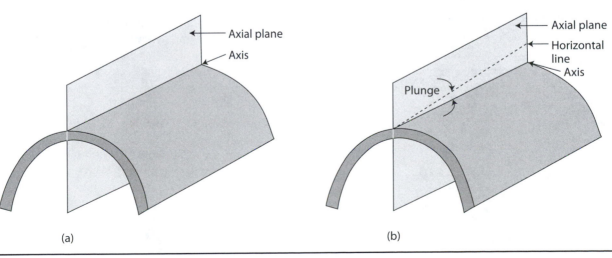

Axial plane

Axis

Axial plane

Horizontal line

Axis

Plunge

(a)   (b)

Figure 11.3   Anticline, with axial plane and axis: (a) nonplunging and (b) plunging to the rear.

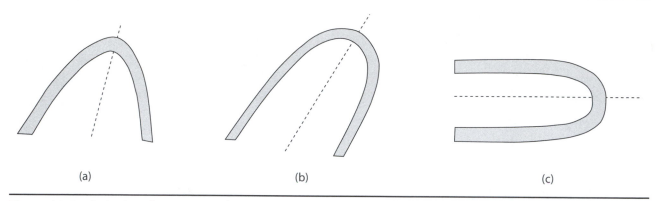

   (a)   (b)   (c)

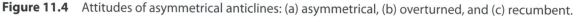

Figure 11.4   Attitudes of asymmetrical anticlines: (a) asymmetrical, (b) overturned, and (c) recumbent.

Strike and dip directions are mutually perpendicular so, for example, a bed dipping to the east will have a north–south strike, usually designated simply as north. By convention, strike and dip directions are measured from true north so that a bed dipping either to the north or to the south would have an east–west strike, which would be designated as N90°E or N90°W. Strike cannot be defined for a horizontal plane; that is, for a plane that has zero dip. Strike and dip are measured in the field by using a Brunton compass and are used to determine the attitudes of structural features. Anticlines and synclines are extensions of the inclined plane concept. By mental construction, segments of planes inclined at different angles can be assembled to yield the fold system.

Strike and Dip Symbols

Symbols are used on a geologic map to designate the strike and dip of beds. The strike and dip symbol looks similar to a T, with the stem indicating the dip direction and the crossbar indicating the strike. The amount of dip is sometimes given in degrees opposite the stem portion. Figure 11.6 shows these symbols for a system of folds in both cross section (side) and in plan (map) views. The lowest numbered bed is always the oldest in these diagrams.

Bed Width Relative to Dip

The width of a given bed can be different in the two limbs of a fold. For example, in section C of Figure 11.7 the width of bed 3 is not the same on opposite sides of the axial plane. On the left side, where the bed is less steep, the width is greater than on the right side where the bed is steeper. This is illustrated in more detail in Figure 11.8. Notice that width $W = t/\sin\theta$, where W is the bed width of the outcrop, t is the thickness of the bed, and θ is the angle of dip (always measured from the horizontal). For vertical beds $W = t$, its minimum width, and for horizontal beds, W becomes infinite. Hence, symmetrical folds will have equal bed widths on opposite sides of the axial plane because the opposing dip angles are the same, whereas asymmetrical folds will have unequal widths.

Apparent and True Dip

A final point concerning the dip of beds involves the variation in the dip angle depending on the line of sight of the observer with respect to the strike direction, as illustrated in Figure 11.9 on page 224. For line of sight A, which is sighting parallel to the strike, the maximum dip or true dip of the bed is observed. For line of sight B, which is sighting perpendicular to the strike, the apparent dip of the bed is zero. As the line of sight increases from 0 to 90°, as measured from the strike direction, the apparent dip reduces from the maximum value to zero. The dip angle θ is shown in

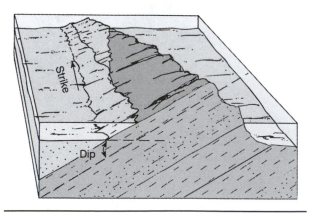

Figure 11.5   Strike and dip of an inclined bed.

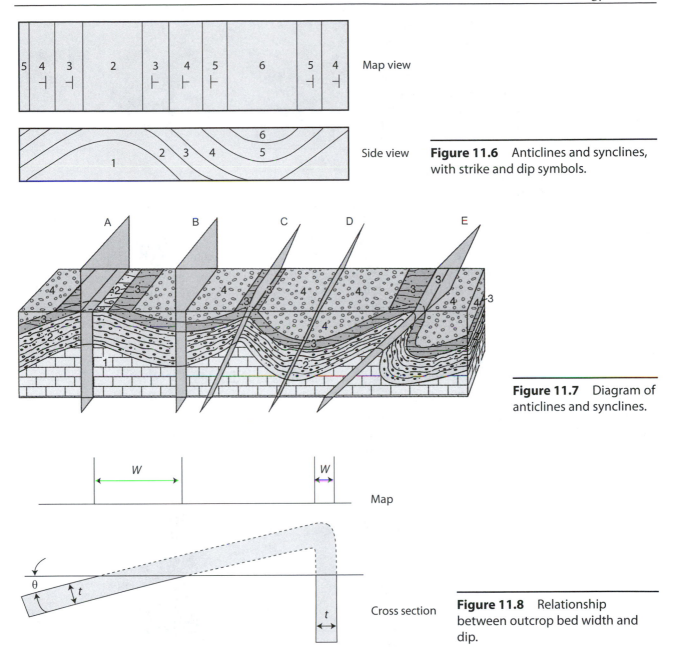

Figure 11.6   Anticlines and synclines, with strike and dip symbols.

Map view

Side view

Figure 11.7   Diagram of anticlines and synclines.

Map

Cross section

Figure 11.8   Relationship between outcrop bed width and dip.

the diagram. The value of the apparent dip in terms of α is given by $\theta_{apparent} = \theta_{true} \cos\alpha$. As shown in Figure 11.9, cosα is measured from the strike direction.

Anticlines and Synclines

First Rule of Anticlines and Synclines

Careful examination of the map view portion of Figure 11.6 provides information that leads to the first rule of anticlines and synclines. The rule states that for the map view of an anticline, the oldest beds are in the center and the beds become progressively younger in each direction. The opposite situation is true for synclines, where the youngest bed is in the center and the beds get progressively older in each direction (Davis et al., 2012; Fossen, 2016). Note that bed 2 is bounded on both sides by bed 3, which is bounded in turn by bed 4, indicating repetition of beds when folding is involved. This rule makes it possible to distinguish between anticlines and synclines in map view, if the relative age of the beds is known, even when strike and dip symbols are lacking.

Figure 11.7 presents a review of some of the information presented on folded rocks. Axial planes are included for reference purposes. Viewing the top portion of the fold system alone in Figure 11.7, the inclination of the axial planes indicates the extent of symmetry of either the anticline or syncline. Also, the age of the beds indicates whether an anticline or syncline is present.

Plunging Folds

Plunging is the term used to indicate the loss of elevation of one or more limbs of the syncline or anticline. Figures 11.10a and 11.10b show the intersections of a plunging syncline and a plunging anticline with a horizontal plane, respectively. The shape and direction of the intersection is important. In Figure 11.10a, the syncline is plunging (losing elevation) to the right and the angle of plunge is shown.

Second Rule of Anticlines and Synclines

The second rule of anticlines and synclines is gained from Figure 11.10. This rule states that, for a plunging anticline, the nose formed by the intersec-

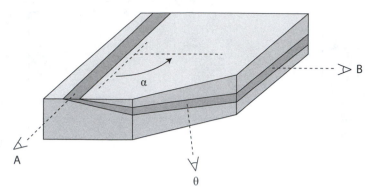

Figure 11.9   Difference between true dip and apparent dip.

tion of the fold system with a horizontal plane points in the same direction as the plunge. The nose of a syncline, therefore, being opposite to an anticline, points in the direction opposite to that of the plunge (Davis et al., 2012). Figure 11.11 further illustrates these details.

Domes and Basins

Three-dimensional fold features also occur in nature. A structural dome is a fold in which the beds slope away from the center in all directions. It is, in essence, a three-dimensional anticline. A structural

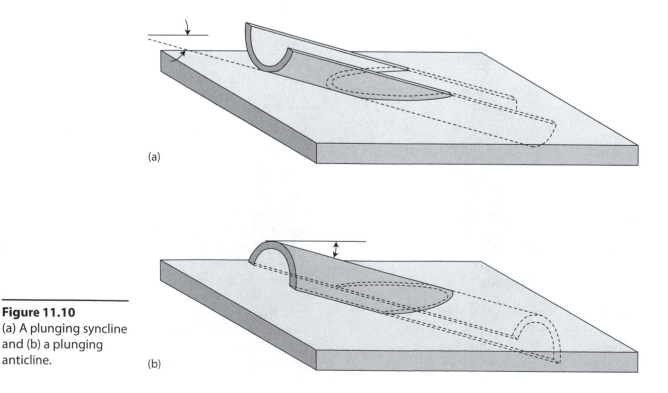

(a)

Figure 11.10
(a) A plunging syncline and (b) a plunging anticline.

(b)

basin has beds that slope inward in all directions. It is a three-dimensional syncline.

Structural Geology Symbols

Included in Figure 11.11a are the standard symbols depicting anticlinal and synclinal axes, along with arrows showing the direction of plunge. Note that the first rule of anticlines still holds for plunging folds. The standard structural symbols used on geologic maps to depict folds, faults, and joints are shown in Figure 11.12 on the following page.

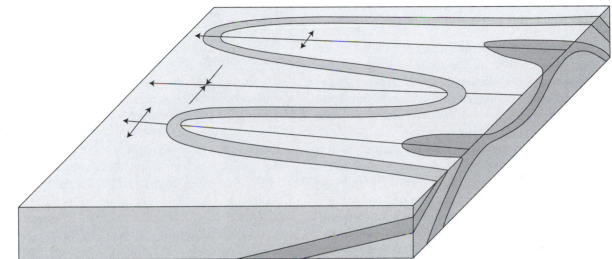

(a)

(b)

Figure 11.11   (a) Block diagram of a plunging anticline and a syncline. (b) Aerial view of an anticline plunging to the left, the same direction as the nose points. An entrenched stream has cut a water gap through the fold. Northern Territories, Central Australia. (Photo by Suzanne Long/Alamy Stock Photo.)

Types of Rock Fractures

Fractures in rock are narrow openings along which the rock mass has lost grain-to-grain contact. Relative movement of the rock blocks on opposite sides of the fracture may or may not have occurred. Based on the extent of movement, three major types of fractures are distinguished: joints, shear zones, and faults.

Joints

Joints are rock fractures along which no movement has occurred parallel to the joint surface. Some displacement perpendicular to the joint may develop because of frost wedging or gravity effects. Similar to bedding planes, strike and dip are used to describe the attitudes of joint planes.

There is a tendency for joints to occur in sets of parallel fractures rather than as a single isolated plane. Therefore, the strike and dip of the joint set is reported along with the joint spacing. Some joints form on release of confining pressures, for example, along the valley wall of a stream or along other steep exposures of rock. These joints are referred to as the valley stress relief joints (Ferguson and Hamel, 1981) and contribute to numerous slope instability problems. Horizontally bedded sedimentary rocks exhibit joint sets that are usually perpendicular to each other, referred to as orthogonal joints. If a horizontal stress regime has been active in a region, two vertical sets of joints, each about 30° from the direction of maximum principal stress, may occur. This is discussed further in a subsequent section of this chapter on the orientation of failure planes in a stress field.

Shear Zones

Shear zones, or shears, are fractures in rock along which some, but not a great deal of, movement has occurred. Most movement is due to the rock slipping to compensate for distortions caused by folding. Typically only a few centimeters (1 in) of movement are involved but some geologists would allow up to 1 m (3.3 ft) of movement and still term the feature a shear zone. Shear zones are typically weak, ground-up areas of rock and soil-like material. They are sites of groundwater flow, are of low strength, and are undesirable areas relative to tunneling, slope excavation,

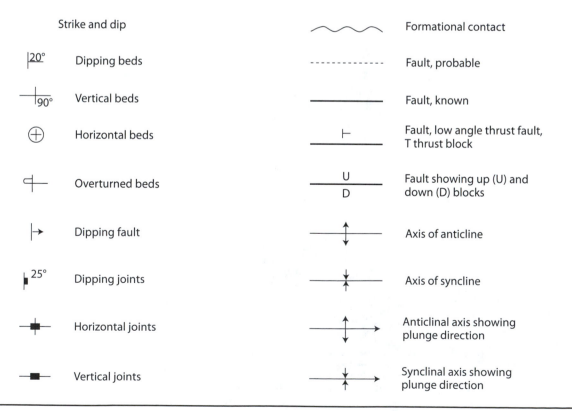

Figure 11.12   Standard structural symbols used on geologic maps.

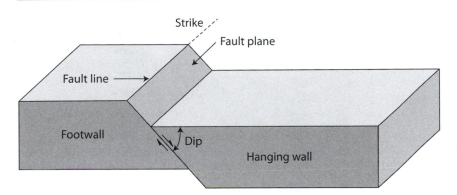

Figure 11.13 Normal fault showing fault terminology.

and foundations for engineering works. They do not appear in persistent sets the way joints do but they tend to be more prominent in certain locations within a mountainous region.

Faults

Faults are fractures along which significant movement has occurred. Displacement may be measured in meters or kilometers (or in feet and miles), typically with a minimum movement of 1 m (3.3 ft) to qualify as a fault.

The attitude of faults is described in terms of strike and dip. The various portions of a fault are labeled in Figure 11.13. The block above the fault plane is called the hanging wall and the portion below it is the footwall. The fault line is the intersection of the fault plane with the surface of the Earth.

Movement along the fault plane can be either (1) primarily along the dip direction (dip-slip faults), (2) primarily along the strike direction (strike-slip faults), (3) a combination of movement in both the strike and dip directions (oblique slip faults), or (4) rotational, in which one block rotates relative to the other (Davis et al., 2012; Marshak, 2013). The fourth possibility is not discussed in this introductory presentation on structural geology.

Dip-Slip Faults

Dip-slip faults can be one of three types: normal faults, reverse faults, and thrust faults. A normal dip-slip fault is one in which the hanging wall has moved down relative to the footwall (see Figure 11.13), whereas a reverse dip-slip fault is one where the hanging wall moves up relative to the footwall (Figure 11.14). Either a vertical compressive force or a horizontal tensile force can cause a normal fault. A low angle (dip angle)

reverse fault is called a thrust fault. A reverse fault can occur because of a horizontal compressive force or a vertical tensile force. Normal faults involve a lengthening of the Earth's crust, whereas reverse faults indicate a shortening of the crust.

Strike-Slip Faults

Strike-slip faults (or wrench faults) occur when the relative movement is, essentially, all horizontal so that the blocks are displaced along the strike direction. Two types of these faults are right lateral and left lateral strike-slip faults. Figure 11.15a on the following page shows a right lateral strike-slip fault. Standing on the front block of the fault and facing the fault plane, one can observe that the opposite side (the hanging wall) has had a relative movement to the right. Figure 11.15b shows a left lateral strike-slip fault. Standing on the front block and facing the fault plane, one sees that the opposite side has a relative displacement to the left.

Oblique Slip Faults

The third type of fault movement discussed herein is the oblique slip faults, also known as translational

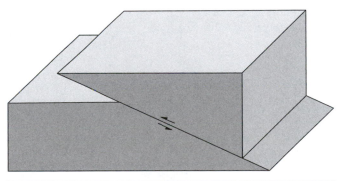

Figure 11.14 Reverse dip-slip fault.

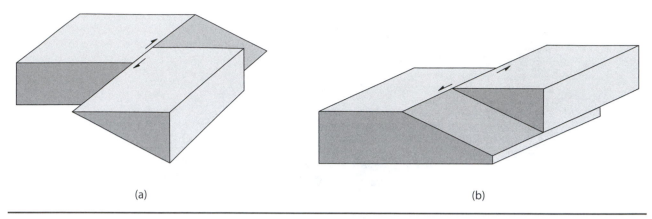

Figure 11.15 Strike-slip faults: (a) right lateral and (b) left lateral.

faults. These have both strike-slip and dip-slip components. In Figure 11.16, one can see that movement has occurred obliquely, moving the blocks apart from point A to point B. This distance is known as the net slip. No rotation of the blocks is involved because lines that were parallel before movement remain parallel after movement. Dip slip and strike slip are the components of the net slip in the strike and dip directions, respectively. Vertical slip (or throw) is the vertical component of the net slip (or of the dip slip), whereas horizontal slip is the horizontal component of the net slip. It is also known as the heave.

For strike-slip faults, it is rather easy to determine the relative movement involved if a marker bed (a well-recognized, persistent layer of rock, such as a coal seam) is present in the fault block. An example is shown in Figure 11.17. Strike-slip faults in nature are more likely to be right lateral than left lateral. The famous San Andreas Fault of the California coastal area is a right lateral fault.

In dip-slip faults it is more difficult to decipher the relative movements because of differential erosion, subsequent to the faulting. Typically the uppermost block is eroded much faster than the lower block, yielding a horizontal surface across them. In such situations, the dip of the marker bed is used to determine the relative movement. Note in Figure 11.18 that the bed dips to the rear of the blocks. If horizontal planes were passed through the block, planes at successively lower elevations would cut the dipping bed further to the rear of the block (because it dips in that direction). Therefore, studying Figure 11.18b, we see that the left block (the footwall) has come from a lower elevation than the right block (the hanging wall) because the bed is displaced further to the rear of the left block. This means that the left block moved up relative to the right block and, consequently, this is a normal dip-slip fault.

Block Faulting

Some additional details concerning fault movements require special consideration. In some areas of the world, for example in the Basin and Range Province of the southwestern United States, *block faulting* is predominant. This involves the movement of three-dimensional blocks of rock upward and downward along steeply dipping fault planes. The upward displaced blocks, known as horsts, provide the ranges and the downthrown blocks, or grabens, yield basins (geographic basins, not structural basins) (Figure 11.19).

AB = net slip
AC = strike slip
CB or AD = dip slip
AE = vertical slip or throw
ED = horizontal slip or heave

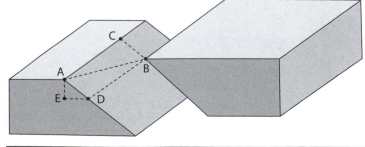

Figure 11.16 Oblique slip or translational fault.

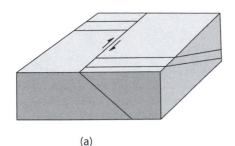

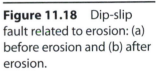

Figure 11.17 Strike-slip movement identification aided by the marker beds: (a) right lateral and (b) left lateral.

(a)   (b)

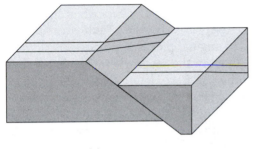

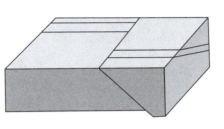

(a)   (b)

Figure 11.18 Dip-slip fault related to erosion: (a) before erosion and (b) after erosion.

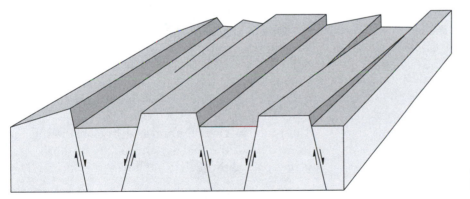

Figure 11.19 Block faulting.

Recognizing Faulting in the Field

Recognizing faults in the field is of great importance because faults are of major concern in engineering construction. The existence of faults should be considered early in the site selection investigation. Surface exposures must be interpreted correctly to discern the presence of faults.

Details about faults are used to recognize them in the field. In general categories, they are as follows: (1) landform features associated with the fault line, (2) abnormal stratigraphic sequences caused by faulting, and (3) features of the fault plane itself.

One of the landform features depicting the presence of faults is the offset or truncation of geologic structures. This can be the offset of a mountain chain or a valley or the sudden cessation of a mountainous feature. Marked changes in elevation are another clue to faulting. These changes might mark the fault line scarp not yet reduced by erosion. The steep eastern face of the Sierra Nevada along the Nevada–California border is an extensive fault line scarp of major proportions. It significantly retarded the westward migration of early settlers and, except for access along a few, narrow passes through the Sierras, prevented transportation by wagons.

Other features that develop along a fault scarp are triangular facets. These occur as a consequence of surface drainage on the fault scarp, yielding a channel that widens in the downhill direction. The area between the drainages yields a triangular shape when viewed looking toward the mountain (Figure 11.20).

Finally, there is a landform feature of strike-slip faults that disrupts the normal, smooth profile of a stream valley. It occurs where the fault intersects the valley to form a sag pond, a swampy area that interrupts the stream profile.

An abnormal stratigraphic sequence related to faulting usually involves the repetition or the omission of the normal beds within that sequence. This is illustrated in the map view presented in Figure 11.21. It shows a series of rocks with a normal sequence of folded sedimentary rocks, disrupted at two locations by faulting (note the F designations).

In Figure 11.21 a repetition of beds is shown on the left side of the diagram and an omission of beds on the right side. The structure is a syncline, as indicated by both the age of the beds (note the numbering system) and the strike-dip symbols. In this example, no change in attitude of the beds occurs at the two fault lines (F designation). In some situations, a change in dip might also occur at the fault line.

Features along the fault plane itself, seen in plan view or in cross section, can help establish the existence of the fault. *Slickensides* are polished rock surfaces, striated in the direction of movement, that occur as a consequence of movement along the fault plane (Marshak, 2013; Fossen, 2016). In some rocks,

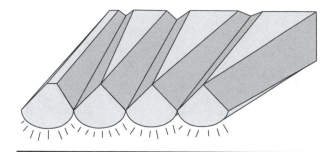

Figure 11.20   Triangular facets depicting fault scarps.

fault displacement causes *gouge* to form. This is finely ground rock dust that occurs when hard rock blocks move against each other. Fault breccia consists of broken angular pieces of massive brittle rock that break off from the mass when the blocks slide past each other and are, later, cemented together. Fault breccia and gouge may be many meters (tens of feet) thick, yielding a fault zone. When marker beds are present within the fault blocks, a drag of the beds at the fault plane may occur, which indicates fault movement (Figure 11.22).

Folds and Faults Combined

One final analysis remains for this discussion on structural geology. It combines plunging folds with dip-slip faults and supplies a final challenge to the student's ability to visualize the relative movements involved. Figure 11.23 shows these two in combination and the caption provides a detailed description. The problem can be unraveled by considering the front view of the fold system relative to horizontal planes passed through it, as shown in Figure 11.24.

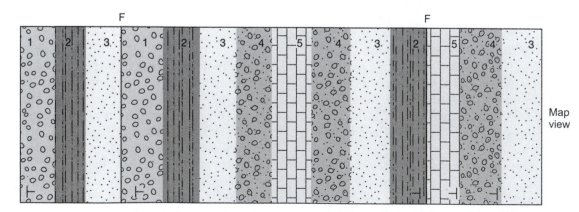

Figure 11.21   Faulting indicated by repetition and omission of beds.

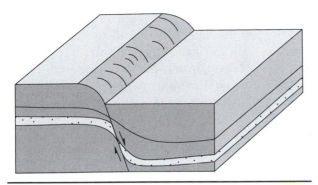

Figure 11.22   Drag of beds along a fault plane.

For the anticline, observe that, as you proceed lower in the cross section, the outcrop of the dipping bed will move further apart, that is, it diverges. Therefore, referring to the top view of Figure 11.23, on the back block the bed is closer together than it is on the front block. Therefore, the bed had to have originated at a higher elevation and, consequently, the back block has moved downward.

The direction of plunge (toward the front), shown in the side view, is a consequence of the second rule of anticlines (on plunging folds). Note that the nose for the anticline points in the same direction as the plunge. The symbols on the top and side views are examples of information supplied for folds and faults on geologic maps.

Figure 11.25 on page 232 is an example of a nonplunging syncline and it completes the details for combinations of folds and dip-slip faults. Note that the beds of the syncline converge with depth. Therefore, because the beds are closer together in the front block, that block had to have originated at a lower depth. Consequently, the front block has moved upward relative to the rear block, yielding a reverse fault.

Direction of Stress and Fault Orientation

The steepness of dip for the fault plane is a consequence of the stress directions that led to faulting. For this purpose, a triaxial stress field can be assumed with the maximum, minimum, and intermediate principal stresses acting mutually perpendicular to each other. This can be shown in a two-dimensional view,

as presented in Figure 11.26 on page 232. The maximum principal stress σ_1 is in the vertical position; σ_3, the minimum principal stress, is horizontal; and σ_2, the intermediate principal stress, lies perpendicular to the page.

The dotted lines represent conjugate planes of failure that will occur at approximately 30° from the direction of maximum principal stress (σ_1) for isotropic rocks (physical and engineering properties the same in all directions) without obvious planes of weaknesses, allowing for an analysis of normal and reverse faults.

For normal faults, the direction of maximum principal stress is close to vertical because excess overburden stresses give rise to this type of faulting. This yields a failure plane 30° from the vertical or 60° from the horizontal. For this reason, most normal faults occur as high-angle faults; that is, faults typically steeper than 45° as measured from the horizontal.

Reverse faults are usually low-angle faults because nearly horizontal compressional stresses (σ_1) are typically the cause of such faults. A high-angle reverse fault suggests that an uplift force, oriented at a steep angle, is the cause of such faulting.

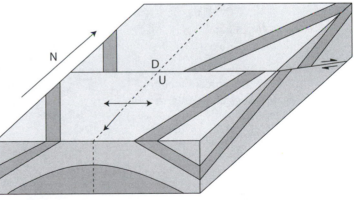

Figure 11.23   South plunging symmetrical anticline cut by an east–west striking normal fault dipping northward.

Figure 11.24   Front view of anticline from Figure 11.23, with imaginary horizontal planes.

A word of caution is in order here, relative to the solutions to exercises on faults at the end of this chapter. It is inappropriate to assume that the type of dip-slip fault (normal versus reverse) can be recognized simply on the basis of the steepness of the dip angle of the fault plane. This would render it unnecessary to work out the details of the dipping beds. Specific details shown on the block diagrams, in keeping with the discussions on bed displacements in this chapter, should be used to determine block movement.

When rocks are anisotropic, the orientation of the weakness planes can influence the direction of breakage or faulting. These weaknesses are known as *s* planes, which consist of prominent bedding in sedimentary rocks, foliation in metamorphic rocks, or flow banding in extrusive igneous rocks.

Laboratory studies have shown that if the *s* planes are oriented within 60° of σ_1, failure will occur along, or be controlled to some extent by, the *s* planes. For *s* planes that are oriented more than 60° from σ_1, the *s* planes have little influence on the direction of failure and, indeed, the failure plane will occur about 30° from the load (σ_1) axis. This is illustrated in Figure 11.27. This can apply to rock faulting, as well. Thinly bedded, horizontal, sedimentary rocks can develop bedding plane faults when earth processes impose horizontal, compressive stresses on them. Low-grade metamorphic rocks, like argillites, may undergo considerable bedding plane slippage during deformation of adjacent mountainous regions.

This suggests that major forces from engineering constructions should not transfer to layered rocks in a direction parallel or nearly parallel to these layers, as slippage and failure may result. Therefore, for example, the thrust of an arch dam should not align in a direction parallel or subparallel to a weakness zone in the rock. Steeply dipping rock strata are also subject to slippage when gravity loads are placed on them. This subject will be examined in greater detail in Chapter 15, Slope Stability and Ground Subsidence.

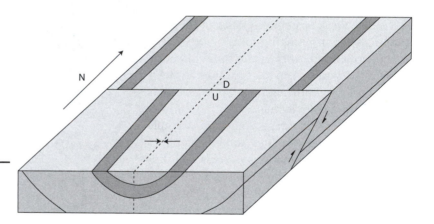

Figure 11.25 Symmetrical syncline cut by an east–west striking reverse fault dipping southward.

σ_1 = Maximum principal stress

σ_3 = Minimum principal stress

σ_2 = Intermediate principal stress (acting into the page)

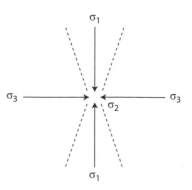

Figure 11.26 Directions of principal stress.

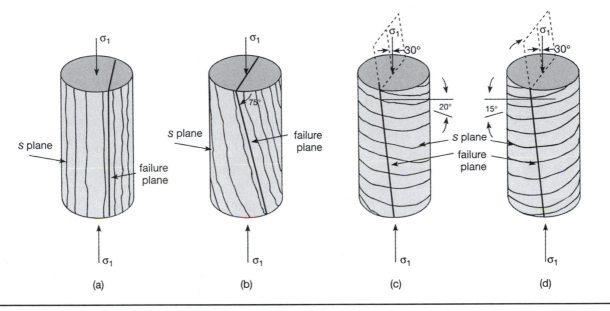

Figure 11.27   Influence of *s* planes orientation on failure plane location in test specimens.

Engineering Considerations of Structural Geology

1. A basic requirement of any geological study related to engineering construction is to determine the specific geologic features for the study area. In addition to soil and rock descriptions, detailed information about the existing geologic structures is required at both local and regional scales. Chapter 20, Physiographic Provinces, provides a detailed description of the different regional geologic structures of the United States.

2. Bedding planes, foliation, joints, faults, and shear zones, jointly referred to as discontinuities, represent planes of weakness in a rock mass. Failure in a rock mass tends to occur along these planes of weakness.

3. The orientation of discontinuities, as defined by strike and dip, has a major effect on the stability of engineering structures and rock excavations. Discontinuities dipping into an excavation at an angle steeper than the friction angle along the discontinuity can promote sliding failures.

4. Rocks deformed by compressive forces yield a system of folds called anticlines and synclines. Strike, dip, and plunge angle are used to describe their position in space. Depending upon their size and attitude with respect to an engineering structure, folds can contribute to both instability and drainage problems. For example, water tends to accumulate toward the center of a syncline; if an underground excavation crosses such a syncline, significant groundwater seepage problems may arise.

5. A combination of plunging anticlines and synclines cut by a dip-slip fault provides a challenging problem to decipher. Considering the relative movements of blocks should prove helpful.

6. Identifying the presence of faults at a construction site is of major importance. Faults can be recognized in the field by landform features indicating the occurrence of displacement. Other features that can help identify faults include the presence of slickensides, gouge, and breccia along the fault plane. On a regional scale, aerial photographs, LiDAR imagery, or unmanned aerial vehicle (UAV) scans can be very helpful in fault identification.

7. After identifying a fault at a construction site, the next step is to determine if the fault is active or not and whether there has been displacement within the last 10,000 years. This is an important consideration for projects like dams, tunnels, and nuclear power plants.

Exercises on Structural Geology

In the following exercises, only strike-slip faults and dip-slip faults are considered. Oblique slip faults are not included. This apparent simplification is justified because most faults in nature show predominantly strike-slip or dip-slip movement. Exercise Figure 11.1 provides a reference of the map view and a block diagram, which will assist in your completion of the exercises.

1. *Folds.* See Exercise Figure 11.2 on pages 236 and 237. Complete the map and front views for (a) through (l). Locate an axial *plane* and *describe* the fold. Bed 1 is the oldest. *Be neat.*

2. *Faults.* See Exercise Figure 11.3 on page 238. Complete the front views for (a) through (f). Identify the structures. Assume dip-slip faults unless otherwise noted.

3. *Dipping Beds.* See Exercise Figure 11.4, page 239.

   a. What is the true thickness of bed Q?

   b. How deep would you have to drill to reach the top of the bed at A?

   c. What is the apparent thickness of the bed that would be found by drilling at A?

4. A hill slopes 20° to the west. At an elevation of 303 m (1000 ft), the top of a sandstone bed, which strikes north–south and dips 30° to the east, is found. Fifteen meters (50 feet) down the slope from the 303-m (1000-ft) elevation, the bottom of the sandstone is found. What is the true thickness of the sandstone? At the 394-m (1300-ft) elevation, how far would you have to drill to reach the sandstone? Sketch the problem using a scale of 2.5 cm (1 in) = 61 m (200 ft). Calculate using trigonometry.

5. *Folds and Faults Combined.* See Exercise Figure 11.5 on page 240. Complete the views and describe completely.

6. Why are sedimentary rock sequences particularly useful in determining the structural geologic history for an area? Relate to the principle of original horizontality.

7. Why would rocks be less brittle when loaded at high temperatures and pressures over long periods of time than when a fast-loading rate is used in the laboratory at room temperatures and pressures? Explain.

8. Distinguish between a hogback and a cuesta.

9. Give the first and second rules of anticlines. Why is it easy to remember how these apply to synclines?

10. How are faults recognized in the field? Give the three general categories and provide examples of each.

11. A coarse gneiss is being tested in uniaxial compression. If the foliation is 10° from the load axis, where is failure likely to occur? If it is 80° from the load axis, where is failure likely to occur? Which of these two tests is likely to provide a greater unconfined compression value? Explain.

12. The Basin and Range Province in the southwestern United States has undergone extensive block faulting. Relate the basins and ranges to the terms *horst* and *graben.* This is a low rainfall area of the United States. What kind of weathering is likely to prevail here? Why?

13. Refer to Figure 11.9 on the true and apparent dip of beds. If a bed strikes N30°E and has a dip of 58°, what would be the apparent dip along a line of sight of N45°E? Give the equation and provide the numerical answer.

14. An anticline contains a bed that is 24 m (80 ft) thick. The dip angle to the east is 30° and the dip angle to the west is 75°.

   a. What type of fold is this? Symmetrical, asymmetrical, overturned, or recumbent?

   b. What is the width of outcrops along a horizontal plane on the east side? On the west side? Show calculations.

15. A coal seam strikes N80°W and dips 5° to the southwest. Depth to the top of the coal is 18 m (60 ft) at point X.

   a. Assuming a horizontal land surface, how far along the direction perpendicular to the strike will you need to travel before the depth to the coal seam is 30 m (100 ft)? A sketch should prove helpful.

   b. What is this direction perpendicular to the strike from point X?

REFERENCES

Davis, G. H., Reynolds, S. J., and Kluth, C. F. 2012. *Structural Geology of Rocks and Regions* (3rd ed.). New York: John Wiley.

Ferguson, H. F., and Hamel, J. V. 1981. Valley Stress Relief in Flat Lying Sedimentary Rocks (pp. 1235–1240). Proceedings for the International Symposium on Weak Rock, September 21–24, Tokyo, Japan.

Fossen, H. 2016. *Structural Geology* (2nd ed.). New York: Cambridge University Press.

Marshak, S. 2013. *Essentials of Geology*. New York: W. W. Norton.

van der Pluijm, B. A., and Marshak, S. 2004. *Earth Structure* (2nd ed.). New York: W.W. Norton.

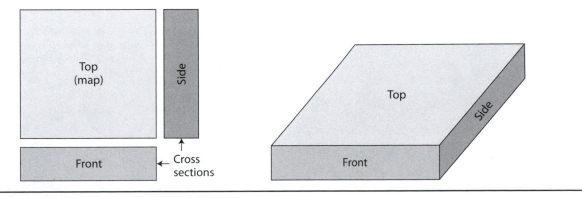

Exercise Figure 11.1

(a)

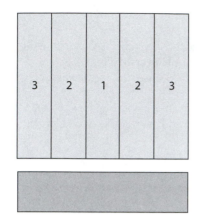

(b)

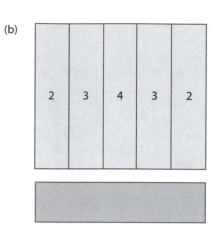

(c)

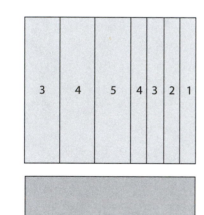

(d)

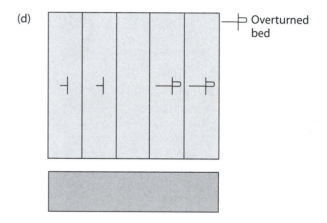

Overturned bed

(e)

No plunge.

(f)

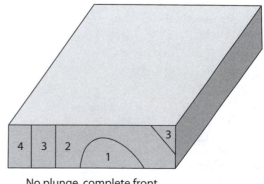

No plunge, complete front and side view.

Exercise Figure 11.2

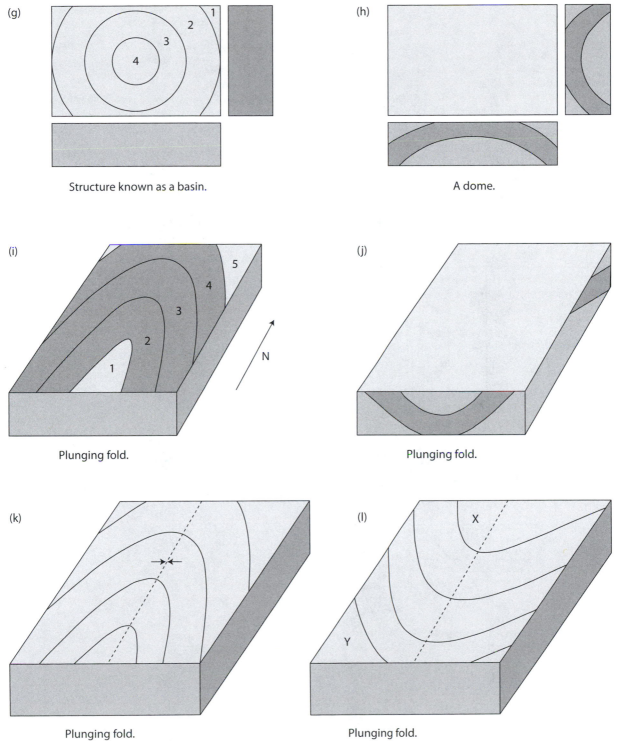

(g) Structure known as a basin.

(h) A dome.

(i) Plunging fold.

N

(j) Plunging fold.

(k) Plunging fold.

(l) Plunging fold.
Bed X is older than bed Y.

X

Y

Exercise Figure 11.2 (cont.)

(a)

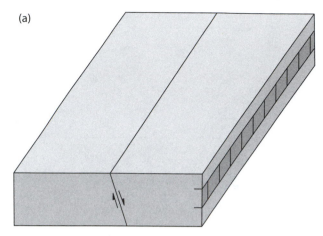

Bed is horizontal. Draw bed
on both sides of front view.

(b)

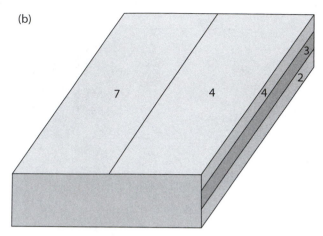

Beds are horizontal,
reverse fault.

(c)

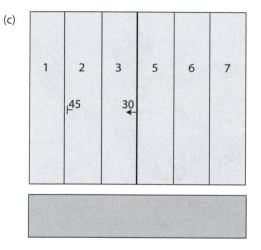

(d)

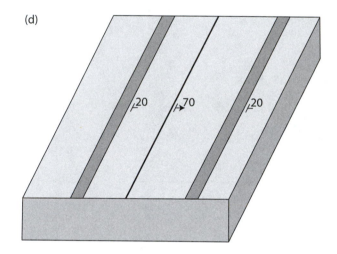

(e)

(f)

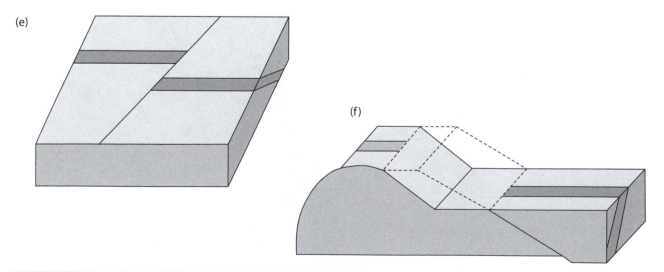

Exercise Figure 11.3

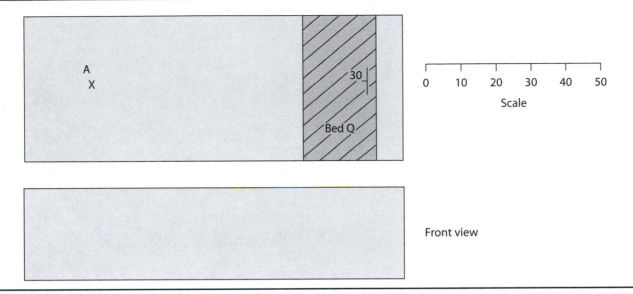

A
X

30

Bed Q

0   10   20   30   40   50

Scale

Front view

Exercise Figure 11.4

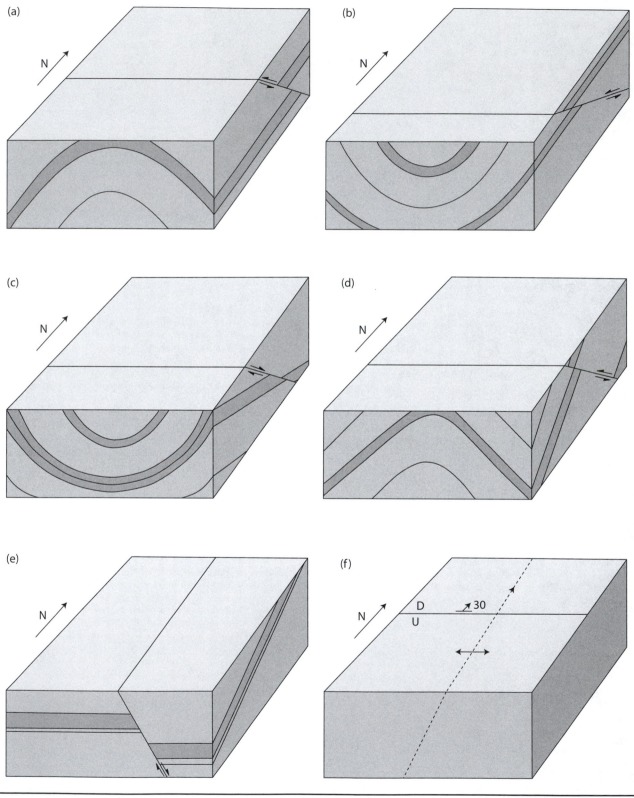

Exercise Figure 11.5

RUNNING WATER AND RIVER SYSTEMS

Chapter Outline

Hydrologic Cycle and Rainfall Equation

Following rainfall or snow melt, water runs from slopes in thin sheets and then collects into rills, gullies, streams, and rivers to arrive, finally, at the ocean. From the sea, it returns to the atmosphere by evaporation and transpiration (evaporation through plants) to begin the pattern once more. This process is known as the hydrologic cycle. The hydrologic cycle, illustrated in Figure 12.1 on the following page, also involves water infiltration into the Earth and groundwater movement which, eventually, returns the water to surface streams or the ocean. The rainfall equation relates these factors in the following way:

$$\text{Rainfall} = \text{Evaporation} + \text{Transpiration} +$$
$$\text{Infiltration} + \text{Runoff}$$

$$(\text{Eq. 12-1})$$

Under natural conditions, the greatest part of rainfall is returned directly to the atmosphere by evaporation and transpiration, jointly referred to as evapotranspiration. A smaller amount flows away on the surface and is known as runoff; the remainder, the smallest portion, soaks into the soil by infiltration.

For humid nonurban regions in the United States from the Mississippi River eastward, 60 to 80% of precipitation returns to the atmosphere by evapotranspiration. In arid to semiarid regions of the western United States, this may reach an excess of 95%. Some 10 to 25% runs off the land surface into stream systems in humid areas with as little as 2 to 5% ending up as runoff in arid regions. The remainder, about 10 to 20% in humid regions, but as little as 1% in arid regions, infiltrates into the ground to replenish the groundwater system (Hamblin and Christiansen, 2003).

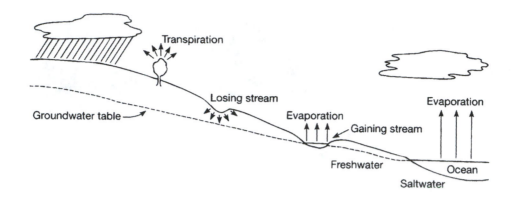

Figure 12.1   The hydrologic cycle.

Factors Influencing Runoff

Several factors influence the percentage of precipitation that becomes runoff in a local area: (1) relief, (2) surface infiltration, (3) vegetative cover, and (4) surface storage (Charlton, 2007). Relief is the difference in elevation between the highest and lowest points within an area of interest. Areas with higher relief have steeper terrain, which encourages more runoff and less infiltration. In steep terrains, the excess runoff can be controlled best by increasing the vegetative cover.

Surface infiltration is related to the permeability of the surface material, whether it is soil or bedrock, and to the pattern of precipitation. Snow versus rainfall is one consideration because snow may melt slowly, allowing a greater opportunity for infiltration. The pattern of rainfall in a storm also has an effect. The infiltration rate is higher in a dry soil so that early in a storm, greater water uptake occurs. As rainfall continues, the pore system becomes temporarily saturated and greater amounts of water must run off. Intense storms will drop more water than can be absorbed directly and, therefore, more runoff occurs. A low intensity, long duration storm will result in a higher amount of initial infiltration compared to a high intensity storm of equal water volume, but will eventually experience a reduced infiltration rate because of soil saturation. The permeability of the soil is determined by the size and nature of its constituent particles. Coarser materials, such as sands and gravels, have a higher permeability than does silt, which, in turn, has a considerably higher value than clay. Unweathered bedrock or rock with a thin soil cover has a low permeability and

infiltration that may be close to zero. The condition of the soil affects permeability and, therefore, infiltration. Soil saturated by previous rains can absorb only small amounts of additional water. Frozen soil is essentially impermeable and major flooding occurs when snow melts rapidly during a sudden warming trend prior to thawing of the soil.

Vegetative cover promotes infiltration and transpiration and reduces runoff. Soil erosion is a consequence of runoff, so vegetative cover should be maintained to prevent loss of topsoil and the subsequent pollution of streams by eroded sediment. In grading operations on earth-moving projects, only the minimum area should be denuded of vegetation at any one time. This includes cuts for road construction and site preparation for housing tracts, shopping areas, or industrial development.

Surface storage or ponding allows water to collect for a time rather than running off immediately. This reduces the peak level of flow and the erosional effects of runoff. It also allows more time for evaporation and infiltration, both of which reduce the total amount of runoff that occurs.

Before leaving the rainfall equation and its various relationships, the contrast between natural conditions and human-induced situations requires attention. This involves the subject of urban hydrology (Baker et al., 2008). Cities contain many hard surfaces including streets, sidewalks, rooftops, and parking lots and, as a consequence of their construction, both vegetative cover and infiltration are significantly reduced or eliminated. Runoff increases from a value of about 15% for a humid area east of the Mississippi River to approximately 60 to 70% in an urban area. Although there may be few soil areas exposed and erosion may not be significant, great quantities of runoff must be accommodated during times of heavy rainfall. Many cities have inherited a sewer system that collects both stormwater runoff and sewage. Sewage treatment plants typically cannot accommodate this extreme volume of water during heavy rainfall periods and, consequently, overflow points in the sewer system are used to reduce the volume of flow. The overflow water mixed with sewage is dumped directly into streams without treatment and this accounts for one of the major contributions to stream pollution in large cities.

The Chicago Deep Tunnel Project, known also as the Tunnel and Reservoir Project (TARP), addresses this problem. This system of collectors and tunnels brings the sewer overflow water from the city and nearby suburban areas into an enlarged sewage storage facility, which can accommodate the excess runoff during times of peak rainfall. The large volume of water is pumped steadily to the sewage treatment plant for treatment, allowing for maximum efficiency rather than the normal, daily or hourly fluctuation of sewage flow.

The construction for Phase I of this project, consisting of 176 km (109.4 mi) of tunnels, ranging from 2.7 to 10 m (9 to 33 ft) in diameter and 46 to 106 m (150 to 350 ft) in depth, started in 1975 and was completed in 2006. It can store up to 2.3 billion gallons of sewage. For Phase 2, the reservoir part, an additional 3.1 billion gallons of sewage can be stored in the Thornton Quarry. Another reservoir near O'Hare Airport and the McCook Reservoir are anticipated to become operative by 2020. Upon completion around 2030, the TARP system will have a storage capacity of 17.5 billion gallons of sewage. Figure 12.2 illustrates TARP.

Types of Runoff

The various types of runoff include: (1) laminar flow, (2) turbulent flow, and (3) jet or shooting flow.

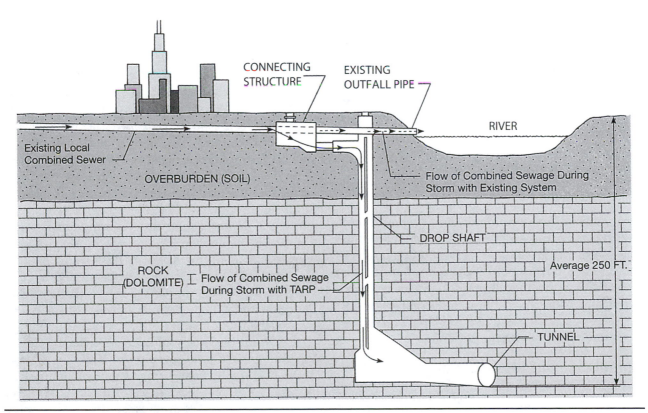

Figure 12.2   Diagram for Chicago Deep Tunnel Project for water reclamation from combined sewers.

Laminar Flow

When water first accumulates on the Earth's surface, it flows in thin layers across a sizable surface in the form of sheet runoff. The water moves slowly in discrete layers flowing parallel to the base. This type of movement is known as laminar flow because water particles move in nearly straight lines. Sheet runoff occurs on cultivated fields where row crops cover only a portion of the soil. Where local irregularities occur in the surface, a surge of flow develops and erosion takes place to remove soil particles. This erosion is called sheet wash and it accounts for the loss of soil over large areas when bare soil is exposed. Erosion of topsoil from farm fields has been a problem of major concern in many areas of the United States. A developing farm practice being encouraged by the Natural Resources Conservation Service (NRCS) is no-till farming. As suggested by the name, traditional plowing no longer occurs; instead, seeds are planted through the stubble of past harvests. First introduced for fields on steeper slopes, it is now practiced on an increasing percentage of farmland in the United States.

Turbulent Flow

Within a short distance, sheet runoff becomes concentrated into a confined channel. First, it forms rills then gullies, creeks, streams, and, finally, rivers. The rate of flow increases in channels and, because of this increase and the roughness of the channel boundary, the straight-line flow is replaced by swirls and eddies. This is called turbulent flow, which is effective in eroding and transporting soil.

Jet or Shooting Flow

Extremely high water velocities are associated with rapids and waterfalls. Called jet or shooting flow, these high velocities are extremely erosive, even in bedrock channels. Although not by design, jet flow may occur in spillways for dams when large volumes of water pass through the spillway gates. Rapid erosion of concrete in the stilling basin of the dam can be a consequence.

Intermittent and Perennial Streams

Not all stream channels carry water year-round. Many smaller streams, termed *intermittent,* carry water only during the wet season or immediately after significant rainfall. In arid regions many of the stream channels run dry, including, in some cases, major streams. In humid regions, the rivers flow year-round, as do many of their primary tributaries, the larger creeks. This occurs even if several weeks separate significant storms because the groundwater table feeds these streams. Streams flowing year-round are known as permanent or *perennial* streams and are designated on topographic maps by solid blue lines. Intermittent streams are shown by dashed blue lines.

Permanent streams are classified as gaining streams because they gain water from the groundwater reservoir. Intermittent streams provide water to the groundwater reservoir as they flow along their channels; consequently, they are termed losing streams. This relationship is illustrated in Figure 12.1.

Stream Flow Terms

A list of the most significant stream flow terms includes the following: velocity, gradient, discharge, flooding, stream hydrograph, load, stream views or profiles, and stream economy (Brutsaert, 2005; Hendriks, 2010).

Velocity

Stream velocity is usually measured in meters per second (or feet per second) because of the convenient values that are obtained using these units. Stream velocity can range from a low of 0.15 m/sec (0.5 ft/sec) to a high of 9.1 m/sec (30 ft/sec). The low velocities are typical of large rivers flowing under low gradients, whereas high velocities are typical of mountain streams flowing under steep gradients and probably in bedrock or boulder strewn channels.

The velocity of a stream is affected by a number of factors, including (1) gradient or slope of the water surface, (2) the nature of the stream bed, that is, its roughness and erodibility, (3) the discharge or volume of flow per unit time, and (4) the load, that is, the material transported by the stream (Brutsaert, 2005; Hendriks, 2010).

Streams do not maintain a constant velocity throughout their cross sections, but flow is fastest where the frictional drag is at a minimum. In a straight portion or reach of a stream, the water flows fastest in the center of the cross section just below the water surface. The surface velocity is approximately

0.8 times the average velocity of the cross section. Figure 12.3a illustrates this point.

Figure 12.3b shows the map view of a river meander. Velocity is greater near the outside of the bend and at a minimum on the inside of the bend. Therefore, active erosion occurs on the outside of the bend and deposition occurs on the inside portion.

Gradient

Gradient is the slope of the water surface or, in general, the slope of the stream bed. Measurements show that the gradient of a stream decreases proceeding downstream from the headwaters toward the mouth. This yields a long profile, which is concave up or concave to the sky.

Stream gradients are expressed in meters per kilometer (feet per mile) and may be as high as 38 to 57 m/km (200 to 300 ft/mi) in the upper reaches of a mountainous terrain to as low as 0.02 to 0.1 m/km (0.10 to 0.5 ft/mi) for large rivers flowing on wide alluvial floodplains. The Missouri River, for example, has a gradient of 3.8 m/km (20 ft/mi) near its headwaters but only 0.2 m/km (1 ft/mi) near its confluence with the Mississippi River at St. Louis. The gradient is determined using a topographic map by locating two points where contour lines cross the stream, finding the vertical difference by subtraction, and measuring the distance of travel between the two points. The drop in elevation divided by the travel distance yields the gradient.

Discharge

Discharge, the quantity of stream flow, is measured in cubic meters per second (cubic feet per second [cfs]) for streams but for water supply and sewage treatment purposes it is typically measured in millions of gallons per day (MGD) (0.028 m^3/sec = 1 ft^3/sec = 0.646 MGD). A small stream may vary from a minimum of several m^3/sec (cfs) to a maximum of several hundred m^3/sec (cfs) during periods of flooding. A major river may have an average of 42.4 m^3/sec (1500 cfs) discharge or more. During and after construction of a dam across a river, discharge equal to or greater than the established minimum discharge for that stream (determined as an average over a number of years) must continue. The minimum discharge is also of interest regarding sewage treatment facilities because it indicates the low-flow conditions and, consequently, the minimum amount

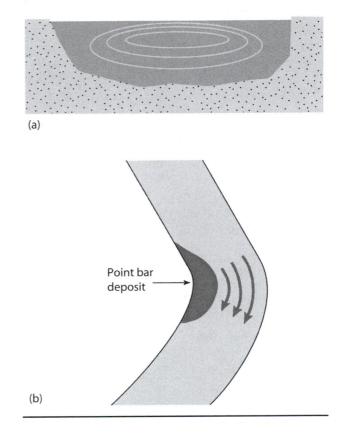

(a)

(b)

Figure 12.3 Variations in stream velocity: (a) cross section showing velocity contours, with velocity increasing toward the center and (b) map view of stream velocity variation around a river bend with length of the arrow indicating the magnitude of velocity.

of water available during the dry season for diluting treated effluent. The maximum discharge, of course, is of interest relative to flooding problems.

For permanent streams, discharge increases downstream due to the additional water supplied by tributary streams and the input of water from the water table.

Flooding

Flooding is one of the major geologic hazards that occurs in nature. A flood marks the highest stage of stream flow. During periods of intense and repeated rainfall, streams swell to bank full levels and then may expand across their floodplains.

Recurrence interval is the time interval during which a flood of a given magnitude has the probability of occurring only once. The 50-, 100-, and

500-year recurrence intervals for flood levels are established from historical data. They indicate the recurrence interval for a flood elevation with one occurrence in 50 years, one in 100 years, and one in 500 years, respectively. The 100-year flood elevation is commonly used as a basis for minimum elevations for building grades and major highways and are incorporated into building codes. Bridges and railroad grades may be established at the 500-year flood elevation or higher.

The flood of record is another item of information recorded for major rivers at designated locations along the river's course. This includes the date and highest elevation given as a level above the local flood stage (Herschy, 2008). In the Midwestern areas of Ohio and Indiana, the floods of record occurred in 1913 or in 1937, when the maximum flood levels were reached. These situations occur during periods of catastrophic precipitation when river levels continue to rise as the extended period of rainfall persists.

Record Flood of 1993

The Mississippi and Missouri River systems at St. Louis, Missouri, had a record flood in summer 1993. The Missouri River has its confluence with the Mississippi River only a few miles north of St. Louis. Intense rainfall on the upper Mississippi River drainage area near Minneapolis continued with an extended period of rainfall in Iowa and western Missouri on the Missouri River system. Flooding began when the water reached 9 m (30 ft). Flood crests from both the Mississippi and Missouri Rivers arrived in St. Louis about the same time. On August 1, 1993, the Mississippi River reached its highest record level in St. Louis, the 15.1-m (49.6-ft) flood stage. The prior record, 13.2 m (43.2 ft), occurred on April 28, 1973.

The floodwall, encompassing a major part of the city along the west bank of the river, extends to 15.8 m (52.0 ft) above flood stage. Therefore, the river came to within 0.73 m (2.4 ft) of topping the floodwall. Many other cities were not as fortunate; for example, Des Moines, Iowa; Alton, Illinois; and Ste. Genevieve, Missouri, suffered major flood damage when levees were overtopped by the flooding rivers.

Floodwater backed up major drainages along the Mississippi River in St. Louis. Many roads, including a number of federal highways, flooded. Bridges along both the Mississippi and Missouri Rivers were closed by flooding, which greatly complicated commuter traffic across the rivers. Locks along the Mississippi River were opened to equalize the water level and barge traffic was curtailed for an extended period. Fortunately, the interstate highway bridges connecting St. Louis and Illinois stood above the flood level so that these busy arteries continued to carry traffic.

The River des Peres, a tributary to the Mississippi River in south St. Louis, became bank full then overflowed onto residential neighborhoods nearby. Several major city streets with bridges across the River des Peres were closed by floodwaters. Volunteers built extensive sandbag barriers and the National Guard was called out to prevent vandalism and looting. Propane tanks from a flooded industrial plant floated into the floodwaters, causing the area to be evacuated and utilities were shut off in nearby neighborhoods to prevent any casualties should an explosion occur. Fortunately, authorities were able to secure the tanks and no additional problems ensued. As the water receded, the clean up of mud and debris provided a major challenge. The river stood above the 9.1-m (30-ft) flood stage for 121 days, breaking the 1973 record of 77 days. Figure 12.4 shows images of the area around St. Louis, Missouri, in 1991 and during summer 1993.

Stream Hydrograph

A stream hydrograph shows the variation of stream discharge with time. The peak discharge represents a flood. Figure 12.5 shows that urbanization in an area can increase the runoff significantly and decrease the lag time (time interval between precipitation and buildup of the flood stage), affecting stream velocity, discharge, and channel characteristics.

Load

Load is both solid and dissolved material transported by streams. The ability to erode material and to transport load increases as velocity increases. This is discussed in detail in a later section of this chapter.

Three Views of a Stream

Three views can be observed for a single stream channel: the channel pattern, the long profile, and the cross profile. The channel pattern involves the map view of the stream. A single stream channel can be straight, meandering, or braided. Figure 12.6 on page 248, a photograph taken by an astronaut from the International Space Station, shows features of a

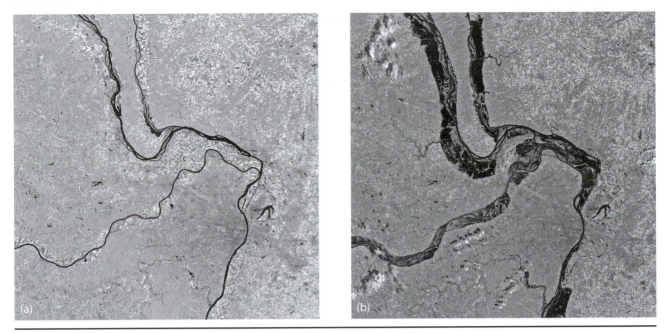

Figure 12.4 (a) St. Louis, Missouri, area, August 1991. (b) St. Louis, Missouri, area, August 1993. (Earth Observatory, NASA, Great Flood of the Mississippi River, 1993.)

meandering stream. Figure 12.7, also on page 248. is a photograph of a braided stream. These features are discussed later in this chapter.

The long profile is a cross section in which the slope of the stream is seen in its profile, or the upstream to downstream view. The cross profile is a cross section taken at right angles to the direction of flow showing the configuration of the stream at a specific point along the bank. Figure 12.3a is a cross profile.

The Economy of Stream Flow

A relationship exists among the parameters that act to establish a stream's cross section. It is not as simple a relationship as that involving, for example, flow in a man-made open channel, constructed of concrete, which provides the width and gradient for the water course. In a natural stream, the banks and bottom can be eroded to widen and deepen the cross section and the gradient can increase or decrease, to some extent, by downcutting or by deposition of load, respectively. For flow in natural channels, the following relationship exists among the parameters involved in stream flow (Brutsaert, 2005; Charlton, 2007; Hendriks, 2010).

$$\text{Discharge } (L^3/T) = \text{width } (L) \times \text{depth } (L) \times \text{velocity } (L/T) \qquad \text{(Eq. 12-2)}$$

Therefore, increased discharge can be accommodated by a larger cross section and increased velocity or some combination of width, depth, and velocity that provides the needed increase.

When the cross profile of a stream is viewed without vertical or horizontal exaggeration, the width is shown to be significantly greater than the depth. It is also clear that resistance to water flow is provided mostly by the frictional drag with the bottom and sides of the stream boundary, or the wetted perimeter. When a stream increases in depth,

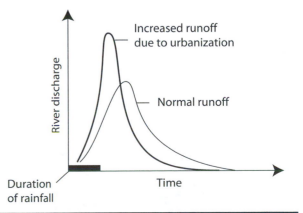

Figure 12.5 Stream hydrographs before and after urbanization.

Figure 12.6 Photograph of landform features of a meandering stream. (Earth Observatory, NASA, Rio Negro Floodplain, Patagonia, Argentina.)

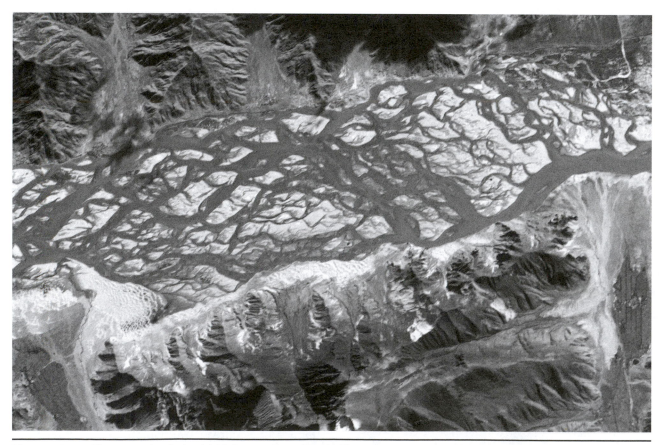

Figure 12.7   Photograph of a braided river system. (Earth Observatory, NASA, Tibetan Braid.)

the additional wetted perimeter is that portion added along the upper banks, as shown in Figure 12.8. This increase in percentage of wetted perimeter is considerably less than the increase in percentage of the cross-sectional area.

In Figure 12.8b the increase in wetted perimeter is about 16% and the increase in cross-sectional area is about 100%. This shows that velocity will increase as a consequence of increased depth. This relationship is indicated by the Manning equation:

$$v = \frac{1.5}{n} d^{2/3} s^{1/2} \qquad \text{(Eq. 12-3)}$$

where:

v = velocity
n = coefficient of bed roughness
d = depth
s = slope or gradient

Erosive power increases markedly with increasing velocity; actually, it is directly proportional to the square of the velocity, that is, $E\alpha v^2$. From this, a sequence of events can be deduced. When discharge increases, the depth increases and, hence, the velocity increases. With an increase in velocity, erosive power increases so that the cross section is widened and deepened by removal of material. This increases the wetted perimeter until the stream readjusts to the increased discharge. Subsequently, when discharge decreases, water level drops again, the velocity decreases, and material is deposited in the stream bed. This sequence of events is depicted in Figure 12.9.

Figure 12.9 highlights a significant problem of river channels. The scour, or removal of bed material, at the base of the stream during flood conditions is extensive. In fact, the depth of scour can range from one to four times the amount of increase in height of the water surface. This has an obvious effect on bridge foundations, loading facilities, and other engineering structures located in the stream channel. Undermining of bridge footings is a common

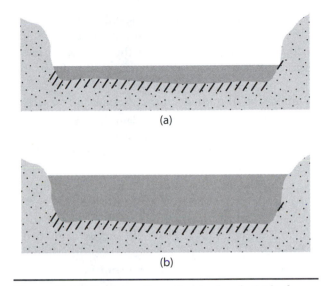

(a)

(b)

Figure 12.8   Cross section of water levels: (a) before and (b) after increased discharge showing increased wetted perimeter.

consequence of scour when this factor is not properly considered during foundation design.

All of this leads to a relationship between discharge, depth, width, bank erodibility, and gradient. Equilibrium is reached between these parameters and holds as long as the discharge remains the same. When discharge changes, the other parameters adjust, in an attempt to seek new equilibrium conditions.

Changes in these parameters along the stream length can be examined next. It is known that proceeding downstream, discharge increases whereas gradient decreases. Measurements at stream gauging stations indicate that the width and depth of streams increase downstream as well (Kondolf and Piégay, 2005; Herschy, 2008). Competency, a measure of the stream's ability to transport material, also decreases downstream. *Competency* is defined as the largest diameter particle that a stream can transport (Charlton, 2007). This decrease is illustrated by the

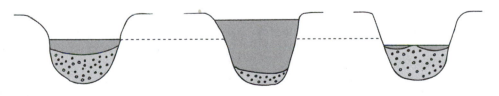

Figure 12.9   Cross sections showing changes in channel cross section with changes in stream discharge.

fact that progressively smaller sized particles occur in the river bed as one proceeds downstream along a major river. As an example, for the Ohio River in its passage along the southern boundary of Indiana, the size of recoverable gravel in the river bed decreases significantly. The effect is that gravel large enough for construction purposes disappears somewhere between Louisville, Kentucky, near the center portion of southern Indiana, and Evansville, Indiana, in its southwestern corner.

Considering this decrease in competency, it is surprising to learn that stream velocity actually increases slightly downstream. It would seem that a decrease in competency would suggest a decrease in velocity, because it is the velocity that dictates the size of particles that will be transported by the stream. The matter may best be explained by the concept of average velocity versus turbulent velocity, the latter being the velocity of turbulence whirls within the stream. Apparently, turbulent velocity decreases as the stream becomes more tranquil, flowing on a lower gradient and, consequently, the ability to transport larger pieces is reduced. This yields a decrease in competency. At the same time, the average velocity increases slightly downstream.

Work of Streams

Water flowing in a stream channel performs three primary functions: (1) eroding the stream bed, (2) transporting debris, and (3) depositing sediment (Charlton, 2007; Hendriks, 2010). The potential energy of the stream, as a result of its elevation, is converted to kinetic energy through the mechanism of flow. This energy is expended mostly in overcoming frictional resistance offered by the stream bed with a lesser amount used to offset the internal friction of the water flow itself, and a small amount is used for erosion and transporting material.

Erosion

Running water has the capacity to erode material. In channelized flow, it removes material from the stream bed and banks. As discussed previously, the erosive power is directly proportional to the square of the velocity. A distinction between the two types of stream beds should be made when considering erosion: (1) alluvial channels are those composed of sediment that may range from clay-sized pieces to pebbles, cobbles, or boulders that were previously deposited by the stream and (2) rocky channels, that is, solid bedrock exposed in the stream bed.

Alluvial Channels

In alluvial channels, the particles of alluvium are eroded by direct uplift from the stream bed. They may remain suspended or be rolled or bounced along by the current. The turbulent flow of water produces eddies and whirls that dislodge particles and lift them into the stream flow. Figure 12.10 shows the basic relationship between stream velocity and dislodgment of particles. This figure indicates the approximate velocity needed to erode clay, silt, sand, and larger particles. It can be seen that medium-sized sand is the easiest grain size to erode because a velocity of only about 0.3 m/sec (1 ft/sec) is needed to dislodge it. With increasing particle size, mass increases, so that increased velocity is needed. The mechanism is different for finer grained soils; with decreasing size, greater cohesion among particles develops so that dislodgment is more difficult. Also, the finer materials lie within the laminar flow regime at the base of a stream, which persists at lower velocities. Higher velocities disrupt this laminar range, allowing the finer sized particles to be eroded.

Figure 12.10 also relates the transportation and deposition of particles in regard to stream velocity. Note that, for fine particles, much less velocity is needed for transportation (to keep particles in suspension) than for dislodgment. Additional details on this subject are presented later in this chapter.

Another view of the water velocity needed for direct uplift of alluvial materials is shown in Figure 12.11. In this graph, the relationship between particle size and critical tractive force is shown. This force is the minimum needed to keep particles moving along the base of the stream. The graph is used to determine the maximum safe velocity for flow in a canal so as to not cause undesirable erosion of the bed material. For example, a fine gravel particle 10 mm in diameter (0.4 in) requires a tractive force of about 9.6 Pa (0.2 lb/ft^2) to move it along the base of a stream channel. For gravel of this size (see Figure 12.10), a water velocity of 100 cm/sec (3.3 ft/sec) would be required.

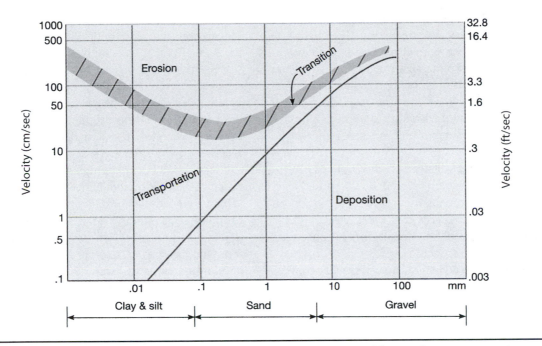

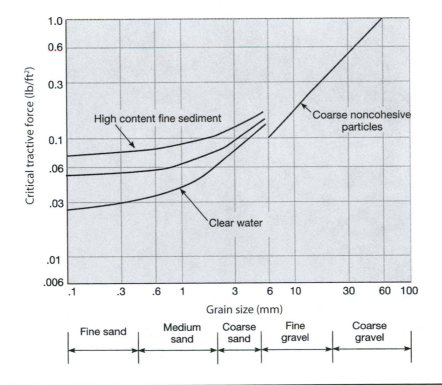

Figure 12.10   Relationship between particle size and stream velocity required for erosion, transportation, and deposition of material.

Figure 12.11   Critical tractive force for particle size movement versus particle size for canal stability considerations.

Bedrock Channels

Other forms of erosion occur mostly in bedrock channels or those strewn with cobble- and boulder-sized particles. Solids suspended in the stream abrade the rock. Potholes, or deep circular depressions in the bedrock, may develop from the swirling action of water charged with abrasive grains. A photograph of potholes formed in quartzite is shown in Figure 12.12.

Smooth, curving surfaces on the rock may form where stream flow is constricted by steep bedrock sidewalls. Plucking, that is, removing blocks of loose rocks along fractures that were weakened by water penetration, can also occur. This is caused by hydraulic action or "hydraulicing."

Some dissolving action of the water will occur as it flows over bedrock or alluvial materials. In most streams, the water carries much smaller amounts of dissolved material than its saturation capacity would allow, simply because insufficient time is provided for greater solution action to occur. Groundwater typically carries a higher concentration of dissolved salts than do streams. When gaining streams receive groundwater, the groundwater is, typically, greatly diluted by the larger volume of surface water. Groundwater volume supplied to a stream becomes its base flow and may constitute the only water it receives during a dry period.

Cavitation occurs at high stream velocities in the range of 7.5 to 9 m/sec (25 to 30 ft/sec). At this high velocity, air bubbles in the stream collapse suddenly, producing an effective loading of up to 13.8 MPa (2000 psi) pressure. This specific erosive process removed more than 0.3 m (1 ft) of concrete in a few days' time from a spillway structure in a major dam. Under natural conditions, cavitation is restricted to rapids and waterfalls.

In summary, stream erosion depends on the following variables: (1) water velocity, (2) kind and

Figure 12.12   Potholes formed in quartzite. Blyde River Canyon, Mpumalanga, South Africa. Pebbles swirling in a fast-flowing stream form these features. (roaldnel, Shutterstock.)

amount of load, (3) nature of the stream bed, that is, the kind of rock or sediment present, and (4) nature, direction, and spacing of rock structures in rocky channels, for example, jointing, bedding, and foliation.

Transportation

A stream dislodges material from its channel and carries it, sometimes intermittently, downstream toward the ocean. The amount of material that a stream transports is called its load. The load of most streams is well below its capacity.

The maximum size particle a stream can carry is its competency. The diameter of the particle varies approximately with the square of the stream's velocity. For example, a medium-sized sand grain, 0.5 mm in diameter, is carried by a velocity of about 10 cm/sec (0.3 ft/sec). If the velocity is doubled to 20 cm/sec (0.6 ft/sec), the particle size transported is four times larger, that is, a large sand grain of 2 mm. Similarly, if the velocity doubles again, to 40 cm/sec (1.2 ft/sec), the particle size transported is a pebble about 8 mm in diameter. Streams are able to transport material in three ways: by solution, suspension, and traction, the last including rolling, sliding, and skipping of particles along the bottom of the stream. The load also consists of three portions: solution load, suspended load, and bedload.

Solution

Most natural streams, especially those in humid areas, consist of extremely fresh water, showing much less hardness than groundwater, which typically contains abundant cations of calcium and magnesium and a lesser amount of iron. On the other hand, surface water is not pure—it does contain some dissolved chemicals. Streams carry measurable amounts of calcium and magnesium cations plus the common anions: chloride, nitrate, sulfate, and silica. Streams deliver these nutrients to the oceans of the world and, thereby, supply the basic needs for the food chain of the sea.

Suspension

Particles of solid material that are suspended in the stream of water as it flows along are designated the suspended load and are said to be carried in suspension. This is distinct from material chemically dissolved by the water or pushed along at its base. As flow occurs, the particles begin to settle, but turbulence

whirls lift them back into the main flow. This provides a relationship between stream velocity and the maximum size of suspended particles because turbulence typically increases with an increase in velocity.

Settlement of solids in water is similar to that in air, the main difference being that the frictional drag of a particle falling in water is considerably greater than that in air. Therefore, just as a projectile fired horizontally through the air from a cannon falls vertically as it moves horizontally, suspended particles fall through the water as the water moves along. They fall at a velocity that increases with increasing particle diameter, assuming that the specific gravity and shape of all particles is essentially the same. If the particles were not lifted again by turbulence whirls, they would settle and the water would clarify.

Figure 12.10, discussed previously, also illustrates the manner in which different size materials are transported. Observe that, for fine-sized particles, much greater stream velocity is needed to dislodge these particles from the alluvial stream base than is needed to keep them suspended in the stream. For coarse particles, this difference narrows considerably.

Because the highest velocity occurs during periods of maximum flow, it should be no surprise to learn that the greatest amount of suspended load is carried during periods of flooding. Because of this, more than three-quarters of the total suspended load transported in a year's time may occur during the short periods of time represented by high-level flooding. This suggests a somewhat catastrophic effect to stream erosion and transportation, rather than a steady regulated effect year-round.

The coarsest material of the suspended load is carried in the lowest part of the stream where the greatest turbulence occurs. This consists of coarse sand pieces for a velocity of about 0.3 m/sec (1 ft/sec). The finer materials, silt- and clay-sized particles, are well distributed throughout the stream depth. Clay particles may remain suspended even after the water has come to rest in a lake or settling basin. Such suspended material is referred to as turbidity. Flocculating agents, for example, aluminum sulfate, are used in water treatment plants to remove the fine sediment for municipal water supplies.

Bedload

The largest pieces transported by flowing water move along the stream bottom as bedload. At high flow rates, as in times of flooding, more than half the total sediment may move along the bed.

The mechanism for bedload movement is traction, which involves rolling, sliding, or saltation (Marshak, 2013). The largest pieces move by rolling or sliding, which may occur in an intermittent fashion, depending on turbulence whirls near the bottom. Saltation involves the jumping and skipping of particles along the bottom. The particle is picked up into the stream flow momentarily, but quickly settles again, only to be picked up once more by the more turbulent flow.

Deposition

When stream velocity decreases owing to reduced depth or gradient, deposition of sediment occurs. The coarsest particles carried in the bedload drop first. Further decreases in velocity allow deposition of the next coarsest material, so that finer and finer materials are deposited as the stream slows down. At a reduction in velocity to 0.3 m/sec (1 ft/sec), pebbles about 4 mm in size are deposited, medium sand is dropped out when the velocity slows to 0.03 m/sec (0.1 ft/sec), and silt will be deposited when velocity drops below about 0.6 cm/sec (0.02 ft/sec).

The ability of a stream to do work is related to its base level. This is defined as the lowest elevation to which a stream can cut its channel. There are several base levels involved. The ocean is the ultimate base level to which the entire length of the stream must adjust, because running water is physically unable to erode below sea level. In the geologic past, sea level stood lower than it does today, so that rivers on the east coast of the United States, such as the Chesapeake, Hudson, Potomac, and Delaware Rivers, have channels that extend well below the present sea level and out into the continental shelf.

The local base level for a tributary stream is controlled by the elevation of the point of confluence with the main stream. Some local base levels may be of a temporary nature, such as a lake that forms because of human activity or a naturally formed dam that crosses the stream. Streams can be dammed by landslides, earthquakes, volcanic eruptions, or glacial action. Within the framework of geologic time, such dams will be destroyed and the previous gradient reestablished.

When the base level is raised, for example, by the construction of a dam, the reduced gradient causes the stream to deposit sediments until the same slope is established, but at a higher elevation. If the base level is lowered suddenly, as by a breaching of the dam, the stream will cut down through its own sediments to reestablish the former gradient.

Landform Features of Streams

Both through erosion and deposition, streams alter the appearance of the land surface. As streams cut down through soil and bedrock, they form a branching network known as the drainage pattern. The actual pattern depends on the nature of the underlying materials and on the history of the stream itself (Kondolf and Piégay, 2005; Charlton, 2007; Marshak, 2013).

Drainage Patterns

Streams exhibit four types of drainage patterns: dendritic, trellis, rectangular, and radial. Figure 12.13 illustrates these drainage patterns.

A stream whose branching habit is like that of the limbs of a deciduous tree is called dendritic (Figure 12.13a). Such a pattern develops on horizontally bedded, sedimentary rocks or various homogeneous materials, for example, massive igneous and metamorphic rocks or thick soil sequences.

Trellis drainage consists of elongated, parallel channels with short, nearly perpendicular tributaries yielding an orthogonal effect, which is similar to a garden trellis (Figure 12.13b). The elongated channels mark the strike valleys of softer, folded rocks with short perpendicular tributaries flowing from the higher, resistant ridges. The folded rocks of the Ridge and Valley Province of the Appalachians and similar portions of the Rockies typically express this drainage pattern.

Rectangular drainage consists of perpendicular segments of streams without the dominant elongation of one orientation as typified by trellis drainage (Figure 12.13c). A combination of strong foliation and joint control usually gives rise to this type of drainage pattern.

Radial drainage is caused by streams radiating from a high central point, such as a volcanic peak (Figure 12.13d). These drainage patterns can be used to deduce information about the geologic structure of an area.

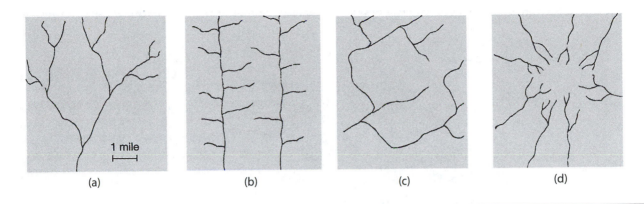

Figure 12.13   Drainage patterns that develop on different landscapes: (a) dendritic, (b) trellis, (c) rectangular, and (d) radial.

Drainage Basins and Divides

The drainage basin, or watershed, is the total area from which a stream and its tributaries obtain runoff. The drainage basin of a major river such as the Ohio or the Missouri is quite extensive—both involve considerable portions of the United States. The Mississippi River includes all of this drainage area because both the Ohio and Missouri Rivers are its tributaries. A drainage divide is the imaginary line connecting the high points between drainage basins. The continental divide separates drainage basins that supply water to the Gulf of Mexico and to the Atlantic Ocean from those yielding water to the Pacific Ocean.

Depositional Forms in Alluvial Valleys

For streams with wide valleys, the features are depositional in nature, except for the valley walls, which mark the sides of the valley. The flat area between the side walls is called the floodplain. Large streams or rivers flowing at low gradients in alluvial channels develop wide valleys. They consist of the landform features discussed next and illustrated in Figure 12.14 on the following page. This figure shows the valley walls and the backswamp deposits. Backswamp areas are poorly drained areas away from the stream channel that accumulate organic materials and are highly compressible when loads are placed on them. Natural levees are the high banks that parallel the river, formed by silt deposition during overbank flow. They form a lip that lies above the general elevation of the floodplain, causing some tributary streams to parallel the main course for a considerable distance before flowing into it. Named the Yazoo effect after the Yazoo River of Mississippi, this feature is illustrated in Figure 12.14.

A meander bend occurs when the channel migrates laterally because of differences in bank erodibility. The outside of a meander loop undergoes erosion, whereas in the inside of the loop, alluvial deposits, known as point bars, form. During periods of high water, a chute cutoff can occur, creating a short cut across the meander bend. Such features may be successful in taking over the primary flow of the river. Meanders cut off from the main stream are called oxbow lakes. These features are shown in Figure 12.14.

Meander scars are high areas, adjacent to the channel, marking the location of a former meander loop. They are the remnants of a natural levee held in place by vegetation, particularly those types of trees with a high water demand. Because a meander bend migrates in the direction of the outside loop, the previous depression will fill with loose, fine sediments. In some locations, a clay core is deposited in the ponded water. This illustrates the diversity of soil materials in an alluvial floodplain. Construction problems associated with floodplains are discussed later in the chapter.

Channel Patterns

The channel patterns of streams flowing in alluvial channels were mentioned briefly earlier in this discussion. A channel can be either straight, meandering, or braided. If a stream is carrying an excess of sediment in relation to its velocity (and indirectly to its discharge and gradient), material will be deposited in the streambed. This yields a complicated network

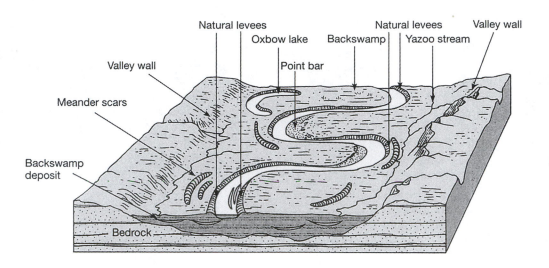

Figure 12.14   Schematic diagram showing landform features of wide alluvial valleys along major rivers.

of small channels choked by sandbars and islands that gives an overall appearance of braided hair (see Figure 12.7). The braided pattern occurs in alluvial fans and glacial outwash deposits as well.

Deltas, Alluvial Fans, and Stream Terraces

Deltas

Other depositional features related to running water include deltas, alluvial fans, and stream terraces. Deltas form where a stream flows into a standing body of water or into one with a much lower velocity and transporting capability. This occurs when the stream reaches the ocean, a lake, or its confluence with a larger stream. The depositional feature that forms takes on a shape similar to the Greek capital delta (Δ).

The delta forms as the transported sediment drops because of decreased velocity. As the debris collects in one location, the channel shifts to a place with lower elevation until that channel builds up and yet another shift occurs. These are called distributary channels. Overall, it yields a deposit, triangular in shape, with the apex pointing upstream. This deposit eventually extends further and further into the body of standing water. Deltas formed in human-made reservoirs accumulate silt from the upper reaches of the stream and, eventually, extend downstream into the main body of the lake. Where a tributary stream flowing on a steeper gradient meets a larger stream, the delta may extend far enough into the channel to deflect the direction of flow of the major stream. If constriction of the channel between the delta and the opposite bank occurs, erosion opposite the delta may develop, thus altering the direction of stream flow.

Some large rivers have extensive deltas where they meet the ocean. The Mississippi River delta is a prime example. However, other rivers form negligible deltas either because they carry no significant amounts of debris or because the ocean currents carry the material away as soon as it is deposited.

Alluvial Fans

An alluvial fan is similar to a delta, but on dry land—or at least where no permanent streams are established. They are found in arid and semiarid regions but may also occur in humid locations if the topography and other conditions are conducive to their formation. Alluvial fans occur where an intermittent stream has a sudden decrease in gradient; for instance, where a mountain stream slopes onto a plain. As with delta formation, the flow channels shift when debris builds the channel base higher than adjacent deposits and, eventually, a semicircular feature sloping toward the plain forms. This is illustrated in Figure 12.15. As the high area recedes because of erosion over geologic time, a sequence of alluvial fans form at higher and higher elevations, close to the mountainous area. Because of the coarse nature of these deposits, alluvial fans are good sources of groundwater. More discussion of this subject is included in Chapter 13 on groundwater.

Stream Terraces

A stream terrace is a flat or gently sloping surface that runs adjacent and parallel to the valley, with a steep bank separating it from the floodplain below or from a lower lying terrace. It indicates the former floodplain location for the stream when located at a higher elevation. This is illustrated in Figure 12.16 on the following page. This feature suggests the following sequence of events: the stream deposited sediments in the valley to the level of the terrace. It aggraded its channel (deposited materials) to this level because of (a) a slow rise in base level, (b) an increase in load, or (c) a decrease in discharge. Then, under equilibrium conditions, it establishes the level floodplain, primarily by meandering of the stream back and forth from valley wall to valley wall. In the next episode, the stream begins a role of active downcutting to establish a channel at a lower elevation. A climatic change could cause a rapid downcutting, yielding either a lowering of the base level or an increase in discharge.

Again, the stream establishes equilibrium conditions and the floodplain forms. This, in turn, may result in another terrace if renewed downcutting occurs, or it may remain as the present-day floodplain. Terraces on opposite sides of the valley are called paired terraces. This feature occurs at Lafayette, Indiana, on opposite sides of the Wabash River. The two terrace levels are depicted schematically in Figure 12.16.

Erosional Features

Water Gaps and Wind Gaps

Several erosional features of stream channels require special attention. Water gaps and wind gaps are of particular interest. When streams are established on a gently sloping surface of uniform resistance to erosion, a channel, oriented parallel to the maximum slope, will develop. As the stream erodes, the channel becomes deeper and the overall land surface is slowly reduced. If this well-established stream

Figure 12.15   Alluvial fans in the arid southwestern United States. (Pi-Lens, Shutterstock.)

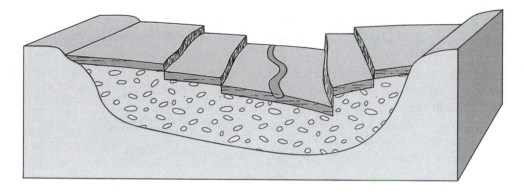

Figure 12.16   Schematic drawing of stream terraces along the Wabash River, Lafayette, Indiana.

intercepts a resistant, folded rock mass at depth, it will continue eroding through the rock material, although more slowly. Eventually the softer rocks at the surface erode more rapidly, leaving the folded, resistant rock standing at a higher elevation than the valley cut by the stream. The depression, or pass, formed where the stream cuts through the steep, resistant beds of the mountain, is called a water gap. Actually, the stream was superimposed on the structure and, through downcutting, it exposed the resistant bed. If the gap is abandoned by the stream because of stream piracy or by some other means, the dry gap remaining is termed a wind gap. Stream piracy occurs when a headward eroding stream with greater erosive activity takes over portions of the drainage area of another stream. Figure 12.17 is a photograph of the Delaware water gap.

Regional Stage

A final detail of erosional features of streams involves the regional dissection obtained in stream development. This is a useful means for describing the appearance of topography and is known as the regional stage. If a flat, gently sloping surface tens of square kilometers (several square miles) in size is uplifted from the sea bottom, runoff would yield a stream channel network in the downslope direction. Early in the development, only a few channels would occur but, as time passes, more tributaries would be added by erosion, as illustrated in Figures 12.18a and 12.18b. The areas between the stream channels would remain flat and, thus, are described as flat, interstream divides. Such an overall condition is known as regional youth. As stream dissection increases, less flat area remains between the stream channels and, at some point, late youth is reached, as pictured in Figure 12.18c. When

the dissection is sufficiently great that only rounded areas exist between channels, regional maturity has occurred. This is shown in Figure 12.18d. With further passage of time, when small rounded areas remain between the channels, late maturity occurs (Figure 12.18e) and, theoretically at least, a point when no interstream divides persist is reached and this is called regional old age, shown in Figure 12.18f.

The concept of a regional stage proves most useful in describing areas that began as flat, planar surfaces. This is true for many continental glacial deposits and for areas such as the Interior Lowlands of the central United States or the Great Plains that lie to the west of them. Youthful topography is typical in much of Indiana, Illinois, and Kansas. Northern Missouri, which was glaciated, shows a mature topography because of the several hundreds of thousands of years since glacial deposition occurred. Central Kentucky and Tennessee depict a mature topography as well, because these areas developed on horizontal sedimentary rocks.

In many texts, the concept of stage is applied to distinguish both individual streams and regions as youthful, mature, and old age. Youthful streams are said to have straight channels with steep gradients and few stream deposits, whereas old-age streams have low gradients, extensive deposits, and a meander pattern. This text prefers to describe the overall area by regional stage and the individual stream by its specific characteristics which, for a mature stream, would be a wide floodplain, low gradient, and meandering pattern. By contrast, youthful streams typically flow in a straight pattern over rocky channels at high gradients.

Figure 12.17   The Delaware water gap. (Jon Bilous, Shutterstock.)

Engineering Considerations of Running Water and River Systems

Many of the engineering concerns related to streams involve the use of wide floodplain areas. The upstream portions of smaller tributaries have engineering concerns but generally to a lesser extent. Both of these are considered in the following discussion.

Problems of Wide Floodplain Areas

1. Wide floodplains contain a diversity of landforms and materials, as shown in Figure 12.14. Most of the soils consist of sands, sandy silts, and clayey silts of relatively low strength. Because of the high groundwater table, these materials are usually saturated as well. All the soils are somewhat compressible, but the organic clays of the backswamp deposits are much more compressible. One major problem is predicting where the

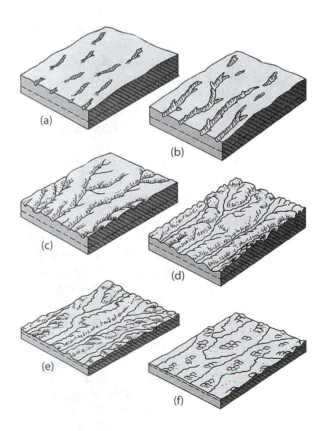

Figure 12.18   Regional stage: (a and b) early youthful topography, (c) late youth with flat interstream divides, (d and e) mature topography with rounded stream divides, and (f) old-age topography.

backswamps are located. As the river meanders and changes elevation, backswamp areas form in different locations. The organic deposits may be 3 m (10 ft) or more below the surface but provide no indication at ground level of their existence.

During construction of interstate highways across wide floodplains, the location of compressible organic soils must be considered. Highway embankments are constructed 3 m (10 ft) or more above the floodplain near the valley wall so as to extend above the flood of record, and even to a greater height near the stream itself. In some low areas of the floodplain and for bridge overpasses, the embankments may be 9 to 15 m (30 to 50 ft) or more high. This leads to settlement of the underlying compressible soils.

Subsurface exploration for pockets of organic soil is difficult. Closely spaced borings, only 7.5, 15, or 30 m (25, 50, or 100 ft) apart, across a wide floodplain are not economically feasible. Instead, probing using a steel rod may be used to determine penetration resistance to a depth of 3.0 to 4.5 m (10 to 15 ft). When compressible soil pockets are located, excavation and replacement is one possible solution, as is constructing a surcharge on the embankment, followed by an extended time interval. A surcharge consists of an extra thickness of soil fill, causing accelerated settlement, which is removed prior to paving. This encourages all the settlement to occur before the pavement is placed. The use of flexible (bituminous) rather than rigid (concrete) pavement, to reduce cracking of the pavement if additional settlement occurs, is also a possibility.

2. A high groundwater table makes deep excavations more difficult. Basements and underground parking garages would not typically be included for most building constructions on floodplains. For structures such as dams, locks, bridge foundations, and power-generating stations, where deep excavation is required, the subsurface must be dewatered prior to and during construction.

3. Because of the availability of groundwater in a floodplain, industrial facilities and communities may elect to obtain their water from the subsurface rather than directly from the river. Groundwater tends to carry more dissolved ions than does surface water but, typically, it has less turbidity and is less likely to be contaminated. Surface water is processed to remove the turbidity and to ensure that pathogens are eliminated, whereas groundwater may be softened to remove nuisance cations; further processing, other than chlorination, may not be needed. Floodplains typically supply the greatest volume of groundwater compared to other geologic situations. Groundwater pollution is a major concern and this subject, along with water supply, is discussed in Chapter 13.

4. Typically, the base of the alluvial soils in a major floodplain lies 30 m (100 ft) or more below the surface. This base may be bedrock or another geologic material, such as glacial deposits, which predate stream development. The thick sequence of alluvial soils tends to consist of loose sands and silts and compressible clays of low shear strength. Consequently, they provide poor support for heavy structures. Deep foundations are required, such as piles, which can transfer the load by end-bearing to a stronger material below the alluvium. This leads to expensive foundation support and economics may dictate that the structure be located away from the floodplain where foundation conditions are better. For some power-generating facilities, the power plant itself is located on the valley wall with only the water intake structures and cooling water ponds positioned directly on the floodplain.

5. The migration of meander bends is a concern. We should always assume that erosion will continue on the outside of a meander bend. Permanent buildings and other structures should not be constructed in these locations because it may be impossible to prevent erosion from occurring and undermining these structures.

6. Flooding is a problem of all floodplain areas. Just a small amount of reflection on the subject makes us realize that the floodplain was formed by the action of the stream through both lateral migration and flooding; the active floodplain marks the location of previous high-volume flows of water. Under the appropriate weather conditions, serious flooding will occur again and the wide floodplain is needed to control this water.

The recurrence interval for flooding may be 10 years or less or it may measure in the hundreds of years.

Constricting floodplains by buildings or flood walls and levees prevents the river from carrying the water volume it possesses during flood time. This transfers the flood effect further upstream by backing up the water and it extends the effect further downstream as well.

The floodway is an area where a stream transmits water by current flow. Adjacent to the active floodway is the floodplain, where water backs up but where active flow does not occur. Construction in a floodway is prohibited by law but encroachment on the floodplain is allowed for structures not subject to property damage. Floodplain encroachment by new construction is nearly nonexistent today because building permits are no longer issued for the floodplain zone. National flood insurance is not available for new homes to be built in a floodplain, so mortgages will not be granted for new home construction.

The question may be raised concerning why people construct homes in areas prone to flooding. Scenic beauty is one factor and individual rights is another. In some mountainous regions, the only flat areas seem to exist in the valleys that are used for agricultural endeavors, home construction, and transportation routes. These locations are particularly prone to flooding because the runoff rate is high and the narrow valleys can carry only relatively small volumes of water.

7. Navigational problems are associated with major waterways of the United States as an aid to shipping. Silt is removed from the channel by dredging. Erosion of the banks is reduced by the placement of riprap where the stream changes direction, and revetments are used to keep the main channel in one specific location rather than allowing it to shift its course. Levee construction and maintenance are necessary for both navigation and flood control purposes.

8. Streams are a primary means for removing liquid wastes from cities and industry. Sewage treatment plants, after removing much of the BOD (biological oxygen demand) from the sewage, dump their effluent into streams. Flowing water has the ability, through its dissolved oxygen (DO), to eliminate the remainder of this BOD and then rejuvenate the oxygen level as it flows along. With greater volumes of sewage have come more advanced sewage treatment plants, so that secondary treatment is mandatory and tertiary treatment is becoming more prevalent. However, in periods of low flow (summer mostly), the dilution capability of the stream is reduced and the DO is lower because of higher water temperature. If the oxygen drops to a critical level, fish kills can occur and an anaerobic condition can develop, yielding a foul-smelling, polluted stream, requiring additional sewage treatment.

9. Urban runoff is a major consideration today. The hard surfaces prevalent in cities yield increased volumes of runoff that the storm sewer systems must accommodate. Because stormwater and sanitary sewer systems are often commingled, they create a greatly increased volume of polluted water. Overflow points occur in the system and at these locations the combined water is dumped directly into the surface stream. In Chicago, the Tunnel and Reservoir Project (TARP) addresses this problem. Other cities in the United States, Milwaukee, Wisconsin, for example, have also constructed deep tunnels to collect the urban runoff.

Problems of Upland Stream Locations

1. The major problems in upland stream areas are erosion, landslides, and flash floods. The upland portion of tributary streams is the steepest and headward erosion is active. These areas are prone to landslides because of steep slopes and undercutting by the stream channel.

2. The upload stream portions, however, are quite scenic and, because of the steep nature of the terrain, may remain forested because clearing for agricultural purposes is not advantageous. These locations make attractive home sites and may require special attention to prevent loss of topsoil and possibly buildings by erosion and landslides.

EXERCISES ON RUNNING WATER AND RIVER SYSTEMS

MAP READING

For these exercises, the US Geological Survey offers TopoView (https://ngmdb.usgs.gov/topoview/). This software will allow you to view and download topographical maps produced from 1880–2010 for locations throughout the United States. Use a map that has a scale of 1:24,000.

SODA CANYON, COLORADO

This map is located in the Colorado Plateau Province. The basic structure is a plateau of horizontal, sedimentary rocks, mostly sandstone.

1. What is the fractional scale, size, and contour interval of the quadrangle?

2. What type of drainage is shown? What does it suggest about the underlying bedrock?

3. What is the regional stage of the quadrangle? Describe the nature of the stream channels—gradients (in general), bed material, valley type, etc.

4. Is the rainfall heavy or light in this area? Explain.

5. What is the origin of the small ponds at the head of Grass and Greasewood Canyons? Their shape should prove helpful in your determination.

6. Sketch on a separate sheet a cross section from the north-central part of Sec. 30, T34N, R14W, southward to Ute Trail. Label the stream locations.

7. What does Mancos Spring tell you about the character of the underlying bedrock?

8. What is the gradient of the lower half of the Mancos River? What is the gradient of Johnson Canyon? Why should it be greater than the gradient of Mancos River?

9. Why does the Mancos River meander? Relate this to its gradient determined in Exercise 8.

10. What engineering problems might you expect in the area (building foundations, road construction, etc.)?

KIMMSWICK, MISSOURI–ILLINOIS

11. What is the fractional scale, size, and contour interval of the quadrangle?

12. What is the regional stage of the area (exclusive of the Mississippi River floodplain)?

13. Sketch on a separate sheet an east–west cross section across the Mississippi River and across Glaze Creek.

14. What are the gradients of the Mississippi River and of Glaze Creek?

15. Describe the nature of both of the streams mentioned in Exercise 14. Be complete.

16. What is Moredock Lake and how did it form?

17. What are the curved or horseshoe-shaped features outlined by contour lines on the flat area east of the Mississippi River?

18. What is the origin of the small depressions that occur between the Meramec and Mississippi Rivers on the upland surface? This is an area of limestone bedrock. Why do some depressions contain water whereas others do not? Where does the surface water drain in these areas?

19. Why does Fountain Creek flow some 10 mi south along the floodplain before entering the Mississippi River? What is this feature called?

20. Why don't the country roads follow the section lines as they do in many parts of northern Indiana?

21. What engineering problems might you expect in this area (building foundations, road construction, etc.)?

LEXINGTON, NEBRASKA

22. What is the fractional scale, size, and contour interval of the quadrangle?

23. What is the regional stage of the area?

24. About how much rainfall is indicated in the area? Explain.

25. Describe the nature of the Platte River (including its gradient). What is the geomorphic name for this channel pattern?

26. Is the Platte River aggrading or degrading its bed? Explain.

27. Would you expect to find bedrock at a shallow depth beneath the town of Lexington? Explain.

28. Would water supply be a problem for Lexington? How would water be obtained?

29. What engineering problems would you expect along the Platte River?

Harrisburg, Pennsylvania

This quadrangle is located in the Ridge and Valley or Folded Appalachian Province. The structure is a folded sequence of sedimentary rocks—sandstones, shales, and limestones primarily.

30. What is the fractional scale, size, and contour interval of the quadrangle?

31. What type of drainage pattern is present in the mountain regions? Why did that type develop rather than some other type?

32. What is the landform (geomorphic) name for the type of pass where the Susquehanna River crosses the mountains? What would this be called if the river abandoned the channel?

33. Describe the configuration of Conodoguinet Creek.

34. Compare engineering problems in eastern and western Harrisburg.

Written Questions

35. If 1% of the rainfall infiltrates into the subsurface in an arid region, how much water per acre is supplied when the annual rainfall equals 10 in? Calculate this in acre feet of water per acre.

36. Explain why a sudden warming trend that melts snow cover is particularly prone to cause flooding with regard to soil infiltration.

37. Explain why urban runoff yields a stream pollution problem. Why is this an expensive problem to solve?

38. Give a low, moderate, and high value for the common parameters with regard to US streams. Parameters should include velocity, gradient, and discharge.

39. The surface velocity of a large stream is 0.8 m/sec (2.6 ft/sec). Approximately what would be the average velocity? If the cross-sectional area of the stream is 42 m^2 (450 ft^2), what value for the discharge is obtained? Give the answer in cubic meters (cubic feet) per second and millions of gallons per day.

40. Distinguish between the three views of a stream. Name them and discuss each in turn.

41. Given the Manning equation,

$$v = \frac{1.5}{n} d^{2/3} s^{1/2}$$

how much will the velocity increase in a stream if the depth increases from 1.5 m (5 ft) to 3.7 m (12 ft),

assuming the other variables do not change? How much does the erosive force increase when this occurs? Show calculations.

42. At zero flood stage, a certain river is 3.7 m (12 ft) deep at its deepest point. If the stream rises 2.4 m (8 ft) during a flood period, how much scour of the stream bottom is likely to occur? What would be the maximum water depth for this situation? Show your calculations.

43. What does the term *competency* mean? Why does competency decrease downstream although the average velocity increases slightly? Explain in terms of turbulent velocity and average velocity. Why is the decrease in competency important with regard to the supply of aggregates for construction?

44. What diameter particle is easiest for a stream to transport? Give the numerical value. Why is this particle easiest to move? Explain.

45. High sediment load in a stream coming from plowed fields or construction sites is considered a type of pollution. Why is this a reasonable conclusion? What effect is it likely to have on fish and other fauna in the stream? How can this be controlled for large construction sites? What farm practice is being implemented to reduce this effect?

46. What is meant by the term *base level*? Why was the ultimate base level lower in the geologic past than it is today? What is the evidence for this? Explain.

47. What is meant by the term *drainage pattern*? How does this differ from *channel pattern*? Explain. Which drainage pattern is indicative of folded sedimentary rocks? Which one is indicative of a resistant central peak?

48. Refer to Figure 12.14. Draw a cross section across the valley showing the various landforms in cross section view rather than in map view. Include some older backswamp deposits in the subsurface.

49. What does a braided channel pattern indicate? What climatic change could cause this to occur?

50. New Orleans, Louisiana, is located on the extensive Mississippi River delta. What engineering construction problems are associated with the city? Base your answer on knowledge gained about problems associated with wide alluvial floodplains.

51. Why would an area adjacent to a water gap be subject to flooding problems? Why would runoff be high in such areas?

52. Northern Missouri has been described as a maturely dissected till plain. What would be the regional stage for this area? Would you expect the secondary roads to run north–south, east–west as they do in northern Indiana and northern Illinois? If not, what orientation would they have?

53. Why is encroachment on a floodplain by buildings a particularly poor policy? If the structures are open to pass water to a height above the 100-year flood, would this solve the problem? Explain.

54. Is the construction of a golf course in a floodway a good use of the land? Explain.

REFERENCES

Baker, D., Pomeroy, C., Annable, W., MacBroom, J., Schwartz, J., and Gracie, J. 2008. Evaluating the Effects of Urbanization on Stream Flow and Channel Stability—State of Practice (pp. 1–10). In R. W. Babcock, Jr. and R. Walton (eds.), World Environmental and Water Resources Congress 2008, Honolulu, Hawaii, American Society of Civil Engineers.

Brutsaert, W. 2005. *Hydrology—An Introduction.* New York: Cambridge University Press.

Charlton, R. 2007. *Fundamentals of Fluvial Geomorphology.* New York: Routledge.

Hamblin, W. K., and Christiansen, E. H. 2003. *Earth's Dynamic System* (10th ed.). New York: Prentice Hall.

Hendriks, M. 2010. *Introduction to Physical Hydrology.* New York: Oxford University Press.

Herschy, R. W. 2008. *Streamflow Measurement* (3rd ed.). Boca Raton, FL: CRC Press.

Kondolf, G. M., and Piégay, H. 2005. *Tools in Fluvial Geomorphology.* New York: John Wiley & Sons.

Marshak, S. 2013. *Essentials of Geology.* New York: W. W. Norton.

GROUNDWATER 13

Chapter Outline

Origin of Groundwater

Distribution of Groundwater

Confined and Unconfined Water

Porosity and Related Properties

Permeability

Springs

Groundwater Movement

Production of Groundwater

Water Well Terminology

Legal Details of Groundwater Ownership

Water Witching

Groundwater Pollution

Formation of Caves in Carbonate Rocks

Engineering Considerations of Groundwater

Groundwater is a subject of major significance in the field of engineering geology because it is a primary source of water for human use; its presence has a pronounced effect on engineering construction and engineering works, including slope stability, surface and subsurface excavation, and foundation support.

The great quantity of water needed for human use is obtained in only two basic ways: from surface supplies (streams, lakes, reservoirs) and from groundwater. The United States uses more than 380 billion liters (100 billion gallons) of water per day, 20% of which comes from groundwater supplies. Each year more than a half million groundwater wells are drilled in the United States alone. The major uses of groundwater are rural, both domestic and farms; public water supply; industry; and irrigation.

Groundwater provides a major storage reservoir for freshwater. Only 5% of the total freshwater supplies occur in the Earth's atmosphere or on its surface. The remaining 95% occurs below the surface where it remains as a largely untapped reservoir.

The use of groundwater has the following advantages compared to surface water:

1. It is commonly free of pathogenic organisms and hence requires no purification for domestic or industrial use.

2. The temperature is nearly constant, which is important for heat exchange purposes.

3. Color and turbidity effects are usually minimal.

4. Chemical composition for a single source is essentially constant.

5. Groundwater reservoirs are generally larger than those for surface water and therefore are not affected by droughts of short duration.

6. Biological and radiological contamination of groundwater is more difficult and less likely.

7. Because groundwater has accumulated in the Earth during long years of recharge, it is sometimes found in areas that do not have significant surface supplies today.

The disadvantages for groundwater development, which apply to only some locations but may negate its use, include:

1. It may not be available in sufficient quantities to match anticipated needs, because some soils and rocks have too low a permeability to transmit much water to a well.

2. In most cases, groundwater has more dissolved solids (e.g., hardness) than surface water in the same region.

3. The cost of developing wells, particularly in humid areas, may be greater than that of impounding reservoirs on small streams for surface supplies.

4. Groundwater contamination by industrial chemicals has become prevalent in some urban areas.

Origin of Groundwater

Most groundwater is of meteoric origin (Karamouz et al., 2011; Fetter, 2018), supplied via the atmosphere by way of the hydrologic cycle (see Figure 12.1). Some water in volcanic areas may be juvenile water, that is, new water supplied from igneous fluids. The hydrologic cycle demonstrates the way in which water circulates from the oceans through the atmosphere and back to the sea by a number of different paths, either over the land surface or underground. These various paths may range from short to long, both in terms of time and distance traveled.

Most moisture in the world evaporates directly from the oceans, an amount estimated to be 328,000 km^3 (80,000 mi^3) per year. Another 61,400 km^3 (15,000 mi^3) of water per year evaporates from the land surface, that is, from lakes, streams, soil, and plants.

Total precipitation from the atmosphere should equal the amount of evaporation supplied to it, so approximately 389,000 km^3 (95,000 mi^3) of water falls on the Earth each year. The continental land-masses receive about 100,000 km^3 (24,000 mi^3)

of freshwater annually. In turn, a portion of this, 61,400 km^3 (15,000 mi^3), returns to the atmosphere with the remainder shared by runoff and infiltration. In this way, the cycle continues from year to year.

Connate Water

Connate water is the water entrapped in sedimentary rocks during their deposition as sediments. For marine sediments, the most common connate water originates as seawater. If the fluids are not flushed out by freshwater after the sediments are uplifted, the water remains as a salt-rich, pore fluid, unsuitable for most uses as a water supply. In the case of oil production, considerable quantities of brine water must be disposed of on the Earth's surface because brine water accompanies the crude oil pumped from the ground. This can be a serious environmental problem.

By definition, an aquifer consists of a formation or strata from which groundwater can be obtained for beneficial use (Karamouz et al., 2011; Fetter, 2018). Formations containing salty, connate water are not considered to be aquifers because the water is not suitable for human consumption. Such formations may be potential sites for disposal of waste fluids if other conditions of containment exist.

Magmatic Water

Magmatic water is another form of groundwater because water is one of the most abundant fluids absorbed by magma moving upward toward the Earth's surface. When lava extrudes onto the Earth's surface, it releases large amounts of steam and other gases into the atmosphere. Consequently, in volcanic areas and areas of high heat flow, it is likely that some of the subsurface water has been released by the magma.

A point sometimes raised by geologists is whether magmatic water is new (juvenile) water. This water could be an entirely new addition to the Earth's supply, implying that such water has not gone through the hydrologic cycle before. It has been determined, in fact, that most magmatic water is acquired from the rocks through which the magma intrudes so that the water is actually meteoric or connate and is not new per se. Many areas of hot springs and geysers, such as Yellowstone Park, are known to spray forth meteoric water or "recycled water" rather than juvenile water.

Distribution of Groundwater

It was first demonstrated, in a convincing way, around 1700 that groundwater and surface water supplies are a direct consequence of precipitation (Fetter, 2018). Prior to this, many mistaken ideas prevailed; for example, that the sea moved underground, lost its salt by distillation or some unexplained process, and reappeared as freshwater in springs, rivers, and lakes. Even after the general public became aware of the direct relationship between precipitation and stream flow, the interdependence between groundwater and surface water remained obscure. This lack of knowledge was reflected in laws concerning water supply, wherein groundwater and surface water were considered to be independent entities. Some of this confusion still prevails in water laws today.

Pierre Perrault (1608–1680) demonstrated during the latter part of his lifetime that rainfall in a drainage basin (watershed) provided the corresponding runoff for that area. He showed that the Seine River system above Arnay-le-Duc, France, had a runoff volume equal to only one-sixth of the precipitation it received during his three-year period of study. This showed that rainfall could easily account for stream flow, water used by plants, and water that infiltrates the ground.

Other contemporary scientists of Perrault were able to verify his findings; one, a fellow Frenchman named Edme Mariotte (1620–1684), suggested that if one-third of the precipitation evaporated and one-third remained in the Earth there would still be enough water to sustain the flow of rivers. Only an extension of this statement was needed to reach the conclusion that groundwater could feed streams to sustain their flow in dry weather and that streams could, in turn, feed the groundwater regime. Despite this situation, the general public did not become cognizant of the interrelationship of surface water and groundwater for more than 200 years.

Subsurface Distribution

Understanding the occurrence and nature of groundwater requires a knowledge of its vertical distribution in the Earth's crust. The outer portion of the Earth, which is fairly porous, is known as the zone of interstitial water and may extend as deep as 9.5 km (6 mi), at least in sedimentary rocks. In this zone, the pores are either partially or completely filled with water. In its deepest portion, pores are isolated from one another so that no water movement occurs. Below this thick interstitial portion, the zone of combined water occurs, a zone where water is attached chemically in the form of hydrated minerals or those with hydroxyl or OH groups. All fractures and open pores in this zone are closed by confining pressure from the overlying rock.

The subdivision of groundwater into various zones is shown in Figure 13.1 on the following page. There are no sharp boundaries between the subdivisions; instead, the changes are more gradual. Soil water, for example, is distinguished from the water directly below it in the intermediate vadose zone by its greater fluctuations in quantity in response to evaporation and transpiration. In a forest environment, trees may have roots that penetrate to depths of 9 m (30 ft) or more. Also, moisture levels retained in the upper few inches, near the ground surface, are affected by the temperature, pressure, and movement of the air.

Vadose Zone

The vadose zone, sometimes referred to as the *zone of aeration* (pores containing some air), is divided into three subzones: (1) soil water zone, (2) intermediate vadose water zone, and (3) the capillary water zone or capillary fringe (Todd and Mays, 2005; Fetter, 2018). In the capillary water zone, capillary rise prevails, yielding saturation closer to the water table. Here the term aeration does not apply, so that zone of aeration is not an appropriate term for the total zone. Vadose or suspended water provides a more appropriate term for this zone.

The intermediate vadose zone separates the soil water from the saturated zone below. This intermediate portion may be 300 m (1000 ft) or more thick in arid regions, or absent in some humid environments.

The capillary zone, which occurs at the base of the vadose zone, is a band directly above the water table where capillary size openings in soil lift water upward because of surface tension. The boundary between the intermediate vadose and the capillary zones is abrupt in coarse-grained, cohesionless soils (sands and gravels) but is quite gradational in silts and clays. The upper surface of the capillary zone is irregular and it changes depending on the amount of recharge. Contained in its upper part are pockets of air that slow the movement of water but, in the lower

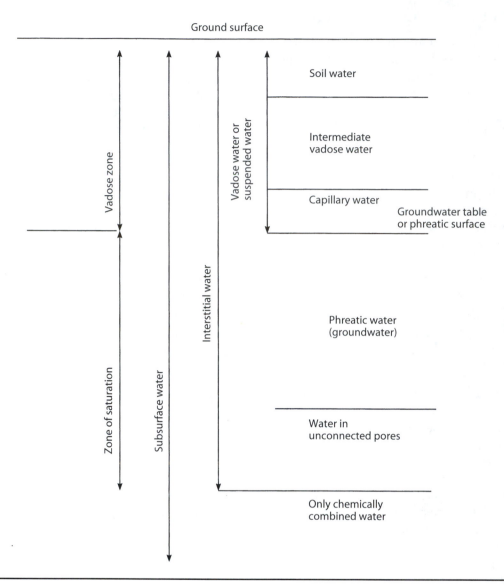

Figure 13.1 Distribution of groundwater within the Earth's crust.

part of the capillary zone, saturation is just as complete as it is below the water table. This movement of water is similar to that below the water table. The capillary zone in silt and clay materials is sometimes as thick as 2.4 m (8 ft), whereas in coarse sand or gravel it may extend upward only about 1 cm (a fraction of an inch) above the water table.

Water Table

The water table, or phreatic surface, separates the phreatic zone (zone of saturation) from the capillary zone. Theoretically, this is the surface depicted by the level of water in wells that penetrate a short distance into the saturated zone (Todd and Mays, 2005; Fetter, 2018). If groundwater flow is essentially horizontal, then the water levels in wells will correspond quite closely to the water table. The shape of the water table is controlled partly by the topography of the land. It tends to follow, in a general way, the land surface so the shape of the water table is commonly considered to be a subdued replica of the Earth's surface.

Common definitions for the water table indicate that it is the upper level of the zone of saturation, or it is the surface of separation between the zone of saturation and the capillary zone. A more precise definition states that the water table is the surface along which the hydrostatic pressure (pressure in the water or pore pressure) is equal to the atmospheric pressure (or zero gauge pressure). Figure 13.2 illustrates this relationship. In the capillary zone the water is in tension rather than in compression.

Figure 13.2 also indicates the positive pore water pressure, which increases with depth below the water table, and the negative pore water pressure, which increases upward into the capillary zone.

Zone of Phreatic Water

In specific geological terms, water below the water table is designated as groundwater and the zone below the water table is called the zone of saturation. Both terms lead to confusion because, in nonscientific terms, groundwater suggests any water below the ground surface and the zone of saturation should include all saturated materials. For this reason, subsurface water is a better general term for all water below the land surface. In addition, the lower part of the capillary fringe is saturated and water there travels at much the same velocity as water just below the water table. Therefore, a substitute for the term *zone of saturation* is needed, with a more acceptable one being *zone of phreatic water*. This is defined as water that will flow freely into a well, thereby indicating the level of the water table. Water in the capillary fringe will not drain freely into a well and will not contribute to the static water level. A pumping well with its intake below the water table, however, will

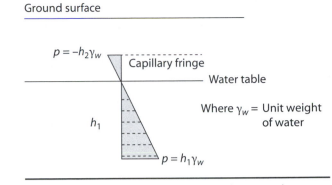

Figure 13.2 Hydrostatic pressure relative to the water table.

pull water from the capillary fringe as it draws down the water table.

Confined and Unconfined Water

Water in direct vertical contact with the atmosphere by way of interconnecting pores in a permeable material is called unconfined water and the permeable material containing unconfined water is called the unconfined aquifer (Figure 13.3) (Todd and Mays, 2005; Fetter, 2018). The water table marks the phreatic surface or top of a saturated zone in an unconfined aquifer. A well penetrating an unconfined aquifer is known as a water well (see Figure 13.3). In many areas the first unconfined water zone encountered in the subsurface lies above the general or regional zone of phreatic water and indicates an isolated body of water. Its position is dictated by permeability differences related mostly

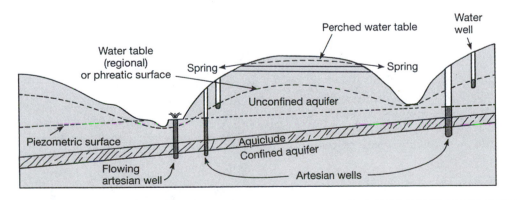

Figure 13.3 Types of aquifers and water wells. The figure also shows the distinction between the water table and the piezometric surface.

to the stratigraphy of the deposit, yielding a permeable zone lying above a small-sized impermeable one. This isolated saturated zone is known as the perched aquifer and its upper surface is called a perched water table (see Figure 13.3).

Confined water is separated from the atmosphere by an impermeable layer that allows for a buildup of pressure in excess of atmospheric pressure at the top of the saturated zone. The saturated soil is called the confined aquifer (Fetter, 2018). Such water, when intercepted by a well, will rise to a level higher than that at which it was intercepted; that is, higher than the top of the saturated aquifer. This type of well is referred to as an artesian well (see Figure 13.3).

The surface to which water in an artesian well will rise is called the piezometric surface (see Figure 13.3), and the rise in meters (feet) above the point of interception (top of the confined aquifer) is the artesian head. If this head is sufficiently large so that the water flows out onto the Earth's surface, the well is known as a flowing artesian well (see Figure 13.3). The term *artesian well*, in the early days of study, referred specifically to a flowing artesian well but, in current usage, *artesian* is synonymous with confined water and flowing artesian wells are those special cases in which the water reaches the surface.

The conversion of water pressure in the confined aquifer to meters (feet) of head in the well or standpipe is related to Bernoulli's equation, which states:

> Total head = pressure head +
> velocity head + elevation head   (Eq. 13-1)

or

$$H_t = \frac{P}{\gamma} + \frac{V^2}{2g} + H \qquad \text{(Eq. 13-2)}$$

where:
- H_t = total head in m (ft)
- P = water pressure in kg/m^2 (lb/ft^2)
- γ = unit weight of water in kg/m^3 (lb/ft^3)
- V = velocity of water in m/sec (ft/sec)
- g = acceleration due to gravity in m/sec^2 (ft/sec^2)
- H = elevation head in m (ft) with respect to datum plane

Because the velocity of subsurface water is very low, unless the water moves through open conduits in caverns or lava tubes, the velocity head is assumed to be equal to zero. When the confined aquifer is intercepted, the pressure head P/γ is converted to elevation head and the water rises in the well. For example, if a confined aquifer has a pressure of 14,061 kg/m^2 (20 psi) and it is intercepted by a well, the water will rise by:

$$\frac{14{,}061 \ kg/m^2}{1000 \ kg/m^3} = 14.1 \ m$$

or

$$\frac{20(144) \ lb/ft^2}{62.4 \ lb/ft^3} = 46.2 \ ft$$

This yields an artesian well. If the Earth's surface is reached within 14.1 m (46.2 ft), a flowing artesian well results.

The material overlying a confined aquifer, referred to as an aquiclude in Figure 13.3, may be semipermeable, allowing some water movement upward under high pressures. Under these conditions, high pressures are not sustained and water will rarely rise more than 3 m (10 ft) or so above the aquifer. Most artesian water in recent alluvial deposits would be semiconfined.

A number of other geologic conditions can lead to confined aquifers: stabilized sand dunes displaying impervious deposits in the interdune area (Figure 13.4a), alternating beds of dipping sandstones and shales, fault and fracture zones in crystalline igneous rocks (Figure 13.4b), folded beds with alternating layers, recharged along fractures (Figure 13.4c), horizontal sedimentary rocks involving permeable and impermeable strata and unconformable relationships (Figure 13.4d), and glacial deposits consisting of till and outwash (Figure 13.4e).

Porosity and Related Properties

Porosity is a measure of the total void space in a soil or rock and it equals the void volume divided by the total volume. It indicates the amount of water that may be present in an aquifer compared to its total volume. The porosity of a soil or rock depends on (1) the shape and arrangement of the particles, (2) the gradation or range of grain size, (3) the degree of compaction and cementation, and (4) the portion, if any, of soluble rock removed by solution. Table 13.1 on page 272 presents a list of porosities for common soils and rock.

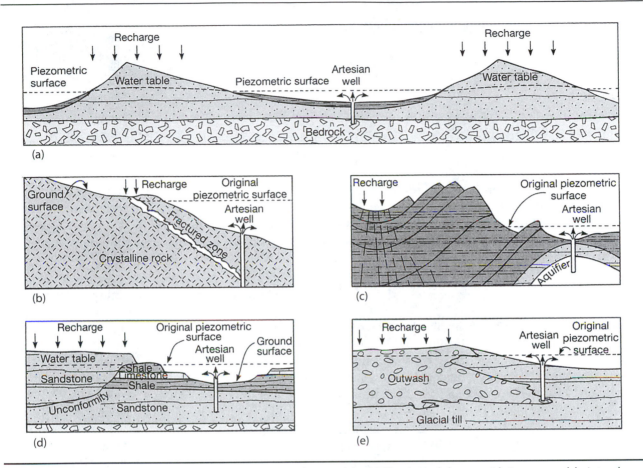

Figure 13.4 Geologic settings yielding artesian conditions: (a) stabilized sand dunes with impermeable interdune areas, (b) crystalline rock, (c) complexly folded and fractured sedimentary rocks, (d) horizontal sedimentary rocks, and (e) glacial deposits.

Other properties related to porosity, such as void ratio, water content, and degree of saturation, can be determined using phase relations discussed in Chapter 7, Elements of Soil Mechanics.

Permeability

Permeability is the rate at which water is able to flow through soil or rock and is of major significance in the fields of construction, civil engineering, and engineering geology. The flow rate is a major factor in situations such as: (1) the rate and volume of water supplied by pumping wells, (2) the rate of leakage through or beneath dams, (3) the ease with which soils can be dewatered, and (4) the rate of consolidation of a saturated soil under load.

Darcy's Law

In 1856, Henry Darcy, a French engineer, developed the following equation, known as Darcy's law, for flow through a porous medium:

$$q = kiA \qquad \text{(Eq. 13-3)}$$

where:

q = quantity of flow through a given cross-sectional area
k = coefficient of permeability
i = hydraulic gradient, a dimensionless number equal to the loss of head over a given flow distance (h/L)
A = cross-sectional area of the soil through which the water flows

Figure 13.5 on the next page illustrates the definition of hydraulic gradient. Practically, it is the slope of the water table. The velocity of flow is determined by the following equation:

$$v = \frac{q}{A} \qquad \text{(Eq. 13-4)}$$

Substituting the value of q in Equation 13-4 yields the following equation:

$$v = ki \qquad \text{(Eq. 13-5)}$$

where v and k have the same units of L/t (or length/time), typically in cm/sec for engineering geology and geotechnical engineering. Similarly, the following equation can be used to obtain the value of permeability:

$$k = \frac{q}{iA} \qquad \text{(Eq. 13-6)}$$

Several assumptions are involved for Darcy's law with regard to groundwater flow: (1) the flow is laminar, (2) the temperature is constant and is approximately 60°F (15.5°C), and (3) steady-state flow under saturated conditions prevails. Laminar

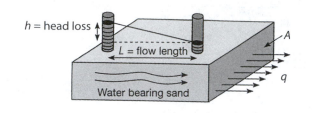

Figure 13.5   Head loss and hydraulic gradient.

flow conditions are violated only in clean, coarse gravels where the actual permeability would be somewhat less than the k value would suggest because of the turbulent flow conditions.

Permeability Units

A different unit of permeability, known as the Meinzer unit after the hydrogeologist who proposed it (Meinzer, 1923), has been used in the past in the United States. Rather than cm/sec, the Meinzer unit measures permeability in gal/ft^2/day. In recent years, because of the emphasis on metric units in scientific literature in the United States, m/day has been employed for hydraulic conductivity in hydrogeology studies.

At first, the gal/ft^2/day unit may not seem compatible with cm/sec but a more careful inspection will show that both have units of L/t, or length divided by time. In hydrogeology the term *hydraulic conductivity* rather than *coefficient of permeability* is commonly used, which tends to underscore the fact that permeability relative to water is involved. The conversion is

$$1 \text{ cm/sec} = 864 \text{ m/day} = 21.2 \times 10^3 \text{ Meinzers}$$

Another unit of permeability used in petroleum engineering is "darcy." A darcy is defined as the flow of 1 cm^3 of fluid having a viscosity of 1 centipose in 1 second under a pressure drop of 1 atmosphere over a length of 1 cm through a porous medium of 1 cm^2 in cross-sectional area.

In the petroleum engineering specialty, the flow of crude oil through porous media under reasonably high pressures (high gradients) is involved. The viscosity of crude oil is not only different from that of water, but varies from one type of crude oil to another and temperatures are usually well above 60°F (15.5°C). For these reasons, the direct application of hydraulic conductivity to the flow of crude oil is not appropriate.

Table 13.1   Porosities of Soil and Rock.

Material	Maximum Porosity (%)
Soil, loam	Up to 60
Sand and gravel	25 to 30
Chalk	Up to 50
Sandstone	10 to 25
Limestone	5 to 30
Chalky limestone	30
Oolitic limestone	10 to 20
Compact limestone	5
Dolomite	5 to 30
Fossiliferous, reef dolomite	30
Crystalline, porous	10
Interlocking mosaic	5
Shale	2 to 25
Siliceous shale	2
Marble	5
Slate	4
Granite	1.5
Other dense rocks	0.5

Additionally, in petroleum engineering, intrinsic permeability is used rather than hydraulic conductivity. Intrinsic permeability is a measure of the medium (rock or soil) alone, not related to a specific fluid, such as water or oil. A darcy has the units of length squared or L^2. Specifically,

$$1 \text{ darcy} = 0.987 \times 10^{-8} \text{ cm}^2 = 1.062 \times 10^{-11} \text{ ft}^2$$

Because the density and viscosity of water are functions of temperature, one can convert from darcies to Meinzers, to cm/sec, or to m/day. Water at 60°F (15.5°C) is typically the conversion point used. Table 13.2 provides values of permeability for common soils and rocks in different units.

Measuring Permeability

Several different procedures are available to measure or estimate permeability. Some test procedures are conducted on laboratory samples, whereas other tests are performed in the field. Those performed in the field consist of two types: (1) the field pumping test with monitoring of the drawdown in observation wells nearby and (2) open hole tests of several different varieties without observation wells (Todd and Mays, 2005; Fetter, 2018). Table 13.3 on the following page contains information on the various types of soil permeability tests. Permeability values are shown in cm/sec on a log scale. Five specific tests are included in the table and can be grouped collectively as field tests on in situ material or laboratory tests on prepared samples. Under the laboratory category, the following tests are included: constant head permeameter, falling head permeameter, grain size distribution (Hazen's approximation), and consolidation.

Field Tests

Field Pumping Test

This test measures the horizontal permeability rather than the vertical permeability in a deposit. This is desirable in most cases because seepage under dams and levees and flow to pumping wells both are horizontal.

Two conditions should be met to apply the equation for the field pumping test. First, steady-state conditions should be established. To accomplish this, prior to conducting the test, the well is pumped for 24 hours or longer at a constant rate. Steady-state conditions are determined by measuring the drawdown values in the monitoring wells at specific time intervals to determine when water levels stabilize. Two observation wells are needed in addition to the pumping well. In practice, the distance that these wells are located from the pumping well is important because, in permeable aquifers, the drawdown may be so small

Table 13.2 Permeability Values for Common Rocks and Soils in Different Units.

Material	k (cm/sec)	k (gal/ft²/day or Meinzers)	k (darcies)
1. Ranges of Values			
Gravel	1–10^2	10^4–10^6	10^3–10^5
Clean sands (good aquifers)	10^{-3}–1	10–10^4	1–10^3
Clayey sands, fine sands (poor aquifers)	10^{-6}–10^{-3}	10^{-2}–10	10^{-3}–1
2. Specific Values			
Argillaceous limestone, 2% porosity	8.6×10^{-8}	1.8×10^{-3}	10^{-4}
Limestone, 16% porosity	1.2×10^{-4}	2.5	1.4×10^{-1}
Sandstone, silty, 12% porosity	2.23×10^{-6}	4.74×10^{-2}	2.6×10^{-3}
Sandstone, coarse, 12% porosity	9.4×10^{-4}	19.9	1.1
Sandstone, 29% porosity	2.1×10^{-3}	43.6	2.4
Very fine sand, very well sorted	8.4×10^{-3}	1.8×10^2	9.9
Medium sand, very well sorted	2.23×10^{-1}	4.7×10^3	2.6×10^2
Coarse sand, very well sorted	3.69×10^1	7.83×10^5	4.3×10^4
Montmorillonite clay	$\approx 10^{-8}$	$\approx 10^{-4}$	$\approx 10^{-5}$
Kaolinite clay	$\approx 10^{-6}$	$\approx 1.0^{-2}$	$\approx 10^{-3}$

3. Equivalencies

1 darcy = 18.2 Meinzer units for water at 60°F, or 8.58×10^{-4} cm/sec for water at 60°F

1 Meinzer = 0.134 ft/day = 4.72×10^{-5} cm/sec = 5.49×10^{-2} darcies for water at 60°F

1 cm/sec = 1.165×10^3 darcies for water at 60°F = 21.2×10^3 Meinzers

10^{-6} cm/sec = 1.165 millidarcies for water at 60°F

1 cm/sec = $1.03 \; 10^6$ ft/yr = 864 m/day

1 millidarcy = 0.001 darcy = 0.858×10^{-6} cm/sec for water at 60°F

Table 13.3   Permeability and Drainage Characteristics of Soils.

Coefficient of Permeability k (cm/sec, log scale)

	10^2	10^1	1.0	10^{-1}	10^{-2}	10^{-3}	10^{-4}	10^{-5}	10^{-6}	10^{-7}	10^{-8}	10^{-9}
Drainage	Good						Poor			Practically Impervious		
Soil types	Clean gravel		Clean sands, Clean sand and gravel mixture				Very fine sands, Organic and inorganic silts, Sand / silt / clay mixtures, Glacial till, Stratified clay deposits			"Impervious" soils, e.g., homogeneous clays below zone of weathering		
							"Impervious" soils modified by effects of vegetation and weathering					
Direct determination of k			Direct testing of soil in its original position–pumping tests; reliable if properly conducted; considerable experience required									
			Constant-head permeameter; little experience required									
						Falling-head permeameter; reliable; little experience required		Falling-head permeameter; unreliable; much experience required		Falling-head permeameter; fairly reliable; considerable experience required		
Indirect determination of k					Computation from grain-size distribution. Applicable only to clean cohesionless sands and gravels					Computation based on results of consolidation tests; reliable; considerable experience required		

Source: After Terzaghi and Peck, 2010.

that it cannot be detected as close as 30 m (100 ft) away. In such cases, observation wells at 3- and 7.6-m (10- and 25-ft) distances should prove satisfactory.

Second, the aquifer should be fully penetrated, that is, the well should extend to the bottom of the aquifer with the well screen placed in the lower one-half to one-third of the aquifer. This ensures that the flow is radial to the well, which is assumed in the analysis and derivation of the equation. Figure 13.6 shows a field pumping test for an unconfined or water table aquifer.

The equation for the water table aquifer, with full penetration of the section, is

$$q = kiA \text{ (Darcy's law)}$$

where:

k = coefficient of permeability
i = dy/dx
A = $2\pi xy$
q = $2\pi k\, xy\, dy/dx$

Separating variables,

$$\frac{dx}{x} = \frac{2\pi k}{q}\, y\, dy$$

Integrating between the limits of two points on the drawdown curve (or at two observation well locations) as shown in Figure 13.6:

$$\int_r^R \frac{dx}{x} = \frac{2\pi k}{q} \int_h^H y\, dy$$

$$\ln \frac{R}{r} = \frac{\pi k}{q}\left(H^2 - h^2\right)$$

where ln is the natural logarithm or

$$\log \frac{R}{r} = \frac{\pi k}{2.3\, q}\left(H^2 - h^2\right)$$

where log is $\log_{10}$ or the common log

$$k = \frac{2.3\, q\, \log R/r}{\pi\left(H^2 - h^2\right)} \qquad \text{(Eq. 13-7)}$$

or for Meinzer units

$$P = \frac{1055\, q\, \log R/r}{\left(H^2 - h^2\right)} \qquad \text{(Eq. 13-8)}$$

where:

k = m/sec (ft/sec)
q = m^3/sec (ft^3/sec)
P = permeability in gal/ft^2/day with q in gpm
r = distance to nearest well (m or ft)
R = distance to farthest well (m or ft)
H = saturated thickness, farthest well (m or ft)
h = saturated thickness, nearest well (m or ft)

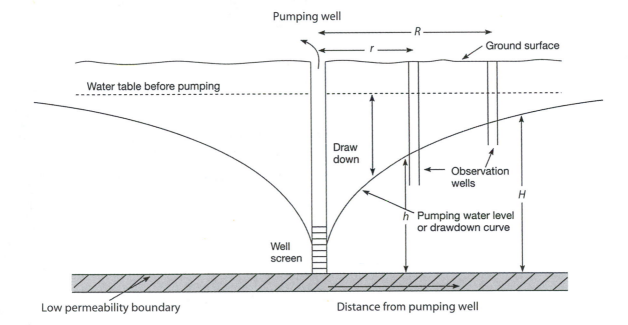

Figure 13.6   Cross section of cone of depression and field pumping test layout.

In a similar analysis for a confined or an artesian aquifer the equation is

$$k = \frac{2.3\, q \log R/r}{2\pi\, m(H-h)}$$

$$= \frac{0.366\, q \log R/r}{m(H-h)} \qquad \text{(Eq. 13-9)}$$

$$P = \frac{528\, q \log R/r}{m(H-h)} \qquad \text{(Eq. 13-10)}$$

with all units consistent in m or ft. Except H and $h =$ heights for the piezometric surface above the base of the aquifer during pumping, $m =$ aquifer thickness, and the same units for P in gal/ft^2/day as in the water table case.

Also note for the artesian case that $T = km$ or transmissibility = permeability multiplied by aquifer thickness. Therefore,

$$T = km$$

$$= \frac{0.366\, q \log R/r}{(H-h)} \qquad \text{(Eq. 13-11)}$$

with all units consistent in m or ft. Refer to Eq. 13-22 for more detail on transmissibility.

It is possible to calculate the coefficient of permeability in a field pumping test even though steady-state conditions (or equilibrium conditions) have not been reached. This is accomplished by applying the Theis equation (Theis, 1935). This concept is beyond the scope of this text but can be found in reference books on groundwater and groundwater seepage. A nonequilibrium well shows a change in drawdown with time rather than reaching a stable steady-state condition.

Example Problem 13.1

A sandy, unconfined aquifer is 15.24 m (50 ft) thick with hard, glacial till below. The depth to the groundwater table is 0.91 m (3 ft). Under steady-state conditions, the well discharge was 1.817 m^3/min (480 gal/min), drawdown in a monitoring well 15 m (50 ft) away from the pumping well was 0.46 m (1.5 ft), and at a monitoring well 30 m (100 ft) away was 0.06 m (0.2 ft). The well fully penetrated the aquifer thickness. Calculate the coefficient of permeability and the transmissibility of the aquifer.

Answers

The equation for this test is

$$k = \frac{2.3q \log R/r}{\pi\left(H^2 - h^2\right)}$$

Using SI units:

- The saturated thickness before pumping is 15.24 – 0.91 = 14.33 m.
- At $r = 15$ m the drawdown is 0.46 m and, therefore, $H = 14.33 - 0.46 = 13.87$ m.
- At $R = 30$ m, the drawdown is 0.06 m and, therefore, $h = 14.33 - 0.06 = 14.27$ m.
- $q = 1.817$ m^3/min $= 0.03$ m/sec
- $H_{sat} = 14.3$ m

$$k = \frac{2.3(0.03)\log\dfrac{30}{15}}{3.14\left(14.27^2 - 13.87^2\right)}$$

$$= 0.0005984 \text{ m/sec} = 5.98 \times 10^{-2} \text{ cm/sec}$$

$$T = kH_{sat}$$

$$= 0.0005984 \times 14.3 = 0.00856 \text{ m}^2/\text{sec} = 739 \text{ m}^2/\text{day}$$

Using British engineering units:

- The saturated thickness before pumping is $50 - 3 = 47$ ft.

- At $r = 50$ ft the drawdown is 1.5 ft and, therefore, $H = 47 - 1.5 = 45.5$ ft.

- At $R = 100$ ft, the drawdown is 0.2 ft and, therefore, $h = 47 - 0.2 = 46.8$ ft.

- $q = 1.069$ ft^3/sec $= 480$ gal/min

- $H_{sat} = 47$ ft

$$k = \frac{2.3(1.069)\log\dfrac{100}{50}}{\pi\left(46.8^2 - 45.5^2\right)}$$

$$= \frac{0.7401}{376.9} = 0.00196 \text{ ft/sec}$$

$$T = kH_{sat}$$

$$= 1.96 \times 10^{-3} \text{ ft/sec (47 ft)}$$

$$= 9.21 \times 10^{-2} \text{ ft}^2/\text{sec} = 7960 \text{ ft}^2/\text{day}$$

This transmissivity is adequate for an industrial supply (> 1350 ft^2/day).

Open Hole Permeability Tests

Open hole permeability tests, those in which only a single well is monitored, can be performed in soils and weakly cemented or porous rocks with reasonably accurate results. These are small-scale, in situ tests as compared to the large-scale tests involving a pumping well and adjacent monitoring wells. The three variations of the open hole tests are: (1) the open auger hole method for shallow depths, (2) the open tube method, and (3) the piezometer method. The auger method is performed in a shallow uncased hole below the water table. The open tube and piezometer methods are performed in cased holes, which are open at the bottom. The piezometer method has a sand-filled zone around the piezometer itself. The three methods are illustrated in Figure 13.7. As shown, the auger hole method and piezometer methods measure primarily horizontal flow, whereas the open tube method measures vertical or effective flow.

For the three test methods, depending on soil permeability, a constant, falling, or rising head can be accommodated in the well. The appropriate equations, derived for similar laboratory tests, can be applied except that the geometry of the "sample" differs; that is, it has a different shape factor. The dimensional details of the three open hole tests are shown in Figure 13.8 on the next page.

Referring to the open auger hole test, water is pumped from a shallow open hole, which lowers the surface to a depth h below the water table, and a rate of inflow q equivalent to the rate of change in h is measured immediately, that is, without waiting for steady-state conditions to develop. In a similar manner, as in the laboratory falling head test

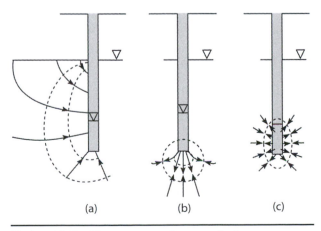

Figure 13.7 Flow patterns for open hole permeability tests: (a) open auger hole, (b) open tube, and (c) piezometer.

$$k = \frac{\pi a}{(A/a)h}\frac{dh}{dt} \qquad \text{(Eq. 13-12)}$$

where a is the radius of the well, h is the drawdown, and A is the shape factor, a function of the borehole and the boundaries of the flow medium.

Integrating,

$$k = \frac{2.3\pi a}{(A/a)t} \log H/h \qquad \text{(Eq. 13-13)}$$

Where horizontal flow and $D \approx 0$, values for the shape factor are:

$$A/a = 0.75\left(\frac{H}{a} + 10\right)\left(2 - \frac{h}{H}\right) \qquad \text{(Eq. 13-14)}$$

and for $D = \infty$, allowing for considerable vertical flow

$$A/a = 0.68\left(\frac{H}{a} + 20\right)\left(2 - \frac{h}{H}\right) \qquad \text{(Eq. 13-15)}$$

The second relationship holds for intermediate values of D to as low as $D/H = 1$ and for high H/a values. This technique can best be illustrated by an example problem.

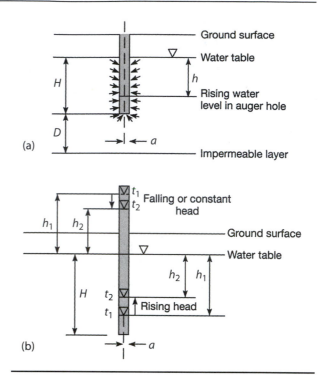

Figure 13.8   Dimension details of open hole permeability tests: (a) auger hole test and (b) open tube or piezometer permeability test.

Example Problem 13.2

In an open hole 10 cm (4 in) in diameter, the water level dropped 3 m (10 ft) in 1 min of pumping (Figure 13.9). Determine the k value for the material.

Answers

$$\frac{D}{H} = \frac{15}{15} = 1$$

Use

$$A/a = 0.68\left(\frac{H}{a} + 20\right)\left(2 - \frac{h}{H}\right)$$

Using SI units:

where: $H = 457.2$ cm
$h = 304.8$ cm
$a = 5.1$ cm
$t = 60$ sec

$$A/a = 0.68\left(\frac{457.2}{5.1} + 20\right)\left(2 - \frac{304.8}{457.2}\right) = 99.7$$

$$k = \frac{2.3\pi(5.1)}{99.7 \times 60} \log\frac{457.2}{304.8}$$

$$= 1.06 \times 10^{-3} \text{ cm/sec}$$

Using British engineering units:

where: $H = 15$ ft

$h = 10$ ft

$a = 2$ in

$t = 60$ sec

$$A/a = 0.68\left(\frac{15}{0.1667} + 20\right)\left(2 - \frac{10}{15}\right) = 99.7$$

$$k = \frac{2.3\pi(1/6)}{99.7 \times 1 \text{ min}} \log\frac{15}{10}$$

$$= 0.0021 \text{ ft/min} = 3 \times 10^{-5} \text{ ft/sec}$$

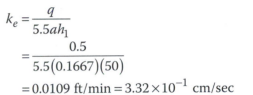

Figure 13.9   Open auger in-hole pump test. Note: 10 ft = 3 m.

For the open tube and piezometer methods the shape factors have also been developed. They are particularly complicated for the piezometer method and are not included here. The reader is referred to a paper by Hvorslev (1949) for these details.

An open tube (also a piezometer) can be tested either under constant head or changing-head conditions during pumping. For the constant head situation, the water surface is raised to a level h above the water table and held there during pumping. A large volume source of water is used to supply the input. The equation for the constant head case is where q is the pumping rate, h_1 is the height above the water table, and A is the shape factor.

For the changing water table case

$$k = \frac{q}{Ah_1} \qquad \text{(Eq. 13-16)}$$

$$k = \frac{2.3\pi a}{A}\frac{\log\dfrac{h_1}{h_2}}{t_2 - t_1} \qquad \text{(Eq. 13-17)}$$

where a is the diameter of the well, h_1 is the initial distance between the water table and the level measured at time t_1, and h_2 is the final distance between the water table and the level measured at time t_2. The shape factor for the open tube case is shown in Table 13.4 on the following page. This can be represented by solved problems. Given the situations shown in Figures 13.10 and 13.11 on the following page, the solutions to these problems are provided here.

Based on Figure 13.10, $q = 0.5$ ft³/min; $2a = 4$ in; a = radius of well = 2 in = 0.1667 ft.

$$k_e = \frac{q}{5.5ah_1}$$

$$= \frac{0.5}{5.5(0.1667)(50)}$$

$$= 0.0109 \text{ ft/min} = 3.32 \times 10^{-1} \text{ cm/sec}$$

Also, based on Figure 13.11, $2a = 4$ in, a = radius of well = 2 in = 0.1667 ft. For one minute of pumping, $h_1 = 50$ ft and $h_2 = 40$ ft.

$$k_e = \frac{2.3\pi a}{4(t_2 - t_1)} \log\frac{h_1}{h_2}$$

$$= \frac{2.3\pi(0.1667)}{4(1)} \log\frac{50}{40}$$

$$= 0.02918 \text{ ft/min} = 8.89 \times 10^{-1} \text{ cm/sec}$$

The above two problems can be solved in equivalent metric units using q in cm³ and other dimensions in cm. Note: 1 in = 2.54 cm and 1 ft = 30 cm.

Field Testing of Rock Permeability

Permeability of rock is controlled, primarily, by fractures. A common in-hole method used to test for permeability of bedrock in situ is the packer permeability test (Fetter, 2018). In this test, water is injected by way of a borehole into a zone of rock through a perforated tube located between two inflated packers. For testing the bottom section of the hole, a single packer can be used. An advantage of this test is that a selected portion of rock can be checked for permeability.

Permeability values for packer tests are commonly expressed in lugeon units. One lugeon is equal to a loss of 1 liter of water per minute per meter length of the borehole at a pressure gradient of 10 kg/cm² (1 MPa). It is equivalent to a permeability coefficient of 10^{-7} m/sec or 10^{-5} cm/sec. Packer tests can be applied both above and below the water table, but greater success is obtained below the water table. The length of the hole tested should be at least five times the hole diameter.

Rock-Fracture Aperture, Frequency, and Permeability

In massive rock, the permeability of fractures is many times greater than that of the solid rock between them. In tunnel and dam design, it is desirable to determine the actual spacing of fractures and their apertures (widths of opening). This relates to rock mass strength, permeability, and ease of grouting the fractures. Spacing of fractures can usually be determined from site investigation, including field mapping and a subsurface drilling program.

Permeability is related directly to the cube of the fracture aperture and to the frequency of fractures. Therefore, fracture frequency has much less effect on permeability than does fracture aperture and, in the same way, fracture frequency has a relatively small effect on fracture aperture for a given permeability. Consequently, a reasonably accurate fracture frequency survey, coupled with good rock mass permeability measurements, can predict, fairly accurately, the fracture aperture of the rock mass. The equation for this relationship is:

$$d^3 = \frac{12\eta k_r}{N\rho_w} \qquad \text{(Eq. 13-18)}$$

Table 13.4   Shape Factors for Open Tube Method of Permeability Determinations.

Drawdown Condition	Open Tube Situation	
	Penetrates Full Aquifer	Within Infinite Aquifer
Constant head	$A = 4a$	$A = 5.5a$
Variable head	$A = 4a$	$A = 5.5a$

Equations Developed:

Constant head	$k_e = \dfrac{q}{4ah_1}$	$k_e = \dfrac{q}{5.5ah_1}$
Variable head	$k_e = \dfrac{2.3\pi a}{4(t_2 - t_1)}\log\dfrac{h_1}{h_2}$	$k_e = \dfrac{2.3\pi a}{5.5(t_2 - t_1)}\log\dfrac{h_1}{h_2}$

where d = fracture aperture (mm), η = coefficient of viscosity of water = 1 centipose = 0.01 dyne/sec/ cm^2, k_r = radial permeability (m/sec), N = frequency of fractures in number/m, and ρ_w = density of water. Substitution of these values yields the following equation:

$$d^3 \cong 1520\frac{k_r}{N} \qquad \text{(Eq. 13-19)}$$

$$d \cong 11.5 \sqrt[3]{\frac{k_r}{N}} \qquad \text{(Eq. 13-20)}$$

If the fracture frequency and aperture are known, Equation 13-20 can be used to determine fracture permeability.

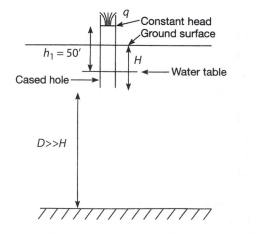

Figure 13.10   Open tube, constant head test, $D = \infty$.

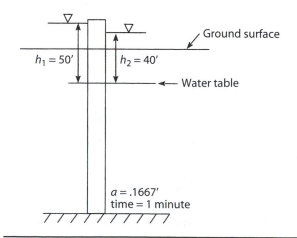

Figure 13.11   Open tube, constant head test, $D = 0$.

Example Problem 13.3

The fracture spacing in a massive rock unit was found to be 0.481 m (1.58 ft) and the rock mass permeability is 5×10^{-8} m/sec. Calculate the fracture aperture for this rock mass.

Answers

$$d \cong 11.5 \sqrt[3]{\frac{k_r}{N}}$$

$$\cong 11.5 \sqrt[3]{5 \times 10^{-8} \times \frac{0.481}{1}}$$

$$\cong 0.072 \text{ mm} = 0.0023 \text{ in}$$

Laboratory Tests

Constant Head and
Falling Head Permeability Tests

In the laboratory, permeability is measured using an apparatus called a permeameter. Two variations of the test are used: the constant head permeameter test (ASTM D2434) and the falling head permeameter test (ASTM D2435). Both tests are based on Darcy's law. Figures 13.12 and 13.13 on the next page illustrate the schematic setups for the constant head and the falling head permeability tests, respectively. The constant head test is used for coarse-grained soils (sands and gravels $k > 10^{-4}$ cm/sec), whereas the falling head test is performed on fine-grained soils (silts and clays). In the constant head test, the volume of water, Q, passing through the saturated sample is collected in time t. In the falling head test, the drop in head is measured in a given time.

Q = total quantity of flow
A = cross-sectional area of conduit
t = elapsed time
k = coefficient of permeability
i = hydraulic gradient (h/L)

For the constant head permeameter test calculation (see Figure 13.12):

$$Q = kiAt$$
$$= \frac{khAt}{L}$$

$$k = \frac{QL}{hAt}$$

For the falling head permeameter test calculation (see Figure 13.13):

$$Q = kiAt$$

$$= kAt\frac{h}{L}$$

$$dQ = k\frac{h}{L}A dt$$

but $dQ = -adh$

$$-adh = k\frac{h}{L}A dt$$

separating variables

$$-\frac{dh}{h} = \frac{kA}{aL} dt$$

$$-\int_{h_1}^{h_2} \frac{dh}{h} = \frac{kA}{aL} \int_0^t dt$$

$$-\ln\frac{h_2}{h_1} = \frac{kA}{aL}t$$

$$k = \frac{aL}{At} \ln\frac{h_1}{h_2}$$

$$= \frac{2.3aL}{At} \log\frac{h_1}{h_2}$$

In the falling head permeameter test, the volume of flow exiting the sample is not measured because fine-grained soils allow very little water to pass through. Instead, the drop of head in the standpipe is measured over a period of time. The permeability can be determined from the elapsed time, the head drop, dimensions of the soil sample, and diameter of the standpipe.

Hazen's Approximation

Generally, at least some of the voids in a particulate system are interconnected and continuous. This holds true for most soils and, as a consequence, the smaller the soil particles, the smaller the water-conducting passageways. Therefore, the coefficient of permeability, k, decreases with decreasing particle size for soils. Based on this principle, Hazen (1892) proposed the following empirical equation for estimating permeability:

$$k = C(D_{10})^2 \qquad \text{(Eq. 13-21)}$$

where C is Hazen's empirical coefficient with values between 1 and 1.5, average being 1; D_{10} is the effective particle size in mm (10% of soil, by weight, being finer than that size); and k is in cm/sec. This approximation works reasonably well for clean sands and gravels, but is less applicable for fine-grained soils. Table 13.5 presents the permeability and the corresponding D_{10} values for different soils.

Consolidation Test

Permeability is a factor in the consolidation test (discussed in Chapter 7) because settlement of the soil is related to the drainage of the sample under load. The permeability is calculated from the coefficient of consolidation C_c, which is obtained during consolidation testing. For further details on that subject the reader is referred to textbooks on geotechnical engineering (Das, 2002; Holtz et al., 2011).

Springs

Springs occur where the water table intersects the ground surface (Marshak, 2013). Another situation for spring formation is where water above the water table moves downward toward the saturated zone. If its downward movement is restricted by layers of low permeability, water will flow down gradient along the interface. Eventually this contact between the permeable layer above and the impermeable layer below will intersect a valley wall or other vertical exposure, and water will exit from the

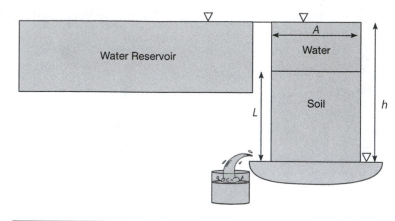

Figure 13.12   Constant head permeameter test setup.

slope. If there is enough volume to show noticeable water movement, such supplies are called springs. Less discharge than this yields surface seepage or, simply, a seep.

The geologic conditions yielding permeable layers that overlie impermeable layers are quite common so that scenarios are numerous. Sequences of (1) sandstone above shale, (2) sand above silt or clay, and (3) broken (talus) or weathered rock lying above massive bedrock are examples of stratigraphic sequences leading to spring development. Other geologic situations that are conducive to moving water toward the ground surface and forming a spring are where a fault zone intercepts permeable strata, concentration of fractures in a fault zone, and where a deep fracture intercepts a confined aquifer (Kresic and Stevanovic, 2009; Marshak, 2013).

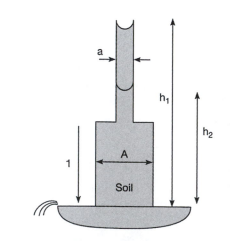

Figure 13.13   Falling head permeameter test setup.

Spring water has a reputation, perpetuated by various advertisements, of being pure or therapeutic or of having some other attribute not available to surface water or well water. Facts do not substantiate those claims but spring water often does have more dissolved solids and, consequently, a stronger taste and can be polluted easily. Pollution occurs because springs flow near the Earth's surface and the flow is in open conduits in the ground without the benefit of filtering through soils.

Springs can be classified according to origin or according to size, with size based on the amount of discharge (Meinzer, 1923; Kresic and Stevanovic, 2009). First-magnitude springs, those with the greatest flow, discharge more than 2.83 m³/sec (100 ft³/sec). Big Spring, Missouri, located in the Missouri Ozarks about 160 km (100 mi) southwest of St. Louis, is a first-magnitude spring with an average discharge of 250 MGD or 10.8 m³/sec (380 ft³/sec) (Figure 13.14).

Second-magnitude springs have a discharge ranging from 0.28 to 2.83 m³/sec (10 to 100 ft³/sec). These designations extend down to eighth-magnitude springs at less than 7.9 ml/sec (1 pint/min) (see Table 13.6).

At best, only a few hundred first-magnitude springs exist throughout the world. This fact is a result of the rare combination needed to produce such high-volume flows: large amounts of infiltration, a large drainage area, and favorable geologic conditions to localize the discharge. Almost all first-magnitude springs are located in lava, limestone,

boulder, or gravel aquifers with flow from cavern-like openings. The common aquifers of sandstone, conglomerate, and sand generally lack the extremely high permeability and hence the enormous discharge to produce first- or even second-magnitude springs.

Small springs may occur in all types of soil and rock. Loess, dolomite, graywacke, gypsum, and serpentine may contain springs of seventh or eighth order. Shales may contain the smallest of springs, issuing from joints or small layers of more silty or sandy rock.

Table 13.5 Permeability and the Corresponding Effective Particle Size Values for Different Soils.

Material	k (cm/sec)	Effective Size, D_{10} (mm)
Uniform coarse sand	0.4	0.6
Uniform medium sand	0.1	0.3
Clean, well-graded sand and gravel	0.01	0.1
Uniform, fine sand	4×10^{-3}	0.06
Well-graded, silty sand and gravel	4×10^{-4}	0.02
Silty sand	10^{-4}	0.01
Uniform silt	5×10^{-5}	0.006
Sandy clay	5×10^{-6}	0.002
Silty clay	10^{-6}	0.0015
Clay (30% to 50% clay size)	10^{-7}	0.0008
Colloidal clay (minus 2 μm ≥ 50%)	10^{-9}	40 Å = 4×10^{-6}

Figure 13.14 Big Spring, located in the Missouri Ozarks. (Missouri Department of Natural Resources.)

Table 13.6 Meinzer's Classification of Springs According to Discharge.

Magnitude	SI Units	British Engineering Units
1st	>2.83 m³/sec	> 100 ft³/sec
2nd	0.283 to 2.83	10 to 100
3rd	28.3 to 283 l/sec	1 to 10
4th	6.31 to 28.3	100 gal/min to 1 ft³/sec
5th	0.631 to 6.31	10 to 100
6th	63.1 to 631 ml/sec	1 to 10
7th	7.9 to 63.1	1 pint/min to 1 gal/min
8th	< 7.9 ml/sec	< l pint/min

Source: Meinzer, 1923.

Groundwater Movement

Groundwater can either provide water to surface streams or derive water from them. A gaining stream is one that is recharged by groundwater; such streams tend to be permanent, that is, they flow year-round. Typically, they occupy well-entrenched valleys located in well-developed floodplains. A losing stream is one that supplies water to the groundwater reservoir. Quite commonly they dry up when the rainy season is past. They are usually small, fairly steep tributaries to the main stream. Older terminology, no longer preferred by most geologists or geological organizations, includes the terms *effluent stream* for a gaining stream and *influent stream* for a losing stream. Older groundwater literature makes extensive use of these terms.

The term *aquifer* was introduced previously to describe a unit that can yield significant quantities of usable water. A rock that neither transmits nor stores water is called an *aquifuge* and one that stores water but does not transmit significant quantities of it is called an *aquiclude*. The term *aquitard* is sometimes used to mean a rock unit that transmits enough water to have a regional effect on flow but not enough to supply a well.

The level of the water table is of special interest to the engineering geologist because it shows the elevation of saturation of soil or rock. When the elevation of the land surface is known, the depth to the water table is easily determined. Maps depicting the water table contours are used to estimate the direction of flow. Movement occurs at right angles to the water table contours in the downslope direction.

Figure 13.15 is a water table map showing both surface contours and water table contours. Observe that the stream flows southwestward with a tributary stream entering from the north. The water table intercepts the main stream, showing it as a gaining stream, whereas the tributary stream loses water to the water table. The pond also intercepts the water table in part and obtains some water from it. The pumping well, by contrast, has depressed the water table surface in the form of a cone, referred to as the cone of depression. In Figure 13.15, the water table contours around the well define the cone of depression.

Similar to the water table, the piezometric surface of a confined aquifer can be depicted on a topographic map. The contours of the piezometric surface show the levels to which the water would rise, not the depths at which it would be intercepted. Piezometric surfaces exist for each confined aquifer and, sometimes, mis-

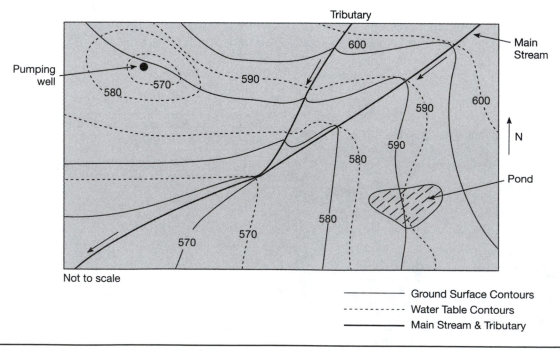

Figure 13.15   Contour map of a water table.

leading data are obtained when two aquifers are intercepted by the same well, which experiences pressure effects from both. Well construction needs to take into account the contribution of multiple layers of artesian head in the subsurface. Maps of the piezometric surface can be used to estimate where pump intakes should be placed in an artesian well and how much lift will be required to bring the water to the surface.

The velocity of groundwater movement is determined based on Darcy's law, using the relationship $v_{avg} = ki$, where v is the average velocity through the cross section and k and i are the coefficient of permeability and hydraulic gradient, respectively. The velocity of flow along the path of seepage, v_s, is appropriately called the seepage velocity, where $v_s = v/n$ with n equal to the porosity represented as a fraction. When the seepage velocity is known, the distance or time of travel through the subsurface can be determined. Quantities of flow can be calculated, also. This is illustrated by Example Problem 13.4.

Example Problem 13.4

Two ponds of water are 305 m (1000 ft) apart with the water surface in one pond 46 m (150 ft) higher than in the other. Both ponds have a square shape, measuring 61 m (200 ft) on a side. A 0.6-m (2-ft) thick sandy layer connects the ponds and the sand has the permeability $k = 5 \times 10^{-2}$ cm/sec with a porosity of 25%. How long does it take for water to travel from the upper pond to the lower pond? How much water seeps from the upper pond into the lower pond on a yearly basis?

Answers

Using SI units:

$$v_{avg} = ki$$

$$= 5 \times 10^{-2} \text{ cm/sec} \left(\frac{46}{305} \right)$$

$$= 7.5 \times 10^{-3} \text{ cm/sec}$$

$$v_s = \frac{v_{avg}}{n}$$

$$= \frac{7.5 \times 10^{-3} \text{ cm/sec}}{0.25}$$

$$= 30 \times 10^{-3} \text{ cm/sec} = 25.92 \text{ m/day}$$

$$t = \frac{\text{distance}}{v_s}$$

$$= \frac{305 \text{ m}}{25.92 \text{ m/day}}$$

$$= 11.76 \text{ days}$$

$$A = 61 \text{ m} \times 0.6 \text{ m} = 36.6 \text{ m}^2 = 36.6 \times 10^4 \text{ cm}^2$$

$$Q = kiAt$$

$$= (7.5 \times 10^{-3} \text{ cm/sec})(36.6 \times 10^4 \text{ cm}^2)$$

$$= 2745 \text{ cm}^3 / \text{sec} = 86,566 \text{ m}^3/\text{year}$$

(continued)

Using British engineering units:

$$v_{avg} = ki$$

$$= 5 \times 10^{-2} \text{ cm/sec} \left(\frac{150}{1000} \right)$$

$$= 7.5 \times 10^{-3} \text{ cm/sec}$$

Knowing 0.134 ft/day = 4.72 × 10⁻⁵ cm/sec (conversion equivalent), then

$$v_{avg} = \frac{7.5 \times 10^{-3}}{4.72 \times 10^{-5}} (0.134 \text{ ft/day})$$

$$= 21.29 \text{ ft/day}$$

$$v_s = \frac{v_{avg}}{n}$$

$$= \frac{21.29}{0.25} = 85.16 \text{ ft/day}$$

$$t = \frac{\text{distance}}{v_s}$$

$$= \frac{1000 \text{ ft}}{85.17 \text{ ft/day}}$$

$$= 11.74 \text{ days}$$

$$A = 200 \text{ ft} \times 2 \text{ ft} = 400 \text{ ft}^2$$

$$Q = kiAt$$

$$= 21.29(400)(365)$$

$$= 3,108,340 \text{ ft}^3 \text{ of water}$$

$$= 3,108,340 \times 7.48 \text{ gal/ft}^3 = 23.2 \times 10^6 \text{ gal/yr}$$

Note: The slight difference in answers is due to conversions of units.

Production of Groundwater

Two criteria must be met to obtain a usable water supply from the subsurface: (1) The soil or rock must be saturated and (2) it must have sufficient permeability to deliver water to the well. If both conditions are not met simultaneously, a dry well will result. A well in dry, coarse gravel or one in saturated clay will produce no flow of water.

A related question is how much water production is needed for a groundwater source to be called an aquifer. The lower limit of production is about 3.8 liters/min or 5472 liters/day (1 gal/min or 1440 gal/day) to be termed an aquifer but, in areas of very low production, water-bearing zones producing 0.27 liters/min or 389 liters/day (0.07 gal/min or 100⁺ gal/day) have been developed to provide a domestic water supply. Typically, a minimum of 190 to 380 liters/person/day (50 to 100 gal/person/day) is required for domestic water use.

The distinction between an aquifer and a water-bearing zone has taken on added significance with regard to groundwater contamination. For aquifers it is typically required that they be cleaned up to background groundwater parameters (Fetter, 2018); the requirement is less stringent for water-bearing zones that cannot support well production.

In most discussions, two types of aquifers are considered: those in soil materials and those in bedrock. Soil or unconsolidated materials (geological definition) offer sites for wells that will have the greatest rate of production.

Production from Unconsolidated Materials

Soils that produce meaningful amounts of groundwater for supply purposes must be coarse grained. Gravel, sandy gravel, gravelly sand, and coarse sand are the usual producers. Fine sand may be too fine in size to be screened out effectively during pumping, which can lead to problems with the pump and the well. In any case, the search for groundwater in unconsolidated deposits is really a quest for sand and gravel. An outline of the geomorphic (landform) features associated with coarse, cohesionless material is presented in Table 13.7.

Landforms Composed of Sand and Gravel

Stream-related features provide the major supply of groundwater. These include items 1, 2, and 3 in Table 13.7 and generally 4 and 5 as well. Outwash plains and valley train deposits occur in conjunction with stream deposits; they form during meltwater flow from glacial ice. So, despite their glacier-related origin, outwash plains and valley trains are associated with major stream development, at least for many of the major deposits related to continental glaciation.

Outwash plains of a smaller areal extent may form during melting episodes of a continental glacier. Such plains may exist at the current surface of the Earth in glaciated terrain or they may have been buried by readvancing and subsequently retreating ice to provide a sand and gravel layer at depth within a deposit of till. Small valley trains formed under similar circumstances and, today, can be found as channels of permeable material within the mass of till.

Alluvial fans form along an intermittent stream course where a sudden reduction in gradient causes the deposition of coarse-grained sediments. Alluvial fans most commonly occur at the base of mountains or of sufficiently high hills. Due to continued erosion, they will be covered by successive deposition during the retreat of the mountain front. Water supplies are obtained from wells drilled in the basins associated with retreating mountain fronts.

Kame terraces form, during glaciation of steep valleys, by alpine glaciers or by continental glaciers

Table 13.7   Groundwater Aquifers, Unconsolidated Materials by Origin.

Stream-Related Features
1. Floodplains
2. River terraces
3. Alluvial fans

Glacial (Glacial-Fluvial) Features
Most reliable
4. Outwash plains
5. Valley train deposits
6. Kame terraces

Less reliable
7. Eskers and kames
8. Selected portions of end moraines
9. Beach ridges

Coastal Features
10. Coastal plains
11. Sea terraces

that partially fill such valleys. The permeable layers form as terraces on opposite sides of the valley by meltwater flowing at the edge of the glacier along the valley wall.

Eskers and kames provide isolated deposits of sand and gravel and local supplies of water. Perched-water aquifers may be the consequence, in which case the associated reservoir is severely limited in volume. Such water supplies run short during periods of drought.

End moraines are dotted with glacial deposits of sorted, coarse-grained material. These features are usually identified easily in the field or from aerial photographs. They too tend to provide limited supplies of water because of their isolated nature.

Beach ridges occur in major lacustrine deposits where they mark the shoreline of the retreating lake water. They are linear features that stand above the terrain and, typically, contain medium to fine sand. Some can supply limited amounts of water.

Coastal plains commonly form areas of wide extent, which mark the flat ocean bottom after retreat or off-lap of the ocean. They are prevalent along the Atlantic Ocean and Gulf of Mexico in the United States, where uplift of the land surface during the late Cenozoic age exposed large land areas. This landform slopes gently toward the sea and major supplies of water are contained within its sandy layers. Excessive drawdown from over pumping and drainage is a serious problem and is considered in a later section of this chapter under the discussion on water supply problems.

Sea terraces are found overlooking the ocean where fairly recent uplift has elevated the beach area some distance above sea level. Such terraces are prevalent along the California coast where active uplift of the Pacific Coast Range continues today. Freshwater occurs at depth but typically at an elevation only a few feet above sea level.

Production from Bedrock

Groundwater can be obtained from bedrock also but, typically, in lesser quantities than from unconsolidated materials, because of the lower permeability of bedrock (see Table 13.2). The permeability of bedrock can be primary or secondary in nature. Table 13.8 provides information on the permeability relationships of bedrock.

Primary Permeability

Primary permeability is intrinsic to the rock and developed when the rock formed. The best bedrock aquifers are those with primary permeability. Sandstones, particularly the poorly cemented, friable ones, provide the most water in the sedimentary rock areas east of the Rocky Mountains. Reef limestones and dolomites are also good aquifers and occur frequently in the central United States. Sandstones and carbonate rocks also comprise the oil producers in the oil fields of the world. For basaltic and andesitic rocks, the flow tops of the lavas provide the major supplies of groundwater. There are only a few areas of the world where these rocks predominate, one being the Columbia Plateau in the northwestern United States. In this area, which is semiarid in nature, the withdrawal of water for domestic supplies and irrigation far exceeds the recharge of water into the aquifer. To this extent, the removal of water is much like mining a valuable mineral from the subsurface.

Secondary Permeability

Secondary permeability is due to fractures or solution openings in the rock and develops after the rock forms. Secondary permeability tends to be sporadic in nature and, therefore, more difficult to characterize in the subsurface and less reliable for

Table 13.8   Bedrock Material and Conditions Yielding Adequate Water Supplies.

Primary Permeability
- Sandstone
- Reef limestones and reef dolomites
- Flow tops of lavas

Secondary Permeability
- Jointing in massive rocks; includes limestone, dolomite, sandstone, intrusive igneous rocks, extrusive lava-type rocks, gneiss, marble, quartzite, massive slates and schist, plus other similarly massive units
- Fault zones and deeply weathered zones
- Caverns and solution channels in limestone, dolomite, and gypsum

water supply purposes. Wells drawing water associated with secondary permeability yield low volumes of flow; typically only a few liters/min (gal/min). Intersecting fractures at depth severely limit the reservoir capacity, thus yielding low volume. Extended dry periods will also cause such wells to run dry or to be seriously reduced. Pollution is a major concern because the few localized joints and fractures tend to be collection points for septic tank effluent and infiltration of runoff from animal waste areas, as well as the location of groundwater recharge.

Water Well Terminology

The permeability of a saturated zone is not the only factor determining the quantity of groundwater available. A 15-m (50-ft) thick layer of saturated coarse sand should be able to deliver to a well more water than a similar layer only 3 m (10 ft) thick.

The coefficient of transmissibility T, or transmissibility, is a measure of the ability to deliver water to a well. It is the rate at which water will flow through a vertical strip of the aquifer 0.3 m (1 ft) wide extending through the full saturated thickness under a hydraulic gradient of 1. Written in equation form:

$$T = kHi = kH \qquad \text{(Eq. 13-22)}$$

where i, the hydraulic gradient, equals 1 and k is given in gal/ft^2/day (Meinzers) or m/day (ft/day) and H in m (ft).

The transmissibility value for an aquifer will indicate its water supply capabilities. Values of T range from 12.4 to 12,400 m²/day (135 to 135,000 ft²/day) or 1000 to 1,000,000 gal/ft/day. Transmissibility of less than 124 m²/day (< 1350 ft²/day) or 1,000 gal/ft/day is good for domestic supplies only, whereas transmissibility values greater than 124 m²/day (> 1350 ft²/day) or of 10,000 gal/ft/day or more is adequate for industrial, municipal, or irrigation purposes. This can be illustrated with an example problem.

Example Problem 13.5

A sandy unconfined aquifer is 12.2 m (40 ft) thick with the water table located 2.44 m (8 ft) below the ground surface. The coefficient of permeability is $k = 4.3 \times 10^{-2}$ cm/sec. What is the transmissibility of the aquifer and is this a good candidate for an industrial water supply?

Answers

$$T = kHi$$

Using SI units:

$k = 4.3 \times 10^{-2}$ cm/sec $= 4.3 \times 10^{-4}$ m/sec $= 37.1$ m/day $= 1.41 \times 10^{-3}$ ft/sec

$H =$ saturated thickness $= 12.2$ m $- 2.44$ m $= 9.76$ m

$i = 1$

$T = 37.1$ m/day (9.76 m) $= 362$ m²/day

Using British engineering units:

$k = 122$ ft/day $= 910$ gal/ft²/day

$H =$ saturated thickness $= 40$ ft $- 8$ ft $= 32$ ft

$i = 1$

$T = 122$ ft/day (32 ft) $= 3904$ ft²/day

$= 910$ (32 ft) $= 29,120$ gal/ft/day

T is > 124 m²/day (> 1350 ft²/day) or $>10,000$ gal/ft/day. Therefore, the aquifer is adequate for an industrial water supply.

When pumping starts in a well, the water level in the nearby aquifer is lowered. The amount the water surface drops is called *drawdown* and the largest drop occurs immediately near the well. Drawdown will be less at greater distances from the well until, at some distance, a point occurs where the water level is, essentially, unchanged. The slope of the unwatered zone increases toward the well and, as the water flows inward from all directions, a cone of depression forms around the pumping well (see Figure 13.6).

As pumping continues, the cone of depression continues to move outward from the well. At some point, the cone of depression stops expanding, although the pumping rate holds constant. When this occurs, the condition of equilibrium occurs and no further drawdown develops as pumping continues. In some wells, equilibrium develops within a few hours after pumping begins. In others, however, it does not occur even after an extended period of pumping. The distance from the well to the outer circle of zero drawdown is known as the radius of influence. In a water table aquifer the radius of influence is related to the transmissibility. Assuming equal pumping rates, a well with very high transmissibility (1240 m²/day or 100,000 gal/ft/day) will have a small drawdown and a small radius of influence as compared to a well with moderate transmissibility (124 m²/day or 10,000 gal/ft/day), which will show a considerably greater drawdown and also a much greater radius of influence. Hence, a highly permeable material has a gentler drawdown curve and smaller drawdown than does a lower permeability material. In addition, wells in artesian aquifers have a much greater radius of influence (in the range of about 1500 m or 5000 ft) than do water table aquifers (a 120 m or 400 ft extent is more common).

Another useful term regarding groundwater supply is the *specific yield*. This is the quantity of water that a unit volume of the aquifer will give up when drained by gravity. It is expressed as a portion of the porosity. The remainder of the water, that portion not given up under gravity drainage, is called the *specific retention*. The specific yield plus the specific retention is equal to the porosity. For example, if 0.0042 m^3 (0.15 ft^3) of water is drained from 0.028 m^3 (1 ft^3) of saturated coarse sand, the specific yield of the sand is 0.15 or 15%. If the porosity of the sand is 35%, its specific retention is 0.20 or 20%.

The coefficient of storage, S, of an aquifer is the volume of water released from storage (or taken into storage) per unit of surface area of the aquifer, per unit change in head. This is the volume of water removed per unit volume of material dewatered during pumping. In water table aquifers, S is the same as specific yield of material in regard to pumping. In artesian aquifers, S is somewhat more complicated. It is the result of two elastic effects, compression of the aquifer and expansion of the contained water when the pressure (or head) is reduced during pumping. As with porosity and specific yield, S is a dimensionless term. For water table aquifers, S ranges from 0.01 to 0.35 and for artesian aquifers the range is from 1×10^{-5} to 1×10^{-3}. The coefficient of storage is used, along with the transmissibility, to determine the permeability of an aquifer that has not reached steady-state conditions. This aspect of groundwater supply is beyond the scope of this text. For further details on this subject the reader is referred to *Groundwater and Wells* by Driscoll (1986).

Legal Details of Groundwater Ownership

Groundwater ownership in the United States can best be understood by considering it in relation to the legal details of surface water. Usually, legal rights to surface water are obtained by either land ownership or by appropriation of specific water resources that are located close to, or a short distance from, the place of use.

Groundwater Rights Based on Land Ownership

Originally, under English common law, by virtue of land ownership a landowner had a riparian right to unlimited use of water from a stream or lake that bordered the owner's land. In regard to groundwater, the doctrine of capture was applied, which stated that anyone who owned the overlying land also owned the water beneath it. As the United States expanded westward and increased its use of natural resources, English common law and the doctrine of capture were no longer practical. A modification known as the American rule developed in many states regarding riparian rights. Under this rule, all landowners whose properties adjoin a body of water have the right to make reasonable use of it as it flows through or over their properties. Also, groundwater must not be removed for a malicious or wasteful reason, but that some positive purpose must exist for pumping water from the subsurface.

Riparian rights for surface water and groundwater ownership prevail in much of the eastern half of the United States. The western states follow the doctrine of prior appropriation. This region is generally to the west of the 100th meridian. Simply worded, the doctrine states that the first person or group of people to appropriate the water gets first rights to use the water. That is, the order of priority is based on the order in which people have established the right to the supply. Reductions start with the last allocation and work back from there. The appropriation can be for water on the owner's land or it can be for water located hundreds of miles away from where it will be used. In the western states where appropriative rights prevail, surface water and groundwater sources are well integrated under the law to protect the prior appropriations.

Development of Water Rights and Conservation

The early practices described here were simply principles on who had the right to use water, and how much they were allowed to use. They did not necessarily consider the preservation or conservation of water as a natural resource. For example, regarding groundwater, common law does not recognize that adjacent owners (and sometimes distant owners) take water from the same aquifer and, thereby, it encourages individual use rather than some plan of sharing. It also encourages increased usage by virtue of individual, unlimited use rather than any form of conservation. Under the American rule, defining the reasonable use of water for an acceptable purpose does not ensure the conservation of water; it only reduces water use by a small amount. Dewatering for construction or for surface and subsurface mining could be considered an acceptable purpose, however, the environmental impacts should be considered.

Groundwater laws have developed in response to the needs and demands of citizens in individual states of the United States. For example, in northwestern Indiana, a problem developed during the 1980s in the Kankakee Outwash and Lacustrine Plain (see Chapter 20), where groundwater is obtained from an artesian bedrock aquifer. Known as the Fair Oaks Farms case, it involved a high rate of pumping for irrigation purposes that drastically lowered the piezometric surface on the adjacent farmer's land. A drop of more than 15 m (50 ft) was noted close to Fair Oaks Farms and the cone of pressure relief with more than 0.3 m (1 ft) of drawdown extended for a radius of more than 16 km (10 mi). Consequently, the pumping affected an area of about one county in the state of Indiana.

Neighbors' wells went dry if the drawdown extended below the depth of their pump intakes. This problem was solved by installing in-hole pumps to lift water more than 7.5 m (25 ft). The piezometric surface, prior to the pumping, was only about 3 m (10 ft) below the ground surface. A suction pump can lift water only by 9.75 m (32 ft), which reduces to about 7.5 m (25 ft) based on pump efficiency. Some older, small diameter wells would not accept in-hole pumps, which necessitated drilling new, larger diameter wells and purchasing new pumps. This was an expensive endeavor that the adjacent landowners felt was unjustified. They appealed to the Indiana legislature and initiated litigation against Fair Oaks Farms.

The result of this controversy was that Fair Oaks Farms settled out of court for damages to the landowners and the Indiana legislatiure changed the state law, replacing prior principles that allowed Fair Oaks Farms to obtain as much water as they desired as long as it was not wasted. Now all large-capacity wells (more than 1 m^3/min or 250 gal/min) must be registered with the Indiana Department of Natural Resources and, if the piezometer surface is drawn down more than 7.5 m (25 ft) in an area, pumpage is reduced by all parties to maintain minimum levels.

Legal Complications

Complications do develop in the application of groundwater laws. It is difficult to apply laws developed for surface water to the groundwater regime. For example, it is extremely difficult to define the boundaries of groundwater supplies, or "groundwater streams" as one would define surface drainage basins, to yield distinct regions to which the doctrine of riparian rights can be strictly applied.

Another problem is related to the large number of people affected by groundwater use in a region. An aquifer may extend beneath the property of hundreds to many thousands of individual owners and include parts of several states. Water removed from one part of the aquifer must eventually affect the water level and possibly the pressure in other parts of the aquifer. An equitable division of water rights under such an involved situation is virtually impossible.

Safe Yield

The total amount of water to be removed from the subsurface under the doctrine of prior appropriation is another difficulty. The annual safe yield, which is determined by the amount of groundwater recharge, is the maximum water withdrawal that can be sustained for extended periods of time without an ultimate shortage. Pumpage greater than this amount is simply the mining of water. In California, safe annual yield is used to determine the amount of water to be appropriated. In other western states this is not the case and, with an ever-increasing population in the southwestern Sun Belt, this problem of safe yield must soon be faced.

Industry and Municipal Use

In many states with riparian rights for water supply, individuals have legal rights that are not extended to industry or municipalities seeking groundwater supplies. As previously stated, individuals can withdraw as much groundwater as they wish as long as it is removed for a positive purpose (American rule). A municipal water supply, however, does not have that same right, without certain conditions. For example, a water supply company or utility could not purchase a small lot in a housing area or subdivision and compete with surrounding landowners for subsurface water by pumping without limit. A landowner accepts certain implied risks when buying property with regard to groundwater supply, but the possibility that a major water user, such as a water works or industrial plant, will locate itself nearby is beyond the normal level of expected risk. Therefore, the industrial plant or municipal water works would have to buy a sufficiently large piece of property so that the water supply of its neighbors would not be altered by groundwater pumping.

Another consideration would come into play regarding the use of subsurface water by the municipal water company. Implied in the common law relationship is that the water would be used on the property from which it is obtained. If used off-site, as in the water works example, the common law rule would no longer apply. The rights of the water works in this situation would not be covered by the rule.

Saltwater Intrusion

Along the eastern seaboard, saltwater intrusion from the ocean into freshwater aquifers is a major concern. The technical details of this subject are discussed in a later section. Here we will provide a brief example of legal complications. In some mining operations, typically those for potash and phosphates, the area adjacent to the ocean is dewatered to allow for a dry surface-mining procedure. Such dewatering accelerates saltwater encroachment. Under previous principles, this practice would be allowed because the water is removed for a productive reason: a more economical mining of a necessary raw material. However, in order to control this problem, a law limiting pumping or permanent lowering of the water table had to be enacted. The apparent technical solution is to return the water to the subsurface in an area located between the operating mine and the ocean, but to continue dewatering operations at the mine.

Appropriate Rights

In the western United States, it is possible that an owner with second appropriative rights can cause a lowering of the water table that affects the owner with first appropriative rights. The net effect is that the first owner must deepen his or her well, which costs money. The second person should compensate the first for this expense but not for all of it because part of the depletion is due to the first person's

pumping as well as that of the second. This percentage of compensation would be determined legally or by mutual agreement.

In many states that operate under riparian rights, special legislation is required to move water from one drainage basin to another. This applies to both groundwater and surface water and involves major consequences, even for states that apparently have more than adequate water supplies, i.e., those in the eastern half of the United States. For the western states, where appropriative rights prevail and water supply is acute, even greater consequences are involved. Typically, federal and state legislation is involved in projects such as the Fryingpan–Arkansas Project, which brings west-slope water from Colorado east of the Rockies into the Arkansas River.

The present-day arrangement is to vest the power for groundwater regulation in governmental agencies at either the district, state, regional, or federal level. In many cases, the individual states have appropriate agencies that have the authority to act in this capacity. Other states have local agencies that set limits of groundwater production and levy taxes on owners of large wells to pay for the cost of recharging groundwater aquifers.

Water Witching

Water witching, divining, or dowsing is a nonscientific method used by some individuals to search for water in the subsurface by means of a forked stick, coat hanger, or metal rod. They assume that some unknown force attracts the stick or other object toward the source of groundwater. However, these individuals lack the training and knowledge to evaluate this phenomenon.

One explanation is that an electromagnetic force is generated by flowing water, which attracts the stick toward water. This, however, has no basis in fact, as groundwater does not flow through the ground at anything approaching a velocity needed to generate an electromagnetic force, nor does it flow rapidly in veins or narrow conduits in the ground. According to Darcy's law, groundwater flow is laminar with velocities well below 10^{-2} cm/sec (0.04 in/sec), or considerably less than 25 cm/hr (1 ft/hr). Many hydrogeologists have witnessed water diving and concluded that a success in finding water was only a coincidence or that the aquifer that was discovered was so extensive that water could be found anywhere below the ground surface. Yet water witching is a persistent myth that many laypeople continue to believe.

Groundwater Pollution

Groundwater pollution can occur in many different ways and has an equal number of causes. Contamination by heavy metal cations; organic chemicals in the form of pesticides, herbicides, and solvents; and biological constituents can occur. If the dissolved solids or suspended solids become too great, thereby lowering water quality to an unacceptable level, then water pollution has occurred. The problems to consider here are the following sources of pollution: (1) septic tanks, (2) sanitary landfills, (3) miscellaneous causes, and (4) saltwater encroachment (Karamouz et al., 2011; Fetter, 2018).

Septic Tank Fields

A septic tank system is a simple sewage treatment method used by individual households or a small group of homes or apartments. Sewage is held in the septic tank, under anaerobic conditions, for a period of time, after which the processed liquid is transferred to the ground through an underground tile system. Figure 13.16 provides the details of a septic system.

Septic tank systems must have free drainage into the subsurface (not onto the Earth's surface) if they are to operate on a long-term basis without causing pollution problems. Two features of the soil can prevent a septic tank from operating properly: (1) too low a permeability at the contact area of the septic tank field to allow drainage into the soil and (2) a perched water table, which prevents gravity flow of water from the tiles in the septic field into the subsurface.

Soil areas can be rated relative to septic tank use based on two different approaches: (1) soil percolation and (2) details from agricultural soils maps, which indicate relative permeability and depth to the highest seasonal water table.

Percolation Tests

Four to six 10-cm (4-in) diameter auger holes are drilled to the depth of the proposed septic tank field at the site. Five cm (2 in) of coarse sand or fine gravel is placed on the bottom of each hole. The holes are filled with water to at least 30 cm (12 in) above the

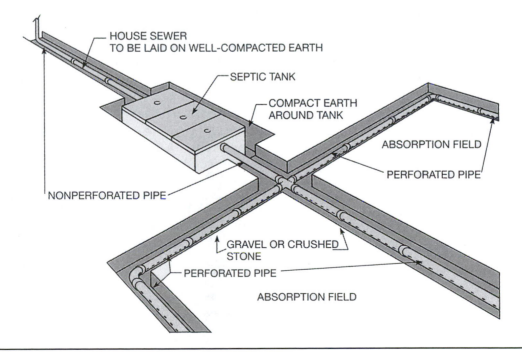

Figure 13.16   Septic tank sewage disposal system.

gravel. Water is left in the holes overnight. Before the test begins the next day, water is added to reach 15 cm (6 in) above the bottom of each hole and the amount of drop is measured at 30-min intervals for 4 hr, refilling to the 15-cm (6-in) level above the bottom as needed. The drop that occurs in the final 30-min period is used to determine the percolation rate.

In sandy or other soils, where the water drains away in less than 30 min, time intervals between readings should be 10 min with the test run for 1 hr. The drop occurring in the last 10 min is used to calculate the percolation rate in this situation. The final reading is used to calculate the percolation rate in terms of the number of minutes required for the water surface to fall 2.5 cm (1 in).

The size of the septic tank leach field is determined on the basis of the area necessary for the base of the septic tank trench or base of a seepage pit. The trench area is related, in a seemingly strange manner, to the number of bedrooms in the planned house and the percolation rate of the soil. Table 13.9, for example, indicates that a percolation rate of 10 min/in requires 165 ft^2 per bedroom as the area for the bottom of the septic tile field trench. If a three-bedroom home is involved, $3 \times 165 = 495$ ft^2 of trench bottom is required. If the trench is to be dug 2 ft wide, then 495/2 = 248 ft of trench length will be needed.

Table 13.10 indicates the restrictions for distances between the septic tank field and other portions of the property. Table 13.11 deals with selecting the size of the septic tank.

The relationship between the *number of bedrooms* versus the size of the septic tank field and the size of the septic tank needs further discussion. Obviously, the more sewage, the larger the septic tank should be and, in turn, the larger the absorption field. The size of the house or the number of occupants is related to this, but so is the number of appliances, such as dishwashers and garbage disposals, and this is related to the general affluence or standard of living of the residents. One measure of affluence is the number of bedrooms, because houses with more bedrooms are generally more expensive and contain more water-using appliances. Therefore, the number of bedrooms is a reasonable measure of house size and affluence and is more appropriate than simply using the square footage of the house. Also this information is usually easily accessible.

Contrary to popular belief, septic tanks do not accomplish a high degree of bacteria removal. Although some treatment occurs, not all infectious material is removed. Also, the effluent released from the septic system may be more objectionable (more odiferous) than was the sewage entering it. This does not detract from the primary, valuable service

Table 13.9   Percolation Rate versus Absorption Rate.

Time Required for Water to Fall 2.5 cm (1 in) per hr	Required Absorption Rate	
	m^2 per bedroom	ft^2 per bedroom
≤ 1	6.5	70
2	7.9	85
3	9.3	100
4	10.7	115
5	11.6	125
10	15.3	165
15	17.7	190
30	23.3[a]	250[a]
45	27.9	300
60	30.7[b]	330[b]

Notes: Rates are based on standard trench and seepage pits.

a Unsuitable for seepage pits if less than 30.

b Unsuitable for leaching systems if less than 60.

Table 13.10   Distance between Septic Tank Field and Property Aspects.

Property Aspects	Minimum Distance to Closest Part of Septic Tank System	
	(m)	(ft)
Water well	30	100
Stream, lake, or other water course	15	50
Dwelling or property line	3	10

Table 13.11   Capacity of Septic Tanks versus Size of House.

Number of Bedrooms	Recommended Minimum Tank Capacity	
	(gal)	(liters)
2 or less	750	2839
3	900	3406
4	1000	3785
Each additional bedroom	add 250	add 946

performed by the septic tank, which is to condition the sewage so that less clogging of the disposal field will occur. Further treatment of the effluent, including removal of pathogens, is accomplished by percolation through the soil. The disease-producing bacteria die in time (one to three months) as a result of the unfavorable conditions prevailing in the soil. An example problem on septic tank absorption field design is in order, followed by a related problem concerning soil retention time.

Example Problem 13.6

The average of six auger holes indicated that 10 min were required for the water level to fall 2.5 cm (1 in) in the hole. The house will have four bedrooms. What size septic tank absorption field is required?

Answers

A reading of 10 min/2.5 cm (1 in) of drop requires 15.3 m² (165 ft²) per bedroom for the standard trench. For a four-bedroom house, 4 × 15.3 m² = 61.2 m² or 4 × 165 ft² = 660 ft² is required. Using a trench 0.6 m (2 ft) wide, the trench should be 61.2 m²/0.6 m = 100 m long or 660 ft²/2 ft = 330 ft long. Three "fingers" 33.5 m (110 ft) long each, would be workable. See the sketch in Figure 13.17.

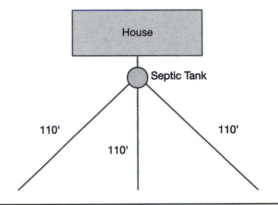

Figure 13.17   Plan view of septic tank absorption field construction.

Example Problem 13.7

Assuming that k for the soil is 1×10^{-3} cm/sec and the gradient of the water table is 1%, how long does it take for water to travel 15 and 30 m (50 and 100 ft) through this soil?

Answers

$$v_{avg} = ki$$

$$= 1 \times 10^{-3} \text{ cm/sec} \times 0.01 = 1 \times 10^{-5} \text{ cm/sec}$$

$$v_s = \frac{v_{avg}}{n_e}$$

where n_e is the effective porosity for the soil. For a silty material, assume $n_e = 15\%$. Therefore,

Using SI units:

$$v_s = \frac{1 \times 10^{-5} \text{ cm/sec}}{0.15}$$

$$= 6.66 \times 10^{-5} \text{ cm/sec} = 6.66 \times 10^{-7} \text{ m/sec}$$

(continued)

$$d = v_s t$$

$$t = \frac{d}{v_s}$$

$$= \frac{15 \text{ m}}{6.66 \times 10^{-7} \text{ m/sec}}$$

$$= 22.25 \times 10^{-6} \text{ sec} = 0.71 \text{ yr}$$

For 15 m, $t = 0.71$ yr.
For 30 m, $t = 1.42$ yr.

Using British engineering units:

$$v_s = \frac{1 \times 10^{-5} \text{ cm/sec}}{0.15}$$

$$= 6.66 \times 10^{-5} \text{ cm/sec}$$

$$= 2.187 \times 10^{-6} \text{ ft/sec}$$

$$d = v_s t$$

$$t = \frac{d}{v_s}$$

$$= \frac{50 \text{ ft}}{2.187 \times 10^{-6} \text{ ft/sec}}$$

$$= 22.88 \times 10^{6} \text{ sec} = 0.725 \text{ yr}$$

For 50 ft, $t = 0.725$ yr.
For 100 ft, $t = 1.45$ yr.

Note: The small difference in answers between British engineering units and SI units is due to round off aspects (e.g., 15 m = 49.2 ft).

Agricultural Soils Maps

Agricultural soils maps and USDA soil classification were discussed in Chapter 4. These maps subdivide ground slope into the following ranges:

- 0 to 2% = level to nearly level

- 2 to 6% = gently sloping (undulating)

- 6 to 12% = sloping (rolling)

- 12 to 20% = strongly sloping (hilly)

- 20 to 35% = steep

- 35+% = very steep

Agricultural soils maps also rate different soil subclasses relative to engineering uses, such as building foundations, ponds, home sites, roadways, septic tank fields, cemeteries, basements, sanitary landfills, and excavations. The various subclasses are rated good, moderate, and severe relative to these uses. Regarding septic tank absorption fields, the factors of slope, permeability, and depth to a seasonal perched water table are the important determinants for the rating.

Subclasses with too steep a slope (76%), too low a permeability, or too high a seasonal perched water table (30 cm [12 in] or less from the surface during the wet season) are rated severe regarding septic tank fields. When these categories are less extreme, a moderate rating is given, and a gently sloping, permeable soil (sands and silts) with no perched water table receives a rating of good.

In some counties of the Midwest (primarily good agricultural producing areas) construction permits for houses with septic tanks are granted on the basis of the soils map or soil subclass, rather than on the basis of percolation tests. There is a reason to question percolation test results in some cases, because the tests may be run in the dry season when the perched water table is absent and high permeability readings will prevail because the soil is cracked from drying out. Percolation tests should illustrate the average, or perhaps the worst, condition rather than the best. Building permits may be denied on the basis of the agricultural soils class alone in some areas.

Rejecting sites for septic tank use focuses attention on a critical concern about the procedure for rating soils relative to specific engineering use. This rating practice causes certain areas to be bypassed as housing sites when the real need is not to prevent construction in marginal areas but, instead, to ensure that proper construction procedures are used as needed. With proper design and construction, severe sites can be acceptable. This may require drainage of perched areas by field tile, the addition of compacted soil at the site to bring the surface above the perched water, or dual septic tank fields with alternate use, allowing a time for each to recover. The proposed solution is to require that construction in severe areas be approved by a competent professional engineering geologist, civil engineer, or agronomist who has established expertise in this specialty. Simple exclusion is not the answer; proper design and construction offer a better approach.

Sanitary Landfills

A sanitary landfill is a method of disposing of refuse or solid waste on the land surface in a manner that protects the public health and environment. Municipal solid waste (MSW) is placed in layers or cells, densified by heavy compacters, and covered daily by a layer of soil. Current landfills have a low permeability base, leachate and methane collection systems, and a clay or flexible membrane cap to minimize infiltration.

The quality of sanitary landfills has improved significantly since the 1970s, when all types of solid waste were disposed of in any preexisting excavation, regardless of the geologic conditions. Commonly, solid waste was placed in abandoned gravel pits, some of which contained standing water. These were appropriately called open dumps. They attracted rodents and birds and were subject to smoldering fires that were difficult to extinguish. An early improvement was to cover the trash with a daily layer of 15 cm (6 in) of soil to keep vectors

(birds and rodents) away and prevent fires. Geologic conditions were considered next, requiring either a natural or man-made dry clay base. These landfills are referred to as attenuation landfills based on the assumption that the clay minerals neutralized the heavy metal cations in the leachate that percolated from the solid waste into the clay base. Leachate is any water that comes in contact with the trash and it has to be disposed of because of its potential to cause groundwater contamination. It was soon realized that attenuation landfills did not function well and leachate pooled at the base of the landfill and commonly seeped out along its base. From this it was clear that the leachate had to be collected, and basal collection systems were installed in the next phase of landfills.

To prevent leachate from moving through the clay base, geomembranes are required at the base and a clay cap at the top of the fill to reduce infiltration and leachate generation (Qian et al., 2002; Karamouz et al., 2011; Fetter, 2018). A geomembrane is an impermeable liquid or vapor barrier made from flexible, polymeric sheets. They are similar to the geosynthetics used extensively in civil engineering construction. The final addition was the installation of methane collection wells, penetrating the solid waste to collect this greenhouse gas. The methane can be used on-site for heating purposes or sold to neighboring industrial facilities. If all the methane is not utilized, the remainder is flared rather than venting it to the atmosphere.

In the past, two different construction methods were used, the trench method and the area method (see Figure 13.18 on the following page). In the trench method a vertical cut 4.5- to 6-m (15- to 20-ft) deep is made, the refuse placed in the trench, and the soil excavated from the advancing trench is used for cover material. In the area method the MSW is placed on level ground or the land surface is excavated to a designated depth, up to 12 m (40 ft), and the excavated soil is used for daily cover. In this way a greater capacity or "air space" is generated for trash disposal.

The top of the completed landfill is sloped steeply enough to encourage runoff but not too steeply as to cause increased erosion. Slopes of 3 to 6% on the top and 20 to 33% on the sides are generally considered. Runoff from the surface of the landfill is collected in settling ponds and kept separate from leachate, which requires treatment before disposal. Hazardous waste landfills typically require double liner systems with two leachate drainage layers between the low permeability layers. The second drainage layer ensures that leachate from the hazardous waste does not enter the subsurface. Figure 13.19 is a cross section of a municipal solid waste landfill showing all the aspects of current operations and standards.

Table 13-12 on page 300 lists the geologic considerations for proper siting of landfills in the Midwestern areas of the United States.

Miscellaneous Groundwater Pollution

There are numerous other means by which groundwater pollution can occur. Industrial wastes can cause groundwater pollution if the wastes are not properly contained or treated. In some cases, toxic heavy metal cations can be transmitted to the groundwater. Corroded containers of industrial wastes supply high concentrations of cadmium, chromium, lead, zinc, and other cations to the groundwater.

(a)

(b)

Figure 13.18 Methods for constructing sanitary landfills: (a) the trench method and (b) the area method.

There is also a concern about chloride contamination. Large supplies of $CaCl_2$ are stored in the summer and fall for use during the winter to deice snow-covered, icy roads. Typically, some of the stockpile remains year-round unless supplies are depleted during a severe winter. In the past, such piles were left outdoors, uncovered. Rainfall dissolved the salt and runoff carried heavy concentrations of Ca^{2+} and Cl^- into surface streams and the groundwater regime. Today, permanent, tall, pointed structures are built to house the salt stockpiles. These structures prevent rainwater accumulation and the subsequent salt-charged runoff.

Groundwater pollution caused by open fissures and caves in limestone and dolomite, and to a lesser extent in gypsum and rock salt, provides nearly an endless list of examples. The high permeability of fractures and caves is the basis for the problem, which, rather than yielding water velocities of approximately 100 m (328 ft) per year, can be hundreds of meters (hundreds of feet) per minute. No

exchange with the soil or rock mass occurs during such rapid transport so there is virtually no reduction in toxicity, bacteria count, or suspended solids during the process.

Countless cases of groundwater pollution by livestock in limestone terrain have been recorded and pollution of distant wells from industrial waste through cavernous limestones is far too commonplace. The flow of septic tank effluent to an adjacent water supply well can be increased by the presence of open fissures, yielding pathogenic bacteria to the water supply. Special care to keep pollution away from open fissures must be practiced in areas with karst topography.

Joint-controlled groundwater flow in massive rocks of all kinds must be carefully scrutinized. In massive quartzites, granites, gneisses, schists, and similar rocks, preferred joint directions may focus much of the flow through a major bedrock region. Joint intersections and shear zones may carry much of the water flow for a bedrock mass measuring many

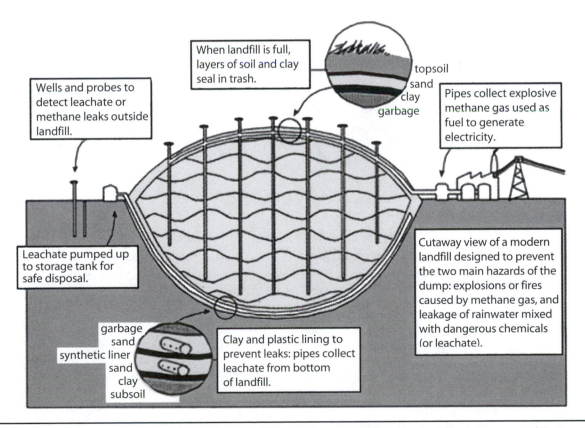

Figure 13.19 Cross section of an MSW sanitary landfill meeting current standards. (US Environmental Protection Agency, 2016.)

Table 13.12   Geologic Considerations for Proper Siting of Landfills in the Midwestern United States.

Type, Nature, and Stratigraphy of Soils

- Glacial till comprises most of ground moraine, which is mostly clayey silt and silty clay with occasional sand lenses. Place monitoring wells in water-bearing sand lenses.
- End moraine features—contain glacial till plus thicker sand lenses in more contorted shapes than ground moraine. Need to determine geologic detail before placing monitoring wells in these sands.

- Outwash sands—challenging areas to locate landfills. May have layers of low permeability materials that subdivide the cross section into separate water-bearing zones. Need to determine the subsurface details before placing monitoring wells.
- River floodplains—similar to outwash sands.
- Residual soils—weathered material above bedrock. Determine the thickness and nature of this material.

Thickness of Unconsolidated Material

- Trench method of landfill—excavate material before placing trash. Must have adequate thickness of soil after excavation, 6 m (20 ft) or more needed between base of trench and an aquifer.

- Concept—want to prevent movement of leachate through the water-bearing zone.
- Monitoring wells need to be placed in the appropriate water-bearing zones.

Type and Nature of Bedrock Below Unconsolidated Material

- Sandstone—permeable.
- Limestone, particularly weathered surface of bedrock—permeable.
- Shale—nonpermeable.

- Other areas outside of the Midwest may consist of igneous or metamorphic rocks, which have different permeabilities.
- Concept—may need monitoring well in bedrock aquifer if overburden soil thickness is not very great or if movement through bedrock aquifer is a concern.

Groundwater Supplies in Vicinity

- Protection of existing water supply is crucial.
- Must study local groundwater supplies before placing monitoring wells. Where do neighbors obtain their water? At what depth? What geologic material is present? What are artesian pressures?

- Geologic detail is extremely important. There may be no connection between water-bearing zones at landfill site and for adjacent areas. Need to study landforms and subsurface information.
- Pressure relationships are important. Water flows from high pressure to low pressure through permeable materials.

Topography of Site Including Flood Potential

- Upland area, away from tributary streams, minimizes the effects of runoff from landfills. This relates to regional water flow as well. Consider during monitoring program.

- Flood potential—generally need to stay away from 100-yr flood elevation. Site can be protected by levee built to this elevation.

Groundwater Table Location and Water-Bearing Zones at Site

- Water table in permeable material—easy to observe during drilling program.
- Water table in low permeability soil—more difficult to determine. Check water levels in borings without sand layers after extended periods of time.
- Need to locate monitoring wells in water-bearing zones that can be affected by leachate migration.

- Need to determine geologic details before placing monitoring wells. Perform subsurface investigation first, using split spoon sampling and Shelby tube sampling. Develop geologic cross sections for site. Return to site and place monitoring wells after evaluating data.
- Perform geologic study first, then place monitoring wells based on geologic findings.

tens of meters (tens of feet) in dimension. Water that supplies domestic wells along with water from septic systems may commonly be directed through this dominant flow system. If the fissures are fairly open and the travel time is short, pollution problems can develop in such terrains.

As a case in point, the city of Ottawa, Ontario, experienced problems with pollution of groundwater from an aquifer in jointed quartzite, which is situated at some depth below the city. The problem was traced to the farm belt, a number of kilometers (miles) away, where the jointed quartzite crops out. Runoff from animal feed lots was getting direct access to the open quartzite joints and little or no improvement in water quality had occurred during the miles of travel underground.

Saltwater Encroachment

Along coastal areas freshwater can be obtained from the subsurface because its lower density allows it to float above saltwater. This balance can be upset by over pumping the groundwater. When that occurs, saltwater encroaches on, or moves into, the freshwater aquifer (Marshak, 2013).

Saltwater encroachment has become a major problem for the populated areas along the eastern seaboard of the United States. Some of the areas, such as Long Island, New York, are heavily populated year-round, and others are popular tourist areas, which experience large increases in population during the tourist season. Atlantic City, New Jersey, is an example of a coastal area that typically experiences heavy increases in population during the summer months and had to allow for such needs in their water supplies. With the advent of casino gambling there, an increase in tourism continues over much of the year, putting a further strain on their water supply. Some coastal areas of California and Florida have also undergone problems with saltwater encroachment.

Recharge of an aquifer after encroachment of saltwater is not easy to accomplish. In lowering the water table, some permeability is lost because of compaction and consolidation so that greater pressures are required to put the freshwater back into the aquifer. This collapsing of the aquifer prevents adequate recovery of the freshwater aquifer even when high pump pressures are used.

The effect of groundwater withdrawal on saltwater intrusion is illustrated for an island, observed in cross section, in Figure 13.20. Since pressure under the ocean must equal the pressure under the island,

$$d(1.03) = H(1.00)$$

but

$$H(1.00) = (h + d)(1.00)$$

$$d(1.03) = (h + d)(1.00)$$

$$d = 33h$$

Hence if h is decreased by 0.3 m (1 ft), that is, if the water table is lowered by 0.3 m (1 ft), d decreases by 10 m (33 ft) or, indeed, the interface moves up 10 m (33 ft) as a consequence of the 0.3-m (1-ft) drop in the water table.

This illustrates why it is so critical that a permanent lowering of the water table not be allowed in coastal areas, either by overuse or by dewatering the sediments to allow mining. As previously stated, this can be prevented in mining areas if the water is recharged into the ground between the mining site and the seacoast, thus replenishing the saturated zone in that location.

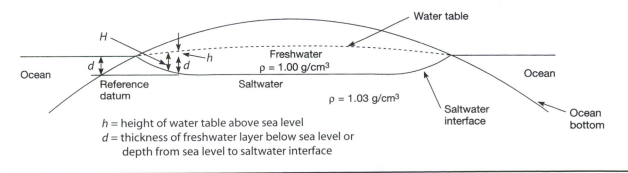

Figure 13.20   Island with saltwater intrusion.

Formation of Caves in Carbonate Rocks

The origin of caves in limestone and dolomite is a subject of much discussion in the field of geology. A limestone cave is shown in Figure 13.21. Stalagmites extend from the ground up and stalactites extend from the ceiling down.

The primary elements of discussion are the ways in which solutioning occurs, how structural features of the rock affect dissolution, and where in the groundwater regime solutioning develops (Ford and Williams, 2007).

Karst topography is common in a limestone or dolomite region and sinkholes, rounded surface depressions, are prominent. Figure 13.22 on page 304 is a photograph of a karst area.

Calcite ($CaCO_3$), the primary constituent in limestone and, to a lesser extent, dolomite [$CaMg(CO_3)_2$], are soluble in dilute carbonic acid, which occurs in subsurface water. This is shown in the following equation:

$$H_2O + CO_2 \rightarrow H^+ + \underset{\text{carbonic acid}}{HCO_3^-} \qquad \text{(Eq. 13-23)}$$

$$CaCO_3 + 2H^+ \rightarrow Ca^{2+} + H_2O + CO_2 \uparrow \quad \text{(Eq. 13-24)}$$

The CO_2 gas yields the sudden effervescence of bubbles that occurs when HCl is dropped on a piece of limestone. Dolomite "fizzes" much less vigorously and it may require that the rock be powdered before any action is observed.

Water movement in massive rocks, such as carbonates, is controlled by the discontinuities: the joints and bedding planes. Near the surface, the joints are open and water moves freely into and along them. Some dissolution undoubtedly occurs. Between the joints, high pinnacles of unweathered rock remain, in contrast to the vertical or nearly vertical joints where the carbonate rock has dissolved. The insoluble residues of chert, iron oxide, and clay minerals remain. This yields an extremely uneven bedrock surface in limestone terrain with pinnacles and pits located in adjacent positions. Excavation and foundation support problems can result from this uneven bedrock surface.

The dissolution occurs over long durations of time (by human standards), typically hundreds of thousands to millions of years. This is a rather short time geologically, but it does illustrate that short-term solutioning of carbonate rocks is seldom an engineering problem. The problem occurs, instead, where the solution channels already exist.

Dissolution of carbonate rocks is controlled mostly by the jointing. The two features of joints involved in this regard are (1) the attitude or orientation of the joints and (2) the continuity or extent to which the joints are through-going in the formation.

Typically, there are two prominent vertical or near-vertical joint sets in gently dipping sedimentary rocks. These are the strike joints, those running parallel to the strike of the rocks, and the dip joints, those parallel to the direction of dip. The two sets are, therefore, mutually perpendicular (orthogonal), or nearly so. Water will move along these joints, intersect a prominent bedding plane, and then continue along this nearly horizontal direction, at right angles to the joint. Hence, vertical and nearly horizontal openings form by enlargement along the rock discontinuity. Figure 13.23 on page 304 shows the joint sets in limestone that give rise to solution action and cave formation.

Not all joints run entirely through a geologic formation. Many terminate at prominent bedding planes. This can be observed in both outcrops and underground openings in rock. Truncated joints not only provide increased rock mass strength but also disrupt the primary conduits for water movement through the mass.

One set of joints is typically more extensive than the other. The strike joints, for example, may be more continuous, actually penetrating the entire formation. If this is the case, water movement will be concentrated along this vertical joint direction.

Dissolution occurs as the water moves through the joint. Also, the purity of the carbonate comes into play after an open joint and conduit system develops. For argillaceous (clayey) limestones and dolomites, less carbonate must be dissolved per unit volume of rock than for purer carbonates, and the clay can be carried away by the moving water. Therefore, a combination of carbonate purity, jointing pattern, availability of water, and topography determines the extent of solution development.

The location within the groundwater regime where the major dissolution occurs is a primary consideration. Various hypotheses have been proposed suggesting that a major cave system forms (1) above the water table, (2) at the water table, or (3) below the water table.

The argument that dissolution and erosional effects to the limestone (or dolomite) would be greater for water flowing under pressure suggests an origin for the major cave system either right at or somewhat below the water table. Above the water table, in the vadose zone, water moves downward under the direct influence of gravity, but no fluid pressure (or positive pore pressure) develops. The dissolution and erosional capabilities would seemingly be less because of this situation.

Some distance below the water table, the volume of flow would be less than that at, and immediately below, this surface. The flow and pressure relationships are shown in Figure 13.24 (page 305). Depicted are flow lines (solid, with arrows) and equipotential lines (dashed and numbered). Between each pair of flow lines an equal volume of water flows. This shows that the upper part of the phreatic zone (immediately below the water table) carries the major portion of flow, and a progressively greater cross section is needed for the same flow as the depth increases.

For the reasons just discussed and based on field evidence, Bretz (1953) proposed the following sequence for cave system formation.

1. Cave system formed at water table when the major stream system was moderately well entrenched and located adjacent to the limestone in the uplands.

2. Deeper entrenchment of the stream lowered the water table and drained the cave system.

3. Caves filled with red mud carried by downward movement of vadose water from the residual soil above.

4. Mud excavated from caves by water from surface flows when erosion works its way down toward cave level.

5. Dripstone, stalactites, stalagmites, etc., formed by precipitation of calcium carbonate in an air-filled cave.

Perhaps the only controversial sequence of this outline is the filling and excavating of the red mud. It does not seem likely that all cave systems would necessarily go through that sequence. The other details of the formational process seem quite reasonable.

Thornbury (1954) suggested that some cave systems in Indiana and Kentucky formed, in part, above the water table. He cites evidence relating sinking streams and subsurface passageways to support this conclusion.

Figure 13.21   Limestone cave showing cave deposits or drip stone. Stalactites from the ceiling and stalagmites from the floor can join near the center to form a column as shown. Missouri Cave connected to Onondaga Cave. (Photo by Pierre Studio, Sullivan, Missouri.)

Figure 13.22   Photograph of a karst area on limestone bedrock. (Vladislav Gajic, Shutterstock.)

Based on these discussions, it is generally accepted that most cave development occurs at or relatively close to the water table when the major stream entrenchment nearby is situated at the proper level to establish caves in the limestone unit in question. The base level elevation for the stream must persist for a sufficient time to develop the cave system at a specific elevation or narrow range of elevations. This requires a special geologic setting.

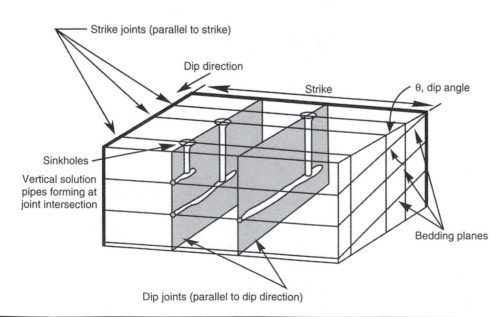

Figure 13.23   Diagram of strike joints and dip joints in limestone that provide pathways for solution channels.

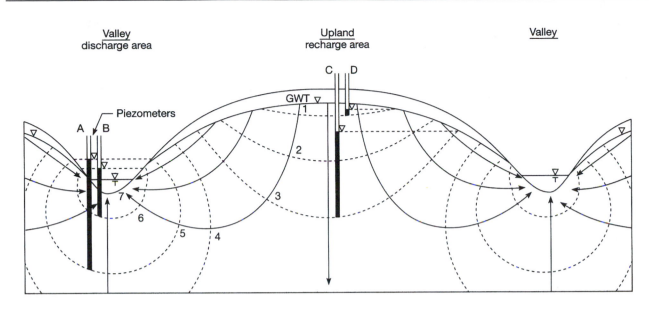

Figure 13.24   Simplified regional flow system in uniformly permeable material. (Adapted from Hubbert, 1940.)

Carbonate units not associated with entrenched streams have little likelihood of developing caves of this type. Consequently, a regional study is necessary to determine the likelihood of cave development. Aided by available data from published reports, maps, and well records on dissolution details of the strata, a determination can be made regarding the likelihood of cave development in a region containing carbonate rocks. Generally, this information is well established for areas in which formations are cavernous.

The question of cave development at depth, that is, below the bottom of the entrenched stream, is sometimes raised. Referring again to Figure 13.24, we can see that at least one flow line in the drawing extends through an area well below the stream. In a uniformly permeable system, most of the water would flow above this level, but in areas of nonuniform materials, water flow may be concentrated in lower zones of higher permeability. Therefore, it is concluded that caves can form, in some cases, not only well below the water table, but well below the base of the entrenching stream that dictates the water table level. This would, by nature of the list of conditions required for it to occur, seem to be an exceptional case rather than a common one.

Engineering Considerations of Groundwater

1. Groundwater is related to many facets of civil engineering and construction. Water supply is a major area of interest. Many of the smaller cities and towns in both the eastern and western United States rely on groundwater as their only water source. Treatment of groundwater typically involves only chlorination to kill bacteria.

2. In addition to water supply and sanitation, groundwater considerations also involve land drainage, irrigation, seepage related to dams and levees, control of water during soil and rock excavation, foundation support, slope stability, and settlement. The need to determine groundwater conditions for a site is one of the major reasons for performing a subsurface investigation. Numerous complications can occur during construction projects when groundwater levels are misinterpreted or when that phase of the investigation is overlooked. Dewatering systems, or other methods of groundwater control, must be designed to provide dry conditions for construction. Settlement, in the form of subsidence, caused by the withdrawal of water from the subsurface must be anticipated or provisions made to minimize subsidence by maintaining the same water level by recharging the aquifer.

3. Slope failures related to pore water pressure are a particularly prevalent problem.

4. Detailed information about caves and cave forming processes is crucial in the study of large dam sites in carbonate sedimentary rock areas. The extent of solution cavities has major significance for dam foundations and water impoundment, and much information must be obtained well before a decision is made to proceed with subsurface investigation.

EXERCISES ON GROUNDWATER

1. Information supplied in this chapter indicates that the continental landmass of the Earth receives 100,000 km^3 (24,000 mi^3) of precipitation per year. What average value, in centimeters (inches) per year, for rainfall on the continental landmasses is suggested by this? Recall that the Earth is essentially a sphere 12,735 km (7915 mi) in diameter, and the continents constitute about 21% of the surface area.

2. The following information pertains to the subsurface at a specific location:

   • The soil is 9.1 m (30 ft) thick and it is a silt.

   • A thick sandstone formation 305-m (1000-ft) thick lies below the soil.

   • The depth to the water table is 5.5 m (18 ft).

   • The silt has a capillary fringe 1.5 m (5 ft) thick.

   Based on this information, show the subdivisions of this cross section relative to the subdivisions of subsurface water. Draw an appropriate hydrostatic pressure diagram for the cross section in keeping with Figure 13.2.

3. a. If an artesian aquifer is intercepted by a well drilled 36.6 m (120 ft) deep and the water rises in a standpipe to a height of 3 m (10 ft) above the ground surface, how many meters (feet) of artesian pressure exist for this well?

   b. What type of well is it?

   c. What was the pressure in kg/m^2 (psf and psi) for the water in the aquifer before entering the well?

   d. If the piezometric pressure in the aquifer is reduced 4.6 m (15 ft) for every 1.6 km (1 mi) of horizontal distance, what is the hydraulic gradient for the aquifer?

   e. If the hydraulic conductivity of the sandstone is 4.7 m/day (15.3 ft/day), what is the average velocity of flow through the sandstone?

   f. Assuming a porosity of 18% for the sandstone, what is the seepage velocity for that unit?

   g. How long does it take for water to travel 1.6 km (1 mi) in the sandstone aquifer assuming the effects of pumping are minimal?

4. A farmhouse is located above a perched aquifer that has a volume of 3058 m^3 (4000 yd^3). The sandy aquifer has a porosity of 28% and a specific yield of 12%.

   a. What is the specific retention of the sandy aquifer?

   b. What is the maximum number of liters (gallons) of water available from the 3058 m^3 (4000 yd^3) aquifer if no recharge occurs because of a drought and all the aquifer is saturated at the start?

   c. If this four-person household uses 2271 liters (600 gal) of water per day (567.75 liters/person/day [150 gal/person/day]) how many days can the family go before all available water is used?

5. An oil-bearing sand is known to have an intrinsic permeability of 120 millidarcies. What is the equivalent hydraulic conductivity for this sand? What would be a typical description for such a sand based on this hydraulic conductivity value?

6. Big Spring in the Ozarks of Missouri has a discharge of about 662,375 liters (175,000 gal/min).

   a. How many m^3/sec (ft^3/sec) does this equal? How many MGD? (*Hint:* 1 m^3 = 264.172 gallons; 1 cfs = 0.646 MGD.)

   b. What magnitude spring would this qualify for in Meinzer's classification (see Table 13.6)?

   c. The Missouri Ozarks consists of sedimentary rocks that are Lower Paleozoic in age. In what specific rock type is the spring most likely to occur?

7. What are the five most productive landforms relative to groundwater supply? What makes them so productive? What problems occur in wells in fine sand and how can they be minimized?

8. Why are shales typically very poor water-producing formations? Consider both concepts of primary permeability and secondary permeability in your answer.

9. A field pumping test is to be performed in an unconfined aquifer of gravely sand. This aquifer is 9.1 m (30 ft) deep with a hard clay zone below it. The water table occurs at a depth of 1.2 m (4 ft).

   a. How deep should the pumping well be drilled and where should the well screen be placed?

   b. How long should the well be pumped in an attempt to reach steady-state conditions?

   c. What is meant by *steady-state conditions* in this regard?

   d. Calculate *k* for the following test situations. For the pumping test assume that steady-state conditions exist and that the pumping well and observation wells are properly completed.

   q = 946 liters/min (250 gal/min)

   r_1 = 7.6 m (25 ft) where drawdown = 1.5 m (5 ft)

   r_2 = 30 m (100 ft) where drawdown = 0.46 m (1.5 ft)

10. A sandy gravel with 15 m (50 ft) of saturated thickness has a permeability of 5×10^{-1} cm/sec.

    a. Calculate the transmissibility of the aquifer.

    b. Rate this transmissibility relative to the needs for domestic or industrial use.

11. An artesian aquifer 30 m (100 ft) thick was evaluated using a field pumping test. The artesian head for the aquifer prior to initiating the test was 36.6 m (120 ft). At a distance of 3048 m (1000 ft), the drawdown was 12.2 m (40 ft), and at a distance of 1524 m (5000 ft) it was 6.1 m (20 ft). Find the permeability of the aquifer assuming that steady-state conditions prevail. The pumping well was sustained at a rate of 1892.5 m/min (500 gal/min) during the test. What is the permeability of the aquifer in m/sec, ft/sec, cm/sec, Meinzers, ft/day, and darcies?

12. A constant head permeameter test was run on a sand sample that had a diameter of 7.6 cm (3 in) and a length of 10.2 cm (4 in). The head loss through the sample was 30.5 cm (12 in). The quantity of water flow in 1 min was 300 cm³. Calculate the permeability in cm/sec, m/sec, and ft/sec.

13. A falling head permeameter test was performed on a clayey silt sample. The sample had a diameter of 5.1 cm (2 in) and a length of 10.2 cm (4 in). The diameter of the standpipe is 0.64 cm (0.25 in). At the beginning of the test, the head was 100 cm (39.4 in) and at the end, 30 hr later, it was 54 cm (21.3 in). Find the permeability in cm/sec.

14. Use the open auger hole procedure to determine a numerical value of permeability for a percolation test. Given that for the 10.2-cm (4-in) diameter hole it took 5 min for the water surface to drop 2.5 cm (1 in), what is the permeability of the soil in cm/sec? If it took 30 min to fall 2.5 cm (1 in), what would be the permeability?

15. What is the main difference between the American rule and English common law, as applied to riparian rights? Why are individuals protected by excessive withdrawal of water from the subsurface by industry or by a municipal water company but not by a typical neighbor's use of water? Explain.

16. A cottage is to be constructed on a lakefront lot and a well and septic tank system are needed. The cottage is to be built on a half-acre lot, having rectangular dimensions and 30 m (100 ft) of lake frontage. The lot slopes gently toward the lake. The dimensions of the cottage will be 15 × 15 m (50 × 50 ft) and it will have the equivalent of three bedrooms relative to use of water and production of sewage. Results of six percolation tests indicated that it took 13 min for the water level to drop 2.5 cm (1 in) in the test hole. Based on this information, determine the size of the septic tank needed. It is desirable to place the cottage as close to the lake as is possible. Show calculations and the layout of the house, septic tank, septic tank absorption field, and well on a plan map of the lot. State assumptions and reasons for various decisions made. Assume the trench in the percolation field is 0.61 m (2 ft) wide.

17. When the gases formed in a landfill are collected, water vapor plus what two gases are the principal ones obtained? How must this gaseous mixture be processed to get a highly combustible product for an energy source?

18. Why are those landforms that are the best for obtaining groundwater supplies also likely to be the worst as a location for the placement of a sanitary landfill? What types of unconsolidated materials and types of bedrock are most favorable as disposal sites for refuse?

19. What are the two advantages of a clayey soil in retarding leachate movement as compared to a sandy soil? Explain in detail.

20. What is meant by the term *saltwater encroachment* and where in the United States is it particularly a problem?

21. A small island on the seacoast has an area of 24.6 km^2 (9.5 mi^2). Because of extensive development of vacation homes, infiltration, which was 15% for the island in its natural state, was reduced to 8% after construction. The annual rainfall for the island averages 76 cm (30 in).

    a. What is the maximum annual safe withdrawal of water for the island in acre-ft, m^3, ft^3, and gallons if no depletion of the groundwater is to occur?

    b. How many tourists can this island support assuming the water consumption is 581.3 liters/person/day (150 gal/person/day)? Calculate this in tourists/week and in tourists/yr.

    c. How many tourists/km^2 (tourists/mi^2) does this maximum number of tourists involve?

    d. If the water table on the island is lowered 2.5 cm (1 in) by over pumping, how much water does this represent? What percent increase over the maximum allowable "safe" number of tourists/yr does this represent?

    e. If the water table is lowered 2.5 cm (1 in), how much higher is the freshwater-saltwater interface raised?

22. Why is movement of water in open joints in a rock a much greater threat for pollution than is water movement through a soil?

23. If dissolution of limestone and dolomite is a slow, geologic process, why are humans concerned about placement of dams and surface reservoirs in such terrains? How are potential problems in these terrains determined?

24. Cave formation is most prevalent near the water table, but some caves develop at much lower elevations. Why is this important relative to leakage from reservoirs? How can the presence of such deep-seated caves be determined?

References

Bretz, J. H. 1953. Genetic relations of cover to peneplains and big springs in the Ozarks. *American Journal of Science* 251:1–24.

Darcy, H. 1856. *Les Fontaines Publiques de la Ville de Dijon* (The Public Fountains of the City of Dijon). Paris: Dalmont.

Das, B. M. 2002. *Principles of Geotechnical Engineering* (5th ed.). New York: Brooks/Cole, Thompson Learning.

Driscoll, F. G. 1986. *Groundwater and Wells* (2nd ed.). St. Paul, MN: Johnson Division.

Fetter, C. W. 2018. *Contaminant Hydrogeology* (3rd ed.). Long Grove, IL: Waveland Press.

Ford, D., and Williams, P. 2007. *Karst Hydrogeology and Geomorphology* (2nd ed.). New York: John Wiley.

Hazen, A. 1892. Some Physical Properties of Sands and Gravels, with Special Reference to Their Use in Filtration (pp. 539–556). In Massachusetts State Board of Health, 24th Annual Report, Publication No. 34.

Holtz, R. D., Kovacs, W. D., and Sheahan, T. C. 2011. *An Introduction to Geotechnical Engineering* (2nd ed.). New York: Pearson.

Hubbert, M. K. 1940. Theory of groundwater motion. *Journal of Geology* 48:785–944.

Hvorslev, M. J. 1949. *Subsurface Exploration and Sampling of Soils for Civil Engineering Purposes*. Vicksburg, MS: US Army Engineer Waterways Experiment Station.

Karamouz, M., Ahmadi, A., and Akhbari, M. 2011. *Groundwater Hydrology: Engineering, Planning, and Management*. New York: CRC Press.

Kresic, N., and Stevanovic, Z. 2009. *Groundwater Hydrology of Springs: Engineering, Theory, Management, and Sustainability*. Butterworth-Heinemann.

Marshak, S. 2013. *Essentials of Geology*. New York: W. W. Norton.

Meinzer, O. E. 1923. *Outline of Groundwater Hydrology, with Definitions* (Water-Supply Paper 494). Washington, DC: US Geological Survey.

Qian, X., Koerner, R. M., and Gray, D. H. 2002. *Geotechnical Aspects of Landfill Design and Construction*. Upper Saddle River, NJ: Pearson.

Terzaghi, K., and Peck, R. B. 2010. *Soil Mechanics in Engineering Practice* (3rd ed.). New York: John Wiley.

Theis, C. V. 1935. The relation between the lowering of the piezometric surface and the rate and duration of discharge of a well using groundwater storage. *American Geophysical Union Transactions* 16:519–524.

Thornbury, W. D. 1954. *Principles of Geomorphology*. New York: John Wiley.

Todd, D. K., and Mays, L. W. 2005. *Groundwater Hydrology* (3rd ed.). New York: John Wiley.

US Environmental Protection Agency. 2016. Municipal Solid Waste Landfills (http://www.epa.gov/landfills/municipal-solid-waste-landfills).

THE WORK OF GLACIERS   14

Chapter Outline

Glacial Ice

Types of Glaciers

Results of Glaciation

Glacial History of the United States

Engineering Considerations of The Work of Glaciers

Glacial Ice

A glacier is a mass of ice formed by recrystallization of snow that may flow across the land surface. Years of accumulation and burial are required to convert snow to glacial ice. Glaciers cover about 10% of the Earth's land surface, some 17.9×10^6 km^2 (6,875,000 mi^2), most of which is in Antarctica and Greenland. They form in regions of the world where snowfall in winter exceeds snowmelt in summer. When the ice volume becomes sufficiently large, glaciers flow, under the influence of gravity, outward from the accumulation area.

The specific gravity of newly fallen snow lies between 0.08 and 0.10, indicating a high degree of porosity. Meteorologists estimate that 25 cm (10 in) of snowfall is equivalent to 2.5 cm (1 in) of rainfall. After the snow settles, its density increases significantly to a specific gravity between 0.2 and 0.3. On further densification, aided by melting and refreezing plus the effects of sublimation, snow turns to firn, a granular material with a specific gravity greater than 0.4 but generally less than 0.8. Firn changes to glacial ice when, through densification, the specific gravity increases to 0.83, but it may reach a value of 0.91 in deeper portions of the glacier. The natural density of 0.917 g/cm^3 for ice frozen directly from water cannot be achieved for glacial ice because air voids in the snow continue to persist, to some degree, in the glacial ice. Glacial ice behaves like a metamorphic rock. In the metamorphic process, the crystal size increases from 0.1 mm for snow to 4 mm or more for the ice.

The boundary between firn and glacial ice occurs at considerable depth within a glacier. Its location depends on ice temperature because colder ice retards the growth of crystals. For valley glaciers this depth may be about 30 m (100 ft) but in the Greenland glacier, the boundary is closer to 300 m (1000 ft) deep.

The upper part of a glacier (3 to 60 m [10 to 200 ft]) is rigid and is referred to as the zone of fracture, where cracks known as crevasses can develop, whereas the lower part is plastic and is the zone of flow.

Types of Glaciers

The three primary types of glaciers are (1) valley glaciers, (2) piedmont glaciers, and (3) continental glaciers. Valley glaciers are rivers of ice that flow down existing stream valleys from mountainous regions above. They are also referred to as mountain or alpine glaciers because of their origin in the mountains. The term *valley glacier* seems more descriptive because the glacier fills a steep valley that determines its direction of flow.

Piedmont glaciers occur when two or more valley glaciers coalesce on the plains below the alpine region where glaciers originate. Piedmont glaciers are not of major significance and a detailed evaluation is not warranted herein.

Continental glaciers are great mound-like volumes of glacial ice that advance from the accumulation area under their own weight. They cover the entire landscape, not just valleys and adjacent plains, as do other glaciers. Their thickness may exceed 3000 m (10,000 ft) in the central portions and are, perhaps, half that near the advancing ice front.

Modern glaciers move slowly, only a few centimeters (1 in) to a meter (3 ft) per day. They move by flow within the ice, much like stream flow, with the central portion moving the fastest and the boundary zones moving the slowest. Much of the movement occurs within the individual ice crystals along atomic slip planes, often resulting in crystals growing from the 4 mm (fraction of an inch) size of newly formed ice to several centimeters (more than an inch) in diameter by the time they reach the snout or furthest extent of a valley glacier.

Results of Glaciation

The effects of present-day valley glaciers can be examined to learn the manner in which they erode, transport, and deposit materials. Many similarities exist between these glaciers and their continental counterparts and, through comparison, considerable detail can be learned about continental glaciers as well.

Valley Glaciers

Erosional Features of Valley Glaciers

Valley glaciers sharpen or steepen the topography by erosive processes because they are concentrated in stream valleys and in snow accumulation areas. They obtain debris through frost action on the adjacent rock and from landslides or avalanches onto the ice itself. This debris, embedded in the ice, abrades the valley walls and bottom. They also pluck blocks of rock from the jointed bedrock surface at their base and the sides of the valley by scouring. Figure 14.1 shows a block diagram of the erosional and depositional features of valley glaciation. The erosional features of valley glaciers include the following (Bennett and Glasser, 2009; Benn and Evans, 2010; Singh et al., 2011):

- *Cirque* (pronounced "sirk"): A bowl-shaped feature or amphitheater developed at the head of the

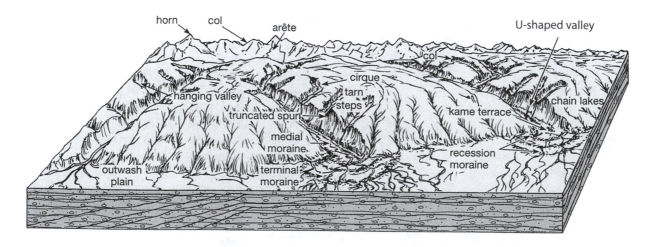

Figure 14.1   Erosional and depositional features of valley glaciation.

Figure 14.2   A cirque filled with glacial ice. (Photo by Nate Saraceno, http://www.adventuresingeology.com [https://creativecommons.org/licenses/by-nc-sa/3.0/].)

valley glacier carved from the rock by plucking and frost action (Figure 14.2).

- *Bergschrund*: A major crevasse developed where the ice and bedrock meet at the head of the glacier; it is a consequence of the ice moving, by gravity, away from the headwall.

- *Rock Flour*: Finely ground rock particles carried by meltwater streams flowing away from the front of the glacier. This gives a grayish color to the water and demonstrates the grinding power of the ice. Glacial lakes, which contain rock flour in suspension, take on a turquoise color caused by the refraction of light.

- *Polish, Striations, and Grooves*: A glacier drags materials of all sizes (ranging from boulders to clay) along its base. These materials abrade the bedrock surface, yielding a smooth polish on fine-grained massive rocks or forming scratch marks, called striations, in other instances. The striations can develop on both the bedrock and the rocks held by the glacier that provide the grinding action. After glacial melting these striated rocks provide evidence of glaciation (Figure 14.3a, on

the following page). Continental glaciers may also abrade deep trough-like features, known as grooves (Figure 14.3b), on a bedrock surface. Both striations and grooves show the bearing of ice movement but they do not indicate the direction from which the ice came.

- *Arête*: An arête is the saw-toothed or serrated ridge between two adjacent cirques.

- *Horn*: A horn is a steep, pyramidal peak left at the intersection of three cirques. It is bounded by the sidewalls of these cirques. The Matterhorn of Switzerland (Figure 14.4, page 313) is the most famous example of this feature.

- *Col*: A gap that forms when two cirques on opposite sides of a ridge intersect.

- *Rock Basins*: Rock basins are depressions of variable size, formed by rock plucking or selective quarrying of a valley floor by moving ice. The rock basins at the base of a cirque, when filled with water, are called tarns, whereas a series of lakes filling rock basins at a lower elevation are known as chain lakes.

Figure 14.3   (a) Bedrock striations from Alaska (Jeff Holcombe, Shutterstock) and (b) glacial grooves from Kelleys Island, Ohio, caused by glacier movement (Kelleys Island Chamber of Commerce).

- *U-Shaped Valleys*: Glaciated valleys have a characteristic U-shaped or trough-shaped cross profile compared to the characteristically V-shaped narrow mountain valleys formed by stream action alone. When the ice moves down the preexisting stream valley, it broadens the base of the valley by eroding the weathered and fractured rock and accumulations of soil, yielding a trough shape (Figure 14.5). The regular gradient of the stream valley is disrupted; the main glaciated valley, long after the ice has melted, typically shows ponded and swampy areas adjacent to the newly established stream channels.

- *Hanging Valleys*: Hanging valleys form where main valleys erode faster and deeper due to glaciation compared to tributary valleys. Consequently, tributary valleys are left hanging at higher elevations. Some hanging valleys are locations for waterfalls such as those observed in Yosemite National Park in the glaciated Sierra Nevada.

- *Fjords*: Fjords occur in coastal areas where bedrock valleys were extensively glaciated and are now covered by the sea. The valleys were glaciated when sea level was considerably lower than today, but glaciers can erode solid bedrock even when it is below sea level. The fjords of Norway, Greenland, Labrador, Alaska, and New Zealand are partially filled by the sea, yielding steep-sided, deep-water passages.

Depositional Features of Valley Glaciers

A number of the depositional features of valley glaciers are similar to the features formed during continental glaciation. Figure 14.1 on page 310 illustrates the depositional features of valley glaciers. Brief descriptions of these features are given below (Bennett and Glasser, 2009; Benn and Evans, 2010; Singh et al., 2011).

- *Drift*: The debris transported by a glacier is deposited when the ice melts. Debris may be deposited directly upon ice-melt or be carried some distance away by the meltwater from the ice. The general term *drift*, or *glacial drift*, includes both modes of deposition. Direct deposition by melting glacier results in unstratified drift, whereas meltwater deposition results in stratified drift. Unstratified drift is known as *till* and consists of a mixture of materials ranging in size from clay to large boulders weighing more than 900 kg (2000 lb). These deposits are prominent in the north-central United States (including Ohio, Indiana, Illinois, Wisconsin, and Michigan) where continental glaciation prevails. In parts of New England and in the state of Washington, there are extensive deposits consisting of large rock fragments and boulders, known as boulder tills.

- *Moraines*: Till is deposited by the receding glacier to yield landforms known as moraines. Depending on their location and configuration, different types of moraines are distinguished. For valley glaciers, the sides of the glacier adjacent to the valley wall are locations for *lateral moraines*. Here, the material that falls from the valley walls accumulates, yielding a ridge along each side of the valley when the ice melts (Figure 14.6). When two glaciers join, adjacent lateral moraines unite to form a

Figure 14.4 The Matterhorn, The Alps, Switzerland. (Vaclav+Volrab, Shutterstock.)

Figure 14.5 A broad U-shaped glaciated valley, Alaska. (akphotoc, Shutterstock.)

medial moraine near the center of the widened glacial flow. These are marked as dark zones in the active glaciers but they are usually reworked and disappear when the ice melts (Figure 14.6). Water running along the valley wall of a melting glacier may deposit coarse material to build up the lateral moraines. After melting, these stratified deposits stand above the valley floor and are called *kame terraces*. They are good sources of aggregate materials for construction.

Common to deposition by both valley glaciers and continental glaciers are *end moraines* and *ground moraines*. An end moraine is a ridge of till that marks the place where the ice stagnated for some time while the advance equaled the retreat caused by melting. Debris accumulates in a high, arc-shaped landform with the concave portion pointing in the direction from which the ice advanced. The terminal moraine is that individual end moraine marking the greatest extent of

ice advance. Other end moraines are recessional moraines, which develop when the ice recedes to another stagnation point as it melts headward. Much of the debris is laid down directly below the ice as melting occurs to provide a rolling terrain along the valley floor. This landform is called ground moraine and it may be hundreds of meters (feet) thick in deposits from continental glaciers.

- *Outwash Sand and Gravel*: These are deposited by meltwater flowing away from the face of a melting glacier. Streams of water loaded with material eroded from the till form braided channel patterns on the land surface beyond the glacier. As these streams lose their velocity, they drop the transported material, largest particles first, to form a wide apron of stratified material. This is called an *outwash plain* (Figure 14.7). If the deposit continues down a narrower valley beyond the outwash plain, a *valley train* deposit is the consequence as more material is dropped by the flowing water.

Figure 14.6 Lateral and medial moraines of a valley glacier, Wrangell–St. Elias National Park, Alaska. (Photo by James W. Frank, US National Park Service.)

Continental Glaciers

Continental glaciers completely override the terrain and therefore tend to smooth the land surface rather than sharpen it in the fashion of valley glaciers. Both the erosional and depositional features of continental glaciers provide a general smoothing effect because the glaciers tend to level off the ridges and the deposits tend to fill in stream valleys (Bennett and Glasser, 2009; Benn and Evans, 2010; Singh et al., 2011). Some less common exceptions to this are the erosive effects and enlargement of stream valleys by the ice sheet, as evidenced by the Great Lakes and by the Finger Lakes of New York State. These zones of less resistant rocks were selectively quarried by the glaciers to yield closed depressions now filled with water. The Great Lakes are discussed more fully in a later section.

Erosional Features of Continental Glaciers

- Roche Moutonnée: Since continental glaciers tend to smooth the land surface, erosional features are rare. However, continental glaciers may erode exposed bedrock hills to yield a unique knob-form known as a *roche moutonnée* (Figure 14.8 on the following page). This is a sculptured hill having a gently sloping face in the direction toward the advancing glacier and a steep face on the opposite slope. The ice rides along a bedding plane or other weakness zone and plucks the rock off a joint plane that is nearly perpendicular to the bedding. Consequently, the gentle slope marks the direction from which the ice advanced.

- *Underfit Stream*: Another consequence of continental glaciation is the enlargement of stream valleys that lie immediately beyond the glacial margin. Water from the melting glacier erodes the stream channel, both widening and deepening it. When glacial melting ceases, the discharge dwindles and a small volume stream is found within a larger valley. Such a feature is called an *underfit stream*. A prime example of this is Mill Creek Valley located in the center of Cincinnati, Ohio. The valley is quite wide and deep but the stream (Mill Creek) today is only of small size.

Figure 14.7   An outwash plain in the foreground of Exit Glacier, Alaska. (Randy Yarbrough, Shutterstock.)

Depositional Features of Continental Glaciers

Continental glaciers yield depositional features somewhat similar to those of valley glaciers. Figure 14.9 is a block diagram of the depositional features of continental glaciation. Table 14.1 provides a comparison between the features of the two types of glaciation.

- *Erratics*: Erratics are boulders or cobbles of a certain rock type carried by glaciers a sufficient distance from their place of origin so that they are foreign to the type of bedrock where they are deposited. They may be found imbedded in till, lying directly on the bedrock surface, or located within outwash formed from glacial meltwater. In the north-central United States, erratics typically consist of igneous or metamorphic rocks from the Canadian Shield. As such, they contrast greatly with the sedimentary bedrock in the area of deposition.

- *Kettles*: Kettles are pits or depressions in the drift. They can occur in outwash plains or in ground moraines as well as end moraines. Formed by the delayed melting of ice blocks in the drift, these depressions range from 10 m to several km (30 ft to more than a mi) in diameter and may be from 1 to 30 m (3 to 100 ft) deep. Outwash deposits marked intensely by kettles are called pitted outwash plains. Kettles may fill with water to form the lakes so common in the northern United States and Canada or may fill with organic deposits to yield extensive peat and muck deposits.

- *Eskers*: Eskers are sinuous ridges of stratified sand and gravel formed by meltwater streams at the base of stagnant ice near the margin of the glacier. They are steep sided, from 3 to 15 m (10 to 50 ft) high, and may extend for 1 to 10 km (0.5 to 6 mi) in length. These ridges of sand and gravel usually occur as discontinuous features.

- *Kames*: A common feature superimposed on ground moraines of continental origin is the small, conical-shaped hills known as kames. They consist of stratified drift collected in the crevasse

Figure 14.8   *Roche moutonnée*, or glacier-carved bedrock hill. (Gestalt Imagery, Shutterstock.)

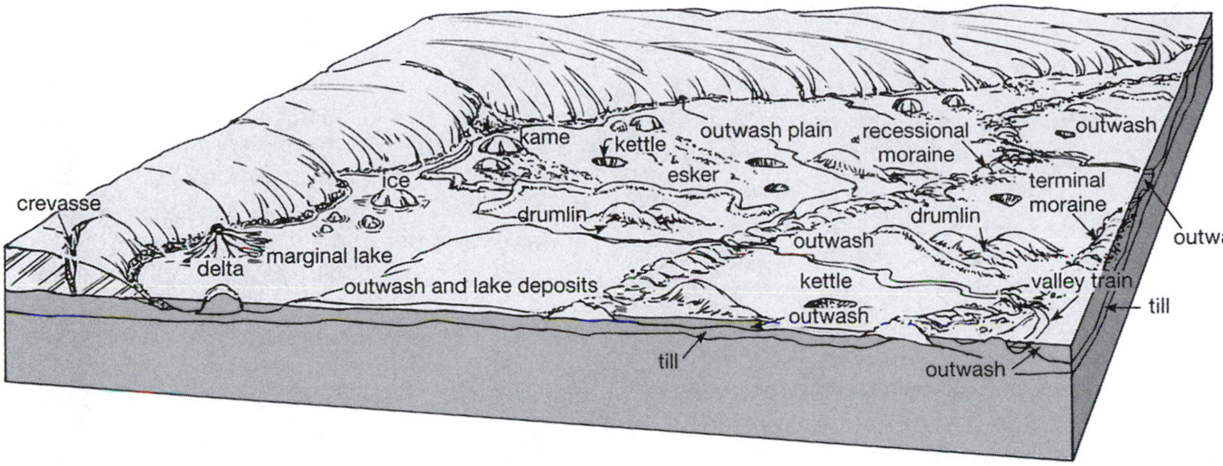

Figure 14.9   Depositional features of continental glaciation.

openings of the melting ice. They may appear as clusters or in isolated mounds and contain a wide range of sand- and gravel-size materials.

- *Drumlins*: Drumlins are cigar-shaped hills, typically composed of till, that were streamlined by the readvance of glacial ice over a rolling till plain. The blunt end indicates the direction from which the ice came and the narrow end shows the direction of ice movement. Many drumlins found together make a drumlin field (Figure 14.10 on the following page).

- *Kame Terraces*: Kame terraces form when ice moves down a stream valley then melts back to yield stratified terraces against the valley wall. They occur in the lower reaches of valleys traversed by valley glaciers and in valleys partially filled by continental glaciers.

- *Lake Beds*: Lake beds form over a period of years as bottom sediments consisting mostly of silt. Because many marginal lakes form as a consequence of glacial melting, lake beds are a common variety of stratified drift deposits. Sandy beach ridges showing the location of a previous shoreline are associated with lake beds. They are collectively referred to as lacustrine deposits.

- *Varves*: Many lake beds consist of pairs of thin sedimentary layers, one consisting of silt and the other of clay. Each pair is known as a varve. The two layers represent a single year's deposition with the darker, clayey layer deposited during

Table 14.1   Comparison of Features for Valley and Continental Glaciation.

Feature	Valley	Continental
Till	Common	Common
Ground moraine	Common	Common
Terminal moraine	Common	Common
Recessional moraine	Common	Common
Lateral moraine	Common	Common
Medial moraine	Common but quickly eroded	Absent
Drumlins	Rare or absent	Locally common
Stratified Drift	Common	Common
Outwash plain	Common	Common
Valley train	Common	Common
Kames	Common	Common
Kame terraces	Common	Hilly country only
Eskers	Rare	Common
Lake beds	Rare	Common
Erosional Aspects		
Striations, polish	Common	Common
Cirques	Common	Absent
Horns, arêtes, cols	Common	Absent
U-shaped valleys, hanging valleys	Common	Rare
Fjords	Common	Absent
Roche moutonnée	Absent	Common
Loess	Rare	Common
Glacial erratic	Common	Common

winter. A varve is usually less than 2.5 cm (1 in) thick but may reach more than 5 cm (several inches) in thickness under some circumstances.

- *Loess*: Associated with many depositional features related to continental glaciation are silt deposits known as loess. These buff-colored materials in North America consist of small angular grains and are generally considered to be wind deposited and linked to glaciation. Glaciologists believe that blowing wind, typically from west to east across newly formed, broad outwash areas, picked up silt-sized particles and deposited them some distance downwind. Shells from air-breathing snails, which abound in the loess, suggest that loess was not deposited in water.

Five major areas of loess accumulation occur in the United States: lower Mississippi River, central interior lowlands, Great Plains, Snake River plain, and Columbia River (or Palouse Area).

Table 14.2 provides details on composition and topography of deposits formed by continental glaciation. Figure 14.11 on page 320 shows a till plain or a ground moraine, typical of the Midwestern United States, on which many depositional landforms of continental glaciation (erratics, kettles, eskers, kames, and drumlins) are superimposed.

Glacial History of the United States

Glacial Stages

In North America, four major advances of continental glaciers occurred during the Great Ice Age of the Pleistocene epoch (Bennett and Glasser, 2009;

Figure 14.10   Drumlin field. (© David Purchase. [https://creativecommons.org/licenses/by-sa/2.0/].)

Menzies and van der Meer, 2016). These four glacial stages are named after the Midwestern states where deposits are extensive or were first studied. In chronological order, starting with the oldest, they are the Nebraskan, Kansan, Illinoian, and Wisconsinan glacial stages. Each major glacial advance is estimated to have persisted about 100,000 years with interglacial episodes of warming lasting for several hundred thousand years. Figure 14.12 at the bottom of the next page is a schematic diagram showing ice advance versus time.

Figure 14.13 on page 321 shows the distribution of glacial deposits according to geologic age, with the surface materials indicated by different symbols. Distinctions are made between Kansan, Illinoian, and Wisconsinan deposits. Isolated patches of glaciation

Table 14.2   Deposits Associated with Continental Glaciation.

Type of Deposit	Lithology	Topographic Form	Special Features and Remarks
Till	Unsorted, nonstratified, heterogeneous. Contains angular rock fragments ranging from clay to boulders. Layer of till deposited by single advance of ice called "till sheet." Larger rock fragments may contain glacial striations or polish.	Ground moraine	Gently rolling: 1.5- to 15-m (5- to 50-ft) relief. When flat, called till plain. When rolling, called "swell and swale." Drumlins (stream-lined hills) characterize some ground moraines. Formed beneath the ice.
		End moraine	Varies from extremely rugged to subdued forms. Commonly called "knob and kettle" topography. Usually occurs as a belt or range of hills marking the edge or margin of a former glacier lobe. Commonly given local geographic name.
Outwash	Stratified, sometimes cross-bedded, sand, gravel, and less frequently silt. Wide range of rock types and mineral suites. Grains and pebbles more rounded than till cobbles due to water transport.	Outwash plain	Flat to irregular, depending on number of stagnant ice masses present when formed. Highly "pitted" outwash plains may appear as rugged as some end moraines.
		Valley train	Preglacial valley filled with extensive glacial outwash. May extend beyond glacial boundary.
		Esker	Subglacial stream deposit, winding ridges, sometimes compound, often discontinuous. Cross-bedded and slumped along sides where former ice walls collapsed.
		Kame	Irregular-shaped knoll believed to form in depressions in stagnant ice or near ice margins. Commonly associated with end moraines.
Lacustrine	Sands and gravel	Beach ridges	Gentle ridges of low relief that mark former shorelines of extinct lakes. May be accentuated by dune deposits. Can be traced for miles in many Great Lakes states.
	Silt and clay	Lake beds	Flat, featureless plains, covered with varying thicknesses of stratified silt and clay. Common in the Great Lakes region. Includes varves and varve-like deposits; fine sands also.
Eolian	Sandy	Sand dunes	Associated with beach sands, either ancient or modern; also found in areas of extensive outwash. Many Pleistocene dunes are now stabilized with vegetation.
	Silty	Loess	Nonstratified, buff-colored, calcareous, up to 90% silt fragments; believed by many to have been derived from valley train deposits associated with advancing or retreating ice front.

Figure 14.11   Till plain or ground moraine feature of the Midwestern United States. (Photo by Pollinator, Wikipedia [https://creativecommons.org/licenses/by-sa/3.0/deed.en].)

(not shown) also occurred in the mountainous regions of the western United States, where valley glaciers prevailed during the Wisconsinan glacial stage. In Figure 14.13 an island of unglaciated terrain is shown in an area of southwestern Wisconsin at the Illinois, Iowa, and Minnesota boundaries, which is known as the "Driftless Area." This territory of resistant Cambrian and Ordovician sandstones marked the extent of ice advance for several glacial stages and substages in the upper Midwest. Never completely encircled by ice at any time, the area experienced an ice advance from the north and west, which then retreated without covering the area. In a later stage, glaciers came

from the north and east also to fall short of covering the area before melting back. Consequently all the land surrounding the Driftless Area was glaciated at different times but leaving it as an untouched island in the glaciated terrain.

The general extent of continental glaciation in the United States can be observed by viewing Figure 14.13. In a general way, we can state that the boundary showing maximum extent of glacial advance extends from Long Island, New York, across northern Pennsylvania to Pittsburgh, down along the Ohio River to its confluence with the Mississippi River, up the river to St. Louis and the confluence of the Missouri River,

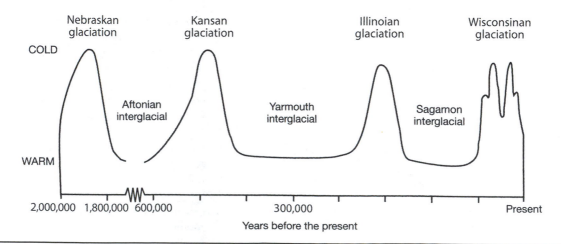

Figure 14.12   Pleistocene-age glacial advances of the United States.

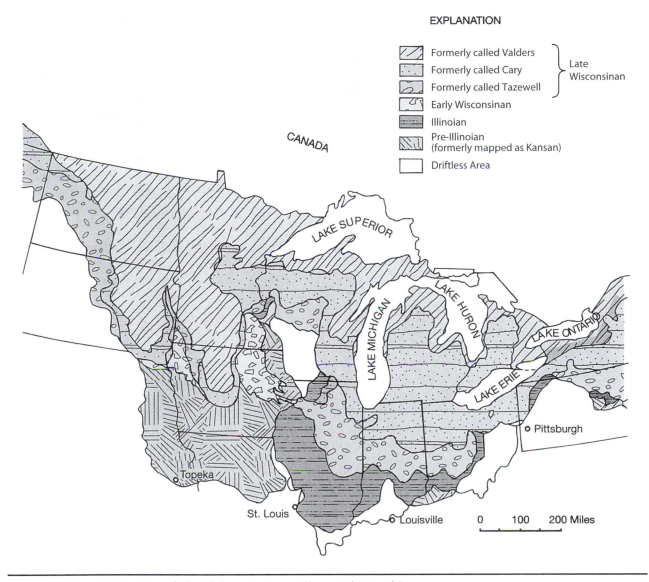

EXPLANATION

- Formerly called Valders ⎫
- Formerly called Cary ⎬ Late Wisconsinan
- Formerly called Tazewell ⎭
- Early Wisconsinan
- Illinoian
- Pre-Illinoian (formerly mapped as Kansan)
- Driftless Area

Figure 14.13   Distribution of glacial deposits in north-central United States.

and then up the Missouri River into North Dakota and straight across the country to Seattle, Washington. There are exceptions to this general boundary, such as southeastern Ohio, south-central Indiana, and the southern tip of Illinois, which were not glaciated. However, this easily remembered boundary serves as an approximate indication for those sections of the United States that were overrun by continental glaciers during the Pleistocene epoch.

The elevation of the ocean's surface is intimately associated with the amount of glacial ice that covers the land surface. Today, if all glaciers were to suddenly melt, it is estimated that sea level would rise 30 to 60 m (100 to 200 ft). Such a rise would, obviously, have a devastating effect on numerous coastal cities throughout the world. Indications are that glacial ice has been retreating in the last century with an attendant increase in sea level of 10+ centimeters (several inches).

During the glacial epochs of the Pleistocene, water locked up in glacial ice was much more than with the glaciers today. Therefore, during glaciation, sea level stood at a much lower elevation. In like manner, during the interglacial periods, sea level was significantly higher than it is today.

It is of interest to estimate how low sea level dropped during the maximum ice advance of Wisconsinan glaciation. To calculate the volume of ice involved, both the area of ice coverage and average thickness of the ice are needed. The area of the Earth's surface covered by ice can be estimated with reasonable accuracy, but determining the average ice thickness is much more difficult. Thickness obviously decreased near the margins. Also the pressure of the ice load can be established for some locations. In all, an estimate of up to 100 m (330 ft) has been suggested for the drop in sea level that occurred during the maximum advance of the Wisconsinan stage.

History of the Great Lakes

Geological evidence indicates that the Great Lakes formed in late Wisconsinan time by the deepening of weak rock lowlands that, in preglacial time, contained streams draining eastward into the St. Lawrence River. The Great Lakes formed in the time period from about 14,000 to 2500 years BP (before the present) by glacial scour yielding ice marginal water bodies when the ice front receded. Lake evolution was influenced by five contributing factors: (1) oscillating ice fronts, (2) topographic irregularities uncovered by retreating ice, (3) variations in directions of retreat and advance of the ice, (4) lowering of lake outlets by erosion, and (5) differential uplift of the land adjacent to the ice following glacial melting.

There were no preglacial (or pre-Pleistocene) Great Lakes, but there would surely have been sequences of lake evolution during the final phases of the Illinoian, Kansan, and Nebraskan glacial stages. No evidence of a pre-Wisconsinan lake system of the magnitude of the Great Lakes has been recognized to date.

The history of the Great Lakes has been determined primarily by tracing topographic and geologic features showing former shorelines and outlets, along with radiocarbon dating of specific deposits. During certain phases of glacial melting, the lake levels stood at considerably higher elevations than the present-day shorelines of the Great Lakes. This yielded lacustrine deposits that extend well beyond the confines of the current lakes. These lacustrine plains, along with the lake outlets that drained the high-level water bodies, are shown in Figure 14.14. The area of marine subsidence along the northeastern end of Lake Ontario is also depicted.

Engineering Considerations of the Work of Glaciers

Glacial deposits are used both as (1) construction materials and (2) for foundation support of engineering structures. Because of the great variety of glacial materials, the contrast in value or usefulness in the former and associated problems in the latter are quite considerable.

1. **Construction Materials**: Sand and gravel are used for many construction purposes, including aggregates for concrete, base course materials under roads, aggregates for bituminous pavements and overlays, cohesionless backfill material, and filter blankets for drainage. Silt and clay are used in engineering applications such as hydraulic barriers where low permeability materials are desired. A brief evaluation of various types of glacial deposits for construction purposes follows.

Outwash Plains and Valley Trains: These yield good supplies of aggregate. They are usually associated with stream valleys and wide plains adjacent to drainage ways. The maximum size and distribution of sizes may vary considerably throughout the deposit, depending on the velocity of the water carrying the material. Typically, the amount of sand and fine pebbles exceeds the

Figure 14.14   Lacustrine areas surrounding the Great Lakes.

amount of coarse gravel needed for aggregates used in concrete and bituminous mixes. Deposits with less than a 20 to 25% gravel fraction are not economical to process because there is little market for the excess sand.

Eskers, Kames, and Kame Terraces: These features are composed of stratified drift but the range of sizes is usually extreme. A range from large boulders to silt may occur with fine sand and silt sizes predominating. These landforms are also prone to extreme chemical weathering of the boulders and cobbles because these hills lie above the general terrain and are affected by the downward seepage of water through time. These deposits can be used for bank run sand and gravel for general-purpose fills or base courses for secondary roads. Consequently, such features often disappear from the landscape because of surface mining soon after urbanization approaches a region.

The economic value of gravel in a glaciated area is determined by its location relative to the point of use. If it must be transported a significant distance, the cost to provide the gravel increases considerably. Such material is said to have a high "place value" because location (or place) is so important in determining its value. High-volume, low-unit cost materials (such as gravel) are particularly subject to such considerations. In glaciated areas, the effect of place value for gravel is magnified, owing to the common occurrence of gravel deposits in these areas. Generally, the market for gravel must be located within about 25 km (15 mi) of the source if its price is to be competitive.

Ground Moraines and Drumlins: These are not good sources for sand and gravel because the unstratified drift consists primarily of silt and clay. If a kame is associated with an end moraine, some sand and gravel may be found within. Careful study of the landform is needed to ensure that coarse materials are contained within the feature of interest.

Loess Deposits: Loess, the wind-deposited material consisting primarily of silt, has special engineering characteristics that warrant consideration. It is yellow to buff in color with a uniform particle size, about 0.01 to 0.005 mm in diameter. The material is without obvious stratification and consists mainly of fresh, angular grains of quartz and feldspar. When it is not saturated with water, loess has a strong tendency to split along vertical joints and to maintain vertical faces when excavated.

Loess is deposited in the windward direction from major valley trains of glaciers; that is, glacial sluiceways. It is thickest adjacent to these features and rapidly thins out from the source. Loess blankets the entire landscape—hills, slopes, valleys, and plains alike. Its origin is linked to a retreating phase of continental glaciation.

Like all silty materials, loess dries quickly when exposed to the air; thus, loess is very sensitive to changes in moisture during earthmoving operations when the material is being compacted in an earth-fill. In the past, slopes were cut vertically in loess deposits because of its ability to maintain this configuration when dry. However, deeper cuts, when related to interstate highway construction, were found to be unstable in a vertical excavation and slumping was extensive. Groundwater seepage contributed greatly to this instability. In such situations, the loess must be cut at a much gentler slope (as low as 3 horizontal to 1 vertical) and trees planted on the slopes to reduce seepage effects and to stabilize the slope.

2. **Foundation Problems**: Engineering properties of glacial deposits vary greatly with regard to the problems associated with foundation support for structures. With this in mind, engineering applications for the following categories are examined: (1) sandy outwash plains and valley trains, (2) ground moraines or till plains, (3) lacustrine plains, and (4) peat and muck deposits.

Outwash Plains: Sandy outwash plains and valley trains typically contain high groundwater tables and cohesionless materials that will settle under static loads and as a result of vibrations. Problems include settlement in sand because of footing contact pressures and vibrations during pile driving. Dewatering of the sands to accomplish dry working conditions may also lead to settlements and loss of ground on adjacent properties. Layers of compressible silts and clays may occur within the outwash materials so that settlements due to consolidation must be considered.

Ground Moraines or Till Plains: Ground moraine in many areas of the Midwest is extremely dense because it was consolidated by the great thickness of the overriding ice. Some layers are referred to locally as "hard pan" because of the difficulty experienced in drilling or in driving a split spoon sampler through the material. The soil sampler is used in the Standard Penetration Test (SPT) where the number of blows per foot to drive the sampler is recorded (ASTM D1586-11). This exploration technique is discussed in Chapter 19, Subsurface Investigations.

These tills provide quite suitable support for most building foundations. Glacial till also provides an excellent material for construction of small dams of uniform cross section. These impound water for farm ponds under the Natural Resources Conservation Service (NRCS) program. For larger dams, glacial till may provide the core material for the dam, whereas the shell material may be somewhat coarser in size.

Lacustrine Plains: A lacustrine plain is the location of stratified drift formed at the base of a glacial lake. These layers of silt and clay are highly compressible. They have low shear strengths and consolidate under imposed loads. The extent of lacustrine deposits associated with the Great Lakes is shown in Figure 14.14. Extensive areas occur in Chicago and at the southern end of Lake Michigan, southeast of Lake Erie in the Toledo, Ohio, area extending along the lake to Cleveland, at the southern end of Lake Huron in the Detroit area, and in Buffalo, New York, and eastward on the south side of Lake Ontario. These are all populous areas and problems associated with the compressible clays have contributed greatly to the study of their geotechnical properties and applications.

Underground construction in compressible lake beds leads to other difficulties. The low shear strength of the clays causes loss of ground into tunnels during excavations by tunneling shields. The Chicago subway system built in the 1940s was constructed in the lake clays of the Chicago Loop area. The problem of weak shear strength and high compressibility had to be dealt with during this construction.

Peat and Muck Deposits: Peat and muck deposits form in glacial depressions that fill with plant remains as vegetation grows in shallow lakes or in swamps. Kettles provide these depressions and they can occur in ground moraines, end moraines, and outwash plains. Peat deposits may comprise an area of approximately 7 km^2 (4 mi^2) or can occur in small patches of less than 0.4 hectares (or 1 acre) (0.004 km^2) in size.

The vertical extent of a peat deposit is generally 3 m (10 ft) thick or less but some large deposits may extend for depths greater than 7.5 m (25 ft). For construction purposes, it is more expedient to avoid peat deposits by relocating buildings or transportation routes and power transmission lines than to deal with the problems they pose. Avoidance is not always possible, however, and for road construction, excavation of the material or bridging over it with a displacement fill may be the only feasible solution. Pile foundations through the muck and peat may be used to support structures that need to be located at the particular site. As more construction and housing developments occur in the United States, fewer options of changing the alignment of roads or the location of proposed buildings will prevail. The solution of difficult problems for supporting buildings and other engineering structures on challenging geologic materials will be required of trained specialists with increasing frequency in the future.

EXERCISES ON THE WORK OF GLACIERS

MAP READING

For these exercises, the US Geological Survey offers TopoView (https://ngmdb.usgs.gov/topoview/). This software will allow you to view and download topographical maps produced from 1880–2010 for locations throughout the United States. Use a map that has a scale of 1:24,000.

MT. RAINIER NATIONAL PARK, WASHINGTON

This map shows mountain or valley glaciers on Mt. Rainier.

1. On which side of Mt. Rainier are the glaciers the largest? Where do they extend to the lowest elevations? How do you account for this? (*Hint*: Consider weather patterns.)

2. Locate on the map examples of the following features. Provide location data points and any identifying information given on the map.

   a. tarn

   b. medial moraine (glacier name)

   c. end moraine (glacier name)

   d. cirque

   e. U-shaped valley

   f. hanging valley

3. Describe the shape of Cathedral Rocks just southeast of the summit of Mt. Rainier. Explain the shape you have just described. What is this feature called?

4. What do the brown dots on Cowlitz Glacier represent? Does this material contribute significantly to the transport of debris by a glacier? Explain.

5. Locate a possible source of glacial gravels for highway construction. What type of glacial deposit have you located?

WATERLOO, WISCONSIN

6. One inch on the map equals how many miles in the field?

7. What is the regional stage of this quadrangle? During which glacial stage did the surface material form?

8. What direction of ice movement can be inferred from the alignment of the drumlins of this quadrangle?

9. What is the average length, height, and width of drumlins in this area? Describe their overall shape. What type of material probably would be found in the drumlins?

10. What does this general area consist of (e.g., outwash plain, till plain, terminal moraine)?

11. Describe the nature of the natural drainage in the area.

12. What engineering problems would the Chicago and North Western Railway be expected to encounter in this area?

OTTERBEIN, INDIANA

The northern part of this quadrangle, which is west of Lafayette, is ground moraine. Note the location of High Bridge and its topographic setting. Also note the Granville Bridge setting.

13. What are the bridges' respective distances from the Wabash River at West Lafayette?

14. How would you describe the topography of the northern part of this quadrangle? (Include its regional stage.)

15. How well drained is the northern part of this map? Has this area undergone much stream dissection to date?

16. What is the geologic age of the glacial drift in this area? Can you tell this by the amount of stream dissection? Explain.

17. Why is Little Pine Creek (near High Bridge) flowing in such a deep, wide valley in spite of the limited dissection of the uplands? What is such an oversized valley called?

18. Why does Lost Creek terminate before reaching the Wabash River? What kind of deposits would you expect to find along the river?

FLORA, ILLINOIS

19. What surficial features shown by the topography suggest an older age of glaciation than that shown on the Otterbein quadrangle?

20. From what glacial stage would you expect the drift in this area to have originated? Explain.

21. What is the regional stage of development for the quadrangle?

22. Account for the meandering of the Little Wabash River.

23. Are there any evidences of drumlins, eskers, and recessional moraines? What glacial landform is present in this area?

24. What do the little circles indicate? (*Hint*: See the key of the map.)

25. Suggest a source for the sand and gravel found in the area.

26. What engineering problems might be expected in this area? Explain your answer.

OLD SPECK MOUNTAIN, MAINE

27. What is the landform name (or physiographic province) for this general location of the United States? How does it differ from the Appalachian Mountains in Pennsylvania and Virginia?

28. What evidence of glaciation appears on the map? (*Hint*: Consider details of surface erosion to arrive at your answer.)

29. If the region has been glaciated, what explanation accounts for the smoothly rounded contours of the hills?

KINGSTON, RHODE ISLAND

This area is located in the Appalachian orogenic belt. Structurally it consists of intrusive igneous rocks overlain in many places by glacial drift.

30. What evidence of glaciation appears in the highland area in the northern portion of the quadrangle?

31. Name the type of landform for the irregular belt of topography just south of Worden's Pond.

32. How is Worden's Pond related to this irregular landform?

WRITTEN QUESTIONS

33. A sample of frozen water, obtained from an Antarctic glacier, measured 0.03 m³ (1 ft³), weighed 13.7 kg (30.1 lb), and had a crystal size of 3.1 mm. What is its specific gravity and what name best applies to the sample? Explain. What would be its approximate air content?

34. Why are the terms *valley glacier* and *mountain glacier* used interchangeably for one of the major types of glaciers? Where in the United States can the effects of valley glaciers be observed?

35. If sea level was 100 m (328 ft) lower during the maximum glacial advance, what volume of water (km³ or mi³) from the oceans does this represent? What volume of ice does this represent (assume a SpG of ice = 0.90)? If the glacial ice averaged 2100 m (6890 ft) thick on the land surface, how much area (km² or mi²) would this volume of ice cover?

36. Approximately how long did the Yarmouth interglacial period last? What would be the effect of this extensive time interval on the glacial deposits of Kansan age? Explain.

37. What is the significance of the Wisconsin Driftless Area? Why was the area not overridden by glaciers? About how large an area is involved?

38. Was the Pittsburgh, Pennsylvania, area glaciated? Was Cincinnati, Ohio? Was Rapid City, South Dakota? What is the approximate percentage of Indiana that is unglaciated?

39. In what areas located around the Great Lakes in the United States do lacustrine deposits prevail? What major cities are located in these areas?

40. What glacial deposits are best exploited for sand and gravel to be used as construction materials? Are these stratified or unstratified deposits? Why?

41. What is meant by the phrase *place value* with regard to geologic deposits? If transportation costs for gravel are $0.25 for the first ton-mile and $0.05 per ton-mile thereafter, how much does it cost to transport gravel 15 miles? What percentage increase in cost does this represent for gravel that sells at $10.00 per ton at the processing plant?

42. What types of concrete deterioration can occur when glacial gravels are used as coarse aggregates? Refer to Chapter 9, Evaluating Construction Materials for details.

43. What are the five major areas of loess accumulation in the United States? In which of these areas is Vicksburg, Mississippi, located? Walla Walla, Washington? East St. Louis, Illinois?

44. In the United States, why would deep loess cuts on the north side of an interstate highway be more stable than cuts on the south side of the highway? Consider climatic factors in your answer.

REFERENCES

Benn, D. I., and Evans, D. J. A. 2010. *Glaciers and Glaciation* (2nd ed.). New York: Routledge, Taylor & Francis Group.

Bennett, M. R., and Glasser, N. F. 2009. *Glacial Geology—Ice Sheets and Landforms*. Hoboken, NJ: Blackwell-Wiley.

Menzies, J., and van der Meer, J. (eds.). 2016. *Modern and Past Glacial Environments*. New York: Elsevier.

Singh, V. P., Singh, P., and Haritashya, U. K. (eds.). 2011. *Encyclopedia of Snow, Ice, and Glaciers*. The Netherlands: Springer.

SLOPE STABILITY AND GROUND SUBSIDENCE 15

Chapter Outline

Classification of Mass Movement

Types of Slope Movement

Causes of Slope Movements

Prevention and Correction of Slope Movements

Ground Subsidence

Engineering Considerations of Slope Stability
 and Ground Subsidence

Mass movement is the bulk transfer of soil or rock debris downslope under the direct influence of gravity. Slope movement and ground subsidence are the main forms of mass movement. Water facilitates mass movement.

Running water and mass movement work hand in hand to erode the land surface. Running water establishes the channels and continues to downcut the terrain over time. Mass movement accepts the local base level set by channel erosion and transports the bulk of soil and rock toward this lower elevation under gravity's influence. The Grand Canyon, Arizona, is an example of this twofold erosion process. Mass movement and slope movement are similar terms but mass movement carries a geologic emphasis implying that slope failure is part of the natural geologic process of gradation, which continues to occur. Slope movement and slope stability usually involve a shorter term consideration. Slope movement studies involve both materials and the mechanisms of failure, with the knowledge that a better understanding of these

factors should reduce the effects of future slope problems. Slope stability is concerned with prevention of slope failure or of extensive movement.

Classification of Mass Movement

Sharpe's (1938) classification, presented in Table 15.1, provides a valuable guide to the overall nature of mass movement, including a historical review of slope movement. Based on the nature of movement, mass movement is subdivided into two categories: the first involving downward and outward displacement, forming a new surface or free side along the failure boundary, and the second, known as subsidence, consisting mostly of downward movement of the land surface, without a new surface.

The first category is further subdivided on the basis of the two types of failure mechanism involved: flow and slip. Flow involves failure of the mass along countless internal planes of slippage, much in the fashion of flow in a viscous liquid such as molasses or hot taffy.

Table 15.1   Classification of Mass Movement.

Movement			Material				
Direction of Movement	**Type**	**Rate**	**Chiefly Ice**	**Ice Plus Earth or Rock**	**Earth or Rock**	**Earth or Rock Plus Water**	**Water**
With free side (downward and outward movement)	FLOW	Imperceptible (usually)	Glacial transport	Rock glacier solifluction	Rock creep Talus creep Soil creep	Solifluction Earth flow Mud flow	Stream flow
		Perceptible, slow, rapid					
	SLIP	Perceptible, very slow to rapid, rapid		Debris avalanche	Slump Debris slide Debris fall Rock slide Rock fall	Debris avalanche	
No free side (primarily vertical movement)	S U	B S I D E N C E					

Source: Sharpe, 1938.

Slip involves the movement of a solid mass downslope along a single failure plane or a well-defined failure zone. In slip, the center of gravity of the mass translates downslope under the force of gravity. This type of failure lends itself to quantitative analysis using the limit equilibrium approach, which states that, at failure, the driving forces along the failure surface are equal to the resisting forces. Prior to failure, the resisting forces must be greater than the driving forces. As shown in Table 15.1, the types of mass movement that occur by slip are slump, debris slide, debris fall, rock slide, and rock fall. Note that although Sharpe's classification includes debris fall and rock fall in the category of movement by slip, they do not involve an appreciable amount of slip nor are they subject to quantitative analysis (Cruden and Varnes, 1996).

Sharpe (1938) used the rate of movement to further subdivide flow and slip types of mass movement. Creep failures of soil and rock occur slowly, whereas avalanches are rapid failures. Slip failures can also be classified according to failure rates as shown in the classification.

The materials carried by mass movement provide the final basis for subdivisions within Sharpe's (1938) classification. The range of constituent materials includes chiefly ice; earth, rock, and ice; earth or rock; earth, rock, and water; and chiefly water. The last category is stream flow and does not qualify as a part of mass wasting per se. Its inclusion in the classification is meant to illustrate the transitional relationship from mass movement to stream flow, which occurs as water becomes the predominant constituent.

Types of Slope Movement

Table 15.2 presents the classification of slope movements proposed by Cruden and Varnes (1996). This classification, widely used by engineering firms, is based on two main criteria: the type of material and the type of movement. The material involved can be either bedrock or engineering soil, with the soil being further subdivided into predominantly coarse-grained and predominantly fine-grained. The coarse-grained soil consists of mostly debris (broken rock) mixed with some soil whereas the fine-grained soil includes sand, silt, and clay. The classification divides slope movements into five major types: falls, topples, slides, spreads, and flows. Figure 15.1 shows examples of slope movements; some of them are not included in Table 15.2 but are discussed in the chapter. The following sections provide detailed descriptions of these five types of slope movement.

Table 15.2   Classification of Slope Movements.

Type of Movement	Type of Material		
		Engineering Soils	
	Bedrock	Predominantly Coarse	Predominantly Fine
Fall	Rock fall	Debris fall	Earth fall
Topple	Rock topple	Debris topple	Earth topple
Slide	Rock slide	Debris slide	Earth slide
Spread	Rock spread	Debris spread	Earth spread
Flow	Rock flow	Debris flow	Earth flow

Source: Cruden and Varnes, 1996.

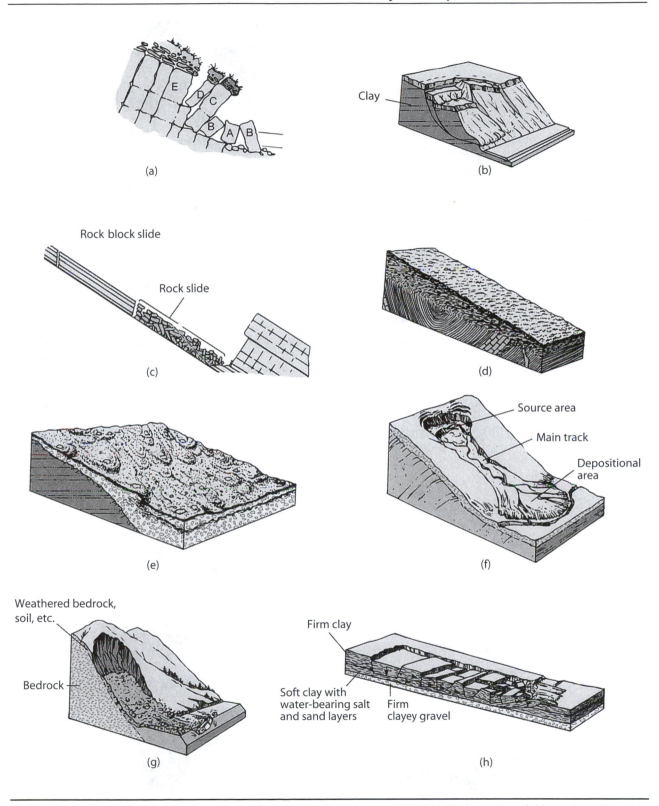

Figure 15.1   Examples of slope movement: (a) rock topple, (b) rotational slide (slump), (c) rock block slide, (d) soil creep, (e) solifluction, (f) earth flow, (g) debris avalanche, and (h) earth spread. (Schuster and Krizek, 1978.)

(a)

(b)

Figure 15.2   Examples of rock falls (a) caused by steeply dipping discontinuities (Igor Dymov, Shutterstock) and (b) caused by undercutting.

Falls

In a fall, the mass of soil or rock detaches along a steeply dipping discontinuity, along which little or no shear displacement occurs, and travels most of the distance through the air. It involves free fall, which is sometimes followed by bouncing and rolling of fragments. Bedrock containing steeply dipping weakness planes that dip into an opening or valley are particularly prone to this problem (Higgins and Andrew, 2012). Weathered bedrock areas may be most susceptible to rock falls, as weathering opens joints and bedding planes. In a series of alternating harder and softer rocks, where differential weathering of rock units results in numerous overhangs, undercutting-induced rock falls are particularly prevalent (Shakoor, 1995; Admassu and Shakoor, 2015). Figure 15.2 shows examples of rock falls. Debris and earth masses on steep slopes may also fail as a unit by falling into the opening. They would be less likely to bounce and roll after dislodging than would the intact blocks of rock.

Rock falls are usually quite hazardous because of their speed and suddenness of occurrence. Because of this, a variety of rock fall hazard rating systems have been developed over the past few decades by different states (Pierson, 2012). Several computer programs, such as Colorado Rockfall Simulation Program by the Colorado Geological Survey and RocFall by Rocscience, are available to determine the trajectories and bounce heights of rock falls of varying sizes. The trajectories and bounce heights are used as input parameters for rock fall hazard rating systems, design of catchment ditches, and design of rock fall barriers. The mechanics of rock fall behavior, in terms of trajectories and bounce heights, can be found in Turner and Duffy (2012).

Topples

Topples involve rotational movement of rock blocks and are a result of an overturning moment about a pivot point below the center of gravity of the block. Toppling occurs when discontinuities dip steeply into the slope. The opening of joint planes, perpendicular to the bedding or to another weakness plane that dips into an opening, is the primary cause of toppling in bedrock. Figure 15.3 shows a field situation conducive to toppling failure. Blocks likely to topple during the service life of a slope should be removed from slopes during construction.

Toppling is a function of the height (h) and width (b) of rock blocks. For toppling to occur, the center of gravity of the block or the weight vector should fall outside the base of the block, as shown in Figure 15.4. This requirement is met when $b/h < \tan\theta$, where θ is the inclination of the plane on which the block rests. Also,

$$\text{factor of safety (FS)} = \frac{b}{h\tan\theta}$$

The following example problem illustrates this requirement.

Figure 15.3   Example of a field situation conducive to toppling failure.

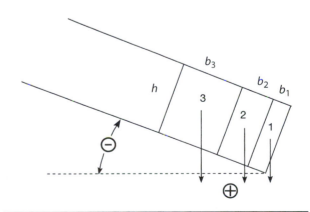

Figure 15.4   Schematic diagram showing the requirement for toppling of rock blocks.

Example Problem 15.1

Bedding planes in a dense limestone dip at 20° into a road cut. The open bedding planes are spaced 2 m (6.5 ft) apart (Example Figure 15.1).
  a. What joint spacing perpendicular to bedding will cause toppling to occur?
  b. For that joint spacing, what would be the FS against toppling if the bedding planes were 3 m (9.6 ft) apart?

Answers

a. $\dfrac{b}{h} = \tan\theta$

Example Figure 15.1

$b = 2\tan 20° = 0.73$ m (2.3 ft)

If the joints are spaced at closer than 0.73 m (2.3 ft) the blocks will topple into the opening.

b. $FS = \dfrac{b}{h\tan\theta}$

$= \dfrac{2}{3\tan 20°}$

$= 1.8$

Slides

In slides, movement occurs along one distinct surface of failure or within a relatively narrow zone of failure. The rock or soil unit fails as a mass by moving down the failure surface. This configuration is well suited to analysis by the limit equilibrium approach, equating the resisting forces (shear strength) to the driving forces (shear stress) acting along the failure surface. The ratio of the two types of forces is expressed as the factor of safety of the slope.

$$FS = \frac{\text{shear strength}}{\text{shear stress}}$$

$$= \frac{c + \sigma\tan\phi}{\text{shear stress}} \qquad \text{(Eq. 15-1)}$$

where:
  c = cohesion
  ϕ = angle of internal friction
  σ = normal stress on the slip surface

If FS > 1, the slope is stable; if FS = 1, the slope is at the verge of failure; and if FS < 1, the slope is unstable. Values for c and ϕ are measured in the field, determined in the laboratory, or simply estimated.

Slides can be either rotational or translational in nature. Rotational slides have a circular failure surface whereas translational slides have a planar failure surface.

Rotational Slides

Rotational slides, commonly referred to as slumps, are extremely common in humid areas. They occur on natural slopes, cut slopes, and embankment slopes comprised of cohesive soils or weak rocks. A buildup of pore pressure in the slope is commonly involved with such failures so that the prime time for rotational slides to occur is shortly after the spring thaw and attendant rainfall. Figure 15.5 illustrates the various parts of a rotational slide and Figure 15.6 on page 334 shows a field example.

Rotational slides are analyzed by using the method of moments, which compares the resisting moment (due to shear strength force along the failure surface) to the overturning moment (due to the driving force). A cross section of the slope, along with the failure surface, is drawn according to scale. The mass of the soil, bounded by the slope and the failure surface, is divided into a series of slices of nearly equal width. These slices are further subdivided into rectangles and triangles (Figure 15.7, page 335). Conventionally, a unit length of slope perpendicular to the plane of drawing is considered. The purpose of

MAIN SCARP—A steep surface on the undisturbed ground around the periphery of the slide, caused by the movement of slide material away from undisturbed ground. The projection of the scarp surface under the displaced material becomes the surface of rupture.

MINOR SCARP—A steep surface on the displaced material produced by differential movements within the sliding mass.

HEAD—The upper parts of the slide material along the contact between the displaced material and the main scarp.

TOP—The highest point of contact between the displaced material and the main scarp.

TOE OF SURFACE OF RUPTURE—The intersection (sometimes buried) between the lower part of the surface of rupture and the original ground surface.

TOE—The margin of displaced material most distant from the main scarp.

TIP—The point on the toe most distant from the top of the slide.

FOOT—That portion of the displaced material that lies downslope from the toe of the surface of rupture.

MAIN BODY—That part of the displaced material that overlies the surface of rupture between the main scarp and toe of the surface of rupture.

FLANK—The side of the landside.

CROWN—The material that is still in place, practically undisplaced and adjacent to the highest parts of the main scarp.

ORIGINAL GROUND SURFACE—The slope that existed before the movement of interest occurred. If this is the surface of an older landslide, that fact should be stated.

LEFT AND RIGHT—Compass directions are preferable in describing a slide, but if left and right are used they refer to the slide as viewed from the crown.

SURFACE OF SEPARATION—The surface separating displaced material from stable material, but not known to be a surface of failure.

DISPLACED MATERIAL—The material that has been moved from its original position on the slope. It may be in a deformed or undeformed state.

ZONE OF DEPLETION—The area within which the displaced material lies below the original ground surface.

ZONE OF ACCUMULATION—The area within which the displaced material lies above the original ground surface.

VC—Vertical component of slump.

HC—Horizontal component of slump.

L—Length of displaced zone along the slope.

LC—Length of slump along the slope.

D—Depth to rupture surface.

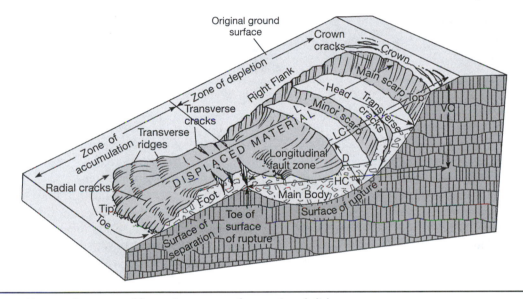

Figure 15.5   Nomenclature used for various parts of a rotational slide.

dividing the slices into rectangles and triangles is to determine the volumes and centers of gravity of each geometrical figure. Knowing the soil/rock density and volumes of triangles and rectangles, the weights of triangles and rectangles can be determined. Knowing the centers of gravity for various rectangles and triangles, their moment arms with respect to the center of rotation, and, hence, the moments can be calculated. The commonly used methods of limit equilibrium analysis for rotational slides, considering the above stated procedure or some variation of it, include the ordinary method of slices (Fellenius, 1927), Bishop's (1955) method, Morgenstern and Price's (1965) method, Janbu's (1968) simplified method, and the modified Swedish method (US Army Corps of Engineers, 1970). Details of these methods are beyond the scope of this text but can be found in books on slope stability (Turner and Schuster, 1996; Abramson et al., 2002; Conforth, 2005).

For the application of limit equilibrium analysis to cut slopes and embankment slopes, it is necessary to determine the location of the critical surface along which the failure is likely to occur. This is done by finding the factor of safety for numerous potential surfaces of failure and choosing the one with the minimum value. These calculations can be accomplished by computer programs such as STABL by Purdue University or Slide by Rocscience.

Figure 15.6   Example of a rotational slide. (R. Couture, Geological Survey of Canada.)

There can be three possible locations of the failure surface with reference to slope geometry: slope failure, toe failure, and base failure (Figure 15.8). Slope failure is a common feature that is limited to the near-surface materials. These failures may involve mostly corrective maintenance for highways and pits. Toe failures are more deeply seated and, hence, involve greater volumes of material. They sometimes develop when the slope is extended by additional excavation. Base failures involve extremely large volumes of material and, because of the considerable depth of the failure plane, it may not be intercepted by borings during exploration prior to construction or during analysis after slope failure. Typically, it involves a weak layer at depth along which failure is concentrated to yield a flat-bottomed shape to the failure surface. Toe and base failures require extensive corrective measures.

Translational Slides

Translational slides, also referred to as rock block slides, occur along planar surfaces, with the rock block sliding down the inclined plane into an opening or valley. This failure is commonly associated with movement along weakness planes in a bedrock mass, referred to as discontinuities.

Rock block slides are of particular interest within the overall category of slides. Many of the large-scale and destructive slope failure events of the past have involved rock block slides. Several of the more spectacular slope failures that have occurred are discussed later in this chapter, but as an introductory statement it is significant that three major slides, the Gros Ventre, Wyoming, failure (1925), the Frank, Alberta, slide (1903), and the Vaiont, Italy, slide (1963) all involved rock block sliding.

Another form of translational slide is the debris slide (see Table 15.2). For the debris slide, the surface of failure is commonly along the debris-bedrock interface, where groundwater accumulates because of the contrasting permeability values of the bedrock (much less) and the debris above. Figure 15.9 on page 336 shows examples of situations prone to rock block sliding and debris sliding.

Figure 15.10, page 337, illustrates the requirements for rock block sliding versus toppling (Wyllie and Mah, 2004): if the slope angle θ is less than the angle of sliding resistance, ϕ, the block will not slide down the inclined plane (zones A and C); if θ is greater than ϕ sliding will occur (zones B and D); if $b/h < \tan\theta$, toppling will occur (zones C and D). The ϕ angle for many competent rocks is about 30 to 35°. For weak rocks such as shale, however, the ϕ

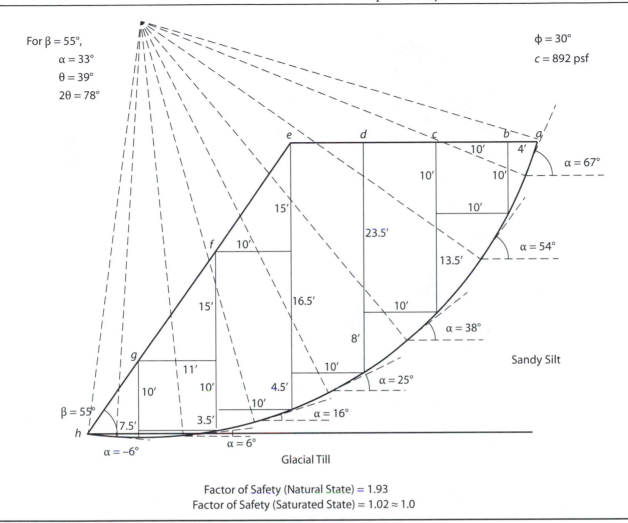

For β = 55°,
α = 33°
θ = 39°
2θ = 78°

ϕ = 30°
c = 892 psf

Factor of Safety (Natural State) = 1.93
Factor of Safety (Saturated State) = 1.02 ≈ 1.0

Figure 15.7   Example of analysis of rotational slides by the method of slices.

value may be only 15° or lower. For Figure 15.10, ϕ is assumed to be equal to 35°. Summarizing Figure 15.10, zone A is stable with respect to both sliding and toppling, zone B shows conditions favorable for sliding but not for toppling, and zone C indicates toppling but no sliding. For zone D, both sliding and toppling will occur because $\theta > \phi$ and $b/h < \tan\theta$.

Similar to rotational slides, the limit equilibrium approach is used to analyze rock block slides. Figure 15.11 on page 338 shows a schematic diagram of rock block sliding, the forces acting on the block, and the Mohr–Coulomb diagram for determining the shear strength parameters (cohesion and angle of internal friction). The normal force, F_n, on the sliding plane is given by the following equation:

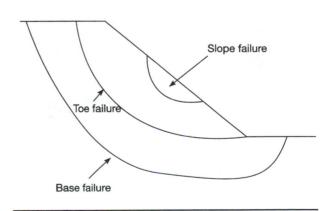

Figure 15.8   Locations of failure surfaces for rotational slides.

(a)

(b)

Figure 15.9   Examples of translational landslides: (a) situation prone to rock block sliding (David Giles, University of Portsmouth) and (b) debris slide (LIUSHENGFILM, Shutterstock).

$$F_n = W \cos\theta \qquad \text{(Eq. 15-2)}$$

The resisting force, F_r, on the failure plane is equal to:

$$F_r = \tau A = cA + W\cos\theta\tan\phi \qquad \text{(Eq. 15-3)}$$

Figure 15.11 defines the parameters involved in Equations 15-2 and 15-3.

Referring again to Figure 15.11, the driving force, F_d, due to the weight of the block, is given by:

$$F_d = W\sin\theta \qquad \text{(Eq. 15-4)}$$

Therefore, the FS is given by:

$$FS = \frac{\text{resisting force}}{\text{driving force}}$$

$$= \frac{cA + W\cos\theta\tan\phi}{W\sin\theta} \qquad \text{(Eq. 15-5)}$$

If cA is considered to be zero, as is assumed in certain cases, and canceling W above and below,

$$FS = \frac{\cos\theta\tan\phi}{\sin\theta}$$
$$= \frac{\tan\phi}{\tan\theta} \qquad \text{(Eq. 15-6)}$$

A final consideration can be made by allowing the ϕ angle to equal the slope angle θ; then

$$FS = \frac{\tan\phi}{\tan\theta} = 1 \qquad \text{(Eq. 15-7)}$$

From this it follows that if $\phi > \theta$ and neglecting any contribution from the cohesion (also assuming dry slopes or zero pore pressure), the FS will be greater than 1, yielding a stable condition. This is the basis for the sliding stability considerations given in Figure 15.10. For most discontinuities, the cohesive force is relatively small, ranging from less than 4.9 to 38.9 kPa (< 100 to 800 psf). Therefore, in many situations, the contribution of cohesion is neglected in the analysis.

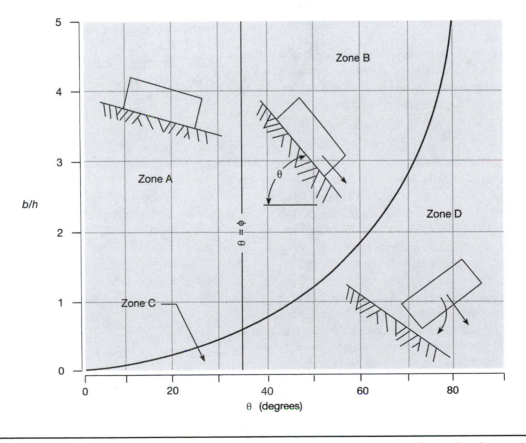

Figure 15.10   Requirements for rock block sliding versus toppling. (Modified from Wyllie and Mah, 2004.)

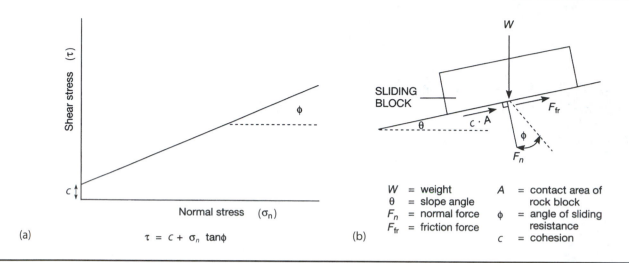

$$\tau = c + \sigma_n \tan\phi$$

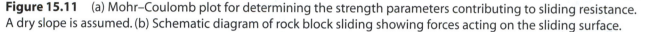

(a) (b)

W	= weight	A	= contact area of rock block
θ	= slope angle		
F_n	= normal force	ϕ	= angle of sliding resistance
F_{fr}	= friction force		
		c	= cohesion

Figure 15.11 (a) Mohr–Coulomb plot for determining the strength parameters contributing to sliding resistance. A dry slope is assumed. (b) Schematic diagram of rock block sliding showing forces acting on the sliding surface.

This conservative approach provides an extra margin of safety against sliding.

The ϕ value for rock sliding on rock is primarily a function of the mineralogy of the constituent materials. Table 15.3 shows the relationship between rock type and ϕ value.

Commonly, an accurate stability evaluation of a previously intact slope can best be made shortly after it has failed. This is accomplished by back calculating the shear strength parameters for a factor of safety equal to unity, assuming the original slope conditions can be determined.

Discontinuities in Rock

Discontinuities are the planes of weakness in a rock mass. The stability of a rock mass depends on the strength along the discontinuities, not of the intact rock. This is in contrast to soil, for which the mass strength is determined by the basic material strength itself. Consequently, laboratory tests on small soil samples, if they approximate field conditions, can be used to indicate field behavior. For a rock mass, however, material strength and mass strength are very different. Relative to the slope stability of a rock mass, shear strength along the discontinuities determines the resistance to movement.

If the compressive strength of a material is greater than about 700 kPa (100 psi), the material behaves as a rock; that is, its mass strength is a function of the discontinuities present. Below 700 kPa (100 psi) the material behaves as a soil, with soil parameters con-

trolling the soil mass behavior. Table 15.4 contains a list of common rock discontinuities.

Significant Aspects of Discontinuities

The nature of discontinuities in the rock mass, that is, their physical properties and characteristics, is of major importance to rock slope stability (Park et al., 2005). The significant aspects of discontinuities are: (1) geometry, (2) continuity, (3) spacing, (4) surface irregularities, (5) physical properties of the adjacent rock, (6) infilling material, and (7) groundwater.

Geometry: The geometry of a discontinuity involves its orientation with respect to slope face. The steepness of discontinuities and how they intercept the slope are critical factors. The strike and dip of the weakness planes and of the slope faces are used to define geometry. Blocks can slide more easily on steeper weakness planes, but these planes must dip toward the slope face (the opening) to be critical. Weakness planes are said to "daylight" if they intersect the slope face by dipping at an angle less than the slope angle but still in the slope direction.

Table 15.3 Angles of Sliding Resistance According to Rock Type, Smooth Contacts.

Rock Name or Varieties	Approximate ϕ Value
Quartzite and quartz-rich rocks	30°
Schists and related rocks	20–25°
Carbonates	30–35°
Clays and clay shales	8–20°

Table 15.4   Types of Rock Discontinuities Commonly Related to Slope Stability.

Joint	A fracture or parting in a rock, without displacement; the joint surface is usually planar and commonly repeats itself in a parallel fashion to form a joint set.
Fault	A surface or zone of rock fracture along which there has been displacement from a few centimeters to a few kilometers (about an inch to a few miles) in scale.
Shear zone	A tabular zone of rock that has been crushed and brecciated by many parallel fractures owing to shear strain.
Bedding surface	A surface, usually conspicuous, within a mass of stratified rock representing an original surface of deposition. It is the interface between two adjacent beds of sedimentary rock. If the surface is more or less regular or nearly planar, it is called a bedding plane.
Foliation	A general term for the planar arrangement of textural or structural features in metamorphic rocks, such as rock cleavage in a slate, schistosity in a schist, and gneissic structure in a gneiss.

Hence, the worst situation occurs when the strike of the weakness plane parallels the strike of the slope and the two dip in the same direction. That is why roadways should be oriented at right angles or at the highest angle possible to the strike of steeply dipping beds in order to minimize potential slope failures.

Continuity: Continuity refers to the extent of the weakness planes and is of considerable importance. Major discontinuities, such as faults, extend for considerable distances but smaller features, such as joints, likely extend for much shorter distances. Some joint sets will extend through a number of beds or even a formation, whereas other sets will terminate after several bed thicknesses or even a single bed. If failure is to occur in a rock mass that contains discontinuities that terminate within the mass, some intact rock must fail in order to propagate the fracture. This involves considerably larger forces for failure to occur than for simple sliding along through-going discontinuities. The shear strength of intact rock will be many times that of the fractured rock. It is sometimes possible to estimate the length of intact rock breakage needed based on the existing discontinuities in a rock mass. This information can be used as a further refinement of rock slope stabil-

ity. Typically, the presence of only 1 or 2% of intact rock along a boundary can nearly double the factor of safety against sliding (West, 1996).

Spacing: The spacing of discontinuities influences rock mass strength. Joints develop in parallel sets or combinations of sets with a more or less regular spacing between the joint planes. In addition, bedding planes have regular spacing, which should be included as part of the descriptive information provided in an engineering geology field report. Spacing ranges from closely spaced, a few centimeters apart (about an inch), to widely spaced or massive with a separation of 1 m (3 ft) or so. The spacing of discontinuities is directly related to the amount of intact rock that must break before through-going fractures can develop. Closely spaced discontinuities, although each, individually, may not fully traverse the rock mass, lead to a weaker situation than do widely spaced discontinuities that also, presumably, do not fully cut the mass. Simply stated, closely spaced discontinuities, with a more continuous fracture surface, indicate a weak rock mass.

Surface Irregularities: Surface irregularities along a discontinuity in rock can provide additional resistance compared to a smooth surface. Two mechanisms allow these irregularities to contribute to increased strength. That is, failure requires: (1) overriding the irregularities or (2) fracturing through them (Patton, 1966). Figure 15.12 on the next page illustrates these two possibilities.

Figure 15.13 on page 341 illustrates the Mohr–Coulomb plots, and the corresponding strength equations, for both overriding irregularities and shearing through mechanisms. Overriding the irregularities, in effect, increases the angle of sliding resistance ϕ by an amount equal to the irregularity angle i (Patton, 1966). This yields the equation:

$$\tau = \sigma_n \tan(\phi + i) \qquad \text{(Eq. 15-8)}$$

Here the small contribution of cohesion along the fracture surface has been neglected or assumed equal to zero. The angle i usually ranges from 10 to 15° in actual practice, although field and laboratory measurements may suggest higher values.

The equation for shearing through irregularities (see Figure 15.13) is:

$$\tau = c_r + \sigma_n \tan\phi_r \qquad \text{(Eq. 15-9)}$$

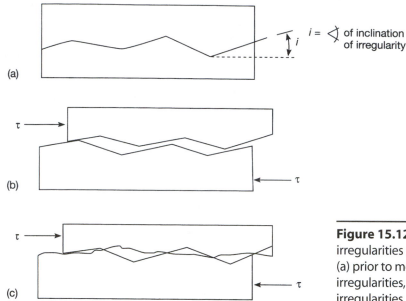

(a)

(b)

(c)

Figure 15.12   Cross-section view of surface irregularities related to movement along a discontinuity: (a) prior to movement, (b) movement by riding over irregularities, and (c) movement by shearing through irregularities.

c_r is the cohesive strength of intact rock and $\tan\phi_r$ is the angle of internal friction for intact rock (approximately 50° for massive rocks). The value of c_r is equivalent to the S_0 values, obtained from triaxial tests for rock, listed in Table 8.1. These values range from 7000 to 70,000 kPa (1000 to 10,000 psi) and are sufficiently large that c_r completely dominates the shear strength equation at depths where the normal stress (σ_n) exceeds the value indicated by the change in the slope of the solid line in Figure 15.13 (West, 1996). The dashed line in Figure 15.13 shows normal stress (σ_n) values where that mode would not control slope failure because the other failure mode occurs at a lower stress level and, thus, prevails.

The lowest plot in Figure 15.13 is for continued movement along a surface of prior displacement. The shear strength equation for this case is:

$$\tau = \sigma_n \tan\phi_{jr} \qquad \text{(Eq. 15-10)}$$

The value of ϕ_{jr} is somewhat less than the ϕ_j value that controlled when slope movement first occurred. For a massive limestone, a typical value for ϕ_j would be 35°, but the ϕ_{jr} value, after block displacement, would be about 25°.

From the preceding discussion, we can conclude that surface irregularities increase the shearing resistance along failure surfaces. These features must be noted in the field to allow for recognition of their contribution to resisting force (Barton, 1971, 1973).

Example Problem 15.2

A series of quartzite beds dips into an open pit mine. How steep can the beds dip if they have an irregularity value of $i = 12°$, assuming a dry slope, no cohesion, and a required factor of safety of 1.2?

Answers

For a dry slope

$$FS = \frac{cA + W\cos\theta\,\tan(\phi + i)}{W\sin\theta}$$

If $c = 0$:

$$FS = \frac{\tan(\phi_j + i)}{\tan\theta}$$

The value of ϕ_j for quartzite is estimated at 30° (see Table 15.3):

$$\tan\theta = \frac{\tan(\phi_j + i)}{FS}$$
$$= \frac{\tan(30 + 12)}{1.2}$$
$$= 0.7503$$
$$\theta = 36.9°$$

Physical Properties of Adjacent Rock: The next aspect of discontinuities for consideration with regard to the resistance of rock block sliding relates to the physical properties of the adjacent rock. Different rock types and their alteration products develop different shearing resistances. This is suggested by the range in ϕ values shown in Table 15.3 and the range in cohesion values (< 4.0 to 38.9 kPa [< 100 to 800 psf]), as indicated previously. For irregular contacts between rock units of different strengths, failure

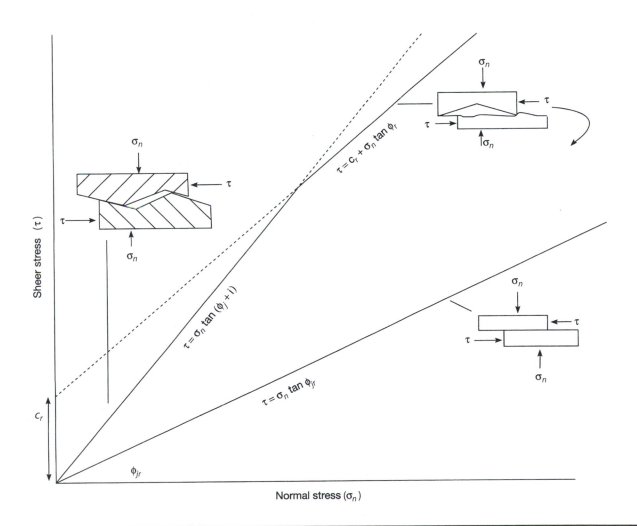

Figure 15.13 Mechanisms of failure on irregular discontinuities in rock and the associated shear strength.

through the irregularities, if it were to occur, should involve more of the weaker rock than of the stronger. Consequently, the value of c_r to include in the equation should represent that of the weaker rock.

Infilling Material: The infilling material of a discontinuity includes all soil-like material that occurs between the walls of the fracture. This includes gouge developed previously during movement between the walls of the discontinuity as well as any sediment that has accumulated in the opening. The resistance to shearing along the discontinuity is related to the properties and thickness of the infilling material as well as the shear strength of the rock walls (Barton, 1974).

Four possible situations exist with regard to the infilled discontinuity:

1. Failure surface is contained entirely within the infilling material and shearing resistance is dependent on the properties of the infilling material alone.

2. Failure surface passes partially through both the infilling and the intact wall rock. Shearing resistance is complex and both materials will contribute to strength.

3. Infilling is present but very thin so that the properties of infilling material will only modify the shearing resistance developed between the rock walls of the discontinuity.

4. Infilling is absent and the shearing resistance is dependent only on properties of the wall rock.

In practice, it is extremely difficult to determine the nature and extent of infilling material for a large rock mass volume even if surface exposures are good. One concern is that infilling material may be washed from rock cores or ground up during the drilling operation. Therefore, it is impossible to obtain truly quantitative information for the total rock mass so that values may have to be extrapolated over considerable distances. Data on infilling material may be used simply to adjust slightly those calculations made assuming the existence of smooth, clean discontinuities.

Another factor concerned with infilling of discontinuities may greatly increase rock slope strength: secondary mineralization, in the form of quartz, calcite, and possibly iron oxide, can greatly increase the rock mass strength. In some cases the joint planes become as strong as or stronger than the intact rock.

This is another reason why careful examination of rock cores and rock exposures is necessary for engineering applications.

Groundwater: The final consideration concerning important features of discontinuities and rock slope stability is the presence of groundwater. The considerations related to groundwater are explored first and then a quantitative evaluation is presented.

Groundwater contributes to the instability of rock slopes in the following ways:

1. A reduction in shear strength occurs because of fluid pressure (pore water pressure) acting along potential failure surfaces. Along discontinuities, water causes an uplift pressure perpendicular to the weakness planes and, in effect, reduces the downward normal stress. When discontinuities are numerous, with diverse orientations, the pore water pressure increases with depth, yielding hydrostatic pressure ($h\gamma_w$) like that occurring in soils. When discontinuities have only a few preferred orientations and the spacing between them is large, abnormal distributions of water pressure (usually higher than hydrostatic) can result.

2. The presence of water increases the bulk unit weight of the rock mass (W) and thus increases the driving force ($W \sin\theta$). The term W is present in the resisting force also, but the reduction in normal stress caused by pore water pressure more than offsets the increase in W caused by the addition of that same water.

3. Groundwater freezing in discontinuities can force open these planes, which increases water access to the slope and causes downslope displacement because of the volume increase when water turns into ice. Another feature sometimes overlooked is that the freezing of water along the downslope surface can block drainage to the surface from the rock mass, thus increasing the water pressure within the slope itself.

4. As a result of high velocity groundwater flow, erosion of infilling material can open up discontinuities. Removal of these materials may reduce the shearing resistance along the discontinuity. Increases in pressure caused by impounding reservoirs or raising the groundwater level by other means may lead to this condition.

The major concerns relative to groundwater effects can be best considered by turning to the basic equation shown in Figure 15.14. Resisting force = cA + F_n tanϕ = cA + W cosθ tanϕ and the driving force = W sinθ. Figure 15.14 illustrates the role of water in decreasing the resisting force and increasing the driving force.

For purposes of an example, assume

L = 3 m (10 ft)
h = 1.2 m (4 ft)
γ_{rock} = 26.7 kN/m³ (170 lb/ft³)
c = 9.6 kN/m² (200 lb/ft²)
ϕ = 35°
θ = 20°
ρ_w = 9.81 kN/m³ (62.4 lb/ft³)

Assume first that the slope is dry. The U and V in Figure 15.14 would equal zero in this situation.

$$FS = \frac{cA + W\cos\theta\tan\phi}{W\sin\theta}$$
$$= \frac{9.6(3)(1) + (26.7)(3)(1.2)(1)\cos 20°\tan 35°}{26.7(3)(1.2)(1)\sin 20°}$$
$$= 2.78$$

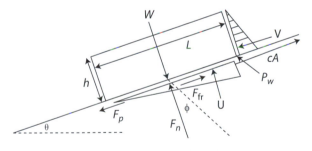

F_p = component force parallel to plane = W sinθ

U = uplift force due to pore water pressure

V = downslope force due to pore water pressure

$$FS = \frac{cL + \left(hL\gamma_{rock}\cos\theta - \dfrac{hL\rho_w\cos\theta}{2}\right)\tan\phi}{hL\gamma_{rock}\sin\theta + \dfrac{h^2\rho_w\cos\theta}{2}}$$

Figure 15.14 Stability analysis of a rock block slide, pore pressures included.

Using British engineering units:

$$FS = \frac{cA + W\cos\theta\tan\phi}{W\sin\theta}$$
$$= \frac{200(10)(1) + (170)(10)(4)(1)\cos 20°\tan 35°}{170(10)(4)(1)\sin 20°}$$
$$= \frac{2000 + 4474.26}{2325.7} = 2.78$$

Next, the pore pressure situation is considered.

The pressure in the water at the bottom of the fracture is P_w. It is equal to the vertical height of the water column, h cosθ, multiplied by the density of water ρ_w, which equals 9.81 kN/m². Therefore, $P_w = h$ cosθ ρ_w = 3 cos20°(9.81) = 26.7 kN/m². The U and V are triangular pressure diagrams, which equal hP_w × 1/2 and LP_w × 1/2, respectively. This yields the equation:

$$FS = \left\langle \frac{\{9.6(3)(1) + [(26.7)(3)(1.2)(1)]\cos 20°\tan 35°\}}{1.2(3)(26.7)\sin 20°} \right.$$
$$\left. \frac{-\dfrac{1.2(3)(9.8)\cos 20°}{2}\tan 35°}{+\dfrac{(1.2)^2(9.8)\cos 20°}{2}} \right\rangle$$
$$= 2.02$$

Using British engineering units:

$$FS = \left\langle \frac{\{200(10) + [(4)(10)(170)]\cos 20°\}}{4(10)(170)\sin 20°} \right.$$
$$\left. \frac{-\dfrac{4(10)(62.4)\cos 20°}{2}\tan 35°}{+\dfrac{(4^2)(62.4)\cos 20°}{2}} \right\rangle$$
$$= 2.02$$

The factor of safety is greater than 1 but substantially less than it was for the same slope when dry (FS = 2.78). The effect of the pore pressure buildup in a slope is quite significant and therefore the slope should be drained or one must consider the water pressure effect in the slope design.

Figure 15.15 shows a final detail for groundwater and rock slope stability. In this diagram, downslope drainage has been blocked by ice on the slope face.

The factor of safety for this condition is now reduced to 0.86. Therefore, from the dry case to the one with the crack full of water under free drainage, the FS is reduced from 2.78 to 2.02. It is further reduced when drainage is blocked, to a value of 0.86, yielding a failure condition. Although $\phi \gg \theta$ (35° versus 20°), failure will occur under these high, pore water pressure conditions.

Figure 15.16 shows additional photos of rock slides, a dominant and usually destructive type of slope movement.

The above discussion of translational or rock block slides focused on sliding occurring along a single discontinuity. It should be noted, however, that translational movement along two intersecting discontinuities, forming a wedge-shaped block, is also a very common occurrence. Figure 15.17 on page 346 shows an example of a wedge failure.

Lateral Spreads

Lateral spreads are the next category of slope failures found in the classification by Cruden and Varnes (1996). Lateral spreads were recognized as a significant form of slope movement only after the Anchorage, Alaska, earthquake of 1964 where the mechanism of lateral spreading was the dominant mode of failure.

The most significant detail of lateral spreading involves fracturing and extension of intact material, either bedrock or soil, because of liquefaction or plastic flow of material at a shallow depth below the surface. Liquefaction is the rapid loss of shear strength that occurs in saturated fine sands and silts because of a sudden increase in pore water pressure and the attendant decrease in effective stress as a result of earthquake shocks. The coherent upper units may subside, slide, rotate, disintegrate, or simply liquefy and flow downhill. The process may involve rotation, translation, and flow and, therefore, lateral spreading failures could be quite complex in nature. Because they are such a distinctive group, however, and they are related to specific geologic conditions, a special designation for them is warranted.

Laterally spreading slope failures occur in fine-grained earth materials on shallow slopes, particularly in sensitive silts and clays that lose most or all of their shear strength when disturbed or remolded. Failure is commonly progressive; it starts locally and spreads from that point into the slope. The initial

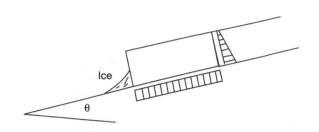

Figure 15.15 Stability analysis of a rock block slide, pore pressures with blocked drainage.

failure may be a rotational slide along a stream bank or along a shoreline with the failure progressing further and further into the bank. The primary surface movement is translation, not rotation. Movement commonly begins suddenly, without warning, and it proceeds with rapid to very rapid velocity from the initial point of failure.

If the underlying mobile zone is thick, blocks at the head of the failure typically drop downward as grabens, which may or may not involve a backward rotation. Upward and outward extrusion and flow may occur at the toe of the failure. Lateral spreads seem to include a gradational series of movement in surficial material, ranging from block slides at one extreme when the mobile zone is very thin, to earth flows or completely liquefied mud flows at the other extreme in which the zone comprises the entire mass.

Most spreading failures in the western United States involve less than total liquefaction and are mobilized by seismic shock. This holds true for the failures during the San Fernando Valley earthquake of 1971 and in the failure of the Bootlegger Cove Clay below the Turnagain Heights residential district of Anchorage, Alaska, in the major earthquake of 1964. It also occurred in some areas during the San Francisco earthquake of 1906 where spreading failures caused direct damage to structures and severed water supply lines, which reduced fire-fighting capabilities. In the Loma Prieta earthquake of October 17, 1989, lateral spreading in the Marina District of San Francisco caused major damage to existing structures. Lateral spreads were negligible during the Los Angeles earthquake of January 17, 1994, because of the dry nature of the valley-fill soils in the city of Northridge.

For lateral spreading failures in Pleistocene-age glacial and marine sediments, several common characteristics of the failed material have been recognized:

(a)

(b)

Figure 15.16   (a) Rock slides along bedding planes (trekandshoot, Shutterstock), and (b) rock slides forming a landslide dam, Thistle, Utah.

Movement commonly occurs without apparent reason, failure is generally sudden, even gentle slopes are unstable, dominant movement is by translation, the materials are sensitive, and pore water pressure is involved in the stability.

Flows

Slope failure by the mechanism of flow occurs by internal slippage along many planes of failure within the slope mass, similar to downslope movement of a viscous liquid. The surfaces of shear in a flow are temporary and closely spaced (Cruden and Varnes, 1996). Rate of movement can be different within different parts of a flow.

The materials affected by flow failure can include ice, as well as a wide variety of soils and bedrock. Various types of flows (see Table 15.2) include rock flow, debris flow, and earth flow, which could change into mud flow with the addition of water. These terms are discussed briefly in the following paragraphs.

Rock Flow

Rock flow or rock creep is the movement of bedrock downslope by the slow-flow failure mechanism of deep-seated creep (Soeters and Van Westen, 1996). It occurs in bedrock that dips downslope by slow slippage along bedding planes, and in rock units that dip into the slope by creep along joint planes that parallel the slope. An extreme case of rock flow into an existing valley can lead to a valley bulge or cambering—the inward movement of the valley sidewalls. This type of flow movement has occurred in England where massive dolomite overlies marine shales, causing bulging of the dolomite units into the valley.

Recently, a flow phenomenon in massive bedrock, manifested by the folding, bending, or bulging of rock, has been recognized during field studies. These features are found in many areas of high relief throughout the world. The term *sackung* is used for the ridgetop depression that develops because of large-scale creep toward the valley walls.

Debris Flow

Debris flows are generally restricted to channels and result from unusually heavy rainfall or from rapid thawing of snow. Debris flows have a considerable runout (Soeters and Van Westen, 1996) zone beyond the base of the slope, a fact that must be made known to planners, architects, and contractors who may attempt to initiate construction in

Figure 15.17 Example of a wedge failure. (National Highway Institute.)

such destruction-prone areas. Figure 15.18 shows an example of a debris flow.

Debris Avalanche

A debris avalanche is an extremely rapid failure of broken rock and soil that is sometimes accompanied by ice. It may be relatively dry but, in some instances, contains enough water to saturate the debris. In general, they are long and narrow with characteristic V-shaped scars tapering uphill to the head. Usually they occupy a preexisting stream channel and may advance well beyond the foot of the slope in an extensive runout zone, an area with a much lower gradient that extends for hundreds or thousands of meters (feet) beyond the foot of the slope, over which the debris expends its final energy.

Mud Flow

Mud flows are rapid failures that are common in arid and semiarid regions and are confined to a definite channel. Mud flows are distinct from debris flows in that debris flows involve material containing a high percentage of coarse fragments, whereas a mud flow consists of 50% or more sand, silt, and clay and the material must be wet enough to flow rapidly. In mud flows, soil and vegetative debris accumulates at the heads of dry channels until an infrequent rainfall event triggers the high runoff typical of these dry, poorly vegetated areas, and the material becomes saturated. With increased moisture comes decreased shear strength and the wet mass flows rapidly down the channel, obliterating any trees or structures in its path. Active mud flow zones can be recognized during field studies or from aerial photographs, and old mud flow deposits can be identified by their diagnostic landforms. Active mud flow areas should be properly zoned to prevent construction of housing and industrial buildings, but golf courses or parks without permanent structures may be quite suitable for these locations (see Figure 15.19, page 348).

A particularly serious set of conditions periodically gives rise to damaging mud flows in California. During a prolonged drought, brush fires burn off the vegetative cover. Then, if a period of intense rainfall occurs, runoff is particularly extreme because of the denuded nature of the landscape, and it triggers the onset of numerous mud flows. This and the lack of restrictive building codes in the past allowed for widespread destruction of homes by mud flows. Proper zoning of mud flow areas and a program to remove excess debris from the heads of mud flow-prone channels can greatly reduce the potential financial loss associated with these slope failures.

Earth Flow

An earth flow is a transitional form of a slow moving earth slide. In an earth flow, most movement occurs on, or is adjacent to, the boundary (Keefer and Johnson, 1983) but the moving mass can be internally deformed to varying degrees. Thus, the movement may range from a slow moving earth flow to a slow moving earth slide. Usually, earth flows can be distinguished from rotational slides by their slow rate of movement and from debris flows and mud flows by a much lower water content. Figure 15.20 on page 349 shows an earth flow in a silty soil.

Figure 15.18   Example of a debris flow. (US Geological Survey.)

Creep and Other Flow Type Movements

Talus Creep and Soil Creep

Talus creep and soil creep refer to the slow movement downslope of either talus (accumulations of broken rock along the base of mountain slopes) or of topsoil and subsoil. Good indicators of soil creep are tilted fence posts, bent trees, and other shallowburied markers that move progressively downslope. Creep can lead to surveying problems and property disputes if the fence line also serves as the property line.

Solifluction

Solifluction involves soil flow down gentle slopes, mostly in arctic and subarctic areas, because of surface melting of frozen soil. Water is unable to penetrate the frozen soil below and the mixture of water, soil, and sometimes ice will flow downslope at angles of as low as 2 to 3°. This process develops flat tundra areas and is thought to have caused the flat terrain of till plains in the years following glacial retreat.

Block Streams

Block streams are narrow channels of rock debris on steep slopes that move extremely slowly and are typically fed by talus accumulations at the head. Rain wash removes the fines from the surface so the resulting appearance is coarser than the overall nature of the deposit.

Wet and Dry Sand Flows

Wet sand or silt flow is a particular problem in dunes along the shores of large lakes or the ocean. Rapid failures can occur during changes from partial saturation to full saturation. Partially saturated sand has an apparent cohesion because of capillary forces caused by numerous water-air interfaces in

(a)

(b)

Figure 15.19   (a) Schematic diagram of a mud flow and (b) area damaged by a mud flow (Harald Schmidt, Shutterstock).

Figure 15.20 Example of an earth flow.

the pores. With saturation, not only is this apparent cohesion lost, but pore water pressure further reduces the effective stress and rapid slope failure can occur. Dry sand flows are common along shores or embankments underlain by dry granular material. They occur as both channelized flows and as sheet-like features.

Dry Loess Flow

Dry loess flow failures are induced by earthquakes on slopes that are generally quite steep. Loess is able to stabilize on a near vertical slope but earthquake shaking provides the vibration needed to mobilize the grains of silt in the porous loess. Apparently, the internal structure is destroyed by the shaking and the loess becomes a fluid suspension of silt in air, which flows down into the valleys, filling them and overwhelming anything in its path.

Loess flows mobilized by earthquake shocks are associated with two major instances of extreme loss of life in the last century. In Gansu Province, China, following the 1920 earthquake, about 100,000 lives were lost when loess flowed into the valleys, burying villages. On July 10, 1949, in Tajikistan, south-central Asia, earthquake-triggered loess flows buried or destroyed 33 villages as the flow covered the valleys to depths of several tens of meters (30 to 50 ft) for a distance of many kilometers (many miles).

Complex Slope Movements

Complex slope failures involve a combination of two or more types of failures (see Table 15.2). Complex failures are no longer included in the classification proposed by Cruden and Varnes (1996), but, because many slope movements are complex in nature, they are included here for completeness. They include:

- Rock fall-debris flow (or rock fall avalanche)
- Slump and topple
- Rock slide-rock fall
- Cambering spreading plus bending overflow
- Slump-earth flow
- Mud slide (or sliding mud flow)

Rock fall-debris flows or, similarly, rock slide-debris flows are most common in rugged mountainous regions. The Elm, Switzerland, failure of 1881, which took 115 lives, started with a small rock slide adjacent to a quarry on the mountainside but quickly became a flow of high velocity. About 10 million m^3 (13 million yd^3) of rock descended an average of 470 m (1540 ft) vertically in an elapsed time of 55 sec. The much publicized Frank, Alberta, slide of 1903 began as a rock slide and turned into an avalanche, which killed about 70 people in the outlying portions of the town. Yet these failures were minor by comparison to the rock fall-avalanche that followed the May 31, 1970, earthquake in Peru, which buried the city of Yungay and part of Ranrahirca, causing a loss of life greater than 18,000. The movement started at an elevation of 6400 m (21,100 ft), coming to rest 3660 m (12,000 ft) below, and involved 50 to 100 million m^3 (65 to 130 million yd^3) of rock, ice, snow, and soil that traveled 14.5 km (9 mi) from its source to Yungay at velocities between 280 to 335 km/hr (175 to 210 mph).

The combined flows of extreme magnitude described seem to be lubricated, at least in part, by a cushion of air trapped beneath the debris. The enormous size of these flows is required to develop this means of transfer, which, fortunately, does not occur in the more commonplace smaller slope failures. The large prehistoric Blackhawk landslide in the San Bernardino Mountains, California, which measures 3.2 km (2 mi) across, is thought to have slid on a layer of compressed air.

The specific combinations of slump and topple, rock slide-rock fall, and slump-earth flow are fairly straightforward relationships between the two failure modes involved. Cambering and valley bulging were described previously under flow failures.

Causes of Slope Movements

Slope failures occur as short-lived phenomena that have a profound effect on the land surface and human activities. The stress applied to a slope must exceed a certain critical level before the equilibrium conditions, which may have endured for many years, are upset and failure proceeds. There are three primary causes of major slope movements: (1) excessive precipitation, (2) earthquakes, and (3) human activities (Alexander, 1992; Wieczorek, 1996).

Excessive Precipitation

Existing moisture conditions in a slope mass determine the extent of rainfall necessary to trigger a landslide. If the slope contains significant moisture from prior rainfall, the contribution of water from a new storm can be relatively small but still induce slope failure. Other factors being equal, the intensity and duration of the storm are the significant factors that contribute to instability of the slope.

Addition of groundwater to the slope increases the total downward-acting force on potential failure surfaces and reduces the frictional resistance along such surfaces by increasing the pore water pressure (with a resulting decrease in effective stress). The geologic conditions of the slope mass have a great effect on the means and degree in which water enters the slope and, in turn, drains from it. High permeability of soils or bedrock provides easy ingress into the slope, except under frozen conditions. Low permeability, or a frozen surface, prevents the exit of water at the base of the slope, yielding an increase in pore water pressure and greater instability.

Exceptionally heavy rains have caused massive sliding to occur in various parts of the world. During 1966 to 1967, very heavy rainstorms in Brazil initiated massive landslides and flooding, which killed more than 2700 people. The landslides included a slump-earth flow, debris slides and avalanches, debris flows and mud flows, and rock slides.

Japan is subject to numerous landslides related to high-intensity storms because of the high relief of the islands and the high annual rainfall. Kobe was hit in 1938 by rainstorms that produced rock and mud flows killing 461 people and destroying 100,000 houses. In 1945 the Makurazaki typhoon hit Kure and produced mud flows causing 1154 deaths. The Kanogawa typhoon of 1958 produced the heaviest rainfall ever recorded for Tokyo: 39.25 cm (15.4 in) in 24 hr. This caused more than 1000 landslides and killed 61 people.

In June 1966 Hong Kong received a series of rainstorms that amounted to nearly 15 times the normal intensity. Near mid-month, 40 cm (15.7 in) of rain occurred in 24 hr, which was added to the previous 31.4 cm (12.3 in) from earlier that month. This was the trigger for extensive landsliding, the most disastrous and spectacular being large-scale, deep-seated rotational slides. Also included were debris avalanches, boulder falls, and rock slides. Bedrock in Hong Kong is predominantly volcanic and intrusive rocks on which a deep residual soil has formed by weathering.

The United States has also suffered extensive slope movements associated with high-intensity storms. In August 1969 Hurricane Camille produced an 8-hr deluge of 71 cm (30 in) in central Virginia. In Nelson County alone, property damage was more than $116 million and 150 people were killed by the combination of debris avalanches and flooding. The debris avalanches followed preexisting drainages on hillsides steeper than 35°; produced head scars at the steepest part of the hill; were prevalent on north-, northeast-, and east-facing slopes; caused rapid surges of water and sediments into stream channels with devastating effects; and typically left scars from 60 to 240 m (200 to 800 ft) in length, 7.5 to 23 m (25 to 75 ft) wide, and 0.3 to 0.9 m (1 to 3 ft) deep. In California, many landslides have been linked to the effects of rainfall. The distribution, amount, and pattern of rainfall strongly influence slope failure. Heavy rainfall that occurs when the slopes are still wet from previous storms causes the largest number of landslides. On this basis, the pattern of rainfall during the rainy season is more important than is the total amount received.

Earthquakes

History suggests that the most catastrophic landslides that occur are related to earthquakes. Earthquake shocks have profound effects on loose and sensitive soils and even on bedrock to the extent that major terrain changes are associated with the world's most severe earthquakes.

The Lisbon earthquake of 1755 produced widespread effects throughout Portugal, causing many failures along sea cliffs and in the valley walls of rivers. The massive Indian earthquake of 1897 profoundly changed and scarred valleys for distances of 36 km (20 mi). The shock was felt as a strong motion in an area of 416,000 km^2 (160,000 mi^2) and was noticeable in a region more than 10 times as great. The 2005 Kashmir, Pakistan, earthquake triggered thousands of slope movements, mainly rock falls and rock slides. The largest movement, an avalanche consisting of mixed sandstone, limestone, mudstone, and shale, measuring 80 million m^3 (105 million yd^3) in volume, killed 1000 people.

In discussions concerning major earthquakes and related landslides in the United States, the specific earthquakes commonly considered are (1) New Madrid, Missouri, 1811; (2) San Francisco, 1906; (3) Madison Canyon, Montana, 1959; (4) Anchorage, Alaska, 1964; and (5) San Fernando Valley, California, 1971.

The New Madrid, Missouri, earthquake caused widespread landslides in the loess river bluffs as far north as the Ohio River. On the east side of the Mississippi River, a continuous strip of landslides prevailed for 56 km (35 mi). Large depressions 30 m (100 ft) in diameter and 30 m (100 ft) deep were still visible as late as 1869.

The San Francisco earthquake, which occurred at 5:12 AM on April 18, 1906, was felt over a land area of 453,000 km^2 (175,000 mi^2). The landslides that occurred were divided into four categories based on field studies conducted soon after the earthquake. The categories were earth avalanches, earth flows, earth slumps, and earth lurches (presumably liquefaction or lateral spreading failures). Most earth avalanches occurred along the sea cliffs of the coastal areas. The largest single landslide scar measured about 1.6 km (1 mi) wide and 0.9 km (0.5 mi) long. One of the largest earth slumps (rotational slides) had a width of 450 m (1500 ft), extending 120 m (400 ft) downslope with a 15-m (50-ft) high scarp. The largest earth flow was 810 m (2700 ft) long, 30 m (100 ft) wide, and 1 m (3 ft) deep involving about 67,000 m^3 (90,000 yd^3) of material.

In the Madison Canyon of Montana, an earthquake triggered a rock slide-rock avalanche that killed 28 campers in August 1959. The rocks were dry because there had been no rain for six weeks prior to failure, but the earthquake shocks were sufficient to weaken the rock mass. Paleozoic dolomites, dipping 40° into the valley, overlying a Precambrian basement complex of gneiss and schist, provided the structure that led to instability. In addition to the main slope failure, numerous other landslides including rock falls, earth flows, slumps, debris slides, and debris avalanches were initiated by the earthquake. Many involved weathered bedrock atop the steep cliffs. Several were debris slides into nearby Hebgen Lake, which may have been helped, in part, by water in the slopes around the lake. Others involved massive slumps of colluvium.

The Alaska earthquake of March 27, 1964, caused significant damage over a 129,000-km^2 (50,000-mi^2) area, killing 114 people. It had a Richter magnitude of 9.2 and the main portion of the earthquake lasted up to 7 minutes. Aftershocks followed the main earthquake for 69 days with some as high as 6.7 on the Richter scale. The greatest damage of property and loss of life occurred in Anchorage where 14% of the city (284 ha [700 ac]) was destroyed by lateral spreading failures. The failures occurred in the Bootlegger Cove Clay, a glacial estuary-marine deposit that underlies much of the Anchorage area.

Another type of failure induced by the Alaska earthquake was the rock avalanche known as the Sherman landslide. Rock blocks broke loose along bedding planes that dipped 40° into the valley. The sandstone-argillite sequence was strongly jointed, which also lead to instability.

The Alaska earthquake also initiated extensive subaqueous slides in the adjacent harbor areas. In the seaport of Seward, the waterfront was altered when deltaic sediments slid oceanward, increasing the depth of the pier area from an original depth of 6 to 9 m (20 to 30 ft) to 39 to 54 m (130 to 180 ft). Subaqueous slides in the Passage Canal near Whittier, Alaska, created 31-m (104-ft) waves. One 15.6-m (52-ft) wave struck Whittier, destroying the harbor, pier, docks, associated installations, and nearby homes.

The San Fernando Valley earthquake of 1971, with its Richter magnitude of 6.5, produced more than 1000 landslides in an area of 260 km^2 (100 mi^2) in the mountainous region nearby. Landslides included rock falls, soil falls, debris slides, debris avalanches, and slumps. Landslides also occurred in human-made structures such as earth embankments for dams and roadway embankments.

Two other severe earthquakes that induced major slope failures occurred in China and Peru. The 1920 failure in Gansu Province, China, that killed from 100,000 to 200,000 people owing to the failure of loess slopes into the stream valley marks the greatest loss of life involving any single slope failure event. An area of 76,000 km^2 (30,000 mi^2) was devastated. The other staggering catastrophe was associated with the Peru earthquake of May 1970. The earthquake (7.7 Richter magnitude) initiated a rock fall-debris flow that took more than 18,000 lives when it destroyed the city of Yungay and parts of Ranrahirca.

Human Activities

An increase in the number of slope failures has occurred since the beginning of the twentieth century because of human activities (Alexander, 1992). The problem is twofold in nature: (1) The construction of engineering structures composed of earth materials to create embankments for roads, bridge abutments, dams, railroads, and tunnel grades, which may encroach upon the limits of the equilibrium conditions for slope stability. This includes stacking refuse material following mining or other excavations and impounding water behind some of these structures,

changing the groundwater regime of adjacent areas. (2) The alteration of natural slopes by excavation and urbanization. Excavation includes steepening slopes, removing the slope base, changing surface drainage conditions, removing vegetation, and disrupting sub-surface drainage characteristics. Urbanization can include construction in snow avalanche and debris avalanche-debris flow areas, on active mud flows, on slopes with a history of previous movement, even simply encroaching on an active area of erosion. The presence of humans can continue to alter the natural environment to some extent with roads, water wells, buildings, sewage disposal, and cultivation.

This is not to say that humans must not encroach on the natural setting. Preservation of nature in an untouched fashion is not the answer to overpopulation and urbanization; instead, the current definition of conservation serves best. It involves using good geology and engineering practices for the management of the land and the greatest benefit to humankind.

The real challenge to the engineering geologist and construction engineer is to "design with nature," taking advantage of the natural setting and minimizing the impact of human activities on it. Humans have the capability of building structures that advance society demands, yet induce few effects to the land-scape, particularly regarding slope stability.

Countless specific examples of slope stability failures related to human activity could be presented here to demonstrate various concerns. These would include failures in dam embankments, in reservoir areas, or natural slopes; embankments or cut slopes due to mining or oil production; and embankments and cut slopes for highways. Only a few specific cases are presented below to illustrate the overall diversity of these slope failures and to provide discussion on some of the most noted, human-related landslides.

Gros Ventre Slide, Wyoming

This is an example of the effects of natural slope failures on human activities through secondary involvement. A rock slide involving 38 million m^3 (50 million yd^3) occurred on June 23, 1925. The slide debris moved 2.5 km (1.5 mi) down the dip slope of Sheep Mountain in northwestern Wyoming, across the Gros Ventre River and 107 m (305 ft) up the opposite slope. The mass, mostly Tensleep Sand-stone, slid along the 18 to 21° bedding, probably at the top of a saturated clay shale unit. The rock slide formed a dam 69 to 76 m (225 to 250 ft) high which, in turn, created a lake nearly 8 km (5 mi) long. Two years later, in May 1927, the lake overtopped the dam, causing serious flooding downstream.

Frank Slide, Alberta

For many years it was thought that coal mining played an important role in the initiation of this slide but more recent mapping and calculations suggest that other factors such as structural control were more dominant. On April 29, 1903, about 30 million m^3 (40 million yd^3) of rock broke loose from Turtle Mountain in southern Alberta, rushing down the mountain side to destroy much of the outlying portion of the town of Frank and kill 70 people.

The Frank slide was considered to be a classic example of a joint-controlled rock slide with the mass moving along joint planes parallel to the slope and perpendicular to the bedding, which dipped steeply into the mountain. Heavy spring rains and settlement due to mining of a coal seam at the base of the mountain adjacent to the town were thought to have triggered the slide.

Recent mapping in the area has shown that the movement was primarily along bedding planes. The mountain had an anticlinal feature prior to failure with the axial plane running through the mountain toward the town. Fractures parallel to the axial plane could have provided ingress for the heavy spring rains which, in turn, caused a buildup of excess pore pressure in the slope and subsequent sliding along the bedding planes.

Vaiont Reservoir Slide, Italy

This example illustrates the extreme need to prevent large-volume landslides from occurring along the rim of a reservoir. On October 9, 1963, the most disastrous landslide in European history, the Vaiont Reservoir slide in northeastern Italy, occurred. A volume of rock of about 250 million m^3 (325 million yd^3) suddenly slid into the reservoir, displacing most of the water and sending a wave 260 m (800 ft) up the opposite slope and at least 100 m (330 ft) over the top of the dam into the river valley below. This wall of water destroyed five villages and killed between 2000 and 3000 people. A remarkable fact is that the thin arch dam, the world's largest at the time of construction, sustained little structural damage. The dam remains intact today, as a monument to this catastrophic event, sitting astride the valley at the base of the empty reservoir.

The area was known to be landslide-prone well before the dam was constructed and landslides were experienced in the reservoir some years before the major failure. The slopes were well instrumented, it was thought by the designers, so that the reservoir could be lowered if slope movement became excessive, and that slope failures of the size anticipated could be accommodated without undue risk. In 1960 a large slide mass (700,000 m^3 [900,000 yd^3]) slid into the reservoir a short distance upstream from the dam but this was only about 3% of the volume of the massive slide that ultimately occurred in 1963. It would appear, in retrospect, that the major error was underestimating the likely volume of a single slide mass.

An adverse set of structural features led to the slide. The Vaiont gorge forms the axis of a syncline with the bedding planes dipping inward from both sides of the valley. The sliding surface was largely along bedding surfaces that also contained well-defined joints parallel to the beds. The failure occurred after a period of sustained, heavy rainfall, so extreme weather conditions should have been considered along with the design details.

Aberfan, Wales

In 1966 a pile of coal mine spoil from a coal processing plant failed by moving downslope to kill 144 people, mostly children, in the town below. The mine spoil had become saturated at depth previous to failure and the slope angle was approximately 30°. Intense rainfall added the needed driving force to cause a flow failure within the material.

Buffalo Creek Dam, Saunders, West Virginia

In 1972, heavy rains induced failure of three coal-refuse impounding structures. The major debris flow, consisting of water, coal wastes, and sludge, traveled 24 km (15 mi) downstream, killing 125 people and leaving 4000 homeless.

Oso Landslide, Washington

The Oso landslide, in Snohomish County, Washington, occurred on March 22, 2014, following a three-week period of high precipitation. The landslide occurred on an approximately 200-m (655-ft) high slope consisting of unconsolidated glacial and colluvial deposits that had experienced previous landslide activity (Keaton et al., 2014). It quickly changed into a debris flow/mudslide, burying approximately 40 homes in the community of Steelhead Haven (US Geological Survey, 2014). It also resulted in 43 fatalities, making it the deadliest landslide event in the United States (Keaton et al., 2014). The Oso landslide displaced approximately 7.6 million m³ (~ 270 million ft³) of material (sand, silt, and clay) that places it among the large landslides in the state of Washington in recent times. The prior activity at the same location took place in 2006 when a landslide, known as the Hazel landslide, occurred and blocked the North Fork Stillaguamish River.

In addition to loss of life, the Oso landslide additionally caused significant economic losses, estimated to be more than $50 million by Washington state officials. The landslide completely destroyed the Steelhead Haven neighborhood and covered approximately 600 m (~ 2,000 ft) of State Route 530 with up to 6 m (20 ft) of debris, closing the highway for more than 2 months.

The lessons learned from the Oso landslide disaster include: (1) zoning laws should not allow the establishment of communities on potentially unstable slopes where evidence of prior slope movement has been documented; (2) the history and behavior of past landslides and associated colluvial soil masses should be carefully investigated for zoning purposes; (3) the risk of landslides to people and property should be communicated clearly and regularly to the public; (4) monitoring and warning systems to reduce the impacts of landslides to people and property should be installed at potentially unstable slopes; and (5) new methods to identify and delineate potential landslide runout zones should be developed (Keaton et al., 2014).

Although human activity was not the major cause of the Oso landslide, it is included here to illustrate the devastating consequences of building human communities on slopes that have known histories of previous movements.

Table 15.5 lists the major slope failure disasters of the world since 1500—a number of the examples are discussed in the text. Figure 15.21 is an photograph of a landslide that occurred in Attabad, Pakistan in 2010.

Figure 15.21   A major and secondary landslide occurred in the Hunza River Valley of northern Pakistan, burying the village of Attabad and damming the Hunza River. (Inayat Ali Shimshal, *Pamir Times* [https://creativecommons.org/licenses/by-nc-nd/2.0/be/].)

Prevention and Correction of Slope Movements

The prevention of slope movements or the correction of the slope after failure involves two basic concepts: increase the resisting force or reduce the driving force or both, to the extent that the slope becomes stable. The specifics of how this is accomplished will depend on the site conditions, budget, and engineer's judgment and experience. The details of individual techniques can be found in Turner and Schuster (1996), Abramson et al. (2002), and Conforth (2005). The general principles can be outlined in a descriptive way, however, as shown in Table 15.6 (page 357). Additional information is supplied in Figure 15.22 and Table 15.7 on slope design procedures. Figure 15.22 (page 357) pertains to slopes in rock, and Table 15.7 (pages 358–359) pertains to slopes in soil.

Ground Subsidence

Subsidence is the lowering of the land surface by a vertical, downward movement or downwarping. Typically a subsidence hollow develops, that is, a dish-shaped feature with the greatest displacement in the center. Subdivisions of subsidence according to the mechanisms of failure are (1) compaction, (2) consolidation, (3) plastic outflow of weak layers,

Table 15.5 Landslide Disasters Since 1500.

Year	Place	Fatalities	Comments
1512	Brenno Valley, Switzerland	600	Rock slide dammed valley; 2 years later dam broke causing destruction.
1584	Tour d'Ai, Switzerland	300	Landslide devastated village of Yvorne in Rhone Valley.
1618	Mount Conto, Switzerland	2430	Rock slide.
1806	Goldau, Switzerland	457	Landslide destroyed village.
1881	Elm, Switzerland	115	Rock avalanche demolished 83 houses.
1893	Trondheim, Norway	111	Liquefaction flow in marine clays.
1903	Frank, Alberta, Canada	70	Rock avalanche destroyed most of town.
1911	Rushon District, Tajikistan	54	Earthquake-induced landslide dammed Murgab River, impounding 65-km long Lake Sarez.
1919	Java, Indonesia	5110	104 villages destroyed or damaged.
1920	Gansu Province, China	200,000	Earthquake caused loess flows.
	Ningxia, China	100,000	675 large loess landslides destroyed many villages, created more than 40 lakes.
1921	Almaty, Kazakhstan	500	Debris flow in valley of Almatinka River.
1933	Sichuan, China	9300	6800 killed by landslides; largest landslide formed a 255-m high dam on Min River, 2500 drowned when landslide dam failed.
1936	Nordfjord, Norway	73	Rock fall created 74-m wave.
1938	Kobe, Japan	461	Rocky mud flows.
1939	Hyogo, Japan	505	130,000 homes destroyed or badly damaged by a major typhoon; 50 to 90% of the impact of Japanese typhoons is caused by mass movements.
1945	Kure, Japan	1154	Rocky mud flows.
1949	Hoit District, Tajikistan	12,000–20,000	Rock slide transformed into large loess and granite debris avalanche; 33 villages destroyed.
1953	Wakayama, Japan	460	4772 homes destroyed by mass movements caused by a major typhoon.
1953	Kyoto, Japan	336	5122 homes destroyed by mass movement and floods.
1958	Shizuoka, Japan	1094	19,754 homes destroyed or badly damaged by mass movement and floods.
1959	Madison, Montana	28	Rock avalanche buried campers.
1962	Ancash, Peru	4000–5000	Major debris avalanche from Nevado Huascarán; average velocity 170 km/hr, destroyed much of the village of Ranrahirca.
1963	Vaiont, Italy	2000	High velocity rock slide into Vaiont Reservoir caused 100-m waves to overtop Vaiont Dam, badly damaging the city of Longarone.
1964	Anchorage, Alaska	114	Major landslide caused by earthquake; damaged cities of Anchorage, Valdez, Whittier, and Seward. Damage estimated to be $280 million in 1964 dollars.
1965	Yunnan, China	444	High speed mass movement destroyed or damaged four villages.
1966	Aberfan, Wales	144	Human activity, mining spoil hill; landslide buried mostly children.
1966–67	Brazil	2700	Combined total from debris avalanches and flood.
	Rio de Janeiro, Brazil	1000	Many landslides in Rio de Janeiro and environs.
	Serra das Araras, Brazil	1700	Many landslides in mountains SW of Rio de Janeiro.
1969	Nelson County, Virginia	150	Combined total from debris avalanches and floods caused by Hurricane Camille.
1970	Huascarán area, Peru	21,000	Combined rock avalanche and debris flow buried two cities.

Table 15.5 **Landslide Disasters Since 1500. (cont.)**

Year	Place	Fatalities	Comments
1970	Ancash, Peru	18,000	Debris avalanche from same peak as in 1962; average velocity of 280 km/hr. Town of Yungay destroyed; Ranrahirca partially destroyed.
1974	Huancavelica, Peru	450	Debris avalanche with average velocity of 140 km/hr. Dammed Mantaro River. 150-m-high dam failed, causing major flooding downstream. Village of Mayunmarca destroyed.
1980	Mt. St. Helens, Washington	5–10	Evacuation saved lives. Began as rock slide; deteriorated into 23-km long debris avalanche with average velocity of 125 km/hr; surface remobilized into 95-km long debris flow. Major destruction to homes, highways.
1983	Thistle, Utah	0	Destroyed major railroad and highways; dammed Spanish Fork, flooding town of Thistle; no deaths but total losses of $600 million in 1983 dollars.
1983	Gansu, China	237	Loess landslide buried four villages; filled two reservoirs.
1985	Tolima, Colombia	20,000	Lagunillas River valley, city of Armero, four towns and villages destroyed. Death toll unnecessarily large because hazard warnings not passed to residents.
1986	Papua, New Guinea	0	Debris avalanche formed 210-m high dam that impounded a 50 million m^3 lake; dam failed, causing 100-m deep debris flow-flood downstream. Village of Bairaman destroyed. Evacuation prevented casualties.
1987	Napo, Ecuador	1000	Landsliding in saturated, residual soils on steep slopes mobilized into debris flows. Many kilometers of trans-Ecuadorian oil pipeline and highways destroyed; total losses $1 billion in 1987 dollars.
1994	Cauca, Colombia	271	Several villages partially destroyed.
1998	Honduras, Guatemala, Nicaragua, El Salvador	10,000+	180-mi/hr winds and torrential rains, at 4 in/hr, caused by Hurricane Mitch, affected Honduras especially, causing large landslides in Tegucigalpa and elsewhere.
1998	Kelso, Washington	0	Slow-moving landslide, which resulted in the condemnation of 137 houses and $40 million in damage.
1999	Vargas, Venezuela	30,000	Caused by a heavy storm that deposited 911 mm of rain in a few days.
2004	Mt. Bawakaraeng, Indonesia	32	Landslide caused by collapse of caldera wall.
2005	La Conchita, California	10	Remobilization of colluvium from 1995 slide into a debris flow.
2006	Southern Leyte, Philippines	1126	Rock-debris avalanche triggered by 10-day period of heavy rain.
2007	Chittagong, Bangladesh	123	Series of landslides caused by illegal hillside cutting and monsoon rains.
2008	Cairo, Egypt	119	Rock fall from cliffs, individual boulders up to 70 tons.
2009	Kaohsiung, Taiwan	439–600	Mudslide resulted from Typhoon Morakot.
2010	Attabad, Pakistan	20	Formed Attabad Lake by damming Hunza River, blocked Karakoram Highway.
2010	Madeira Island, Portugal	42	Floods and mudslides.
2010	Gansu, China	1287	Mudslides.
2012	Seti River, Nepal	70	Rock slope failure turned into a rock avalanche that entered the upper gorge of the Seti River and became a debris flow causing massive destruction.
2014	Oso, Washington	43	49 structures destroyed or affected.
2014	Badakhshan, Afghanistan	350–500	Mudslides, 4000 people displaced.
2014	Hiroshima, Japan	50+	38+ missing, deadliest landslides in Japan in 42 years.
2014	Badulla District, Sri Lanka	100+	Badulla landslide. 300+ feared dead. Hundreds more missing.
2017	Bali Province, Indonesia	12	Landslides triggered by heavy rain buried 5 houses in Songan Village.

Sources: Modified from Schuster, 1966 and Wikipedia.

Table 15.6   General Considerations for Preventing or Correcting Slope Failures.

1. Weight at the head of a slope is a driving force. Never load the head unnecessarily. Relieve the head by removing material. Tie the surface to stable material deeper within the slope if needed.
2. Weight at the toe is a resisting force. Never excavate material at the toe without being sure that the slope will be stable afterward. Load the toe with berms or support the toe with piles, retaining walls, cribbing, etc., if more support is required.
3. Keep surface water off the slope. Infiltration may occur or erosion and gullying may develop depending on the nature of the slope. Water should be channeled and drained away from the slope.
4. Provide internal drainage for the slope when needed. This reduces the pore water pressure, increasing the frictional component of shear strength. It also increases cohesion of plastic materials by lowering the water content. Horizontal drains perpendicular to the slope direction may be required.
5. Reduce the slope angle as needed. This will decrease the driving force but greatly increase excavation costs. The slope should be graded to obtain a stable surface profile.

and (4) collapse of underground openings. Subdivision based on the location where the subsidence originates yields two varieties: shallow subsidence and deep subsidence. These types of subsidence are discussed in the following paragraphs.

Compaction and Consolidation

With regard to engineering construction, the downward movement of the ground surface is termed *settlement*. Geologically speaking, this downward movement is called subsidence. Within the soil mass, settlement or subsidence is typically of two types: compaction and consolidation.

Compaction, in both a construction sense (geotechnical engineering) and an engineering geology sense, is the reduction in volume of a soil mass under load caused by reorientation and realignment of soil particles to form a denser packing, but without drainage of water from the soil. Compressive loads (rollers used on earth-moving projects), vibration (vibrocompactors, earthquakes, pile-driving operations), or the downward movement of water through a dry, loose soil can cause such reorientation.

Compaction in a strictly geologic sense is the volume reduction and loss of water that occurs when sediments (soil) become more rock-like

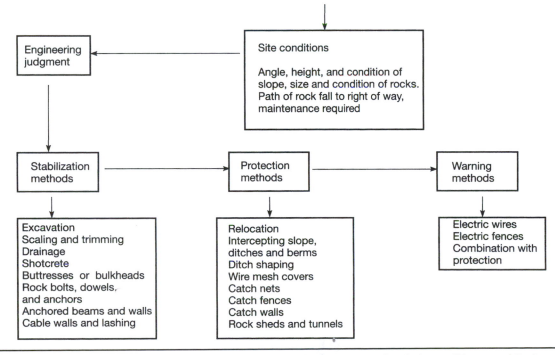

Figure 15.22   Order of slope treatment methods for design and construction of rock slopes. (Piteau and Peckover, 1978.)

Table 15.7   Summary of Slope Design Procedures for Soils.

Category	Procedure	Best Application	Limitation	Remarks
Avoid problem	Relocate highway	As an alternative anywhere.	Has none if studied during planning phase; has large cost if location is selected and design is complete; also has large cost if reconstruction is required.	Detailed studies of proposed relocation should ensure improved conditions.
	Completely or partially remove unstable materials	Where small volumes of excavation are involved and where poor soils are encountered at shallow depths.	May be costly to control excavation; may not be best alternative for large slides; may not be feasible because of right-of-way requirements.	Analytical studies must be performed; depth of excavation must be sufficient to ensure firm support.
	Bridge	At sidehill locations with shallow-depth soil movements.	May be costly and not provide adequate support capacity for lateral thrust.	Analysis must be performed for anticipated loadings as well as structural capability to restrain landslide mass.
Reduce driving forces	Change line or grade	During preliminary design phase of project.	Will affect sections of roadway adjacent to slide area.	Slope vegetation should be considered in all cases.
	Drain surface	In any design scheme; must also be part of any remedial design.	Will only correct surface infiltration or seepage due to surface infiltration.	
	Drain subsurface	On any slope where lowering of groundwater table will affect or aid slope stability.	Cannot be used effectively when sliding mass is impervious.	Stability analysis should include consideration of seepage forces.
	Reduce weight	At any existing or potential slide.	Requires lightweight materials that are costly and may be unavailable; may have excavation waste that creates problems; requires consideration of availability of right-of-way.	Stability analysis must be performed to ensure proper use and placement area of lightweight materials.
Increase resisting forces	Drain subsurface	At any slide where water table is above shear plane.	Requires experienced personnel to install and ensure effective operation.	
	Use buttress and counterweight fills	At an existing slide, in combination with other methods.	May not be effective on deep-seated slides; must be founded on a firm base.	
	Install piles	To prevent movement or strain before excavation.	Will not withstand large strains; must penetrate well below sliding surface.	Stability analysis is required to determine soil-pile force system for safe design.

Table 15.7 Summary of Slope Design Procedures for Soils. (cont.)

Category	Procedure	Best Application	Limitation	Remarks
	Install anchors	Where rights-of-way adjacent to highway are limited.	Involves depth control based on ability of foundation soils to resist shear forces from anchor tension.	Study must be made of in situ soil shear strength; economics of method is function of anchor depth and frequency.
	Treat chemically	Where sliding surface is well defined and soil reacts positively to treatment.	May be reversible action; has not had long-term effectiveness evaluated.	Laboratory study of soil-chemical treatment must precede field installation.
	Use electroosmosis	To relieve excess pore pressures at desirable construction rate.	Requires constant direct current power supply and maintenance.	
	Treat thermally	To reduce sensitivity of clay soils to action of water.	Requires expensive and carefully designed system to dry out subsoils artificially.	Methods are experimental and costly.

Source: Gedney and Weber, 1978.

under geostatic loading. It includes the engineering aspects of both compaction and consolidation.

Consolidation, in the geotechnical sense and engineering geology sense, is the volume reduction under load that occurs as water flows from the soil, and is related to the reduction in excess pore water pressure. Therefore, saturated, compressible clays consolidate under load whereas all types of soils, cohesive and noncohesive alike, can be compacted by moving the grains closer together. Consolidation in the geologic sense is the process by which sediments become rock-like or become lithified. It may include consolidation and compaction in the engineering sense and, in some cases, cementation. Geotechnical engineering details of consolidation and compaction are discussed in Chapter 7.

One feature of compaction that is of considerable interest is the collapse that occurs when water flows through loosely packed, dry soils. It is sometimes called hydrocompaction. The motion of the water provides sufficient energy to move the soil particles to a tighter packing arrangement. This occurs to some soils when water is first impounded in a reservoir or when water leaks from conduits or canals into the loosely packed material below. An example of interest is related to the construction of the California State Water Project, which supplies water to southern California from the northern part of the state. Mud flow deposits in the San Joaquin Valley are particularly susceptible to such settlement. Water leaking through cracks in the concrete canal lining caused large-scale settlement below the canal and the subsequent collapse of the lining. Prewetting of susceptible soils prior to construction is a viable means for preventing the subsequent failure of the project.

Consolidation occurs when heavy loads are placed on compressible soils such as plastic clays and organic materials. The foundation loads from footings, piles, and mats can initiate such settlement by inducing consolidation. These loads increase the pore pressure or neutral stress on the soil, causing water to flow from it. Hence, as the effective stress increases, the soil reduces in volume.

Another cause of consolidation is the removal of the subsurface fluids (water or oil), usually because of human activities. Lowering the water table reduces the buoyancy force contributed by the water and, in effect, increases the downward pressure on the soil at the former level of the water table. This initiates consolidation and settlement. Since it takes time for water to flow through a fine-grained soil, consolidation is time dependent. This is in contrast to compaction, which occurs much more quickly in response to the applied load.

Consolidation commonly occurs in oil or groundwater supply areas. Interbedded sands and clays provide the conditions needed for the consolidation process. Drainage of the sand layers leads to greater loads on the clays, which consolidate. This problem has occurred in Mexico City where groundwater removal from the volcanic ash beds and clays has been extensive. It also occurred in Long Beach, California, where great volumes of oil and brackish water were removed, and in the Houston, Texas, area where groundwater removal from the coastal plain sediments has caused extensive consolidation. In some Houston suburbs, surface areas are now below the high-tide elevation and are subject to ocean encroachment at high tide.

A bit of reflection on this subject will bring the concept of shallow and deep subsidence into perspective. Compaction due to water movement, vibration, or applied load typically occurs near the land surface because the soil is loosest there and surface loads have the greatest effect. Therefore, shallow subsidence is typically related to compaction.

Consolidation, however, can occur hundreds or thousands of feet below the land surface if fluids are removed from these depths. Settlement is transferred upward from a considerable depth; therefore, it is termed deep subsidence. Most groundwater pumping for irrigation causes deep subsidence, whereas shallow subsidence is provided by initial wetting near the surface.

Subsidence above underground openings is particularly destructive to surface structures because of differential settlement, which occurs because the center settles much more than the edges of the trough or dish. In addition, it provides tensile stresses near the edge of the trough, which can be more destructive to buildings than are compressive stresses. Soil tunneling, which yields a subsidence trough at the surface, can be extremely destructive to buildings, particularly relatively stiff buildings such as some old masonry, wall-bearing structures. This is why the loss of ground into a tunnel must be minimized in areas of urban development. Loose, cohesionless soils or weak, cohesive soils are particularly prone to loss of ground during tunneling when a tunnel shield is used. Compressed air, dewatering, and possibly grouting ahead of the advancing face may be needed to strengthen the soil to minimize soil movement into the tunnel.

Plastic Outflow

Plastic outflow occurs when an organic or silty layer near the surface is subjected to loading by a heavy structure and the compressible layer exits to the surface causing subsidence under the loaded area. Before final site selection is made, this problem can be alleviated by exploring the subsurface to determine the presence of such layers close to the surface.

Underground Opening Collapse

The collapse of underground openings can involve natural openings or those due to human activities. Natural openings subject to collapse include caverns in soluble rock such as limestone, dolomite, gypsum, and rock salt, and lava tubes and tunnels in extrusive igneous rocks. When the roof of the opening collapses, a stable arch may form if the cavern width is sufficiently small or the rock above the cavern is massive, containing only widely spaced joints. Under these conditions, the void will not migrate upward but instead form a cathedral shape extending upward to a point above the opening. Stable openings as large as 30 m (100 ft) wide have been found in large limestone caves with a stable roof of massive rock. The value for the inward sloping angle is a function of the rock and its joint spacing.

When the roof opening is wide compared to the strength of the rock mass, roof failure will occur as blocks of material fall into the opening. The side boundaries of these blocks tend to slope outward from the vertical so that a wedge of rock moves downward from above. The angle measured from the horizontal is known as the angle of break. The relationship between the angle of break and the subsidence trough at the surface is shown in Figure 15.23.

The angle of break, α, can be estimated for soils and rocks using ϕ, the angle of internal friction, and a relationship based on the Mohr circle representation for the state of stress on the plane of failure. This yields $\alpha = \phi/2 + 45°$, which is derived in Chapter 7. Table 15.8 presents a listing of ϕ values, calculated α values, and observed α values for various soil and rock types.

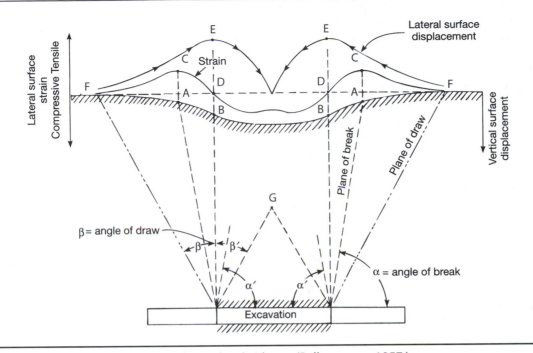

Figure 15.23   Idealized representation of trough subsidence. (Rellensmann, 1957.)

In Figure 15.23, the plane of draw and angle of draw, β, are also shown. β is measured from the vertical whereas α is measured from the horizontal. The angle β is related to the boundary on the surface of the subsidence hollow and, typically as a matter of convenience, this boundary is taken at a point where only 5% of the maximum vertical displacement occurs. A similar designation must be made or the bounds of the trough will become a function of surveying accuracy rather than actual subsidence. The plane of draw always extends further away from the underground opening than does the plane of break [that is to say, $\beta > (90 - \alpha)$].

Note in Figure 15.23 that point B lies vertically above the boundary of the opening. Point B is the inflection point on the subsidence curve, the point where the shape changes from concave up (in the center of the trough) to concave down (on the two limbs of the subsidence hollow). When the subsidence trough shape is compared to a normal probability curve (the Gaussian curve used in statistics), the inflection point occurs about 1/2.5 of the distance from the center of the trough to the edge, again the edge being at the 5% vertical displacement point. If the half width is designated w, point B occurs at $w/2.5$.

The plane of break on Figure 15.23 intersects the ground surface at A and the strain curve at C. It can

Table 15.8   Measured Angles of Internal Friction, Calculated Angles of Break, and Observed Angles of Break for Various Soils and Rocks.

Material	Measured Angles of Internal Friction, ϕ (degrees)	Calculated Angle of Break, α (degrees)	Observed Angle of Break, α (degrees)
Clay	15–20	52.5–55	
Plastic strata			60–80
Clays and shales			60
Sand	35–45	62.5–67.5	40
Unconsolidated strata			40–60
Soft shale	37	63.5	
Hard shale	45	67.5	
Sandstone	50–70	70–80	
Coal (average)	45	67.5	
Rocky strata			80–90
Hard sandstone			85

be seen that C is the point of maximum strain and, therefore, the likely place for the break to occur. In the drawing, the plane of break occurs about one-third of the distance between D and the edge of the trough, or AB ≅ BF × 1/3. Recall that Figure 15.23 is an idealized representation of the subsidence relationship so values for angles of draw, obtained from angles of break using the geometry of this drawing, will be, by their very nature, only approximate.

Values for the angle of draw are measured from actual cases of subsidence, both for tunnels driven through soil and for underground mining in rock. Values of β as low as 9° and as high as 58° have been determined for different soils and geologic conditions. For massive rocks, β values are in the range of 11 to 20°. Table 15.9 provides a list of estimated angles of draw based on field observations.

Referring again to Figure 15.23, point G occurs at the intersection of lines inclined inward from the vertical at an angle of β or β', the angle of draw. As the excavation is widened equally on both sides (outward along the darkened bed), point G moves up vertically toward the surface. As long as G remains below the ground surface, the subsidence is termed *subcritical*, with the critical point occurring when G reaches the surface. This marks the situation when maximum subsidence occurs at the center of the trough. As further widening ensues, G rises above the surface and the trough will widen but no longer deepen. Movement will cease in the center of the trough. When G lies above the surface, the subsidence is termed *supercritical*. Total vertical movement in the center of the trough will be from 70 to 90% of the height of the mine, that is, of the thickness of the extracted zone.

The cathedral effect that forms above some small cavities is related to the location of point G and to the strength of the rock mass. If G lies too far above the excavation, progressive subsidence will occur and eventually reach the surface. Below that distance, a stable arch can develop and no further subsidence ensues. The width of the opening and the angle of draw influence the location of G. This holds true for underground mines as well as caverns. For coal mines, the critical width may be 3 to 5 m (10 to 15 ft), whereas for massive carbonate rocks (limestones, dolomites) this width may be 12 m (40 ft) or greater.

Mining Activities

Human-made openings underground are the result of mining activities and of engineering construction, such as tunnels and underground power plants. Mining activities include extraction of coal,

Table 15.9   Estimated Angles of Draw for Various Soils and Rock.

Material	Angle of Draw β (degrees)
Rock, hard clay, sands above the groundwater table	11–26
Stiff to soft clays	26–50
Sands below the groundwater table	< 50

limestone, gypsum, salt, or metallic ores such as iron, copper, lead, zinc, gold, silver, and molybdenum. Mining methods include (1) room and pillar, (2) stoping, (3) block caving, and (4) longwall methods. In block caving and longwall methods, most or all support is eventually removed so that subsidence is imminent. In the stoping and the room and pillar methods, sufficient support is commonly left behind to prevent roof failure and general subsidence. However, in room and pillar mines when a higher extraction ratio is desired and subsidence is acceptable, the pillars are removed (robbed) during a retreating operation from the far reaches of the mine section working back toward the permanent haulage ways. Surface subsidence is almost immediate when the support pillars are removed.

Solution mining of salt (NaCl or halite) leads to complicated subsidence problems. In this mining technique, two boreholes into the salt body are connected by hydrofracturing, that is, by fracturing between the holes using water pressures that exceed the overburden stresses. Hot water is then circulated down one of the holes and brine is taken out of the other. On the surface, this water is allowed to evaporate to yield the salt for further processing. Cavities thus formed are commonly asymmetrical in shape, because the flow paths take on preferred orientations and remove salt along these directions. When a sufficiently large cavity forms, subsidence occurs in the rock layers above the salt and propagates to the surface. Areas near Detroit, Michigan; Windsor, Ontario; Rochester, New York; and central Kansas have major subsidence problems related to such salt extraction.

Sinkholes

A final point for consideration of subsidence has to do with the development of sinkholes. In this context, the term *sinkhole* refers to the conical-shaped surface depressions that occur in carbonate and other soluble rock terrains (usually limestone, dolomite, and marble, sometimes gypsum and salt). The depression marks a point above a vertical joint, enlarged by solutioning, which leads to the underground drainage system. Soil and weathered rock collapse into the enlarged joints to form the sinkhole. These features are shown on topographic maps as circular closed contours with dashed lines pointing inward.

Sinkholes are part of the natural, geologic, erosional process. In a natural setting it takes a considerable span of time for downward migrating waters moving through narrow openings to dissolve away the rock volume needed to develop sinkholes. Terrains exposed during the Pleistocene age (1 to 2 million years) had sufficient time to develop sinkholes. However, this natural development is not expected to occur in a human lifetime or during the life of an engineering structure. Therefore, when sinkholes suddenly occurred at the rate of many per year in areas of Alabama and Florida, human involvement in the situation was logically suspected.

One reason for rapid sinkhole development is the lowering of the groundwater table. This commonly occurs as a consequence of human activities such as removal for water supply, dewatering for construction, and construction of permanent road cuts or surface mines—all of which bring down the groundwater table. Figure 15.24 illustrates one of the possible situations. In some areas, deep weathering of previously exposed limestone terrain is covered with sand or clay deposits from a later stage of development.

Such deposits serve as a cap for the vertical openings below. When the water table is lowered, the buoyancy force is lost, yielding an increase in effective stress. This stress acts downward on the clay and may force it to fail (punch) into the opening below. When this occurs, a sinkhole forms quite suddenly on the surface. An example of rapid sinkhole failure was the formation of "December Giant," which occurred overnight (December 1972 in Alabama). The sinkhole measured 120 × 90 m (400 × 300 ft) across and 46 m (150 ft) deep. Fortunately, it occurred in an undeveloped, wooded area adjacent to an expanding residential tract. The groundwater table had been dropping in this general region for several years. A photograph of this large sinkhole is shown in Figure 15.25 on the following page.

A question for consideration is could this sinkhole have occurred suddenly because of purely natural (nonhuman-related) causes? Obviously the groundwater table could lower for strictly geologic reasons alone. The answer involves the rate of change, not its magnitude. For the groundwater table to lower such a significant amount to trigger this failure, many years of reduced rainfall would be required. A long-term climatic change would be needed to cause the failure. But for the widespread development of new sinkholes in a short period of time in areas occupied by homes or other human activities, the problem is placed squarely on human involvement. Lowering of the water table because of increased pumping is the most logical answer.

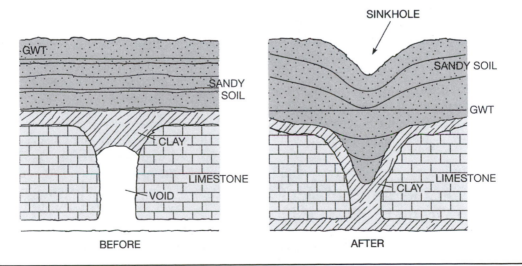

Figure 15.24   Possible mechanism for rapid sinkhole formation, human-induced failure.

Figure 15.25   Collapse of large sinkhole in Alabama. (US Geological Survey.)

This is not to say that humans should stay away from all sensitive areas. Instead, cognizance of the problems is required and development of proper solutions to prevent such catastrophic failures from occurring is needed. If the water table level is maintained by balancing withdrawal and recharge, the problems will be alleviated. Other situations require different solutions but humankind must engineer these solutions to fit the potential problem.

Engineering Considerations of Slope Stability and Ground Subsidence

1. Slope movements and ground subsidence are of major concern to engineering geologists, geotechnical engineers, and planners because of their adverse impacts on design, construction, and maintenance of infrastructure, on the environment, and on socioeconomic conditions.

2. Slope movements in the United States alone result in 25 to 50 fatalities and approximately $2 billion in property damage, annually.

3. Slope movements result in reduced real estate values; loss of tax revenues from affected properties; loss of tourist revenues because of damage to facilities or interruption of transportation systems; loss of industrial, agricultural, and forest productivity because of damaged land; and loss of human productivity because of injury or death.

4. The stability of oil- and gas-carrying pipelines depends on subaerial and submarine slope movements that can result in major changes in

topography and morphology. Therefore, slope stability studies are very important for the location of pipelines.

5. Engineering geologists and geotechnical engineers are routinely involved in slope stability evaluations when designing and constructing such infrastructure as dams, highways, tunnels, bridges, and airports.

6. Sediment from slope movements adversely impacts the quality of water in rivers and streams, changing the habitats of native fauna.

7. Ground subsidence, caused by mining, tunneling, and fluid withdrawal, can impact areas ranging from very small to very large. In several states (e.g., Ohio, Pennsylvania, West Virginia,

Illinois) subsidence due to coal mining is widespread, impacting the design and construction of infrastructure, especially highways, residential communities, and commercial buildings. In other states with excessive withdrawal of fluids, both water and petroleum, large areas are affected by ground subsidence. Texas, California, Oklahoma, and Nevada are some of the states that have experienced severe ground subsidence problems. In Texas, for example, many waterfront properties have been abandoned because of ground subsidence. Engineering geologists and geotechnical engineers should carefully investigate the ground subsidence problems and their influence on design and construction, where relevant.

EXERCISES ON SLOPE STABILITY AND GROUND SUBSIDENCE

1. Discuss the various ways in which Sharpe's classification of mass movement (see Table 15.1) differs from Cruden and Varnes' classification of slope movement (see Table 15.2).

2. Explain how an earth flow would be distinguished from a rotational slide in the field. What is the difference between a mud flow and a debris flow?

3. Stability against toppling of rock blocks is maintained when $b/h > \tan\theta$ (see Figure 15.4). For a bed dipping at 30° how closely spaced can those joints perpendicular to bedding be and still maintain stability if the beds are 0.3 m (1 ft) apart?

4. The term *landslide* by common usage typically includes slope failures by fall or topple, flow and lateral spreading, and slide or slip. How did this usage develop? Why is it important to distinguish between flow failures and slip failures?

5. Why is it important that the potential failure surface of a rotational slide be intercepted during subsurface exploration? Describe two geological settings in which a deep-seated slump could occur. A labeled cross section should prove helpful in this regard.

6. Why is the continuity of fractures in a rock mass, related to the spacing of these weakness planes, relative to slope stability? How are surface irregularities of bedding planes related to slope stability? Explain.

7. Refer to Figure 15.13 in the text. Note the point where the upper solid line changes slope.

   a. Calculate the σ_n value for this point given the following information for a rock mass of quartzite: $c_r = 42,000$ kPa (6000 psi), $\phi_r = 40°$, $\phi_j = 32°$, and $i = 15°$.

   b. Having calculated σ_n, assuming this to be a vertical stress at depth in a rock mass, calculate the depth required to obtain this σ_n value given a unit weight or unit density for quartzite of 2720 kg/m^3 (170 lb/ft^3).

8. A block of limestone, generally similar to that illustrated in Figure 15.14, is resting on an inclined, smooth bedding plane. The block measures 2.4 m (8 ft) long by 2.7 m (9 ft) high by 3 m (10 ft) thick (into the slope). The mass is dry. The unit weight of the rock is 2.56 g/cm^3 (160 lb/ft^3). Cohesion equals 14.7 kPa (300 lb/ft^2) and $\theta = 32°$.

   a. Calculate the factor of safety if the slope angle $\theta = 35°$.

   b. Here is a more difficult question requiring iterative calculations: What is the maximum slope angle that the block can have to yield a factor of safety against sliding of 1.2?

9. Assume conditions in Exercise 8a except that the fracture behind the block is completely filled with water and free drainage occurs at the downslope end of the block (see Figure 15.14).

   a. Calculate the factor of safety.

b. Assume next that the fracture is completely filled and drainage is blocked at the downslope end (see Figure 15.15). Calculate the factor of safety.

10. If those slopes that fail by flow (earth flows and lateral spreads in particular) cannot be analyzed by the limit equilibrium method, how is stability for these slopes determined?

11. How is a mud flow recognized in the field and on aerial photographs? How can an ancient mud flow be told from an active one? How is shallow subsidence related to mud flows? How can the damage caused by mud flows to residential areas be minimized?

12. What is meant by the term *liquefaction?* Why would this be of concern for earth dams constructed over alluvium in earthquake-prone areas? Explain.

13. If a major earthquake occurred in Evansville, Indiana, what areas in the state would be most prone to landslide failures? See a map of Indiana provided in Figure 20.3 for reference purposes. Explain. (*Hint:* Alluvial areas are most prone to earthquake effects.)

14. Examine Figure 15.22 on slope treatment for rock slopes. What is the purpose of berms (or benches) in a cut slope? Consider construction and maintenance primarily in your analysis.

15. What is the purpose for drilling horizontal drains back into a rock slope? Does this seem to be a difficult operation to accomplish? Explain.

16. In contour strip mining today, mining companies are required to put all stripped rock and soil back on the mining bench after coal has been removed. What slope stability problems might develop under such circumstances? If a failure on such a slope occurs, what methods could be used to explore the failed area so that a back calculation of stability could be made? What problems do you anticipate in the exploration?

17. Discuss the difference between the terms *compaction* and *consolidation* as used by civil engineers and by traditional geologists (not including engineering geologists). What geologic materials would be particularly subject to consolidation (in the engineering sense)?

18. What factors lead to shallow subsidence? To deep subsidence? How can the two types be alleviated or greatly reduced?

19. The load on a rock pillar for a room and pillar mine is defined by the following equation:

$$\text{Stress on pillar} = \frac{h\gamma_{rock}}{1 - \text{fraction of material removed}}$$

where h is the height of overburden and γ_{rock} is the unit weight of the rock. If the pillar strength is determined by its compressive strength, the factor of safety is

$$\text{FS} = \frac{\text{compressive strength}}{\text{pillar stress}}.$$

Show this in terms of the equation given for pillar stress. If a factor of safety of 3 is required and the pillars have a compressive strength of 34,475 kPa (5000 psi) (for good coal), develop, in graph form, the relationship between depth and allowable extraction ratio (or fraction of material removed). Show the equation for depth versus extraction ratio and plot it on Exercise Figure 15.1.

20. Why do some sinkholes hold water whereas others are free draining? What problems might develop if sinkholes are simply filled with earth and buildings placed on the leveled land surfaces? How might the fill be properly constructed to prevent such problems?

21. Regarding construction of interstate highways in mountainous areas (such as I-70 west of Denver), sometimes it is more economical to make cuts and fill embankments that might fail locally rather than to prevent all such failures from occurring by more costly construction procedures. What are the economical considerations involved in this decision? Other than economics, what other factors should be considered?

22. In a room and pillar limestone mine, the pillar dimensions are 6 m × 6 m (20 ft × 20 ft) and the distance between the pillars is 9 m (30 ft). They are offset in the manner shown in Exercise Figure 15.2. The pillars are numbered for reference purposes. When pillars 8 and 9 are completely removed, a subsidence trough develops above the opening, with its long dimension located between pillars 7 and 10 and its shorter dimension between 3 and 15.

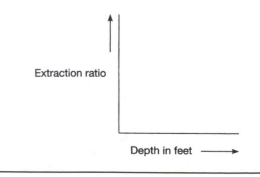

Exercise Figure 15.1

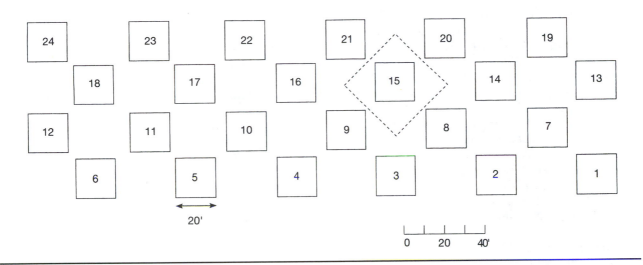

Exercise Figure 15.2

a. What is the extraction ratio for this mine? (That is, the volume removed divided by the volume of the mine for the level in question.)

b. If the limestone weighs 25.6 kN/m³ (2563 kg/m³ or 160 lb/ft³), its compressive strength is 103,400 kN/m² (15,000 psi), and the mine depth is 153 m (500 ft), what is the factor of safety against crushing for the pillars?

c. Give a representative angle of break for this limestone. Give a representative angle of draw for it. Make a cross section from pillars 3 to 15 extending to the ground surface showing these reference angles.

d. Would the subsidence be subcritical, critical, or supercritical? Explain what this means and show calculations.

e. How wide would the subsidence hollow be if it reached the surface?

f. What is the factor of safety now against crushing for pillars 2, 3, 4, 7, 10, 14, 15, and 16? As an aid for this problem draw in the new areas of influence for the remaining pillars on the mine plan given in Exercise Figure 15.2. Are additional pillar failures likely to occur? Show calculations.

REFERENCES

Abramson, L. W., Lee, T. S., Sharma, S., and Boyce, G. M. 2002. *Slope Stability and Stabilization Methods* (2nd ed.). New York: John Wiley & Sons.

Admassu, Y., and Shakoor, A. 2015. Cut slope design for stratigraphic sequences subject to differential weathering. *Environmental & Engineering Geoscience* XXI(4):311–324.

Alexander, D. 1992. On the causes of landslides: Human activities, precipitation, and natural processes. *Environmental Geology and Water Sciences* 20(3):165–179.

Barton, N. R. 1971. A Relationship between Joint Roughness and Joint Shear Strength. Proceedings, International Symposium on Rock Mechanics (pp. 1–8), Nancy, France.

Barton, N. R. 1973. Review of a new shear strength criterion for rock joints. *Engineering Geology* 7:287–332.

Barton, N. R. 1974. *A Review of Shear Strength of Filled Discontinuities in Rock* (Publication No. 105). Oslo: Norwegian Geotechnical Institute.

Bishop, A. W. 1955. The use of the slip circle in the stability analysis of slopes. *Geotechnique* 5(1):7–17.

Conforth, D. H. 2005. *Landslides in Practice: Investigation, Analysis, and Remedial/Preventative Options in Soils.* New York: John Wiley & Sons.

Cruden, D. M., and Varnes, D. J. 1996. Landslide types and processes. In A. K. Turner and R. L. Schuster (eds.), *Landslides—Investigation and Mitigation* (Special Report 247, pp. 36–71). Washington, DC: Transportation Research Board, National Academy of Sciences.

Fellenius, W. 1927. *Erdstatische Berechnungen mit Reibung und Kohasion.* Berlin: Ernst.

Gedney, D. S., and Weber, Jr., W. G. 1978. Design and construction of soil slopes. In R. L. Schuster and R.

J. Krizek (eds.), *Landslides: Analysis and Control* (Special Report 176, pp. 172–191). Washington, DC: Transportation Research Board, National Academy of Sciences.

Higgins, J. D., and Andrew, R. D. 2012. Rockfall types and causes. In A. K. Turner and R. L. Schuster (eds.), *Rockfall: Characterization and Control* (pp. 21–55). Washington, DC: Transportation Research Board, National Academy of Sciences.

Janbu, N. 1968. *Slope Stability Computations* (Soil Mechanics and Foundation Engineering Report). Trondheim: Technical University of Norway.

Jones, C. L., Higgins, J. D., and Andrew, R. D. 2000. Colorado Rockfall Simulation Program (Version 4.0), User's Manual. Denver: Colorado Department of Transportation.

Keaton, J. R., Wartman, J., Anderson, S., Benoît, J., deLaChapelle, J., Gilbert, R., and Montgomery, D. R. 2014. *The 22 March 2014 Oso Landslide* (Report No. GEER-036). Washington, DC: Geotechnical Extreme Event Reconnaissance Association.

Keefer, D. K., and Johnson, A. M. 1983. *Earth Flows: Morphology, Mobilization, and Movement* (Professional Paper 1262). Washington, DC: US Geological Survey.

Morgenstern, N. R., and Price, V. E. 1965. The analysis of the stability of general slip surfaces. *Geotechnique* 15(1):79–93.

Park, H. J., West, T. R., and Woo, I. 2005. Probabilitistic analysis of rock slope stability and random properties of discontinuity parameters, Interstate Highway 40, western North Carolina, U.S.A. *Engineering Geology* 79(3–4):230–250.

Patton, F. D. 1966. Multiple Modes of Shear Failure in Rock. Proceedings, First International Congress of Rock Mechanics (Volume 1, pp. 509–513). Lisbon, Portugal.

Pierson, L. A. 2012. Rockfall hazard rating systems. In A. K. Turner and R. L. Schuster (eds.), *Rockfall: Characterization and Control* (pp. 56–71). Washington, DC: Transportation Research Board, National Academy of Sciences.

Piteau, D. R., and Peckover, F. L. 1978. Engineering of rock slopes. In R. L. Schuster and R. J. Krizek (eds.), *Landslides: Analysis and Control* (Special Report 176, pp. 173–228). Washington, DC: Transportation Research Board, National Academy of Sciences.

Rellensmann, O. 1957. Rock Mechanics in Regard to Static Loading Caused by Mining Excavation. Second Symposium on Rock Mechanics, Quarterly, Colorado School of Mines, Vol. 52.

Rocscience Inc., 2015, RocFall, Version 5, University of Toronto, Ontario, Canada.

Rocscience Inc., 2016, Slide, Slide, Version 7, University of Toronto, Ontario, Canada.

Schuster, R. L. 1996. The 25 most catastrophic landslides of the 20th century. In J. Chacon, C. Irigaray, and T. Fernandez (eds.), *Landslides*. Proceedings of the 8th International Conference and Field Trip on Landslides, Granada, Spain. Rotterdam: Balkema.

Schuster, R. L., and Krizek, R. J. (eds.). 1978. *Landslides: Analysis and Control* (Special Report 176). Washington, DC: Transportation Research Board, National Academy of Sciences.

Shakoor, A. 1995. Slope stability considerations in differentially weathered mudrocks. In W. C. Haneberg and S. A. Anderson (eds.), *Clay and Shale Slope Instability, Reviews in Engineering Geology* (Vol. X, pp. 131–138). Boulder, CO: Geological Society of America.

Sharpe, C. F. S. 1938. *Landslides and Related Phenomena: A Study of Mass Movements of Soil and Rock.* New York: Columbia University Press.

Soeters, R., and Van Westen, C. J. 1996. Slope instability—recognition, analysis, and zonation. In A. K. Turner and R. L. Schuster (eds.), *Landslides: Investigation and Mitigation* (Special Report 247, pp. 129–177). Washington, DC: Transportation Research Board, National Academy of Sciences.

Turner, A. K., and Duffy, J. D. 2012. Evaluation of rockfall mechanics. In A. K. Turner and R. L. Schuster (eds.), *Rockfall: Characterization and Control* (pp. 21–55). Washington, DC: Transportation Research Board, National Academy of Sciences.

Turner, A. K., and Schuster, R. L. (eds.). 1996. *Landslides: Investigation and Mitigation* (Special Report 247). Washington, DC: Transportation Research Board, National Academy of Sciences.

US Army Corps of Engineers. 1970. *Engineering and Design—Stability of Earth and Rockfill Dams* (Engineer Manual EM 1110-2-1902). Washington, DC: Department of the Army Corps of Engineers, Office of the Chief of Engineers.

US Geological Survey. 2014. *Preliminary Interpretation of pre-2014 Landslide Deposits in the Vicinity of Oso, Washington* (Open-File Report 2014-1065). Washington, DC: US Geological Survey.

West, T. R. 1996. The effects of positive pore pressure on sliding and toppling of rock blocks with some considerations of intact rock effects. *Environmental & Engineering Geoscience* II(3):339–353.

Wieczorek, G. F. 1996. Landslide triggering mechanisms. In A. K. Turner and R. L. Schuster (eds.), *Landslides: Investigation and Mitigation* (Special Report 247, pp. 36–71). Washington, DC: Transportation Research Board, National Academy of Sciences.

Wyllie, D. C., and Mah, C. W. 2004. *Rock Slope Engineering* (4th ed.). New York: CRC Press.

COASTAL PROCESSES 16

Chapter Outline

Wave Motion

Tides

Effects of Severe Storms

Erosion in Coastal Areas

Deposition in Coastal Areas

Estuaries

Classification of Coastlines

Sensitivity of Shoreline Features to Human Activities

Engineering Considerations:
  Shoreline Protection Structures

The zone along the coast of oceans or large lakes, such as the Great Lakes, is a site of active geomorphic processes. Wave action and the relative level of the water surface provide the erosive force that modifies and rebuilds coastal landforms (Marshak, 2013; Nance and Murphy, 2016). Waves are the undulations on the water surface, usually set in motion by the frictional drag of wind. The size of waves depends on wind speed, wind direction, and the length of open water (Bird, 2008).

Only the water surface and its near subsurface are disturbed by the wind. The area over which wind acts to generate and promote waves is known as the fetch. This area is extremely large for the oceans and the Great Lakes and is much of the reason why storm activity can be so devastating to adjacent coastal areas. For smaller lakes and reservoirs, the fetch is smaller. Engineers designing earth dams try to minimize the fetch where wind builds waves that will, ultimately, strike the dam. Protection of the upstream slope of an earth dam can be reduced when the fetch for that shoreline is minimized.

Wave Motion

The motion of waves in the ocean is described by the terms used to describe wave behavior. Figure 16.1 on the next page provides this terminology. Wavelength L is the horizontal distance between successive repeat points on the wave trace, that is, between adjacent crests or adjacent troughs. Wave height H, or amplitude, is the vertical distance between the crest and the trough. The period T is the time elapsed between successive crests or troughs. Two additional terms associated with wave motion are *frequency* and *velocity*. Frequency, N, is the number of wavelets or cycles that pass a point in a unit of time and velocity, V, is the rate of movement of the waveform. Frequency is the reciprocal of the period (or $N = 1/T$), so it can be observed that $V = LN = L/T$.

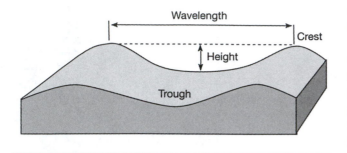

Figure 16.1 Terminology associated with waveforms.

The two primary types of ocean waves are oscillatory waves, which occur in the open ocean away from the coasts, and translational waves, which develop a short distance offshore (Bird, 2008; Bush and Young, 2009).

Oscillatory Waves

For oscillatory waves only the waveform advances, not the individual water molecules themselves. This is similar to stalks of grain bending in the wind as a breeze ripples across a field. Just as the stalk sways back and forth, yielding a circular motion for a point on the stalk, so is a circular motion in the water generated. The circles decrease in size with depth, as depicted in Figure 16.2. At the surface, the circular orbit has a diameter equal to the wave height with particles completing the circle in time T (10 to 20 seconds). Beneath the surface, the motion decreases rapidly (exponentially) and, at a depth of about one-half the wavelength ($L/2$), it is less than

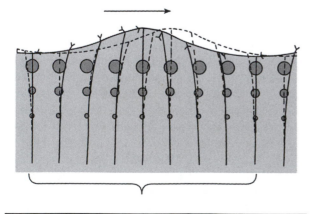

Figure 16.2 Movement of water molecules at depth as the wave travels along the surface.

5% of the velocity at the surface. The level at which negligible motion occurs is known as the *wave base*.

The velocity of the waveform, presented previously in the equation $V = L/T$, is not the same velocity as that of the water particles. Instead, it is the rate of propagation for the surface wave itself. Water particles moving along the circular path travel at a considerably lower velocity with depth. For example, for a wave with a wavelength of 100 m (330 ft) and a height of 5 m (16 ft), traveling at 45 km/hr (28 mi/hr), the orbital velocity of the surface particles is about 7.0 km/hr (4.4 mi/hr). At a depth of 20 m (65 ft), the orbital motion is only about 2 km/hr (1.2 mi/hr), and at 100 m (330 ft), movement is negligible. By contrast, for a wave with a wavelength equaling 380 m (1250 ft) and a height of 10 m (33 ft), traveling at 90 km/hr (56 mi/hr), the surface particles have an orbital velocity of 7.4 km/hr (4.6 mi/hr), but at a depth of 100 m (330 ft) the orbital motion is still nearly 1.6 km/hr (1 mi/hr).

Wind Velocity versus Wave Velocity

Wind supplies the energy to waves in two ways: (1) by pushing against the wave crests and (2) by friction of the air against the water surface. Wind blowing at 50 km/hr (31 mi/hr) will produce a wave 6.7 m (22 ft) high with a 76 m (250 ft) wavelength traveling at 40 km/hr (25 mi/hr). By contrast, wind at 110 km/hr (69 mi/hr) provides a wave 14.5 m (48 ft) high, with a wavelength of 376 m (1230 ft) traveling at 88 km/hr (55 mi/hr). The energy of the resultant waves increases very rapidly with increasing velocity of the generating wind. It appears to vary as the fourth or fifth power of the wind velocity.

Waves from storm winds are transferred into waves called *swells* and interference of these waveforms yields larger waves. These typically have periods in the range of 8 to 12 seconds. Sea captains have estimated wave heights at more than 24 m (80 ft) during hurricanes and sailors aboard US naval ships have sighted waves with heights of about 35 m (115 ft) during these storms.

Tsunami

Waves with greater periods are generated by phenomena other than wind. Short-period waves have lower wave heights and less wave energy. Tsunami (from the Japanese, "tsu" for harbor and "nami" for wave), or seismic sea waves, are long-period waves, ranging from 800 to 3000 seconds. Tsunami are generated by fault displacement in the ocean basins some distance offshore (Nance and Murphy, 2016). Tsunami have long wavelengths of from 55 to 200 km (35 to 125 mi) and travel across the ocean at speeds between 500 to 900 km/hr (310 to 560 mi/hr). These waves differ from wind-generated waves because their energy is transferred to the water from the seafloor so that the entire water column is involved in the wave motion. In the open ocean, the wave height of a tsunami is only 20 to 60 cm, so that ships at sea may not perceive their passing. However, when the tsunami approaches the shore, significant changes occur. Energy distributed in a deep column of water now becomes concentrated in an increasingly shorter column of water, resulting in a rapid increase in wave height. This is why the waves increase to 15 m (50 ft) in height and, in rare situations, to more than 30 m (100 ft). Such waves can cause enormous damage to low-lying coastal areas. Nearly all tsunami originate in the Pacific Ocean within active earthquake zones. Total evacuation of low-lying areas in the path of the tsunami is the only effective defense. Elaborate warning systems for the coastal areas of the Pacific Ocean have been devised to alert people in time to move to high ground.

Waves in Shallow Water

In shallow water, oscillatory waves change their mode to become translational waves. The waves begin to drag or "feel the bottom" when the water depth decreases to about one-half of the wavelength and the wave base is contacted. When this occurs, the wave height increases, the wavelength grows smaller, and the velocity of the waveform decreases. Progressively, the wave front becomes steeper until the wave finally breaks. After the wave breaks, a translational wave takes over and water expends its energy running ashore and up the beach as wave runup or swash (Masselink et al., 2011). Figure 16.3 shows this process of wave movement in shallow water.

Figure 16.3 also shows an empirical relationship between wavelength, water depth, and wave height. The wave reaches its peak at about $d = 2H$ where d and H are water depth and wave height, respectively. This occurs at about a water depth of $d = L/2$ with L being the wavelength. The wave breaks at about $d = 1.3H$ and either a predominantly plunging wave or a spilling wave develops, depending on the configuration of the ocean bottom. A plunging wave forms on steep, smooth bottoms where the crest is propelled ahead of the wave. A spilling wave occurs on smooth, flat bottoms with the crest spilling down in front of the wave. Most breakers actually fall somewhere within these two extreme cases.

The force of waves is an important agent for coastal erosion and sediment transport. Its magnitude is primarily a function of wave height and frequency, with winter storms typically yielding the greatest force. Such forces against a vertical wall have been measured for the New England area. The study showed that hydrostatic pressure against a vertical wall occurs as a consequence of an upward surge of water. A smaller, reflected wave moves away from the vertical wall. In this situation, the water is sufficiently deep in front of the wall that no breakers develop prior to striking it, which, of course, would expend some energy. Values for this dynamic impact indicate a maximum destructive force of about 287 kPa (6000 lb/ft^2) with averages of about 96 kPa (2000 lb/ft^2) in the winter and 29 kPa (600 lb/ft^2) in the summer.

Wave Paths and Erosive Forces

Wave Refraction

When waves approach the shore, they are *refracted*, or bent, in a natural effort to strike per-

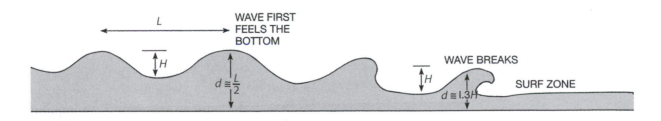

Figure 16.3   Waves breaking in shallow water.

pendicular to the shoreline. In this regard, waves can be depicted either by their crest line or by their ray path, the latter indicating the direction of propagation. Hence, the crest line tends to parallel the coast as refraction occurs (Figure 16.4).

Wave refraction concentrates energy on the headlands and directs it away from the bay areas between them. As the wave moves toward the shore, the portion offshore from the headlands (see Figure 16.4) encounters or "feels" the bottom first and slows down. Meanwhile, the rest of the wave continues forward without velocity reduction to be refracted later when it drags the bottom. This causes progressive bending of waves toward the headlands, which finally receive the concentrated force of the breakers. As a result, the erosive force in the bays is minimized and sediment is deposited there to form beaches. In this way, coastlines are straightened as the promontories reduce by erosion and deposition builds up in the bays. Figure 16.5 shows wave refraction and the longshore drift resulting in beach formation.

Longshore Drift

Although wave direction changes by refraction as the waves drag the bottom, they typically strike the shoreline at an angle to the perpendicular. This provides a component of movement parallel to or in the "downcoast" direction. Longshore drift (or littoral drift) of sediment is generated as waves strike the shore obliquely. Figure 16.4 shows the details of this process.

As the wave strikes the shore at an angle, water and sediment are propelled obliquely up the beach. When the wave is spent, water and sediment return down the beach, perpendicular to the shore. Successive waves move the sediment obliquely up the beach and the backwash returns it. In a series of steps, the net transport is parallel to the beach in the component direction. This is known as beach drift and the current that occurs in the breaker zone is the longshore current, or littoral current. These two zones of movement provide transport parallel to the coast; the first along the beach, as described, and the other in the surf and breaker zone where material is transported by suspension and saltation. The combined action of these two comprises the longshore drift (Komar, 1983; Masselink et al., 2011).

A great volume of sediment can be transported by the longshore drift and, for this reason, beaches have been described as rivers of sand. As long as a

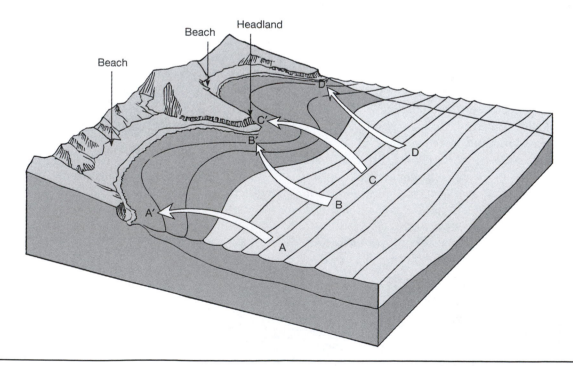

Figure 16.4 Wave refraction along an irregular shoreline, showing ray paths (A, B, C, D) and crest lines.

Figure 16.5   Wave refraction and longshore drift. (Gavan Leary, Shutterstock.)

continuous supply of sediment occurs from the up-coast direction, the beach may show little change within a season. Increased or decreased sediment supply, however, typically brought about by human activities, may greatly disrupt this balance. From one season to another, or because of differences in direction between prevailing winds and storms, waves may approach the shore at different angles. This will change the intensity of the longshore drift or even reverse its direction. Longshore currents build their strength with increased distance down the shoreline. Ultimately, they return seaward through the breakers in a narrow zone known as a rip current.

In a general sense, the longshore drift for coastal areas of the United States moves southward. This holds true for the Pacific, Atlantic, and Gulf Coasts and for Lake Michigan. This is a consequence of the prevailing winds and storms, which come from a northwest or northeast direction. Locally, the direction of longshore drift may differ from this because of shoreline configuration, tributary streams, offshore islands, or other complications.

For Lake Michigan, the longshore current moves southward on both the western side of the lake (the Wisconsin—Illinois shorelines) and the eastern side

(the Michigan shoreline). Indiana Dunes National Lakeshore is located at the southern end of Lake Michigan where the migrating sand has accumulated in large dunes along the shoreline (see Figure 16.6 on the next page).

Wave energy and wave direction both affect shoreline erosion. High-energy waves that are very steep (that is, they have a high ratio of height to length) cause beach erosion, whereas less steep, smaller waves provide beach replenishment. Intense seasonal storms have resulted in the concept of a winter beach, where the beach is eroded, narrow, and deficient in sand, as opposed to a summer beach with a wide, sandy expanse. These extreme conditions do not occur along all beaches because of the different intensities of winter storms from one coastal area to another.

Tides

Location and Frequency of Tides

Tides are daily fluctuations in the level of the sea caused by the combined effects of the gravitational

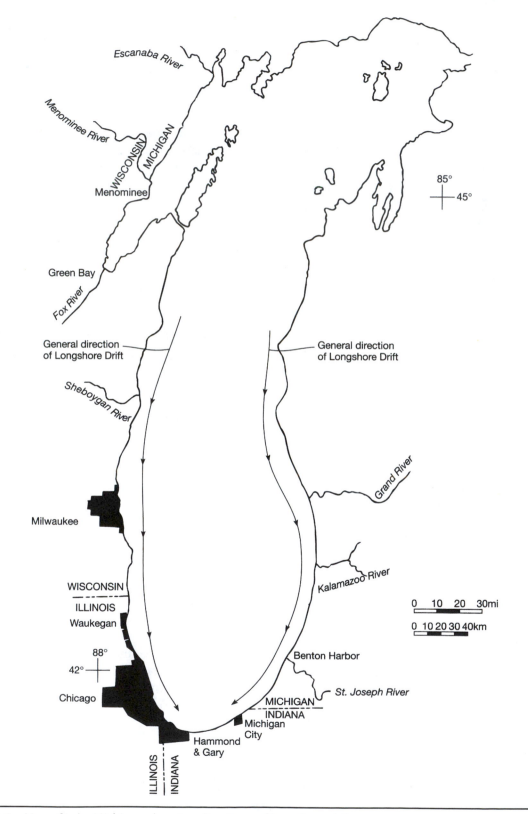

Figure 16.6   Map of Lake Michigan showing directions of longshore drift.

forces exerted by the Moon, the Sun, and the rotation of the Earth. Both the side of Earth facing the Moon and that directly opposite it will experience a high tide. As the Moon revolves around Earth, these bulges in the water surface follow along. At the intermediate locations, 90° away, or one-fourth of the Earth's circumference, the coastal areas experience low tide. Because of this relationship, in a period of one day, typically two high tides and two low tides are experienced. More accurately, the double cycle repeats itself once every 24 hr and 50 min (not every 24 hours) as the Moon advances eastward in its orbit. The bulging of the tides relative to the position of the Moon is depicted in Figure 16.7.

Causes of Tidal Fluctuations

The bulge of water on the opposite side of Earth from the Moon can be explained in two ways. First, because the center of Earth is closer to the Moon, it is pulled more strongly toward the Moon than is the water on the opposite side of Earth. Hence, Earth, in effect, is pulled away from its water body on the side opposite the Moon. The other explanation is that Earth's centrifugal force, caused by its rotation, exerts the greatest influence on the opposite side of Earth as it pulls against the Moon's gravitational force at that point.

The Moon exerts more influence on tidal fluctuations than does the Sun, despite the Sun's much greater mass and greater attractive force, because Earth is closer to the moon than to the Sun, and Earth's diameter (12,800 km [8000 mi]) is more significant relative to the distance between the Earth and the Moon (320,000 km [200,000 mi]) than when compared to its distance from the Sun (149 × 10⁶ km [93 × 10⁶ mi]). The net result is that the Sun's effect on the tides is just less than half (0.46) that of the Moon.

Spring Tides and Neap Tides

Twice each lunar month, at new moon and full moon, the Sun and the Moon lie in a straight line and their influence on Earth is additive. These conditions cause large tides, called the spring tides. In like fashion, twice each lunar month, at first and third quarters, the Sun and the Moon are in opposite positions (at 90° to each other when viewed from the Earth), and the tides are smaller. These are known as the neap tides.

Details on Tidal Extremes

The tides are not as simple a phenomenon as the previous discussion would suggest because (1) many parts of the world have only one tide per lunar day, (2) in other areas the high tide lags many hours behind the time when the Moon passes overhead, and (3) in still other locations the two daily tides are of greatly different heights.

In reality, several components comprise the tidal effect, and their relative contribution is a function of Earth latitude. The semidiurnal tide effect mentioned previously has its maximum contribution at the equator and a minimal one at the poles. Another diurnal effect has its maximum contribution at 45° latitude and none at the equator.

Theoretically, the maximum fluctuation of the tides should be only about 50 cm (20 in), but values of more than 10 m (33 ft) are experienced in some areas of the world. Tides are affected by the geometry of the coastline, the location of the landmasses, the variable water depth of the oceans, and the Coriolis force, which deflects tidal currents. This force, a result of Earth's rotation, is an apparent force experienced by moving objects, which trends to the right in the Northern Hemisphere and to the left in the Southern. Because of these complications, tide charts for the world are prepared based on empirical relationships involving past observations of tidal effects.

Locations with Maximum Tidal Effects

Mouths of estuaries and straights between bodies of water whose maximum water levels are not

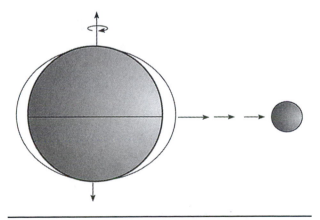

Figure 16.7 Earth and Moon system showing the bulges in the oceans toward and away from the Moon.

in phase with the local high tide experience strong tidal currents. In the English Channel, between England and France and into the North Sea, this effect is quite pronounced and tends to increase the extremes of high and low tides. In another extreme example, water flowing under the Golden Gate Bridge in San Francisco has a daily flow of 6.5×10^4 m³/sec (2.3×10^6 ft³/sec), or 3.5 times the discharge of the Mississippi River.

In some rivers, the tide enters as a rushing wave of water called a bore. The bore of the Qiantang River in Hangzhou Bay in China can be as great as 3 m (16 ft) with speeds up the river at 25 km/hr (16 mi/hr). Smaller tidal bores occur at the head of the Bay of Fundy between New Brunswick and Nova Scotia.

As previously mentioned, the configuration of the coastline has a pronounced effect on the tides. Irregular coastlines, with long estuaries extending inland, provide the highest tides. As the tide advances, the wall of water funnels into the estuary and, due to the decreasing cross section, the water height must increase to accommodate a constant water volume. In long estuaries, the tidal rise can be multiplied tenfold or twentyfold by this aspect. Table 16.1 provides a list of the tidal differentials for a few selected areas of the world.

The Bay of Fundy has the largest tide range on Earth, 16 m (54 ft) as a spring tide. It is situated at the head of an estuary 100 km (62 mi) long, and one consequence is a reversal in the flow direction of tributary streams between high and low tide. Tidal currents traveling 15 km/hr (9 mi/hr) during the rise

and fall of the tides have scoured basins 45 m (150 ft) deep in the bay.

In Brittany, France, an estuary of the Rance River that is 100 km (62 mi) long has been developed for tidal power generation. At this location, the maximum tide range is 13.7 m (45 ft) and currents reach 13 km/hr (8 mi/hr) during rise and fall of the tides. A concrete dam 366 m (1200 ft) long, 38 m (125 ft) wide at the base, and 26 m (85 ft) high above the foundation level has been constructed. It consists of 24 concrete bays, with each bay containing a 10,000 kW, reversible pump turbine. These turbines, combined, deliver some 624 million kW-hr/yr. With the incoming tide, the water runs through the turbines into the head of the bay to generate electricity. When the water level on the ocean side is only about a meter (a few feet) above that of the bay, generation is curtailed, waiting for the outgoing tide. Water is also pumped from the ocean side to the bay side as the tide retreats, thus building a greater differential head. When the tide has ebbed sufficiently, the turbines are reversed and water flowing from the bay side to the ocean side again generates electricity. Obviously, a considerable time interval occurs within each cycle of the tides in which no power can be generated. The tidal power station is tied into a large, power-generating network so that other generating plants can supply the load when the Rance River is not producing.

An interesting difference in tidal fluctuations occurs for the Panama Canal. On the Atlantic (Caribbean) side, the tides are less than 0.3 m (1 ft), whereas for the Pacific side they are 4 m (13 ft).

Effects of Severe Storms

Two distinct processes act to modify the coastline. The first provides a gradual change, resulting in slow, continuous erosion and deposition. The tides and longshore currents provide this regular, continuous action. The second causes rapid changes and acts in an irregular, intermittent fashion. These are the catastrophic storm events that cause flooding and severe erosion of coastal areas when they occur. Severe storms, hurricanes, and typhoons involve very high wave energies that develop as a consequence of low pressure cells in the atmosphere. They produce four related effects: (1) storm surge, (2) high wave energy, (3) extreme rainfall, and (4) high winds.

Table 16.1   Maximum Tide Range for Selected Locations of the World.

Location	Range (m)	Range (ft)
Bay of Fundy, Nova Scotia	16	54
Rance River, Brittany, France	14	45
London, England	6	20
Boston, Massachusetts	2.7	9
New York City, New York	2	6
Key West, Florida, and Galveston, Texas	0.6	2
Cook Inlet, Anchorage, Alaska	11	35
Honolulu, Hawaii	0.5	1.5
Seattle, Washington	5	16

Storm surge manifests by a rise in elevation of the water surface because of low pressure conditions. This surge affects the shoreline long before the direct storm activity hits. As the storm approaches, the wind onshore increases in velocity and the wave heights increase as well. As the storm moves inland, the previously flooded coastal area is subjected to further flooding from heavy rainfall. Erosion of the beach and dunes can be excessive during such a storm. In some instances the storm waves will breach the foredune and flood the inland area.

Erosion in Coastal Areas

Coast is a general term that designates the broad area adjacent to the sea. It typically consists of a combination of features that may include steep cliffs, low-lying beaches, bays, tidal flats, or marshes. The topography that develops is a function of the uplift or subsidence of the coastline relative to sea level, the geologic materials involved, the processes of erosion and deposition by the sea, and time elapsed since relative movement between land and sea has occurred (Dean and Dalrymple, 2004; Bird, 2008).

Factors in Wave Erosion

Waves erode the coasts in two primary ways: (1) by impact and hydraulic pressure and (2) by abrasion—the grinding action of sand, gravel, and cobbles on the cliffs or across the foreshore. The massive impact of water against a sea cliff can exert pressures up to 144 kPa (3000 psf) during major storms with pressures of only one-quarter that from smaller storms. Water is also driven into every opening and fracture in the rock, compressing the air against the solid boundary. This acts as a wedge to widen fractures and to loosen blocks of rock.

Landform Features in Massive Rock

Features that form by this wave action include sea cliffs, wave-built terraces, and wave-cut beaches, illustrated in Figure 16.8. In coastal areas composed of steeply sloping resistant rocks, the wave action cuts a horizontal notch into the rock at sea level. By subsequent erosion this may produce sea caves, sea arches, or sea stacks. Sea arches occur when notches encroaching from two sides of a rocky promontory connect to form a passageway through the rock.

When the arch eventually collapses, an isolated pinnacle, known as a sea stack, remains. In these same areas, water spouts and potholes in the rocky cliffs commonly provide pleasant tourist attractions.

Undercutting of the cliff causes overhanging rocks to collapse, with the fallen debris soon carried away by wave action. With a new cliff face exposed, the process is repeated and by this progressive action the sea cliff retreats. As a consequence, a wave-cut bench is produced, typically with the upper portion visible at low tide. Sediment transported by longshore currents can accumulate in deeper waters to form a wave-built terrace.

Equilibrium Conditions

As the bench is widened, the waves break farther and farther from the shoreline. Their energy dissipates by friction as they travel across the shallow water on the beach. This greatly reduces the wave action on the cliffs, and beaches can develop. The cliff face is now subject only to weathering and mass wasting. In effect, an equilibrium condition develops between the length of wave-cut bench, the sea cliff, and the wave action of the sea for the established elevation difference between the landmass and sea level. If his relationship should change because of tectonic uplift of the land or a worldwide alteration in sea level, a renewed erosional sequence would be set in motion.

Impact on Human Activities

The extreme effects of coastal erosion are of particular concern to human activity because property and human activity are affected by landform change. Humans strive diligently to minimize the alterations that natural processes bring about. The profound effect of wave action onshore protection structures is well documented. At Wick, in northern Scotland,

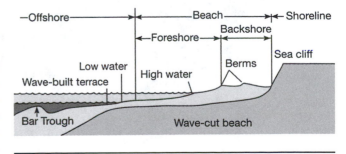

Figure 16.8 Cross section of features along a shoreline.

storm waves have eroded away about 2300 metric tons (2500 tons) of concrete from breakwaters built for shore protection. At Ymuiden, Holland, a 6.3 metric ton (7 ton) block in the breakwater was moved nearly a meter (3 ft) by the sea. The White Cliffs of Dover are so rapidly undercut by the sea that, occasionally, large landslides occur. A major landslide in 1810 caused earthquake vibrations felt in the town of Dover, which is located several miles away.

The erosion rate of poorly consolidated materials by wave action of the oceans can be extreme, with an average retreat rate of 1.5 to 2 m (5 to 6.5 ft) per year. Along parts of England's coastline, the shore has retreated more than 5 km (3 mi) since Roman times so that a number of coastal villages and landmarks have disappeared. The effect of erosion along the coastline of Lake Michigan during the mid-1970s was also profound. In the suburban areas north of Chicago several meters (5 to 10 ft) of erosion occurred within a 2-year period. Many homes, beach cottages, and recreational areas were damaged severely during this period when the lake level was higher than average. The Lake Erie shoreline experienced similar levels of erosion and bluff retreat during the mid-1980s.

Deposition in Coastal Areas

Longshore currents and waves transport sediment along the coast until it is deposited in areas of low energy. These landforms include beaches, spits, tombolos, bars, and barrier islands (Bird, 2008; Nance and Murphy, 2016), as illustrated in Figure 16.9. Small beaches may develop during the active erosional sequence in the bays between prominent rocky headlands, but the other depositional landforms are more common on smooth or gently curved coastlines. These coastlines occur when shoreline equilibrium occurs, as previously discussed.

Beaches

A beach is the gently sloping portion of a shore consisting of unconsolidated sediment. Typically,

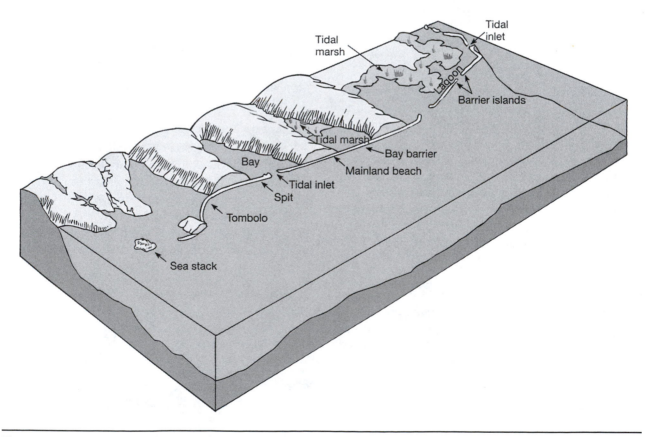

Figure 16.9 Depositional features along a coastline.

sand will dominate but some beaches are composed of cobbles and boulders, and others are silt and clay. In general, beaches have a similar morphology, as illustrated in Figure 16.8. The characteristics of a beach, its slope, composition, and size, depend on wave energy and the supply of sediments. Beaches composed of fine-grained sediments are typically flatter than those made of coarse sand and gravel.

Rivers that carry sediments from the landmass are the primary source of material for beach formation. This sediment is transported by the longshore current down the coast. If the source of sediment is curtailed, by damming the river or by shore protection devices, serious changes may occur downcoast. This problem is addressed in a later section.

The beach itself consists of two parts, the fore-shore or the strand, which lies between low tide and high tide, and the back beach, which is formed by two berms (see Figure 16.8). Both the lower and upper berms are the consequence of storm activity and mark the boundary of the beach. Landward are found cliffs, dunes, or limiting features that denote the extent of the erosive process of the sea.

Sea Terraces

Sea terraces occur in regions with emergent coastlines; that is, areas where the coast has recently moved upward relative to the sea. This marks the location of the beach and wave-cut bench, which formed when that portion of the landmass stood at sea level. When such features are observed and the differential movement between the ocean and the land is noted, one must decide whether the land moved up while the sea level remained stationary or if the land remained fixed and the sea level went down. Of course, if the sea retreated, worldwide evidence of this event should be present. Evidence indicates that emergent coasts typically involve tectonic uplift of the landmass rather than a drop in sea level.

Spits and Sand Bars

Spits are extensions of the beach across indentations in the shoreline, such as bays or estuaries (see Figure 16.9). Spits may continue to grow out into the bay as materials accumulate at the end. With continued growth, the spit may completely close the front of the bay. Such features are called *bar barriers* or *baymouth bars*.

An extensive spit may acquire a claw-shape tip at its end by either further wave refraction or storm activity. This feature is termed a *recurved spit* or a *hook*. A *tombolo* is a spit that connects the mainland to an island (see Figure 16.9).

A bar is an offshore, submerged, elongated body of sand, built by wave action or by longshore currents. Bars may be partially exposed at low tide and, eventually, may develop into a barrier beach.

Barrier Islands

Barrier islands are prominent landforms that occur along the Atlantic Coast from New Jersey to Florida and along the Gulf Coast from Texas to Florida. Also called barrier offshore beaches, they parallel the shoreline, consist of elongated masses of sand, and are located some distance offshore. Typically, they are separated from the mainland by a lagoon and most are cut by one or more tidal inlets. Individual islands may range from 100 m (330 ft) to more than 1600 km (1000 mi) long and, generally, have a width of less than 4 km (2.5 mi) with an average of only 0.4 to 1.6 km (0.25 to 1 mi). The islands usually lie less than 3 m (10 ft) above sea level, although sand dunes may rise as high as 13 m (40 ft).

Barrier islands are depositional features formed by detrital sediments (sand and gravel) and are, therefore, distinguished from organic formations such as reefs. They are transitory because they are continuously undergoing changes as a result of natural, dynamic processes that have been complicated by human activities. These islands are thought to have originated on submergent shorelines through either the emergence of offshore bars, the continual growth of complex spits, the submergence of coastal beach ridges, or some combination of these occurrences. Humans are greatly interested in preserving the barrier islands because of their natural beauty and commercial value so beach erosion prevention is crucial in these areas.

The barrier islands forming the Outer Banks of North Carolina are shown in Figure 16.10 on the next page.

Dune Ridges

In many areas, a dune ridge, or dune line, is associated with the sand beach along an ocean or major inland lake. In a natural state, the dune line migrates forward and back in response to beach erosion and deposition, in keeping with the continuing processes of tides and longshore currents and the intermittent processes associated with severe storms. The dune system may take the form of a primary and secondary dune with a trough area between them. Figure 16.11 illustrates primary dunes, secondary dunes, backdunes, bayshore, and bay areas.

Human Activity in Dunal Areas

When humans develop a coastal area for housing and recreation, the first tendency is to stabilize the dune line. This protects the local residences and provides a roadbed for the primary transportation route. Eventually, small towns and villages develop in these recreational areas and the natural changes, such as tidal inlet openings, dune migration, and beach overwash, are no longer acceptable.

Stabilizing the dune line, however, stops the processes of migration behind the dunes. As the sea level rises because of storm surges and other phenomena, the beach zone shifts landward, but because the dunes no longer migrate with this shift, the width of the beach and berm is reduced and wave energy is concentrated over a smaller and smaller runup area. This throws the system out of equilibrium and leads to long-term dune erosion.

Under natural conditions, the migrating dunes give up their sand to the local system. That sand is redeposited nearby. Dune and beach replenishment occurs when the lower water level is reestablished. However, for stabilized dunes, the sands are continually eroded and taken from the system by longshore currents carrying it downcoast. Eventually the wave energy will dissipate directly against the dune

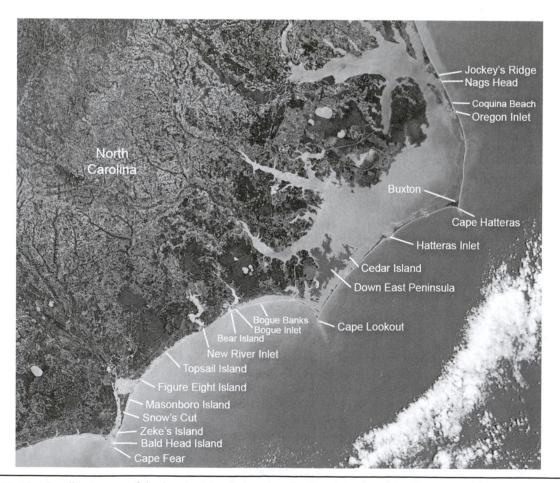

Figure 16.10   Satellite image of the barrier islands forming the Outer Banks of North Carolina. (Image by Jacques Descloitres, MODIS Land Rapid Response Team, NASA and the Science Education Resource Center at Carleton College.)

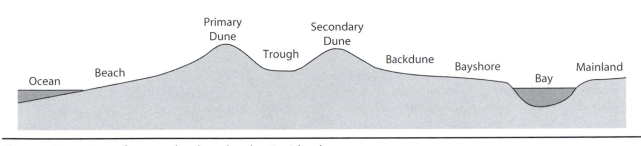

Figure 16.11   Dune features developed on barrier islands.

line. Vegetative cover provides erosion protection to the dune surface but does not prevent undercutting. Breaching and overwash will eventually occur, both of which greatly concern local residents. The response is to plug the gap as soon as possible and rebuild the dune. Closing the breach by simple sand replenishment is not often a successful long-term solution. Commonly, the next storm breaches the dune in the same area. A more successful procedure is to encourage dune growth with slotted fences (snow fence), which have a better chance for permanently reestablishing the dune. In any event, when the dune line is held stationary, the likelihood of breaching the dune is markedly increased.

The dune ridge on a barrier island is affected in yet another way when a continuous dune line is established by humans in order to prevent outwash and tidal delta development. Because of the exclusion of the sea at high tide, the interior portions of the islands, including marshy areas in the lagoons, become progressively lower relative to the rising sea. Closing the natural inlets reduces sediment supply to the marshes and slows the accumulation of organic matter. The results are twofold: (1) the marsh areas subside, yielding, in effect, a thinner island, and (2) the barrier loses the buffering effect of the marshes against wave energy dispelled through the inlets. Therefore, the islands erode from both sides and are more susceptible to flooding from lower level surges. With no outlet through the dune to the ocean, surges from the lagoon pile up against the inside of the dune and flood extensive areas.

The rotating winds and low atmospheric pressures of hurricanes crossing barrier islands will attack the dune line from both sides. The level of the interior will be raised by overwash through breaching of the dune as new sand is added to the backshore and marsh areas. Seaward, the dune line will suffer severe erosion and some beach areas will become nearly devoid of sand. In built-up areas, large sums of money will be required to rebuild dunes and beaches and to remove sand deposited on roads and in residential areas. Flood damage caused by surges from the back bay will also be extensive. In natural areas, the damage, measured monetarily, will be less as the dune and backdune area should repair themselves in time through the reestablishment and migration of the dune system.

Estuaries

Estuaries are large volume streams emptying into the ocean, which flow on very low gradients and experience the ebb and flow of the tides. They are common to drowned coastlines such as the Atlantic seaboard from New England to Washington, DC. Here the land has submerged relative to the sea and submarine canyons mark the location of the downstream end of channels formed during lower ocean levels in the geologic past.

Uses of Estuaries

Estuaries provide excellent harbors because of the shelter of the land. New York City, Boston, Washington, DC, Baltimore, and Norfolk are examples of harbors in estuaries in heavily populated areas. They occur along major rivers, for example, the Chesapeake, Potomac, Delaware, and Hudson Rivers. Estuaries are the center of activity because of the population density and the multiple-use capabilities of the waterways: shipping, fishing, mining, industry, farming, and recreation.

Pollution of Estuaries

Pollution of estuaries is a serious environmental problem. Major contaminants include municipal wastes, industrial wastes, heat, oil, toxic chemicals, pesticides, insecticides, nutrients, flesh-tainting substances, and sediments. Major sources of the contaminants include municipal sewage treatment plants, industry, power-generating stations, farming, construction, and shipping. Pollution must be controlled if the multiple uses of the estuary are to be maximized. Table 16.2 lists contaminants and their sources that cause pollution in estuaries.

Some interactions involving these pollutants and the various uses of estuaries require further discussion. Flesh-tainting substances are organic liquids contributed by industrial wastes and oil from shipping. They discolor both finfish and shellfish, making them unsuitable or unappetizing for human consumption. This can be minimized by proper enforcement of sewage treatment specifications and cessation of the practice by ships of discarding oil and other nondesirable liquids just prior to docking.

Pesticides, insecticides, and nutrients used in farming affect both fishing and recreation. Use of these pollutants should be kept at the lowest level possible. Construction, farming, and mining also contribute sediments to estuaries that, on deposition, cause a negative effect on shipping, fishing, and recreation. Suitable construction, mining, and farming practices can minimize the amount of contaminants transported by streams to estuaries.

Heat is supplied by power-generating stations, from both fossil fuel and nuclear power plants. Heat loss into streams should be minimized by cooling ponds with sufficient retention times to preclude the supply of warm water to surface streams. The proper use of cooling towers and mechanical heat exchangers can also lower the temperature of the effluent to an acceptable level.

To a certain extent, several uses of estuaries are potentially at odds with other uses. Mining coexisting with fishing and recreation seems difficult to attain, as is shipping with fishing and recreation. For this reason, it has been proposed that certain areas of a large estuary system be zoned for a specific group of uses, much as a county is zoned according to certain types of buildings and purposes: residential, light industrial, heavy industrial, and farming. This arrangement should minimize the conflicts between the various uses and environmental impacts for estuaries.

Table 16.2   Contaminants and Their Sources in Estuaries.

Major Sources	Contaminants
Municipal wastes	Pathogenic organisms
	Organic matter
	Nutrients
Agriculture	Pesticides
	Herbicides
Industry and shipping	Heavy metals
	Oil
	Flesh-tainting substances
	Toxic chemicals
Power plants	Heat
Agriculture and construction	Sediments

Classification of Coastlines

Coastlines can be classified in a variety of ways depending on the purpose or intended use of the classification (Bird, 2008). Details of origin and/or physical characteristics are generally involved to some extent. In some cases, it is advantageous to apply the classification based on recent emergence from the sea, or submergence, of the landmass. Table 16.3 provides this classification. One of its advantages is the easy identification from a topographic map of the features used for classification. Table 16.4 presents two classifications, one based on lithology and the other on geomorphic processes. Both classifications in Table 16.4 have shortcomings because neither includes all types

Table 16.3   Classification of Coastlines Relative to Ocean Level.

Submergent Coastline
- Drowned shoreline
- Rivers become estuaries
- Irregular coastline
  — Active erosion with sea cliffs
  — Numerous rocky islands

Emergent Coastline
- Terraces
- Few bedrock islands
- Straight coastline with baymouth bars, spits, and large beaches

Compound Coastlines
- Contain features of both submergent and emergent types or have no distinguishing characteristics

Table 16.4   Classifications of Coastlines Based on Lithology and Geomorphic Processes.

Lithology of Shoreline
- Igneous and metamorphic rocks (massive rocks)
- Sedimentary rocks (less massive rocks)
    — Durable sandstones
    — Less resistant: carbonates and weakly cemented clastics

Geomorphic Process of Formation
- Glacial
- Littoral
- Tectonic
- Reef rocks, carbonates

of coasts. They do, however, provide a valuable basis for comparison of coastal features.

Sensitivity of Shoreline Features to Human Activities

When it comes to the sensitivity of shoreline features, environmental regulations and local zoning codes for coastal areas should reflect not only the needs of the community, but also the needs of the feature to maintain a stable environment. As an example, we will consider each portion of the dune system in regard to its proper use (see Figure 16.11). First, the ocean is tolerant of intensive recreation when subject to pollution controls. The beach is tolerant of intensive recreation but no permanent buildings should be constructed there. Both the primary dune and the secondary dune are truly intolerant. No passage, breaching, or building should be allowed. The natural grasses prevent deflation and pathways across the dune destroy this grass. If a passage to the beach is needed, a tunnel should be constructed through the dune or a bridge over it.

The trough area between dunes is relatively tolerant. Limited recreation and construction are possible there. It is an acceptable site for parks and picnic areas with picnic shelters and other day-use facilities. If permanent housing is constructed, it should be widely distributed.

The backdune region is the most tolerant and is suitable for development of housing and other permanent structures. This is also the best location for a highway running parallel to the sea and the dunes. If sufficiently elevated, it will provide a view of the sea and tend to negate the need to travel on the primary dune. If dredging is needed to elevate the roadbed, the sand should be taken from the ocean, not the bay. The beach and ocean area are not very rich biological environments, whereas the bay is extremely rich and varied. These biological colonies should be preserved.

The bayshore is intolerant and no filling of the low-lying areas should be allowed. It is best left untouched. The silts of the bayshore are not suitable for septic tank systems because of their low permeability. Pollution of groundwater is a concern for areas with greater permeability. A sewer system and sewage treatment facility are necessary if any degree of population density develops.

Bay areas are tolerant to intense recreation, much as is the ocean. Fishing, boating, and shelling are favorite pastimes. If the coastal zone is used carefully—respecting the sensitive environments—then the benefits of this attractive area can be enjoyed without the underlying risk of undue degradation of these landforms.

Figure 16.12 on othe following page shows an outline for solving coastal engineering problems. It includes many of the protective structures and details discussed in this chapter.

Engineering Considerations: Shoreline Protection Structures

Coastal processes result in erosion and deposition. Several types of shoreline protection structures are used to prevent or reduce the effects of erosion and deposition of sand along a shoreline. These include groins, jetties, breakwaters, seawalls, bulkheads, and revetments (Thorne, 2002; Dean and Dalrymple, 2004; Brose et al., 2013). Groins and jetties are built perpendicular to the shoreline extending into the breaker zone; breakwaters are built some distance offshore for harbor protection; and seawalls, bulkheads, and revetments are placed parallel to the shoreline separating the land from the water and providing protection against wave attack. Figure 16.13 on the following page illustrates these structures and Figure 16.14 on page 386 shows photographs of various shoreline protection structures. A brief discussion of erosion protection structures follows.

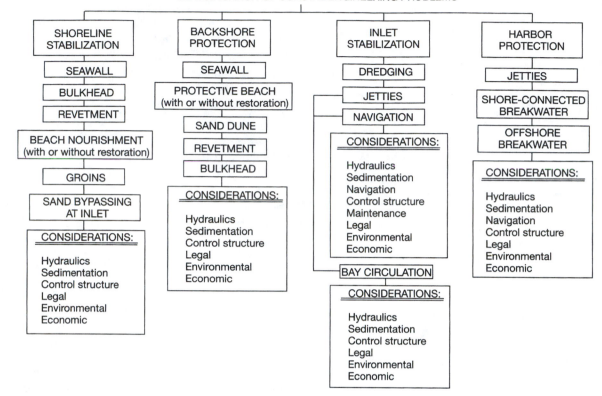

Figure 16.12   General classification of coastal engineering problems.

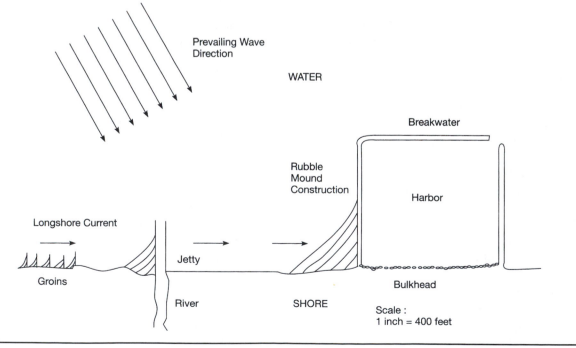

Figure 16.13   Shoreline protection devices.

1. Groins and Jetties

Groins and jetties are structures used to trap or modify the longshore transport of sands. This sand is supplied to a coastal area by two distinct means: (1) rivers transporting sand produced by weathering of quartz-rich rocks from upstream areas and (2) erosion of the upcoast region of the shoreline by longshore currents. In the first situation, sand enters the longshore drift at the mouths of rivers. If dams are constructed on the upstream reaches of such rivers, the fluvial sediments (sands) are trapped behind the dams and the reduced sediment supply encourages erosion downcoast from the river's mouth.

The manner of construction of groins and jetties is quite similar. The distinction exists in their intended use and placement along the shoreline. A groin is used to extend a beach area or to retard erosion by trapping longshore drift. A jetty, by contrast, extends into the water to direct or confine river or tidal flow in a channel and to prevent or reduce shoaling (sediment buildup) of the extended channel by longshore drift.

Groins are generally placed along continuous sections of shoreline, whereas jetties are built to extend river banks into the river mouth, tidal inlet, or harbor entrance. Jetties at the entrance of a channel also protect the channel from wave action and crosscurrents.

Groins and jetties are both constructed perpendicular to the shoreline and vary in length between 30 to 90 m (100 to 300 ft). Beach extension will equal about 50% of the structure's length with the structures spaced at one to three times that length apart. Wider spacing is commonly associated with a continuous, straight shoreline. A limiting factor to groin length (and hence for the additional beach width they provide) is that groins extend only to the breaker zone or to a water depth of about 2 m (6.5 ft), because they are not effective beyond that point. Groins may be classed as high or low, long or short, fixed or adjustable, and permeable or impermeable; they may be constructed of timber, steel, stone, or concrete.

Groins and jetties usually induce erosion in the downdrift region because of their interruption of sediment transport. Sand passing these structures will not reach the shoreline for a distance of three

to five times the structure's length. To alleviate sediment loss, sand can be pumped from the updrift side of the jetty or groin to the sand-starved beach area below, bypassing inlets where present. Studies should be made to determine the effects of a groin or jetty design for the specific beach in question.

2. Breakwaters

Breakwaters are structures used to protect shore areas from wave attack by reducing or eliminating wave action. They cause the waves to break some distance offshore, expending their energy in that location rather than immediately adjacent to the shore. These structures may be shore-connected or offshore breakwaters and may be constructed of a rubble mound or a cellular steel-sheet pile structure. The rubble-mound variety is the stronger of the two structures, is adaptable to almost any water depth, and can be designed to withstand severe wave action. It may contain armour stone rock pieces as heavy as 10 metric tons (10 tons) each. The cellular steel-sheet pile combination structure is used for lower energy conditions. Jetties and groins can be used in combination with various breakwater designs and configurations for shoreline structures.

With the elimination of wave action by breakwaters, longshore transport is greatly reduced and the downdrift beaches are deprived of their normal supply of sediment. In the case of an offshore breakwater, sand is deposited in the quiet area between the structure and the shore. When breakwaters are built updrift of inlets, they impound sand, preventing the shoaling of the inlet channel. Shelter is also provided for dredging operations from which sand is pumped either across the breakwater into the longshore drift or onto the beach.

3. Seawalls, Bulkheads, and Revetments

Generally, seawalls, bulkheads, and revetments are placed parallel to the shoreline to separate the land from the water. Their primary purpose is to provide protection against wave attack. Seawalls are the most massive and they are designed to resist the full force of the waves. Bulkheads are next in capacity and their function is to hold fill in place, but they are not exposed to severe wave action. Bulkheads are used to provide slope stability for the soil behind the structure. Revetments, the lightest of the three

Figure 16.14   Shoreline protection structures: (a) groins to reduce longshore drift (Loma Roberts, Shutterstock), (b) breakwater consisting of armour stone slabs (photo by J. C. Rock, Ltd.), and (c) seawall with revetments (photo by Oikos-team at English Wikipedia),

structures, are designed to protect the shoreline against erosion from currents and light wave action.

Seawalls may be of curved, stepped, curved and stepped, or rubble-mound construction. The curved face and rubble-mound types are used for higher intensity wave action than are stepped structures. An armour of large stone is needed for the curved face structure to reduce the scouring effect of the waves. Sheet piles are used in conjunction with the curved face structure to prevent loss of foundation materials and to reduce uplift and seepage forces.

Bulkheads can be constructed of concrete, steel, or timber. They consist of a thin wall structure embedded into the sediment, with sand backfill behind the wall and a height of water in front. Tie rods and restraining weights at intervals along the wall hold the flexible sections in place.

Revetments are sloping structures placed against the shoreline and may consist of ridged, cast-in-place concrete, flexible articulated concrete, or a riprap structure. In all cases, relief of the hydrostatic uplift pressure generated by wave action is provided by a gravel or crushed stone filter beneath the total length of the structure. Selection of the specific type of structure for shore protection (seawall, bulkhead, or revetment) is based on foundation conditions, exposure to wave action, availability of materials, and cost analysis.

4. **Nonstructural Methods of Beach Protection**

Beach protection, like that for floods along major rivers, can be accomplished by either structural or nonstructural means. Structural means for beach protection have been discussed with regard to jetties, groins, breakwaters, and seawalls. Nonstructural means, by contrast, entail the use of the natural sand for beach and dune control and repair.

Beach nourishment requires the addition of sand at the upcoast end of the longshore drift system in order for it to be deposited downcoast in an eroded section. Dune reconstruction is facilitated by storing extra sand in dune areas so that, following storm wave erosion, the dune ridge can be rebuilt. Building regulations and zoning of the coastal area are other nonstructural methods for beach and dune line protection.

EXERCISES ON COASTAL PROCESSES

MAP READING

For these exercises, the US Geological Survey offers TopoView (https://ngmdb.usgs.gov/topoview/). This software will allow you to view and download topographical maps produced from 1880–2010 for locations throughout the United States. Use a map that has a scale of 1:24,000.

BOOTHBAY, MAINE

1. What type of shoreline is shown, based on the classification of coastlines relative to ocean level? Give two reasons to support your answer.

2. Why are there swamps in the highlands? (A geologic process not included in the shoreline classification is involved.)

3. By latitude and longitude to the nearest minute, locate a tidal flat.

4. List two highway construction or maintenance problems for this area and indicate how they might be solved.

OCEANSIDE, CALIFORNIA

5. What type of shoreline is shown based on the classification of coastlines relative to ocean level? Give two reasons to support your answer.

6. Why doesn't the San Luis River outlet to the ocean? What happens to the water?

7. Why is there no tidal marsh in Canyon de las Encinas?

8. What are the relatively flat areas parallel and adjacent to the seacoast? How did they form?

TAMALPAIS, CALIFORNIA

9. Locate by name or general location:

   a. tidal flats

   b. a baymouth bar

   c. a raised terrace (between what elevations?)

   d. headland

   e. sea stack(s)

   f. a cove

10. How did Lake Lagunitas form?

SANDY HOOK, NEW JERSEY
(USE THE ENGINEERING MAP EDITION)

11. What is Sandy Hook?

12. In view of your answer to Exercise 11, how did Spermaceti Cove form? What caused the interesting "stairstep" pattern in the vicinity of Normandie (SE part of map)?

13. Why is there an offset (indention) of the east coast at the end of the piling in the vicinity of C tower (base of Sandy Hook)? What is this structure called?

14. What direction is the littoral sand drift from the town of Highlands to East Keansburg? On what information did you base your decision?

15. Why is the seawall necessary at Atlantic Highlands (along railroad) when there is abundant sand at Water Witch and Atlantic Beach Park?

16. Is there any apparent reason to control the topography and shorelines of the Atlantic Highlands and Belford areas? Explain.

17. What are the straight dark blue lines in the Belford Marsh area called? What is their function?

18. One centimeter on the map equals how many kilometers on the ground? One inch on the map equals how many miles on the ground? Is this a large-scale or small-scale map? Explain.

OBERLIN, OHIO

19. What are North Ridge, Middle Ridge, and Butternut Ridge? Did they form at the same time or did they form at different times? Defend your answer and, if you decide they formed at different times, what was the order of formation?

20. What kind of material would you expect to find in the ridges mentioned in Exercise 19?

21. Are there good bathing beaches in this region? Why or why not?

WRITTEN QUESTIONS

22. If the tides in a certain coastal area have a period of 8000 seconds, what is the frequency of the wave in wavelets per day? What is its period in hours? At this rate, approximately how many of these long wavelengths occur during the time from high tide to low tide?

23. If an oscillatory wave begins to feel the bottom at a depth of 3 m (10 ft) as it approaches the shore and it has a period of 10 sec, what are its wavelength and

wave velocity? What is the wave height at the wave's peak? At what water depth does the wave break?

24. This question refers to the example on orbital particle velocity of waves discussed at the beginning of the chapter. The example wave had L = 100 m (330 ft), H = 5 m (16 ft), velocity of the wave trace = 45 km/hr (28 mi/hr), and the orbital particle velocity = 7 km/hr (4.4 mi/hr). The text indicates that the circular orbit with a diameter of H is traveled in the time T. This yields for V_p the particle velocity:

$$V_p = \frac{\pi H}{T}$$

   a. Calculate T from the equation $V = L/T$.

   b. Calculate V_p from the equation $V_p = \pi H/T$. Check this with the value given in the problem.

   c. For a second example wave, make a similar calculation using L = 380 m (1247 ft), H = 10 m (33 ft), V = 90 km/hr (56 mi/hr), and V_p = 7.4 km/hr (4.6 mi/hr).

25. The following equation relates wave velocity to water depth and wavelength:

$$V^2 = \frac{gL}{2\pi} \tanh \frac{2\pi d}{L}$$

   where g, the acceleration of gravity, is 9.81 m/sec^2. For shallow water, or $d = L/20$, it reduces to $V^2 = gd$ and for deep water, $d = L/2$, it reduces to $V^2 = gL/2$.

   a. If d = 1 m (3.3 ft) and L = 30 m (98 ft), calculate V using the full equation and the reduced equation. Compare answers.

   b. If d = 20 m (66 ft) and L = 30 m (98 ft), calculate V using the full equation and the reduced equation. Compare answers.

26. Why are tides on the Earth, which, theoretically, should vary only about 50 cm (20 in), much more complicated than this? How is this indicated? What factors on Earth affect the tidal variation?

27. What complications related to the tides would be involved if a sea level canal were built in the general vicinity of the Panama Canal? Would the two levels eventually equalize? Explain.

28. a. Calculate the relative attractive force between the Sun and Earth and between the Earth and the Moon. Use the equation:

$$F = G\frac{m_1 m_2}{d^2}$$

where G is Newton's gravitational constant = 6.67 × 10^{-8} in cgs units. For the Earth–Moon system use d equal to 320,000 km (200,000 mi) and for the Sun–Earth system 150 × 10^6 km (94 × 10^6 mi). For mass use mass of the Earth, M_E = 1, mass of the Moon, M_M = 1/82 M_E, and mass of the Sun, M_S = 3.32 × 10^5 M_E. Is the value of G actually needed to calculate the relative attractive forces? Explain.

   b. What is the difference in the force for each if 12,800 km (8,000 mi) (Earth's diameter) is added to the distance between the celestial bodies in question? Explain.

29. For many years, tidal power generation was considered for the Bay of Fundy and eventually a small tidal power generating facility was constructed. Severe winter conditions made construction slow and difficult. A large capacity facility is now operational. What is the latitude for the Bay of Fundy? How does it compare to the northern coast of France? Explain the difference in the weather conditions for the two.

30. Give the names of the estuaries associated with New York City, Boston, Washington, DC, and Norfolk, respectively. What is the relationship between estuaries, harbors, and tide differential? Explain. What type of mining is likely to be associated with estuary areas? Explain.

31. In a beach area, the groins are spaced 60 m (197 ft) apart. How much beach widening is possible if the groin extends outward 45 m (148 ft) but goes 8 m (26 ft) beyond the breakers at low tide?

32. Erosion occurs beyond the downbeach side of jetties. If a jetty extends 150 m (492 ft) into the water, for what distance would beach erosion likely occur down beach? How can this be alleviated or minimized?

33. Breakwaters are constructed of rubble mounds with sizable pieces of armour stone for outside protection. What are the dimensions of a 10 metric ton (11 ton) cube of granite? Why is it difficult to find proper armour stone today for Lake Michigan's breakwaters? Explain.

34. The terms *structure*, *process*, and *stage* are used relative to the development of landscapes or physiographic provinces. Indicate how this relates to the development of shoreline equilibrium discussed in the text.

35. Refer to the classification of coastlines relative to ocean level (see Table 16.3). Is it likely that some certain combinations of the first and second groups could occur in the same areas, such as active sea cliffs

and sea terraces? What about irregular coastlines and numerous spits? Discuss in detail.

36. Sometimes breakwaters are blamed for causing beach erosion in areas adjacent to their location. What effect would the breakwater shown in Exercise Figure 16.1 have on downbeach erosion? Explain. What distances might be involved?

37. What is meant by beach protection using nonstructural methods? What is beach nourishment?

38. Why is breaching of the primary dune such a poor practice relative to beach preservation? How can traffic to the beach be accomplished otherwise?

39. Why must filling of the bayshore area be prevented? What is the origin of the compressible silts of the bayshore area?

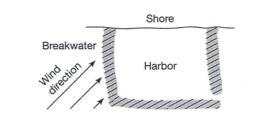

Exercise Figure 16.1

40. In the erosion of Lake Michigan's shoreline in the mid-1970s, some of the affected areas consisted of glacial till with interbedded outwash sands. Why would this sequence be particularly prone to failures by undercutting and landslides? Explain.

References

Bird, E. C. F. 2008. *Coastal Geomorphology: An Introduction* (2nd ed.). New York: John Wiley & Sons.

Brose, B., Clark, G., and Cox, J. 2013. *Great Lakes Coastal Shore Protection Structures and Their Effects on Coastal Processes*. Superior: University of Wisconsin Sea Grant Institute.

Bush, D. M., and Young, R. 2009. Coastal features and processes. In R. Young and L. Norby (eds.), *Geological Monitoring* (pp. 47–67). Boulder, CO: Geological Society of America.

Dean, R. G., and Dalrymple, R. A. 2004, *Coastal Processes with Engineering Applications*. New York: Cambridge University Press.

Komar, P. D. 1983. *Handbook of Coastal Processes and Erosion*. Boca Raton, FL: CRC Press.

Marshak, S. 2013. *Essentials of Geology*. New York: W. W. Norton.

Masselink, G., Hughes, M., and Knight, J. 2011. *Introduction to Coastal Processes and Geomorphology* (2nd ed.). Florence, KY: Routledge.

Nance, D., and Murphy, B. 2016. *Physical Geology Today*. New York: Oxford University Press.

Thorne, C. R. 2002. *River, Coastal and Shoreline Protection: Erosion Control Using Riprap and Armourstone*. New York: John Wiley & Sons.

ARID ENVIRONMENTS AND WIND

17

Chapter Outline

Arid Climates

Geologic Processes in Arid Regions

Wind

Engineering Considerations of Arid Environments and Wind

Arid Climates

The climate in a region has a significant influence on the geologic processes that prevail and on the appearance of the resulting landscape (Hill, 2002; Cooke et al., 2006; Goudie, 2013). The dominance of chemical over mechanical weathering is a consequence of abundant, unfrozen water, whereas the nature of streams or the presence of glaciers is a direct result of rainfall and temperature, that is, of climate.

In arid regions, the action of streams is entirely different from that of humid areas. Yet the effects of running water are more profound in arid areas because of the scarcity of vegetation, greater erodibility of the soil, and localized heavy rainfall (French and Miller, 2011).

In arid areas, wind is also an important geologic agent because of its contribution to erosion and deposition (Fernandez-Bernal and De La Rosa, 2009). Wind was once thought to be the dominant agent in shaping desert landforms but, today, only secondary details of landscape are attributed to its role. Nevertheless, the effects of wind are highly visible in arid environments.

Deserts, or arid lands, in low to temperate latitudes are typified by low rainfall, high temperatures (at least seasonally), and a rather higher level of evaporation than of precipitation (Parsons and Abrahams, 2009). Deserts receive less than 25 cm (10 in) of rainfall per year and can experience daytime temperatures well in excess of 38°C (100°F) during the summer months. Death Valley, in the Basin and Range Province of southeastern California, has the record in the United States for these climatic extremes with an average of less than 5 cm (2 in) of rainfall per year and a maximum recorded temperature in excess of 44°C (112°F). Features known as cool deserts are also found throughout the world. They occur at high latitudes and/or high elevations and, typically, have low rainfall but much lower temperatures than those occurring in low to temperate latitudes. The Atacama Desert in Chile and the Gobi Desert in northern China and southern Mongolia are examples of cool deserts.

Most deserts also experience frequent high-velocity winds due to the convection that occurs as the heated air rises and other air moves in as a replacement. Convection also provides much of the precipitation for deserts. As the rising air cools, it becomes saturated and delivers its moisture as localized, possibly heavy, rainfall (Fernandez-Bernal and De La Rosa, 2009). Such cloudbursts are a regular event in desert areas and the geologist performing field work in arid and semiarid regions is advised to watch for the buildup of such storms. Dry, scorched, drainage basins can become a torrent of runoff on short notice.

Vegetation is scant, at best, in arid regions but the actual amount is a direct function of the available rainfall (Parsons and Abrahams, 2009). Sagebrush will grow in the semi-desert areas of the United States, as will juniper trees, but not when the rainfall is below 25 cm (10 in) per year. Cactus, Joshua trees, yucca, and small hardy bushes may be the only vegetation in these drier regions. The low bushes grow some distance apart, leaving extensive, bare areas between them. This setting is conducive to erosion by both wind and running water. Figure 17.1 depicts desert vegetation and topographic features in the Mojave Desert of California.

Geologic Processes in Arid Regions

Geologic processes operative in arid regions produce landform features that are different from those in humid areas. Table 17.1 presents the landform features characteristic of arid regions. In arid climates, the unconsolidated material (regolith) is thin and coarse-textured and is primarily a product of mechanical weathering (Goudie, 2013). Slopes are generally steeper than in humid regions and, as discussed in Chapter 4 on rock weathering, sandstones and limestones form cliffs. Rock units tend to break along joints, yielding rugged cliffs, known as buttes, and mesas in flat-lying sedimentary rock terrains.

Figure 17.1 Desert vegetation and topographic features, Mojave Desert, California. A Joshua tree is in the foreground. (S. Borisov, Shutterstock.)

Table 17.1   Landform Features of Arid Regions.

Primal Agent	Depositional Features	Erosional Features
Running water	Alluvial fan	Mesa
	Bajada	Butte
	Playa	Box canyon
		Pediment
Wind	Dunes	Deflation basin
	Loess	Wind caves and arches
	Tephra deposits	Deflation armor
		Ventifacts
		Desert varnish

In humid areas, the regolith is relatively fine textured, formed mostly by chemical weathering, and transported downslope by creep. It is usually covered by vegetation and sandstones are the only prominent cliff-formers in sedimentary terrains. Figure 17.2 illustrates the contrast between slope development in humid and arid regions.

Deserts are characterized by well-entrenched stream channels that are dry much of the time. The water that flows in them shortly after heavy rainfall soon disappears by infiltration or evaporation and is unable to reach the ocean (Stephenson, 2004). This condition is known as internal drainage. The Colorado River is an exception to this situation because the discharge from its source in the mountains is so great that the river continues to flow despite the losses under desert and semi-desert conditions. Groundwater contributes to the discharge in the Colorado River in some areas as well.

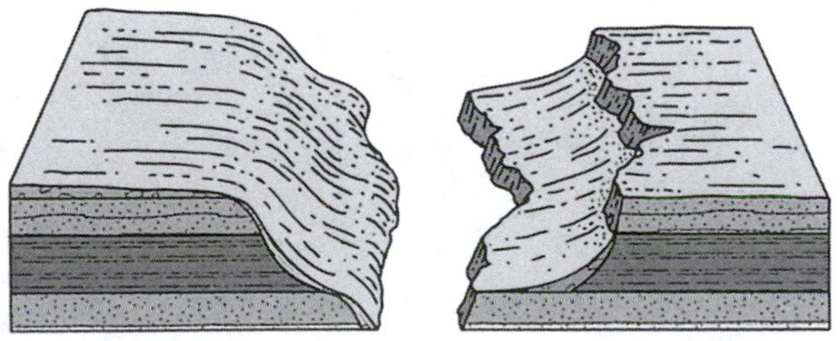

Figure 17.2 Effects of climate on slope weathering: (a) humid climate and (b) arid climate. Rock sequence for both diagrams, top to bottom, consists of sandstone, shale, and sandstone.

Massive erosion of alluvial floodplains occurs when heavy rains result in flash floods (French and Miller, 2011). Banks may be undercut and slopes may cave in to form steep-sided channels, called box canyons. Erosional and depositional effects may cause considerable damage to alluvial areas; bridge abutments and center span footings may be undermined while transverse roads are buried by debris. A flash flood east of San Diego, California, in 1978 covered a major highway in the Imperial Valley with 1.5 m (5 ft) of deposits, requiring the construction of a new section of road at a higher elevation.

Clay-rich units, with their attendant low permeability, provide a terrain where little infiltration occurs and, therefore, runoff is abundant. In the absence of vegetation in arid climates, very closely dissected topography develops, yielding landforms referred to as badlands. The Painted Desert of Arizona and the Badlands of South Dakota are well-known examples of this feature. Figure 17.3 on the following page shows the South Dakota Badlands.

A somewhat similar landscape has developed in the old strip mine areas of the Midwest where unreclaimed coal mine refuse is closely dissected. Vegetation is curtailed by the high acidity of the soil, caused by the weathering of pyrite from shale units. These strip mines predate the enactment of state reclamation laws, which took effect in the mid-1950s or early 1960s. Figure 17.4 on the next page shows an abandoned mine lands area in Indiana.

In the deserts of the southwestern United States, basins formed by block faulting are prevalent. This is the Basin and Range Province described in Chapter 20. Streams flowing from the highlands (ranges) seldom persist until they reach the center of the adjacent basin. After high-intensity storms, however, water may accumulate in the basin to form a shallow lake that persists for a few days or weeks. These are known as playa lakes; the lake bed, when dry, is called a *playa* (Stephenson, 2004). Figure 17.5 (page 395) shows a playa located in Nevada.

The Great Salt Lake in Utah is a playa lake today, although the basin itself has a complex history of block faulting and subsequent glacial deposition. Because of internal drainage, dissolved salts are not carried away to the sea but are concentrated in the playa lakes instead. On evaporation, salts accumulate on the surface of the playa or, if the water infiltrates, brackish water is carried downward into the subsurface.

Figure 17.3   South Dakota Badlands. (Phil Lowe, Shutterstock.)

The nature of the salts deposited is a function of the composition of rocks in the adjacent ranges. Deposits may be rich in sodium, chloride, alkalis (sodium and potassium carbonates), bittern salts (sulfates), or borates (borax and related minerals). The rocks of the Basin and Range Province have a complex history of igneous and metamorphic origin so that all of the ions for the salts just mentioned are available in one area or another.

A prominent depositional feature in a desert basin is the alluvial fan, a triangle-shaped deposit of debris that forms near the change in slope between

Figure 17.4   Erosion developed on abandoned mine lands area in Indiana. (Indiana Department of Natural Resources, Division of Reclamation.)

Figure 17.5　Playa of the Black Rock Desert, Nevada. (Robert Stolting, Shutterstock.)

the sloping basin and the highland. As erosion proceeds, the highlands recede and additional alluvial fans form. The nearly flat area, downslope from an active alluvial fan consisting of coalescing fans, is called a *bajada* (Goudie, 2013).

Another erosional feature of desert terrain is a sloping rock surface at the base of the highland.

Cutting across bedrock, this landform, a *pediment*, generally has a concave up shape, much like the long profile of a stream (Goudie, 2013). At the downslope extreme of the pediment, alluvium covers the rock surface. Figure 17.6 depicts the various landforms discussed here.

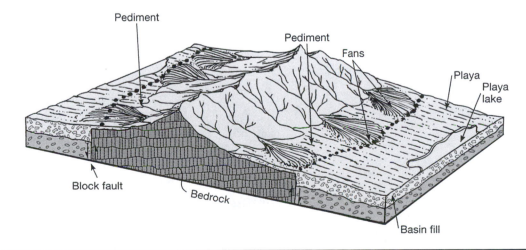

Figure 17.6　Landform features in arid climates; block faulted highlands and adjacent basin.

Wind

Action of the Wind

Winds in arid regions cause both erosional and depositional features (Fernandez-Bernal and De La Rosa, 2009) (see Table 17.1). While the effects of running water are most significant in shaping desert topography, wind plays an important secondary role. Researchers working in deserts have observed that two layers of particles are transported in the air when strong winds persist (Fernandez-Bernal and De La Rosa, 2009). The lower layer contains sand grains and extends about 10 to 45 cm (4 to 18 in) above the ground. The upper layer carries silt and clay particles and it can extend to significant heights—100 m (330 ft) or higher. The two layers are equivalent to the bedload and suspended load of a stream.

Sand grains move by saltation, in a series of long jumps. These grains usually rise only about 10 cm (4 in) off the ground, as indicated by wind tunnel experiments. Observations in deserts show that 45 cm (18 in) is about the highest level to which telephone poles are sandblasted. Diameters of sand grains transported by the wind as bedload range from 0.15 to 0.3 mm, a medium-size sand. Air velocities of 5 m/sec (16 ft/sec) or 18 km/hr (11 mi/hr) are needed to lift sand grains from the ground.

Most sediment finer than sand remains at low altitudes. Turbulent eddies in air (as in water) move in all directions. On average, the velocities of upward eddies are about one-fifth the forward velocity. Therefore, a wind at 5 m/sec (16 ft/sec), which is required to lift sand grains from the ground, would keep silt and finer sizes in suspension.

Gravity pulls these particles downward at velocities greater than particles in water because of water's greater density and viscosity. Figure 17.7 shows a comparison of water versus air as the transporting medium. This indicates that a coarse silt particle has a settling velocity of 1 m/sec (3.3 ft/sec) in air and 0.250 cm/sec (0.1 in) in water. An upward velocity of 1 m/sec (3.3 ft/sec), therefore, is necessary to maintain this coarse silt particle in suspension.

A dead air layer persists immediately above the ground. Its thickness is about one-thirtieth that of an obstruction, and the most common obstruction on a soil surface is a sand grain about 1 mm in diameter. This yields a dead air layer about 0.033 mm thick, which is equivalent to coarse silt size. Silt and finer particles are protected by this dead air layer and will not be picked up by the wind until saltating sand grains, or some other disturbance, breaks the smooth surface. A dry gravel road in the country serves as an example. The wind will generate little or no dust from the road until a car comes by; this breaks the smooth surface and allows the silt particles to become airborne. Wetting the road will reduce dust from traffic by increasing the cohesion of the soil which, in turn, resists the uplifting force of the wind. This practice is common on construction sites where water trucks wet down the roads to control dust generation.

Most silt- and clay-sized particles settle within several tens of kilometers (tens of miles) from the source. Loess

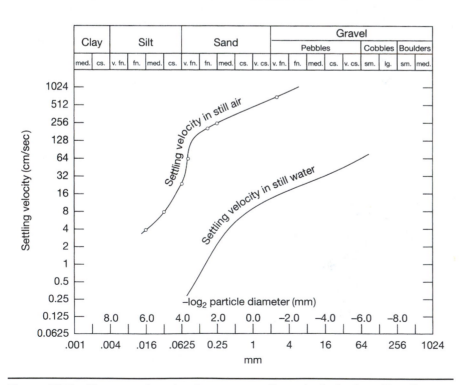

Figure 17.7   Diameter of sand grains versus settling velocity in air and in still water.

or wind-deposited silt forms in this way. Both average grain size and thickness of the deposit decrease in the downwind direction. This is discussed in more detail in a later section.

In some instances, fine particles are carried long distances by travel through the stratosphere. Updrafts in the atmosphere lift the material and downdrafts, related to major air circulation, deposit the particles a great distance away. Falls of tephra from distant volcanoes are of this nature and the average diameter of particles carried does not necessarily decrease downwind.

Eruption of the Hekla Volcano in Iceland in 1947 provides an example. Some tephra was raised 9000 m (29,500 ft) into the atmosphere and traveled northeastward to Finland some 3000 km (1875 mi) away. The 1980 eruption of Mount Saint Helens, in the Cascade Range of Washington, deposited tephra in five adjoining states. Dust accumulations in the stratosphere were noted as far east as the Ohio River.

Wind Erosion

The wind erodes surface materials in two major ways: by deflation and by abrasion (Fernandez-Bernal and De La Rosa, 2009). Deflation is the removal of fine, loose particles by the wind and accounts for most of the volume of eroded material. Abrasion is the natural sand-blasting effect of the wind (rubbing action involving friction), which cuts and polishes bedrock surfaces and pieces of rock pebble size and larger.

Deflation can completely lower the surface or can work in localized areas to yield deflation basins. In arid regions, when vegetation is absent, which occurs during periods of extended drought, large-scale deflation will develop. Nearly a meter (several feet) of the land surface may be removed over a period of a few years under such conditions.

Deflation basins occur in semi-arid regions of the United States and Canada, such as the Great Plains. They are typically less than 1.5 km (1 mi) long and have depths of a meter or less (a few feet). In wet years they are covered with grass and may contain shallow lakes. In the past, such features were known as buffalo wallows. Some, in western Nebraska, have been mined for potash or borax.

Deflation of a soil mass, consisting of sand, silt, and pebbles, will eventually form a protective coating, which limits further deflation. Figure 17.8 shows the stages of development. The sand and silt are car-

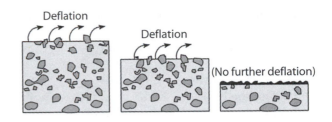

Figure 17.8 Three stages in the development of deflation armour, known also as desert pavement or lag deposit.

ried away by the wind and the pebbles remain. When a surface consists of pebbles fitted together tightly, the surface will stabilize. This feature is referred to by several terms: *deflation armour*, *desert pavement*, or *lag deposit*. It is defined as a surface of coarse particles primarily developed by deflation. Figure 17.9 shows a photograph of a desert pavement.

Abrasion in desert areas is concentrated on rock at the surface that is not removed by deflation. A *ventifact* is a rock fragment cut and faceted by wind action. Ventifacts are characterized by polished, often pitted or fluted surfaces, with facets separated from each other by sharp edges. Most of the cutting and polishing occurs by saltating sand grains. Studies have shown that the facet being created (abraded) faces the wind (Parsons and Abrahams, 2009; Goudie, 2013). If the rock is rotated by undermining or shifts

Figure 17.9 A desert pavement or lag deposit. (Richard Trible, Shutterstock.)

for some other reason, two or three facets can be cut on the rock. Sometimes, the German terms *eink-anter*, *zweikanter*, and *dreikanter* are applied to these rocks; *kanter* means "edge" or "face." Figure 17.10 shows the development of ventifacts.

Fine-grained massive rocks that undergo abrasion by the wind also take on a dark brown to purple color known as desert varnish. This coating is thought to consist of iron oxide or manganese oxide imparted by the sand-blasting effects. Desert varnish tends to reduce the contrasting appearance between different rock types in outcrop in the desert. Care must be taken to break a fresh surface on a rock cliff or hand sample before examining the lithologies involved.

Wind Deposits

Wind-deposited landforms are of two basic types: dunes of sand formed close to the source and blankets of silt (loess) or clay formed much farther away (Livingstone et al., 2007).

Sand Dunes

A dune is a mound of sand piled up by the wind. Generally it is initiated by an obstacle that disrupts the flow of air. Sand dunes are not symmetrical in shape but have a steep lee slope with a gentler slope on the windward side. Sand grains roll or push up the windward slope and drop over the crest of the dune to the lee slope, coming to rest at their own angle of repose. For loose sand, this repose angle is about 34°. As more sand is pushed over the crest, the sand slides or slips downward, maintaining the angle of repose. Hence, the lee slope of a dune is sometimes referred to as the slip face. The angle for the windward slope is determined primarily by the wind velocity and it ranges from about 5 to 15°. Figure 17.11 shows windward and lee slopes of a sand dune.

Sand dunes may reach heights of 30 to 90 m (100 to 300 ft) and, in some extreme cases, they may be up to 180 m (600 ft) high. Movement of sand from the windward to the lee side of a nonvegetated dune can cause this landform to migrate slowly downwind. Rates of migration of desert dunes may be as high as 10 to 20 m/yr (35 to 70 ft/yr). Adjacent to sandy beaches associated with the oceans or the Great Lakes, migration of dunes can bury houses, cover roads, and, generally, disrupt inhabited areas. Encroachment is best curtailed by establishing vegetation that can

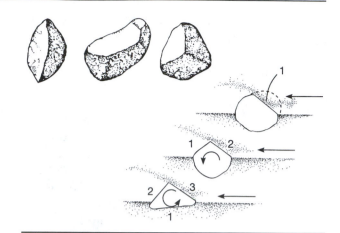

Figure 17.10   Ventifacts, pebbles shaped and polished by the wind.

thrive in the dry sandy soil. Beach grasses also prevent deflation by holding the sand grains in place.

The stratification of sand in dunes becomes fairly complicated as migration occurs. Originally, the cross-strata, formed on the lee side, yield a series of beds truncated by the windward face of the dune (Figure 17.12). With shifting wind direction and velocity, dune migration changes direction. Deposition may replace erosion, causing a truncation of one series of cross beds and establishment of another (Figure 17.12). In any event, the steep beds indicate the lee side of the dune during the migration of a specific episode, with the wind coming from the direction opposite the dip direction of the steep beds.

The shapes of dunes or dune fields are quite diverse relative to both geometric form and orientation with regard to the wind. Availability of sand, wind velocity, constancy of wind direction, general terrain, and vegetative cover are factors that determine the nature of the dunes (Livingstone et al., 2007). Sand dunes can develop in a variety of geologic settings that have a supply of sandy materials at the surface. Beaches, coastal plains, floodplains, outwash

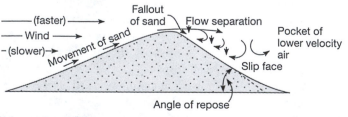

Figure 17.11   The windward and lee slopes of a bare sand dune.

plains, and extensive outcrops of sandstone can provide the sand needed to generate dunes.

Most sand dunes consist of quartz grains, primarily because of the abundance and weathering resistance of that mineral. These grains may have experienced several cycles of lithification and erosion before coming to rest in a present-day dune. In certain areas of the world, other minerals are as abundant as sand-sized pieces that accumulate into dune forms.

On the island of Bermuda in the Atlantic Ocean, the dunes consist of calcite grains derived from the limestone that forms the island. In New Mexico, near Alamogordo, at White Sands National Monument, the area is covered by dunes consisting of gypsum grains. These derive from the rock gypsum exposed at the surface in this arid environment.

It is usually possible to distinguish sandstones of water-deposited origin from those formed by lithification of wind-deposited dunes. The eolian (wind-derived) deposits are larger and have more extensive cross bedding. They also have frosted, more rounded grains than water-deposited sands. Water cushions the grains, making them less subject to rounding by abrasion. Frosting results from the mutual impact of sand grains during saltation by the wind. Complications can occur in such analysis, however, because some water-deposited sands derive from wind-deposited sandstones. In this case, the frosted surface of grains may persist in the second sandstone. However, not all of the grains would be frosted because, undoubtedly, some of the sand would be derived from other sources.

Loess Deposits

Loess is wind-deposited silt, usually accompanied by some clay and fine sand (Liu, 1988; Muhs, 2007). Typically it is not stratified because of the small range of grain sizes deposited together and the contribution of plant roots and worms, which work on the sediment during and after its deposition.

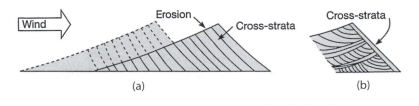

Figure 17.12 Cross section view of migrating sand dune: (a) simple parallel layers and (b) complex cross-strata caused by variations in wind direction.

Loess will usually stand in a vertical fashion when exposed. This is thought to be because the fine particles present, the silt grains, are surrounded by a clay surface coating that provides cohesiveness to the material. A loess cut near Vicksburg, Mississippi, is shown in Figure 17.13. Yet porosity and permeability are high because of the angular nature of the silt and the open network of pores between them. For this reason, loess holds water, which is available to plants, thereby yielding a productive agricultural soil.

Loess is typically yellow in color and consists mostly of quartz, feldspar, mica, and calcite (Muhs, 2007). The color is caused by slight oxidation of iron present in minor amounts. It closely resembles the fine product known as *rock flour*, which is formed by glacial action. In the United States, loess is found

Figure 17.13 Loess cut near Vicksburg, Mississippi, showing vertical excavation. (Mark A. Wilson, Department of Geology, College of Wooster.)

immediately leeward of major glacial outwash and valley train deposits. Loess was deposited, in glacial times, in areas just beyond the glacial ice. At this location, the cold, windy climate and denuded valleys choked with glacial outwash provided the source areas for the silt. Deflation in the valleys moved sand grains and provided silt particles for the suspended load of the wind. The silt settled, forming thick blankets up to 30 m (100 ft) thick on the leeward side (eastern side mostly) of major stream valleys and trailing out to thicknesses of 1.5 m (5 ft) or less within about 80 km (50 mi).

The silt came to rest within the zone and was not transmitted further downwind. This is probably because the fine size was not as easily picked up again in the absence of sand grains to initiate its suspension and because obstructions protected it from the wind. Grasslands and forested areas were likely present on the upland areas where the loess came to rest.

Five major areas of loess deposits exist in the United States: (1) the lower Mississippi River section of the Gulf Embayment, (2) the central area of the United States in the Interior Lowlands, including parts of eastern Nebraska and Kansas, Iowa, South Dakota, Missouri, Minnesota, Wisconsin, Illinois, and Indiana, (3) the plains of eastern Colorado and western Nebraska and Kansas, (4) the Snake River Plains of southern Idaho, and (5) the Palouse area of eastern Washington and north-central Oregon. These areas are shown in Figure 17.14. Although these deposits are not the product of an arid environment, they are discussed in detail in this chapter because of their deposition by wind.

Loess in the lower Mississippi River section extends along the east bank of the river floodplain from New Orleans to the Ohio River. The silt deposit is thickest near the river, 30 m (100 ft), and gradually thins within 80 km (50 mi) or so. The northern

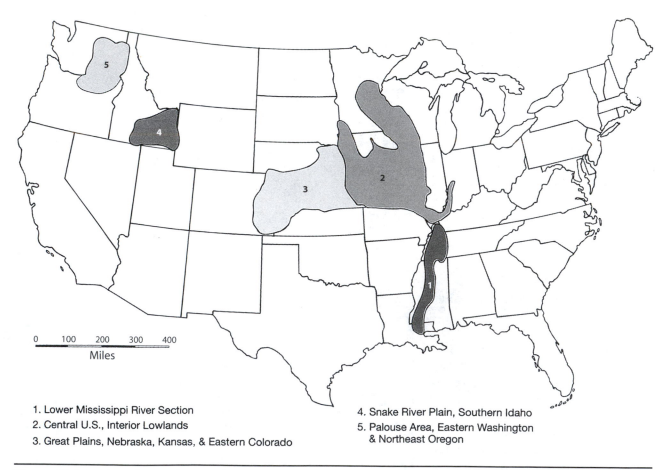

0   100   200   300   400
Miles

1. Lower Mississippi River Section
2. Central U.S., Interior Lowlands
3. Great Plains, Nebraska, Kansas, & Eastern Colorado
4. Snake River Plain, Southern Idaho
5. Palouse Area, Eastern Washington & Northeast Oregon

Figure 17.14   Map of United States showing the five major areas of loess deposition.

portion of the deposit is shown in Figure 17.15 and depicts the area of confluence of the Missouri, Mississippi, and Ohio Rivers. Shown also is Crowley's Ridge of southern Missouri and northern Arkansas, which is capped with loess and lies west of the river. Floodplain sediments lie east of the ridge, which were likely the origin of that loess deposit. Rainfall in this region is about 127 cm/yr (50 in/yr).

Loess in the Interior Lowlands is generally underlain by glacial drift which, in many cases, is Illinoian or Kansan in age. A portion of the eastern segment of this loess deposit is shown in Figure 17.15. On that drawing, loess blankets the eastern part of Illinois, covering Wisconsinan drift on the north and Illinoian drift on the south. The deposit is thickest on the uplands, east of the Mississippi River, and the Illinois River, another glacial sluiceway, contributes more loess on its leeward side. In Indiana, not included on the map, loess is found on the east side of the Wabash River. It is more prevalent in the southern half of the state between Terre Haute and Evansville. Rainfall in this extensive area ranges from 76 cm/yr (30 in/yr) in eastern Kansas to 102 cm/yr (40 in/yr) at its eastern boundary in Indiana.

The loess of the plains lies on a large, treeless region sloping gently eastward from the Rocky Mountains. In eastern Colorado and western Kansas and Nebraska, the loess overlies Rocky Mountain outwash materials. In central Kansas, it overlies chalky limestones and shales, whereas in eastern Nebraska and Kansas, the underlying material is glacial drift. Because of erosion, this glacial drift is frequently exposed in small isolated areas, particularly along valley slopes. The rainfall ranges from about 51 cm/yr (20 in/yr) in the western portion to nearly 76 cm/yr (30 in/yr) in the eastern part.

In the Snake River Plain of southern Idaho, the loess deposits cover an extensive area of about 51,000 km² (20,000 mi²) and lie at an altitude between 1200 to 1800 m (4000 and 6000 ft). Rainfall is 25 cm/yr (10 in/yr) or less and natural vegetation is primarily desert shrub and sagebrush. Irrigation, from the Snake River or from groundwater supplies for agricultural purposes, is prevalent in the area. The underlying bedrock is Columbia River Basalt, which extends through the loess in some areas, yielding a terrain formed on basalt flows known as *scablands*.

The Palouse area of eastern Washington and north-central Oregon is a loess-covered upland on

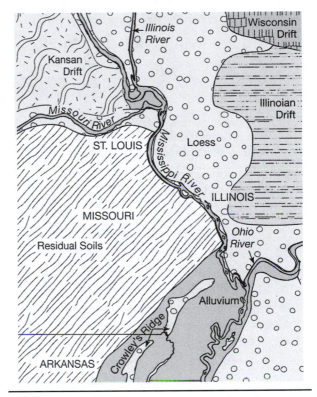

Figure 17.15 Surficial geology, area of confluence, the Missouri, Mississippi, and Ohio Rivers.

the Columbia Plateau with an average elevation of about 600 m (2000 ft). Annual rainfall is 51 cm (20 in) or less and native vegetation is primarily bunch grass. The silt can be 23 m (75 ft) thick or more but basalt exposures can occur abruptly in the otherwise deep loess areas. The Palouse loess is thought to be derived partly by wind from residual soil of the basalt and partly from volcanic ash. This latter contribution seems quite likely in light of the 1980 eruption of Mount Saint Helens, which supplied considerable tephra to the adjacent cities and states of the Northwest. The Palouse loess area is used extensively for farming. In eastern Washington, some areas can supply wheat with a minimum of irrigation. In other areas, irrigation is required for various grain crops.

In some regions of the world, loess is found in large areas that lie in the leeward direction from deserts. Mechanical weathering obviously provides the silty material that accumulates as alluvium on the desert floor adjacent to the mountains. The loess of western China, derived from the great desert basins of central Asia, is up to 70 m (230 ft) thick (Liu, 1988).

Engineering Considerations of Arid Environments and Wind

Engineering and environmental problems encountered in the arid regions, and the associated effects of winds, are as follows:

1. Scarcity of both surface water and groundwater. Because of the scarcity of water it is difficult to maintain growth of trees and bushes in residential and commercial areas.

2. High temperatures are detrimental to asphalt pavements.

3. Flash flooding can cause extensive channel erosion, property damage, and loss of life.

4. Roads can be covered with thick accumulations of sand, posing hazardous driving conditions.

5. In situ dynamic compaction may be required in areas of loose sand before construction of an engineering structure.

6. Loess, being uniform in texture (with 50 to 90% in the silt-size fraction), requires care during compaction because silts have a narrow range of moisture for maximum density and because they dry readily.

7. Loess is subject to cracking in earth dams if compacted on the dry side of the optimum moisture content.

8. Slopes in loess tend to be stable when cut vertically, provided that the loess will remain relatively dry, i.e., well below saturation. When saturated, however, or when the phreatic zone is intercepted in a deep loess cut, the back slope (cut slope) must be reduced markedly to a 2:1 (2 vertical to 1 horizontal) or 3:1. Trees with deep roots and vegetation supplying thick ground cover will aid in drying out loess from a saturated condition.

EXERCISES ON ARID ENVIRONMENTS AND WIND

MAP READING

For these exercises, the US Geological Survey offers TopoView (https://ngmdb.usgs.gov/topoview/). This software will allow you to view and download topographical maps produced from 1880–2010 for locations throughout the United States. Use a map that has a scale of 1:24,000.

FURNACE CREEK, CALIFORNIA

This area represents a youthful stage in the cycle of erosion in an arid region. It lies within the Basin and Range Province, characterized by a series of north–south trending fault-block mountains and valleys.

1. Draw an east–west profile across the map area. Note the different topography of the mountains, the valley sides, and the valley bottom. Sketch in the surface area that is probably bedrock.

2. What is the typical gradient of a stream in the mountains along Trail Canyon, where Salt Creek is in the center of the valley?

3. What do the arc-shaped contour lines adjacent to the mountains outline?

4. Why do some streams end abruptly?

5. What do the numerous nicks on the surface of the alluvial fan contours represent? What does this suggest about local relief?

6. What is the base level of the map area?

7. In what direction is Salt Creek flowing?

8. Drainage in this quadrangle is described as internal. Why?

9. How will the valley shape change with time?

10. Explain the green strips radiating from Furnace Creek Ranch.

11. List engineering problems that might be associated with the area.

ANTELOPE PEAK, ARIZONA

This area represents the erosion cycle in an arid climate. It is also located in the Basin and Range Province.

12. Sketch a profile from Antelope Peak to the northeast corner of the map. Sketch the bedrock surface.

13. Contrast topographic features on this map with those on the Furnace Creek Quadrangle.

14. What do the large, gently sloping areas on this map represent?

15. Why are alluvial fans on this map not as prominent as on the Furnace Creek Quadrangle?

16. What is the dominant agent of erosion in this area? Cite evidence.

Written Questions

17. The Colorado Plateau Province in the southwestern United States consists primarily of flat-lying sedimentary rocks of the Paleozoic age. Which of the sedimentary rock types would likely be cliff-formers and which would be slope-formers in this arid environment? What is the significance, in this regard, of the formation named Redwall Limestone, which crops out in the Grand Canyon, Arizona?

18. The Appalachian Plateau of West Virginia consists primarily of flat-lying sedimentary rocks. Which of these would be cliff-formers and which slope-formers in this humid area?

19. How does the vegetative cover differ between the Colorado Plateau and the Appalachian Plateau area considered in Exercises 17 and 18? In what way does this affect the rock exposures? Explain.

20. Why is runoff and stream erosion in a desert region credited with a greater extent of erosion than is wind action, despite the infrequent rainfall that occurs in such regions? What is the effect of vegetation in this regard?

21. Outcrops of coarse-grained granite in the desert because of mechanical disintegration do not develop desert varnish to the extent that basalts or fine gneisses would. Why is this likely to be the case? Explain.

22. The settling velocity of sand in air is given as 1 m/sec (3.3 ft/sec). For how long would the sand be airborne if it fell from a height of 45 cm (18 in)? This would require that wind gusts of what velocity persist for that period of time to sandblast posts in the desert up to 45 cm (18 in) high?

23. In the past, road oil has been placed on gravel roads to prevent dust from billowing as automobiles pass. Why is such a practice effective in reducing dust? Explain.

24. What is the origin of sand grains that form a sandstone? Why is it likely that several cycles of weathering and lithification could take place if the quartz formed during the Precambrian era?

The following questions pertain to Figure 17.15.

25. Note that the northern portion of St. Louis, Missouri, has loess covering the surface but the southern part of the city has residual soils instead. Much of this soil was derived from limestone.

    a. Why is the deposit of loess in northern St. Louis located on the west side of the Mississippi River? Note that it is not in a valley area but on an upland surface.

    b. What compaction problems might occur when constructing fills or embankments on loess material? Why? What should be done to alleviate this?

    c. What problems might occur during construction of buildings in the residual soils in southern St. Louis? Explain.

26. Note the three different ages of glacial drift. Arrange them in chronological order from younger to older. How would the topography differ in the three areas of differently aged drift? Explain.

27. Note that the loess extends farther eastward from the Mississippi River in the area of the Illinois River than it does to the south from there. Give a likely reason why this is so.

References

Cooke, R. U., Warren, A., and Goudie, A. S. 2006. *Desert Geomorphology*. Boca Raton, FL: CRC Press.

Fernandez-Bernal, A., and De La Rosa, M. A. (eds.). 2009. *Arid Environments and Wind Erosion*. New York: Nova Science Publishers.

French, R. H., and Miller, J. J. 2011. *Flood Hazard Identification and Mitigation in Semi- and Arid Environments*. Singapore: World Scientific.

Goudie, A. S. 2013. *Arid and Semi-Arid Geomorphology*. Cambridge, UK: Cambridge University Press.

Hill, M. 2002. *Arid and Semi-Arid Environments* (2nd ed.). London: Hodder & Stoughton.

Liu, T. 1988. *Loess in China* (2nd ed.). Berlin: Springer-Verlag.

Livingstone, I., Wiggs, G. F. S., and Weaver, C. M. 2007. Geomorphology of desert sand dunes: A review of recent progress. *Earth-Science Reviews* 80(3–4):239–257.

Muhs, D. R. 2007. Loess deposits, origins, and properties. In S. Elias (ed.), *The Encyclopedia of Quaternary Sciences* (pp. 1405–1418). Amsterdam: Elsevier.

Parsons, A. J., and Abrahams, A. D. 2009. *Geomorphology of Desert Environments*. New York: Springer.

Stephenson, D. 2004. *Water Resources of Arid Areas*. New York: Taylor and Francis.

EARTHQUAKES AND GEOPHYSICS

18

Chapter Outline

Earthquake Studies

The Earth's Interior

Engineering Geophysics

Engineering Considerations of Earthquakes and Geophysics

Earthquakes are shock waves that travel through the Earth. They occur apparently as a result of the elastic rebound of rocks when strain energy is suddenly released. The energy builds slowly in a fault zone until the stored stresses become greater than the rocks can withstand and the earthquake occurs (Marshak, 2013; Nance and Murphy, 2016).

Geophysics involves the application of physical laws to study the Earth and earth materials. It is used in purely scientific studies but also in subsurface exploration to search for valuable minerals or to determine subsurface characteristics. The connection between earthquakes and geophysics lies in the common concern for waves propagating through the Earth and what they reveal about the nature of the propagating medium.

Earthquake Studies

Types of Waves

There are two types of earthquake waves: body waves and surface waves. Body waves travel through the volume of the mass or body, and surface waves occur at the boundary of the mass. Body waves are of two types, primary (*P*) or longitudinal waves and secondary (*S*) or shear waves. Primary waves are compressional in nature; they travel by a series of compressions and rarefactions of the waveform in the solid. They are a push-pull-type wave. Sound waves are compressional waves. These waves vibrate parallel to the direction of propagation and have the greatest velocity of the earthquake waves. Consequently, they are the first waves to arrive at the Earth's surface from the source and hence are termed *first arrivals*.

Secondary waves are transverse in nature and vibrate perpendicular to the direction of propagation. They push out into the medium and that medium must push back in order to keep the wave moving. For this to occur, the material must possess some shear strength, or the wave will not propagate. Materials without shear strength, such as liquids, cannot supply the return push and *S* waves will not travel through them. Secondary waves do not penetrate the Earth from one side to another. This provides the basis for the conclusion that the outer core of the Earth is a liquid. Secondary waves travel slower than primary waves but have a greater velocity than surface waves.

When body waves contact the Earth's surface, they cause new waves, called surface waves. There are several types of surface waves, including Love waves and Rayleigh waves, but all are characterized by long wavelengths and high amplitudes. Collectively, surface waves are referred to as L waves (for long wavelength). Their velocities are slower than S waves, so they are the third wave to arrive at a location following an earthquake. The velocities can be summarized as follows:

$$V_P > V_S > V_L \qquad \text{(Eq. 18-1)}$$

The primary and secondary wave velocities can be defined in terms of several material properties of rocks. The properties of interest are (Reynolds, 2011):

- E, the modulus of elasticity or Young's modulus

- K, the bulk modulus

- S, the shear modulus

- ρ, the density

- μ, Poisson's ratio

Modulus of elasticity E is the familiar compressive stress/longitudinal strain relationship for a solid (longitudinal strain = $\Delta L/L$, where ΔL is the change in length of a sample under the application stress and L is the original length of sample).

$$E = \frac{\sigma}{\varepsilon} \qquad \text{(Eq. 18-2)}$$

and K is the stress divided by volumetric strain (volumetric strain is change in volume of a sample under the application of stress divided by the original volume of sample).

$$K = \frac{\sigma}{\dfrac{\Delta v}{v}} \qquad \text{(Eq. 18-3)}$$

The shear modulus S for most materials is numerically somewhat less than half that of E.

The shear modulus provides the relationship between shear stress and deformation in that

$$S = \frac{\tau}{\delta} \qquad \text{(Eq. 18-4)}$$

where:
  S = shear modulus
  τ = shear stress
  δ = deformation

Density ρ is the mass per unit volume and Poisson's ratio is the lateral unit strain divided by the vertical unit strain under load:

$$\mu = \frac{\dfrac{\Delta B}{B}}{\dfrac{\Delta L}{L}} \qquad \text{(Eq. 18-5)}$$

where ΔB is change in diameter of a sample under stress application and B is the original diameter. Using established relationships:

$$V_P = \sqrt{\frac{K + \dfrac{4}{3S}}{\rho}} \qquad \text{(Eq. 18-6)}$$

$$V_S = \sqrt{\frac{S}{\rho}} \qquad \text{(Eq. 18-7)}$$

By inspecting these equations we observe that when S, the shear modulus, is zero or when τ, the shear stress, is zero, V_S is zero. This is in agreement with the understanding that liquids do not propagate shear waves.

It is also well established that

$$S = \frac{E}{2(1+\mu)} \qquad \text{(Eq. 18-8)}$$

$$K = \frac{E}{3(1-2\mu)} \qquad \text{(Eq. 18-9)}$$

Substituting these relationships for K and S in the earlier equation for V_P yields:

$$V_P = \sqrt{\frac{1}{\rho}\left[\frac{E}{3(1-2\mu)} + \frac{\dfrac{4}{3}E}{2(1+\mu)}\right]} \qquad \text{(Eq. 18-10)}$$

$$= \sqrt{\frac{E}{\rho}\frac{1-\mu}{(1-2\mu)(1+\mu)}}$$

In like fashion

$$V_S = \sqrt{\frac{S}{\rho}}$$

$$= \sqrt{\frac{E}{\rho} \frac{1}{2(1+\mu)}}$$

(Eq. 18-11)

Therefore,

$$\frac{V_P}{V_S} = \sqrt{\frac{K}{S} + \frac{4}{3}}$$

$$= \sqrt{\frac{1-\mu}{\frac{1}{2}-\mu}} > 1$$

(Eq. 18-12)

so $V_P > V_S$. This shows that the velocity of the primary wave is always greater than that of the secondary wave.

Several other details are also of interest. One of these is the so-called "density paradox" for V_P.

The equation

$$V_P = \sqrt{\frac{K + \frac{4}{3}S}{\rho}}$$

(Eq. 18-13)

suggests that as ρ increases, V_P decreases. This is not actually the case because as ρ increases, K also increases but at a rate greater than that of ρ. Therefore, in general, as the density increases, so does the velocity of the primary wave. Hence the paradox.

The equations also show that the shear wave velocity can be determined from E, ρ, and μ, parameters that are easily determined in the laboratory by static loading of a sample. Values under dynamic loading differ somewhat (Reynolds, 2011). For this reason, in studies of earthquake resistance, for many major structures the shear wave velocity and shear modulus are typically measured under field conditions to obtain dynamic values.

Example Problem 18.1

a. A rock has a Poisson's ratio of $\mu = 0.25$. What is the ratio of V_P/V_S?
b. If $E = 44.8 \times 10^6$ kN/m², what are the values of S and K?
c. If V_P is 4420 m/sec, what is V_S in km/sec and ft/sec?

Answers

a. $$\frac{V_P}{V_S} = \sqrt{\frac{1-\mu}{\frac{1}{2}-\mu}}$$

$$= \sqrt{\frac{1-0.25}{0.50-0.25}}$$

$$= \sqrt{3} = 1.732$$

b. $$S = \frac{E}{2(1+\mu)}$$

$$= \frac{44.8 \times 10^6 \text{ kN/m}^2}{2(1+0.25)}$$

$$= 17.9 \times 10^6 \text{ kN/m}^2$$

$$K = \frac{E}{3(1-2\mu)}$$

$$= \frac{44.8 \times 10^6 \text{ kN/m}^2}{3(1-0.50)}$$

$$= 17.9 \times 10^6 \text{ kN/m}^2$$

c. $V_P = 4420$ m/sec $= 4.42$ km/sec

$$V_S = \frac{V_P}{1.732}$$

$$= \frac{4.42 \text{ km/sec}}{1.732}$$

$$= 2.552 \text{ km/sec}$$

$$= 2552 \text{ m/sec} \times 3.28 = 8371 \text{ ft/sec}$$

Earthquake Scales

Two distinct scales are used to measure the level of earthquakes. These are the magnitude scale, a measure related to the amount of energy released, and the intensity scale, which measures the damage the earthquake causes (Marshak, 2013).

Magnitude Scale

The Richter scale (M_L) is used today to measure the magnitude of an earthquake (Richter, 1935, 1956). Presented in the current form in 1957, it is based on the maximum displacement or amplitude of the earthquake wave, measured on a standard seismograph 100 km from the epicenter. That reference point, the epicenter, lies on the Earth's surface directly above the focus or place of origin for the earthquake. The Richter value is determined by measuring the maximum amplitude in thousandths of a millimeter (or in microns = micrometers = 10^{-6} m) and taking the common logarithm of this number. Hence, a 0.100-mm displacement would yield a Richter value of 2 and 1.000 mm would yield a value of 3. In like manner, 0.500 mm becomes 2.698. A factor of 10 times the displacement is involved from one integer value to another. This shows that a Richter value of 6 has a displacement 1000 times greater than a value of 3.

Because the Richter scale is based on a linear measurement of length, there is no upper limit to the scale as far as physical measurements are concerned. A common misconception creeps in, however, when the news media or others, on occasion, allude to some maximum, usually 10, for this scale. This is simply not the case. The Earth does, however, seem to impose its own limit of 9.0 to 9.5; this appears to be about the maximum of equivalent stress that the Earth can store before releasing its load. The Chilean earthquake of 1960 has the highest Richter magnitude (9.5) ever assigned to an earthquake (Marshak, 2013). Earthquakes above 6 are important and those of 7 or more are major earthquakes. The truly massive, devastating earthquakes in history have been in the range of 8 or more and, fortunately, have occurred quite infrequently in the recorded past. Some of the major earthquakes in recent history are listed in Table 18.1.

In recent years additional studies of earthquake magnitudes have led to modifications to the Richter scale. The designation used for the local magnitude

is M_L, the surface wave magnitude is M_s, the body wave magnitude is M_b, duration magnitude M_D, and moment magnitude, M_w.

A detailed discussion of these other magnitude scales is beyond the scope of this text but it is important to realize that M_L is not an adequate measure for all levels of magnitudes of earthquakes ranging from less than 3 to greater than 7.5. Because M_L scale is not particularly sensitive to earthquakes that are well below magnitude 3 and it approaches a saturation point at about magnitude 7, it does not distinguish very well between earthquakes with magnitudes below 3 and above 7. Therefore, several scales are suggested for various ranges of magnitude:

- M_D or M_L magnitudes less than 3
- M_L or M_b magnitudes between 3 and 7
- M_s magnitudes between 5 and 7.5
- M_w magnitudes greater than 7.5

An empirical relationship between the amount of energy released and the Richter magnitude scale has been developed:

$$\log E = 11.4 + 1.5 M_L \qquad \text{(Eq. 18-14)}$$

with E in ergs and M_L on the Richter scale. The two constants 11.4 and 1.5 are subject to modification as more data are acquired. Previously a value of 11.8 was used for the intercept value. Table 18.2 shows the energies released, expressed in tons of TNT, for earthquakes of different magnitudes.

The average number of earthquakes per year, including those that can be felt near their epicenters, is estimated to be more than 150,000. The total number, including those imperceptible to humans, is probably about 1 million per year. Despite this large number, much of the total energy released in a given year will be associated with just a small number of major earthquakes. A single earthquake of magnitude 8.4 can release almost as much energy as the total energy released each year by low to moderate magnitude earthquakes. Table 18.3 on page 410 illustrates information on earthquake frequency and magnitude worldwide.

Intensity Scale

Earthquake intensity is a measure of the damage done to human-made structures that occurs as a result of the ground shaking. It describes the damaging effect that the earthquake motions inflict on

Table 18.1   Richter Magnitudes of Selected Significant and Major Earthquakes in the World.

Year	Location	Magnitude	Year	Location	Magnitude
1755	Lisbon, Portugal	8.7 est.	1983	Japan	7.7
1811–12	New Madrid, MO	7.8–8.1 est.	1983	Turkey	6.9
1886	Charleston, SC	7.3 est.	1985	Chile	7.8
1899	Alaska	8.6	1985	Mexico City, Mexico	8.0
1906	Colombia	8.6	1986	El Salvador	5.4
1906	San Francisco, CA	7.7–7.8 est.	1987	Columbia–Ecuador	7.0
1906	Chile	8.4	1988	Burma–China	7.0
1911	Soviet Union–China	8.4	1988	Armenia	7.0
1914	Quebec, Canada	5.5	1989	Loma Prieta, CA	7.0
1920	China	8.5	1989	Australia	5.6
1925	Santa Barbara, CA	6.5	1990	Iran	7.7
1933	Long Beach, CA	6.3	1990	Philippines	7.8
1933	Japan	8.5	1992	Landers, CA	7.5
1940	El Centro, CA	7.1	1994	Los Angeles (Northridge), CA	6.6
1950	Pakistan–Tibet–Burma	8.6	2005	Kashmir, Pakistan	7.6
1952	Kern County, CA	7.7	2010	Haiti	7.0
1960	Chile	8.3	2010–11	New Zealand	7.1, 7.0
1964	Anchorage, AK	8.4	2011	Chile	8.8
1965	Puget Sound, WA	6.5	2011	Tohoku, Japan	9.0
1970	Peru	7.8	2012	Aceh, Indonesia	8.6
1971	San Fernando, CA	6.6	2013	Balochistan, Pakistan	7.7
1977	Indonesia	8.0	2014	Iquique, Chile	8.2
1979	El Centro, CA	6.5	2015	Nepal	7.8
1980	Italy	7.2	2016	Ecuador	7.8
1981	Iran	7.3			

buildings and other engineering works. It is not measured by instruments, but is, instead, determined from reports supplied by trained observers of damage effects. Because intensity is based on actual observations of earthquake damage at specific locations, the evaluation is perhaps more meaningful to the layman than is the magnitude. Considerable time, weeks or months, may be required to assemble an intensity map for a particular earthquake. Intensities for earthquakes that occurred several hundred years ago, well before seismographs monitored such phenomena, can be constructed on the basis of damage reports provided by historians, whereas an accurate measure of the magnitude is more difficult to compile.

The first scale involving earthquake intensities was based on work by M. S. de Rossi of Italy and F. A. Forel of Switzerland in 1883. This scale, with values from I to X, was used for about 20 years before a more refined measure was needed. With advances in earthquake studies, in 1902, Giuseppe Mercalli,

Table 18.2   Energies of Some Major Earthquakes in the Western United States.

Location	Date	Richter Magnitude	Energy in Tons of TNT
San Francisco, California	1957	5.3	500
Long Beach, California	1933	6.3	15,800
El Centro, California	1940	7.1	250,500
Kern County, California	1952	7.7	1,990,000
San Francisco, California	1906	8.2	12,000,000
Anchorage, Alaska	1964	9.2	410,000,000

Note: One ton of TNT = 10^9 calories; 1 calorie = 4.186×10^7 ergs.

an Italian seismologist, developed a new scale with a I to XII range. In 1931, the Mercalli scale was modified by American scientists Harry O. Wood and Frank Neumann (1931) to incorporate the modern structural features of buildings. In use in the United States today, the modified Mercalli (MM) intensity scale is shown in an abbreviated form in Table 18.4.

There is always some interest in comparing the intensity scale to the magnitude scale. Approximate correlations exist between the two scales, but difficulties are involved because the intensity depends on several factors other than earthquake energy. Geologic conditions and specifics concerning the structures are also involved. These factors are discussed in the next section. The ground acceleration induced by the earthquake, designated in terms of the equivalent acceleration of gravity, is also

Table 18.3   Annual Number of Earthquakes Relative to Magnitude Worldwide.

Class	Magnitude[a]	Average Number	Release of Energy (Approximate Explosive Equivalent)
Great	8 or more	2	50,000 1 megaton (hydrogen) bombs
Major	7.0–7.9	12	
Strong	6.0–6.9	108	
Moderate	5.0–5.9	800	1 megaton bomb or 1 small atom bomb (20,000 kg of TNT)
Light	4.0–4.9	6200	
Minor	3.0–3.9	49,000	
	2.5–2.9	100,000	
	2.5	700,000	0.5 kg of TNT

[a] For a magnitude of 5.0 or less, the average number of occurrences is based on estimates by sampling special regions.

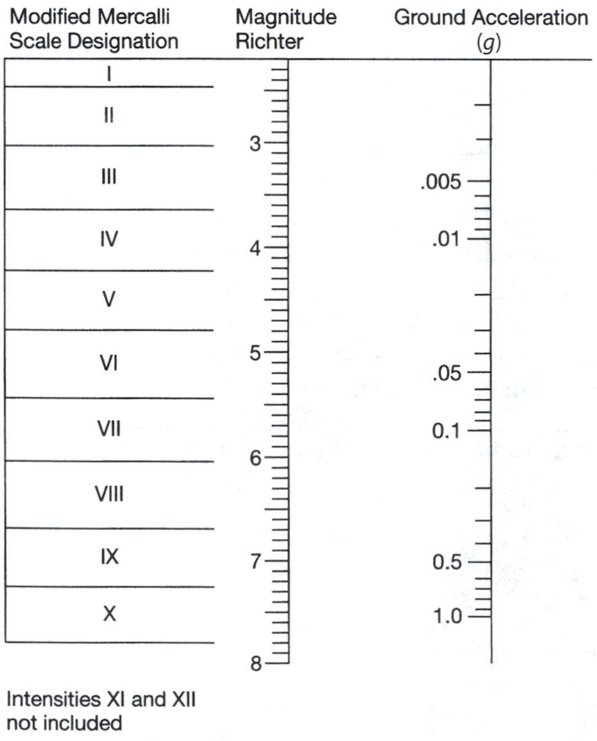

of interest because structural engineers commonly work in these units when analyzing building safety. Figure 18.1 shows the general correlation among the three measures associated with earthquakes.

Earthquake Damage

Earthquake damage to human-made structures is influenced by five principal elements (Sharma, 1997; Reynolds, 2011; Marshak, 2013):

1. Strength of the earthquake wave,

2. Length of earthquake motion (and effects of aftershocks),

3. Proximity to a fault zone,

4. Type of geologic foundation, and

5. Building design.

The strength of the earthquake at the location in question is obviously important. This is indicated by the magnitude of the event and the distance from the epicenter. The ground motion falls off rapidly with distance as seen in Figure 18.2 where the distance is related to the maximum acceleration for various magnitude earthquakes.

The magnitude of a potential earthquake is related to the length of rupture of the triggering fault. Obviously the greater the rupture length, the greater the energy produced and the higher the magnitude of the earthquake event. The length of rupture is estimated for active faults in order to evaluate

Figure 18.1   MM intensity scale and approximate relationship with magnitude and ground acceleration.

Table 18.4   Earthquake Intensity or Modified Mercalli Intensity Scale.

Intensity	Characteristics
I	Not felt; animals uneasy.
II	Barely felt by people at rest.
III	Felt indoors; like a passing truck; some hanging objects swing.
IV	Like passing heavy trucks; hanging objects swing; autos rock; windows, dishes, etc., rattle; walls creak.
V	Felt outdoors; sleepers awakened; small objects upset; doors swing open or close.
VI	Felt by all; people frightened and run outdoors; dishes and glassware broken. Furniture moved, overturned; weak masonry cracked; small bells ring.
VII	Difficult to stand; noticed in moving autos; furniture and masonry broken; chimneys break at roof line; ponds become turbid.
VIII	Auto steering affected; masonry damaged, some collapse; frame houses move on foundations; twisting and falling of large chimneys; branches broken from trees.
IX	General panic; masonry damaged or destroyed; frame buildings cracked, underground pipes broken; sand boils, soil cracks.
X	Most masonry buildings destroyed; wood frame buildings destroyed; landslides; railroad rails bent.
XI	Railroad rails badly bent; underground piping destroyed.
XII	Total damage; objects thrown into air.

Source: Wood and Neumann, 1931.

the potential for occurrence of a major earthquake. Table 18.5 shows the relationship between earthquake magnitude and fault rupture length.

The length of earthquake motion is significant because structural cracking may propagate while shaking continues so that weakened members may eventually fail completely. For the most significant earthquakes, the motion persists for about 10 sec to 1 min. In the Anchorage earthquake of 1964, however, reports indicate that the shaking lasted for 4.5 min. Thus, the high magnitude (9.2) and the poor geological conditions account for the extensive destruction of that major earthquake.

Table 18.5   Earthquake Magnitude versus Fault Rupture Length.

Magnitude (Richter)	Fault Rupture Length	
	(km)	(mi)
5.5	5–10	3–6
6.0	10–15	6–11
6.5	15–30	11–19
7.0	30–60	19–38
7.5	60–100	38–62
8.0	100–200	62–125
8.5	200–400	125–250

Aftershocks are also important. These occur for a period of months after the major earthquake as the Earth readjusts following the sudden stress release. For example, the Kern County, California, earthquake ($M_L = 7.7$) of 1952 had an aftershock of 5.8 about one month later. The aftershock caused more damage to Bakersfield than did the main earthquake because the aftershock caused previously damaged, yet still standing, structures to collapse. This points to a concern for civil engineers; damaged public

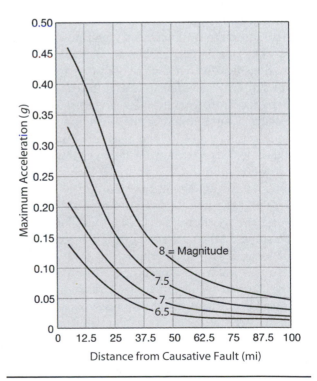

Figure 18.2 Relationship between peak bedrock acceleration, earthquake magnitude, and fault distance.

buildings should be examined immediately and closed to the public following the main shock, if justified, to ensure that additional effects are minimized if aftershocks develop.

Proximity to a fault zone is important because the energy dissipation seems to concentrate in a fault area, yielding more shaking of the buildings than for adjacent areas. Therefore, if a building is on one fault zone and an earthquake occurs along another fault zone some distance away, the damage is likely to be greater along the nonconnected fault zone than in adjacent nonfaulted material. However, in an area where very strong ground movement occurs, the presence of the fault may be less important than the actual nature of the foundation material.

The geologic materials on which a building is constructed play a primary role with regard to earthquake damage. Massive bedrock provides the best foundation because it passes wave motions on with minimum attenuation, resulting in less vibration to the structure. Soil or unconsolidated material tends to amplify rock acceleration, and soft soil may increase this level by several times. Low-density or poorly compacted soils will fare the worst because they supply much greater ground motion to the structure.

In addition to this increased vibration level, low-density soils yield other problems. They may be subject to densification, yielding a volume reduction and loss of support. Low-density, unsaturated sands will settle by reorienting their grains as the ground shaking occurs. Saturated fine sands and silts are subject to liquefaction caused when pore pressures build, thus drastically reducing the effective stress. When in a loose condition (N = 10 to 15 blows/foot; see Chapter 19 for details on the standard penetration test), these saturated fine soils are prone to liquefaction and the soil flows downhill, except for on very gentle slopes. Alluvial areas along major streams are prime locations for such failures.

Human-made fills, not properly compacted, are also subject to settlement during shaking. Graded lots on a hillside can permit major residential failures if the fills are not properly compacted. As shown in Figure 18.3, at the site soil and weathered rock were pushed over the edge of the hillside during construction and compacted only by overburden weight and that of a bulldozer. Hence, the footing on the right is founded on rock and the footing on the left on poorly compacted soil. Major settlement of the downhill

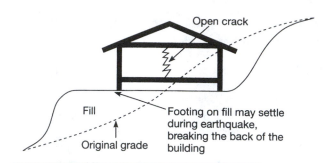

Figure 18.3 Effect of earthquakes on structures built on poorly compacted side hill cuts.

footing results in failure of the house. Also shown in place is the original topsoil, left at the base of the fill. This should have been removed prior to fill placement because this material can cause slope failure.

Building design is the last of the five factors that influence damage. For residential buildings, the object is to build a structure that can overcome inertia and move as a unit with ground motion, without the individual parts moving independently of each other. This requires integrity of the structure and is obtained by adequate bracing, with secure anchoring, and bonding of all elements. The whole building—foundation, frame, walls, floors, and roof—must be balanced and well-tied together.

Conventional wood frame construction with stud supports and wood veneer or aluminum siding has a high degree of earthquake resistance, provided sound workmanship is involved. Adequate lateral bracing is supplied by such elements as the frame, roof, wall sheathing, interior walls, and ceiling panels. Lateral supports can be improved if the sill is bolted to the foundation, all studs are toe-nailed to the sill from both sides, and cross-wall bracing is included. This bracing should be installed in opposing pairs either in an X or V shape. Figure 18.4 shows these aspects in detail.

Ordinary masonry structures may be particularly vulnerable to earthquake shaking. This is a consequence of their greater mass than wood frame houses and, commonly, their inability to move as a unit. If good brick or concrete block is laid with strong Portland cement mortar, properly reinforced with steel rods through the masonry, and tied in well with the foundation and inner walls, masonry buildings can resist shaking quite well. Concrete slabs on well-compacted soil and secure footings should per-

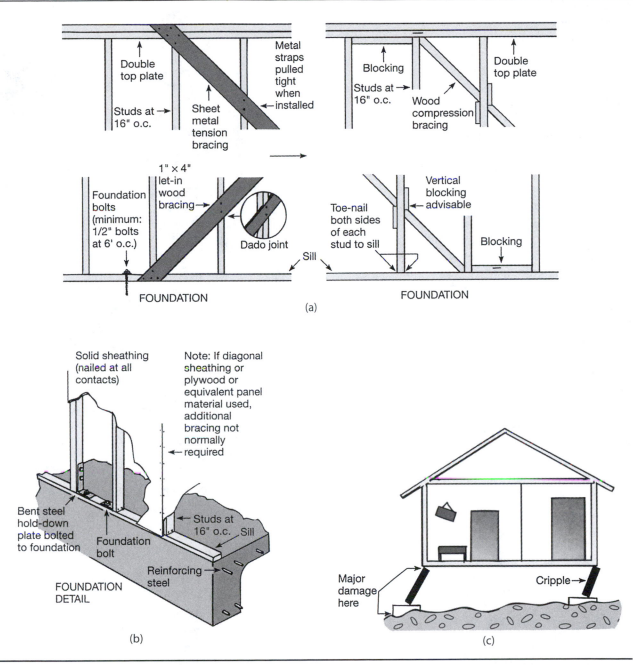

Figure 18.4   (a and b) Construction techniques to improve earthquake resistance. (c) Structural damage to residential construction during earthquakes.

form without undue problems. Nonreinforced bearing walls of adobe or hollow concrete block will be seriously damaged in even moderate earthquakes. This is a problem of large and small buildings alike and for block walls and masonry veneer for buildings where reinforcing is omitted.

Surface waves also generate horizontal movement and contribute to earthquake damage. Tall buildings must be resistant to transverse motion and to the frequency at which the building vibrates during earthquakes. If a resonant frequency for the building develops, severe damage will be induced.

This is known as *site resonance* and involves the relation between the natural period of the ground and the natural period of the structure (Reynolds, 2011). Other design details are involved, but these are discussed in a review of the San Fernando earthquake of 1975 presented in a later section.

The secondary wave is typically the most destructive body wave of an earthquake because it induces transverse or horizontal movement in a structure. Increased lateral stiffening of the structure is sometimes required to provide sufficient resistance to transverse shaking. Vertical movement or forces can be more easily resisted because these act in the same direction as gravity or the weight of the building. Surface waves with a horizontal component are also destructive to buildings.

Field measurements of the secondary wave velocities of the soil, V_S, commonly provide valuable information concerning earth shaking. Figure 18.5 shows how secondary waves are generated and measured in the field. A sledgehammer is used to strike the end of a wooden plank held against the ground by the weight of a field vehicle. As hammering continues, the sonde is lowered down the hole and V_S is determined at various depths in the soil column.

The period of the soil column P is calculated from the equation

$$P = \frac{4T}{V_S} \qquad \text{(Eq. 18-15)}$$

where T is the thickness of the soil column and V_S is the secondary wave velocity. For example, if V_S = 250 m/sec and the soil or unconsolidated column is 32 m thick, then

$$P = \frac{4(32)}{250}$$

$$= 0.51 \text{ sec for the period of the soil column}$$

A critical condition develops when the period of the soil column matches the period of vibration of a building founded on the soil. These two periods should not coincide because, if they do, the building will likely be destroyed during a major earthquake. The rule of thumb for buildings is shown by the following equation:

$$P_{\text{bldg}} = \frac{N}{10} \qquad \text{(Eq. 18-16)}$$

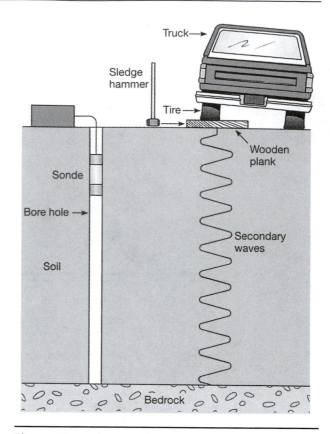

Figure 18.5   Generating and measuring secondary wave velocity in the field.

where N is the number of floors in the building. So, for a five-story building,

$$P_{\text{bldg}} = \frac{5}{10} = 0.5 \text{ sec}$$

Also, some bridges have a period of about 0.5 sec. A comparison of the 0.51 value to the 0.5 sec for a five-story building leads us to conclude that ground shaking would cause severe damage to this five-story building.

Geologic Investigations of Earthquake Prone Areas

Geologic investigation, by necessity, is extensive in earthquake-prone areas where new construction is planned or the safety of an existing structure is to be determined. The following are typically required in an engineering geology investigation (California Geological Survey, 1990):

1. A structural geology map of the region indicating recent tectonic movements.

2. A compilation of active faults including the nature of displacements. Fieldwork will likely be necessary and would involve trenching across suspected fault zones. Geologic criteria showing fault movement in the Holocene time (the past 10,000 years) are important; these include displacements in recent soils, dating of organic materials by hydrocarbon methods, and evaluation of landforms to determine relative ages of formation.

3. Mapping of the structural geology for the site, including scarps in the bedrock, differential erosion, and offsets in sedimentary deposits. Included should be rock type, surface features, and local faults. The latter should include probable length, continuity, and sense of movement.

4. Use of geophysical exploration on through-going faults near the site to define recent fault ruptures. Techniques include electrical resistivity and gravity measurements in a direction normal to the fault. Examine evidence for segmentation of the fault, looking for stepover of fault stands and changes in strike.

5. Evaluation of reports involving landslides, major settlements, or effects of flooding at the site.

6. Investigation of groundwater barriers in the area that are associated with faults or may affect the soil response to earthquake shaking.

7. Borings, trenches, excavation, and sampling to locate and describe the presence of sand layers, which may be subject to liquefaction.

8. Measurement of the physical properties of the soil (density, water content, shear strength, behavior under cyclic loading, attenuation) in situ or in borehole samples.

9. Determination of P and S wave velocities and attenuation values in the soil layers using geophysical exploration methods.

This summary addresses particularly critical construction in populated areas. Both the density of the population and the nature of the proposed engineering structure impact the level of risk involved in an earthquake-prone area.

Example Problem 18.2

If the time interval between the first arrival of the P wave and the S wave at a seismic station is 250 sec, the velocity of the P wave is 7.89 km/sec, and that of the S wave is 4.33 km/sec, what is the distance from the station to the epicenter?

Answers

Assume x = distance from seismic station to the epicenter in kilometers. Then x/V_P = the travel time for the P wave (t_P) and x/V_S = the travel time for the S wave (t_S).

$$t_P = \frac{x}{7.89 \text{ km/sec}}$$

$$t_S = \frac{x}{4.33 \text{ km/sec}}$$

Therefore, the time interval between the first arrival of the P and S waves or $\Delta t = t_S - t_P = 250$ sec.

$$\frac{x}{4.33} - \frac{x}{7.89} = 250$$

$$7.89x - 4.33x = 250(7.89)(4.33)$$

$$3.56x = 8541$$

$$x = 2400 \text{ km}$$

Locating Earthquakes Using Seismology

Seismograph stations in the world network receive the wave trace from a strong earthquake starting with the arrival of the P waves and followed, subsequently, by the S waves and then the L waves. This is illustrated in Figure 18.6. If the velocities of the P and S waves are known, the distance to the epicenter can be calculated based on the time delay between the first arrivals of the P wave and S wave. This is somewhat of an oversimplification, however, because the P and S wave velocities are not constant with distance from the epicenter; both increase in velocity with greater distances. An example problem serves to illustrate this ability to calculate the distance.

An accurate plot of travel time versus distance for earthquake waves is shown in Figure 18.7. The slope of these curves (its first derivative) equals $1/v$ where v is the velocity. Observe that the slope decreases or the velocity increases with distance. This occurs because the waves travelling longer distances have wave paths located deeper in the Earth than do waves traveling shorter distances. The seismic velocity increases with increasing depth because of increasing density.

Also shown in the plot of Figure 18.7 is the cessation of the S wave (corresponding to a depth of 2900 km) and an offset for the P wave. This occurs at the boundary between the mantle and outer core, where S waves are attenuated and the P waves are refracted. These topics are discussed further in a later section concerning the nature of the Earth's interior.

In actual practice, the distance from the seismic station to the epicenter is determined from precise plots of travel-time curves of the nature shown in Figure 18.7. The travel time difference between the P and S waves is compared to the two sloping curves until a fit is obtained. The distance is indicated by the distance values designated along the abscissa.

After the distance between the epicenter and seismic station is determined, a circle with that radius can be drawn using the station as the center. Three such stations are required to locate the earthquake, and the triple intersection point marks the location of the epicenter. Details are shown in Figure 18.8.

The depth of focus for the earthquake can also be determined. If the focus is only a few kilometers deep, then the P and S waves arrive at the epicenter

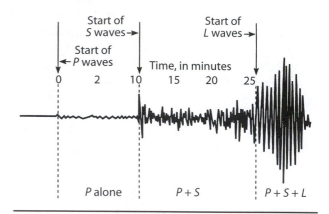

Figure 18.6   Earthquake wave traces on a seismogram.

sooner than they reach points 100 km (62.5 mi) or so away. By contrast, if the focus is 150 to 300 km (94 to 188 mi) deep, the waves arrive almost simultaneously at all points within a large circle around the epicenter. Therefore, the surface area in which the shock is felt simultaneously increases rapidly as the depth of focus increases. The delays in arrival times for the S and L waves relative to the P wave are also indicative of the depth of focus and can be compared from a number of seismic stations.

Based on these data, earthquakes can be classified according to depth and are subdivided into shallow-focus earthquakes, intermediate-focus earthquakes, and deep-focus earthquakes. Information concerning the depth of focus is presented in Table 18.6.

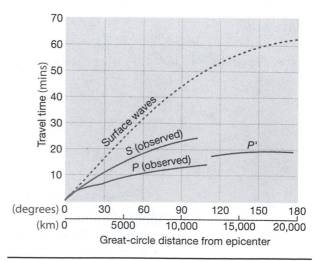

Figure 18.7   Travel-time curves for seismic waves from earthquakes.

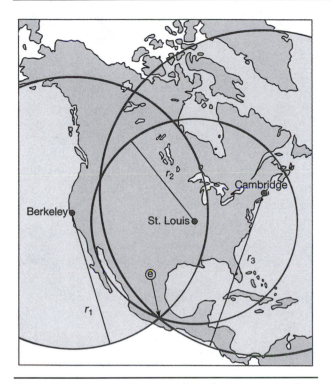

Figure 18.8 Locating an earthquake epicenter using three seismic stations.

In the middle of the twentieth century, the low-velocity zone (LVZ) of the upper mantle (from about 40 to 300 km deep) was discovered through seismological studies (Gutenberg, 1959). In this zone, the velocity of the P wave is reduced by as much as 10% and the S wave about 2%. It is now thought that many shallow-focus earthquakes occur above the low-velocity zone, many of the intermediate-focus variety within it, and deep-focus earthquakes below it.

Causes of Earthquakes

Following studies of the San Francisco earthquake of 1906, it was soon generally accepted that earthquakes were caused by faulting, as a result of rock rupture in the Earth's crust. This is known as the elastic rebound theory. It was based on observations that the San Andreas Fault, associated with the 1906 earthquake, had shown distortion in years prior to the event and considerable displacement along the fault afterward. This suggested a sudden release in stress during rupture, followed by elastic recovery (Reid, 1910).

Today this mechanism is questioned for earthquakes that originate at a depth of 20 to 30 km (12.4 to 18.6 mi) or more. The confining pressures of the overburden load at that depth prevent sudden extensive ruptures from occurring; instead, the rocks deform plastically. During this deformation some buildup of stress is possible because of rock viscosity so that sudden jerks or ruptures could interrupt the continuous plastic deformation and cause an earthquake. This mechanism seems more acceptable in light of current knowledge about the Earth's interior.

Another piece of evidence in support of plastic deformation is the fact that in some cases surface fault movement occurs after an earthquake rather than during it. In the El Centro earthquake of 1979, creep movement across a bituminous road continued for several weeks following the earthquake. The surface fault displacement was apparently a result of the earthquake-causing mechanism, not the cause of the strong ground motions. This suggests that such faulting is a consequence of earthquakes, not a cause.

Earthquake Prediction

The details of earthquake prediction that receive the greatest interest are location, magnitude, and time of occurrence. For many interested citizens, earthquake prediction simply means predicting the time of occurrence of a major earthquake. In reality, the ability to predict the specifics of the strong ground motion at a specific site are more important for mitigation of earthquake hazards. Strong motion data are now used for building design, evaluating existing structures, and zoning for new construction (Cicerone et al., 2009; Hayakawa, 2015).

In 1990, there was great concern about a major earthquake occurring in the lower Midwest of the

Table 18.6 Classification of Earthquakes by Depth of Focus.

Type	Range of Depth (km)	Range of Depth (mi)	Relative Abundance
Shallow	70	44	75% of earthquakes occur in this range.
Intermediate	70–300	44–188	Most of the remaining earthquakes occur in this range.
Deep	300	188	About 3% of all earthquakes; no earthquakes below about 700 km ever recorded.

United States. This concern, which is more typical in certain areas of California and Alaska, was elevated to fear—and near panic in the minds of some citizens—to the extent that classes in primary and secondary schools were cancelled on the day of the predicted earthquake.

Iben Browning, a climatologist from New Mexico, predicted that on December 3, 1990, a major earthquake would occur on the New Madrid Fault located in eastern Tennessee and southern Missouri (Spence et al., 1993). This was the location of a series of major earthquakes in 1811–1812 that had an estimated Richter magnitude of as high as 8.1, causing major soil ruptures and liquefaction, and vibrations that were felt up to 1600 km (1000 mi) away (Johnston and Schweig, 1996; Hough, 2009). No major earthquakes have occurred on the New Madrid Fault since 1811–1812, and the repeat interval for a major event has been estimated to be between 200 and 500 years.

Browning's prediction was apparently based on the fact that the gravitational pull on the Earth would be maximized during this particular spring tide event. As discussed in Chapter 16 on coastal processes, a spring tide occurs twice each lunar month during the full and new moons when the Sun and Moon lie in a straight line and their pull on the Earth is additive. December 3, 1990, was a time of full moon, but another celestial condition prevailed, making this a rare occurrence. This event was called a triple junction. In addition to the Sun and Moon being aligned (spring tide situation), the Moon was in perigee, or at a point in its orbit when closest to the Earth. This triple occurrence happens only once about every 61 years.

Selection of December 3, 1990, by Browning as a rare celestial occurrence can certainly be affirmed, but the conclusion he reached (or jumped to) that this would trigger a major earthquake on the New Madrid Fault system, was highly questionable. Other researchers in the past have attempted to correlate earthquake occurrence with celestial gravitational attraction and have found no consistent relationship to exist.

Because of Browning's prediction, people of the lower Midwest have become more conscious of the possibility of a major earthquake sometime in the not too distant future. Certainly people in such cities as Memphis, Tennessee, St. Louis, Missouri, and Evansville, Indiana, have become more concerned about earthquake damage. This greater awareness is a posi-

tive contribution to what was otherwise an unwarranted scare. The news media treated Browning's theory as a hot news item, which helped to elevate the concern to fear and near panic in the affected areas.

A number of methods for predicting the time and size of an earthquake have been tried (Cicerone et al., 2009; Hayakawa, 2015). Earthquake locations, of course, are related to the presence of existing, active faults and overall seismic activity. As discussed previously, fault rupture length is used to provide a rough estimate of the size or magnitude of the expected earthquake.

Physical clues used for predicting the time of occurrence are:

1. Increased seismicity (foreshocks) and animal restlessness;

2. Ground uplift;

3. Reduction in P wave velocity;

4. Increased radon emission;

5. Decreased electrical resistivity;

6. Temporal and spatial gaps in the seismicity of a region, suggesting that a portion of a fault zone has locked-in stress, which eventually will release yielding an earthquake;

7. Determination of a periodicity of earthquake occurrence; and

8. On monitored fault zones, changes in the groundwater level and fault slippage or displacement.

All of these physical determinations have encountered complications when applied to a series of earthquake situations. Continued application of these procedures to monitored fault zones should show which, if any, will prove to be reliable prediction techniques for the time of earthquake occurrence.

Earthquake Distribution

No portion of the Earth's surface is completely free from earthquakes, yet earthquake occurrence and magnitude are, by no means, evenly distributed. For most areas, only infrequent shocks of small or moderate intensity have occurred. By contrast some zones of the Earth are subject to frequent earthquakes that range from weak tremors to strong motion events. These comprise the major earthquake belts of the world.

The most prominent seismic areas are as follows: (1) the Circum-Pacific Belt, (2) the Mediterranean and Trans-Asiatic Belt, (3) the Mid-Ocean Ridges of the Atlantic and Indian Oceans, and (4) the East African Rift Zone (Marshak, 2013).

The Circum-Pacific Belt accounts for some 80% of the earthquakes each year. It extends from the tip of South America at Cape Horn along the Andes Mountains to Central America and along the western area of North America to Alaska, across the Aleutians to Asia and southward down the Asian coast to the island arcs extending southeastward to New Zealand. This area is also known as the "Ring of Fire" because volcanoes are commonly associated with this zone. This extensive belt marks the location where many of the plates of the Earth are actively opposing one another (refer to the discussion on plate tectonics in Chapter 1 for more details).

Approximately 15% of all earthquakes occur in the Mediterranean and Trans-Asiatic Belt. It extends from Gibraltar on the west end of the Mediterranean Sea across Italy and Greece through the mountainous regions of the Middle East to the high mountain ranges of southern Asia. Many major earthquakes with severe levels of damage have occurred along this belt during historic times.

The mid-ocean belts follow the mid-ocean ridges, which mark the centers from which seafloor spreading originates. High rates of heat flow are also associated with these highs in the ocean basins.

In East Africa, the fault zone lies along a line from the Red Sea on the north through Lakes Turkana, Victoria, Tanganyika, and Malawi to the south. This area is known as the Rift Valley and, in some places, is marked by a prominent fault escarpment along the earthquake zone.

Today we also recognize, through studies of plate tectonics, that most earthquakes occur along belts marking the boundaries of crustal plates. Earthquakes occur where these plates collide, separate, or slip past each other. They coincide with chains of volcanoes, zones of crustal disturbance, active strike-slip faults, deep-sea trenches along continental borders, young mountain belts, and mid-ocean ridges. These are the areas of greatest geologic activity on Earth. Of the four major earthquake belts described here, only the East African Rift Zone does not lie along a plate boundary.

Earthquakes of the United States

In 1967 and 1968, S. T. Algermissen of the US Coast and Geodetic Survey studied approximately 28,000 earthquakes that occurred in the contiguous United States from 1900 to 1965. Prior studies used frequency of occurrence to help determine a seismic probability map for areas throughout the United States. In this study, MM maps and strain release were used. Strain release is assumed to be proportional to the square root of the energy released and this, in turn, involves the Richter magnitude.

Based on a strain release rate, the United States was divided into four 10,000-km^2 (3861-mi^2) areas and a rate of occurrence of equivalent magnitude 4 earthquakes for each block was applied. This map is presented in Figure 18.9 on the next page. It is useful in that it shows a relative rate of tectonic activity that would not normally show on a maximum intensity map.

Algermissen also developed a seismic risk map of the United States, which has been revised repeatedly. However, new probabilistic hazard maps complied by the US Geological Survey are derived from seismic hazard curves calculated on a grid of sites across the United States. They display earthquake ground motions for various probability levels (Figure 18.10, next page).

Midcontinent Region

For the Midwest, several locations are designated as high strain-release zones. In Figure 18.9, note that an area in western Ohio has 64 to 256 equivalent magnitude 4 earthquakes, whereas southern Illinois has from 16 to 64. In Figure 18.10, the top three highest hazard zones are centered around an area of southern Illinois, the southeast corner of Missouri, and the far western portions of Kentucky and Tennessee. This hazard area is also the site for the massive earthquakes of 1811–1812 in New Madrid, Missouri.

On November 9, 1968, in south-central Illinois, an earthquake occurred. It was felt in an area of approximately 1,485,000 km^2 (580,000 mi^2) with a maximum intensity of VII and a Richter magnitude equal to 5.5. An isoseismal map of this earthquake area is shown in Figure 18.11 on page 421. An isoseismal map uses contour lines to depict locations with equal seismic intensity. As expected, the intensity decreases with distance from the epicenter, but there are isolated pockets of lower or higher values. Also, contours travel along the stream valleys of the Missouri,

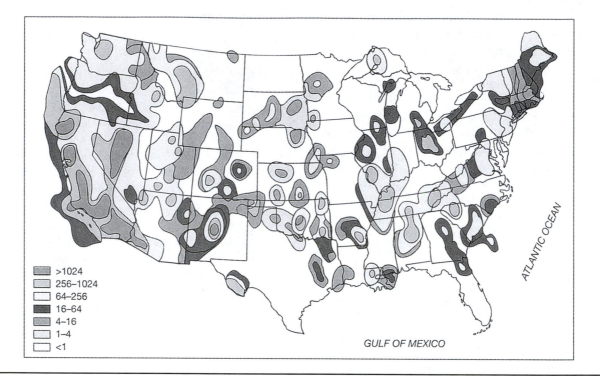

Figure 18.9   Strain release in the United States, 1900 to 1965, expressed as equivalent number of magnitude for earthquakes. (Algermissen, 1969.)

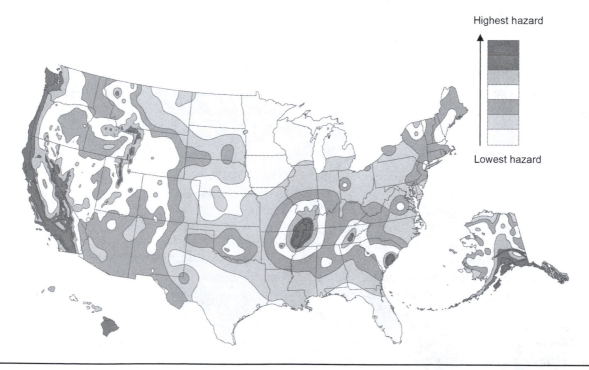

Figure 18.10   National Seismic Hazard Map. (US Geological Survey, Earthquake Hazards Program [https://earthquake.usgs.gov/hazards/hazmaps/conterminous/index.php#2014].)

Mississippi, Ohio, and Wabash Rivers. This indicates how the higher intensities are associated with unconsolidated materials of major stream valleys.

Using information from the 1968 earthquake and other shocks that have occurred in the Midwest, we can develop a map of maximum historical intensity. Such a map for Indiana and contiguous states is shown in Figure 18.12. In general, the values increase toward the southwest with a maximum of IX occurring near the confluence of the Ohio and Mississippi Rivers, again near the historic site of the New Madrid earthquakes of 1811–1812. A potential earthquake

problem in some areas is the liquefaction of loose, saturated fine sands and silts. This would occur most likely in the alluvial areas in the zones IX and VIII, and possibly in zone VII, found in Figure 18.12. Homes, industrial and commercial building, bridges, and fill embankments would likely be affected.

It is not warranted in the midwestern United States to build structures that will not sustain some damage in a severe earthquake. The cost to provide this is simply too great, relative to the level of risk involved. A structure should be strengthened to prevent collapse and loss of life but funds are unwisely

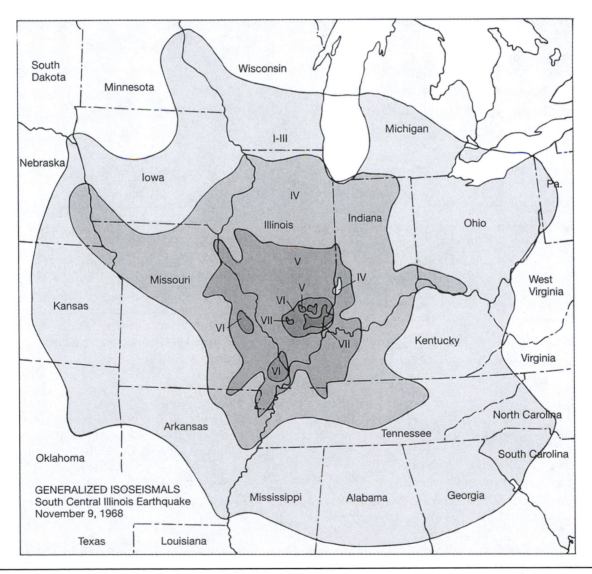

Figure 18.11   Generalized isoseismal map of the south-central Illinois earthquake, November 9, 1968. Roman numerals refer to the MM intensity scale.

spent if used for total prevention with expenditures significantly greater than the cost to repair damage that would occur.

San Fernando, California, Earthquake

A number of lessons can be learned from analyzing the damage in the San Fernando earthquake of 1971 (M = 6.5). This earthquake caused 65 deaths and more than half a billion dollars in property loss, but was not called a "major" earthquake ($M > 7$). Prior to its occurrence, the Los Angeles area had 40 years of seismic building codes regulating development. Of course, codes are intended primarily to prevent structural collapse during an earthquake; they do not guarantee the usefulness of the structure after the shock.

Highway bridges and embankments were particularly prone to damage in the San Fernando earthquake. Design weaknesses included the following:

1. Concrete columns did not withstand twisting and grinding; increased strength and closer spiral reinforcing was needed.

2. Expansion joints did not hold bridge decks in place. Movement could have been reduced by using tension retaining rods.

3. Inadequate bond where concrete shattered; addition of bars at top of block footings and bottom of column caps was needed.

4. Weak connection between pile caps and footings; anchor system was needed that could accommodate tension.

5. Shear keys did not provide lateral restraint of bridge abutments; larger shear keys and longitudinal ties were needed.

6. Inclusion of dynamic earthquake forces in the analysis of a structure is needed rather than using "equivalent" static forces.

Loma Prieta, California, Earthquake

On October 17, 1989, at 5:04 PM, Pacific Daylight Time, a magnitude 7.1 earthquake occurred on the San Andreas Fault, 16 km (10 mi) northeast of Santa Cruz, California (McNutt and Sydnor, 1990; Schiff, 1998). This major event takes its name from the Loma Prieta mountain peak located less than 10 km (6 mi) from the epicenter. The epicentral location is shown on Figure 18.13, with San Jose 33 km (21 mi)

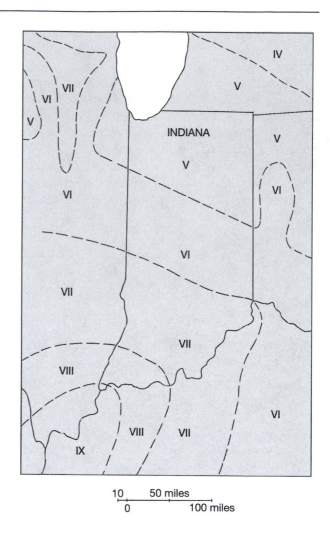

Figure 18.12   Maximum historical intensity map of Indiana, 1811 to 1974. Roman numerals refer to the MM intensity scale.

to the northeast and the cities of San Francisco and Oakland about 90 km (55 mi) to the northwest.

The focal depth of the earthquake was 17.6 km (11 mi), thus qualifying it as a shallow-focus earthquake. It registered the following magnitude values: M_s = 7.1, M_b = 6.5, M_L = 7.0, and M_w = 6.9 (refer to the previous discussion in this chapter for the distinction between these different earthquake magnitude measurements).

Intensities from the earthquake reached VIII on the MM intensity scale throughout much of the epicentral area. Isolated values of MM IX were assigned to the collapse of the elevated section of Interstate

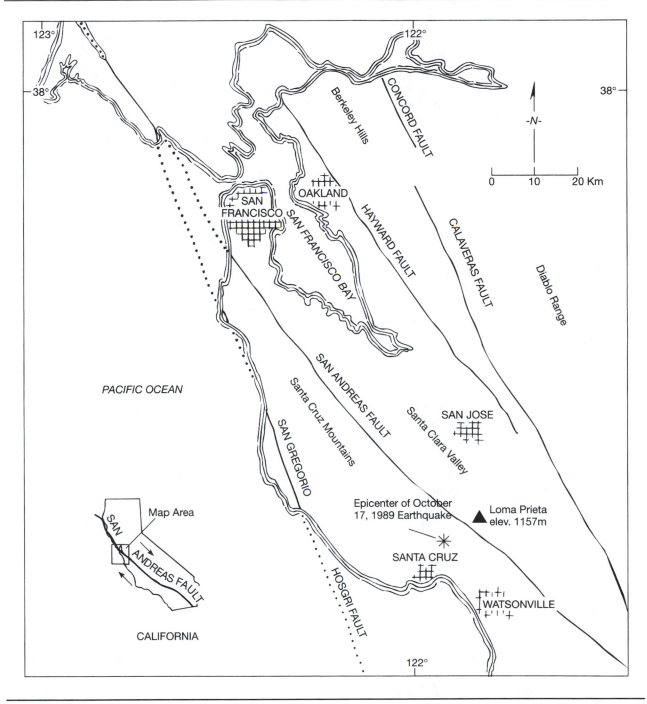

Figure 18.13   Map of affected area, Loma Prieta earthquake, October 17, 1989.

880 (Nimitz Freeway) in Oakland, to the damaged Highway 480 (Embarcadero Freeway), and to the Marina District in San Francisco. Some areas of San Francisco underlain by thick deposits of Quaternary bay mud and sand dunes experienced intensities of 1 to 2 MM units higher than the central part of the city underlain by bedrock. The area shaken at intensity VII or greater (shaking that causes significant structural damage) measured 4300 km² (1860 mi²) on land. As a comparison, the area of MM VII or greater in the 1906 San Francisco earthquake was 48,000 km² (18,750 mi²) on land, or about 11 times greater. Most

people in the earthquake's vicinity experienced about 10 to 15 seconds of ground shaking stronger than 0.1g. Felt by millions of people, the P wave was experienced first, followed by the strong shaking associated with the S waves and surface waves. The strongest ground shaking recorded in an area independent of structures (which amplify the motion) was 0.64g horizontal. It yielded little damage, whereas damage nearby, particularly on ridges, was extensive. This suggests that shaking on the ridges and in much of the epicentral area was stronger than 0.64g.

Numerous ground failures occurred over an area 100 km long by 40 km wide (62 mi by 25 mi), extending from San Gregorio to Hollister. Ground failures included several types of landslides, most commonly rock falls, rock slides, soil falls, and soil slides, of less than 100 m³ (130 yd³). Additional coastal bluff landslides occurred along the coast to Marin County north of San Francisco. Liquefaction failures were widespread; liquefaction of unconsolidated, saturated, sandy deposits occurred along alluvial plains and near the coast from Oakland to Salinas, and produced numerous sand boils and mud volcanoes. A dramatic example of damage was the collapse of a portion of Highway 1 spanning Struve Slough where strong shaking caused liquefaction of the subgrade soils.

Damage caused by the Loma Prieta earthquake can be considered under two categories: (1) damage near the epicenter and (2) damage located a far distance from it (more than 81 km [50 mi]). The area near the epicenter includes the Santa Cruz Mountains plus the cities of Santa Cruz and Watsonville.

Severe ground shaking in the Santa Cruz Mountains caused major damage—local peak accelerations may have approached 1.0g. Eyewitness reports indicate household items flew across the room and water tanks became airborne. Nearly 5000 homes in unincorporated areas of Santa Cruz County experienced earthquake-related damage and some houses were thrown off their foundations. A massive landslide closed the northbound lanes of Highway 17 for 33 days and large slumps severely damaged dozens of homes in several housing subdivisions.

In flatter terrains near the Santa Cruz Mountains, ground shaking displaced sidewalks and curbs. In the town of Los Gatos, about 31 commercial buildings and 318 residences experienced $240 million of damage. Much of the damage was localized, involving masonry structures or older residences. On the

Stanford University campus near Palo Alto, 25 of the 60 damaged buildings required structural reinforcement or repair for an estimated cost of $160 million.

In Santa Cruz, damage to commercial properties, residential structures, and municipal properties was about $51 million, $50 million, and $12 million, respectively. On the nearby University of California campus, about $5.8 million was incurred, of which about $4.8 million was structural damage. In the downtown area, Pacific Garden Mall and the adjacent residential blocks of mostly older homes experienced the greatest damage. These areas are underlain by recent alluvium of the San Lorenzo River and ground shaking was more damaging than in neighboring areas underlain by older, more consolidated marine terrace deposits. Older brick buildings were damaged extensively and several people were killed when buried under rubble. About 35,000 m² (375,000 ft²) of commercial property was demolished. Residential areas experienced collapsed chimneys and porches as well as foundation failures. One building burned down, presumably because of a gas line rupture. Some residences, recently upgraded structurally, showed little earthquake damage whereas many houses that had not been retrofitted were severely damaged. Similar damage occurred in the town of Watsonville. Brick buildings were damaged seriously and many older houses suffered partial collapse from foundation failure.

The Loma Prieta earthquake also caused damage far from the epicenter, about 100 km (60 mi) away in San Francisco and Oakland. The intensity of ground shaking typically decreases with distance from the epicenter for sites underlain by bedrock or well-consolidated soil. In contrast, sites on bay mud can experience high levels of shaking because of amplification of the wave energy by the soft soil. Consequently, areas underlain by bay mud and soft artificial fill near San Francisco underwent liquefaction. Most of the $2 billion damage for the city occurred in the Marina District underlain by artificial fill. For the city as a whole, 22 buildings suffered so much damage that they had to be demolished, 363 were declared unsafe, and 1,250 were suitable for only limited access pending repairs.

The hard hit Marina District is located on an old lagoon filled in to accommodate construction of the Panama–Pacific Exhibition, which celebrated the rebuilding of the city after the devastating 1906 earthquake and fire. Sand boils and evidence of lateral spreading indicate that liquefaction of the sandy fill led to extensive damage. Many of the affected structures were three- and four-story apartments with parking garages on the first floor. The limited shearing resistance of the open, first floor makes these buildings particularly prone to collapse during shaking or lateral spread of the underlying soil. Therefore, a combination of ground deformation and susceptible building design led to the collapse. Lateral spreading of fills also apparently damaged water mains, hampering initial efforts to fight fires resulting from the earthquake. Many unreinforced masonry buildings were damaged severely and six fatalities occurred when a brick structure collapsed.

Across the bay in Oakland, damage was concentrated in areas underlain by bay mud and sandy alluvium. The 2.4-km (1.5-mi) long collapsed portion of Interstate 880 (Cypress Structure) was underlain by thin bay mud and artificial fill, whereas the portion that did not fail was underlain by older alluvium. Locally, enhanced ground motion likely contributed to the disastrous pancaking of the freeway. Similarly, sections of Highway 280 and Highway 480 (Embarcadero Freeway) in San Francisco were closed because of earthquake damage. Other areas of the East Bay margins underlain by sandy fill and bay mud showed damage caused by liquefaction and lateral spreading. The Port of Oakland was damaged by lateral spreading. Liquefaction of sandy fill caused damage to the approaches to the San Francisco–Oakland Bay Bridge, at the Oakland International Airport, and at the Almeda Naval Air Station.

In Oakland, about 1400 residential, 200 commercial, and 12 public buildings sustained damages of about $29 million, $750 million, and $250 million, respectively. In addition, 13 commercial buildings were destroyed, with an estimated loss of $234 million, yielding an estimated total of about $1.3 billion. This does not include the demolition and reconstruction costs of the Cypress Structure. Also, significant structural damage occurred in areas underlain by older alluvium where presumably the ground shaking was less than for bay mud and artificial fill areas. The buildings in West Oakland, which partly collapsed or were thrown off their foundations, were either old Victorian-style or masonry structures that had not been seismically upgraded. In all, much of the damage in downtown Oakland, where the structures are underlain by older alluvium, occurred to masonry buildings.

The nature and distribution of the damage caused by the Loma Prieta earthquake illustrates that local geology and building construction details greatly affect the extent of structural damage during seismic shaking. Bay mud and loose, sandy, human-made fills yield increased seismic hazards for structures built on them. Also, the fatalities in San Francisco and Santa Cruz mostly occurred in unreinforced, masonry buildings. A modification of land use practices to avoid failure-prone areas and reinforcement of existing susceptible buildings is required to reduce future earthquake damage effects.

Human-Induced Earthquakes

Earthquakes are induced in a number of ways by human activity (Ehlen et al., 2005). There is, of course, the obvious contribution of shock waves by rock blasting for quarries and construction and the major effects of underground nuclear explosions. In the latter case, sensitive seismographs can sense the unique patterns of such blasts. Details of construction blasting are considered in Chapter 21.

In addition, many disturbances at the Earth's surface generate small seismic waves. Included are the wind, auto and truck traffic, railroad trains, and vibrating industrial machinery. These waves can be measured by seismographs and, collectively, provide part of the microseisms that supply the background displacements on a seismograph.

Earthquakes can be induced by other human activities that have a less obvious connection. The filling of reservoirs after dam construction can yield low-magnitude earthquakes and there is always concern that rapid filling may provide a disruptive shock. We also know that disposal of liquids can increase pore pressures along incipient fault zones. The Rocky Mountain Arsenal in Denver, Colorado, began experiencing this phenomenon in 1961. It was later established that low-magnitude earthquakes in an area not previously prone to earthquakes (see the seismic hazard map in Figure 18.10 on page 420) were directly related to injections of disposal liquids in the deep well. This deep-level disposal has been discontinued since the relationship was proven by a local Denver geologist. Injection-induced earthquakes have also been on the rise in Oklahoma (Hough and Page, 2015).

This suggests a mechanism for releasing smaller magnitude earthquakes to prevent the buildup of strain that would ultimately produce a major earthquake. Water injection into a fault zone might trigger the release of a magnitude 5 earthquake when the strain builds to that level. Also the length of a fault rupture might be reduced by pumping water from a fault zone at specific locations, thereby breaking the fault into segments. To do this work with an acceptable level of certainty of the outcome, more research is required leading to specific knowledge about pore pressures and groundwater movements in fault zones.

There is also the question of liability for damage caused by a 5.0 magnitude earthquake that is triggered by a governmental agency or government contractor. The government could be held responsible for this damage. Also it may be difficult to prove that a natural earthquake of greater magnitude would have eventually occurred to yield greater damage than that caused by the induced earthquake.

The Earth's Interior

Studies of travel paths and wave velocities in the Earth have revealed much information concerning its interior structure. Wave velocities increase in denser materials, waves are refracted and reflected at boundaries, and S waves are not transmitted beyond a depth of 2900 km (1802 mi). This latter fact leads to the shadow zone on the side of the Earth opposite from an earthquake, a zone where no S waves are received.

The refraction of waves reveals major discontinuities or changes in physical properties in the Earth's interior. From this, four major zones within the Earth can be distinguished: the crust, mantle, outer core, and inner core (Figure 18.14).

Crust

The crust is the thin, outermost layer of the Earth, often referred to as the skin of the Earth. The boundary between the crust and mantle is the Mohorovicic discontinuity, known more commonly as the "Moho" or "M" discontinuity. It was recognized first by the Croatian seismologist Andrija Mohorovicic, after whom it is named, in the study of a 1909 earthquake in Croatia.

Depth to the Moho is quite different below the continents than below the ocean basins. On the landmass, the crust averages about 33 km (20.5 mi) thick but it varies between 20 and 60 km (12.5 and 25 mi), with the higher value occurring under mountain ranges. Under the oceans, the crust averages about 5 km (3 mi) thick (Figure 18.15).

The average velocity of waves on the upper portion of the continental crust is about 6 km/sec (3.7 mi/sec) for the P wave and 3.5 km/sec (2.1 mi/sec) for the S wave. This is the typical velocity for a rock similar in composition to granite, granodiorite, or granite gneiss—rocks rich in silica and aluminum. A transitional layer, likely of intermediate composition, occurs below this zone and, at the base of the crust, a zone of basaltic composition occurs, as indicated by the seismic velocity of the layer. Representative values for these three zones of the crust, designated as layers 1, 2, and 3, respectively, are given in Table 18.7 for the crust in the New England region.

The basaltic layer of the crust apparently extends below the ocean basins as shown in Figure 18.15. The P wave velocity is somewhat lower for this zone below the ocean basins, 6.7 km/sec (4.1 mi/sec), as compared to the 7.0 to 7.2 km/sec (4.3 to 4.5 mi/sec) for the zone under the continents.

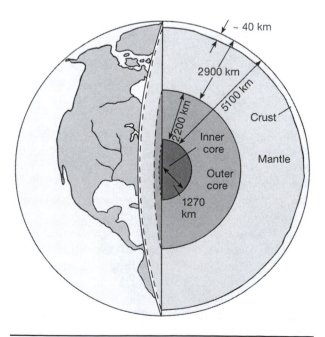

Figure 18.14   Major zones in the Earth's interior.

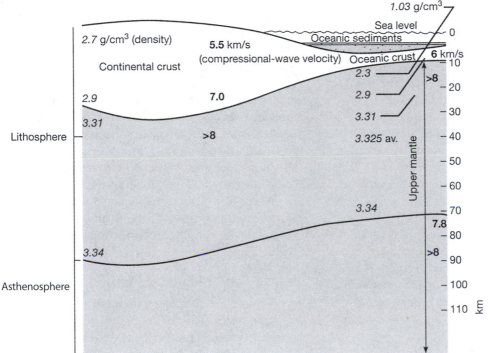

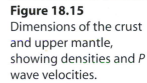

Figure 18.15
Dimensions of the crust and upper mantle, showing densities and *P* wave velocities.

Mantle

The zone between the crust and the outer core, about 2900 km (1800 mi) thick, is known as the mantle. The mantle transmits both *P* and *S* waves and, therefore, must consist of solid materials. It also comprises about 80% of the Earth's volume.

Recent studies have shown that a low-velocity zone exists within the upper mantle. The upper boundary of the low-velocity zone occurs at a depth of 70 to 80 km (43.4 to 49.2 mi) below the oceans and about 100 km (62 mi) below the continents; the low-velocity zone extends to a depth of nearly 300 km (186 mi) (see Figure 18.15). The upper boundary of the low-velocity zone forms the base of the lithosphere, with the asthenosphere occurring below. In plate tectonics discussions, the plates comprise the lithosphere, which moves slowly over the asthenosphere. Plates consist of crust and upper mantle although they are commonly referred to simply as crustal plates. The upper mantle is in motion, carrying the oceanic and continental crusts above it. The lower boundary of the low-velocity zone marks the contact between the upper, colder portion of the mantle and the lower, hotter part.

Core

The last major zone of the Earth is the core, which extends from a depth of 2900 km (1800 mi) to the center of the Earth at 6370 km (3950 mi). Analysis of earthquake waves that have traveled a great distance shows that two parts of the core exist, the outer core, which is 2200 km (1364 mi) thick, and the inner core with a radius of 1270 km (788 mi). Below the discontinuity at the base of the mantle, *S* waves

Table 18.7   Continental Crust Thickness, New England Region.

Zone	Thickness (km)	(mi)	Velocity (km/sec) P Wave	S Wave	Rock Composition
Layer 1	16	10	6.1	3.5	Granitic
Layer 2 (transition)	13	8	6.8	3.9	Intermediate
Layer 3	7	4	7.2	4.3	Basaltic
Moho					
Top of mantle			8.4	4.6	

no longer propagate and P waves undergo a reduction in velocity. The most widely accepted explanation for this drastic change is that the outer core is liquid. The P waves markedly increase in velocity at a depth of about 5150 km (3193 mi). This marks the boundary with the inner core and the velocity increase suggests that the inner core is solid.

As the Earth revolves around the Sun, it behaves in a manner consistent with a sphere of specific gravity of 5.5. Crustal rocks have a specific gravity of only 2.7 for the granitic zone and perhaps 2.9 in the basaltic layer. Rocks of similar composition, compressed under the extreme pressures at 2900 km (1800 mi) depth or more, would increase the specific gravity to about 5.7. Therefore, to produce the average specific gravity of 5.5, the core must measure about 15.0. To accomplish this, the core must be composed primarily of iron with perhaps about 8% nickel and some cobalt, in a manner similar to the proportions shown in metallic meteorites.

Engineering Geophysics

In the study of geophysics, the application of physical laws is used to discern details about the nature of the Earth. This procedure may apply to studies of basic science or to subsurface investigation, including mineral exploration or engineering construction. In engineering geology, the objective is typically related to construction but may involve mineral exploration as well. The application of geophysics to engineering works or construction projects is often referred to as engineering geophysics.

Six primary groups of geophysical methods are considered when applied to geologic studies. These include seismic, gravity, magnetic, electrical, ground penetrating radar, and well logging (Reynolds, 2011). The following sections present a summary of these methods, followed by the details of seismic refraction and electrical resistivity.

Seismic Methods

Seismic methods involve the propagation of waves through earth materials and are a direct extension of earthquake wave studies. Two specific seismic methods are available for exploration purposes: refraction and reflection. The refraction seismic method, because of its widespread application in engineering, is discussed in greater detail later on.

In the reflection seismic method, waves are reflected from rock unit boundaries in the subsurface and returned to the Earth's surface. In this situation, the angle of incidence, measured from the vertical, is equal to the angle of reflection. This technique is used by oil company seismic crews to explore for oil and gas. Actually, their objective is to look for useful geologic structures or "oil traps" that could accumulate petroleum. This method has the advantage of penetrating thousands of meters into the subsurface, but it also involves lengths of seismic cable measured in kilometers and numerous geophones to cover the distance along which the signal is returned. It is a major effort and expense for oil companies to support a seismic exploration crew in the field for extended periods of time.

Gravity and Magnetic Methods

The gravity and magnetic methods deal with the strength of the fields of gravity or magnetism generated between a mass of rock and the Earth. From any point on or above the Earth's surface the gravity or magnetic field is measured and compared to that of an adjacent area. Changes in values, known as anomalies, that supply information about the size, nature, and location of a high- or low-gravity or magnetic source within the Earth are recorded. This provides information concerning mineral exploration or details of the subsurface of interest in construction. The relationships have the form:

$$F = G \frac{m_1 m_2}{r^2} \qquad \text{(Eq. 18-17)}$$

for the gravity force. In the equation, m_1 and m_2 are masses, m_1 usually being that of the Earth concentrated at the Earth's center and m_2 that of the instrument; r is the distance between them; and G is the universal gravity constant = 6.670×10^{-8} in cgs units. The measure of F or gravity force is the "gal," named after Galileo and equal to 1 cm/sec^2, with the Earth's gravity force approximately equal to 980 gals.

For the magnetic method:

$$F = \frac{1}{\mu} \frac{P_0 P}{r^2} \qquad \text{(Eq. 18-18)}$$

where F is the magnetic attraction or repulsion depending on the sign, P_0 and P are poles separated by distance r, and μ is the magnetic permeability, which depends on the magnetic properties of the medium in which the poles are located. The unit of magnetic field strength used for geophysical studies is the gamma, defined as 10^{-5} oersted. The total magnetic field of the Earth is normally about 0.5 oersted.

Both the gravity and magnetic methods have been used in engineering geology studies, but they are usually applied in rather specialized situations. Also, they are typically performed by geophysicists rather than by engineering geologists, geological engineers, or civil engineers.

Electrical Methods

Electrical methods involve measuring the electrical properties of earth materials. In general, they consist of two varieties: the measurement of natural earth currents and the resistance to induced electrical flow. Natural earth current flow is generated under some geologic conditions in which an anode and cathode develop naturally in the subsurface and electron flow ensues. Measurement of the strength and extent of this current is useful in establishing the nature of the geologic conditions.

Electrical resistivity is related to the resistance to electrical flow through earth materials. In this method, a current is induced into the Earth, the resistivity is measured, and information about a basic property of the material under study is thereby determined. Electrical resistivity is the resistance between opposite faces of a unit cube of material. For a conducting cylinder with length L and cross-sectional area A, the resistivity ρ is:

$$\rho = \frac{RA}{L} \qquad \text{(Eq. 18-19)}$$

with R in ohms, A in cm^2, and L in cm to yield ρ in ohm-cm. The conductivity of a material, $K = 1/\rho$, has units in reciprocal ohm-cm. Because electrical resistivity is a primary method of interest in subsurface exploration for engineering construction, it is discussed in more detail in a later section.

Ground Penetrating Radar

Ground penetrating radar (GPR) is a geophysical technique that uses radar pulses to image the subsurface (Jol, 2008). It is a nondestructive method that utilizes electromagnetic radiation in the microwave band of the radio spectrum, a range of 10 MHz to 2.6 GHz, detecting reflected signals from subsurface structures. It has applications in the following media: rock, soil, ice, fresh water, pavements, and structures. Figure 18.16a shows a GPR survey being conducted in an open field in the Midwest and Figure 18.16b shows an anomaly (a buried object) indicated by such a survey.

GPR uses high frequency (usually polarized) radio waves to emit electromagnetic energy into the ground. When the energy encounters a buried object or boundary between materials, it may be reflected or refracted back to the surface. A receiving antenna then records the variations in the return signal.

The greatest depth of penetration is achieved in ice where the depth of exploration is up to several thousand meters. In Greenland, using low GPR frequencies, the depth of investigation can extend to the bedrock. Dry sandy soils or massive dry materials, including granite, limestone, and concrete, tend to be resistive rather than conductive, yielding a depth of penetration up to 15 m (50 ft). In moist and clay-rich soils and similar materials that have a high electrical

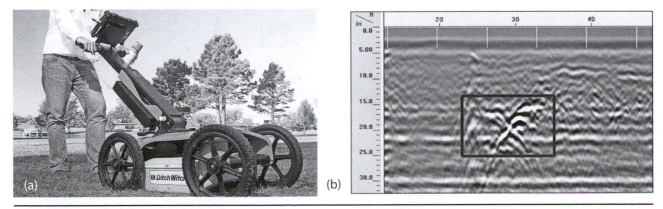

Figure 18.16　(a) GPR survey in an open field in the Midwest (Ditch Witch, Wikipedia [https://creativecommons.org/licenses/by-sa/2.0/deed.en].) (b) GPR image showing an anomaly. (Geophysical Survey Systems, Inc.)

conductivity, penetration may be as low as 15 cm (6 in). Table 18.8 shows the antenna frequency and soil penetration depth for various soil conditions.

GPR has many applications in several fields, including earth sciences. It has also been applied to prospecting for gold nuggets and diamonds in alluvial gravel beds by locating natural traps in buried stream beds where heavier particles accumulate. Engineering applications involve nondestructive testing of structures and pavements, locating buried objects and utility lines, studying soils and bedrock, and investigating ice and groundwater. For environmental remediation, GPR is used to study landfills (contaminant plumes) and other remediation sites. In archeology it is used to map archeological features and cemeteries whereas GPR is used in law enforcement to locate clandestine graves and buried evidence. Military use includes detecting mines, unexploded ordnance, and tunnels.

Well Logging Methods

Well logging methods include a variety of techniques that involve lowering instruments into a well or boring and generating data on subsurface lithologies as the instrument is pulled from the hole (Lau, 2000). These techniques were developed for oil exploration and, for 50 years or more, have been standard methods for obtaining subsurface information from exploration wells drilled for oil. Well logging is of particular significance today because these methods are now used in exploration for major engineering construction. Much information can be obtained through the transfer of expertise from standard oil exploration procedures to applications in engineering construction. When a borehole has already been drilled for sampling purposes,

the additional cost of well logging with geophysical equipment is quite small compared to the total exploration costs.

In well logging techniques, various probes or sondes are lowered down a completed hole to obtain information on rock units and rock boundaries. These methods are useful to confirm those lithologic boundaries, particularly in areas where core recovery was poor. The methods are also used to obtain data in rock zones that were not cored; this is a common occurrence in deep holes drilled for oil exploration.

The common borehole geophysical methods are summarized in Table 18.9. Electrical resistivity well logging is an essential part of the geophysical analysis performed on wells drilled by oil companies. The electric log shows the apparent resistivity of the subsurface formation, along with the spontaneous potential (SP) curves. Shallow wells in freshwater zones show featureless SP curves, whereas brackish and saltwater zones have a negative SP deflection in contrast to clay-rich zones. These show permeable strata in the potential oil-bearing zones in the subsurface.

Gamma-ray logging is performed in soil borings and in bedrock. In this procedure, the natural radiation of gamma rays that emanates from certain radioactive elements in the subsurface is measured. Materials containing higher concentrations of radioactive elements, such as uranium, thorium, and radioactive isotopes of potassium, yield greater gamma radiation. Typically, clay and shale contain more radioactive materials than do limestone, sandstone, or sand. Therefore, gamma-ray logs are used to indicate clay versus sand units in soils and the clayey and shaley beds versus limestone or sandstone in bedrock. Figure 18.17 shows the relative response of different types of geological intervals to gamma-ray activity.

Table 18.8   Antenna Frequency and Soil Penetration Depth of GPR.

Antenna	Approximate Penetration in Dense Wet Clay	Approximate Penetration in Clean Dry Sand	Example of Smallest Visible Object
100 MHz	6.0 m (20 ft)	18.0+ m (60+ ft)	Tunnel at 18 m (60 ft) depth; 60 cm (2 ft) diameter pipe at 6 m (20 ft)
250 MHz	4.0 m (13 ft)	12.0 m (40 ft)	90 cm (3 ft) diameter pipe at 12 m (40 ft) 15 cm (6 in) diameter pipe at 4 m (13 ft)
500 MHz	1.8 m (6 ft)	4.4 m (14.5 ft)	10 cm (4 in) diameter pipe at 4 m (13 ft) 0.5 cm (3/16 in) hose at 1.8 m (6 ft) or less
1000 MHz	0.9 m (3 ft)	1.8 m (6 ft)	0.5 cm (3/16 in) hose at 0.9 m (3 ft) Wire mesh, shallow
2000 MHz	0.15 m (0.5 ft)	0.6 m (2 ft)	Monofilament fishing line

Table 18.9   Common Borehole Geophysical Methods.

Method	Description and Uses	Conditions Needed
Electric Logging		
Electric resistivity	Several methods with electrode spacing from 0.4 m (16 in) to 5.8 m (19 ft) apart. Measures resistivity between beds. Rock types and effective porosity can be determined.	Fluid-filled uncased hole. Fresh drilling mud usually required.
Spontaneous potential	Potential difference between single electrode in hole and another at ground surface. Values remain relatively constant against shale, reduced values for non-shaley layers.	Fluid-filled uncased hole. Fresh mud.
Radiation Logging		
Gamma ray	Measures inherent radioactivity of rock units. Shale yields higher activity than sandstone.	Fluid-filled or dry, cased or uncased hole.
Neutron	Measures hydrogen-rich aspect of rock or presence of water. Indicates filled porosity of rock. Discerns oil-filled pores from water-filled ones.	Fluid-filled or dry, cased or uncased hole.
Sonic logging	Measures seismic velocity, can correlate with specific lithologies.	Not affected by fluid, hole size, or mud.
Temperature Logging	Determines thermal gradient, presence of cement behind casing, zones of active gas flow or of lost circulation. In uncased holes, locates fissures and solution openings.	Cased or uncased hole. If no fluid present, logged at slow speed. Fluid should be undisturbed.
Caliper Survey (section gauge)	Determines hole or casing diameters. Locates fractures, solution openings, and other cavities. Correlation of formations, used to select zone for packer placement, placement of gravel pack, evaluation of explosive efficiencies, finding details of abandoned wells.	Fluid-filled or dry, cased or uncased hole.

Gamma-ray logging is commonly performed on borings in glacial material in the Midwest and in other areas. Logging is performed through the casing, which can be composed of either steel or PVC. Borings for foundation studies, water supplies, or environmental studies can be logged in this fashion. The method has proved particularly useful in studies for sanitary landfill sitings, general stratigraphic studies, and analysis of earthquake potential.

It is often important to identify sandy zones versus silts or clays in the subsurface (Figure 18.18 on the following page). In addition, a thick sequence can provide a diagnostic signature curve that can be used to correlate unconsolidated sequences from one area to another. For example, the presence of sand at the base of a unit with higher amounts of clay above it can provide a diagnostic curve for a unit that has a designation of fining upward (or getting finer in grain size in the upward direction). By con-

trast, increasing amounts of sand yield a sequence that is getting coarser upward.

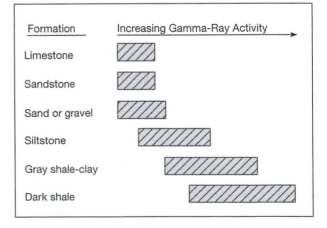

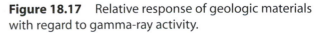

Figure 18.17   Relative response of geologic materials with regard to gamma-ray activity.

Gamma-ray logging of soil sections adjacent to existing sanitary landfills provides details on the migration of contaminant plumes. Water-bearing zones only a foot or so in thickness can be easily missed during subsurface drilling and sampling, yet migrating groundwater contamination is transmitted by these zones because they control the pathways of leachate migration.

Additional details on geophysical logging methods can be obtained in the literature (Lau, 2000; Reynolds, 2011). Cross-hole seismic techniques are also available. In that method, a shock wave is generated in one hole and its travel time to an adjacent hole is measured. Seismic velocities are determined based on distances between the boreholes.

Refraction Seismic Method

The refraction seismic method is used for exploration at shallower depths than the reflection seismic method and, therefore, is more appropriate for construction-related projects. A depth of 30 to 60 m (100 to 200 ft) is the usual limit for this method (Reynolds, 2011; Sjogren, 2013).

At major lithologic boundaries, seismic waves are refracted, or bent, depending on the contrast in velocity between the two layers. This refracted wave travels to the surface before direct waves, traveling just below the surface, arrive. Primary (or *P*) waves are the fastest seismic waves, i.e., arrive first, and are used in refraction seismic studies. The refraction seismic method provides two types of information: seismic velocities and depth to those lithologic boundaries where the velocities change suddenly.

To initiate primary waves in the ground, energy must be released or generated. This can be accomplished by detonating a blasting cap or larger explosive in the ground or by hitting a sledgehammer against a metal plate on the Earth's surface. The depth of penetration desired for the shock wave, the sensitivity of the seismograph, and the background vibration noise will dictate how much energy must be supplied to obtain the needed detail.

When an explosion is detonated, a wide range of frequency waves is initiated, up to 1000 Hz (or cycles per second), but the Earth soon filters out the high-frequency portion, allowing the lower frequencies, from 15 to 70 Hz, to propagate. Hence, the seismograph measuring the *P* wave arrivals should be sensitive to this range of frequencies.

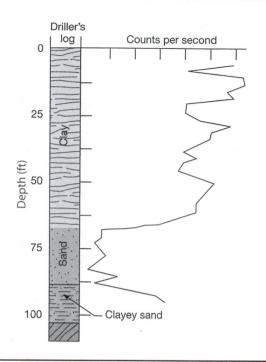

Figure 18.18 Gamma-ray log versus driller's log for unconsolidated materials.

A seismograph for engineering geophysics applications involves an accurate timing device to note both the instant of energy generation and of the first wave arrival at a specific location. A geophone, consisting of a coil in a magnetic field, is inserted into the soil below the root zone. When the shock wave arrives, the coil moves, generating a small electrical current that registers as an offset on the recording device.

Seismographs can be single-channel or multiple-channel devices. For a 10-channel seismograph, for example, a string of 10 geophones is used to record a single energy-release event. Travel times and distances to each geophone are recorded and become the basis for velocity measurements and depths for lithologic layers. Smaller equipment consists of a single-channel, single-geophone arrangement and the energy is usually input by sledgehammer on a metal plate. The geophone is located a specific distance away and the travel time is determined. Then the geophone is moved to the next station at a specific distance and the procedure is repeated. This continues until the specified total distance is covered.

The seismic velocity (*P* wave velocity) has major importance in seismic refraction work. Table 18.10

Table 18.10   Seismic Velocities of Different Materials for Engineering-Related Studies.

Soil to Rock, Nonsaturated	Seismic Velocities	
	m/sec	ft/sec
Sand	200–2000	650–6500
Loess	300–600	1000–2000
Alluvium	500–2000	1650–6500
Loam	800–1800	2600–5900
Clay	1000–2800	3300–9200
Marl	1800–3800	2600–12,500
Saturated soil	1520	5000
Sandstone	1500–4300	5000–14,000
Limestone	1700–6400	5600–21,000
Slate and shale	1800–4600	6000–15,100
Granite	4000–5700	13,000–18,700
Quartzite	6100	20,000

provides information on seismic velocities. From this table we can observe that saturated soil below the groundwater table can be discerned from the nonsaturated soil above the water table. Bedrock, whether above or below the water table, typically has a seismic velocity above 1520 m/sec (5000 ft/sec) and its degree of saturation is not directly involved in determining the velocity. In general, as bedrock becomes denser with depth, the seismic velocity increases. Velocity also increases with reduced porosity or increased cementation of sedimentary rocks. Porous, poorly cemented sandstones have low velocities but strongly cemented sandstones have velocities near that of quartzite. Calcareous shale (cementation shale) will have higher velocities than other varieties (compaction shale). Changes in jointing, dipping beds, and overall cementation will affect the seismic velocity as well.

Seismic velocities have been used for many years to predict the ease of excavation of earth materials. Some bedrock varieties require blasting, whereas others can be ripped using a bulldozer with a rock-ripper attachment. With this hydraulic equipment, ripper teeth are forced into the rock and the bulldozer is pulled forward. Thinly bedded or foliated rocks, shales, claystones, and mudstones can be removed relatively easily by this method and usually is cheaper than blasting the rock. Figure 18.19 on the following page provides specific information based on the ability of a Caterpillar D-9 dozer to remove rock materials by ripping. The rule of thumb that rocks can be ripped using a D-9 dozer if the seismic velocity is below 2150 m/sec (7000 ft/sec), should be applied with reasonable caution.

To obtain information on depths to lithologic boundaries, a consideration of wave refraction must be included. This is governed by Snell's law. Figure 18.20 (next page) illustrates refraction for two cases: (1) a higher velocity layer below the surface layer and (2) a lower velocity layer below the surface layer. When energy is released in the ground, waves move out in all directions from that point. This can be indicated in the form of a ray or as a wavefront.

Snell's law indicates that:

$$\frac{\sin i}{\sin r} = \frac{V_1}{V_2} = n \qquad \text{(Eq. 18-20)}$$

where: n is the index of refraction, i is the angle of incidence, and r is the angle of refraction (Figure 18.20).

In Case 1, where there is a higher velocity layer below the surface layer, the ray is bent away from the normal and toward the interface. This can generate a wave that will return to the Earth's surface. For Case 2, where there is a lower velocity layer below the surface layer, the wave is bent downward and no opportunity exists for it to return to the surface. Therefore, lithologic units must increase in velocity with depth to ensure that a refracted wave can return to the Earth's surface.

Referring to Case 1 again, as i values increase for different rays, r will increase until a value of $r = 90°$ is reached. When $r = 90°$, $\sin r = 1$ and as $\sin i = V_1 \sin r / V_2$,

$$\sin i_{cr} = \frac{V_1}{V_2}(1) = \frac{V_1}{V_2} \qquad \text{(Eq. 18-21)}$$

where i_{cr} is the critical incident angle. This provides a ray that moves along the interface of the two layers at a velocity of V_2. As the ray proceeds, it serves as a point source for new primary waves, which emanate from the boundary as the wave proceeds. This is illustrated in Figure 18.21 on page 435.

In Figure 18.22 (page 435), waves move back through Layer 1 at V_1 velocity after being carried along the interface at V_2 velocity. Eventually, a wave refracted along the interface will initiate a second-generation wave that reaches the surface before a direct wave can travel immediately below the ground surface. It will travel back along a wave direction, also related to i, as shown in Figure 18.22. Also shown is a

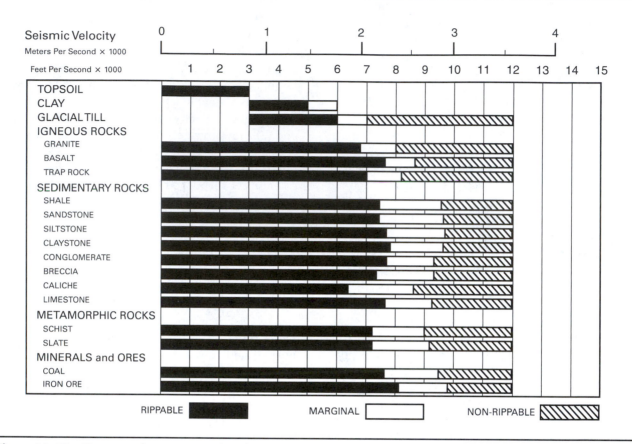

Figure 18.19   Ripper performance as related to seismic wave velocities. (Caterpillar. January, 2014. Track-type tractors. *Caterpillar Performance Handbook* (44th ed.). San Diego, CA: Caterpillar.)

time-distance curve that indicates when the V_2 velocity begins to exert its influence on the first arrivals. The refracted wave arrives first and will continue as the first arrival. In the same fashion, if a third layer exists with V_3 velocity, such that $V_3 > V_2 > V_1$, eventually the wave from Layer 3 will control the first arrivals (see Figure 18.22).

The velocities of the three layers are determined from the slopes of the time-distance plots. Because velocity = distance/time, velocities are the reciprocals of the slopes of these straight lines. The relationships

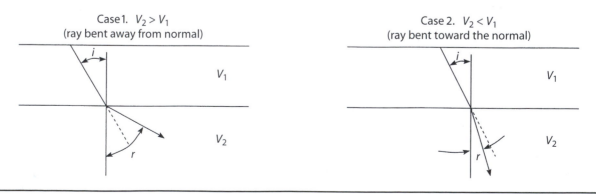

Figure 18.20   Illustration of Snell's law of wave refraction.

for the thickness of the lithologic layers can be calculated based on the geometry involved for the ray paths.

For the Z_1 depth, the relationship is as follows:

$$Z_1 = \frac{X_{12}}{2} \sqrt{\frac{V_2 - V_1}{V_2 + V_1}} \qquad \text{(Eq. 18-22)}$$

where X_{12} = the critical distance for the arrival of the wave influenced by V_2; or

$$Z_1 = \frac{T_2 V_1}{2 \cos i_{12}} \qquad \text{(Eq. 18-23)}$$

where T_2 is the time intercept for the $1/V_2$ slope and $\cos i_{12}$ is the cos of the critical angle where $\sin i_{12} = V_1/V_2$.

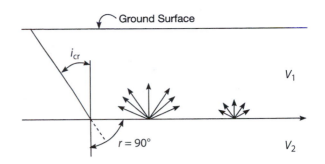

Figure 18.21 Refraction and critical wave propagation.

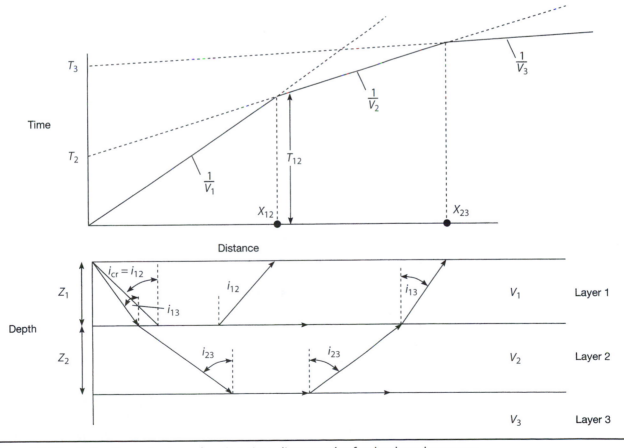

Figure 18.22 Refraction seismic exploration, time-distance plot for the three-layer case.

For the Z_2 thickness

$$Z_2 = \frac{X_{23}\left(\sin i_{12} - \sin i_{13}\right) - Z_1 \cdot 2\left(\cos i_{13} - \cos i_{12}\right)}{2 \sin i_{12} \cos i_{13}}$$

(Eq. 18-24)

where X_{23} is the critical distance for the wave influenced by V_3 and the angles are indicated in Figure 18.22. Also

$$Z_2 = \left(T_3 - \frac{2Z_1 \cos i_{13}}{V_1}\right)\frac{V_2}{2 \cos i_{12}}$$

(Eq. 18-25)

where T_3 is the time intercept for the $1/V_3$ slope.

Several approximations have been proposed for the two-layer and three-layer thickness calculations. They are as follows:

$$\frac{2Z_1}{V_1} + \frac{X_{12}}{V_2} \cong T_{12} \text{ (two layer)}$$

(Eq. 18-26)

$$Z_1 + Z_2 \cong 0.85 Z_1 + \frac{X_{23}}{2}\sqrt{\frac{V_3 - V_2}{V_3 + V_2}} \text{ (three layer)}$$

(Eq. 18-27)

Example Problem 18.3

Use the time-distance plot shown in Figure 18.23 to calculate Z_1 and Z_2. See also Figure 18.22 for a similar three-layer cross section. First, use both the critical distance method and the time intercept method to find both depths. Then use the approximate methods for comparison.

Answers

Where:

$$i_{12} = 15.5°$$
$$i_{13} = 5.4°$$
$$i_{23} = 20.65°$$

$$\sin i_{12} = \frac{V_1}{V_2} = \frac{244}{915} = 0.2666$$

$$\cos i_{12} = \cos 15.5° = 0.9636$$

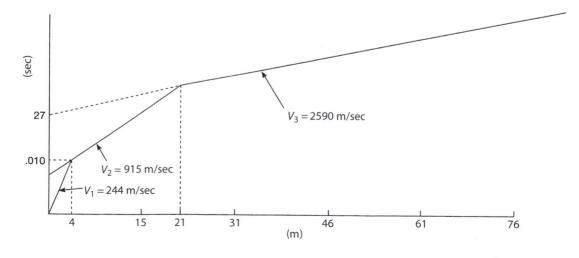

Figure 18.23   Refraction seismic exploration, time-distance plot for Example Problem 18.3.

a. $Z_1 = \dfrac{3.6}{2}\sqrt{\dfrac{915-244}{915+244}}$

$= 1.8(.76) = 1.37 \text{ m}$

b. $Z_1 = \dfrac{T_2 V_1}{2\cos i_{12}}$

$= \dfrac{0.01(244)}{2(0.9636)} = 1.27 \text{ m}$

c. $Z_2 = \dfrac{70\left(\sin\dfrac{V_1}{V_2}-\sin\dfrac{V_1}{V_3}\right)-Z_1\cdot 2(\cos i_{13}-\cos i_{12})}{2\sin i_{12}\cos i_{13}}$

$\sin i_{13} = \dfrac{V_1}{V_3} = 0.0941, \quad i_{13}\, 5.4°, \quad \cos 5.4° = 0.9956$

$\sin i_{23} = \dfrac{V_2}{V_3} = \dfrac{915}{2590}$

$= 0.3529, \quad i_{23}\, 20.65°, \quad \cos 20.65° = 0.9357$

$Z_2 = \dfrac{21.3(0.266-0.09411)-1.27(2)(0.9956-0.9636)}{2(0.2666)(0.9357)}$

$= \dfrac{3.674-0.08128}{0.4989} = 7.21 \text{ m}$

d. $Z_2 = \left(T_3 - \dfrac{2Z_1\cos i_{13}}{V_1}\right)\dfrac{V_2}{2\cos i_{23}}$

$= \left[0.027 - \dfrac{2(4.1)(0.9956)}{244}\right]\dfrac{915}{2(0.9357)} = 8.2 \text{ m}$

e. $\dfrac{2Z_1}{V_1} + \dfrac{X_{12}}{V_2} \cong T_{12}, \quad T_{13} \cong 0.010 \text{ sec}$

$\dfrac{2Z_1}{244} + \dfrac{3.6}{915} \cong 0.010$

$2Z_1 + \dfrac{3.6}{3.71} \cong 2.44$

$Z_1 \cong 2.44 - 0.97 \cong 1.47 \text{ m}$

f. $Z_1 + Z_2 \cong 0.85 Z_1 + \dfrac{X_{23}}{2}\sqrt{\dfrac{V_3-V_2}{V_3+V_2}}$

$\cong 0.85(1.27) + \dfrac{21.3}{2}\sqrt{\dfrac{2590-915}{2590+915}}$

$\cong 8.44 \text{ m}$

$\cong 8.44 - Z_1 \cong 8.44 - 1.27 \cong 7.17 \text{ m}$

Earth Resistivity Method

Derivation

The earth resistivity method, based on the apparent resistivity of earth materials, is used to discern details about the subsurface. It, like the seismic method, is an indirect procedure and the data should be compared to boring logs or to an outcrop description to ensure that the lithology is properly identified. Many mistakes have been made in the application of geophysics to engineering exploration that could have been precluded if this simple rule were followed.

To begin developing this method, a review of the relationship for resistivity ρ is required. Recall that:

$$\rho = \dfrac{RA}{L} \qquad \text{(Eq. 18-28)}$$

where R can be expressed in terms of the Ohm's law relationship, $E = IR$, where $E = V =$ electromotive force or voltage, $I =$ current flow in amps. So $R = E/I$ and, therefore,

$$\rho = \dfrac{EA}{IL} = \dfrac{VA}{IL} \quad \text{and} \quad V = \dfrac{I\rho L}{A} \qquad \text{(Eq. 18-29)}$$

Figure 18.24 shows a diagram of the earth resistivity array. If materials of constant resistivity are assumed, with current electrodes located at Points A and B, plus potential electrodes at C and D, the resistivity can be determined. Because of this configuration and the volume of material through which the current flows, the voltage or potential of an electrode at Point C will be:

$$V_C = \dfrac{I\rho}{2\pi}\left(\dfrac{1}{r_1} - \dfrac{1}{r_2}\right) \qquad \text{(Eq. 18-30)}$$

and, in a similar fashion, the potential of an electrode at Point D will be:

$$V_D = \frac{I\rho}{2\pi}\left(\frac{1}{r_3} - \frac{1}{r_4}\right) \qquad \text{(Eq. 18-31)}$$

Referring to Figure 18.24, the potential difference V measured between C and D is $V_C - V_D$. Subtracting these and solving in terms of ρ yields the following:

$$\rho = \frac{2\pi V}{I}\frac{1}{\left(\dfrac{1}{r_1} - \dfrac{1}{r_2}\right) - \left(\dfrac{1}{r_3} - \dfrac{1}{r_4}\right)} \qquad \text{(Eq. 18-32)}$$

which is the fundamental equation for the resistivity method.

To this point, uniform resistivity has been assumed for the material under study. When this is the case, the resistivity value will remain the same, independent of the volume of material measured. However, for natural earth materials the resistivity is not constant throughout the cross section, so a change in the volume of material measured will likely yield a change in the resistivity value. In the case of nonuniform materials, the value of ρ is called the *apparent resistivity*.

For most engineering projects in the United States, the Wenner configuration of electrode spacing is used. In this arrangement the four electrodes are spaced equally along a line, with the two on the outside being current electrodes and the two on the inside being potential electrodes, as shown in Figure 18.25. The distance between electrodes is designated as a, which is increased as successive readings are taken, keeping the same central point for the spread. With an equal distance between the electrodes (refer to Figure 18.24) the equation for resistivity is simpli-

fied as shown below. In Figure 18.24, $r_1 = r_4 = a$ and $r_2 = r_3 = 2a$. Therefore, for the Wenner configuration:

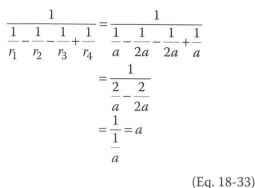

$$\text{(Eq. 18-33)}$$

and

$$\rho = 2\pi a \frac{V}{I} \qquad \text{(Eq. 18-34)}$$

It is also assumed that the depth of penetration of the current flow lines is approximately equal to a, the electrode spacing. Hence, by increasing a, the depth of exploration is increased.

Application of Method

In practice, when using the Wenner configuration, each potential electrode (P_1, P_2) is a distance $a/2$ from the center position and each current electrode (C_1, C_2) is a distance $3a/2$ from the center. Electrode spacing is increased from one reading to the next, yielding data for ρ versus a spacing for the investigation. If we want to increase the electrode spacing (a spacing) by 3 m (10 ft) from one reading to the next, each potential electrode must be moved 1.5 m (5 ft) further from one reading to the next, and each current electrode must be moved 4.5 m (15 ft). This simple procedure for moving electrodes makes the work proceed faster in the field.

Several suggestions have been proposed for electrode spacing in electrical resistivity surveys, or *soundings* as they are sometimes called. For rapid reconnaissance, the "leap-frog" sequence is quite useful. With this procedure, the first reading is taken at the largest electrode separation that will be required or is anticipated; refer to this as a_0. It may be, for example, equal to 27.5 m (90 ft). The next spacing will be $a_0/3$ and third $a_0/9$. The location of the potential

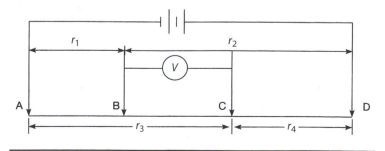

Figure 18.24   Elevation view, electrical resistivity survey.

electrodes of one reading becomes the location of the current electrode for the next reading. Usually four or five successive readings will cover the range of *a* spacing desired, as illustrated in Figure 18.26. For our example, the first *a* spacing would be 27 m (90 ft), then 9 (30), 3 (10), 1 (3.33), and 0.303 (1). Intermediate points from 9 (30) to 27 (90) and 3 (10) to 9 (30) would be required to complete a detailed survey.

For some interpretations of the data, equal interval values for the *a* spacing make that phase of the study more manageable. These interpretations include the Barnes layer method and the Moore cumulative method. Therefore, increases of 1.5 m (5 ft) may be convenient for detailed work. Use of the Wenner configuration, as indicated before, requires that the potential electrodes each be moved *a*/2 or 0.75 m (2.5 ft) and the current electrodes moved 3*a*/2 or 2.25 m (7.5 ft) between successive readings. Intervals involving 0.15 m (0.5 ft) are manageable but any other fractions of feet or decimal values of meters are too cumbersome to work with conveniently in the field and still allow for reasonable productivity.

For scientific work and some general geologic studies, the optimum electrode spacing would be that which is equally spaced on a logarithmic scale. For analyses, the ρ versus *a* spacing values are plotted on double logarithmic paper where the electrode intervals will appear to be equally spaced.

One disadvantage of the Wenner configuration is that variations laterally in the subsurface can be misinterpreted as being caused by variations in the material with depth. As the *a* spacing increases, so is the depth of penetration, and the lateral extent of the survey is widened as well. If a contrasting material is intercepted laterally, it shows up as a change in ρ which, more than likely, is assumed to be caused by a change in a layer at depth. This confusing detail can be discerned by either taking measurements along two perpendicular lines or by using the Lee electrode modification procedure, shown in Figure 18.27.

In the Lee electrode configuration, an additional electrode P_0 is driven into the ground at the center of the spread. Three readings are taken for each electrode spacing, the first being the total value using normal potential electrodes P_1 and P_2

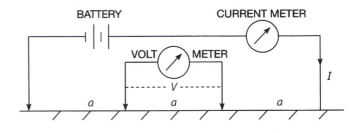

Figure 18.25 Wenner configuration, electrical resistivity survey.

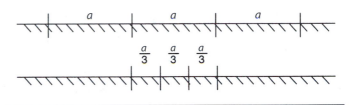

Figure 18.26 Leap-frog sequence of electrical resistivity survey readings.

in the conventional way. For the other two, the voltage is read between P_1 and P_0 and P_2 and P_0 in succession. If the readings are similar for both, i.e., plots overlap, as shown in Figure 18.27b, lateral variations in resistivity are not great. If they differ considerably, i.e., diverge, as shown in Figure 18.27c, the lateral change is significant. A new center point away from the contrasting material may have to be selected or the traverse rotated 180° to obtain a better measure of subsurface changes. Several

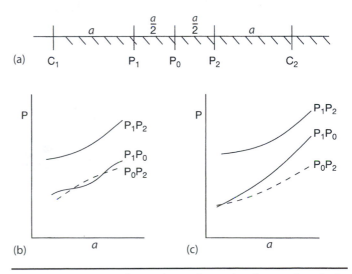

Figure 18.27 Lee electrode configuration for an electrical resistivity survey.

additional guidelines are recommended for electrical resistivity soundings when using the Lee electrode configuration:

1. The largest electrode spacing should be at least 3 times (but preferably 5 to 10 times) the maximum depth of interest.

2. The smallest electrode separation should be less than one-half the minimum depth at which a change in material is expected. However, electrode separations of less than 0.6 m (2 ft) are seldom needed.

3. Reading intervals should be more closely spaced at small electrode separations than at large electrode separations.

This, of course, is in contrast to the use of equally spaced *a* spacings, a convenient procedure for interpreting the Barnes layer method and the Moore cumulative method. For the Lee configuration, to increase reading interval with increasing spacing, readings should be equally spaced according to the log scale of distances. This procedure is mostly appropriate to scientific investigations rather than engineering site investigation studies.

Interpretation

Interpreting apparent resistivity data is the most difficult part of this procedure. A number of different methods are used for this purpose and they are summarized in Table 18.11.

Barnes Layer Method

The Barnes layer method was developed because of a perplexing problem that occurs with earth resistivity studies. If a layer at depth is to

influence the resistivity value when this layer is included in the volume, the resistivity contrast has to be extremely large. The reason for this is illustrated in Figure 18.28. Each time the electrode spacing is increased, a new incremental volume is added to the total volume to be measured. As the number of readings grows, the increment added to the volume represents a smaller and smaller contribution to the total being measured. A small contrasting resistivity near the surface can affect the ρ measurements as much or more than a large contrasting portion at depth. Stated directly, the ability of near-surface materials to overwhelm the effects of deeper lying deposits is an inherent problem with the resistivity method.

The Barnes layer method attempts to delineate the resistivity of layers within the sample volume studied. The thickness of the layer is assumed to equal the increment in the electrode spacing. Equal increases of increments work best if, for no other reason, it allows the Moore cumulative method to be used for

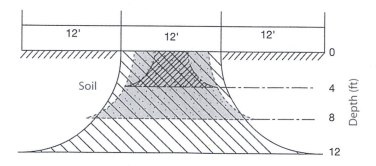

Figure 18.28   The increase in volume of soil measured that occurs with increasing *a* spacing for the Wenner configuration, electrical resistivity survey. The cross-hatched volume was measured using a 4-ft electrode separation; the dotted volume with an 8-ft separation; and the total volume with a 12-ft separation.

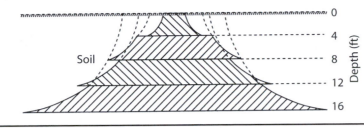

Figure 18.29   Barnes layer method: isolating layers at depth from the total volume of soil measured.

comparison. The Barnes layer method is illustrated in Figure 18.29. The equations are developed as follows:

$$\rho = 2\pi a \frac{E}{I} \qquad \text{(Eq. 18-35)}$$

for the Wenner configuration but $R = E/I$ or $1/R = I/E$:

$$\frac{I}{E} = \frac{2\pi a}{\rho}, \quad \text{or} \quad \frac{1}{R} = 2\pi \frac{a}{\rho} \qquad \text{(Eq. 18-36)}$$

if

$$\frac{1}{R_n} = \frac{1}{\overline{R}_n} - \frac{1}{\overline{R}_{n-1}} \qquad \text{(Eq. 18-37)}$$

where $1/R_n$ is the conductance of a given layer increment when $1/\overline{R}_n$ is the total conductance between the ground surface and the bottom of a given increment, and $1/\overline{R}_{n-1}$ is the total conductance between the ground surface and the bottom of the increment directly above the increment under consideration. Equation 18-35 assumes that the layers act electrically as though they comprise a parallel circuit.

$$\frac{1}{R} = \frac{2\pi a}{\rho} \quad \text{or} \quad \rho = \frac{2\pi a}{1/R} \qquad \text{(Eq. 18-38)}$$

where ρ is in ohm-cm (ohm-ft) and a is in cm (ft).

$$\rho_L = \frac{2\pi \times 2.54 \times 12 A_L}{1/R_n} \qquad \text{(Eq. 18-39)}$$

where ρ_L is the layer resistivity in ohm-cm (ohm-ft).

$$\rho_L = \frac{191 A_L}{1/R_n} \qquad \text{(Eq. 18-40)}$$

where:

  191 = constant, collecting 2π and unit conversions

  A_L = thickness of a given layer in m (ft)

  $1/R_n$ = layer conductance of a given increment

Typical values of resistivity for various earth materials are presented in Table 18.12 on page 444. Note that no hard and fast rules apply between resistivity values and material types. This is indicated, to some extent, by the wide range of values shown. Actually, the changes and contrasts in resistivity at a specific site may be more useful than the absolute values in determining the nature of the materials present. Example Problem 18.4 is provided.

Example Problem 18.4

Data from a hypothetical earth resistivity survey, based on the Wenner configuration, is supplied in Example Table 18.1. This was constructed with the cross section shown in Example Figure 18.1 in mind but not adhered to in a strict fashion. Refer to Figure 18.30 on page 443. Complete data sheet calculations for the Barnes layer method and the cumulative ρ method.

Answers

The data sheet calculations for the Barnes layer method are in Example Table 18.2 (refer also to Figure 18.31 on page 443):

The data sheet calculations for the cumulative ρ method are in Example Table 18.3.

(continued)

Example Table 18.1

ρ (ohm-ft)	*a* Spacing (ft)
40	5
37	10
51	15
63	20
72	25
60	30
47	35
37	40
39	45
41	50
44	55
45	60

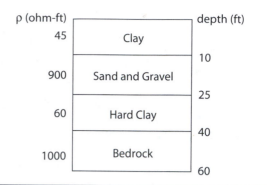

Example Figure 18.1   Cross section sample for Example Problem 18.4.

Example Table 18.2

a (ft)	ρ (ohm-ft)	$2\pi a$	$\dfrac{1}{\overline{R}_n} = \dfrac{2\pi a}{\rho}$	$\dfrac{1}{R_n} = \dfrac{1}{\overline{R}_n} - \dfrac{1}{\overline{R}_{n-1}}$	$\rho_L = \dfrac{2\pi a}{1/R_n}$	Layer Involved (ft)
5	40	31.4	0.78	0.78	40	0–5
10	37	62.8	1.697	0.917	68	5–10
15	51	94.2	1.847	0.150	628	10–15
20	63	125.6	1.994	0.147	854	15–20
25	72	157.1	2.182	0.188	836	20–25
30	60	188.5	3.142	0.960	196	25–30
35	47	219.9	4.679	1.537	143	30–35
40	37	251.3	6.792	2.113	119	35–40
45	39	282.7	7.244	0.452	625	40–45
50	41	314.2	7.656	0.412	763	45–50
55	44	354.4	8.053	0.397	893	50–55
60	45	377.0	8.370	0.317	1189	55–60

Example Table 18.3

a	ρ	Cumulative ρ
5	40	40
10	37	77
15	51	128
20	63	191
25	72	263
30	60	323
35	47	370
40	37	407
45	39	446
50	41	487
55	44	531
60	45	576

Selection of Geophysical Exploration Methods

The two geophysical methods discussed in detail in the previous sections are refraction seismic and electrical resistivity. As explained, these two techniques are most applicable to engineering geology studies because they are the best methods available for exploring shallow depths of the subsurface to 30 m (100 ft) deep or so. Sometimes the two methods are used in combination and the results can augment each other. However, in other instances, one method or the other is preferred because of the specific details of the subsurface.

For the refraction seismic method to provide useful data, the subsurface zones or layers must increase in velocity with depth. This is necessary for waves to be refracted upward to travel along the layer boundary, generating waves that return to the surface. (Refer to Figure 18.21, page 435, for details.) Therefore, if a low-velocity layer lies below a high-velocity layer, the refraction seismic method will not work properly. An example of this would be a sand and gravel layer lying below a clay layer. The gravel has a lower seismic velocity than does the clay, so waves refracted by the boundary between them would not return to the ground surface (see

Figure 18.20, Case 2, page 434). Hence, the refraction seismic method is not useful to explore for sand and gravel deposits within glacial till or clay.

By contrast, the electrical resistivity method should work well in this situation. The contrast in resistivity between the clay (lower resistivity) and the sand and gravel (higher resistivity) should be sufficient that when the *a* spacing, or depth of penetration of the electrical flow path, reaches the sand and gravel layer, a measurable increase in resistivity is encountered.

The refraction seismic method works well in an area where soil overlies bedrock. The seismic velocity of the bedrock is typically much greater than that of the overlying soil. The electrical resistivity method would also likely work well here, because bedrock typically has a considerably higher resistivity than the soil. However, as the depth to bedrock increases, the resistivity contrast must be greater for the bedrock to provide a difference significant enough to influence the collective resistivity value.

In areas of the Midwest where dense glacial till overlies shale bedrock, the contrast in resistivity may not be great enough to yield the necessary difference for the resistivity method to delineate the bedrock surface. However, the seismic velocity of the shale can be twice or more that of the till, and the refraction boundary would be discerned. Yet, if the glacial till is extremely dense, its seismic velocity will approach that of the shale and then it would not be possible for the refraction seismic method to delineate the bedrock surface.

The electrical resistivity method can be used to locate the groundwater table in a sand and gravel deposit. The saturated sand and gravel would have a

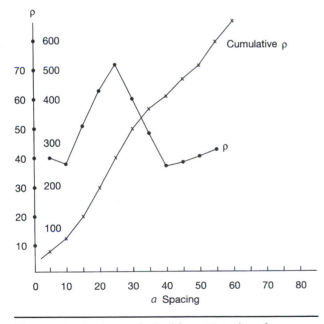

Figure 18.30 Example Problem 18.4, plot of ρ versus *a* spacing and cumulative ρ versus *a* spacing.

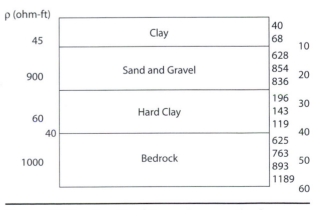

Figure 18.31 Results of Barnes layer method calculation related to geologic cross section.

Table 18.12 Typical Earth Resistivity Values for Selected Geologic Materials.

Material	ohm-cm	ohm-ft	2π ohm-cm for Barnes Method
Soils			
Wet to moist clayey soil	160–320	5–10	1000–2000
Wet to moist silty clay and silty soils	320–15,000	10–50	2000–10,000
Moist to dry silty and sandy soils	1500–15,000	50–500	10,000–95,000
Clay	90–1100	3–35	560–7000
Sand	3000–20,000	100–6500	20,000–1,250,000
Sand and gravel with layers of silt	30,000–250,000	1000–8000	190,000–1,500,000
Coarse dry sand and gravel	250,000+	8000+	1,500,000+
Soft shale	50–1000	2–33	300–6300
Bedrock			
Well fractured to slightly fractured with moist filled cracks		500–1000	
Slightly fractured bedrock with dry soil-filled cracks	30,000–250,000	1000–8000	190,000–1,500,000
Massive bedded and hard bedrock	300,000	8000	
Sedimentary Rocks			
Hard shale	800–60,000		
Sandstone	3000–300,000		
Porous limestone	9000–900,000		
Dense limestone	200,000–1,000,000		
Chattanooga Shale	2×10^3–1.4×10^5		
Michigan shale	2×10^5		
Calumet and Hecla conglomerates	2×10^5–1.3×10^6		
Muschelkalk Sandstone	7×10^3		
Ferrugenous sandstone	7×10^5		
Muschelkalk Limestone	1.8×10^4		
Marl	7×10^3		
Metamorphic Rocks			
Group values	5×10^3–1×10^9		
Garnet gneiss	2×10^7		
Mica schist	1.3×10^5		
Biotite gneiss	10^8–6×10^8		
Slate	6.4×10^4–6.5×10^6		
Igneous Rocks			
Group values	9×10^3–2×10^9		
Granite	5×10^5–10^8		
Diorite	10^6		
Gabbro	10^7–1.4×10^9		
Diabase	3.1×10^5		

lower resistivity, yielding a contrast that the method could measure. In this case, refraction seismic would probably work as well because the saturated sand and gravel layer has a higher seismic velocity than does the nonsaturated material. The contrast of saturated clay or till versus its nonsaturated counterpart would likely not be great enough, however, with regard to either resistivity or seismic velocity, for either method to pick up the saturated zone in a clay deposit.

The electrical resistivity method has been used to locate plumes of contaminated groundwater in a sand and gravel aquifer. The contaminated water contains a higher concentration of dissolved solids, both anions and cations, which decrease its electrical resistivity. Hence, the contaminated plume shows up as a low-resistivity layer. Electrical resistivity surveys, performed periodically, can be used to plot the rate of contaminant transport.

Figure 18.32 shows a series of geologic cross sections that are evaluated in terms of the refraction seismic and electrical resistivity methods.

As a final detail in this introduction to geophysics, it is worth noting that geophysical methods of exploration are indirect. In each method, a property

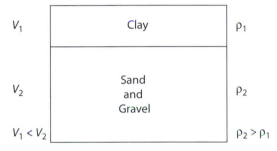

(a) Exploration for sand and gravel.

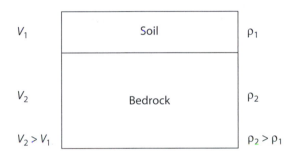

(b) Soil over bedrock, major contrasts.

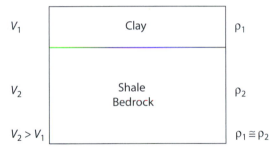

(c) Clay over shale bedrock.

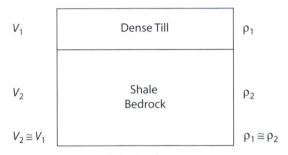

(d) Dense till over shale bedrock.

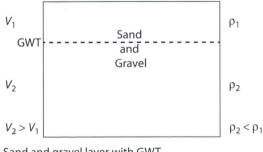

(e) Sand and gravel layer with GWT.

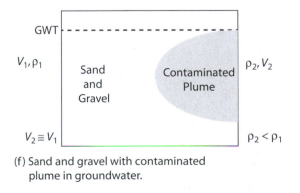

(f) Sand and gravel with contaminated plume in groundwater.

Figure 18.32   Geologic cross sections evaluated using the refraction seismic and electrical resistivity methods.

of the material is measured (i.e., gravitational force, magnetic attraction, seismic velocity, electrical resistivity). However, none of these yields as direct a physical picture as a rock core, rock sample, excavation, or borehole. To infer a rock name or description from geophysical data requires at least one step beyond the basic information obtained, with several assumptions involved. This limitation of the methods in geophysical exploration should always be recognized.

Engineering Considerations of Earthquakes and Geophysics

1. Earthquakes constitute one of the most expensive and deadliest geologic hazards.

2. Evaluating earthquake history is a fundamental requirement in site selection for engineering construction.

3. Earthquakes occur in four major areas of the Earth, the most significant being the Circum-Pacific (or Ring of Fire) Belt.

4. Secondary and surface waves are the most destructive to buildings and other structures.

5. For large engineering projects, such as dams and tunnels, it is essential to investigate if any faults present at the site are active or not.

6. For seismically active areas, engineering structures must be designed to withstand ground motion without major damage.

7. Earthquake damage to buildings is related primarily to magnitude, site geology, and construction details. These factors must be taken into account when designing a structure in a seismically active area.

8. The National Seismic Hazard Map indicates where earthquakes are most frequent. This map can be a useful guide for designing earthquake resistant structures.

9. The Earth has a zoned interior, with the outer portion, the crust, being most significant for engineering purposes.

10. Engineering geophysics includes procedures used to augment drilling and sampling to acquire information. Useful techniques include seismic, gravity, magnetic, electrical, ground penetrating radar, and well logging. Gamma-ray logging of soil borings provides a means to discern sand versus clay and, in groundwater contamination studies, tracing plumes from sanitary landfills.

EXERCISES ON EARTHQUAKES AND GEOPHYSICS

1. Refer to the appropriate equations involving the shear modulus, bulk modulus, and modulus of elasticity. Given a granite with $E = 37.9 \times 10^6$ kN/m^2, $q_u = 152$ MN/m^2, and $\mu = 0.23$, what is the K and S value for the rock? What is likely its τ_0 or shear strength at zero normal stress? What deformation would develop just prior to shearing failure?

2. Given the empirical relationship $\log E = 11.4 + 1.5M_L$, where E is the energy released by an earthquake in ergs and M is the magnitude on the Richter scale, calculate the increase in energy per integer value of the magnitude scale. Collect the data in tabular form in Exercise Table 18.1.

3. Using the relationship $\log E = 11.4 + 1.5M_L$, calculate the energy release for the six earthquakes listed in Table 18.2 of the text. How well does this coincide with the values given for energy release in Table 18.2? One ton of TNT = 10^9 calories and 1 calorie = 4.186×10^7 ergs.

Exercise Table 18.1

Richter Magnitude	$\log E$	E (ergs)	Multiple from Previous Value
1	12.9	8×10^2	—
2			
3			
4			
5			
6			
7			
8			

4. The average energy released worldwide by earthquakes in a year is about equivalent to that of an 8.4 magnitude event. How much energy is this in ergs, tons of TNT, and in megaton bomb equivalents?

5. Refer to Figure 18.1 in the text. If an earthquake registers 6.5 on the Richter scale, what intensities are likely associated with it near the epicenter? How extensive would the damage be? What level of ground acceleration would likely occur near the epicenter? (See also Figure 18.2.) At a distance of 80 km (50 mi), what would be the equivalent magnitude, intensity, and acceleration? Explain as needed. If ρ for the rock = 2.65 gm/cm^3, what are the V_P and V_S values in m/sec, km/sec, and ft/sec?

6. Why is the damage so great for adobe buildings in earthquake-prone areas? How could masonry walls in nuclear power plants be made more resistant to earthquake damage?

7. The value of V_P in the upper layer of the Earth's crust is 6.1 km/sec and V_S is 3.5 km/sec. For a seismograph station 1200 km from the epicenter, what would Δt (or $t_S - t_P$) be equal to at this location using the velocities given?

8. a. The depth to the focus for an earthquake is 20 km. If a value of 6.1 km/sec for V_P is used, how long does it take for the P wave to travel from the focus to the epicenter directly above it? How long does it take for the P wave to travel from the focus to a point on the Earth's surface 100 km away from the epicenter? What is the difference in elapsed time for the two?

   b. For another earthquake, the depth to the focus is 250 km. Using $V_P = 6.1$ km/sec, how long does it take to travel from the focus to the epicenter and how long does it take to travel from the focus to a point on the Earth's surface 100 km from the epicenter? What is the difference in elapsed time for the two?

   c. For both the earthquake with a 20-km depth of focus and the one with a 250-km depth of focus, find the area of the circle located around the epicenter in which the arrival time difference is 1 sec or less. How many times greater is the area for the deep-focused earthquake?

9. At a 30-km depth, what is the overburden pressure on the rock? Assume the density of the overburden to be 2.8 g/cm^3. Calculate this in kg/m^2, kN/m^2, lb/ft^2, and psi.

10. Describe strain release. How did Algermissen apply it to a seismic risk map for the United States? What specific earthquakes caused the concentric effect observed around southeastern Missouri, around eastern South Carolina, and around eastern Massachusetts (see Figure 18.9)? Refer also to Table 18.1 for information.

11. Refer to Figure 18.11. Note that a portion of a large zone IV area extends eastward along the Ohio River. What is the cause for this extension?

12. Describe the seismic hazard level and seismic activity of Denver, Colorado. For more information, refer to Figure 18.10 and the US Geological Survey earthquake website (https://earthquake.usgs.gov/earthquakes/byregion). What is the likelihood and size of earthquakes in this area? What are the chances for damaging shaking?

13. Examine Figure 18.15. How deep under an ocean would a hole have to be drilled to reach the Moho or M discontinuity? Was this ever planned or attempted? Explain.

14. Examine Equation 18-15 for the gravity method. What effect does elevation on the Earth's surface have on the F value obtained? Why is this so? Why would a mass of granite yield a higher value than a mass of sandstone below the Earth's surface, all other factors being equal?

15. What limitations and restrictions are imposed when it is noted that geophysical methods are of an indirect nature? How can geophysical information be checked by direct information in the field? Explain.

16. What two types of information are supplied by the refraction seismic method? What type of wave is used in these studies? Why?

17. Why must a seismograph for engineering geophysics studies be an accurate timing device and be sensitive to frequencies in the 15- to 70-Hz range? How does a geophone work?

18. Examine Table 18.10, which provides seismic velocities for common soils and rocks. What is the seismic velocity listed for granite, measured in m/sec? How does this compare for the velocity in the upper crust (see Table 18.7)?

19. Typically, it is not possible to distinguish the groundwater table in a continuous mass of bedrock using the refraction seismic method. Why is this the case? Why then, in some instances, can the water table be discerned in soil materials?

Exercise Table 18.2

Distance from blast to geophone (ft)	10	20	30	40	50	60	70	80
Travel time of wave (sec)	0.008	0.016	0.018	0.020	0.022	0.023	0.024	0.025

20. Refer to Figure 18.19 on seismic velocity versus ripper performance. Note that both rock type and ripper size are factors in this figure. Why is this significant? What is the lowest value for marginal ripping shown on the chart and what material is it for? Why then is 7000 ft/sec used as the rule of thumb for rippability? Explain.

21. For a horizontally bedded two-layer situation, $V_1 = 45$ m/sec and $V_2 = 2400$ m/sec. What types of materials could make up Layer 1 and Layer 2? Calculate the value of i_{cr} for this sequence.

22. Information from a refraction seismic survey for a three-layer case in which V_3, V_2, and V_1 is involved is given in Exercise Table 18.2. Draw the time-distance curve similar to that in Example Problem 18.3 and in Figure 18.22. Find V_1, V_2, and V_3 based on the slope of the lines in the plot. Calculate Z_1 and Z_2 using both the critical distance method and the time intercept method. Explain any discrepancies in your answers.

23. Relative to the earth resistivity method, what is meant by (a) Wenner configuration, (b) leap-frog sequence, and (c) Lee configuration?

24. Under what geologic conditions could the earth resistivity method give useful results whereas, for the same conditions, the refraction seismic method would not? Explain and then give a geologic cross section to illustrate this situation.

25. Refer to Example Problem 18.4 in the text. Given the a and ρ values, compute the columns for $\log a$ and $\log \rho$ in Exercise Table 18.3.

a. Make a plot of $\log a$ (abscissa) versus $\log \rho$ (ordinate). Compare this plot to that of a versus ρ and a versus cumulative ρ in Example Problem 18.4. For this problem, which curve seems to be most useful? Why is that?

Exercise Table 18.3

a	$\log a$	ρ	$\log \rho$
5	0.70	40	1.60
10	1.00	37	
15	1.18	51	
20	1.30	63	
25	1.40	72	
30	1.48	60	
35	1.54	47	
40	1.60	37	
45	1.65	39	
50		41	
55		44	
60		45	

26. Complete the Barnes layer calculations in Exercise Table 18.4 for another electrical resistivity survey.

a. Make a plot of a versus ρ for the data.

b. Compute cumulative ρ. Make a plot of a versus cumulative ρ.

27. Refer to Table 18.9 on common borehole geophysical methods. For the electrical and radiation logging types, which logs could be run in a cased hole through soil? What useful information could be obtained from these well log data? Explain.

Exercise Table 18.4

a (ft)	ρ (ohm-ft)	$2\pi a$	$\dfrac{1}{R_n} = \dfrac{2\pi a}{\rho}$	$\dfrac{1}{\overline{R}_n} = \dfrac{1}{R_n} - \dfrac{1}{\overline{R}_{n-1}}$	$\rho_L = \dfrac{2\pi a}{1/\overline{R}_n}$
5	60	31.4	0.523	.523	60
10	45	62.8	1.395	.872	72
15	50	94.2	1.884	.489	193
20	60	125.6	2.093	.209	193
25	71	157.1			
30	80				
35	89				
40	98				

References

Algermissen, S. T. 1969. Seismic Risk Studies in the United States. Proceedings, Fourth World Conference on Earthquake Engineering (Vol. I, pp. 14–27), Santiago, Chile.

California Geological Survey. 1990. *The Seismic Hazards Mapping Act*. Public Resources Code, Chapter 7.8, Section 2690–2699.6. Department of Conservation.

Cicerone, R. D., Ebel, J. E., and Britton, J. 2009. A systematic compilation of earthquake precursors. *Tectonophysics* 476(3–4):371–396.

Ehlen, J., Haneberg, W. C., and Larson, R. A. (eds.). 2005. *Humans as Geologic Agents: Reviews in Engineering Geology* (Reviews in Engineering Geology, Vol. 16). Boulder, CO: Geological Society of America.

Gutenberg, B. 1959. *Physics of the Earth's Interior*. New York: Academic Press.

Hayakawa, M. 2015. *Fundamentals of Earthquake Prediction*. Singapore: John Wiley & Sons.

Hough, S. E. 2009. Cataloging the 1811–1812 New Madrid, central U.S., earthquake sequence. *Seismological Research Letters* 80(6):1045–1053.

Hough, S. E., and Page, M. 2015. A century of induced earthquakes in Oklahoma? *Bulletin of the Seismological Society of America* 105(6). doi: 10.1785/0120150109.

Johnston, A. C., and Schweig, E. S. 1996. The enigma of the New Madrid earthquakes of 1811–1812. *Annual Review of Earth and Planetary Sciences* 24:339–384.

Jol, H. M. (ed.). 2008. *Ground Penetrating Radar Theory and Applications*. New York: Elsevier.

Lau, K. C. 2000. *A Review of Downhole Geophysical Methods for Ground Investigation* (Geo Report No. 99). Geotechnical Engineering Office, Civil Engineering Department, The Government of the Hong Kong Special Administrative Region, Hong Kong.

Marshak, S. 2013. *Essentials of Geology* (4th ed.). New York: W. W. Norton.

McNutt, S. R., and Sydnor, R. H. (eds.). 1990. *The Loma Prieta (Santa Cruz Mountains), California, Earthquake of 17 October 1989* (Special Publication 104). Sacramento: California Division of Mines and Geology.

Nance, D., and Murphy, B. 2016. *Physical Geology Today*. New York: Oxford University Press.

Reid, H. F. 1910. The mechanics of the earthquake. In A. C. Lawson (Chairman), *The California Earthquake of April 18, 1906: Report of the State Earthquake Investigation Commission*. Carnegie Institution of Washington Publication 87.

Reynolds, J. M. 2011. *An Introduction to Applied and Environmental Geophysics* (2nd ed.). New York: John Wiley & Sons.

Richter, C. F. 1935. An instrumental earthquake magnitude scale. *Bulletin of the Seismological Society of America* 25(1–2):1–32.

Richter, C. F. 1956. *Elementary Seismology*. San Francisco: W. H. Freeman.

Schiff, A. J. (ed.). 1998. *The Loma Prieta, California, Earthquake of October 17, 1989—Lifelines*. T. L. Holzer (coord.), Performance of the Built Environment (Professional Paper 1552-A, pp. A1–A131). Washington, DC: US Geological Survey.

Sharma, P. V. 1997. *Environmental and Engineering Geophysics*. New York: Cambridge University Press.

Sjogren, B. 2013. *Shallow Refraction Seismics*. Netherlands: Springer.

Spence, W., Herrmann, R. B., Johnston, A. C., and Reagor, G. 1993. *Responses to Iben Browning's Prediction of a 1990 New Madrid, Missouri, Earthquake* (Circular 1083). Washington, DC: US Geological Survey.

Wood, H. O., and Neumann, F. 1931. Modified Mercalli Intensity Scale of 1931. *Seismological Society of America Bulletin* 21(4):277–283.

SUBSURFACE INVESTIGATIONS 19

Chapter Outline

Subsurface Model

Reconnaissance

Subsurface Exploration: Sounding, Drilling, and Sampling

Miscellaneous Considerations

Site Selection

Engineering Considerations of Subsurface Investigations

Subsurface Model

Designing and constructing foundations for engineering structures requires detailed knowledge of the subsurface. Examples of engineering structures include: earth, rock-fill, and concrete dams, tunnels, bridges, heavily loaded and certain lightly loaded buildings, highways, levees, docking facilities, and large underground openings such as power-generating stations and subway stations (Peck et al., 1974; Das, 2015). Information on subsurface conditions is also needed for mining, underground storage of natural gas or liquids, liquid waste disposal, landfill design, and for oil exploration. Specifically, the objectives of a subsurface investigation program are: (1) determine the depth to the water table; (2) determine the depth, thickness, and nature, both lithological and structural, of various soil or rock strata within the zone of influence of the structure; (3) determine the horizontal and vertical variability of soil and rock strata; and (4) determine the engineering properties of various strata for characterization and design purposes (Peck et al., 1974; Adeyeri, 2015; Das, 2015).

In most cases, varying amounts of information about the subsurface are available prior to drilling and sampling, or geophysical exploration. Generally, the depth to bedrock, the type of soil and bedrock, and groundwater conditions are known. Depending upon the site location, some information about soil and rock properties also may be available. Based on the available information, a conceptual model of the subsurface is developed. Without such a model, it is both difficult and expensive to plan a proper subsurface investigation. The purpose of the drilling and sampling phase of subsurface investigation is to confirm or to modify the "model" of the subsurface developed by the individual in charge of subsurface exploration. This individual is typically a soils engineer (geotechnical engineer), engineering geologist, mining geologist or mining engineer, petroleum geologist, or some similarly trained person.

Traditionally, subsurface investigation is conducted in three stages: (1) reconnaissance, (2) drilling and sampling, and (3) laboratory testing of the samples collected during drilling (American Society of Civil Engineers, 1976; Lowe and Zaccheo, 1991; Das, 2015).

Reconnaissance

Reconnaissance is the collection of all available information about the subsurface prior to drilling and sampling. Depending on the engineering organization and on the nature of the project, varying amounts of effort are expended on the reconnaissance stage. This may involve many hours of effort for a major structure, such as a large dam, to relatively little effort for a one-story industrial plant or an office building. Drilling and sampling programs can be planned more effectively if available information is reviewed prior to the initiation of drilling and sampling. This is commonly accomplished in two phases: office reconnaissance and field reconnaissance. For certain projects, such as highway bridges or foundations for large buildings, the distinction between the two is clear-cut, office reconnaissance first, followed by field reconnaissance; then the boring plan is developed and drilling and sampling commence. For larger structures such as major dams, tunnels, power plants, and docking facilities, several stages of office and field reconnaissance may occur, with preliminary drilling between them, prior to the major drilling program associated with the predesign stage. The nature and scope of the project determine the need for alternating office and field activity, but the important fact must be underscored that both office and field reconnaissance are needed prior to drilling and sampling (American Society of Civil Engineers, 1976; Bowles, 1996).

Office Reconnaissance

In the office phase, several tasks can be accomplished as sources of information are reviewed. Not all the sources discussed here are appropriate or necessary for all sites or even for any single site. However, this can be used as a checklist for sources of available information. It should be kept in mind that gaining information by drilling and sampling is expensive and, thus, unwarranted if that same information is already available in the literature. It may be necessary, however, to proceed without such information in remote areas of developing nations or in desolate areas throughout the world.

Review of Design Plans or Preliminary Plans

When the subsurface investigation begins, preliminary, if not final, plans for a structure should be available. These may consist of a series of alternative plans being considered, or a complete building design. The geotechnical engineer or engineering geologist in charge of the foundation investigation should become familiar with the proposed structure through the study of plans and design data. One should determine the approximate magnitude of loads to be transmitted to the foundation as the depth of boreholes depends on these loads; this information is available from the architect or the structural engineer for the project.

Review of Engineering Reports

Engineering reports may provide general information on the soil, rock, and groundwater conditions that are likely to prevail in the study area. Many private companies and public agencies have accumulated information on certain geographic locations and on specific types of projects. A study of previous construction activity may prove helpful for anticipating foundation problems and providing geologic detail. This may involve reviewing previous projects of a similar nature from the files of a private company. Federal agencies and state governmental offices are the primary sources for information of public record.

Review of Published Information or Open File Geologic Reports

Other information pertaining to the site area can be obtained from published maps, records of railroads and utilities, aerial photographs, water well logs, oil well logs, and other similar information filed with the state government. Abundant site data are available from governmental agencies.

As with all available information, there are limits on the accuracy and extent to which data can be applied for the situation under study. Knowing such limitations is a vital part of the analysis procedure.

Topographic Maps

These maps, which depict geographical information as well as surface elevations, are useful in the reconnaissance program. Site location and access to the site can be determined from these maps. Surface elevations for boring locations can also be estimated. Topographic maps provide a general idea of the geology to the experienced engineering geologist, particularly in conjunction with other geologic information. They also serve as base maps for field reconnaissance.

Topographic maps of 7.5-min quadrangles are available from the US Geological Survey (USGS) for most areas of the United States. These maps, with a scale of 1:24,000, are named for the largest city or town contained on the map and are indexed by state according to this quadrangle name. Fifteen-minute topographic maps covering four times the area of the 7.5-min sheet may also be available and will prove useful if a larger area is to be viewed. These are mapped at the smaller scale of 1:62,500. The National Geologic Map Database (http://www.ngmdb.usgs.gov) offers stratigraphic, geologic, and topographic maps for both viewing and download. The state geologic survey for most states stocks a supply of "topomaps" for their own state that are available for sale.

Google Earth Maps

Google Earth maps are a popular and valuable means for evaluating the site conditions, site accessibility, and planning field reconnaissance. Google images can be printed or downloaded at different scales and annotated to show various features. They can be used to show tentative boring locations and demarcate any problematic areas. A variety of apps are available to modify these maps for the intended purpose.

Geologic Maps

Geologic maps may also be available for the area in question. On a geologic map, the top of the bedrock surface is depicted, which is what would be observed if the soil were stripped away from the rock. Geologic units of differing rock type and age are delineated on such maps. The mapping units are typically individual formations. Appendix B presents specifics concerning geologic maps.

Geologic maps are found in a number of sources; for example, The USGS Publications Warehouse (http://www.pubs.er.usgs.gov) provides access to several published series, such as GQ (geologic quadrangle), HA (hydrologic investigation atlas), I (miscellaneous/geologic investigations), and COAL/C- (coal investigations). An index of USGS geologic maps is available in most libraries in the form of a list of published maps. Regional geologic

maps for areas of 1° × 2° may also be available, with a typical scale of 1:125,000. Several counties in the midwestern United States would be included on such a map.

Geology-Related Maps

Other types of geological maps are available from many state geological surveys. Glacial thickness maps typically covering an entire county or several counties and bedrock-surface contour maps for an equal-sized area are available in many locations with glacial cover. These maps prove useful in determining the depth to bedrock when used in conjunction with topographic maps. Environmental geology maps may prove somewhat useful if available for a study site, particularly since they tend to be mapped at rather large scales, say, 1:12,000 to 1:1000. However, some information must be interpreted from these maps, which requires a more extensive knowledge of geology.

Agricultural Soils Maps

These maps, which were discussed in Chapter 4, Rock Weathering and Soils, are available through the US Department of Agriculture (USDA). They cover an entire county and, generally, are available from the county farm agent or the Natural Resources Conservation Service office in the county. Agricultural units are depicted on these maps but engineering significance can be applied to the mapping units as well, as discussed in Chapter 4.

Aerial Photographs

Aerial photographs are obtained from an aircraft that takes successive photos using a vertical view with overlap in the ground image from photo to photo. A 60% overlap from photo to photo with a 20% overlap between successive flight lines is commonly used. This provides two views of the same object from a slightly different location. A stereographic image can be obtained by viewing two photos simultaneously. Most commonly, air photos are taken using black-and-white film, but color and color infrared photography are also used. Black-and-white aerial photographs are available from several governmental agencies. The USDA has photo coverage for much of the primary agricultural lands of the United States, whereas the US Forest Service has photography for much of the western area of the country. These photos can be obtained after consulting the index of available photos and ordering photos by number from the appropriate agency.

Geographic details of the ground surface, such as roads, houses, and farm fields, are recorded on the photo and are useful for purposes of location. The tone and texture of objects and patterns are also captured on the photo and these features are useful in identifying soil types, rock units, geologic structures, and various landforms. Aerial photograph interpretation is an important tool for reconnaissance study of terrain.

Color photography and color infrared photography (false color IR) are also available for analysis. Color photography is of particular value in mapping various soil types because color differences indicate differences in soil types. Color photography also enhances bedrock features. Infrared photography utilizes the near-infrared portion of the spectrum, just beyond the range of the visible spectrum, but in a portion still sensitive to certain types of film. Color IR photography is useful for highlighting green vegetation (or lack thereof) and water bodies. Vegetation takes on a magenta (red) color, whereas water is completely black.

Color and color IR photography, in most cases, must be obtained by scheduling new aerial photography for a site. No large depository of color photos of specific areas is available for sale. If such reconnaissance is needed, an aerial survey company should be contacted to obtain the new photos. Photo scale and other details are designated when ordering the photos.

LiDAR Imagery

These days light detection and ranging (LiDAR) imagery is available online on a statewide basis for most of the states. With some experience, a significant amount of information about site conditions, especially bedrock structure, can be obtained from such imagery for large project sites.

Well Log Data

Thousands of water wells are drilled each year in the United States. In many places, state law requires that drillers file the logs for such wells with a state agency, usually the Department of Natural Resources or some similarly titled agency. Such well data are identified according to location and kept on open file for people to use as needed. Some state agencies have posted these logs online.

Oil well logs are also filed with a state agency, typically the Bureau of Oil and Gas or some similarly titled

agency. These logs commonly begin their detailed description at the bedrock surface, lumping all material above rock as overburden. In areas with a thick glacial cover such collective terminology overlooks valuable information for engineering construction, but it does indicate the depth to bedrock. In areas of residual soils, the well logs provide information from the near-surface downward.

In some cases, the USGS has compiled the water well information into lists of data with an accompanying report. These are published as USGS professional papers and water supply papers. USGS publication lists are available from libraries at major universities or in the libraries of most large cities.

Geologic Reports

Geologic information, including maps, geologic columns, and descriptive material, are published in many sources in the United States. These include the major journals of the geological sciences plus other related scientific and engineering fields. A set of reference volumes titled *Bibliography and Index of North American Geology* is found on the reference shelf of most university libraries and is also available in the GeoRef Database (http://www.americangeosciences.org/georef). Referenced by subjects and by author, the publications pertaining to a specific geographical area can be found. Published articles in reference journals can be retrieved from the library to determine whether the information is helpful in the reconnaissance study.

Formulation of the Boring Plan— Spacing and Depth of Borings

During the final stage of office reconnaissance, a tentative boring plan should be developed for subsequent review during field reconnaissance. Several factors help determine the number, depth, and the layout of borings: the nature of the construction project, the complexity of the subsurface materials, the personal judgment of the project manager, and the policy of the organizations involved, both the organization doing the investigation and the one paying for the work. The purpose is to learn as much as possible about the subsurface and how it will affect or be affected by the construction by doing a minimum amount of drilling and sampling. Therefore, both engineering geology or geotechnical engineering and economics play a major role.

Sometimes, to help determine the number of borings needed, a seismic or an electrical resistivity survey of the site may be performed prior to, or in conjunction with, the boring program. Frequently the subsurface data can be extended between borings by use of these and other geophysical methods. These techniques are discussed in detail in Chapter 18, Earthquakes and Geophysics. Some general considerations concerning the density of borings and geological conditions are as follows. Limestone units may require more borings than other rock types because of the irregular nature of soil-bedrock contact and likelihood of solution cavities. Igneous and metamorphosed-igneous rocks are typically fairly uniform across a small site and may require a minimum number of borings. In alluvial valleys and in glacial deposits, it is difficult to generalize except that inhomogeneity is common. Soils associated with shallow ocean deposits tend to be fairly uniform and fewer borings are required. Keep in mind, however, that these rules of thumb have many exceptions. They must be applied only as a guide for formulating the plan, with changes forthcoming as the study progresses. Below are the commonly used guidelines for selecting the spacing and depth of borings.

Spacing

The spacing of borings at a construction site depends on spatial variability of the geologic materials, type and importance of the structure, and the available budget. Most engineering firms use the guidelines proposed by American Society of Civil Engineers (1976); for important structures, a spacing of one hole per 232 m^2 (2500 ft^2) or one hole per 465 m^2 (5000 ft^2) is used. This gives a spacing of 15 m (50 ft) and 22 m (70 ft), respectively. For relatively uniform soil or rock conditions, a spacing of 30 m (100 ft) can be used. The number of borings indicated by these guidelines is then adjusted in light of the available budget. Also, initially, a larger spacing is used and, based on the findings, the spacing is adjusted. Some engineering geologists or geotechnical engineers prefer to use a random boring plan rather than a grid pattern.

Depth

Depth of holes depends on the loads that are likely to be imposed by the structure. In the case of multistory buildings, the depth will depend on the number of stories (Das, 2015). Three guidelines are commonly used to determine the boring depth:

1. Depth = 2B below the footing bottom; where B = footing width

2. Depth = $\Delta\sigma/\sigma'_o \leq 0.1$; where $\Delta\sigma$ = change in stress at a given depth due to the imposed stress, as determined by the methods discussed in Chapter 7, and σ'_o = effective overburden pressure at the depth at which $\Delta\sigma$ is determined. By iterations, a depth is determined where the above ratio equals 0.1 (10%). For very important structures the ratio is reduced to 0.05 (5%).

3. Depth = bedrock depth, if the overlying soil material is of poor quality.

Field Reconnaissance

After the office reconnaissance has been accomplished the field phase should commence with a visit to the site. Preferably, the project manager should make this field inspection but, if that is not possible, a field representative or the project geologist should be involved. For major projects, the field inspection party may include a group of specialists consisting of the designers, project geologist, project engineer, and the construction inspection people. This allows for constructive discussion of the project in terms of all the factors that will have an impact on the total procedure. Site conditions commonly relate to design limitations and construction details. If these are known and discussed early in the project, many potential problems can be prevented.

Field reconnaissance has two major purposes regardless of the size of the construction project: (1) to allow an experienced observer to view the site and record information that will affect either the foundation design or boring plan and (2) to gather information needed for the drill crew to accomplish the boring program.

There are a number of items to look for in the field during reconnaissance, as summarized below, and field notes should be taken for subsequent reference.

Proposed Location of Structure

The designer's proposed location of the building or structure on the tract of land should be observed on the ground. If the proposed location appears to be in poor material, and better locations are available on the tract of land, a recommendation to shift the location may be in order. Swampy or low marshy areas or unstable slopes are examples of locations that could likely be improved in most situations. A knowledge of the proposed structure (from the review of plans during office reconnaissance) makes it possible to view the needs of the structure and the field location collectively.

Topography and Vegetation

In nearly all cases a topographic map is available for the site, either a USGS topographic quadrangle sheet or a larger scale topographic map prepared specifically for the site. During reconnaissance, the field observer should walk the tract of land while studying the topographic map and notations concerning vegetation and topographic relief should be made. If aerial photographs are available, they should be referred to as the site is traversed.

Notes on topography are useful to provide indirect evidence of the subsurface. For example, narrow, steep stream channels are suggestive that bedrock is close to the surface while wide alluvial channels suggest a much thicker soil cover. The notes on topography are valuable to the drill crew as well, allowing them to plan access to the site and determine the type of equipment needed, such as truck-mounted, skid-mounted, or all-terrain vehicles or winched equipment.

Vegetation, to some extent, is indicative of surface soils and groundwater conditions. Growth is likely to be dark and lush in areas of groundwater seepage. In some parts of the United States, the growth of certain kinds of trees or shrubs is linked to locations of specific rock types or even rock units. Other trees are indicative of slow slope movement and, of course, treeless, scarred areas within a forest indicate a recent snow or rock avalanche or other landslide. Obviously, the use of vegetation to determine detailed information requires much experience on the part of the observer and in the specific part of the country involved.

Surface Soils, Gullying, and Natural Slopes

Surface soils are studied by direct observation of the surface and through the use of a shovel, posthole digger, or hand auger. These soils may be indicative of the specific parent material from which they were derived, either the rock type or even the individual formation.

Alluvial soils supply information about the stream from which they were deposited. Boulders in a small stream suggest a high stream velocity and possibly scour problems during high flow levels.

Glacial soils in upland areas, terrace deposits, wind-deposited soils, or residual soils on bedrock uplands can be investigated during the on-site investigation. A knowledge of geomorphology (the study of landforms) is extremely helpful in this situation.

Any surface depression, slope dissection, or other natural or human-made excavation in the ground can provide valuable information about the subsurface. Commonly, such cuts into the subsurface provide more information than a borehole because of the larger size and three-dimensional observation available. Soil type, depth to bedrock, and type of bedrock may be visible at such locations. Narrow, actively downcutting streams can be good sources of subsurface data, which must be explored during the on-site reconnaissance. Countless examples can be quoted in which a proper examination of the stream valleys would have provided a wealth of information for use in developing the drilling program and for correlation of boring data after the drilling is completed.

Surface and Subsurface Water

The presence of surface or subsurface water is valuable information in designing foundations for structures and in creating the boring plan. All surface flows, surface seeps, and springs should be noted as well as available information on groundwater levels. Natural lake levels may be indicative of the groundwater surface and major streams commonly set the lowest level of the groundwater table for the area.

Geology of the Site

Depending on the size and nature of the project, and on the site itself, different degrees of geologic field investigation are warranted. For a major project such as a large dam, tunnel, or deep highway cut, a detailed, large-scale geologic map should be prepared. This would involve a careful traversing of the site by an engineering geologist, geological engineer, or field geologist during which time the nature of the rock units and their attitudes (strike and dip) would be determined. Maps would be prepared on the topographic base map with assistance from the aerial photographs for the site. Knowledge of field geology and geomorphology are necessary for an accurate mapping program. This map will be used, along with the other information gathered (as described earlier), to plan the drilling program so that samples can be obtained from rocks at depth and their physical character determined. This information is used to modify the subsurface model that is formulated during the reconnaissance program.

Information Needed by the Drill Crew

The drilling crew needs to know the following information: how to get to the site, where to drill, what equipment to take for the project, and what difficulties will likely occur. It is important that most of the drill crew's time be spent drilling, with a minimum of time spent moving from point to point and virtually no time spent searching for the site, specific boring locations, or traveling back to the shop for more equipment. Generally, the following factors are involved.

Finalization of the Boring Plan

The proposed boring locations should be checked for accessibility and for suitability relative to the needs for drilling information to complete the subsurface model. Deletions, relocations, and additions should be made based on the accessibility and need to supply the necessary details.

Types of Equipment Required

During reconnaissance, notes should be taken that provide information concerning which type of drilling equipment is most suitable (rotary, auger, auger type). The most applicable method should be used based on field reconnaissance and accessibility. If a continuous flight auger is to be used, will the soil samples from the auger flights be suitable? If a three-flight auger is used will the hole stay open when the auger is withdrawn? When caving of the hole is anticipated, a rotary drill rig is required with casing to keep the hole open, or drilling mud will need to be used instead. A rotary coring rig will also require a water supply, as do hollow stem augers when certain sandy strata are encountered. The size of pump, length of hose, and similar details must be determined for this situation in addition to the location of a water supply.

Any borings to be drilled in a steam bed require special consideration. The size of barge, its anchorage, size of drill rig to penetrate to the required depth, and whether special drilling permits are needed must be determined prior to sending the drillers to the site. The type of equipment might be influenced by the time of the year in which the borings will be made.

The winter season may be a better time for drilling some sites if frozen ground will facilitate access.

Benchmarks and Locations of Borings

Reconnaissance should determine if benchmarks or other reference points are in place and whether they are adjacent to or on the site. These benchmarks should be properly referenced on the plans. The boring locations may be laid out during reconnaissance if close tolerance location is not critical. If surveying is required to locate borings, a survey crew should be used to accomplish this task.

If a large-scale topographic map is available for the site, boring locations can be made directly from this plot rather than by survey crew. In this case, the boring locations should be laid out (staked) during reconnaissance.

Permission of Property Owners

For drilling to be accomplished on land not yet purchased, or if access through property owned by others is involved, permission to proceed should be obtained by the reconnaissance party. A team of specialists performs this function for major projects. Using the drill crew to accomplish this task is a poor and generally expensive practice.

Location of Utilities

Underground or overhead utilities on the site should be accurately shown on the plans, or their locations staked on the ground, or both. The names of the agencies or people to contact before work begins should be supplied on the plans for the drilling foreman.

The use of site maps alone to locate utilities without field verification is not a good practice. Checking with management personnel of the utilities involved is a recommended procedure, which should be accomplished during reconnaissance. A careful record of the names and job titles for the utility should be included with the information supplied regarding utility line locations.

Pertinent Notes

Notes providing general information for the drill crew should be prepared during reconnaissance. This includes instructions on the best route to the site and routes between borings, as well as other pertinent information. If additional exploration methods, such as geophysical surveys, are applicable to the site, notes on locations and types of exploration techniques desired should be made during reconnaissance. This information, of course, is subject to modification based on the results of the boring program.

Procedure for Reconnaissance

It is difficult for all construction projects to generalize a precise reconnaissance procedure. Obviously the level of concern is greater for a major dam than for a conventional small building, and this is reflected in the extent and detail of the respective reconnaissance programs. For major structures such as large dams, tunnels, or docking facilities, for example, the procedure may include a review of the regional features followed by an analysis of local considerations of the project.

In the regional review for a major project, analysis of geologic details of a significantly large portion of the Earth's surface is involved, perhaps an area the size of a state or physiographic division of the United States. A general view of the broad structural geology features, the geologic column involved, the nature of the rock, general topography, major drainages, climatic details, and other pertinent information is included. This allows the investigator to determine the overall nature of the geologic factors that will affect the total area, making it possible to determine the relationship between the site and the region as a whole.

With this accomplished, the local factors of the site are considered, including geologic structure, rock types, topography, groundwater conditions, soil types, depth to bedrock, and any other relevant data. This makes it possible to evaluate the local conditions in light of the overall regional concerns.

Subsurface Exploration: Sounding, Drilling, and Sampling

For most foundation studies, the reconnaissance program is followed by subsurface exploration. This may be limited to finding the depth to refusal by a probing technique or it may involve sophisticated drilling, sampling, and in-hole measuring procedures. Refusal is the depth at which a probing device is no longer able to penetrate into the subsurface.

Soil exploration can be accomplished in three basic ways: (1) probing or sounding of soil depth

to reach refusal, (2) drilling and sampling of soil at various depths, and (3) excavation of test pits followed by inspection and sampling. Rock exploration is accomplished by analyzing drill cuttings or rock cores, or with in-hole measuring devices. These are described in the following sections.

Soundings

Sounding, probing, or pricking is a means by which an indication of the depth to bedrock (refusal) is obtained without drilling. A steel bar or hand auger is forced manually into the ground until refusal is reached. This method is most appropriate for soft, soil materials of shallow depths that overlay bedrock. In some cases, a minimum soil depth is needed for construction purposes; for example, a gas or oil pipeline, sewer line, or water line. If this minimum depth is reached, then the probing may not extend to refusal but be terminated just beyond the minimum depth. In rough terrain where a drilling rig would have major access problems, probing may be a necessary means of subsurface exploration.

The sounding method has serious shortcomings. No samples are taken and typically no recording of soil type is made, even of near-surface material. Refusal can occur either on bedrock or on any other hard surface including boulders, slightly cemented sands, or strongly consolidated clays. Many assumptions are involved when one concludes that the refusal depth indicates bedrock; this should be checked by a thorough geologic investigation. Soundings may be used to provide information between widely spaced borings. When this is done, the boring information is extended by the soundings but, again, care must be taken to ensure that no false conclusions are reached based on refusal depths.

Drilling Methods

Two distinct steps are involved in obtaining samples from the subsurface: the first involves making a hole to the depth of interest and the second involves obtaining the desired samples.

Cable Tool Well Drilling

The cable tool or churn drill is used primarily for water well drilling and only rarely for engineering exploration. A chisel bit is pounded and churned into the ground by the lifting and dropping action on the hoisting drum of the drill rig. Only an up-and-down motion (no rotation) is used, which allows the chisel bit to pound the soil or moderately hard rock to small pieces. Rock and soil fragments are removed by a bailer (bailing bucket), which is dropped into the well on a cable to collect the fragments and any water that may accumulate. Sampling by this technique is obviously limited.

Rotary Drilling

In this method, the drill rod is rotated around the drill axis (usually vertical) and various drill bits are used to cut through soil or rock. A more elaborate drilling rig than the cable tool rig is required to deliver the rotational power needed to penetrate soil or rock. Special roller bits are available for soil and rock that will cut through the material in question. Cuttings are returned to the surface through the wash water. Sampling at selected intervals can be accomplished for most rotary drilling methods but not for all.

Four primary rotary procedures are used for exploration drilling: auger boring with solid augers, auger boring with hollow-stem augers, rotary drilling with roller bits, and rock or soil coring. A truck-mounted rotary drill rig is shown in Figure 19.1. Hollow-stem augers are being advanced into the ground in this photograph.

Auger Borings with Solid Auger Systems

These augers can be either continuous-flight, three-flight, or bucket augers. Continuous-flight augers appear similar to a wood screw in that the flutes are rotated into the ground and the soil returns to the surface by way of the rotating flutes. Auger cuttings can be taken from the flutes; the soil is disturbed in this process and a delay is involved between the time when a soil zone is penetrated by the auger tip and the excavated soil reaches the surface. These holes are drilled dry and no casing is needed to keep the hole open. As with all augers, drilling is limited to soil.

Three-flight augers and bucket augers must be retracted from the hole when the auger flutes or the bucket is filled with soil. The drill stem is attached to the specific auger in use. For the three-flight augers the auger is rotated quickly after extraction from the hole and the soil spins off in a circular fashion, a short distance from the hole. Sampling can be obtained from these disturbed piles of soil. Holes can be 10 to 38 cm (4 to 15 in) or more in diameter. Typi-

cally the holes are not cased so that drilling is limited to cohesive soil or moist cohesionless soil above the water table.

Holes as large as 90 cm (36 in) in diameter, used for caisson holes, can be drilled with three-flight augers. The base and sidewalls of these large holes can be inspected by lowering a knowledgeable observer (an engineering geologist or geotechnical engineer) into the hole to describe the subsurface conditions. Caissons are cast-in-place concrete piles that are founded on a hard, bearing layer. They are used as column supports for buildings.

Bucket augers have two flights of augers at the base of the stem plus a bucket attached just above the augers. This bucket has an opening at the bottom so that soil will rotate into it but not easily fall out when the stem is pulled to the surface. Disturbed

Figure 19.1 Truck-mounted rotary drill rig using hollow-stem augers. (Photo by Central Mine Equipment Company.)

samples can be obtained from the bucket of soil at the ground surface. Bucket auger holes are usually 15 cm (6 in) or more in diameter and usually are not cased. Therefore, they have the same limitations concerning soil type as do the three-flight augers. Only cohesive soils will not cave into the hole when the auger is extracted.

Hollow-Stem Augers

These are large diameter, usually 20 to 30 cm (8 to 12 in) continuous-flight augers that have a hole, 7.5 cm (3 in) in diameter, inside the augers. These augers are rotated down into the ground and the soil is carried upwards by the flutes. Instead of sampling this returned, disturbed soil, or pulling the augers to the surface to obtain soil collected near the bottom flights, sampling is accomplished through the hollow portion of the auger by a sampling tool. This feature is discussed later. The hollow-stem augers act as a casing for the hole during the sampling process. Because these augers are not extracted from the hole until the exploration is complete, sampling in cohesionless or cohesive soil above or below the groundwater table can be accomplished. These augers cannot penetrate fresh bedrock.

Rotary Drilling with Roller Bits

In this method of drilling, a tricone bit is used to drill through the rock or soil, grinding the material to coarse sand size or smaller particles, which are returned with the wash water. No other sampling is possible than that obtained through analysis of the cuttings. This procedure is used in soil if the soil unit is not of interest or was previously defined, or if drilling to bedrock is desired as quickly as possible. When this method is used in rock, the purpose is to drill down to a depth of interest where rock cores will be taken. For large portions of the deep oil wells drilled for the petroleum industry, only the major zones of interest are cored, with roller bits used for the remainder. This is done in the interest of saving both time and money. Useful information can be obtained by a trained geologist studying the rock cuttings returned with the drilling fluid.

Percussion Drilling

Percussion drilling uses jack hammers or air tracks that advance a hole by percussion effects. This method is used in bedrock, including both soft and hard rocks. The rock is ground to dust and only color

and mineralogy can be determined with any degree of assurance. Typically, only the rate of penetration is recorded, whereby soft seams and cavities can be discerned. The holes are usually drilled for purposes other than exploration, for example, for presplit blasting holes, holes for explosives, and holes for grout placement.

Sampling Methods

To complete an adequate subsurface exploration program, it is necessary to obtain representative samples of soil and rock for identification, classification, and determination of other engineering properties. Both disturbed and undisturbed samples are collected. Disturbed samples are those in which the soil structure has not been maintained and they can be used only for description, soil classification, and laboratory tests in which soil structure is not important. In contrast, relatively undisturbed samples are used for determining wet and dry density, unconfined compressive strength, shear strength parameters, permeability, and consolidation characteristics.

The degree of disturbance of soil samples depends on the nature of the material being sampled, the core barrel or sampling method used, the drilling equipment, and the skill of the driller. Soil samples, on extended exposure to the atmosphere, will become unusable for testing, mostly because of changes in moisture content. Therefore, undisturbed samples must be properly sealed, transported, and stored.

The core samples of weak shales, coals, and other weak rocks should not be exposed to the atmosphere for extended periods prior to logging the core or wrapping the samples for laboratory testing. Commonly, clay-rich shales, claystones, and mudstones will form thin wafers or disks on drying in the field even when placed in a core box. Cores should be logged as soon as possible after they are obtained. Samples should be selected for testing and properly protected at that time.

The simplest procedures for soil sampling have been mentioned in the preceding discussion on drilling. For continuous-flight augers, the disturbed soil is taken from the surface once the auger flutes push it up to the surface. For three-flight augers and bucket augers the soil is obtained when the auger is retracted from the hole when either the bucket or the flutes become filled with soil. In the cable tool method, soil or rock cuttings are returned to the surface by the bailer. Percussion drilling in rock provides only rock dust, which yields little if any specific information about the character of the rock being drilled.

Standard Penetration Test

The most common procedure used for routine soil sampling involves the standard penetration test (SPT) and the split-spoon sampler (ASTM D1586). The sampler is 60 cm (24 in) long, 3.6-cm ID, 5.0-cm OD (1 3/8-in ID, 2-in OD). When disconnected from the drill rod, it splits in half vertically to expose the column of soil obtained during the sample-driving operation. This generally disturbed sample can be described and portions stored in glass jars with screw-top lids for future reference. The soil can be used to determine moisture content, specific gravity, grain size distribution, and Atterberg limits.

The standard penetration test consists of attaching the split-spoon sampler to the drill rod, lowering it to the bottom of the drilled hole, and advancing it into the ground by a series of blows from a drop hammer. The hammer has a standard weight or mass of 64 kg (140 lb) with a height of fall of 0.76 m (30 in). The sampler is advanced in three increments of 15 cm (6 in), with the number of blows recorded

for each increment. The first increment is supposed to seat the sampler in the bottom of the hole and the second and the third increments are used to determine the penetration resistance. The results would be recorded as, for example, 6/7/5, meaning 6 hammer blows for the first 15-cm (6-in) increment to seat the sampler and 7 and 5 blows, respectively, for the second and the third increments. The penetration resistance, designated as the SPT or N value, is the sum of the blows for last two increments, which in this case equals 12. Some engineers prefer to drive the entire sampler into the ground in four 15-cm (6-in) increments and use the blow counts for the middle two increments to determine the N value. Figure 19.2 is a photograph of the standard penetration test in progress. The split-spoon sampler is attached to the drill rod with the standard hammer shown at the top of the photo.

In Table 19.1, N values have been correlated with the relative density of cohesionless soils and with the consistency and unconfined compressive strength of cohesive soils. During the last 50 years an enormous quantity of data has been collected based on the SPT, so that useful comparisons of soil density, consistency, and strength can be made through the use

Figure 19.2   Standard penetration test in progress. (Photo by Handalan Enterprise.)

Table 19.1   Comparison of *N* Values to Density and Compressive Strength of Soil.

Soil	Penetration Resistance (*N*) (blows/30 cm [1 ft])	State	Relative Density (%)	Unconfined Compressive Strength (q_u) (psf)	(kN/m²)
Sand	0–4	Very loose	0–15		
	4–10	Loose	15–35		
	10–30	Medium	35–65		
	30–50	Dense	65–85		
	50	Very dense	85–100		
Clay	2	Very soft		500	25
	2–4	Soft		500–1000	25–50
	4–8	Medium		1000–2000	50–100
	8–15	Stiff		2000–4000	100–200
	15–30	Very stiff		4000–8000	200–400
	30	Hard		8000	400

Note: 1 kN/m² $\cong$ 20.88 psf.

of this test. In addition to the above parameters, N values exhibit significant correlations with the friction angle of sands and undrained shear strength of clays (Bowles, 1996; Das, 2015).

The SPT is a valuable tool in determining soil conditions for the subsurface but the test is not without problems. The N values are subject to variations depending on the free fall of the hammer. The hammer may not fall the full 0.76 m (30 in), or excessive friction on the hammer stem may occur due to too much friction because of caked oil or soil; the hammer may strike the upper stop too sharply on retrieval, pulling the sampler off the bottom between each blow; the boring may not be clean, reducing hammer energy; a rock may partially plug the sampler; in an uncased hole, soft clay may close around the drill rod, causing additional friction; and in deep holes the hammer energy loss may be greater due to the heavy drill stem and rebound within the stem itself.

Standard penetration tests can be performed in conjunction with drilling through hollow-stem augers, or in cased holes drilled by three-flight, bucket, or continuous-flight augers. For these latter three methods, the auger is withdrawn from the hole and the split-spoon sampler is inserted. Soils subject to caving or squeezing will yield problems when the augers are withdrawn. Dry, cohesionless soils and cohesionless soils below the water table are particularly prone to caving. Soft saturated clays are subject to squeezing. Cased holes are needed in these situations.

Hollow-stem augers may develop problems in loose sands below the water table. Such sands tend to push up inside the augers so that the split-spoon sampler cannot be extended to the base of the auger stem. This can result in incorrect N values. Washing of the sand under pressure with the water nozzle can remove such sands and sampling can then proceed. This pushing-up action disrupts the soil below the auger, however, and a truly disturbed sample is obtained. Plugs for the auger opening prevent soil from coming into the hollow opening. The plugs are removed after drilling and before each split-spoon sample is taken. Use of such plugs is not problem free because pebbles can lodge against the plug, making it extremely difficult to disengage the plug after drilling and before sampling.

Pushed Tube Sampling

Pushed barrel or thin-walled tube sampling is used for cohesive soil or soft rock to obtain a relatively undisturbed sample. A thin-walled tube (Shelby tube) is forced into the soil, using the static force from the weight of the drilling rig, or driven into soft rock. The tubes range from 7.5 to 125.0 cm (3 to 49 in) in diameter with 10.0 cm (4 in) being a common size. The tubes are pushed into the soil for a length of 10 to 20 times the tube diameter, with a 1-m (3-ft) sample length common for the 10.0 cm (4.0 in) diameter sample. The soil or soft rock must have sufficient cohesion to remain in the barrel while the sampler is being withdrawn from the hole. For cohesionless materials, a piston sampler is used, which allows the influence of a vacuum effect to aid in keeping the sample in the tube. The piston, which is controlled by a rod, remains stationary while the outer thin-walled tube is forced ahead into the soil. The Osterberg-type sampler activates the piston hydraulically, whereas the Hong-type sampler uses a ratchet mechanism. The common tests performed on thin-walled tube samples include moisture content, mass unit weight, Atterberg limits, permeability, unconfined compression, direct shear, triaxial compression, and consolidation.

Direct Push Sampling

In the direct push sampling method, a machine pushes the sampling device into the ground to create space for the sampler, without drilling. The static weight of the vehicle to which the machine is attached, combined with some percussion action, is used to advance the sampler into the soil. The process involves compression or compaction of soil under static load to create space and advance a continuous string of the sampling tool. Since drilling is not involved, the sampling tool is pushed to the

depth possible. Sampling of sands, gravels, and hard clays usually requires percussion action. The direct push method provides a continuous sample. Figure 19.3 shows the tube sampler along with the lower part of the machine.

Rotary Coring

This method is used to obtain a cylindrical sample of soil, soft rock, swelling clay, or swelling soft rock. An annular hole, or kerf, is cut into the soil or rock by rotating the outer tube of the sampler with the sample sliding up inside of this tube.

Rotary coring of soil or soft rock is accomplished by a core bit with embedded steel teeth that cut through the subsurface. Soft materials are protected by a stationary inner barrel, which accepts the sample as the outer core barrel penetrates the soil or rock. Cuttings are flushed upward by the drilling fluid, which comes down through the core barrel and upward on the outside or perimeter of the hole. Adaptations include the Denison sampler, which has a fixed cutter on the outer barrel, and the Pitcher sampler, which has a spring-loaded outer tube. The Acker core barrel uses air and mud as the drilling fluid to support the hole in soft materials. Relatively

Figure 19.3   Direct push sampling apparatus. (Photo by Geoprobe Systems®.)

Table 19.2   Core Sizes for Conventional Coring in Rock.

Core Designation	Core Barrel Outside Diameter		Approximate Diameter of Core Hole		Approximate Diameter of Core	
	(cm)	(in)	(cm)	(in)	(cm)	(in)
E_x	3.6	1 7/16	3.8	1 1/2	2.2	7/8
A_x	4.4	1 27/32	4.7	1 7/8	3.0	1 3/16
B_x	5.9	2 5/16	6.0	2 3/8	4.1	1 5/8
N_x	7.5	2 15/16	7.6	3	5.4	2 1/8
3 in	10.6	4 3/16	10.8	4 1/4	7.6	3
4 in	13.8	5 7/16	14.0	5 1/2	10.2	4

undisturbed samples, 5 to 20 cm (2 to 8 in) in diameter and 0.3 to 1.5 m (1 to 5 ft) long can be retained in the inner barrel of the appropriate size. Sampling is effective in firm to stiff cohesive soils and in soft but intact rock.

Coring of rock by the rotary method involves a core bit with embedded diamonds attached to a core barrel, which rotates to cut the annular hole in the rock. The core is protected by a stationary inner barrel and cuttings are flushed upward by the drilling fluid, which is commonly water, but can be air or drilling mud. The rock cylinder (core) can range from 2.2 to 10.0 cm (0.9 to 3.9 in) in diameter with perhaps the most common size being 5.4 cm (2 1/8 in), known as N_x size. Table 19.2 lists the other common core diameters with the 10.0-cm size prevailing for some projects. One and a half or 3 m (5 or 10 ft) lengths of core are common, but this can be extended to lengths of 6 m (20 ft) if the rock is extremely sound. Each core barrel full of rock core is referred to as a *core run* or a *run*.

For conventional rock coring the core barrel and the drill bit are attached to the drill rods, which are rotated by the drill rig to advance the hole. The inner barrel does not rotate and it accepts the core as the bit rotates downward into the rock. Typically, rock core holes are not cased. When the core barrel is full, the entire drill stem is lifted from the hole to obtain the core sample. In this procedure, the core barrel is brought to the surface, the core removed, and the barrel returned to the bottom of the hole. This cycle may take a reasonably long time if the hole is fairly deep because only 10 m (30 ft) or so of drill stem can be kept intact without needing to break it down (disconnect the stem) as it is extracted from the bore hole. Further details of this procedure are discussed in a subsequent section.

Core recovery is an item of major concern in any drilling program because a continuous sequence of samples is the objective of coring. Core recovery is indicated as the percent of core obtained compared to the length of hole drilled. Lost core creates questions concerning what caused the core loss. Closely spaced rock fractures, weak rock, shear zones, and changes in lithology are only a few of the common reasons for core loss. The drilling equipment and skill of the driller are also factors. Although 100% recovery is the objective, this may not be physically possible under the geological conditions at the site. An acceptable core recovery, using the most appropriate sampling techniques, should be stipulated by the engineering geologist or geotechnical engineer. This is the performance level to be expected from the driller or specified in the drilling contract.

In addition to core recovery, the *rock quality designation (RQD)* is determined on the core sample. RQD is the ratio of the sum of the lengths of core pieces 10 cm (4 in) or more long to the total length of hole drilled, represented as a percentage (Table 19.3). Cores should be N_x size or larger for this value to be meaningful, and the core breaks should be natural—not those induced by drilling or drying out

Table 19.3   Rock Quality Designation.

RQD (% recovery in 10-cm [4-in] lengths)	Description
10–25%	Very poor
25–50%	Poor
50–75%	Fair
75–90%	Good
90–100%	Excellent

Source: After Deere and Miller, 1966.

of the core. Drilling-induced fractures are fresh and straight whereas natural fractures are weathered and stained because of groundwater flow. A photograph of rock cores placed in a core storage box is shown in Figure 19.4. These are 7.6-cm (3-in) diameter cores obtained from limestone bedrock.

The dip of bedding and of joints is evident in vertical rock cores but the strike cannot be determined from conventional drilling. When complete orientation (both strike and dip) of bedding planes or joints is desired, an oriented rock core can be obtained with special equipment. The method is similar to conventional coring, but, in addition, continuous grooves are scribed on the rock core relative to a specific compass direction. The core is typically about 5.4 cm (2.1 in) in diameter with a 1.5 m (5 ft) run. In highly fractured rock, this method may not work because of core rotation and possible core blockage.

Core drilling using the wire line procedure has become more common in subsurface investigations for engineering purposes. Long used in oil exploration drilling, the method is best suited for deep holes where there is a much faster cycle of core recovery and resumption of drilling than is possible by conventional core drilling. This is possible as the core and stationary inner barrel are retrieved from the outer core barrel by a lifting device suspended on a thin cable or wire line. The lifting device is disengaged from the outer core barrel, and the inner barrel and core are brought to the surface while the outer barrel and core bit remain in place in the hole. Rather than small-diameter drill rods, a large-diam-

eter tube through which the core can be lifted serves as the drill stem for the wire line apparatus. Consequently, the equipment to outfit a wire line operation is different and more expensive than that used for conventional drilling. However, the rate of coring (production) is greatly improved using this method when deep holes (15 m [50 ft]) are involved.

Rock cores 3.8 to 7.9 cm (1 1/2 to 3 1/8 in) in diameter are possible using the wire line method, with the N_Q size (4.8 cm [1 7/8 in]) being most common. Core barrel lengths of 1.5 to 4.6 m (5 to 15 ft) are available. Another advantage of the wire line method, in addition to increased core production (and an attendant lower cost), is a decreased tendency for core loss. In some instances, a third core barrel or inner lining of plastic is used to retain the sample, which yields improved core recovery for soft or weathered materials. This barrel can also be used in the conventional rock coring procedure to improve that recovery as well. Swelling clays or swelling rocks to be sampled commonly require the special triple barrel core sampler to minimize core loss.

Another procedure is available to sample highly fractured or weathered rock. Using rotary coring, an integral or single sample can be obtained by adapting the standard coring method. A central hole, 2.2 cm (7/8 in) in diameter, is drilled through the length of the proposed core and a steel rod is grouted into this hole using a cement slurry. Using conventional rotary equipment, a large-diameter core 10 to 15 cm (4 to 6 in) is drilled around the grouted rod to obtain the core sample.

Preparation of Borehole Logs

An important part of any site exploration program following sampling is the development of the borehole log, which provides the basis for further subsurface evaluation. This log should supply an accurate record of the soil and rock units encountered in the boring along with any other pertinent information obtained during drilling.

Generally, borehole logging is accomplished in the field as the samples are brought to the surface by the drill rig. A more detailed description of the soil or rock samples is provided sometime later in the laboratory but the initial description of the samples on retrieval from the sampler provides valuable

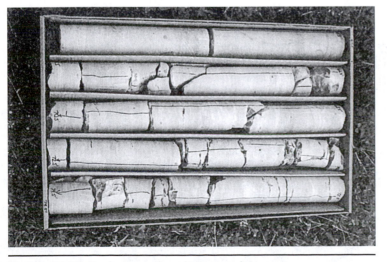

Figure 19.4 Rock cores placed in a core box.

information. The final borehole log is developed from the field and laboratory descriptions. The format of the log and its accuracy depend on the drilling method and on the type of material drilled.

A complete borehole log typically contains much of the following information (Bureau of Reclamation, 2002). Figure 19.5 shows an example boring log.

1. Name of the project, date, and weather conditions during drilling.

2. Borehole identification, elevation, and location relative to project grid or accompanying boring plan.

3. Method of drilling and sampling plus details of equipment as needed to identify drilling specifics.

4. Whether boring is vertical or inclined; if inclined, an angle of inclination and the bearing.

5. Rate of drilling progress either as penetration resistance in soil (N values) or, for rotary drilling in soil or rock, the penetration rate (such as 0.3 m [1 ft] in 15 min) and the rpms during the period.

6. Location of coring runs, samples, and in situ tests, if included; also depths where casing was used and its diameter.

7. Descriptions of groundwater level, changes in standing water level, zones of water loss (when water is used as drilling fluid), or zones of water gain (when air is the drilling fluid).

8. Detailed geologic description of the soil or rock samples. For soils, this involves a description of the moistness, color, texture (grain size), consistency, and organic content, if any, using proper terminology. For rocks, the texture, color, significant mineralogy, type of cementation, if appropriate, rock name, and degree of weathering or alteration are included. A description of the joints, seams, and openings in the rock core is also included, and, for joints, the degree of openness, spacing, inclination, and infilling material is noted.

9. Percent recovery of rock cores or soil samples is included for each run. Core fractures in breaks per meter or per foot are determined for rock, in addition to the RQD. A description of the rock quality can be added, usually in parentheses, to the general description (see Table 19.3) (Deere and Miller, 1966).

10. Unusual occurrences, such as a sudden drop of drill rods, change in color of return wash water, or loss of circulation of wash water, should be noted.

11. A symbolic log is sometimes included at the extreme left or right of the sheet. It is drawn to scale and consists of standard petrographic (or soil) symbols depicting the stratigraphic layer described adjacent to it. The symbols used for soils and rock are given in Figure 19.6 on page 471.

Details of Drilling Procedures

The following paragraphs provide an explanation of the drilling procedures in order to emphasize some of the specifics involved. Both split-spoon sampling and rotary rock coring are considered.

In soil materials, split-spoon samples can be taken continuously, at specific intervals, or at changes of soil type. For interval sampling for a 8-m (25-ft) boring, a common practice is to sample at 0.8-m (2.5-ft) intervals to depths of 3 m (10 ft) or 4.6 m (15 ft) and at 1.5-m (5-ft) intervals for the remaining distance to 8 m (25 ft). The split-spoon sampler is driven 0.46 m (1.5 ft), so to end the sample at a depth of 0.8 m (2.5 ft), sampling is begun at a 0.3-m (1-ft) depth. The sequence, therefore, is as follows, assuming hollow-stem auger equipment:

1. Drill to 0.3-m (1-ft) depth, insert split-spoon sampler;

2. Drive split-spoon to 0.8-m (2.5-ft) depth in three 15-cm (6-in) increments recording the number of blows for each increment;

3. Remove split-spoon, open spoon, describe sample, place sample in a glass jar, and tighten with a screw-top lid;

4. Drill to depth of 1.1 m (3.5 ft); and

5. Put split-spoon sampler back in the hole and drive to depth of 1.5 m (5 ft).

This procedure continues until the full depth of the hole is sampled.

For deep borings in soil, the procedure becomes more complicated. To obtain a sample at 9 m (30 ft), drilling proceeds to 8.7 m (28.5 ft) and the split-spoon sampler is inserted. The sample is driven to the 9-m (30-ft) depth (in three increments) and the sampler is pulled from the hole. This yields a 46-cm (18-in) sample on the end of 9 m (30 ft) of drill rod.

PROJECT NAME ___Coal City___   BORING NO. _____   SHEET NO. ___1___ of ___4___
LOCATION_____   DATE _____   SURFACE ELEV. _____
CONTRACTOR _____   CREW CHIEF_____   WEATHER _____
DRILLING METHOD___HSA & RC___   BOREHOLE DIA ___8", HQ core___   _____
VERTICAL BORING____X____   INCLINED BORING _____   ANGLE AND BEARING_____

DESCRIPTION	Stratum Depth, ft.	Ground Water	Depth Scale, ft.	Sample No.	Blows/6 in. 3–6" increments	% Recovery	Shelby Tube No.	Notes on Drilling Conditions, Etc.
Ground Surface								
Topsoil	0.7							
Brown, moist, medium-dense Clayey sand (SC) with little gravel	2.5			1	1/3/2	75		
Brown, moist, medium-dense Clayey sand (SC) with little gravel			5	2	4/7/7	50	1	
				3	3/3/2	75	2	
loose	8.5							
Brown, moist, medium-dense, well-graded sand and gravel (SW-GW)			10	4	8/11/14	40		Scale Change
			15	5	13/8/10	10		
			20	6	14/19/22	20		
	23.5		25	7	20/29/50	50		
Brown, moist, very dense, well-graded sand (SW) with trace gravel			30	8	22/50/.4	75		
			35	9	41/50/.4	75		
wet	38.5		40	10	22/33/50/.4	75		
Brown, moist, very dense, well-graded sand and gravel (SW-GW)		▽	45	11	8/19/22	75		
Brown, wet, dense, fine silty sand (SP-SM)	48.5		50	12	8/20/26	75		

BORING METHOD
HSA = Hollow Stem Auger
CFA = Continuous Flight Auger
RC = Rock Coring

SAMPLING METHOD
SPT = Standard Penetration Test
ST = Shelby Tube

GROUND WATER
Noted on rods_46.0_
On completion_____
After_____ hours_____

Figure 19.5   Example of a boring log.

(continued)

PROJECT NAME ___Coal City___   BORING NO. _____   SHEET NO. ___2___ of ___4___
LOCATION_____   DATE _____   SURFACE ELEV._____
CONTRACTOR_____   CREW CHIEF_____   WEATHER_____
DRILLING METHOD_____   BOREHOLE DIA _____
VERTICAL BORING_____   INCLINED BORING _____   ANGLE AND BEARING_____

DESCRIPTION	Stratum Depth, ft.	Ground Water	Depth Scale, ft.	Sample No.	Blows/6 in. 3-6" increments	% Recovery	Shelby Tube No.	Notes on Drilling Conditions, Etc.
	53.5							
Brown, wet, dense well-graded sand (SW)			55	13	8/28/42	75		
	58.5							
Brown, wet, dense well-graded sand and gravel (SW-GW)			60	14	6/14/22	75		
	63.5							
Gray, very hard weathered shale with gravel			65	15	50/.4	10		
	67.2			70	16	50/.4	30		
Gray, moist, very hard clayey silt (ML) or highly weathered shale			75	17	50/.3			Cased with HQ casing to 79.5'
	79.5		80		Coring information recorded			
Gray highly to slightly weathered shale	87.0			1	$\frac{7.5}{7.5}$	100		Rock Coring RQD= 60%
			87					
Black shale	90.0		90					
Gray shale	91.5			2	$\frac{10.0}{10.0}$	100		
Black shale	93.0							Run 2
Coal	94.5							RQD = 55%
Gray shale			97					
highly weathered to slightly weathered			100	3	$\frac{10.0}{10.0}$	100		Run 3 RQD = 45%
			107					
			110	4				

BORING METHOD
HSA = Hollow Stem Auger
CFA = Continuous Flight Auger
RC = Rock Coring

SAMPLING METHOD
SPT = Standard Penetration Test
ST = Shelby Tube

GROUND WATER
Noted on rods_____
On completion_____
After_____ hours____

PROJECT NAME ____Coal City____
LOCATION _____
CONTRACTOR _____
DRILLING METHOD_____
VERTICAL BORING_____

BORING NO. _____
DATE _____
CREW CHIEF _____
BOREHOLE DIA _____
INCLINED BORING _____

SHEET NO. ___3___ of ___4___
SURFACE ELEV. _____
WEATHER _____
_____
ANGLE AND BEARING_____

DESCRIPTION	Stratum Depth, ft.	Ground Water	Depth Scale, ft.	Sample No.	Coring Information	% Recovery	Shelby Tube No.	Notes on Drilling Conditions, Etc.
Gray Shale			117	4	$\frac{10.0}{10.0}$	100		Run 4 RQD = 65%
	120.0		120					Run 5, RQD = 20% Coal washed away from 120.5 to 125.0'
Coal	125.0			5	$\frac{5.5}{10.0}$	55		
Gray shale, weathered			127 130					
	133.5			6	$\frac{10.0}{10.0}$	100		Run 6 RQD = 65%
Coal	134.5							
Gray shale, weathered			137					
	141.0		140					
Coal	142.5			7	$\frac{4.0}{10.0}$	40		Run 7 RQD = 20%
Gray shale			147					
	150.0		150					
Sandstone seam	154.0			8	$\frac{7.0}{10.0}$	70		Run 8 RQD = 60%
Gray shale	157.5		157					
Coal, very friable	159.5		160					
Gray shale, highly weathered				9	$\frac{10.0}{10.0}$	100		Run 9 RQD = 70%
			167 170	10				

BORING METHOD
HSA = Hollow Stem Auger
CFA = Continuous Flight Auger
RC = Rock Coring

SAMPLING METHOD
SPT = Standard Penetration Test
ST = Shelby Tube

GROUND WATER
Noted on rods_____
On completion_____
After_____ hours____

(continued)

PROJECT NAME ___Coal City___ BORING NO. _____ SHEET NO. ___4___ of ___4___
LOCATION _____ DATE _____ SURFACE ELEV. _____
CONTRACTOR _____ CREW CHIEF _____ WEATHER _____
DRILLING METHOD _____ BOREHOLE DIA _____
VERTICAL BORING _____ INCLINED BORING _____ ANGLE AND BEARING _____

DESCRIPTION	Stratum Depth, ft.	Ground Water	Depth Scale, ft.	Sample No.	Coring Information	% Recovery	Shelby Tube No.	Notes on Drilling Conditions, Etc.
				10	$\frac{4.8}{8.0}$	60		RQD = 45%
Gray shale, hard 170-175'			175					
	180.5		180		$\frac{8.0}{10.0}$	80		RQD = 70% 5" of coal washed away
Coal	181.5			11				
Sandstone, hard, thinly bedded			185					
Cross-bedded								
			190	12	$\frac{6.0}{10.0}$	60		RQD = 40%
			195					
			200	13	$\frac{4.0}{5.0}$	80		RQD = 60%
Bottom of boring 200.0'								

BORING METHOD
HSA = Hollow Stem Auger
CFA = Continuous Flight Auger
RC = Rock Coring

SAMPLING METHOD
SPT = Standard Penetration Test
ST = Shelby Tube

GROUND WATER
Noted on rods _____
On completion _____
After _____ hours _____

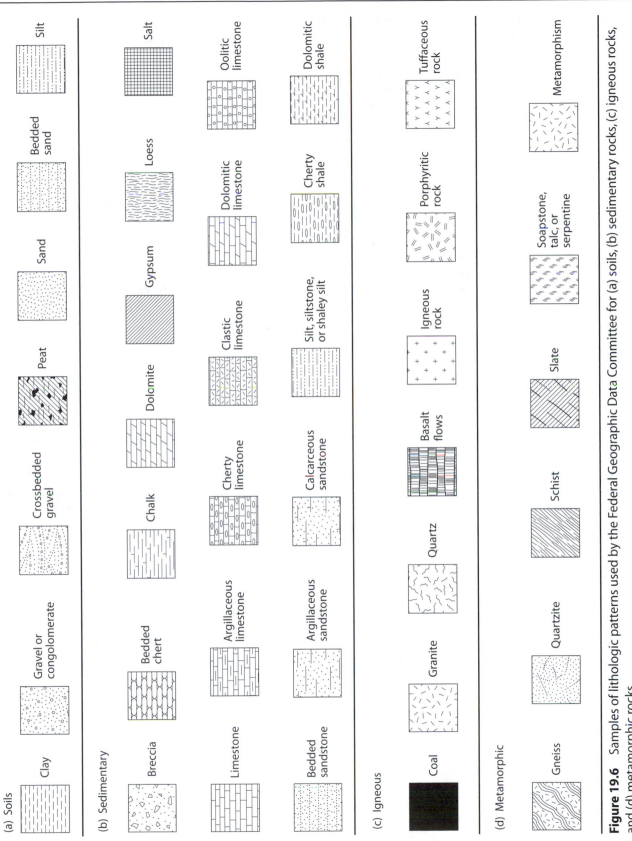

Figure 19.6 Samples of lithologic patterns used by the Federal Geographic Data Committee for (a) soils, (b) sedimentary rocks, (c) igneous rocks, and (d) metamorphic rocks.

The drill rod is hoisted into the air against the derrick of the rig and held in place by a slip ring on the sand line until the split-spoon sampler is removed from the end of the rod. After soil removal, the sampler is screwed into place on the rod and, when drilling is complete, the rod is returned to the hole with 1.5 m (5 ft) of rod added to the top of the drill stem. For depths exceeding 9 m (30 ft), no more rod can be safely held in place by the slip ring and derrick, so 3-m (10-ft) lengths must be dismantled in turn as the rod comes out of the hole until the final 9 m (30 ft) are left and these are pulled into the air as before. Therefore, for a 30-m (100-ft) deep hole, seven 3-m (10-ft) sections must be dismantled in turn as the rod comes out of the hole. This is very time consuming. Some drill rigs utilize two sand lines with slip ring attachments. In this case, for a 30-m (100-ft) depth, four 3-m (10-ft) lengths are dismantled in turn as the rod comes out of the hole. Then, a 9-m (30-ft) section is pulled from the hole and disconnected to be held by the first slip ring and sand line and finally the last 9 m (30 ft) are pulled out and held by the second line while the sampler is removed. As illustrated here, deep sampling takes much longer to accomplish than does shallow sampling and this must be reflected in the price for taking the samples.

In conventional rock core drilling, the core barrel and diamond core bit are connected to the drill rod (about a 3.7 to 6.0 cm [1.5 to 2.4 in] in diameter steel rod), which rotates to cut the core. The most common sizes of core barrels used for subsurface exploration in engineering geology are 1.5 m (5 ft) and 3 m (10 ft). These yield 1.5-m and 3-m (5- and 10-ft) runs of rock core, respectively.

Before rock coring can begin, a casing is set on the bedrock surface and coring proceeds through the casing. Typically, a casing is not placed in the rock section unless that rock is extremely weak and unable to remain open when unsupported. Rock coring can also be accomplished through a hollow-stem auger drilled to refusal on the bedrock surface.

The procedure for rock coring below a thick layer of soil is considered in the following discussion. For a geologic section with 24 m (80 ft) of soil above bedrock, the soil cross section would be sampled using split-spoon sampling likely at 1.5-m (5-ft) intervals, except for the upper 4.5 m (15 ft) or so where 0.8-m (2.5-ft) intervals would prevail. In the last soil sample ending at 24 m (80 ft), the split-spoon sampler would be retrieved by raising the drill rod to the surface. Nine meters (30 ft) of rod could be held by the slip ring and sand line. If a second sand line were available on the drill rig, another 9 m (30 ft) of rod could be held intact by that apparatus. Without the second sand line, the drill string would have to be broken down in 3-m (10-ft) segments until the lower 9-m (30-ft) section was reached. This would involve five 3-m (10-ft) segments. With two sand lines, two 3-m (10-ft) segments would be separated first, then two 9-m (30-ft) sections would be lifted and held on the slip rings and sand lines of the rig. This is obviously a time-consuming operation in both cases but considerably greater when only one sand line is available.

To core rock at the 24-m (80-ft) depth the core barrel and core bit replace the split-spoon sampler at the base of the drill rod stem. The drill stem is put back into the hole and the sections are reassembled to reach the 24-m (80-ft) depth. Typically, a 3-m (10-ft) long core barrel is used so that a 3-m (10-ft) core run is obtained. On completion of the rock coring, the sample is brought back to the surface by breaking down the drill stem in the same manner described before. After the rock core is removed from the core barrel, the barrel is put back into the hole for another 3-m (10-ft) run beginning at the 27-m (90-ft) depth. Obviously, it takes a long time to core a distance of 15 to 30 m (50 to 100 ft) into bedrock under these circumstances. Therefore, for extensive rock coring projects below thick sections of soil, the conventional coring procedure is very time consuming. In fact, coring rock at depths below the ground surface of 15 m (50 ft) or more becomes much slower because of the travel time of going in and out of the hole with samples.

Rock coring at greater depths can be accomplished more quickly using wire line drilling rather than the conventional method. For the same example as before, a 24-m (80-ft) soil section above rock, the hollow-stem auger would be used to drill to refusal on the bedrock surface. For wire line drilling, the drill stem consists not of a small-diameter steel rod but of a tube of sufficient diameter that the core barrel can slip through the inside of it. This drill rod stem or tube must extend from the surface downward to the level of drilling, and this tubing for wire line drilling is more expensive than is the small-diameter drill rod used for conventional drilling. After a

3-m (10-ft) core run is completed, the core barrel (containing the core) is detached from the drill stem and brought to the surface on a wire line that leaves the remaining hardware behind. The core barrel is lifted to the surface quickly by the hoist, the core is removed, and the barrel is returned to the bottom of the hole where the coring is resumed.

In addition to much greater core production, the wire line method produces better core recovery than does the conventional method. For wire line drilling in highly fractured, weathered, or altered rock a triple barrel is used.

When the core barrel is brought to the surface the core is removed and stored in wooden core boxes. These boxes are typically 1.5 m (5 ft) long and made of either wood or cardboard. The core is placed in the core box in columns according to the order in which it is taken from the core barrel. Wooden blocks mark the points where the individual core runs start and end. Depths are indicated on these blocks so that the cores are properly located in the core box. Core recovery percentages, RQD, and fractures per meter or per foot are determined from measurements made of the cores in the core box.

Test Pits and Trenches

Open pits and trenches are a useful means for obtaining both disturbed and undisturbed soil samples from relatively shallow depths. They also provide a way to observe a large exposure of the subsurface in the sidewalls of the excavation, which is not possible by other exploration methods. Details of soil stratigraphy, such as alternating layers of different soil textures, can be fully appreciated by such observation to an extent that could not be discerned by split-spoon samples (even taken continuously) or from pushed thin-walled tube sampling (Shelby tubes).

Open test pits can be dug either by hand or more commonly with backhoes or clamshells on excavating equipment. Typically, such pits range in size of 1.2 m × 1.2 m (4 ft × 4 ft) to about 2 m × 2.5 m (6 ft × 8 ft). In some soils, dewatering and/or bracing would be required to keep the pit open while excavating and sampling. In the past, pits have been dug by hand to depths as great as 30 m (100 ft), particularly in developing nations where hand labor is inexpensive. In other situations, hand-dug pits are restricted to areas inaccessible to heavy equipment. Samples are carefully cut from the sidewalls or from the bottom

and properly preserved for laboratory testing. These samples can be used for uniaxial compression, direct shear, triaxial shear, and consolidation tests. Descriptions of the pit should include the four sidewalls and, possibly, the bottom as excavation proceeds.

Pits should be braced, ventilated, and covered as conditions warrant for safety purposes. In addition, if pits are dug in or near a footing location, the depth should not extend below the footing, to prevent any disturbance or loosening of the bearing material.

Test trenches are used to expose soil materials along a given line or section. When dug by hand they are usually about 1 m (3 ft) deep, whereas a bulldozer can extend such trenches to depths of 4 to 5 m (12 to 15 ft). Using a dragline, test trenches can reach depths of 6 to 10 m (20 to 30 ft). Clamshells and backhoes are also used to dig test trenches. Samples typically are removed from the sidewalls but can be taken from the bottom of the trench during excavation. As with test pits, the total exposure of soil should be described for the trench. Color photographs provide a permanent record of the soil details of the excavation.

Trenching is particularly suited for exploration of moderately steep slopes but is adaptable to flat terrain as well. The excavation may consist of a single slot trench down the slope face or it may include a series of short trenches spaced at short intervals along the slope. Test trenches are used extensively to locate fault traces in alluvial materials in order to date the time of fault movements.

Miscellaneous Considerations

Several other items related to drilling and sampling deserve consideration. One method to gain further information from a borehole is through the use of a borehole camera. This equipment consists of a photographic or television camera that is lowered into an N_x size, 10 cm (4 in) or larger diameter hole to yield a circular photograph or scan of the sidewalls. This is used to view stratification, fractures, and cavities in the walls of the borehole. The method works best above the water table, in a pumped down hole, or when the hole can be stabilized by clear water.

Calyx holes are another means for obtaining subsurface information. These are large holes with diameters of 0.9 m (36 in) or more. Rock cuttings are removed by circulating water. When a section has been drilled the core is extracted by drilling a small hole in the top of the core, an eye bolt is grouted in, and the core is pulled from the hole using the hoist on the drill rig. Calyx holes are particularly useful for firsthand observation of the subsurface conditions by viewing the sidewalls of the hole. A geologist or foundation engineer can be lowered on a bosun's chair into the hole to describe the lithologies, stratification, fractures, and cavities. This may require dewatering of the hole prior to the investigation. Because of the considerable cost of these calyx holes, and because the observations of the sidewalls of the hole are the most rewarding detail, there is a tendency under current conditions to take no actual core but to cut through the rock with a large-diameter roller bit. The cuttings are removed by the wash water.

Exploration drifts and pilot tunnels are sometimes used to obtain direct information about rock conditions. These are typically 1.2 to 2 m (4 to 6 ft) wide and 1.2 to 2.5 m (4 to 8 ft) high. The floor (invert) slopes toward the entrance to facilitate drainage and mucking (rock removal). Descriptions of the tunnel are made and some in situ tests are performed at selected locations.

Geophysical logging of boreholes is becoming increasingly more common in the exploration of major engineering projects. A long-time practice for exploration in the petroleum field, these procedures are now proving helpful in supplying needed detail for engineering studies. Well logging consists of running various measuring devices into the boreholes to determine a number of the physical properties of the rock column that has been drilled. These techniques are discussed in Chapter 18, Earthquakes and Geophysics.

For large construction projects, the drilling and sampling program may consist of several phases rather than a single effort. One phase may involve widely spaced borings to determine the general nature of the site in terms of specific data at these locations. The final design of the project may be accomplished following this preliminary drilling program. Prior to construction, a more detailed drilling program would ensue to provide information for construction details and for bidding information for prospective construction contractors. The plans and specifications for the project would be developed based on all reconnaissance and fieldwork, including this last drilling program.

As a final step in the construction of a major project, the geologic data encountered or learned during construction are recorded along with specific construction details. The purpose is to document the construction record and to provide geologic detail should subsequent problems develop during operation of the project. This makes correction much easier. It also provides a learning process by which planners and designers can improve their techniques through experience gained on previous projects.

Site Selection

Each site for an engineering project, from a small commercial building to a large dam, is unique in terms of its location, accessibility, intended purpose, topography, subsurface geology, hydrogeology, availability of construction materials, economics, and environmental considerations. The site selection criteria and the extent of investigation are, therefore, different for different sites. For structures like dams, highways, and sanitary landfills multiple sites may be considered and the one that provides the best combination of safety and economy may be selected. For a high-rise building in a major city the land may have already been purchased and, therefore, the option of alternate sites is not available. In such cases, the structure needs to be designed according to subsurface conditions, even if the subsurface is less than desirable. Regardless of whether multiple sites or a single site are considered for a given structure, subsurface investigations play a vital role in site selection. The design of a structure, its stability, cost of construction, and long-term performance all depend on subsurface conditions.

Engineering Considerations of Subsurface Investigations

1. Subsurface investigations provide detailed information about the depth and thickness of various soil and rock units, depth to groundwater table, and depth to bedrock. Such information is essential for designing and constructing foundations for all types of engineering structures.

2. There are countless examples in the literature where inadequate subsurface investigations resulted in increased cost, delays in a construction schedule, and, in some cases, failure of the structure.

3. Subsurface investigations, in addition to other considerations, provide the basis for final site selection for projects where alternate sites are considered, such as dams and associated structures.

4. Samples of various soil and rock units obtained during subsurface investigation are used for determining relevant engineering properties required for stability analyses.

5. Detailed information about subsurface conditions is necessary to investigate the causes of any problems that may arise after the construction of an engineering structure. Logs of boreholes, trenches, and pits, along with photographs of the trenches and pits, can be used for this purpose.

6. Records of subsurface investigations are needed for legal cases involving partial or complete failure of an engineering structure.

Exercises on Subsurface Investigations

1. What is the distinction between office reconnaissance and field reconnaissance regarding subsurface investigations for an engineering construction project? List the categories of information to review during office reconnaissance.

2. Regarding published information or open file reports, list the maps and other data that should be reviewed prior to performing a subsurface investigation.

3. Give the two major purposes for field reconnaissance prior to subsurface investigation. List the primary items to be observed during field reconnaissance.

4. List the typical information needed by the drill crew before they can begin subsurface investigation. Discuss the aspects of (a) permission of property owners and (b) location of utilities.

5. What is meant by the term *sounding* when applied to subsurface investigation? What are the limitations of this method?

6. What is the distinction between drilling and sampling? List the eight different methods given in the text for drilling holes. Discuss the hollow-stem auger method.

7. Describe how the standard penetration test (SPT) is accomplished. What is the length of the split-spoon sampler?

8. Two holes are to be drilled to a depth of 15 m (50 ft). Both will involve hollow-stem auger drilling and the SPT method.

   a. In the first hole, SPT tests are to be performed at 1.5-m (5-ft) intervals with the bottom of the sample ending at multiples of 1.5 m (5 ft). Draw this in a cross section. How many SPTs are involved? What percent of the boring is sampled in this case?

   b. In the second hole, SPT tests are to be performed at 0.8-m (2.5-ft) intervals with the bottom of the sample ending at multiples of 0.8 m (2.5 ft). Draw this in a cross section. How many SPTs are involved? What percent of the boring is sampled in this situation?

9. What is a pushed tube soil sample? What are its advantages over a SPT? What are its disadvantages, if any? What laboratory tests can be run on these samples?

10. Which drilling and sampling procedure would most likely be used to investigate an 8-m (25-ft) thick lake bed on which a three-story building was to be constructed?

11. If a standard penetration test yielded the following results: 6/5/7, what would be the N value? If the soil was a sand, what would be the state and relative density? What would these be if the sample was a cohesive clay?

12. Give an example in which probing would be adequate. Give another geologic setting in which it would not be. Explain why.

13. What would be the nature of the reconnaissance for an interstate highway through a mountainous region? Explain in as much detail as possible.

14. What is meant by *rotary coring*? For rock coring, what is the difference between a core bit and a core barrel? What is the diameter of an N_x core? How big is the N_x core hole?

15. What is meant by *core recovery*? Why is it important to have a high core recovery?

16. Distinguish between conventional core drilling and wire line core drilling. What is the major advantage of wire line drilling?

17. What are bore hole logs? Why is it so important that these be made accurately and completely?

18. When logging core how can natural joints in rock be distinguished from fractures that occurred during the drilling process? Which ones are included in determining the RQD? Why?

19. Prepare a drilling log depicting the following information. Include a lithologic section using proper symbols.

    - Vertical hole, 9.5 m (31 ft) of soil, 6 m (20 ft) cored into rock, N_x size. XYZ Project, Boring No. 1. Drilled ten days ago.

    - Split-spoon samples at 0.8-m (2.5-ft) intervals ending on multiples of 0.8 m (2.5 ft). Values were 2/5/4; 3/3/3; 4/3/5; 5/6/7; 6/6/8; 8/7/9; 6/8/8; 12/12/14; 14/14/13; 15/15/16; 15/17/18; 30/30/28.

    - Soils interrupted included 0 to 0.6 m (0 to 2 ft), topsoil; 0.6 to 2.1 m (2 to 7 ft), sand; 2.1 to 5.5 m (7 to 18 ft), silty clay; 5.5 to 8.2 m (18 to 27 ft), clayey silt; 8.2 to 9.5 m (27 to 31 ft), silt with rock fragments.

    - Rock intercepted: 9.5 to 12.8 m (31 to 42 ft), brown friable, quartz sandstone; 12.8 to 15.5 m (42 to 51 ft), gray shale.

    - Condition of core: 9.5 to 12.5 m (31 to 41 ft), 2.9 m (9.4 ft) of core, all pieces greater than 10 cm (4 in) except for 0.3 m (1 ft), 7.6 cm (3 in) total;

12.5 to 15.5 m (41 to 51 ft), 2.7 m (8.9 ft) of core, only 1.8 m (6 ft), 10 cm (4 in) of core consisting of pieces greater than 10 cm (4 in) long.

    - Water on completion and 48 hr later, 3.7 m (12 ft) below the ground surface.

20. What is the rock quality designation (RQD)? How is it determined? Why is it significant?

21. What drilling equipment would be needed to explore along the centerline of a proposed earth dam where it crosses an existing stream? What samples would be taken and at what depths?

22. If rock coring costs $17 per ft for N_x size conventional drilling for borings to a depth of 12 m (40 ft), and $25 per ft for depths of 12 to 23 m (40 to 75 ft), how closely spaced can borings be placed for a 152 m × 152 m (500 ft × 500 ft) area if the borings will be 20 m (65 ft) deep and the drilling bill cannot exceed $15,000? Make a sketch and locate the borings.

23. What is the advantage of test pits and trenches over rock or soil sampling? What are the limitations for pits and trenches? Explain.

24. What is a calyx hole? How large are they? Can they be viewed directly by human inspectors? Why is groundwater a concern in this situation?

25. What would be the purpose of driving a pilot tunnel into a dam abutment prior to dam construction? What information could be supplied?

26. What is meant by *geophysical bore hole logging*? Refer to Chapter 18, Earthquakes and Geophysics for further information.

REFERENCES

Adeyeri, J. B. 2015. *Technology and Practice in Geotechnical Engineering* (Chapter 12, Subsurface Investigation, pp. 744–795). Hershey, PA: IGI Global.

American Society of Civil Engineers. 1976. *Subsurface Investigation for Design and Construction of Foundations of Buildings*. Reston, VA: American Society of Civil Engineers.

Bowles, J. E. 1996. *Foundation Analysis and Design* (Chapter 3, Exploration, Sampling, and In Situ Soil Measurements, pp. 135–210). New York: McGraw-Hill.

Bureau of Reclamation. 2002. *Engineering Geology Field Manual* (2nd ed.). Denver, CO: Bureau of Reclamation, US Department of the Interior.

Das, B. M. 2015. *Principles of Foundation Engineering* (8th ed.). Boston: Cengage Learning.

Deere, D. U., and Miller, R. P. 1966. *Engineering Classification and Index Properties for Intact Rock* (Technical Report No. AFWL-TR-65-116). Urbana: University of Illinois at Urbana–Champaign.

Lowe III, J., and Zaccheo, P. F. 1991. Subsurface explorations and sampling. In H-Y. Fang (ed.), *Foundation Engineering Handbook* (Chapter 1, pp. 1–71). New York: Springer.

Peck, R. B., Hanson, W. E., and Thornburn, T. H. 1974. *Foundation Engineering* (2nd ed.). New York: John Wiley & Sons.

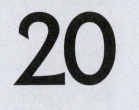

PHYSIOGRAPHIC PROVINCES 20

Chapter Outline

Physiographic Provinces

Factors Affecting Physiography

Subdividing Physiographic Provinces

Engineering Considerations of Physiographic Provinces

Physiography is the study of landform features of the Earth's surface. In this chapter, regional physiography is examined first, followed by the details of geologic history that produced these landscapes. Next, a review of the geologic and physiographic characteristics of two sample locations, Indiana and Oklahoma, is provided. Finally, the problems that are common to certain physiographic regions are considered using Oklahoma as an example.

Physiographic Provinces

Continental areas, such as the United States, have been divided into regions, each having a particular type of landscape. These subdivisions of the continental landmass are called physiographic provinces (Fenneman, 1931, 1938; Lobeck, 1948; Hunt, 1967; DiPietro, 2013).

In the United States, the physiographic provinces are large, ranging from one to several states in size. Within a province, the landscape is generally similar because it has developed under similar climatic and geologic conditions. Figure 20.1 shows the physiographic provinces of the United States. Mountainous provinces, in general, are underlain by orogenic belt structures, that is, by various combinations of folded and faulted sedimentary rocks, intrusive igneous bodies, and regional metamorphic rock masses—all of which are associated with mountain building processes. The lowland and plateau provinces are underlain by stable region structures. These consist of horizontal to gently dipping sedimentary rocks or broadly domed sedimentary features that extend over areas of possibly several states in size. Thick layers of lava flows and volcanic cones form a final group of landscape features. These typically formed at a later time on top of preexisting orogenic belt (mountain forming) structures. Most Precambrian orogenic belts have been reduced to an erosion surface of low relief.

Table 20.1 provides a brief discussion of the major physiographic provinces of the conterminous United States. These descriptions are greatly simplified but provide a general indication of the features that prevail in a particular province. In reality, each province comprises a chapter in books that focus on this subject.

Table 20.1   Descriptions of the Physiographic Provinces of the Conterminous United States.

Canadian Shield Province

This is the foundation on which the North American continent was built. It has undergone many periods of uplift and erosion. Most of the units are hard metamorphic and igneous rocks of Precambrian age, highly resistant to erosion. Surface relief on the Canadian Shield has been greatly reduced by erosion since Precambrian time, from a mountainous terrain to one of gently rolling hills and valleys.

Two parts of the province are located in the United States: the Superior Upland of northern Minnesota and Wisconsin, and the Adirondacks of New York. Both are composed of rocks with extremely complex structures, have been exposed to erosion for nearly two billion years, and have undergone, most recently, extensive glaciation. The soil cover is thin because of glacial erosion and numerous lakes occur in the massive bedrock areas.

Gulf and Atlantic Coastal Plain Province

This province is composed entirely of undisturbed sedimentary strata consisting mostly of loosely consolidated sands and gravels. Much more area than is visible today was exposed during Pleistocene glaciation when sea level stood considerably lower. The plain is relatively flat, sloping gently to the ocean, with some low, hilly sections. Beginning with Long Island, New York, on the north, the coastal plain extends through New Jersey, Delaware, Maryland, Virginia, North Carolina, South Carolina, Georgia, Florida, Mississippi, Louisiana, Arkansas, and Texas.

Ridge and Valley Province

Labeled as Folded Appalachians on Figure 20.1, this province consists of upturned, strongly folded Paleozoic sedimentary rocks extending from the New England Province to Alabama. It is centrally located between the Appalachian Plateau to the west, the Piedmont Plateau to the east, and the Blue Ridge Province to the east. The structure is not nearly as complex as the belt to the east, nor are the mountains as high. The belt consists of parallel, elongated, steep-sided ridges, with valleys between them. Drainage is almost entirely confined to these valleys. The province contains major coal deposits.

Appalachian Plateau Province

This is an area of very gently folded rocks that has been uplifted like a table and then dissected by streams. The northern half was glaciated and contains many lakes and other glacial features. Numerous coal deposits are found here. This province extends from New York through Pennsylvania, Ohio, West Virginia, Kentucky, Tennessee, and Georgia.

Blue Ridge Province

This is a massive ridge complex that extends some 800 km (500 mi) from southern Pennsylvania to northern Georgia. It consists of metamorphosed sedimentary rocks of the Lower Cambrian and Upper Precambrian ages plus Precambrian gneisses and granites. The Great Smoky Mountains occur in the southeastern portion of this province.

Piedmont Plateau Province

This province extends from southern New York State to Alabama and lies as a low plateau above the Atlantic Coastal Plain Province to the east but at a much lower elevation than the Blue Ridge Province to the west. It consists of Precambrian age metamorphic and igneous intrusives; hence the name "Crystalline Appalachians." Deep residual soils known as saprolites occur in the province as well as occasional resistant bedrock domes such as Stone Mountain, Georgia.

New England Province

This is the northern extension of the easternmost mountain belt, that is, of the Piedmont, Blue Ridge, and Ridge and Valley Provinces. The rocks are very complex having been subjected to folding, faulting, and intrusions by igneous rocks, with most episodes in early Paleozoic time. Extensive areas of gneiss, schist, slate, marble, and quartzite indicate the widespread metamorphism. The entire region has been recently glaciated, leaving a landscape dotted with thousands of lakes and numerous streams.

Central Lowlands Province

This extensive province is underlain by limestone, sandstone, and shale of Paleozoic age. It is gently warped into domes and basins. The northern half was recently glaciated and contains many lakes and glacial features. The province embraces almost the entire Mississippi Valley and the Great Lakes area. It is one of the largest food producing areas of the world.

Interior Low Plateaus Province

This is a transition area between the Appalachian Plateau and the Central Lowlands with the Gulf Coastal Plain to the southwest. It consists of gently dipping sedimentary rocks ranging in age from Ordovician to Cretaceous. Extensive erosion has formed a series of rock escarpments in the province.

Ozark and Ouachita Provinces

These are sometimes collectively referred to as the "interior highlands." They are remnants of an extensive mountain range that can be correlated with the Appalachian system. The Ozark Plateau consists of Paleozoic sedimentary rocks, predominantly cherty limestone, whereas the Ouachita Mountains are strongly folded and faulted rocks with sandstone and shale predominating.

Great Plains Province

This province extends from the Rio Grande River on the south into northern Canada with the Central Lowlands to the east and the Rocky Mountains to the west. The rocks are mainly Mesozoic or Cenozoic in age and, at the surface, consist mostly of poorly lithified rock units formed on the weathering of the Rocky Mountains, or of glacial deposits. Beds are mainly horizontal. Only the northern section has undergone continental glaciation.

Southern Rocky Mountains Province

This province consists of a north–south trending mountain system extending from southern New Mexico to southern Wyoming. It is comprised of a series of anticlinal ranges with cores of Precambrian igneous or metamorphic rocks, and yields the highest peaks in the Rockies.

Wyoming Basin Province

An extension of the Great Plains Province, the Wyoming Basin lies between the Southern and Middle Rocky Mountains Provinces. Tertiary age formations occur at the surface, attesting to its relationship to the Great Plains.

Middle Rocky Mountains Province

Separated from the Southern Rockies by the Wyoming Basin, this province consists of two distinct mountain groups. The eastern group includes the Bighorn, Wind River, Bear Tooth, and Owl Creek Mountains, which are similar to the Southern Rockies with their cores of Precambrian igneous and metamorphic rocks. The western group includes the Gros Ventre, Hogback, Wyoming, Snake River, Salt, and Teton Ranges, which were formed mainly by thrust faulting. The origin of these ranges is more closely related to that of the Basin and Range Province than to the Southern Rockies. The Tetons, Jackson Hole, and Yellowstone Park are located in the western group of the Middle Rockies Province.

Northern Rocky Mountains Province

The northern "Rockies" are located in the northern two-thirds of Idaho and the western part of Montana. They are composed of a major "granite batholith," the Idaho Batholith, and prominent folded mountains in western Montana and other prominent structures.

Columbia Plateau Province

This plateau, located adjacent to the Columbia River in Idaho, Oregon, and Washington, consists of an accumulation of thousands of feet of lava flows. The northern portion was glaciated and several lakes formed by this process and by lava flows transecting the rivers. It is an arid region and not heavily populated.

Colorado Plateau Province

Located in the southwestern United States, this plateau system formed on gently dipping sedimentary rocks and includes the Grand Canyon. It is an area where slow uplift has been matched by downcutting by the rivers to form deep canyons. In the southern section, extensive lava flows and episodes of volcanism, some as recent as 900 years ago, have occurred. The sharp topographic expression is due largely to the arid climate of the province.

Basin and Range Province

The geologic history of this region is complex with block faulting forming large basins and ranges, yielding deep, closed basins, surrounded by relatively flat-topped ridges. The area is one of the most arid in the world and poorly populated.

Sierra–Cascades Province

This province consists of the Sierra Nevada, a prominent block-faulted mountain system located along the California–Nevada border and the Cascade Range, a series of volcanic peaks extending from northern California through Oregon and Washington. Included in the Cascades are such prominent peaks as Mount Shasta, Mount Rainier, Mount Lassen, and Mount St. Helens. Mount Whitney, the highest point in the 48 contiguous states of the United States, is located in the Sierra Nevada.

Pacific Coast Ranges Province

These are the westernmost mountains on the continental United States, extending along the coastal area. They are also the most recently formed mountain belt and many of the mountains consist of partially unconsolidated materials. Los Angeles and San Francisco are located in this area. To the north (in Washington), the mountains form offshore islands. In the south, rainfall is sporadic, yielding a semiarid condition. Because of the young nature of these mountains there are many very active faults in the area.

Pacific Troughs Province

This belt of deep valleys divides the Sierra–Cascades from the Pacific Coast Ranges. The troughs are formed by large down-faulted blocks. They have been accumulating sediments from the mountains on both sides and some are very fertile. The San Joaquin Valley lies south of San Francisco and the Sacramento Valley to the north. In Oregon, the Willamette Valley forms the trough and in Washington it is the Puget Sound. In the north the troughs are water filled and form the island passage up the coast to Alaska.

Figure 20.1   Physiographic provinces of the United States. (Adapted from Lobeck, 1957.)

Factors Affecting Physiography

There are three factors that determine the appearance of the land surface or the landscape: (1) rock type and geologic structure, (2) the dominant geologic process that has operated in the region, and (3) the stage or degree to which the geologic process has been able to carve the land surface (Thornbury, 1965; Hunt, 1967; DiPietro, 2013).

Rock Type and Geologic Structure

The type of rock present, its attitude or position in space, and the extent of faulting and folding of the rock mass play a dominant role in shaping the land surface. For example, the horizontal sedimentary rocks of the Central Lowlands yield a different landscape than do the massive igneous intrusives that form the Sierra Nevada. Similarly, in the Ridge and Valley Province, the harder sandstones form the ridges and the weaker shales or other weak rocks form the valleys.

The primary geologic structures of continental regions that strongly influence the appearance of the landscape are shown in Table 20.2 on page 482. There are three subdivisions of these geologic structures: (1) undisturbed, (2) disturbed, and (3) volcanoes and lava flows. The structures in each group owe their general similarities to a similar geologic history.

The undisturbed structures underlie most of the interior areas of the United States, extending from the Appalachian Plateau on the east to the Great Plains on the west. This is the stable interior of essentially horizontal sedimentary rocks, a common feature for all the continents of the world. The rocks have not been deformed by mountain building stresses.

Disturbed structures include folded and faulted sedimentary rocks along with igneous intrusives and complex metamorphic configurations. These are the areas of active mountain building regions and of such locations now worn down through the great expanse of geologic time. The Appalachian Mountain region, the Canadian Shield, the Rocky Mountain system, the Pacific Coast Ranges, and the Basin and Range Province make up the vast area of disturbed structures in the United States.

Volcanoes and lava flows comprise the third major subdivision of disturbed structures. These are extensive areas where volcanic mountain chains dominate the terrain or where great expanses of flood basalts cover large areas of the landscape. The extensive areas in the world where flood basalts occur were discussed in Chapter 3, Igneous Rocks. In the United States the province comprising extensive lava flows is the Columbia Plateau Province, located in the northwest portion of the country. The Cascade system, located in central Washington and Oregon and extending some distance into northern California, is the prominent volcanic mountain system in the United States. Mount St. Helens, which violently erupted in 1980, is located within the Cascades in southern Washington.

Figure 20.2 on the next page illustrates the geologic structures described in Table 20.2. The eight drawings of Figure 20.2 are keyed to the specific subdivisions shown in Table 20.2. The two upper drawings show the A1 structures, coastal plain features in the upper left and horizontal strata in the upper right. The A2 structure of broadly domed sedimentary layers is shown in the drawing directly below.

The disturbed structures of the B subdivision are illustrated in four drawings below the A structures in Figure 20.2. B3 shows the folded sedimentary rocks common in the Ridge and Valley Province of the eastern United States. B4 illustrates major faulting and B5 shows the rounded topography of homogeneous crystalline igneous and massive metamorphic rocks. B6 is a diagram depicting complex combinations of igneous, metamorphic, and folded sedimentary rocks including faulting.

The terrain characterized by volcanic cones is shown in diagram C7. This is typical of the Cascades region. Layered volcanic rocks typical of flood basalts formed by fissure eruptions are not included in Figure 20.2. These have a generally similar appearance to the dissected plateau of horizontal sedimentary rocks shown in A1, upper right drawing.

The various features depicted in Figure 20.2 can be observed on the topographic maps of selected areas of the United States. By observing the configuration of the contour lines, the relief, drainage pattern, gradients, and other factors, details can be ascertained about the geologic structure that

prevails in the area. This information is used in turn to predict the construction problems, concerns, and geologic constraints that should be considered. The interpretation of topographic maps to determine the underlying geologic structure is required in the map exercises presented at the end of this chapter.

Geologic Processes

The geologic processes of major significance relative to the appearance of the land surface are glaciation, weathering and erosion in humid regions, weathering and erosion in arid regions, subsurface solution action in limestone terrains, and development of coastal features along the oceans and the Great Lakes. Although less significant than rock type and structure in determining the overall appearance of the landscape, geologic process is, nonetheless, important. As an example, the New England Province is equivalent, structurally, to the combined area of the Ridge and Valley Province plus the Blue Ridge Province. Extensive continental glaciation in the New England area, as contrasted to no glaciation in the two other provinces, however, provides a major difference in appearance and, therefore, the separate designations.

Additional details of geologic processes are considered in other chapters of this text; running water (Chapter 12), glaciation (Chapter 14), coastal processes (Chapter 16), groundwater (Chapter 13), and wind (Chapter 17). Differences in weathering effects of humid versus arid conditions can be observed for the eastern versus the western areas of the United States, respectively. These differences are discussed in detail in Chapter 4, Rock Weathering and Soils. See Figures 20.3 (page 483) and 20.4 (page 484) for geologic maps of Oklahoma and Indiana, respectively.

Stage

The concept of stage was discussed in a previous chapter on running water and stream development (Chapter 12). Figure 12.18 shows the effect of different stages of development on the appearance of the landscape. The existence of flat interstream divides is a criterion for youthful topography, with mature topography occurring when these flat areas are destroyed.

Stage can be thought of as the extent to which the process has run its full course. Youthful topography in a humid region shows only the initial effects of stream erosion, leaving extensive interstream divides that are flat and untouched by channel development. A con-

Table 20.2   Geologic Structure of the Continents.

A. Undisturbed Structures (Stable Region)

These structures develop on layers of sedimentary rocks, mostly shale, sandstone, and limestone, or on thick glacial deposits. These layers are horizontal, gently dipping, or very broadly warped into domes or basins, and form a relatively thin veneer over the Precambrian basement complex, ranging in thickness from tens of meters to hundreds of meters (a few hundred to a few thousand feet). The broad expression of a stable region structure is that of a plateau if the relief is moderately high, or that of a plain if the relief is low.

1. Horizontal or very gently dipping sedimentary layers, including coastal plains.

2. Broadly domed sedimentary layers.

B. Disturbed Structures (Orogenic Belt)

These structures are formed on sedimentary, metamorphic, and igneous rocks. If layered, these rocks are seen to be folded and faulted from their original positions, and they form a thick, though long and narrow, section above the basement complex. Post-Cambrian orogenic belts are typically expressed as long, narrow mountain chains, having extensive amounts of granitic rock exposed in their central part. Also within the belt, and usually bordering the granitic rocks, are areas of belted metamorphic rocks. Most ancient (Precambrian) orogenic belts have generally been reduced by erosion to a surface of low relief.

3. Folded or tilted sedimentary layers (striped, canoe-shaped, and zigzag patterns).

4. Major faults (may be expressed as an abrupt change in topography along a more or less straight line).

5. Homogeneous crystalline rocks: granite, gneiss, some schists.

6. Complexly related igneous, sedimentary, and metamorphic rocks.

C. Volcanoes and Lava Flows

7. Volcanic cones.

8. Layered flows of lava.

Figure 20.2   Geologic structures related to landscape.

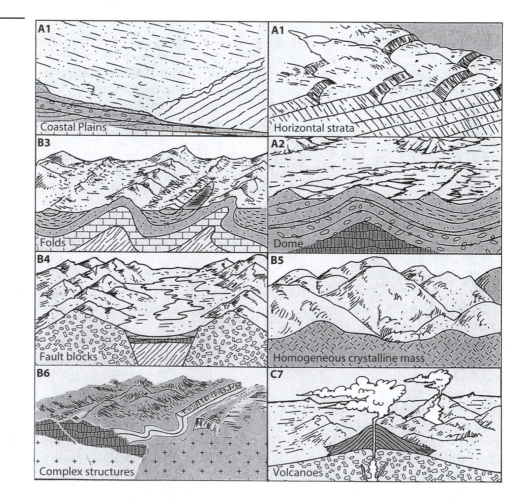

trast can be seen between the appearance of the till plains in central Indiana and the same type of feature in northern Missouri. The till plain of Wisconsinan age in Indiana (only 14,000± years old) is in early youth with only a small network of streams. In Missouri, the topography is mature because the Kansan age till has been undergoing dissection for nearly a half-million years. Obviously, the rolling terrain of Missouri looks different from the Indiana region.

The regional stage for an area is discerned by comparing the appearance of the topographic map to Figure 12.18, which depicts the different types of stage development. Youth, late youth, early maturity, and late maturity are possible designations. If an area did not begin as essentially a flat sloping plane on which stream drainage systematically developed, then the comparison is not applicable.

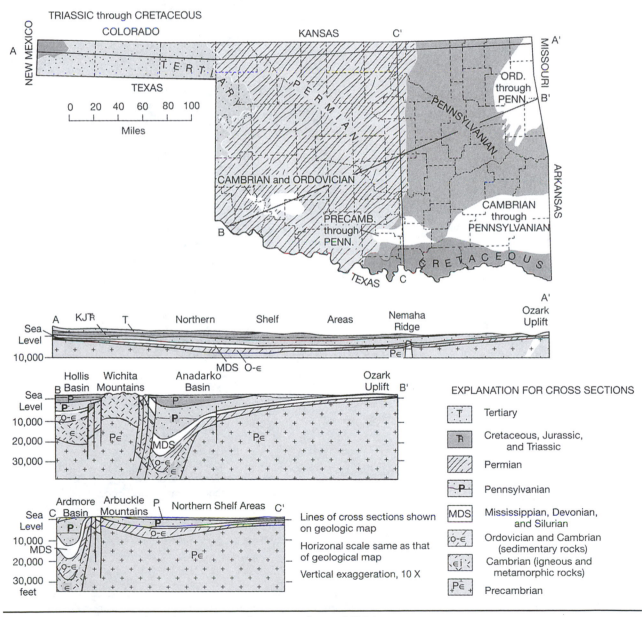

Figure 20.3   Generalized geologic map and cross sections of Oklahoma.

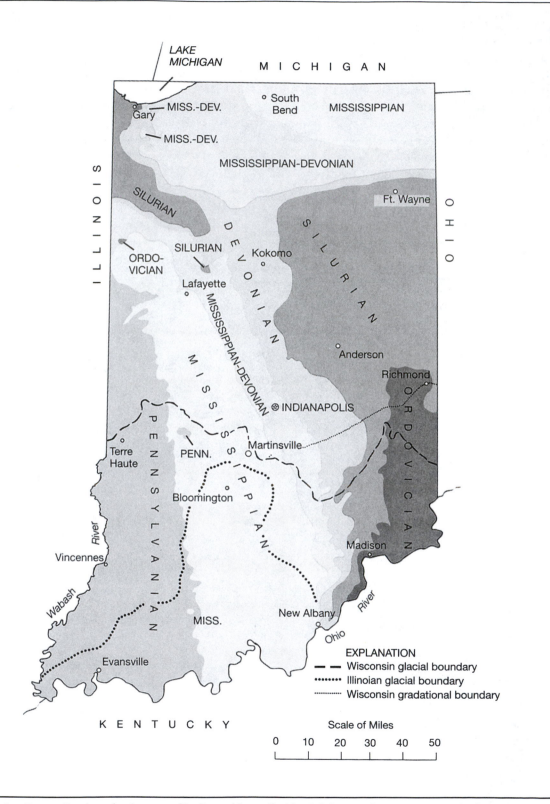

Figure 20.4   Generalized geologic map of Indiana. (Compiled by J. B. Patton, January 1, 1952, Indiana Department of Conservation, Geological Survey.)

Subdividing Physiographic Provinces

Sections

All of the physiographic provinces described in Table 20.1 can be subdivided into smaller portions based on physiography. These smaller units, typically called physiographic sections, have more features in common and a less diverse range of landscape. In the case of provinces consisting of mountainous terrain, the sections may be individual mountain ranges and intermountain basins. This is the case for the Middle Rockies. Another example, the Ozark Province, consists of individual plateau areas and higher mountain terrains, which have distinct structural differences. On a site-specific basis, analysis of the individual sections is a must, preceded by a study of the regional conditions, which is accomplished by a review of one or more physiographic provinces.

Divisions on a State Basis

Another means for subdividing physiographic provinces, or large parts of them, is studying the landscape detail on a state-by-state basis. This proves useful because state geological surveys approach the studies of surface geology in this fashion and the citizens of that state can best appreciate these subdivisions. Many states can be studied using this approach. Subdivisions of a state physiographic map into these areas based on a very similar landscape appearance are referred to as regional physiographic units.

As an example, the map in Figure 20.5 on page 487 shows the physiographic units for the state of Indiana and also provides the extent of Wisconsinan and Illinoian glaciations. The landscape of the units with a north–south elongation, located in the central to southern portion of the state, is influenced by the bedrock characteristics in these areas. For northern Indiana the landforms are dictated primarily by the effects of glacial deposition.

Table 20.3 on the following page provides the descriptions for the physiographic units of Indiana. They are keyed to the map of the state presented in Figure 20.5.

Oklahoma is another example of state-wide physiographic divisions. Oklahoma has been selected to illustrate how physiographic divisions can be related to highway construction problems. This is discussed later in the section on engineering considerations. Figure 20.6 on page 489 shows the physiographic units of the state and Table 20.4 on page 488 provides a description of these physiographic subdivisions.

Table 20.3   Description of Regional Physiographic Units of Indiana.

1. **Calumet Lacustrine Plain**
   Lake sediments of glacial Lake Chicago. Series of beach ridges and sand dunes from former and present shorelines.

2. **Valparaiso Morainal Area**
   Wide terminal moraine of a substage of Wisconsinan glaciation. It is the prominent relief feature in the area.

3. **Kankakee Outwash and Lacustrine Plain**
   Sandy glacial lake deposits developed on outwash from the Valparaiso moraine. Many low scattered sand dunes on the flat lake plain.

4. **Steuben Morainal Lake Area**
   Composed of recessional moraines from the ice lobe that entered the state from the northeast (Saginaw lobe).

5. **Maumee Lacustrine Plain**
   Lake sediments—an extension into Indiana from an extensive lake plain in Ohio.

6. **Tipton Till Plain**
   Flat Wisconsinan age till sheet underlain by locally rugged bedrock topography giving rise to a great range in thickness of glacial deposits.

7. **Wabash Lowland**
   Area of moderate to low relief developed on Pennsylvanian age shale and sandstone.

8. **Crawford Upland**
   Most rugged topography in southern Indiana; results from erosion of alternating massive sandstone, shale, and limestone of Late Mississippian age.

9. **Mitchell Plain**
   Area of moderate to low relief developed by solution of Middle Mississippian limestone, many caves.

10. **Norman Upland**
    Gently westward-sloping surface on resistant sandstones and siltstones of Early Mississippian age. The Knobstone escarpment along its eastern margin rises 120 to 180 m (400 to 600 ft) above the adjoining Scottsburg lowland.

11. **Scottsburg Lowland**
    Narrow lowland area developed on Devonian and Early Mississippian age shales.

12. **Muscatatuck Regional Slope**
    A westward-sloping surface held up in the east by resistant cherty Silurian limestone along the border of the Dearborn upland.

13. **Dearborn Upland**
    Flat upland surface with deeply entrenched valleys developed on Upper Ordovician limestone.

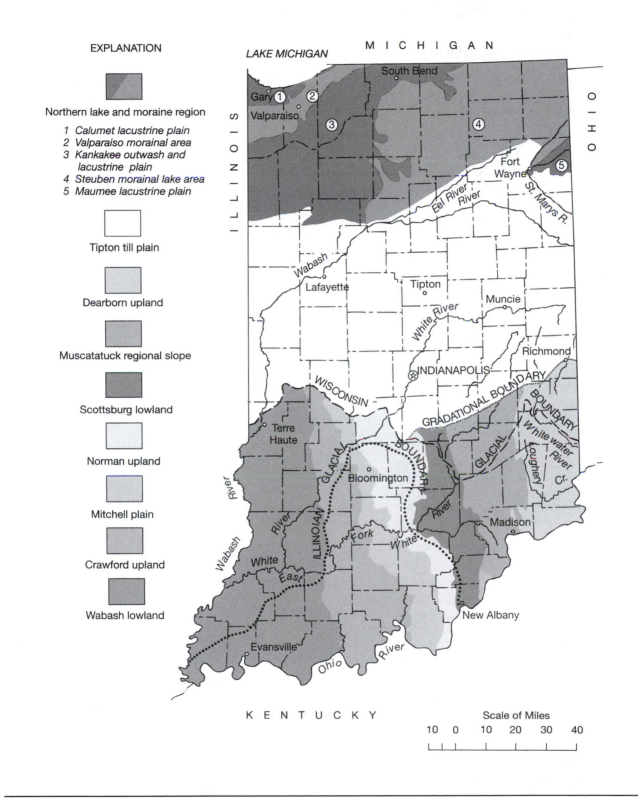

Figure 20.5   Regional physiographic units of Indiana.

Table 20.4   Descriptions of Regional Physiographic Units of Oklahoma.

1. **High Plains**
   Composed principally of nearly flat-lying, weakly to unconsolidated calcareous sandstones and conglomerates with some soft, chalky, sandy limestone present locally. Tertiary in age, these plains were formed essentially by Rocky Mountain outwash. Steep slopes adjacent only to large streams.

2. **Prairie Plains Homocline**
   Permian and Pennsylvanian shale and sandstone that dip westward 2 to 9 m/km (10 to 50 ft/mi). Shale is red to the west (Permian) and gray to the east (Pennsylvanian). Region consists mostly of gently rolling hills and broad, flat plains. Resistant beds of cemented sandstone or of limestone are more common in the Pennsylvanian strata. Hence, relief is greater in the east where long cuestas of east-facing escarpments overlook broad shale plains.

3. **Ozark Uplift**
   A deeply dissected plateau formed on gently dipping (westward) cherty limestones of Mississippian age. Caves, solution cavities, and other karst features are more prevalent than in other sections of Oklahoma.

4. **Anadarko–Hollis–Marietta Basins**
   Flat-lying red Permian shale and sandstone. Few rock units are hard and the topography is mostly gently rolling hills and broad flat plains. Locally 30-m (100-ft) thick beds of gypsum occur along with thin (1.5 m [5 ft]) dolomite beds. Resistant gypsum layers in Anadarko Basin and western Hollis Basin locally cap high escarpment. Dips are mostly gentle (usually less than 3 m/km [15 ft/mi]), although a narrow area 8 km (5 mi) at the southern edge of Anadarko Basin has a dip 50 m/km (250 ft/mi). In some areas thick beds of soft sandstone are deeply dissected into steep-walled canyons. Extensive granular terrace deposits occur adjacent to major streams.

5. **Arkoma Basin**
   Pennsylvanian strata consisting of gray shale and hard, resistant sandstone. Strata are gently folded into a series of many local anticlines and synclines. Dips typically less than 15° and many faults are evident locally. Resistant sandstones yield broad hills and mountains rising 90 to 600 m (300 to 2000 ft) above wide, hilly plains and valleys.

6. **Ardmore Basin**
   Lowland of folded Mississippian and Pennsylvanian shale and sandstone lying between the Arbuckle Mountains and Gulf Coastal Plain. General structure is synclinal, but a large number of anticlines are also present. Dips are steep, particularly along margins of the basin, with angles of 45 to 90° being common.

6. **Ardmore Basin (continued)**
   Dark gray shale units are 30 to more than 600 m (100 to more than 2000 ft) thick and separated by resistant sandstone, limestone, and conglomerate only 3 to 5 m (10 to 15 ft) thick. These latter units form conspicuous, subparallel ridges rising up to 30 m (100 ft) above adjacent lowlands.

7. **Ouachita Mountains**
   Along with the Wichita and Arbuckle Mountains, the Ouachitas comprise Oklahoma's three mountain regions. Located in the southeast, the Ouachitas form an arcuate fold belt consisting mostly of Mississippian and Early Pennsylvanian sandstone and shale locally about 9100 m (30,000 ft) thick, deposited in an elongated trough. During the Pennsylvanian age these strata were folded into broad anticlines and synclines and thrust northward along a series of major thrust faults. Resistant, steeply dipping sandstones form long sinuous ridges and hogbacks towering 300 to 450 m (1000 to 1500 ft) above adjacent shale valleys.

8. **Arbuckle Mountains**
   Located in south-central Oklahoma, this is a small area of low to moderate hills containing 4500 m (15,000 ft) of folded and faulted sedimentary rocks, ranging in age from Cambrian to Pennsylvanian. About 80% of this thickness is comprised of limestone and dolomite with the remainder being shale and sandstone. Rocks were folded and thrust faulted during several mountain building episodes of the Pennsylvanian. Sedimentary rock cover has eroded to expose Precambrian granites in a 400 km^2 (150 mi^2) area in its southeastern portion, the largest exposure of Precambrian rocks in the state.

9. **Wichita Mountains**
   Located in southwest Oklahoma, granite, rhyolite, and gabbro are the dominant rocks, Middle or Early Cambrian in age. These igneous rocks are flanked by scattered outcrops of Cambrian and Ordovician limestone and dolomite, much like the folded rocks of the Arbuckle Mountains. Thrust faulting during several episodes in the Pennsylvanian uplifted the igneous rocks and erosion removed the sedimentary cover. Igneous rocks now form mountains 150 to 300 m (500 to 1000 ft) high, rising above a surrounding plain of Permian red beds.

10. **Gulf Coastal Plain**
    Located in southeast Oklahoma, loose sand, gravel, limestone, and clay, Cretaceous in age, dip gently southward toward the Gulf of Mexico. Sediments, only slightly dissected by streams, form gently rolling hills and plains.

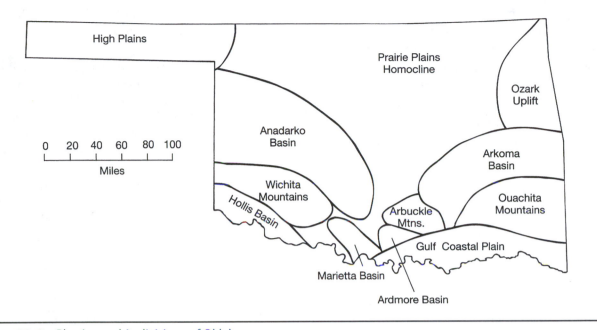

Figure 20.6   Physiographic divisions of Oklahoma.

Engineering Considerations of Physiographic Provinces

In engineering construction a significant advantage accrues from considering geologic features on a region-by-region basis. Construction problems and geologic concerns in one area of the United States will differ from those of another area. One major reason for these differences is the contrast in geologic setting and geographic details that occur in one place versus another. Many examples can be cited where nonlocal construction contractors, coming into an unfamiliar area, overlook a local geologic condition and bid a major job too low. After winning the contract, they subsequently must fight to minimize their losses or attempt to break even on the project.

Physiography, or landform features, can be used as a basis to isolate specific construction problems or concerns for an individual region. The geologic formations that make up an area have a direct effect on construction problems. These formations are a consequence of geologic history, from their origin, through time, and the processes that have acted on them. The concept of physiographic provinces can be applied relative to highways, dams, bridges, and building foundations. Table 20.5 presents a list of possible construction problems or considerations

that could pertain to different parts of the United States and different physiographic provinces.

We now use the state of Oklahoma as an example. Table 20.6 shows the physiography of Oklahoma influences the nature of highway-related problems. Its geologic map and physiographic units are presented in Figures 20.3 and 20.6, respectively. Hayes (1971) listed the following highway construction problems for Oklahoma, in order of the frequency of occurrence rather than by severity of hazard or by economic cost: (1) seepage, (2) landslides and slumps (slope stability), (3) rippability, (4) expansive soils, (5) rock excavation, (6) construction materials location, (7) erosion-sedimentation, (8) sinkholes and cave-ins, (9) corrosion of metals underground, and (10) land use.

Some of these problems are more prevalent and severe in certain parts of the state. For instance, landslides are almost always restricted to the eastern third of the state because of the higher annual rainfall (112+ cm [45+ in]), steep slopes, and thick sequences of shale and/or presence of colluvium. Additional details concerning highway construction problems in Oklahoma include:

- Difficult to locate riprap or coarse aggregate for paving in the High Plains area.
- Severe sinkholes occur in gypsum areas. Gypsum beds thicker than 3 m (10 ft) cannot be ripped.

- Seepage problems occur where the granular terraces overlie less permeable shale.
- Salt springs typically are corrosive to culverts.
- Seepage is a frequent problem on east–west roads that must cross east-facing escarpments.
- Some clay shales are expansive.
- Nonrippable limestones are common; seepage in these areas is also common; underground mine areas yield cave-in problems.
- Acid seepage water is corrosive to culverts.
- Expansive soils are common in the Gulf Coastal Plain.
- Landslide problems can be related to certain rock formations, members, and beds in the state.

Johns Valley Formation in southeast Oklahoma (Ouachita Mountains) almost always results in landslides unless a special design is used.

The two states used in this chapter as examples of physiographic subdivisions, and the relation of these subdivisions to engineering construction, represent an area where glacial deposits and horizontal rocks prevail (Indiana), and where a variety of sedimentary structures and igneous rocks are found (Oklahoma). The Oklahoma example illustrates the relationship between physiography and the anticipated problems during engineering construction. Similar engineering applications can be made to physiographic divisions of the other states.

Table 20.5   Engineering and Geological Factors of Concern for Various Regions of the United States.

1. Tunnels required	12. Embankment construction	23. Storms near seacoast
2. Frost action in soils	13. Water for compaction	24. Beach erosion
3. Water supply	14. Flash floods	25. Soil erosion
4. Permafrost	15. Snow removal, snow avalanches	26. Reservoir leakage
5. Aggregate shortage	16. Accessibility during exploration	27. Saltwater intrusion (coastal)
6. Landslides and slope stability	17. Blowing sand	28. Expansive soils
7. Location—bridges, foundations, etc.	18. Earthquakes	29. Compressible soils
8. Excavation, rock cut	19. Fires	30. Poor bearing capacity
9. Sinkholes	20. Icing of roads	31. Reactivity of concrete aggregates
10. Irregular bedrock surface	21. Water pollution	32. Presence of faults
11. Sanitation	22. High groundwater table	33. Subsidence

Table 20.6   Highway Construction Problems Related to Oklahoma Physiographic Units.

	1 High Plains	2 Anadarko–Hollis–Marietta Basins	3 Prairie Plains Homecline	4 Ozark Uplift	5 Arkoma Basin	6 Ouachita Mountains	7 Gulf Coastal Plain
Seepage	SN, G	SN, G	M, G	S, G	S, G	S, G	M, G
Slumps	SN, G	SN, G	M, L	M, L	S, L	S, L	M, G
Rippability	SN, G	M, L	M, G	S, G	S, G	S, G	SN, G
Expansive soils	SN, L	SN, G	M, G	SN, G		M, G	S, G
Rock excavation	SN, G	M, L	M, G	S, G	M, G	S, G	SN, G
Materials location	S, G	M, G	SN, G	M, G		M, G	SN, G
Erosion	M, G	M, G	M, G	SN, G		M, G	M, L
Sinkholes	SN, L	S, L	SN, G	S, L	M, L	SN, G	SN, G
Corrosion	SN, G	SN, G	SN, G	S, G	M, G		SN, G
Land use	SN, G	SN, G	M, L	SN, G	SN, G	SN, G	SN, G

Notes: SN = slight to none; G = general; L = local; S = severe; M = moderate.

EXERCISES ON PHYSIOGRAPHIC PROVINCES

1. Refer to the physiographic provinces map of Figure 20.1 to complete this exercise. Draw the boundary lines around the various physiographic provinces. It is suggested that this be done in pencil first, checked carefully, and then outlined using red pencil or red pen.

2. Review the brief descriptions of the US physiographic provinces in Table 20.1. Using the list below, indicate for each province whether it is basically a stable region structure (SRS), orogenic belt structure (OBS), a volcanic cone (VC), or a volcanic lava flow structure (VLS). Where does the Canadian Shield fit into this classification since it is not a mountainous region? Why is the Ridge and Valley Province considered to consist of geographic "mountains" but not mountains in a structural geology sense?

Northern Rockies _____

Middle Rockies _____

Wyoming Basin _____

Southern Rockies _____

Basin and Range Province _____

Sierra–Cascades Province _____

Pacific Troughs Province _____

Pacific Coast Ranges Province _____

Atlantic Coastal Plain _____

Gulf Coastal Plain _____

Canadian Shield _____

New England Province _____

Piedmont Plateau Province _____

Blue Ridge Province _____

Ridge and Valley Province _____

Appalachian Plateau _____

Central Lowlands _____

Interior Low Plateaus _____

Ozark Plateau _____

Ouachita Mountains _____

Great Plains _____

Columbia Plateau _____

Colorado Plateau _____

3. Review the list of engineering and geological factors of concern for various regions of the United States in Table 20.5. Compose two lists. For the first list, identify which of the factors in Table 20.5 would be associated with each of the physiographic provinces listed in Exercise 2. In the second list, identify the engineering and geological factors of concern that would be

associated with (a) mountainous regions, (b) wide alluvial flood plains, (c) arid portions of the southwestern United States, and (d) limestone regions in Kentucky and southern Indiana.

MAP READING

The topographic maps in the following exercises relate to the physiographic provinces of the United States. To obtain these maps, the US Geological Survey offers TopoView (https://ngmdb.usgs.gov/topoview/). This software will allow you to view and download topographical maps produced from 1880–2010 for locations throughout the United States. Use a map that has a scale of 1:24,000.

KASSLER, COLORADO

4. To what physiographic province does the eastern part of this quadrangle belong? The western part? Refer to Figure 20.2. Which geologic structure describes each part?

5. What is the regional stage of each part? See Figure 12.18.

6. Give the maximum relief of the eastern part and of the western part (use elevation difference from mountain to valley).

7. In what direction do the hogbacks dip? Explain the difference between a hogback and a cuesta.

8. What evidence indicates that the basement complex rocks underlying the western portion are homogeneous crystalline rocks (rather than folded sedimentary rocks)?

ALLENS CREEK, INDIANA

This quadrangle is underlain by sedimentary rocks (shale, limestone, and sandstone) that dip gently westward and form part of the western flank of a very broad, gentle arch that has its center near the Indiana–Ohio state line.

9. To what physiographic province does this area belong?

10. Allens Creek is located a few miles north of Bloomington; in which Indiana physiographic section is it contained (see Figure 20.5)?

11. What is the approximate maximum relief?

12. What evidence on the map suggests that the beds are nearly horizontal? Is there any evidence of sinkhole formation? How does your answer relate to the nature of the rocks present?

13. What evidence of recent changes in the course of Salt Creek may be seen on the map? Where is Salt Creek currently widening its floodplain?

14. What depth to bedrock is suggested in this area? What type of problem can occur during construction of a sewer line in this area?

ORBISONIA, PENNSYLVANIA

15. In what physiographic province is this quadrangle located? Refer to Figure 20.2. Which geologic structure describes it?

16. What evidence on the map indicates that the rock layers are folded rather than being only tilted?

17. Note the two water gaps and the trellis drainage. These are indirect evidence of folded rocks. How are water gaps formed?

18. What rock type is likely to make up the mountain tops; what kind makes up the valleys?

MOUNT WHITNEY, CALIFORNIA

The scarp forming the western edge of Owens Valley marks the eastern front of the Sierra Nevada, which was formed by a great block of the Earth's crust that was upfaulted and evidently tilted toward the Pacific Ocean. The Sierra Nevada is formed mostly of granite; just east are the Inyo Mountains, which are formed of granite and disturbed sedimentary rocks. Movement along the fault was essentially vertical and did not take place all at one time, but rather in small increments of a meter or several meters (few feet or tens of feet) over a long period of geologic time.

19. In what physiographic province is this quadrangle located? Refer to Figure 20.2. Does it appear similar to diagram B4? Explain.

20. What evidence of faulting appears along the front of the Alabama Hills?

21. If the front of the Sierras is a fault scarp, what is the total vertical movement along the fault, as measured between the east front of the Alabama Hills and the top of Mount Whitney?

22. What evidence indicates that Owens Valley cannot be a stream-cut valley, such as is the case for the Grand Canyon?

23. What engineering problems would be expected to occur in this area?

EWING, KENTUCKY–WEST VIRGINIA

The northern part of this quadrangle is underlain by nearly horizontal sedimentary beds and the southern part by beds that are folded.

24. To what physiographic province does the northern part of this quadrangle belong? The southern part?

25. Why does the rock layer forming Cumberland Mountain yield a ridge, whereas the adjacent layers form valleys? What rock types are involved? What is the strike of these beds?

26. Name one stream on the map that has a trellis drainage pattern.

27. What is the regional stage of the northern part? Why is that area regarded as a plateau?

28. What engineering problems would you suspect in this area?

THOUSAND SPRINGS, IDAHO

This area is underlain by layered flows of basalt that have a total thickness of several thousand feet.

29. To what physiographic province does this quadrangle belong? Refer to Table 20.2. Which geologic structure relates to this area?

30. What is the regional stage in the northeast part of the map?

31. What evidence of layered bedrock structure appears on the map?

32. What properties of basalt would account for the fact that there is an aquifer from which many large springs emerge?

33. What engineering problems would you envision in this area? Consider this also relative to the major dams built on the Columbia River.

PHYSIOGRAPHIC REGIONAL UNITS OF INDIANA

34. In what US physiographic province is Indiana located?

35. Examine the geologic map of Indiana (see Figure 20.4). Describe in a general way the age and configuration of the bedrock in the state.

36. Omitting glacial deposits, to what part of the state would you go to find the oldest rocks? Where would you find the youngest rocks?

37. a. Compare the physiographic map of Indiana (see Figure 20.5) to Figure 14.13 showing the extent of glaciation in the north-central United States.

    b. Refer to Table 20.3. For the first six units, identify the glacial stage it belongs to.

38. In which Indiana physiographic province would you expect to find caves (see Figure 20.5)? Explain why.

39. Describe the possible groundwater problems in the following provinces:

    a. Kankakee Outwash and Lacustrine Plain

    b. Tipton Till Plain

    c. Crawford Upland

    d. Scottsburg Lowland

40. What kind of foundation material (specifically the bedrock or glacial material type) would be found in the following cities?

    a. Gary

    b. Lafayette

    c. Indianapolis

    d. Fort Wayne

    e. Bloomington

    f. Evansville

41. List the regional provinces in which swamps would be common and why.

42. What would you expect to be the most common bedrock type in the state? What bedrock type is most commonly used as a crushed stone aggregate for Portland cement concrete and for bituminous mixes in Indiana? Why would it not be basalt?

43. What highway engineering problems might be expected in the area of Wisconsinan glaciation in Indiana?

44. What highway engineering problems would be expected in the nonglaciated portion of Indiana that would not occur elsewhere in the state?

45. During the 1960s the US Nuclear Regulatory Commission proposed Eagle Creek, Indiana, as a building site for a linear particle accelerator. (The accelerator was subsequently located near Batavia, Illinois.) From your knowledge of Indiana geology, what problems would you expect in the foundation for such a structure in the original proposed location (Eagle Creek is north of Indianapolis)?

Physiographic Regional Units of Oklahoma

46. Review the physiographic divisions and the regional physiographic descriptions of Oklahoma (see Figure 20.6 and Table 20.4). Using the list below, indicate whether each of the units belongs to a stable region structure (SRS) or an orogenic belt structure (OBS).

    High Plains   _____

    Prairie Plains Homocline   _____

    Ozark Uplift   _____

    Anadarko–Hollis–Marietta Basins   _____

    Arkoma Basin   _____

    Ardmore Basin   _____

    Ouachita Mountains   _____

    Arbuckle Mountains   _____

    Wichita Mountains   _____

    Gulf Coastal Plain   _____

47. Only seven physiographic sections are considered in Table 20.6 but 10 are described in Table 20.4. Which three were omitted? Why do you think they were omitted? (*Hint:* See Figure 20.6 and note the size and location of the omitted units.) Which of the highway construction problems given in Table 20.6 is likely to be associated with these other three physiographic sections? Explain why.

48. Examine the geologic map of Oklahoma on Figure 20.3. Where are the Mesozoic age rocks located? With which physiographic division(s) are these associated? (*Hint:* See Figure 20.6.) Are there any Cenozoic age rocks present? Explain.

49. What are the three mountain regions in Oklahoma? Which of the three seems to be the most rugged or highest? Explain your answer.

REFERENCES

DiPietro, J. A. 2013. *Landscape Evolution in the United States: An Introduction to the Geography, Geology, and Natural History.* New York: Elsevier.

Fenneman, N. M. 1931. *Physiography of Western United States.* New York: McGraw-Hill.

Fenneman, N. M. 1938. *Physiography of Eastern United States.* New York: McGraw-Hill.

Hayes, C. J. 1971. Engineering Classification of Highway Geology Problems in Oklahoma. Proceedings, 22nd Highway Geology Symposium, Norman, Oklahoma.

Hunt, C. B. 1967. *Physiography of the United States.* San Francisco: W. H. Freeman.

Lobeck, A. K. 1948. *Physiographic Diagram of North America.* New York: Geographic Press.

Lobeck, A. K. 1957. *Physiographic Diagram of the United States.* Maplewood, NJ: Geographic Press.

Thornbury, W. D. 1965. *Regional Geomorphology of the United States.* New York: John Wiley & Sons.

ENGINEERING GEOLOGY
HIGHWAYS, DAMS, TUNNELS, AND ROCK BLASTING

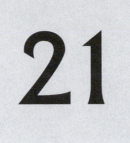

Chapter Outline

Engineering Geology

Engineering Geology of Highways

Engineering Geology of Dams

Engineering Geology of Tunnels

Rock Blasting

Environmental Geology

This chapter presents a brief discussion of the field of engineering geology in relation to highways, dams, tunnels, and rock blasting. Engineering geologists play a vital role in site investigations and other important aspects of these structures.

Engineering Geology

Engineering geology is the branch of geology that deals with the role of geologic factors in the design and construction of engineering structures. The purpose is to ensure an economic and safe structure. As defined by the American Geological Institute (AGI), "Engineering geology is the application of the geological sciences to engineering practice for the purposes of assuring that the geologic factors affecting the location, design, construction, operation, and maintenance of engineering works are recognized and adequately provided for" (Neuendorf et al., 2005).

Engineering geology had its formal beginning in the United States following the failure of the St. Francis Dam in California in 1928. Investigations of the failure showed that weak geologic materials comprising the foundation and the abutments (valley walls) of the dam were overlooked. More recent studies attribute the failure to the reactivation of an ancient landslide under the left abutment (Rogers, 2006a, 2006b). Today, nearly all major construction activities include engineering geology input from the planning through construction phases of the work.

In addition to a sound understanding of geology, an engineering geologist has a strong background in civil and geotechnical engineering and is well aware, through education, training, and experience, of the engineering aspects of construction directly related to geology. An engineering geologist is expected to be well versed in the following subjects: hydraulics and hydrology, soil mechanics, rock mechanics, construction materials, slope stability, and site investigations. In past years, a working knowledge of construction, obtained through on-the-job experience, was considered sufficient for a geologist to become an engineering geologist. Today, it is more efficient and productive to obtain much of this knowledge through course work in civil engineering, engineering geology, and hydrogeology.

The distinction between an engineer and an engineering geologist is clear. The engineer is involved in the design of engineering works and is responsible for their long-term performance, whereas an engineering geologist provides the geologic input for design purposes and ensures that geologic factors are adequately considered through all phases of construction but, usually, is not involved in designing a structure.

Many aspects of physical geology are significant in the field of engineering geology. In this text, the most important aspects of geology have been discussed with the specific purpose of showing how they relate to engineering. From minerals and rocks to subsurface investigation and earthquakes, the application of geology in engineering has been detailed. Table 21.1 lists the subjects of physical geology relating to engineering geology.

The case history approach is used sometimes in the study of engineering geology. In this method of instruction, specific construction projects are reviewed in an attempt to evaluate the procedures used for a specific set of conditions. Various phases of the work can be studied: exploration, design, construction, and various subdivisions of each. Care should be taken in this approach to ensure that the specifics of engineering geology are covered rather than simply broad generalities. One useful method is to discuss the construction procedures in detail first and then examine the case history. For example, for a case history on grouting, discuss types of grouts, mixtures, equipment, procedures, and limitations, then review an article on curtain and foundation grouting for a specific dam.

In many ways, engineering geology encompasses the obscure part of construction, which involves the uncertainties of geologic materials. Examples are bedding planes and joints yielding rock-block failures in an excavation, seeps in clayey deposits caused by sand layers supplying water into an opening, encountering deeply weathered soil or pinnacles of rock when another condition is expected, solution cavities in limestone in a building foundation area, or a methane-bearing zone in a shale tunnel. Many times it is the most challenging aspects of construction that involve engineering geology so it should be no mystery that knowledge of geology and of engineering is needed to anticipate such occurrences and to react in the best way possible when they occur.

Geological engineering is a sister field to engineering geology but, here, the student is an engineer well versed in geological knowledge rather than vice versa. More engineering courses are required and somewhat less geology on the bachelor's degree level. Generally, geological engineering has three distinct specialties: petroleum, mining, and engineering construction. For mining and construction engineering, the required background is similar, calling heavily on civil engineering and on some mining engineering expertise. Petroleum aspects of geological engineering are somewhat different from the others; more involvement with subsurface reservoirs, liquid and gaseous flow in porous media, and oil well systems are prime considerations. Geological engineers in mining and construction are more closely aligned to civil engineering than to oil and gas production.

Examples of Construction Projects Related to Engineering Geology

Table 21.2 provides a list of construction projects related to engineering geology. Each of these would require a chapter to explore the subject fully. Such detail is beyond the scope of this introductory text. As an example of the nature of the subject matter, however, engineering geology considerations for three types of projects are discussed in the following sections: highways, dams, and tunnels.

Table 21.1   Geology Subjects Related to Engineering Geology.

1. Rock description and identification.
2. Engineering properties of rocks and materials for construction.
3. Rock weathering and soil development.
4. Map reading, both topographic and geologic.
5. Structural aspects—bedding planes, joints, faults.
6. Mass movement and landslides.
7. Running water—erosion, flood effects, water impoundment.
8. Groundwater—control during construction, water supply, pollution, subsidence, slope stability.
9. Shoreline erosion and protection—engineering structures, jetties, groins, breakwaters, sea walls; finding rocks for shoreline protection such as riprap and armour stone.
10. Earthquakes, seismic risk, slope failure, liquefaction.
11. Glacial deposits—construction materials, groundwater supply.
12. Arid environments, wind, accelerated erosion.
13. Subsurface geology, stress distribution at depth.

Engineering Geology of Highways

Table 21.3 provides engineering geology considerations pertaining to highways, the details of which are discussed in the following sections.

Table 21.2   Construction Projects Related to Engineering Geology.

1. *Highways*: location, alignment, classification of soil and rock borrow, groundwater control, slope stability, materials inventory and selection.
2. *Dams*: site selection, materials selection, subsurface investigation, geologic mapping, foundation support, groundwater control, slope stability, grouting, instrumentation.
3. *Building foundations*: site selection, subsurface investigation, geologic correlation, rock or soil strength determination, groundwater, construction monitoring.
4. *Tunnels*: Regional geology, site selection, subsurface investigation, geologic correlation, strength determination, estimation of rock behavior during tunneling, blasting, groundwater control, support recommendations, geologic mapping during construction.
5. *Rivers*: erosion of banks, dredging, scour effects, channel control.
6. *Surface mining*: subsurface investigation, geologic correlation, groundwater, slope stability, excavation aspects, blasting, stripping procedures, reclamation.
7. *Underground openings and mines*: subsurface investigation, rock strength, stress conditions, groundwater, orientation of openings, extraction ratios, subsidence.
8. *Sanitary landfills*: site selection, subsurface investigation, geologic correlation, groundwater, soil and rock permeabilities, baseline water quality determination.
9. *Environmental geology and land use*: preparation of large-scale geologic maps and interpretive maps, determination of current land use, development of acceptable future land uses based on geologic, economic, and, to some extent, political restraints.
10. *Evaluation of construction materials*: materials inventory and selection, evaluation of masonry construction materials.
11. *Slope protection*: along shorelines and lakes, beach erosion, engineering structures.
12. *Nuclear and fossil fuel power plants*: subsurface investigation, geologic correlation, structural geology analysis and faulting evaluation, determine seismicity of site, groundwater, and other aspects of building foundations.
13. *Blasting and vibration effects on engineering structures*: particle velocity determination and blast design aspects.
14. *Environmental impact*: measure baseline information, estimate the changes that will occur if a structure is added.

Highway Location

Highway location is dictated, in part, by the topography and nature of geologic materials present in the area (Fwa, 2005). Existing maps are used to select the highway alignment. In glaciated regions, care is taken to avoid compressible soils and other poor foundation materials. River crossing locations are selected carefully and road distances are kept to a minimum, compatible with the other constraints. In bedrock areas, attempts are made to minimize the volume of cuts and fills, thereby reducing the amount of earth and rock to be moved. Where troublesome rock formations exist, road grades and elevations are selected so that they minimize the exposure of the problematic rock units. Swelling shales and claystones and landslide-prone rocks are examples of rock formations to be avoided, minimized, or accounted for in the design.

Subsurface Investigations

Exploration borings are used to obtain subsurface data along the proposed highway centerline and for bridge foundations. Standard penetration tests are performed in the soil and rock cores are taken from bedrock. Borings are made in both the cut and fill sections. Borings should extend to a minimum depth of 2 m (6 ft) below the road grade in cut sections. If a large embankment is to

---

Table 21.3   Engineering Geology Considerations Pertaining to Highways.

1. Highway alignment, locations of right-of-way for the proposed construction.
2. Subsurface exploration along highway centerline and for bridge foundations.
3. Classification of materials for excavation, rock versus common borrow (soil).
4. Cut and fill volumes determination to minimize the need for offsite borrow pits or rock waste areas; volume changes in both soil and rock from the cut to the fill are estimated.
5. Overbreak of rock cuts minimized by presplitting rock faces.
6. Recommend angle of back slope (rock cut slope) based on rock conditions.
7. Groundwater aspects related to construction.
8. Evaluation of areas prone to slope movements.
9. Recognition of compressible soil materials.
10. Legal aspects, highway effects on adjacent landowners.
11. Construction materials, location, and inventory.
12. Increasing problems with highway location because of human activities.

be placed, borings should extend to a depth of two-thirds the height of the embankment or to bedrock, whichever is less. As with the cut section, a 2-m (6-ft) minimum depth below the grade is required. Atterberg limits and grain size distribution tests are performed to characterize the soils. Compaction tests are performed on borrow pit soils used to construct the embankment.

Subsurface investigation for bridge foundations is performed to provide the details needed for the bridge footing design. Appropriate samples are obtained for laboratory testing. Since rocks gradually transition into soils, the criteria for distinguishing between soil and rock must be established for determining the soil-bedrock interface. The cost of soil excavation is a few dollars per m³ (per yd³), whereas the cost for rock excavation is typically $100 per m³ (per yd³) to several times that amount. The soil-rock interface is established during the drilling phase and this information may be augmented using engineering geophysics.

Balancing Cut and Fill Volumes

During the design process, attempts are made to balance cuts and fills so that offsite borrow or rock waste areas are kept to a minimum. Soil borrow pits are common for interstate highway construction because the grade is elevated above low-lying areas to prevent flooding and encourage runoff. Careful calculations are required to ensure that the volumes of cuts and of fills (embankments) are determined accurately. Rock increases in volume from the cut to the fill because excavated rock pieces have much larger void spaces between them compared to the intact rock, whereas soils decrease in volume during this process because their natural void space is reduced upon compaction in the fill area. In general, clayey soils tend to decrease more during the compaction process in the fill area than do sandy soils. By contrast, massive rocks such as well-cemented sandstones, limestones, and hard igneous or metamorphic rocks increase considerably in volume from cut to fill area, whereas weaker rocks such as shales have a smaller increase in volume. This is illustrated by an example problem.

Example Problem 21.1

A cut section is located adjacent to a fill section for an interstate highway project. In the cut, 11,460 m³ (15,000 yd³) of clay and 9168 m³ (12,000 yd³) of sand will be excavated, along with 15,280 m³ (20,000 yd³) of hard shale and 9932 m³ (13,000 yd³) of limestone. The shrinkage factor is 15% for the clay and 5% for the sand; the swell factor for the hard shale is 10% and for the limestone is 20%.
   a. What will be the volume of the fill constructed from these materials?
   b. What volume change from the cut to the fill will occur?

Answers

   Using SI units:
      a. Clay volume in the cut converted to volume in fill:
            $11,460 \times 0.85 = 9741$ m³
         Sand volume in the cut converted to volume in fill:
            $9,168 \times 0.95 = 8710$ m³
         Hard shale volume in the cut converted to volume in fill:
            $15,280$ m³ $\times 1.1 = 16,808$ m³
         Limestone volume in the cut converted to volume in fill:
            $9932 \times 1.20 = 11,918$ m³
      b. Volume increase is:
            $47,177$ m³ $- 45,840$ m³ $= 1337$ m³
         Percent increase in volume from cut to fill area = $(1,337/45,840)(100) = 2.9\%$
         If the fill could accommodate only 45,840 m³, the extra 1337 m³ would have to be disposed of in an offsite waste pile.

Using British engineering units:
   a. Clay volume in the cut converted to volume in fill:
      $15,000 \times 0.85 = 12,750$ yd^3
      Sand volume in the cut converted to volume in fill:
      $12,000 \times 0.95 = 11,400$ yd^3
      Hard shale volume in the cut converted to volume in fill:
      $20,000 \times 1.10 = 22,000$ yd^3
      Limestone volume in the cut converted to volume in fill:
      $13,000 \times 1.20 = 15,600$ yd^3
   b. Volume increase is:
      $61,750 - 60,000 = 1750$ yd^3
      Percent increase in volume from cut to fill area = $(1,750/60,000)(100) = 2.9\%$
      If the fill could accommodate only 60,000 yd^3, the extra 1750 yd^3 would have to be disposed of in an offsite waste pile.

Overbreak and Presplitting

Overbreak is the volume of rock blasted from a rock cut in excess of that designated in the slope design (Fwa, 2005). It occurs during production blasting in highly jointed or fractured rock. Overbreak can be minimized by performing presplitting prior to production blasting. This consists of drilling closely spaced boreholes, 1 to 2 m (3 to 7 ft) apart, along the designated boundary of the rock cut. These holes are loaded lightly with explosives and blasted prior to the production blasting. A plane of fracture is established at the designated rock cut boundary. This fracture prevents blast-wave propagation beyond the cut boundary so that no overbreak occurs during the production blasting operation.

Cut-Slope Angle

In horizontal sedimentary rocks, the angle of the cut slope or back slope depends largely on the durability of the rock. Massive, durable rock such as limestone, dolomite, and well-cemented sandstone can be cut nearly vertical, 1/6:1 (1/6 horizontal, 1 vertical), for example, soft shale may be cut at 1:1 slope, and some claystones (redbeds) may require slopes as gentle as 3:1. Intermediate conditions such as interbedded shale and limestone may be cut on a 1/2:1 slope (Admassu and Shakoor, 2015). There is a major advantage in making the rock slope as steep as safety permits. Steeper slopes require a smaller volume of rock removal and a narrower width for right-of-way at the top of the cut. Both of these considerations reduce the cost involved.

Groundwater Problems

Groundwater problems related to highway construction are of two types: (1) groundwater seepage into cut sections and at the base of fill sections and (2) groundwater supplies adjacent to the highway. Seepage into cut sections is controlled by installing horizontal drains, intercepting most discontinuities, as well as surface drains. This prevents seepage onto the highway and subsequent icing of the road during freezing conditions. Springs are intercepted at the base of the fills and carried by drains under the road. This prevents saturation of the embankment, which could lead to possible settlement problems. In case of a shallow water table, underdrains are installed below the pavement to intercept water.

The water supplies of adjacent landowners may be affected by road cuts that intercept the groundwater table. The water table may be locally depressed in this case. Effects on neighboring wells must be properly documented by preconstruction and postconstruction surveys of these wells. Property owners affected by such road cuts should be compensated by supplying them with an alternate water supply. Water supplies are also needed at regular intervals along the roadside for rest areas for interstate travelers. The water supply demand is considerable for such facilities and a major water supply, in the hundreds of liters (gallons) per minute range, is required. For this reason, locating a source of water supply for a large rest area can be a major geologic challenge.

Areas Prone to Slope Movements

Engineering geologists evaluate landslide-prone areas along the proposed highway right-of-way. They provide geologic information for slope stability analysis through a review of published information, using remote sensing techniques such as aerial photographs, LiDAR, and unmanned aerial vehicles (UAV), field investigation, and, in some cases, subsurface drilling and sampling. Compressible soils are identified during the planning stages and recommendations made to minimize their effects on the completed highway. Prior to construction, avoidance, excavation, and settlement are evaluated.

Legal Aspects

Highway geologists may be called on to provide expert opinions regarding domestic water supplies including wells, springs, and spring-fed ponds. Evaluation of the effects of highway construction on oil well fields, shale pits, stone quarries, and other facilities are commonly required. Testimony as an expert witness in court or in hearings involving these situations may also be necessary.

Construction Materials: Location and Inventory

A primary job of many highway geologists involves the location and inventory of construction materials for highway construction (Fwa, 2005). Aggregates consisting of gravel, crushed stone, or, possibly, blast furnace slag comprise the major constituents of a highway pavement. Figure 21.1 shows an idealized cross section of a highway pavement system. Aggregates are used in all three portions: pavement, base course, and subbase course.

The subgrade consists of compacted soil at the base of the highway section with the grade line marking the top of the subgrade. Compacted aggregate forms the base and subbase courses. Subbase material, typically, is of lower quality than the base course. It is used in the lower portion of flexible pavements where, generally, a greater total thickness is required than for concrete pavements. High-quality aggregate is used in both concrete and asphaltic pavements.

Highway geologists and materials engineers are responsible for maintaining an inventory of aggregate sources, including gravel pits and stone quarries. This includes details of the stratigraphy and nature of the pits and quarries as related to the engineering quality of the aggregate. In the western United States, gravel supplies vary greatly relative to rock composition and overall quality. Because transportation of aggregates is a major expense, quality aggregates must be located as close as possible to the construction site.

Human Activity Considerations

Human activity can generate challenging problems for highway construction. Today, highways cross areas of abandoned or active underground mines, former strip mines, landfill areas, subsiding oil well fields, sliding hillsides, compressible floodplain soils, and other challenging situations. Highway geologists are called on to decipher the details of surface and subsurface conditions that affect highway construction in these complex areas. Avoiding poor areas for highway construction is no longer a choice for many projects today.

Engineering Geology of Dams

Dams are engineering structures built across stream channels that impound water at a considerably higher elevation on the upstream side. A dam holds back a large volume of water that possesses an enormous destructive power if released suddenly. Dams are one of the few engineering structures with a capacity for destruction that extends far beyond

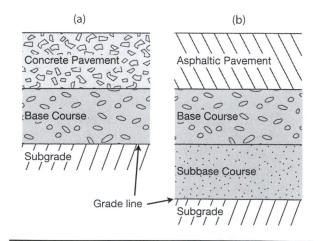

Figure 21.1 Schematic diagram of highway pavements: (a) rigid (concrete) pavement and (b) flexible (asphaltic) pavement.

their location. A simple comparison to highways, buildings, tunnels, mines, and even sanitary landfills points to this conclusion. The role of a dam is to resist the lateral pressure of the impounded water and the associated loads of ice, wind, and silt and transfer these loads to its base (the foundation) and valley walls (the abutments). The stability of dams depends largely on site conditions, particularly on the site geology, perhaps more than any other engineering structure (Jansen, 1988; Fell et al., 2014). Dams are built to serve the following purposes: flood control, water supply, irrigation, hydroelectric power generation, and recreation. Most major dams serve more than one of these functions. In the eastern United States, the US Army Corps of Engineers is responsible for the construction of most large dams. Electric utility companies, the Natural Resources Conservation Service (NRCS), and the Tennessee Valley Authority (TVA) have constructed sizable dams as well. In the western United States, the US Bureau of Reclamation is responsible for constructing dams that provide irrigation for the arid and semiarid regions. State governments and electric utility companies also have constructed large dams in the western United States.

Classification of Dams

Based on the material of construction, dams are classified as concrete dams, earth dams, and rock-fill dams. Earth and rock-fill dams are referred to jointly as embankment dams (Jansen, 1988; Fell et al., 2014). Many dams are combinations of concrete and embankment structures. Temporary dams, of much smaller size, may be built of timber or of steel sheet pile. Site geology and the configuration of the valley greatly influence which of the three major types of dams will be selected (Jansen, 1988; Fell et al., 2014). Concrete dams require stronger foundations than do embankment dams. Also because of the higher unit cost of concrete, these dams are more appropriate in narrow valleys. Embankment dams are selected for wide valleys that extend several hundred meters (thousand or more feet) in length. Earth dams are constructed where sufficient, fine-textured soils are available. In some mountainous areas where clays are in short supply but coarse-sized deposits prevail, a rock-fill dam may prove to be the most economical type to construct.

Concrete Dams

Concrete dams consist of three basic varieties: gravity dams, buttress dams, and arch dams. Gravity dams are roughly trapezoidal in cross section, approaching a triangular shape because of the much greater width at the base than at the crest (top) of the dam. Sound rock is the preferred foundation material for gravity dams but they can be constructed on fractured or variable rock units or in some cases on alluvial fill in the stream valley. In the latter case, potential leakage below the dam is a major problem. Also, intact abutments are required to prevent seepage around the dam but the strength requirements are less stringent than those required for arch dams. Figure 21.2 shows a concrete gravity dam.

Buttress dams consist of a vertical or sloping upstream slab of reinforced concrete, supported by buttresses on the downstream side that transmit the water load from the slab to the foundation. The

Figure 21.2 Concrete gravity dam. The image shows the gated overflow spillway, the chute of the spillway, the stilling basin, and the right abutment training wall.

buttresses are perpendicular to the dam axis. Figure 21.3 shows a buttress dam.

Buttress dams have the advantage of a smaller volume of concrete than a gravity dam and a decreased volume of foundation excavation because of the smaller size of foundation footings. The area between buttresses can also be used to place the outlet works and power house. The various parts of a dam are discussed in a following section.

Buttress dams typically require more labor than gravity dams because of the added detail of the buttress configuration. Hence, there is a trade-off in volume of concrete (less for buttress dams) versus labor (more for buttress dams) that must be weighed in the dam selection process.

Buttress dams also require a good foundation because buttresses transmit concentrated pressures to the foundation. The space between buttresses is not loaded so, in poorer foundations, a buttress may punch through the rock. Gravity dams, by contrast, are not subject to this concentrated loading condition. In general, a buttress dam requires a stronger foundation than a gravity dam. However, like gravity dams, the abutment requirements are less stringent than that needed for arch dams.

An arch dam consists of a curved concrete wall in the plan view, with the convex face pointing upstream. Figure 21.4 shows an arch dam. Part of the water pressure is transmitted to the rock abutments by the arching action of the dam so that strong abutments are required to take this thrust.

Arch dams can be further subdivided based on the ratio of base thickness to height. The following relationship is used:

Description	Base thickness Height
Thick arch	≥ 0.3
Medium thick arch	0.2 to 0.3
Thin arch	< 0.2

Terminology Used for Concrete Dams

Figure 21.5 shows the plan view and the cross section of a concrete gravity dam, illustrating the terms used to designate various parts.

- **Heel and Toe:** The heel is the upstream base of the dam, in contact with the foundation, whereas the toe is its downstream counterpart.

- **Crest:** The crest forms the top of the dam over which a road or walkway is commonly placed.

- **Freeboard:** The freeboard is the vertical distance between the maximum pool level and the top of the dam.

Figure 21.3   Concrete buttress dam showing a gated spillway on the side. (Andy Pernick, Bureau of Reclamation.)

Figure 21.4   Concrete arch dam. (Photo by US Bureau of Reclamation.)

- **Axis**: The axis of the dam is an imaginary line through the center of the crest extending from one abutment to the other.

- **Galleries**: Galleries are openings through the concrete dam, running longitudinally and located just above the dam foundation. They drain seepage water from the face or the foundation and allow access inside the dam for instrumentation or for equipment to drill grouting and drainage holes.

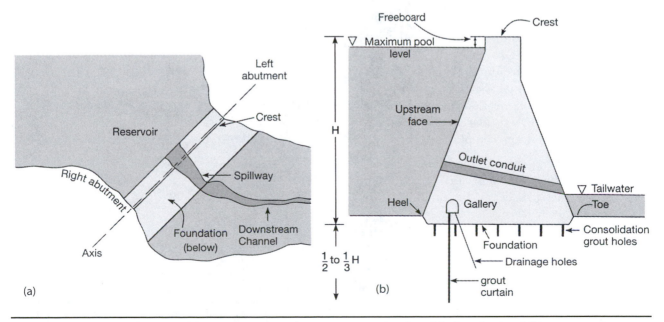

Figure 21.5   Terminology used for various parts of a concrete gravity dam: (a) plan view and (b) cross section.

- **Tail water**: Tail water is the water on the downstream side of the dam, discharged from the spillway, outlet works, or power house.

- **Outlet Works**: Outlet works divert water from the reservoir for irrigation, water supply, or power generation. There are two types of outlets, the conduit type shown in Figure 21.5 or the tunnel type where the water is carried in either lined or unlined tunnels through the abutments. For the power generation case, the outlet works provides water to the hydroelectric power plant. This installation can be aboveground, near the toe of the dam, or be belowground with an extensive opening excavated into bedrock.

- **Spillway**: A spillway is a structure used to discharge floodwater from the upstream to the downstream side of the dam where it is released as tail water. Spillways can be of several types. Figure 21.2 shows the normal, or overflow, spillway with gates to control the amount of discharge. In this case, water flows over the crest of the dam and is carried downstream by a channel or chute. The water flowing down the chute of a spillway enters the stilling basin, which may contain baffles (concrete structures) to dissipate energy. The concrete walls forming the sides of the chute are called *training walls*.

In a side-channel spillway, water flows from the reservoir along a pathway that is perpendicular or at an angle to the dam axis (see Figure 21.3. It flows around the dam in an open channel, through the abutment in an excavation cut perpendicular to the axis. Earth and rock-fill dams commonly employ a side-channel spillway.

Another structure is the *shaft spillway*, also termed a *morning glory* or *glory hole* spillway. This is illustrated in Figure 21.6. Water moves through the gates, into a vertical or oblique shaft that connects with a gently sloping section that usually extends below, rather than through, the dam. A variation of the shaft spillway excavated into rock is a tower

spillway constructed of concrete, which extends upward into the reservoir. It is founded on bedrock and is connected to a gently sloping section that extends below the base of the dam. This type of spillway can be used in conjunction with an earth or rock-fill dam.

- **Gates**: The gates control the amount of water flow allowed to enter the spillway (see Figure 21.2). Gated spillways are known as controlled spillways, whereas an uncontrolled spillway is without gates.

An emergency spillway is an uncontrolled spillway designed to carry floodwater that exceeds the capacity of a normal spillway. They are included in a reservoir design to ensure that an intense rainfall event, occurring when the reservoir is nearly full, will not allow the dam to be overtopped if the normal spillway capacity is exceeded. The elevation of the emergency spillway is located slightly above the maximum pool level, within the freeboard distance, but well below the crest of the dam.

Emergency spillways are located in natural topographic saddles of the reservoir rim and enlarged by excavation. For many emergency spillways, no concrete construction is included and it is understood that extensive erosion of the base and side

Figure 21.6   Shaft or glory hole spillway located in the reservoir area. (Central California Area Office, US Bureau of Reclamation.)

walls of the excavation will occur if the spillway is ever used. This would likely be a once in 500- to 1000-year occurrence but the effects would be catastrophic if, instead, the dam were overtopped by floodwaters.

- **Penstocks**: Penstocks are steel pipes of large diameter, extending from below the dam to the aboveground power station that carry water, under pressure, along the ground surface. They supply water to the electrical generators at the pressure of the upstream elevation.

Embankment Dams

Embankment dams are selected for wide valleys with an absence of firm bedrock. Large volumes of construction materials are required. For example, a 30-m (100-ft) high, 305-m (1000-ft) long earth dam would consist of about 710,000 m³ (1,000,000 yd³) of material for the portion above the original ground surface.

Embankment dams include both earth and rock-fill dams, with similar design concepts (Jansen, 1988; Fell et al., 2014). Earth dams can be homogeneous or zoned. Usually, a homogeneous earth dam is constructed of low plasticity clay with slopes ranging from 2:1 to 3:1, depending on foundation conditions. A filter blanket or a rock toe is installed at the downstream end to control seepage through the dam and foundation and to prevent the seepage line from emerging on the downstream face. The upstream face of a homogeneous dam is protected from wave action by a layer of riprap and the downstream face is covered with either gravel or grass to reduce erosion by rain or snowmelt. Figure 21.7 shows a schematic cross section of a homogeneous earth dam.

Zoned earth dams consist of different soil materials compacted to form a cross section, with the less permeable, lower strength materials in the center to serve as a hydraulic barrier, and the coarser, stronger materials on the outside to increase stability. Zoned sections in the dam provide for the combined requirements of low permeability to reduce seepage and soil strength for stability. Figure 21.8 on the next page shows the cross section of a zoned earth dam with three major zones: a central clay core, the upstream and downstream shells of silty or sandy material, and a sand drain on the downstream side of the core. The sand drain collects and drains seepage, after it has passed through the core, which relieves pore pressures. A 2½:1 slope is shown for the upstream slope and a 2:1 slope for the downstream slope.

The foundation for the dam is shown as permeable material, which is typical of alluvial or stream deposits. The clay core extends through the permeable material to intersect the low permeability material below. This serves as a cutoff wall to minimize

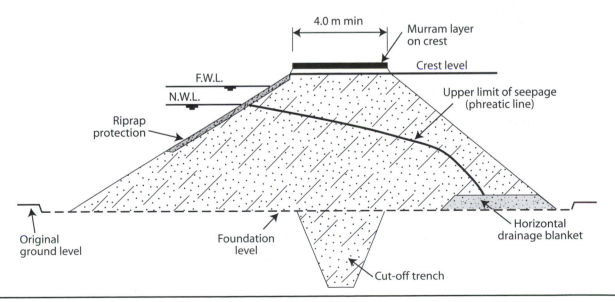

Figure 21.7 Schematic cross section of a homogeneous earth dam. (Ministry of Water and Irrigation. 2015. Practice Manual for Small Dams, Pans and Other Water Conservation Structures in Kenya. State Department for Water, Ministry of Water and Irrigation, Government of Kenya.)

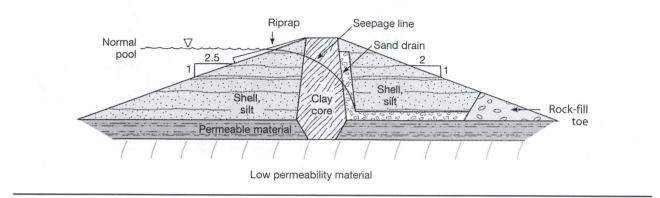

Figure 21.8   Cross section through a zoned earth dam.

seepage below the dam and reduce the uplift pressure under the downstream portion of the dam.

Riprap, consisting of large size blocks of quarry stone, is used along the upstream slope of the dam for wave protection. Rock pieces must be large enough that wave action will not dislodge them and durable enough so the pieces do not reduce in size with time as weathering processes prevail. Along the toe of the dam, a small rock-fill zone is added. This helps support the downstream slope and allows for drainage both from the sand drain and from the shell. It also provides added strength to the slope. Earth berms are sometimes added at the base of the downstream slope. Such berms are typically about 15 m (50 ft) high and about 7.5 m (25 ft) wide at the top. The addition of a downstream berm makes it possible to steepen the downstream slope of the dam and may reduce the total volume of fill required.

Earth dams are constructed by placing different soils in designated zones, followed by compaction with sheep's foot or other suitable rollers to achieve the necessary density. The soil is transported to the fill area by large dump trucks or by large scrapers. Figure 21.9(a) shows an earth dam under construction and Figure 21.9(b) shows an earth dam in operation.

Rock-fill dams are constructed from coarse rock pieces, gravel size and larger, with either an impervious upstream face, usually of bituminous concrete, or an impervious core of low plasticity clay surrounded by filters. Rock-fill dams tend to be more economical when the following conditions are present: silt and clay size materials are in short supply, the foundation is not sound enough for a concrete dam, suitable large-sized rock is available in large quantities, cold weather greatly limits the season

for rolled fill construction, and considerable earthquake hazards exist. Rock-fill dams, because of their inherently flexible nature, are less prone to damage during earth shaking (Jansen, 1988; Fell et al., 2014). California has the largest number of rock-fill dams in the United States.

Materials for Dam Construction

Large quantities of material are required for constructing dams. For concrete, the coarse aggregate, consisting of gravel or crushed stone, must be located near the construction site. Sand, for fine aggregate in the concrete, is also required. Even for earth dams, large quantities of concrete are needed for the spillway and other engineering structures.

Large quantities of clay, silt, and sand are required for an earth dam. The core, shell, sand drain, rock-fill toe, and riprap (see Figure 21.8) consist of materials obtained from the dam site. Considerable geological and engineering investigations are required to locate and select the needed materials. Special planning is required if excavated rock or soil is to be used; rock excavated from the foundation or in stripping back the abutments must be processed for use as aggregate, riprap, or soil borrow material. Rock excavated from spillways, diversion tunnels, or supply tunnels can also provide useful construction materials.

Exploration for Dams

Dam sites are selected using office reconnaissance and geologic field studies. Several sites are evaluated and the prime site is selected on the basis of geology, design, construction, and economic considerations. The details of this major undertaking are beyond the scope of this discussion.

Exploring for a dam site consists of making soil borings and rock corings along the centerline of the proposed dam. Standard penetration tests are performed and Shelby tube samples of the soil are collected. Soil descriptions and detailed cross sections are developed. Laboratory tests are performed on soil samples to determine Atterberg limits, grain size distribution, and strength parameters. Bedrock is cored to the depth of influence of the dam, considering both the foundation loads and the seepage paths of water when the reservoir is impounded. Rock cores are obtained from the foundation and the abutments. Typically, N_x-size rock cores, or larger, are obtained and used to develop detailed stratigraphic sections and determine RQD. During the rock coring process, water pressure testing is conducted in the boreholes. This locates high permeability zones in the bedrock where water would leak from the reservoir. This information also helps determine the zones where grouting will be done, using vertical and angled holes. Rock strength, permeability, and other relevant properties of various rock units are determined in the laboratory from the rock cores.

Grouting

Grouting of foundation and abutments is done prior to and during dam construction. In this operation, holes are drilled and a mixture of cement and water is pumped, under pressure, into the holes. The cement sets up and seals the openings in the rock. It is imperative that pressures are restricted to a level that does not lift the rock column, thereby inducing additional fractures in the rock mass. This would weaken the rock and increase permeability. Two primary types of grouting are used for a dam foundation: consolidation grouting and curtain grouting (Jansen, 1988; Fell et al., 2014). Figure 21.5b shows locations of grouting.

Figure 21.9   (a) An earth dam under construction (photo by US Bureau of Reclamation) and (b) an earth dam in operation.

Consolidation grouting is performed for the contact area of the foundation, to increase the strength and Young's modulus of the rock. This reduces and equalizes the amount of deflection of the foundation under the base of the dam.

For concrete dams, curtain grouting is done from the galleries, upstream of the centerline of the dam, and extends below the dam to form a cutoff to the flow of water below the dam. The reason for performing curtain grouting from the galleries is to be able to use higher grouting pressures. The weight of the dam adds pressure to the foundation and, therefore, higher grouting pressure can be used without lifting the rock column. Cement is forced more thoroughly into rock fractures and bedding planes, which greatly reduces underseepage—the ability of water to seep through the foundation. By contrast, if curtain grouting is needed for an earth dam, it must be done prior to construction of the embankment. Grouting through the embankment may erode the earth, causing voids to develop in the embankment. For embankment dams, consolidation grouting and curtain grouting are performed first and then the dam is constructed above the foundation.

Malpasset Dam Failure—A Case History Illustrating the Role of Geology

Figure 21.10 shows a photograph (a), the plan view (b), and the cross sectional view (c) of the Malpasset Dam, built between 1950 and 1952 on the Reyran River in southern France, only 9 km (15 mi) upstream from the Mediterranean Sea. The dam was approximately 61 m (200 ft) high, with an overflow spillway and a spillway apron downstream. It had a dewatering gate and a water intake for irrigation supply.

On December 2, 1959, during the first filling, the dam failed, killing approximately 340 people downstream. Failure of the dam was attributed to excessive uplift pressure from water flowing through a fault zone in a contorted gneiss under the left side of the dam, 30 m (100 ft) below the foundation (Duffaut, 2013). This fault zone was connected hydraulically to a fracture zone or second fault zone below the reservoir, so that the total reservoir pressure (more than 43 m [140 ft] of water) was applied upward on the foundation and blocks of rock were displaced more than 75 cm (30 in). This caused massive failure of the left side of the dam. In the subsurface investigation prior to construction these faults were not detected. However, it is

questionable that exploration, no matter how thorough, would locate every geological detail.

Other well documented examples of catastrophic failures attributed to geologic conditions include the Vaiont Dam in Italy (Ward and Day, 2011) (see Chapter 15), the St. Francis Dam in California (Rogers, 2006a), and the Austin Dam in Pennsylvania (Martt et al., 2005). These catastrophes show how important are extremely detailed geologic investigations.

Engineering Geology of Tunnels

A brief discussion of tunnels in rock is presented here to provide an introduction to this subject. Advanced books on engineering geology, applied rock mechanics, and tunnel engineering (Hoek and Brown, 1980; Goodman, 1993; Maidl et al., 2013), for example, provide more extensive treatment of this subject. Table 21.2, item 4 provides a brief outline of

Figure 21.10   Malpasset Dam: (a) the original structure (photo by Association of State Dam Safety Officials), (b) the plan view, and (c) cross-sectional view.

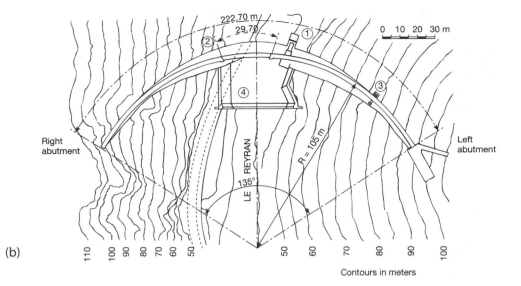

(b)

Contours in meters

the geology of tunnels. This forms the basis for the following discussion.

Site Investigations

Site selection and subsurface investigation for rock tunnels can be an extensive endeavor. Tunneling through poor-quality, weak rock is extremely expensive and time consuming, so it is beneficial to site tunnels in sound rock. Investigations for tunnels involve geologic field studies and extensive rock coring along the proposed tunnel alignment. The borehole logs provide lithologic descriptions of rock cores, water conditions encountered during drilling, percent recovery, and the RQD values. Selected laboratory tests are performed on core samples to determine relevant properties such as density, absorption, slake durability index (in a case of weak rocks), and strength. Geologic mapping is performed during tunnel construction as well, so that any future problems can refer to the geologic details of the existing rock.

Estimating Overburden Pressure

Much of a major rock tunnel usually has hundreds to thousands of meters (feet) of overburden material. The roof of the tunnel, however, does not have to support the total overburden stress because the load is distributed around the tunnel by a condition called the *arching action*. Only the load of the immediate roof is supported by the tunnel and the stronger the rock, the smaller the rock load.

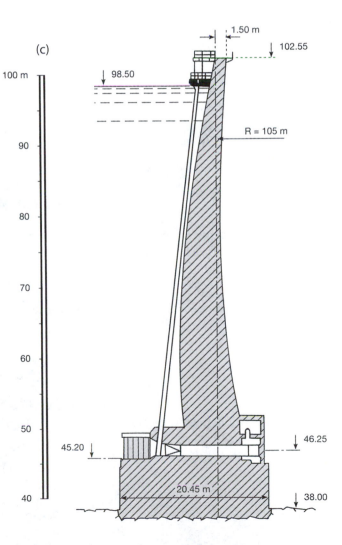

(c)

Figure 21.11 shows the concept of arching. Shown also are the dimensions B for the tunnel width, H_t for the tunnel height, H_p for the height of the rock that exerts load on the tunnel roof, and H for the total overburden height. As with most rock tunnels $H_p \ll H$.

Terzaghi (1946) developed a method for estimating the loads on tunnels based on a description of the rock encountered in tunneling. The better the rock quality, the lower the rock load. Based on quality, Terzaghi (1946) categorized rock mass into different classes and, for each class, expressed the height of the overburden rock exerting load on the tunnel roof, H_p, as a function of the width, or width plus height of the tunnel. The following equations show this relationship:

$$H_p = fB \qquad \text{(Eq. 21-1)}$$

for very sound rock, with f varying from 0 to 0.5, and

$$H_p = f(B + H_t) \qquad \text{(Eq. 21-2)}$$

with f varying from 0.35 to 1.1 for most weak rocks.

For worst case situations, $H_p = H$, the total overburden height. Terzaghi (1946) provided rock descriptions for different rock conditions that occur in tunnels and suggested corresponding f values for each category. For additional details on Terzaghi's method of estimating rock loads, the reader is referred to Hoek and Brown (1980) and Farmer (1983). An example problem is presented to illustrate the concept of rock load.

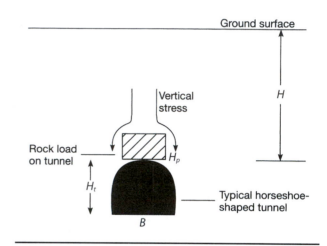

Figure 21.11 Arching action and transfer of rock loads to tunnel sides.

Example Problem 21.2

A horseshoe-shaped tunnel, 3.6 m (12 ft) wide and 4.9 m (16 ft) high, is driven through a fairly massive rock with a unit weight of 26 kN/m³ (165 lb/ft³). The f factor of Terzaghi indicates the following relationship:

$$H_p = 0.3B$$

What is the rock load (pressure), P, that will be carried by the tunnel roof?

Answers

Using SI units:

$$H_p = 0.3B = 0.3(3.6) = 1.1 \text{ m}$$

$$P = H_p \gamma_{rock} = 1.1(26) = 28.6 \text{ kN/m}^2$$

This load would need to be carried by the supports placed in the tunnel. If, for this tunnel, the following relationship prevailed:

$$H_p = 1.1(B + H_t)$$
$$= 1.1(3.6 + 4.9) = 9.4 \text{ m}$$

Then

$$P = 9.4(26) = 244 \text{ kN/m}^2$$

Using British engineering units:

$$H_p = 0.3B = 0.3(12) = 3.6 \text{ ft}$$

$$P = H_p\gamma_{rock} = 3.6(165) = 594 \text{ lb/ft}^2$$

This load would need to be carried by the supports placed in the tunnel. If, for this tunnel, the following relationship prevailed:

$$H_p = 1.1(B + H_t)$$
$$= 1.1(12 + 16) = 30.8 \text{ ft}$$

Then

$$P = 30.8(165) = 5082 \text{ lb/ft}^2$$

Although Terzaghi's method was widely used in the past to estimate rock loads, it is too conservative. Presently, the rock mass rating systems discussed briefly in Chapter 8, Engineering Properties of Rocks, are used to estimate rock loads for design of tunnel support systems. Among these, the rock mass rating (RMR) system developed by Bieniawski (1973, 1989) and rock mass quality or Q-system developed by Barton et al. (1974) are the most widely used methods.

Another method is the new Austrian tunneling method (NATM) (Brown, 1990; Karaku and Fowell, 2004). In this method rock loads in the next tunnel section to be excavated are estimated based on measurements and descriptions of the material already excavated. Continuous adjustments are made for tunnel support in consecutive sections.

Other techniques for estimating rock loads on tunnels are based on instrumentation data collected during tunnel construction. These include using load cells under the steel supports and borehole extensometers installed in the roof above the tunnel.

Standup Time

An important consideration in tunnel construction is known as the standup time, or bridge action time. This is the amount of time that rock (or soil) will stand intact after excavation, before the roof and side walls begin to break apart and fall into the tunnel. As would be expected, sound rock has a longer standup time than poor-quality rock. Therefore, rock mass rating systems can be used to estimate standup time. Obviously, some amount of time is required after mining to erect tunnel supports, so it is very challenging to tunnel through a rock with little or no standup time.

Geologic Hazards during Tunneling

Knowing the major geologic hazards in tunneling is helpful in understanding rock tunnels. These are as follows (Maidl et al., 2013):

1. Sudden increase in rock load as the tunnel advances.

2. Sudden reduction in standup time as the tunnel advances.

3. Rapid inflow of groundwater.

4. Presence of harmful gases.

If the rock load increases suddenly when a new section of the tunnel is mined, the operation for supplying tunnel supports is greatly affected. A brief discussion of tunnel supports occurs later in this section. A sudden reduction in the standup time for the excavated rock also greatly affects the tunnel construction process. Typically, this is associated with the sudden increase in rock load and is a consequence of a major change in the rock quality. A rapid flow of water into the tunnel can be disastrous. It must be pumped from the workings and requires the proper equipment to do so. Heavy water flows are dangerous to tunnel workers and large volume flows may lead to casualties. Rock units with open fissures and conduits, such as cavernous limestones or porous basalts, are possible candidates for this problem. Geologic studies must provide the knowledge that such conditions exist so that water can be drained or controlled before and during tunneling operations. The presence of harmful gas in a tunnel is a very dangerous condition. Harmful gases include CO_2, CO, H_2S, SO_2, and methane. Methane is an explosive gas when mixed in the proper proportion with the air in the tunnel. Methane, CO_2, and CO can occur in organic rich soils and rocks, including black shales and coal bearing strata. H_2S is a product of black shales and of volcanic rocks. SO_2 is also present in volcanic rocks. Of the various rock types, Holocene volcanics (less than 10,000 years old) can be the most treacherous rocks for tunneling because all four of the major geologic hazards can occur in them.

Tunneling Procedures

The two basic procedures for advancing a tunnel in rock are (1) the drill and blast method and (2) the tunnel boring machine (TBM) method (Maidl et al., 2013).

The drill and blast procedure is sometimes termed the conventional method because it preceded the TBM method. The drill and blast method is a six-step operation: (1) drill blast holes, (2) load the holes with explosives, (3) shoot or blast the explosives, (4) ventilate gases from the explosion, (5) muck or remove the blasted rock, and (6) erect tunnel supports. The standup time of the rock must be greater than the time elapsed between steps 3 and 6.

The explosive used in tunnel excavation is ammonium nitrate and fuel oil (ANFO). These ingredients, themselves, are inert, which is a safety advantage over the use of dynamite or similar explosives. Blasting caps are used to detonate the ANFO and a delay shooting array is used. Delay shooting involves millisecond (thousandths of a second) delays between shooting different groups of loaded holes. This allows for more efficient breaking of the rock, reduces overbreak, spreads the amplitude of the waveform from the blast, relative to time, and reduces the blast damage to nearby structures. ANFO can be used only in dry blast holes; otherwise the mix of ammonium nitrate and fuel oil in proper proportions is not possible. In rocks that produce significant amounts of seepage, dynamite or a similar explosive must be substituted for the ANFO.

The second method for advancing a rock tunnel involves a tunnel boring machine. A TBM consists of a rotating cutter head system that rotates against the tunnel face. Rock is sliced and sheared off as the TBM advances. The cutter face rotates at a few revolutions per minute and the muck is collected on a conveyor system that transports it to a muck hauling system. Rock bolts are commonly placed in the roof as the TBM moves forward, exposing the newly cut surface. Figure 21.12 shows a TBM.

TBMs cut circular tunnels rather than the traditional horseshoe shape. They also disturb the rock

(a)

(b)

Figure 21.12   (a) In July 2013, the TBM known as "Bertha" began tunneling work in Seattle, Washington, to replace the SR 99 Alaskan Way Viaduct. (b) The launch pit for Bertha. (Photos by the Washington State Department of Transportation.)

much less than the drilling and blasting method. Typically, the amount of temporary support in a TBM tunnel is considerably less than that of a drill and blast tunnel of comparable size with the same initial rock conditions. The rate of progress with a TBM can equal or exceed that accomplished by drilling and blasting.

However, not all rocks are suitable for TBM excavation. Poor rock conditions lead to many problems with using a TBM. Very hard, strong rocks cannot be excavated economically using a TBM because the bit wear on the rock cutters is so extensive. This causes undue amounts of downtime related to repairs and replacements. Rocks containing minerals with a Mohs hardness of 6 and above cause greater amounts of cutter bit wear as do rocks with a high unconfined compressive strength, above 105 to 140 MPa (15,000 to 20,000 psi). Therefore, softer, low-strength, but intact rocks are the best candidates for tunneling by a TBM. The geologic conditions should be as uniform as possible; the same rock unit with similar overall conditions yields the best situation for the systematic work of the TBM. TBMs up to 12 m (40 ft) in diameter have been used to excavate rock tunnels. The largest TBM in the world, known as Bertha, was built specifically for the Washington State Department of Transportation's (WSDOT) Alaskan Way Viaduct replacement tunnel project in Seattle. The machine, made by Hitachi Zosen Corporation's Sakai Works in Osaka, Japan, had a cutterhead diameter of 17.5 m (57.5 ft) across, was 99 m (325 ft) long, and weighed 6,100 metric tons (6,700 short tons). TBMs are most efficient on large tunneling jobs involving tunnels of a mile or more in length. The high cost of setting up a TBM system requires that the project be large enough to justify the cost involved.

Tunnel Supports

Types of supports for tunnels include (1) steel sets, (2) roof bolts, (3) bolt straps and mesh, (4) shotcrete, and (5) spiling (Maidl et al., 2013). Commonly, a permanent concrete lining is placed over these supports during a later stage of tunnel construction.

Steel sets are H beams or wide flange beams that are deformed into the shape of the tunnel by the manufacturers, according to the specifications. Purchased from the steel mill prior to construction, they are transported to the job site and then underground, where they are assembled to support the rock load (Figure 21.13). Considerable labor is involved in erecting large steel sets in a tunnel. Obviously, the larger the rock load, the more steel support is required.

Roof bolts are placed in holes drilled into the roof of the tunnel. The bolt length depends upon the stratigraphy immediately above the roof and varies from project to project. Bolts are tensioned after placement to provide greater support. Several types of bolts and anchors are used, depending upon geologic conditions. Bolts are placed at regular intervals, such as at 1-m (3- or 4-ft) centers, and

Figure 21.13   Steel set supports in a tunnel under construction in Greece. Notice also the use of shotcrete. (Photo by Paul Marinos and Evert Hoek.)

bolt straps and mesh can be placed against the roof to hold small, loose rocks in place.

Shotcrete is a relatively dry concrete, containing a small size aggregate (1.25 cm [0.5 in]), that is sprayed on the roof and sides of a tunnel to provide strength by holding surface materials together (see Figure 21.13). Layers of shotcrete are 15 cm (6 in) or more in thickness. Rock bolts can be placed in the shotcrete to anchor it into the adjacent rock.

Spiling consists of concrete reinforcing bars (deformed bars or rebars) and larger pieces of steel that are driven at an upward angle into the roof of a tunnel prior to excavation. This method is used in poor quality rock with a very short standup time.

Rock Blasting

Rock blasting, using explosives, is frequently required for excavation of highway cuts, dam foundations, and tunnels. A major concern associated with the use of explosives for rock excavation is the potential damage caused by waves generated during blasting. Two types of damage can occur during blasting: damage by ground vibrations and damage by airblasts. The situation is further complicated by the fact that humans can feel vibrations at levels much below those required to cause damage, and transient vibrations are inherently alarming to them. Hence, the local residents tend to complain even when the blast vibrations are of relatively low levels and safe.

Blast-Damage Criterion

Particle velocity, measured in cm/sec (in/sec), is the commonly used criterion for relating blast vibrations to possible damage. The value of 5 cm/sec (2 in/sec) particle velocity is commonly considered as the lower boundary of onset for possible damage to a sound, residential structure because the plaster starts showing cracking at this velocity (Edwards and Northwood, 1960). Humans can notice transient motions at a velocity as low as 0.15 cm/sec (0.06 in/sec) (Figure 21.14). The response of humans without accompanying noise (or sound effects) is shown in Figure 21.14a. As indicated, vibrations become disturbing at 1.0 cm (0.4 in/sec) and feel severe at 3.0 cm/sec (1.2 in/sec). If the sound of the blast accompanies the vibration, people react with

even greater concern, as shown in Figure 21.14b. For this combined situation, blasting is noticeable at 0.05 cm/sec (0.02 in/sec) and will be judged severe, yielding complaints, at particle velocities as low as 0.5 cm/sec (0.2 in/sec). This is why good communications with the public, educational responses, and blast monitoring are needed when blasting in urban or other populated areas is conducted. A lower criterion that has been developed in some cases to preclude damage and to gain more tolerance from the general public is 2.5 cm/sec (1.0 in/sec) particle velocity.

The frequency of blast waves is also a contributing factor as low-frequency waves are more destructive to structures. This relationship is shown on Figure 21.15. For frequencies less than 3 Hz (cycles per second), less

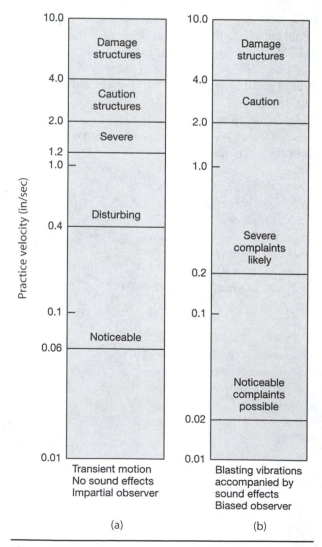

Figure 21.14 Human response to blasting vibrations.

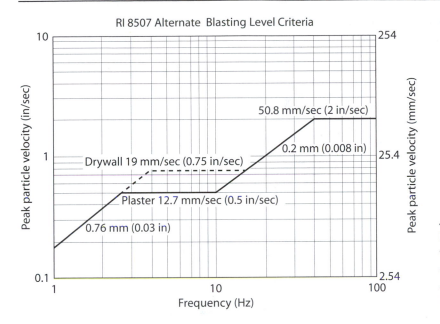

RI 8507 Alternate Blasting Level Criteria

50.8 mm/sec (2 in/sec)

0.2 mm (0.008 in)

Drywall 19 mm/sec (0.75 in/sec)

Plaster 12.7 mm/sec (0.5 in/sec)

0.76 mm (0.03 in)

Peak particle velocity (in/sec)

Peak particle velocity (mm/sec)

Frequency (Hz)

Figure 21.15 Safe levels of blasting vibrations for houses, expressed as a combination of peak particle velocity and frequency of wave vibration. (Siskind et al., 1980.)

than 2 cm/sec (0.75 in/sec) will cause damage. Fortunately, these low frequency waves are filtered out by the ground in a short distance. From 3 to 20 Hz the damage level for drywall is 0.75 Hz and at 30 Hz the 5 cm/sec (2.0 in/sec) threshold is in control. For Indiana the maximum level of vibration for all structures is set at 2.5 cm/sec (1.0 in/sec).

The effects of ground vibration on engineering structures have been studied in detail since the early 1940s (Thoenen and Windes, 1942; Crandell, 1949; Devine, 1966; Oriard, 2005). As one would expect, the level of ground motion needed to cause damage to a structure is not the same for all types of buildings. Also the degree of damage acceptable in one building based on its specific use may not be the same as that accepted in another. In the early work, acceleration a of the ground coupled with wave frequency n was used as the damage criterion. Using conditions of harmonic motion (sinusoidal waves) the energy ratio a^2/n^2 was used to estimate damage effects. It was later found that damage could be correlated with displacement and frequency. Because particle velocity is proportional to a product of displacement and frequency, particle velocity became the main criterion.

Blast-Induced Ground Vibrations

Ground vibrations experienced at a certain point adjacent to a blasting site are dependent on the weight of explosive detonated per delay, the distance between the blast site and the observation point, and the characteristics of the earth material transmitting the vibration. The term delay refers to the short time intervals, measured in milliseconds (0.001 sec). between detonations of different portions of an explosive charge. Up to 10 delays at several millisecond intervals can be involved at one time to prevent the blast vibrations from being superimposed on each other.

The effects of blast vibrations fall off rapidly with distance. The association between weight of explosives and distance has been developed in a form known as the scale distance (SD) relationship. Devine (1966) of the US Bureau of Mines proposed the following relationship:

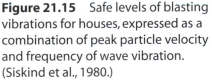

$$V = 121.1 \text{ in/sec} \frac{1(\text{ft/lb})^{1/2}}{R/W^{1/2}(\text{ft/lb})^{1/2}} \qquad \text{(Eq. 21-3)}$$

where:

V = particle velocity (in/sec)
R = distance (ft)
W = weight of explosive (lb)

This equation involves the square root law for blast damage. Since the concept of scale distance was developed in British engineering units, the following discussion uses the original units.

A conservative approach can be used to estimate the allowable maximum charge using Devine's (1966) suggested scale factor of $R/W = 50 \text{ lb/ft}^{1/2}$.

This is included in Table 21.4. For purposes of comparison, a scale factor of 20 lb/ft$^{1/2}$ is shown as well.

A scale distance of 50 lb/ft$^{1/2}$ yields a particle velocity of less than 6 cm/sec (2.4 in/sec) (based on Eq. 21-3). Soft, organic soils yield one of the most destructive vibration levels possible to buildings, as contrasted to massive rock, which yields the least effects for a given blast. If more than a 5 cm/sec (2 in/sec) particle velocity is to be allowed when a seismograph is employed for monitoring, or if a value less than 50 lb/ft$^{1/2}$ is employed for the scale distance without measuring particle velocity directly, it is suggested that a well-documented study of the site conditions be made in advance. Such a study would be accomplished by a vibrations expert or geophysicist.

Table 21.4   Amount of Allowable Explosive for Varying Distances.

Distance R (ft)	Allowable Weight of Explosive per Delay	
	50 lb/ft$^{1/2}$	20 lb/ft$^{1/2}$
100	4	25
500	100	625
1000	400	2500
2000	1600	10,100

Note: This table is based on a scale distance of $R/W^{1/2}$ or $W = (R/SD)^2$.

Additional details on blast damage criteria are available in the literature (Hendron, 1977).

Example Problem 21.3

Blasting for a tunnel excavation will take place 39.6 m (130 ft) away from a building that needs to be protected from blast damage. What is the maximum weight of dynamite charge that can be detonated based on a scale distance of 50 lb/ft$^{1/2}$? Based on 20 lb/ft$^{1/2}$?

Answers

$$SD = R/W^{1/2}$$

- If R = 130 ft and SD = 50,
  $W = (130/50)^2 = 6.76$ lb (3 kg) of dynamite per delay

- If R = 130 ft and SD = 20, then
  $W = (130/20)^2 = 42.2$ lb (19 kg) of dynamite per delay

Effects of Air Blasts

Air blasts occur when surface charges are detonated or when demolition occurs aboveground. An overpressure of 7 kN/m^2 (1 psi) typically will break windows and cause cracks in new plaster. Windows less than 5.5 m^2 (60 ft^2) in area are safe from breakage if the air blast is less than 0.7 kN/m^2 (0.1 psi) overpressure. It turns out that overpressures of 0.7 kN/m^2 (0.1 psi) for air blast damage correspond fairly well with ground vibrations of 5 cm/sec (2 in/sec). The overall effect generally is that air blast damage can be prevented by using a scale factor that will preclude damage from ground vibration. Precautions should be taken nonetheless to reduce or minimize air blast effects by proper stemming of holes and the use of blasting mats or barricades if surface blasting is involved.

Environmental Geology

Environmental geology and engineering geology are interrelated fields. Environmental geology, as a designated specialty, developed in the late 1960s. A concern for the environment affected the geological sciences in much the same way as for the other fields of study, and a specific focus on the effects of humans on the geologic environment became a key issue (Ehlen et al., 2005). The definition by Flawn (1970) describes the essence of the subject, "Environmental geology is a branch of ecology . . . that . . . deals with relationships between man and his geologic habitat; it is concerned with the problems that people have in using the earth and the reaction of the earth to that use."

Design and construction of engineering structures must consider the environmental impacts of the structure. For example, constructing a dam completely changes the habitat of water life, fish in particular, in the reservoir area. That is why a dam design team will have one or more environmental scientists as its members. This necessitates a brief discussion of the subject in this chapter.

Environmental geology typically addresses the subjects listed in Table 21.5. These are usually discussed in terms of statistics and detrimental factors of the problems are underscored.

Environmental geology also includes the biological aspects of the land. This involves the biosphere, or the biological portion of the lithosphere. The effects on mineral, elemental, or chemical cycles are a major concern. These include, for example, the phosphate, nitrogen, carbon, and protein cycles. To ensure that the Earth will continue to support human life, in the near and distant future, we must accomplish the following: (1) regulate the number of humans (unless population stability is achieved, everything else will fail), (2) conserve and recycle the basic materials we use to the greatest extent possible, and (3) ensure that the food supply is adequate for the regulated numbers. It certainly is in the realm of possibility for humans to allow the environment to deteriorate to the extent that their extinction could occur. Climate change is an example of this threat. Far short of that extreme, it seems necessary that the environment is preserved in a state similar to that in which it was inherited, both for aesthetic and simple survival purposes. The answer, however, is not only preservation but also conservation, accomplished by being the best stewards of the land.

Large-Scale Interpretive Maps

One feature of environmental geology involves preparing and using environmental geology maps. These large-scale maps are prepared at a scale of 1:1000 or larger to show maximum detail. They may be strictly geologic maps or interpretive maps derived from geologic maps.

Many different forms of interpretive maps can be developed from geologic maps or geologic data. Maps can be made for depicting the extent of

---

Table 21.5   General Outline of Environmental Geology.

1. **Geologic Hazards**

   Earthquakes
   Landslides
   Floods
   Tsunamis
   Volcanoes
   Subsidence
   Loss of mineral resources by urbanization

2. **Geologic Constraints to Construction**

   Slope stability along actively downcutting streams
   Siting sanitary landfills
   Septic tank percolation fields
   Shortage of natural construction materials
   Excavation in bedrock areas
   Compressible, organic soils
   Loess and lacustrine deposits
   Karst terrain
   Expansive soils
   Presence of faults
   Groundwater complications
   Reactive concrete aggregates
   Soil erosion and siltation
   Floodplains, flooding, and building restrictions

3. **Environmental Health**

   Geologic factors and environmental health
   Trace elements and health
   Chronic diseases related to geologic environment

4. **Human Interaction with the Environment**

   Water supply
   Waste disposal
   Effects of urbanization on the landscape

5. **Mineral Resources and Depletion**

   Mineral resources and population
   Renewable and nonrenewable resources
   Resources versus reserves
   Environmental impact and mineral development
   Recycling of minerals, effects of planned obsolescence

6. **Current Land Use and Land Use Planning**

   Determining current land use
   Landscape analysis and evaluation
   Mineral resource mapping
   Landscape aesthetics
   Land use planning
   Environmental impact

available water in the subsurface, locating construction materials, indicating areas where soils with particular properties (such as swelling potential, rock strength, or modulus of elasticity) occur, indicating conditions of septic tank acceptability, or areas of landslide susceptibility. Smaller scale maps (1:24,000 and less) have been used in engineering geology reports as interpretive maps for many years.

Fault mapping in earthquake-prone areas is a situation in which environmental geology maps prove useful. Fault traces in the subsurface are found by trenching and fault displacements on the Earth's surface are noted for the study area. This information is transferred to a large-scale map (1:500 to 1:1000 or so) where a high degree of detail and accuracy can be depicted. Areas where building cannot be allowed can be designated accurately on these maps so that restrictive, meaningful zoning can be established.

Past experience in Nicaragua, for example, has shown that fault zones designated on small-scale geologic maps are not sufficiently obvious or accurate to indicate, effectively, potential earthquake damage. Although located at an apparently safe distance from the fault zone, based on the map, construction of major buildings directly in an active fault zone occurred. A combination of imprecise location of the fault zone and mapping on too small a scale led to this dangerous situation.

Applying geology to land use planning is another significant criterion for environmental geology. The goal is to introduce geologic factors into the planning process to influence procedures for mineral extraction, development of housing tracts, groundwater development, preservation of agricultural land, and so on.

For major cities and for counties of approximately 1300 km^2 (500 mi^2), an appropriate map scale is 1:24,000 (7.5-minute quadrangle). Topography, surface geology, agricultural soils, bedrock and structural geology, depth to bedrock, and groundwater information are important data that must be compiled. Current land use information for the study area is obtained as well. With these combined data, site selection for those areas suitable as housing tracts, sanitary landfills, quarries and gravel pits, groundwater supply or recharge, septic tank percolation fields, industrial building sites, and prime agricultural land can be outlined. Economic and political restraints are considered and, in the final analysis, these and geological factors can be used for land use planning. Individual maps are made depicting this information.

An environmental study of the type detailed lends itself to a database and computer analysis approach. This is accomplished by establishing a grid pattern for the area, each grid having a unique address, to which values for the significant geologic factors for the different proposed uses of the land can be assigned.

In this way, a value for each unit area of the grid can be totaled, relative to individual proposed uses. Areas registering the highest sum for a particular proposed use are the most suitable in that area for that application (Hasan and West, 1982). Geographic information systems (GIS) have been created by government and research organizations throughout the United States for data storage and retrieval of this type of information.

Environmental impact studies are required for all projects involving expenditures by the federal government. These include details on the current nature of the environment and the anticipated impacts rendered by a proposed project. Significant impacts must be mitigated, or clearance to proceed is not granted. Opportunities are provided for the public to make their views known regarding the proposed project.

EXERCISES ON ENGINEERING GEOLOGY:
HIGHWAYS, DAMS, TUNNELS, AND ROCK BLASTING

MAP READING

Exercises 1 through 5 pertain to the US Geological Survey (1953) publication *Interpreting Geologic Maps for Engineering Purposes: Hollidaysburg Quadrangle, Pennsylvania* (scale 1:62,500). This folio contains a topographic map (map 1), general-purpose geologic map (map 2), foundation and excavation conditions (map 3), construction materials (map 4), water supply (map 5), and site selection for engineering works (map 6) for the Hollidaysburg Quadrangle. It affords a unique opportunity to compare the different maps for one area. Be sure to examine all maps carefully before completing this exercise.

1. Using a road map of Pennsylvania, give the location of Hollidaysburg with regard to major cities in Pennsylvania and indicate its general location in the state. With the aid of a landform map of the United States and the US physiographic map in Chapter 20, indicate the physiographic province in which this area is located.

2. Refer to the explanation found on map 2. Construct a simplified geologic column for the quadrangle based on this information. Use the following degree of detail for the subdivisions. Use abbreviations for periods if needed. Continue from the portion shown here, beginning with the formations of the Middle Devonian:

Quaternary
  Recent
    Alluvium

UNCONFORMITY
  Carboniferous
    Pennsylvanian
      Allegheny Formation
      Pottsville Formation

UNCONFORMITY
  Mississippian
    Mauch Chunk Formation
    Loyalhanna Limestone
    Pocono Formation
  Devonian
    Upper Devonian
      Hampshire Formation
      Chemung Formation
      Brallier Shale
      Harrell Shale
    Middle Devonian

3. a. Sketch a map view of the major structural features of the quadrangle (use map 2). Show anticlinal and synclinal axes, faults, sufficient bedding planes, and geologic symbols to indicate the structural features. Refer to map 1. How are the mountain regions related to the structural features? What rock types comprise the mountain tops in this area?

   b. Name the maps contained in this folio. How are maps 3 through 6 obtained from maps 1 and 2? How is this useful in land use planning?

   c. The following system is used for reference purposes on the Hollidaysburg Quadrangle:

3	2	1
4	5	6
9	8	7

   Why isn't the township and range method used? Why isn't a latitude and longitude designation used?

   d. How are the rock units grouped to obtain map 3? How would such a map be useful to highway construction? Foundations for heavy buildings?

   e. How are the rock units grouped to obtain map 4? Why is this of use to planners in this area?

   f. How are the rock units grouped to obtain map 5? Why is this of use to planners and home builders?

4. a. A tunnel is proposed that will extend from East Sharpsburg to the southern end of Oldtown Run Road. On the topographic cross section in Exercise Figure 21.1 on the following page, draw in the geologic cross section between these two points; include the tunnel grade line on the drawing. A 2.5 cm = 792 m (l″ = 2600′) map scale for the horizontal dimension has been used in the drawing. For the vertical scale on the cross section, a 2.5 cm = 152 m (l″ = 500′) scale is employed. What is the vertical exaggeration? Describe any problems that might occur during rock tunneling.

   If the load on the roof of the tunnel is expressed by the relationship

   $$H_p = f(B + H_t)$$

where:

B = width of tunnel in m (ft)

H_t = height of tunnel in m (ft)

f = factor dependent on rock conditions

What would be the pressure on the roof of the tunnel in the Tuscarora Quartzite if f = 0.25 and the tunnel had a height of 6 m (20 ft) and a width of 4.5 m (15 ft)?

b. A dam has been proposed for Clover Creek near the town of Henrietta (rectangle 7). What is the gradient of the creek at this general location? Describe completely the geology of the site. What would be the maximum height of the dam? Discuss any problems that might occur during and after construction of the dam. Would you recommend this location as a dam site? Explain in detail.

c. Using a piece of 22 cm × 28 cm (8½ in × 11 in) tracing paper and map 4, outline all the areas in rectangle 8 where underground caverns could likely occur. Indicate also where limestone quarries for concrete aggregate might be located in this rectangle. Explain your choice.

5. Environmental geology maps are used today for planning and development. The US Geological Survey has mapped the geology adjoining some large cities with map scales of 2.5 cm = 33 m (1″ = 100′). Locations with complex geology and rapid growth are mapped first: Denver, Los Angeles, San Francisco, and so on.

Why would the Hollidaysburg map fall short in this effort? Explain.

DANVILLE QUADRANGLE

6. Refer to the Geologic Map of the 1° × 2° Danville Quadrangle, Indiana and Illinois, published by the Indiana Geologic Survey. This map includes the area of Lafayette, Indiana. What is the map scale? To what extent does this fall short of the 2.5 cm = 33 m (1″ = 100′) scale? What can be done to improve the situation?

SASKATOON, ALBERTA

7. The environmental geology map of Saskatoon, Alberta, Canada, is one of the maps available in many university libraries. How is this map an improvement over the Hollidaysburg and Lafayette maps? Explain.

ENVIRONMENTAL GEOLOGY, CLINTON COUNTY, INDIANA

For Exercises 8 through 13, several types of maps are provided for Clinton County, Indiana.

8. Using the topography map (Exercise Figure 21.2) and the bedrock geology map (Exercise Figure 21.3), what is the depth to bedrock at the corner where the boundary between T20N and T21N meets the boundary between R2W and R1W? (*Hint:* This is determined in the following way: Find the surface elevation, then find the elevation for the top of the bedrock; the numerical difference is the depth to the bedrock.)

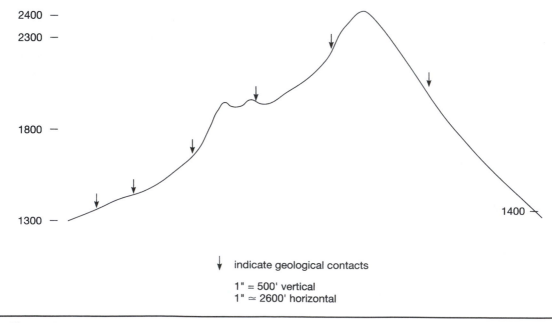

↓   indicate geological contacts

1" = 500' vertical
1" ≈ 2600' horizontal

Exercise Figure 21.1

Check this same point on the surficial geology and glacial thickness map (Exercise Figure 21.4). Are the two values the same? Why or why not?

9. What is the maximum relief in the county based on the topographic map (Exercise Figure 21.2)? Why is the area so flat? Refer also to the generalized glacial geology map of the state (see Exercise Figure 21.7).

10. Refer to the surficial geology map (Exercise Figure 21.4):

    a. What material comprises the primary portion of the map area?

    b. What type of soil texture is this most likely to be? Explain.

    c. What are the other four mapped materials?

    d. Which is (are) most likely to supply gravel materials for construction? Explain.

e. Where is the depth to bedrock least in the county? What would be the bedrock material there? Be specific.

11. Examine the piezometric surface map of Clinton County (Exercise Figure 21.5). This is the elevation to which water would rise in a well after an aquifer was intercepted.

    a. If the piezometric surface is higher than the elevation at which water is intercepted, what type of well is involved?

    b. At the corner where the boundary between T22N and T23N meets the boundary of R2W and R3W, what is the elevation of the piezometric surface? How far below the ground surface is this? At what depth would a pump intake have to be set in this well? Explain. What is the likely direction of groundwater flow in this area? Explain.

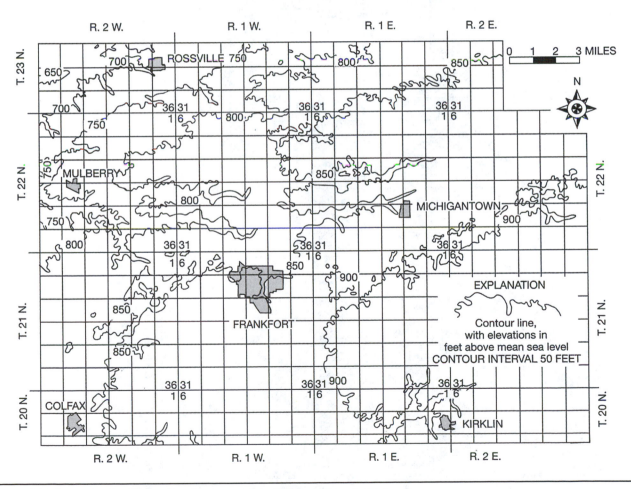

Exercise Figure 21.2   Generalized topography of Clinton County, Indiana.

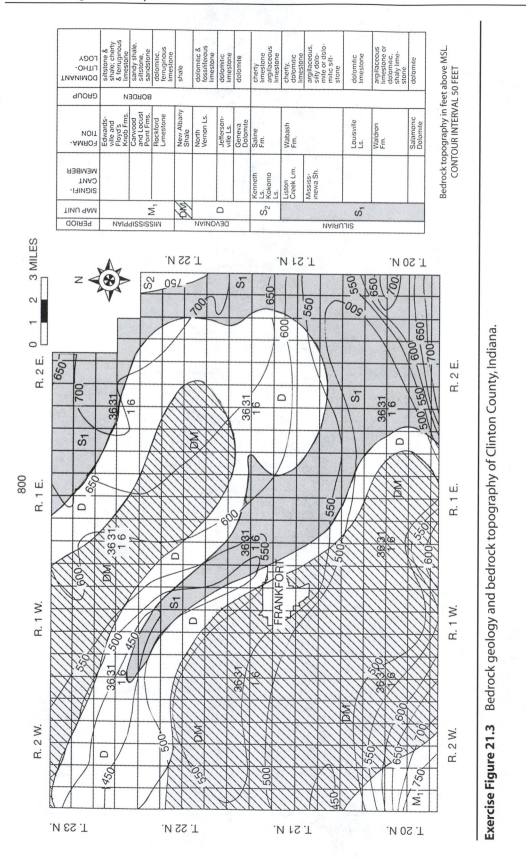

PERIOD	MAP UNIT	SIGNIFI-CANT MEMBER	FORMA-TION	GROUP	DOMINANT LITHO-LOGY
MISSISSIPPIAN	M_1		Edwards-ville and Floyd's Knob Fms.	BORDEN	siltstone & shale; cherty & ferruginous limestone
			Carwood and Locust Point Fms.		sandy shale, siltstone, sandstone
			Rockford Limestone		dolomitic, ferruginous limestone
DEVONIAN	DM		New Albany Shale		shale
	D		North Vernon Ls.		dolomitic & fossiliferous limestone
			Jefferson-ville Ls.		dolomitic limestone
			Geneva Dolomite		dolomite
SILURIAN	S_2	Kenneth Ls. Kokomo Ls.	Saline Fm.		cherty limestone argillaceous limestone
		Liston Creek Lm.	Wabash Fm.		cherty, dolomitic limestone
		Mississ-inewa Sh.			argillaceous, silty dolo-mite or dolo-mitic silt-stone
	S_1		Louisville Ls.		dolomitic limestone
			Waldron Fm.		argillaceous limestone or dolomitic, shaly lime-stone
			Salamonic Dolomite		dolomite

Bedrock topography in feet above MSL.
CONTOUR INTERVAL 50 FEET

Exercise Figure 21.3   Bedrock geology and bedrock topography of Clinton County, Indiana.

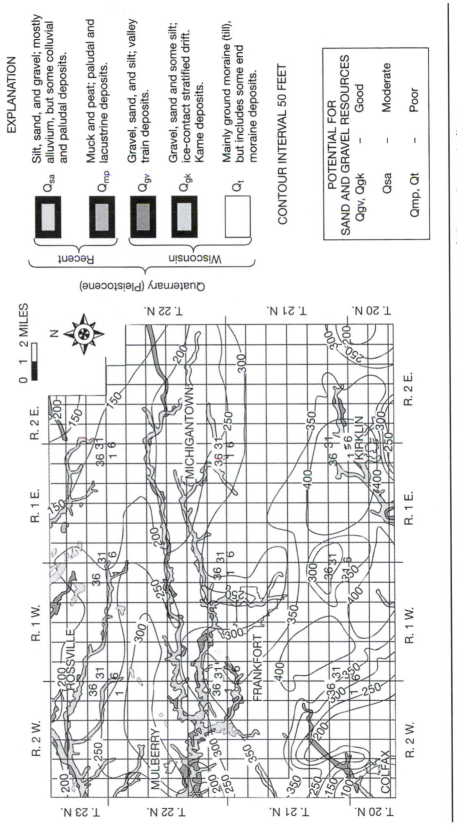

Exercise Figure 21.4   Surficial geology and glacial drift thickness showing sand and gravel resources of Clinton County, Indiana.

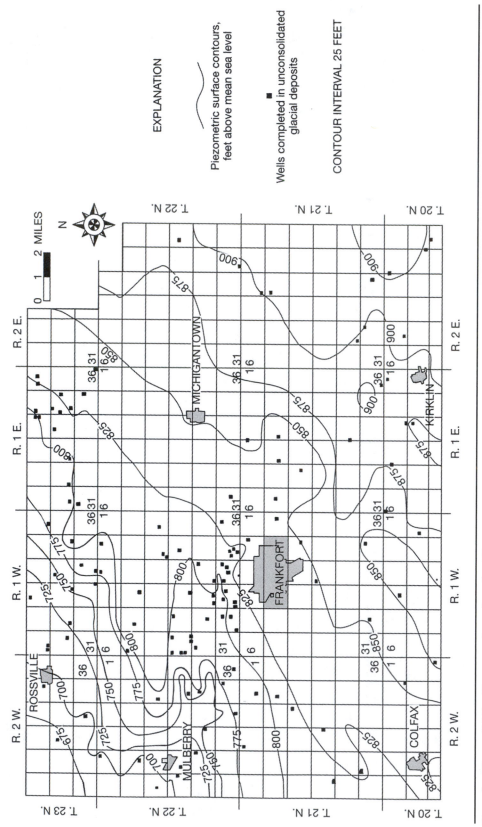

EXPLANATION

Piezometric surface contours,
feet above mean sea level

Wells completed in unconsolidated
glacial deposits

CONTOUR INTERVAL 25 FEET

Exercise Figure 21.5   Piezometric surface for Clinton County, Indiana.

SOIL ASSOCIATIONS

4. Genesee-Shoals-Eel: Nearly level, well drained, loamy Genesee, moderately well drained, loamy Eel, and somewhat poorly drained, loamy Shoals in alluvial deposits.

64. Crosby-Brookston: Nearly level, somewhat poorly drained, clayey Crosby and very poorly drained, loamy Brookston in glacial till.

66. Fincastle-Ragsdale-Brookston: Nearly level, somewhat poorly drained, silty Fincastle in wind-blown silts and glacial till, very poorly drained, silty Ragsdale in wind-blown silts and loamy Brookston in glacial till.

73. Raub-Ragsdale: Nearly level, somewhat poorly drained, silty Raub in wind-blown silts and glacial till and very poorly drained, silty Ragsdale in wind-blown silts.

81. Miami-Russell-Fincastle: Sloping, well drained, loamy Miami in glacial till, silty Russell in wind-blown silts and glacial till and nearly level, somewhat poorly drained, silty Fincastle in wind-blown silts and glacial till.

83. Miami-Crosby: Sloping, well drained, loamy Miami and nearly level, somewhat poorly drained, clayey Crosby in glacial till.

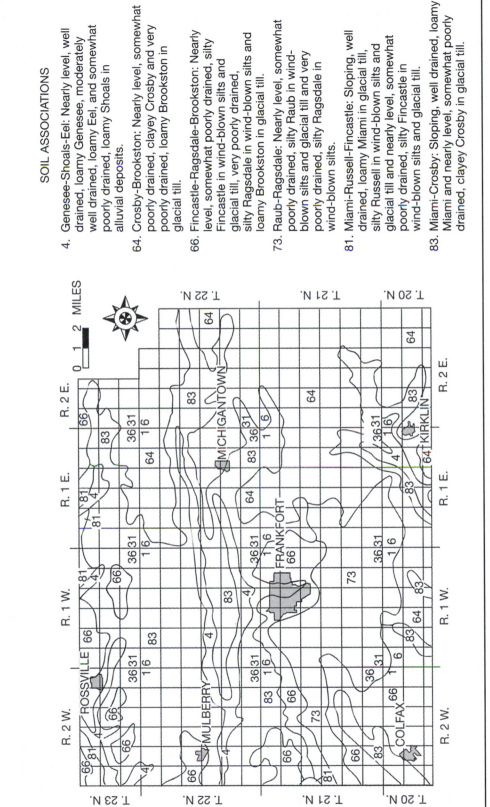

Exercise Figure 21.6   General soil associations of Clinton County, Indiana.

12. Examine the soil associations map of the county (Exercise Figure 21.6). A septic tank system should be located in relatively permeable material. Does soil association 81 qualify for this? Explain. Does association 64 qualify? Discuss.

13. Section 12, T21N, R1W is being considered for a sanitary landfill site. Refer to the surficial geology map (Exercise Figure 21.4) and the soil associations map (Exercise Figure 21.6). Is this a suitable site for a landfill? Explain in terms of information from both maps.

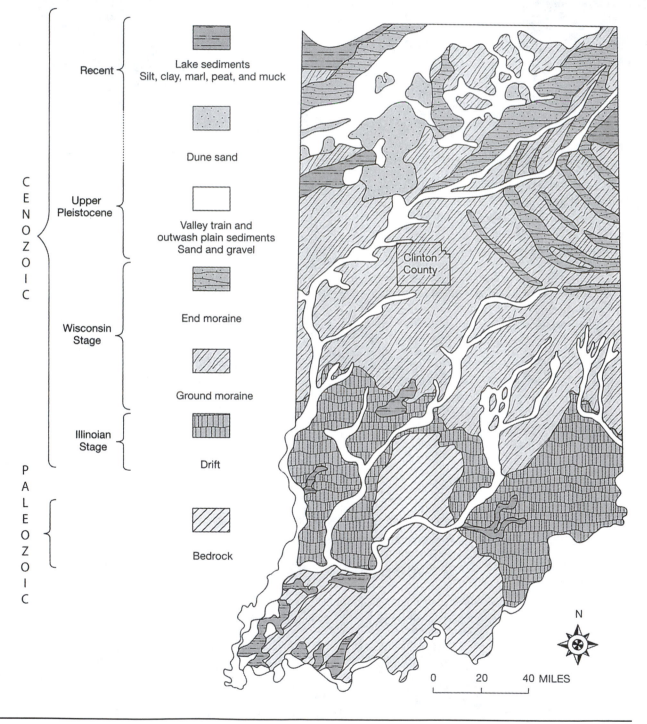

Exercise Figure 21.7   Generalized glacial geology map of Indiana. (From Wayne, 1958.)

WRITTEN QUESTIONS

14. For a highway construction project, a soil cut will be 9 m (30 ft) deep and an adjacent embankment will be 15 m (50 ft) high. The cut area will be used as a borrow site for the embankment. The depth to bedrock is more than 30 m (100 ft). Describe how typical borings should be taken in the cut and in the fill. Include the depth of the boring, drilling technique, and soil sampling. Include also a list of laboratory tests needed to characterize the material in question.

15. The following volume of soil and rock is to be removed from a highway cut: Soil: 9932 m^3 (13,000 yd^3) of silt and 12,988 m^3 (17,000 yd^3) of silty clay; rock: 13,752 m^3 (18,000 yd^3) of limestone and 11,460 m^3 (15,000 yd^3) of hard shale. The shrinkage factor for the silt is 8% and for the silty clay it is 12%; the swell factor for the limestone is 19% and for the hard shale 9%.

    a. What will be the volume of the fill constructed from these materials? What is the overall volume change in cubic yards and percent?

    b. If the volume of the fill is designed to be 44,312 m^3 (58,000 yd^3), what would be the needed volume of offsite rock disposal? Next, assuming the rock dumped offsite would be shale, how much volume does this represent in the cut section?

16. About 12,000 people per day will stop at a rest area along an interstate highway during the maximum tourist period. If the average water use is 13 liters/person (3.5 gal/person) stopping at the rest area, how much water will be consumed per day? How many minutes per day must a well pump if its capacity is 455 liters/min (120 gal/min)? What percent of the time must the well pump?

17. A highway is to be constructed over an old, coal strip mine site that was not reclaimed after mining. It contains elongated spoil piles and a lake, which marks the final strip of coal excavation. List the possible highway construction problems that could occur at the site. State assumptions, if any.

18. What are five primary purposes for dam construction? Where in the United States does the US Army Corps of Engineers have most of its responsibility for dam construction? The US Bureau of Reclamation? Name two other governmental agencies that construct and maintain dams in the United States.

19. List the different types of dams, based on construction materials. Include subclasses of the various types. Under what conditions would a thin arch concrete dam be the most feasible? Explain.

20. a. Why is an emergency spillway a necessary part of a dam construction project, particularly in heavily populated areas?

    b. List some likely advantages and disadvantages of an underground power generation station compared to an aboveground installation. Consider topography and climate in your analysis.

21. a. A major sanitary landfill was located in Indianapolis, Marion County, Indiana, along the sandy terrace deposits of the White River. Refer to Table 13.12 in Chapter 13, Groundwater. Would this site be considered favorable or unfavorable based on that table? Explain.

    b. The White River is immediately downgradient from the landfill and, normally, leachate would migrate from the landfill to the river. Because the White River has a large discharge, dilution of the leachate would occur. Nevertheless, is it good practice to allow leachate to migrate offsite in the groundwater? What are the long-term consequences of such a practice?

    c. Because of heavy pumping of groundwater by industry upgradient from the landfill, contaminated water from the landfill was drawn to the industrial wells. What is required to reverse the gradient in this fashion? Does this seem to be a likely occurrence along major floodplain areas in an urban environment? Explain.

22. The Eagle Creek Reservoir is located just to the west of Indianapolis and it supplies a major portion of the drinking water for the city. The Eagle Creek Dam is an earth structure that is 26 m (85 ft) high and 474 m (1554 ft) long.

    a. What subsurface exploration procedure would be appropriate for a proposed earthen dam structure in this area? Make a sketch in plan view, showing the centerline of the proposed dam, a small existing stream, an adjacent floodplain, and the two valley walls. Locate where borings should be placed along the centerline. What type of borings should be made and approximately how deep should they be?

    b. The cross section of a zoned earth dam is shown in Figure 21.8. The different sections of the dam are labeled. The materials for construction must be located adjacent to the dam prior to construction. The Eagle Creek Dam is located within a Wisconsinan age till plain.

(i) What material would be used for the clay core?

(ii) The silty shell material would likely be obtained from the proposed reservoir area. What is this stream-deposited material called? Why should this not be removed too close to the dam itself? Explain.

(iii) Clean sand is used in the sandy drainage zone. This is to relieve pore pressures after water has passed through the clay core. This material would likely have to be removed, selectively, from deposits in the reservoir area. Why is a clean sand required in this drainage zone? Explain.

(iv) Settlement of the foundation, and of the dam itself, is a concern. Why might the foundation settle? What would be the vertical stress induced on the foundation at the center of the dam, based on the description provided? What test should be run to determine how much settlement of the foundation is likely to occur? (*Hint:* Refer to the last section in Chapter 7, Elements of Soil Mechanics.)

(v) How is settlement of the dam itself (the embankment) kept at a minimum? How is this accomplished during the construction of the dam?

(vi) A concrete spillway is used to control the level of water in the reservoir. What are the materials of construction needed for the spillway? Where, on the construction site, would the concrete aggregates be obtained?

23. A coal-fired, electric-generating power plant is to be constructed in an area of sedimentary rock strata adjacent to a large stream. The foundation for the power plant and a water intake structure in the river are to be designed by a major engineering company. The power plant will be located on the valley wall on bedrock rather than on the floodplain or terrace of the river.

a. Why was the power plant probably located on the valley wall rather than in the floodplain? Assume flooding is not the issue, because an area above flood effects was available on the terrace.

b. Review Table 21.1. Indicate how each of these items would be involved in the design considerations for the project described above. If an item does not apply indicate why this is the case.

24. Why does engineering geology encompass the more obscure parts of construction? Is it to be expected that analytical analysis of geologic phenomenon related to construction does not always provide exact answers? Explain.

25. Geologic hazards generally involve the catastrophic effects related to nature, whereas geologic restraints are more of a long-term effect of geologic agents acting on the landscape. Relative to physiographic units, where do the catastrophic geologic hazards occur and where are the geologic restraints more in control? Consider relative to stable regions (SR) and orogenic belt areas (OB) as presented in Chapter 20, Physiographic Provinces.

26. The Fortville Fault in central Indiana is a high-angle normal fault that displaces the bedrock surface but does not extend through the Pleistocene glacial deposits above. The bedrock surface displaced ranges in age from Mississippian to Silurian.

a. From this description, the fault is younger than which age, Mississippian or Silurian? The fault is older than which age, Mississippian or Silurian?

b. The fault strikes about N30°E and the south portion is the downthrown side. Draw the fault in map view and cross-section view, including the information about the fault provided in this exercise.

c. If the fault does not extend through the glacial deposits to intersect the ground surface, how was the fault's presence determined? Thickness of glacial deposits in this area typically exceeds 30 m (100 ft).

d. There is no increase in earthquakes or microseisms associated with the Fortville Fault in Indiana as compared to the surrounding area, which overall is in a low-intensity earthquake zone. What does this indicate about the activity of this fault system? *Capable faults* are those capable of movement because of earthquake activity. Is the Fortville Fault capable? Explain. Did movement occur along it within the last 10,000 years? Explain.

27. Under the center of a mountain range, a tunnel is located 380 m (1250 ft) below the ground surface.

a. If the average unit weight of the rock is 26 kN/m^3 (165 lb/ft^3), what is the total vertical stress above the tunnel in kN/m^2, psf, and psi?

b. The Terzaghi information for the rock above the tunnel indicates that the equation $H_p = 0.45(B + H_t)$ would apply. The tunnel is 6 m (20 ft) wide and 4.5 m (15 ft) high. Calculate H_p in m (ft) and the roof pressure in kN/m^2 (lb/ft^2).

28. a. List the four major geologic hazards in tunneling. In a fissured limestone, which of these is likely to occur during the tunneling?

    b. What harmful gases can occur in rocks that contain thin coal beds? How can this problem be minimized?

29. What are the six steps of the drill and blast method for tunneling? What is the preferred explosive? How can blast damage to adjacent structures be minimized?

30. Under what conditions are tunnel boring machines most effective and efficient in rock tunneling? Why might a long tunnel in a fine, massive granite not be a good site for a TBM?

31. A quarry owner decided to set off a large blast in the quarry on a weekend when few people were in the vicinity. An equivalent of 9091 kg (10 tons) of dynamite was set off at one time. An office building is located 655 m (2150 ft) away from the quarry. What was the scale distance factor for this blast? Was it likely to cause vibration damage to the office building? Explain.

    a. If the office building has several large windows of 9.3 m^2 (100 ft^2) in size, would air blast damage be a problem if a direct line of sight occurred between the quarry blast and the building? Explain in detail.

REFERENCES

Admassu, Y., and Shakoor, A. 2015. Cut slope design for stratigraphic sequences subject to differential weathering. *Environmental & Engineering Geoscience* XXI(4):311–324.

Barton, N., Lien, R., and Lunde, J. 1974. Engineering classification of rock masses for the design of tunnel support. *Rock Mechanics* 6(4):189–236 .

Bieniawski, Z. T. 1973. Engineering classification of jointed rock masses. *Transactions of the South African Institute of Civil Engineering* 15(12):335–343.

Bieniawski, Z. T. 1989. *Engineering Rock Mass Classifications*. New York: John Wiley & Sons.

Brown, E. T. 1990. Putting the NATM into perspective. *Tunnels & Tunneling*, Special Issue, Vol. 22, pp. 9–13.

Crandell, F. J. 1949. Ground vibrations due to blasting and its effect upon structures. *Journal of the Boston Society of Civil Engineers* 36(2):206–229.

Devine, J. R. 1966. *Avoiding Damage to Residences from Blasting Vibrations* (Highway Research Record No. 135, pp. 35–42). Washington, DC: Highway Research Board, National Academy of Sciences.

Duffaut, P. 2013. The traps behind the failure of Malpasset arch dam, France, in 1959. *Journal of Rock Mechanics and Geotechnical Engineering* 5(5):335–341.

Edwards, A. T., and Northwood, T. D. 1960. Experimental studies on the effects of blasting on structures. *The Engineer* 210.

Ehlen, J., Haneberg, W. C., and Larson, R. A. 2005. *Humans as Geologic Agents*. Denver: Geological Society of America.

Farmer, I. 1983. *Engineering Behavior of Rocks* (2nd ed.). New York: Chapman and Hall.

Fell, R., MacGregor, P., Stapledon, D., Bell, G., and Foster, M. 2014. *Geotechnical Engineering of Dams* (2nd ed.). New York: CRC Press.

Flawn, P. T. 1970. *Environmental Geology: Conservation, Land-Use Planning, and Resource Management*. New York: Harper & Row.

Fwa, T. F. 2005. *The Handbook of Highway Engineering*. New York: CRC Press.

Goodman, R. E. 1993. *Engineering Geology—Rock in Engineering Construction*. New York: John Wiley & Sons.

Hasan, S. E., and West, T. R. 1982. Development of an environmental geology data base for land-use planning. *Bulletin of the Association of Engineering Geologists* XIX(2):177–192.

Hendron, Jr., A. J., 1977. Engineering of rock blasting on civil projects. In W. J. Hall (ed.), *Structural and Geotechnical Mechanics* (pp. 242–277). Englewood Cliffs, NJ: Prentice Hall.

Hoek, E., and Brown, E. T. 1980. *Underground Excavation in Rock*. London: Institution of Mining and Metallurgy.

Jansen, R. B. 1988. *Advanced Dam Engineering for Design, Construction, and Rehabilitation*. New York: Springer.

Karaku, M., and Fowell, R. J. 2004. An Insight into the New Austrian Tunneling Method (NATM). Seventh Regional Rock Mechanics Symposium, Sivas, Turkey.

Maidl, B., Thewes, M., and Maidl, U. 2013. *Handbook of Tunnel Engineering*. New York: John Wiley & Sons.

Martt, D. F., Shakoor, A., and Greene, B. H. 2005. Austin Dam, Pennsylvania: The sliding failure of a concrete gravity dam. *Environmental & Engineering Geoscience* 11(1):61–72.

Neuendorf, K. K. E, Mehl, Jr., J. P., and Jackson, J. A. (eds.). 2005. *Glossary of Geology* (5th Edition). Alexandria, VA: American Geological Institute.

Oriard, L. L. 2005. *Explosives Engineering, Construction Vibrations, and Geotechnology*. Solon, OH: International Society of Explosives Engineers.

Rogers, J. D. 2006a. *The 1928 St. Francis Dam Failure and Its Impact on American Civil Engineering.* Reston, VA: American Society of Civil Engineers.

Rogers, J. D. 2006b. Lessons learned from the St. Francis Dam failure. *GeoStrata* 6(2):14–17.

Siskind, D. E., Skagg, M. S., Kopp, J. W., and Dowding, C. H. 1980. *Structure Response and Damage Produced by Ground Vibration from Surface Mine Blasting* (Report of Investigations 8507). Bureau of Mines, US Department of the Interior, Office of Surface Mining Reclamation and Enforcement.

Terzaghi, K. 1946. Rock defects and loads on tunnel supports. In R. V. Proctor and T. White (eds.), *Rock Tunneling with Steel Supports* (pp. 15–99). Youngstown, OH: Commercial Shearing and Stamping.

Thoenen, S. R., and Windes, S. L. 1942. *Seismic Effects on Quarry Blasting* (Bulletin 442). Washington, DC: Bureau of Mines.

US Geological Survey. 1953. Interpreting Geologic Maps for Engineering Purposes: Hollidaysburg Quadrangle, Pennsylvania. Washington, DC: US Geological Survey.

Ward, S. N., and Day, S. 2011. The 1963 landslide and flood at Vaiont Reservoir, Italy: A tsunami ball simulation. *Italian Journal of Geoscience* 130(1):16–26.

Wayne, W. J. 1958. Glacial Geology of Indiana, Atlas of Mineral Resources in Indiana, Map 10. Bloomington: Indiana Geological Survey.

Appendix A

IDENTIFICATION OF MINERALS AND ROCKS

Appendix Outline

Minerals   Rocks

This appendix contains charts used for identification of minerals and rocks. The purpose of these charts is to supply the means whereby a beginning student in geology can identify common minerals and common rocks based on the physical characteristics of those materials.

Minerals

Table A.1 includes the most common rock forming minerals. The chart is so arranged that three groups of minerals are considered with regard to hardness: (1) those softer than steel, (2) those about the same hardness as steel, and (3) those harder than steel. Hardness is used as the primary subdivision because it is a very diagnostic property of minerals and can be readily discerned by a beginning student.

After hardness is determined the properties of cleavage, luster, color, and streak are used to discern the mineral name. Only 19 mineral species are included in Table A.1 because these are the most common rock forming minerals and they comprise the great volume of materials within the Earth's crust.

Table A.1   Common Mineral Identification Chart.

Hardness	Cleavage or Fracture	Luster	Color	Streak	Remarks	Name and Composition
Group 1: Minerals Softer Than Steel						
1–2.5	Not visible	Dull to earthy	White, light brown, other colors due to impurities	White	SpG of pieces about 2, soft compact earthy masses. Greasy feel that becomes plastic when moistened. Clay-like odor to mass.	Clay mineral
1–5	Uneven fracture	Earthy	Yellow	Yellow, brown to black	Commonly in earthy masses. SpG 3.5–4.	Limonite (Iron oxides)
1–5	Uneven fracture	Earthy	Deep red	Deep red to brownish red	May be granular or earthy. SpG 5.	Hematite (Iron oxides)
2	One plane	Vitreous silky	White to pastels	White	Fine grained, massive. SpG 2.3.	Gypsum (hydrous calcium sulfate)
2–3	Excellent one plane	Pearly to vitreous	Colorless	White	Transparent in thin sheets, yields plates flexible and elastic. SpG 2.8–3.1.	Muscovite (Mg silicate) (Mica)
2.5–3	Excellent one plane	Vitreous	Black to dark brown	Black	Yields plates flexible and elastic. SpG 2.7–3.2.	Biotite (K, Mg, Fe, Al, silicate) (Mica)
3	Three planes, rhomb shape	Vitreous	Colorless or yellow to white	White	SpG 2.7. Reacts with hydrochloric acid.	Calcite (calcium carbonate)
2.5–4	Rhomb shape	Vitreous to dull	Colorless, white to light brown, pink	White	SpG 2.85. Powder effervesces slowly in cold, dilute HCl, coarse crystals do not.	Dolomite (Ca, Mg carbonate)
2.5–4		Metallic	Brass		See pyrite.	Chalcopyrite (Cu, Fe sulfide)

Hardness	Cleavage or Fracture	Luster	Color	Streak	Remarks	Name and Composition
Group 2: Minerals about the Same Hardness as Steel						
5–6	Two planes at 90°	Vitreous to dull	Black to dark green	Gray greenish gray, brownish gray or blackish gray	SpG 3.1–3.5. Cleavage planes more prominent and more shiny in amphiboles than in pyroxenes (Ca, Mg silicate).	Pyroxene (augite) — *Pyroboles*
	Two planes at 56 and 124°				SpG 3–3.3.	Amphibole (hornblende) — *Pyroboles*
6	Uneven fracture	Metallic	Black	Black	Commonly in compact granular masses. Attracted by magnet. SpG 5.2.	Magnetite (iron oxide)
6.5–7	Conchoidal fracture	Vitreous	Light green	White to pale green	Granular, sugary masses common, transparent to translucent. SpG 3.2–3.6.	Olivine (Fe, Mg silicates)
5.6–7.5	Conchoidal or uneven fracture	Vitreous to resinous	Blood red	White	Transparent to opaque, well-formed crystals common. SpG 3.5–4.3.	Garnet group (Ca, Mg, Fe, Mn, Al silicates)
6–6.5	Uneven fracture	Metallic	Brassy yellow	Greenish black	Granular masses and well-formed cubic crystals common. Variety softer than steel nail chalcopyrite. SpG 4.1–5.2.	Pyrite (Fe sulfide)
Group 3: Minerals Harder Than Steel						
6–6.5	Two planes at about 90°	Pearly to vitreous	White, pink	White	Commonly pink. SpG 2.5–2.7	Feldspar group (Na, Ca, K, silicate) / Orthoclase (K feldspar)
			White, gray greenish gray, dark gray		May show striations on cleavage plane. SpG 2.5–2.7. Sodic plagioclase, white to light gray. Calcic plagioclase, dark gray to black.	Plagioclase (Na₂Ca feldspar)
7	Conchoidal fracture	Vitreous or greasy	White, colorless, pink gray, violet	White	Massive or well-formed crystals, transparent to translucent. SpG 2.65.	Quartz (silicon dioxide)
7	Conchoidal fracture	Dull	White to black	White to gray	Opaque, black variety is flint. SpG 2.65.	Chert (silicon dioxide)

Table A.2 is a tabulation sheet for mineral identification. It is so arranged from left to right to prompt the student through the important physical properties that are used to identify a mineral sample. This sheet is used in accordance with Table A.1 for mineral determination.

Table A.3 lists common minerals that have significant economic importance. They can also be considered as ore minerals, or those that when present in sufficient quantities can be mined at a profit. These minerals are arranged alphabetically from top to bottom. Key mineral properties from left to right are supplied. This table can be used for identification of minerals or as a source of information on common economic minerals.

Table A.2   Mineral Tabulation Sheet.

Specimen Number	Hardness	Color	Streak	Type of Cleavage or Fracture	General SpG	Miscellaneous	Mineral Name	General Composition

Table A.3   Common Minerals with Economic Importance.

Mineral	Chemical Composition	Specific Gravity	Hardness	Cleavage or Fracture	Luster	Color	Streak	Form	Other Aspects
Actinolite (an asbestos and an amphibole)	$Ca_2(MgFe)_5$ $Si_8O_{22}OH_2$	3.0–3.3	5–6	See amphibole, Table A.1	Vitreous	White to light	White	Slender crystals usually	Common metamorphic mineral, asbestos
Apatite	$Ca_5(F,Cl)(PO_4)_3$	3.15–3.2	5	Poor cleavage	Vitreous	Green, also brown or red	White	Massive, granular	Important source of fertilizer (P)
Azurite	$CuCO_3$	3.77	4	Fibrous	Vitreous to dull	Dark blue	Blue	Sometimes radial fibers	Minor ore of copper, effervesces in HCl, commonly found with malachite
Bauxite	Al hydroxides, not a mineral	2–3	1–3	Uneven fracture	Dull	Yellow, brown gray	Colorless	Rounded grains or earthy clay-like masses	Ore of aluminum, weathering product in tropical areas, clay-like odor when wet
Bornite	Cu_5FeS_4	2.5	3	Uneven	Metallic	Bronze on fresh surface, tarnishes to purple	Gray-black	Usually massives, rarely as rough cubes	Important copper ore, associated with chalcocite, chalcopyrite, malachite, and other Cu minerals
Carnotite	$K_2(UO_2)_2(VO_4)_2$	4	Very soft	Uneven fracture	Earthy	Bright yellow	Yellow	Earthy powder	An ore of uranium and vanadium, usually occurs as incrustation on sandstone, radioactive
Cassiterite	SnO_2	6.8–7.1	6–7	Conchoidal fracture	Submetallic to adamantine to dull	Brown or black	White to light brown	Massive and granular	Principal ore of tin
Chalcocite	CuS_2	5.5–5.8	2.5–3	Conchoidal fracture	Metallic	Lead-gray to black	Grayish black	Massive and aphanitic	Important copper ore, associated with bornite, chalcopyrite, and malachite

(continued)

Mineral	Chemical Composition	Specific Gravity	Hardness	Cleavage or Fracture	Luster	Color	Streak	Form	Other Aspects
Chalcopyrite	$CuFeS_2$	4.1–4.3	3.5–4	Uneven fracture	Metallic	Brass-yellow	Greenish-black	Usually massive	Softer than steel whereas pyrite is harder than steel, more brittle than gold
Chromite	$FeCr_2O_4$	4.6	5.5	Uneven fracture	Metallic to submetallic	Black to brown-black	Dark brown	Massive, granular to compact	Only ore of chromium
Copper (native)	Cu	8.9	2.5–3	Hacky fracture	Dull because of tarnish	Copper red	Copper red	Hacky	Minor ore of copper, highly malleable
Corundum	Al_2O_3	4.02	9	Basal parting	Adamantine to vitreous	Usually pink, brown, or blue	Colorless	Barrel-shaped crystals or granular	Used as an abrasive, gem forms are ruby (red), sapphire (blue)
Fluorite	CaF_2	3.18	4	Four planes of cleavage yielding octahedron	Vitreous	Bluish, purple, yellow, light green	White	Well-formed cubes or massive	Used as flux in making steel, may show fluorescence
Galena	PbS	7.4–7.6	2.5	Cubic cleavage	Metallic	Lead-gray	Lead-gray	Cube shaped, also massive	Principal ore of lead
Gold (native)	Au	15–19.3	2.5	Uneven	Metallic	Yellow	Yellow	Usually irregular plates or masses	Highly malleable and ductile, this and high specific gravity distinguish it from pyrite, chalcopyrite, and altered mica flakes
Graphite	C	2.3	1	Good in one direction	Metallic or earthy	Black to steel gray	Black to steel gray	Foliated, scaly, radiating or granular	Feels greasy, smudges hands
Gypsum	$CaSO_4 \cdot H_2O$	2.32	2	Good in one direction, flexible	Vitreous, pearly, silky	Colorless, white, gray	Colorless	Prismatic crystals, granular or fibrous	Fibrous variety is satin spar, selenite is transparent, alabaster is massive

Mineral	Chemical Composition	Specific Gravity	Hardness	Cleavage or Fracture	Luster	Color	Streak	Form	Other Aspects
Halite	$NaCl$	2.16	2.5	Perfect cubic cleavage	Glassy to dull	Colorless or white	Colorless	Cubic crystals or massive	Salty taste, table salt
Hematite	Fe_2O_3	5.26	5.5–6.5	Uneven fracture	Metallic to dull	Reddish brown to black	Deep red	Long prismatic crystals or coarse to fine masses	Most important ore of iron
Limonite	Hydrous iron oxide, not a mineral	3.6–4	5–5.5	None	Vitreous to dull	Yellow to dark brown	Yellow-brown	Amorphous, nodular or earthy	Used as a pigment, yellow ocher, found with iron minerals
Magnetite	Fe_3O_4	5.18	6	Some octahedral parting	Metallic	Iron-black	Black	Usually massive granular, aphanitic	Strongly magnetic, primary ore of iron
Malachite	$CuCO_2(OH)_2$	3.7–4.03	3.5–4	Poor	Silky to dull	Green	Green	Radiating fibers and kidney-shaped forms common	Effervesces in HCl, associated with other copper ores
Pyrrohtite	$Fe_{1-x}S$	4.58–4.65	4	Poor	Metallic	Bronze yellow	Black	Massive to granular	Magnetic
Serpentine	$Mg_3Si_2O_5(OH)_4$	2.2–2.65	2–5	Conchoidal fracture	Greasy, waxy, or silky	Variegated shades of green	Colorless	Platy or fibrous	Platy variety is antigorite, fibrous one is crysotile, which is an asbestos
Sphalerite	ZnS	3.9–4.1	3.5–4	Perfect cleavage in six directions	Resinous	Brown to yellow or black	White to yellow brown	Many-sided, distorted crystals common	Most important ore of zinc
Sulfur	S	2.05–2.09	1.5–2.5	Conchoidal to uneven fracture	Resinous	Yellow	Yellow	Irregular masses	Burns with ease, used in chemical industry
Talc	$Mg_3Si_4O_{10}(OH)_2$	2.7–2.8	1	Good cleavage in one direction	Pearly to greasy	Gray, white, apple green	White	Foliated, massive	Greasy feel, metamorphic mineral, used in talcum powder

Rocks

Igneous Rocks

Figure A.1 is a simplified identification chart for igneous rocks. It is based on two primary characteristics of igneous rocks: texture and mineral composition. Texture is designated on the vertical axis on the upper right side. A decrease in grain or crystal size occurs from top to bottom, that is from pegmatite to glassy. Pyroclastic is the igneous texture for rocks whose particles were ejected into the air during volcanic eruptions and form by settling and subsequent lithification.

Mineral composition ranges from high silica (acid) to low silica (basic), which is shown from left to right in the diagram. High-silica igneous rocks are also light in color (white, gray, pink), whereas low-silica igneous rocks are dark in color (dark gray,

black, green) with intermediate compositions generally having an intermediate color between these two.

Only four groups of minerals are included in the composition chart shown: orthoclase, quartz, plagioclase, and ferromagnesians. The latter group are silicates rich in iron and magnesium; they include pyroxenes, amphiboles, and olivine. In practice, the amount of dark mineral is discerned first (ferromagnesians) then the percentage of quartz and orthoclase are estimated. This approximately defines a vertical line somewhere along the lower diagram. Based on texture this is extended vertically to the upper diagram to obtain a rock name.

For example, assume a dark coarse-grained igneous rock had about two-thirds ferromagnesians and one-third dark feldspar or plagioclase. This vertical

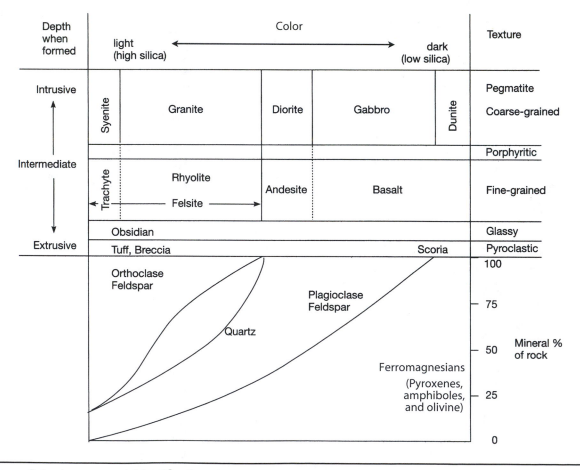

Figure A.1   Igneous rock identification chart.

line lies about one-quarter of the distance from the right side of the diagram. Extending this upward to coarse-grained texture yields the gabbro category. Therefore, the rock is designated as a gabbro.

Table A.4 consists of a tabulation sheet for igneous rock identification. The texture and composition (or color) are determined first by examination. Then, using Figure A.1 the rock name is obtained by relating the composition to the texture in that chart.

Information on origin is provided on the extreme left-hand side of the table. It is related to texture. Extrusive igneous rocks have either pyroclastic or a lava-type origin. Coarse-grained rocks are intrusive in origin, whereas rocks of hypabyssal origin are commonly porphyritic. These concepts are discussed in detail in Chapter 3.

Table A.4   Tabulation Sheet for Igneous Rocks.

Specimen Number	Texture	Composition	Origin	Rock Name

Table A.5   Classification of Sedimentary Rocks.

A. Texture Chiefly Clastic (Composed of Pieces)

Diameter of Pieces	Unconsolidated Sediment	Rock Name	Composition	Descriptive or Special Features
> 2 mm	Gravel	Conglomerate	Quartz or varied	Larger pieces rounded, cement commonly calcite
	Rubble	Breccia	Varied	Angular pieces, poorly sorted
1/16 to 2 mm		Quartz sandstone	Quartz grains, calcite or iron oxide cement	
	Sand	Arkose	Feldspar grains 25–50%	Light color (often pink or red) micaceous
		Graywacke	Dark rock fragments present	Dark color, cement clay or chlorite
		Clastic limestone	Mostly calcite	Fossil fragments, oolites
1/256 to 1/16 mm	Mostly silt, some clay	Siltstone	Mica, clay minerals, quartz	Laminated, flat-shaped rock
	Silt and Clay	Shale	Clay minerals	Laminated, flat-shaped rock

B. Texture: Coarsely Crystalline (Visible, Interlocking Grains)

Limestone	Mostly interlocking grains of calcite	Soluble in cold HCl
Dolomite	Mostly interlocking grains of dolomite, some calcite	May be porous, effervesces slowly in cold HCl
Evaporites		
Anhydrite rock	Mostly anhydrite	Commonly fine grained
Gypsum rock	Mostly gypsum	
Rock salt	Mostly halite	Salty taste
Bittern salts	Mostly Mg and K salts	Bitter taste

C. Texture: Aphanitic, Cryptocrystalline, Dense, or Amorphous

Dense limestone (lithographic)	Mostly calcite	Very fine grained, soluble in cold HCl
Dense dolomite	Mostly dolomite mineral	Very fine grained, effervesces slowly in cold HCl
Chert	Very fine quartz, opal, or chalcedony	White, black, green, or brown; conchoidal fracture; hardness of 7 if unweathered

D. Texture Composed of Whole Fossils or Their Alteration Products

Fossiliferous limestone, shell conglomerates	Mostly calcite	Large calcite fossils
Chalk	Mostly calcite	Microscopic calcite fossils (Foraminifera)
Peat	Matted plant remains	Brown; stringy or spongy
Lignite	Mostly hydrocarbons and carbohydrates	Dark brown to dull black
Bituminous coal	Hydrocarbons, carbon, and ash	Black; cubical fractures

Sedimentary Rocks

Table A.5 is a classification chart for sedimentary rocks. Four subdivisions based on texture are designated in this chart: the clastic, coarsely crystalline, aphanitic, and whole fossil groups. These are determined through careful examination of the rock's texture. Next based on grain size, composition, and other features one ascertains the name of the sedimentary rock.

Table A.6 is a tabulation sheet for sedimentary rock identification. By this chart the student is directed to determine the nature of the texture, specifics of the texture, and the minerals present. Based on this information and Table A.5, the proper rock name can be applied.

Table A.6   Tabulation Sheet for Sedimentary Rocks.

Specimen Number	Texture	Composition	Origin	Rock Name

Metamorphic Rocks

Table A.7 is a simplified classification chart for metamorphic rocks. Two main categories are recognized: rocks that are foliated and those that are not. Determination of the type of foliation and the minerals present leads directly to the name of the metamorphic rock.

Nonfoliated metamorphic rocks are identified according to their mineral composition and their textural detail. Only marble and quartzite are considered in this group of rocks.

Table A.8 is a tabulation sheet for metamorphic rock identification. If the rock is foliated, the type of foliation is discerned. This along with the mineral composition leads directly to the name of the metamorphic rock. If the metamorphic rock is not foliated, the nature of the texture and the mineral composition are used to determine the rock name.

General Considerations

In most courses on rock identification the three genetic groups, igneous, sedimentary, and metamorphic, are studied individually in turn, typically beginning with igneous and finishing with the metamorphics. The student therefore is not concerned with the problem initially of determining to which of these genetic groups the rock specimen belongs. This information is supplied for each group. On completion of the rock study, however, identification of all of these rocks becomes the assignment and a student is required to identify all rock specimens taken collectively, igneous, sedimentary or metamorphic, presented as a combined group.

A significant problem that must be addressed is how to determine the genetic group to ensure that the correct rock name is applied to a specific test specimen. Table 6.1 in the chapter on metamorphic rocks is an important aid in making this determination.

Table A.7   Metamorphic Rock Classification.

Type of Foliation	Descriptive or Special Features	Minerals Present	Name	Probable Original Rock
A. Foliated Rocks (Those with Directional Structure)				
Slaty rock cleavage	Fine-grained, cleavage more pronounced than bedding in shale, mica sheen not obvious	Mica and quartz, but too fine to see	Slate	Shale, tuff
Slaty rock cleavage, surface may have undulations	Cleavage as in slate, mica sheen obvious	Mica and quartz, mica in phyllite larger than in slate and produces sheen	Phyllite	Shale, tuff
Schistosity	Planar directional property due to visible platy or needle-shaped minerals	Muscovite in large, visible sheets or visible needles of hornblende	Schist	Fine-grained igneous and sedimentary rocks
Gneissic	Banding due to segregation of coarse-grained minerals	Quartz, feldspar, and hornblende; lesser amounts of mica than schist	Gneiss	Granite, diorite arkose, graywacke, quartz sandstone
B. Nonfoliated (Without Directional Structure)				
Texture				
Crystalline	Massive, granular; lack of sedimentary features	Calcite or dolomite	Marble	Limestone, dolomite
Crystalline	Massive, granular; grains cannot be removed as easily as with sandstone	Quartz	Quartzite	Quartz sandstone

Mineral composition, texture, structure, and other features are used to determine the genetic rock group. A careful application of the three classification charts is also used to ensure that the rock belongs in that specific category. Experience, of course, is a major factor because the points of confusion and the diagnostic features for identification become more apparent as more time is spent working with the rock specimens.

Table A.8   Tabulation Sheet for Metamorphic Rocks.

Specimen Number	Foliated or Nonfoliated	If Foliated, Type of Foliation; If Not, Describe Texture	Minerals Present	Rock Name

Appendix B

TOPOGRAPHIC MAPS

Appendix Outline

Elements of Topographic Maps

Geologic Maps of Horizontal Rocks

Maps represent a portion of the Earth's surface and provide one of the most useful means for engineers and geologists to record and subsequently present information. They are generally classified into two types, planimetric maps and topographic maps. Planimetric maps depict only the horizontal dimensions showing the location of surface features but not the vertical distance between them. Topographic maps typically show the same features as do planimetric maps, including horizontal measurements, but they also illustrate the third dimension, that is, the configuration of the Earth's surface.

Topography or *relief* can be indicated in three different ways on a flat map: shading, hatchure lines, or contour lines. Shading can provide only relative information indicating higher versus lower elevations but no accurate vertical distances can be obtained from such maps. Hatchures are lines drawn parallel to the maximum slope direction of the land surface but are inaccurate in depicting topography and again provide no information on actual vertical distances between points on a map.

The most common method used on maps in the United States to depict topographic information is the contour line. These consist of numerous, fine, brown lines, which provide extensive topographic information to the trained observer, including stream location and direction of flow, shapes of landforms, and specific information on elevation differences.

The United States Geological Survey (USGS), a division of the Department of the Interior, has been providing standard topographic maps of the United States for over a century. Three common series of topographic quadrangle maps have been published by the USGS. Early maps included 30 min of latitude and 30 min of longitude, which represents a large area, about 2331 km^2 (900 mi$^{2)}$) when located in the middle latitudes (continental United States). Presented on a map approximately 76 cm × 76 cm (30 in × 30 in), this yields a small-scale map that cannot provide much detail. Latitude lines (meridians) form the north and south boundaries and longitude lines (parallels) form the east and west boundaries of the map.

A larger scale map covering 15 min of latitude and longitude was initiated later, representing only one-quarter the area and allowing more detail. At the present time, 7.5-min series maps are produced, which, of course, depict 7.5-min of latitude and longitude. This large-scale map contains one-quarter the area of the 15-min quadrangle and shows considerably more detail. The publication of 7.5-min quadrangles covering previously unmapped areas provides an active program for the USGS. Their goal is to map the entire nation at that scale.

Quadrangle maps carry as their title the name of the largest city or most important feature that lies within the map's boundaries. Reference maps for each state are available that provide the names and locations of individual quadrangles. Maps are typically filed in map libraries according to map series and then by the quadrangle name. State geological surveys commonly offer for sale the quadrangle maps available within their state. The USGS provides complete coverage of available maps for the United States and its possessions.

Three major features are shown on topographic maps: (1) relief (topography), which has been discussed in some detail, (2) water features—streams, lakes, ponds, swamps, and canals, and (3) cultural features showing human activity including roads, railroads, bridges, tunnels, land boundaries, various buildings, and so on. The print color is indicative of this threefold division. Topography is shown in brown, water in blue, and cultural features in black. In addition, interstate highways may be printed in red, vegetation in green, and updated information added since a previous map printing is shown in purple. Figure B.1 illustrates some of the standard topographic map symbols.

Elements of Topographic Maps

Latitude and Longitude

To locate points on the Earth's surface, the system depicting latitude and longitude has been developed. The Earth's surface is divided into units of area by two sets of intersecting lines that form a grid. One set of lines extends in the north–south direction from the true North Pole to the true South Pole. These are longitude lines or meridians of longitude and they are measured up to 180° east and 180° west from 0° or Prime Meridian, which passes through Greenwich, England, southeast of London. The meridian or longitude value is determined by the angle between the principal meridian and the meridian in question, measured at the center of the Earth. In the United States all values are west longitude.

The other set of intersecting lines consists of the parallels or the lines of latitude. These circle the Earth in an east–west direction with zero latitude being the equator. Latitude is numbered north and south from the equator to 90° at the poles and is measured by the central angle between that parallel and the equatorial plane.

An important distinction between the two sets of lines is that longitude lines converge at the poles, whereas latitude lines are always parallel. This yields a rectangular shape for quadrangle maps in the middle latitudes (such as the continental United States) and trapezoidal-shaped maps near the poles where rapid convergence occurs.

The unit of measure for the grid system is the degree, which can be subdivided into minutes and seconds. At the equator, one minute of arc is equal to a nautical mile or about 1853 m (6080 ft) in length. A spot on the Earth is located by giving the latitude first (either north or south) and the longitude (either east or west) both to the degree, minute, and second. One second at the equator is equal to about 31 m (101 ft) of distance.

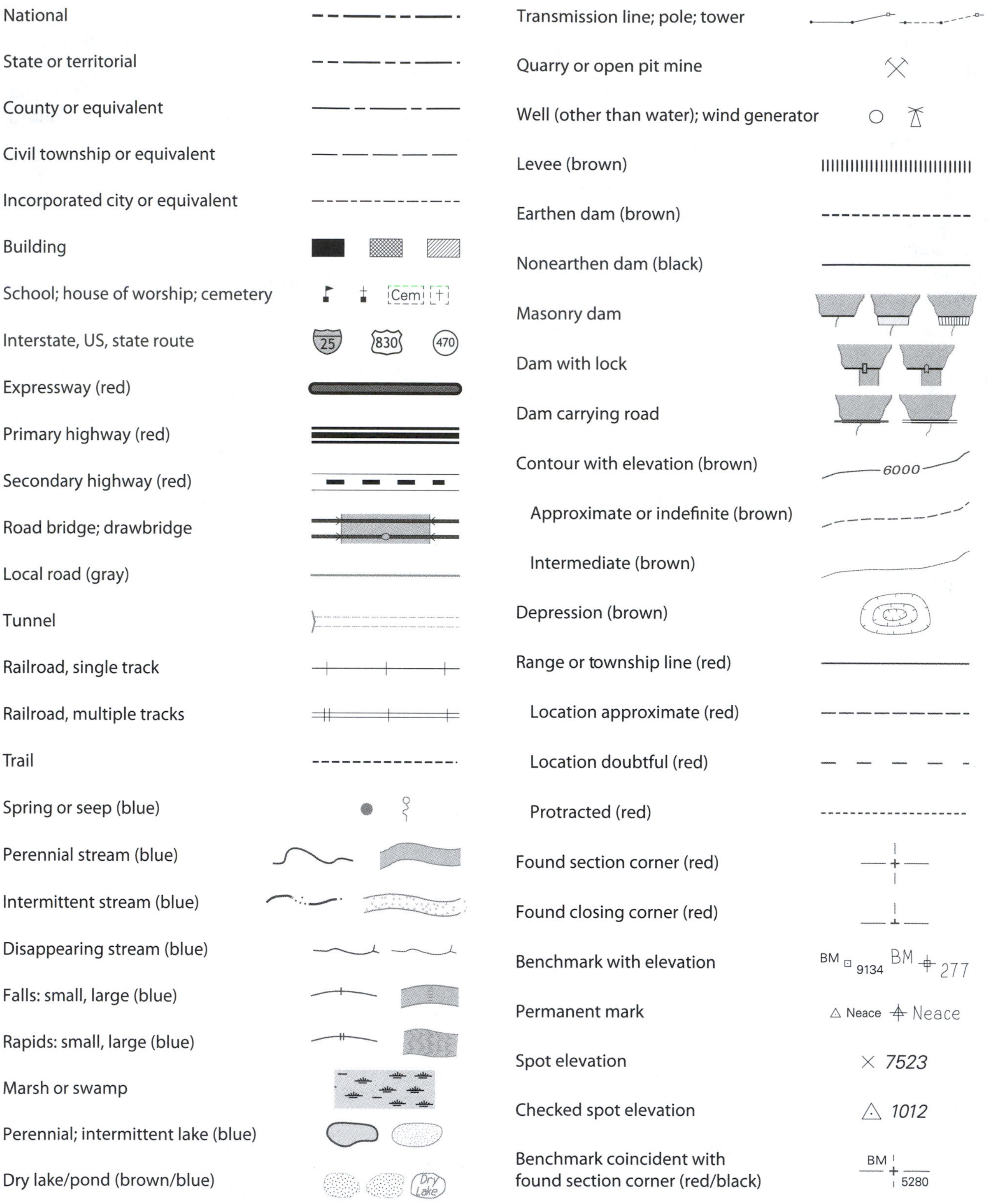

National	— — — — —	Transmission line; pole; tower	
State or territorial	— — — — —	Quarry or open pit mine	
County or equivalent	— — — — —	Well (other than water); wind generator	
Civil township or equivalent	— — — —	Levee (brown)	‖‖‖‖‖‖‖‖‖
Incorporated city or equivalent	— — — — — —	Earthen dam (brown)	— — — — —
Building		Nonearthen dam (black)	———————
School; house of worship; cemetery	Cem ☐	Masonry dam	
Interstate, US, state route	25 830 470	Dam with lock	
Expressway (red)		Dam carrying road	
Primary highway (red)		Contour with elevation (brown)	—6000—
Secondary highway (red)		Approximate or indefinite (brown)	
Road bridge; drawbridge		Intermediate (brown)	
Local road (gray)		Depression (brown)	
Tunnel		Range or township line (red)	———————
Railroad, single track		Location approximate (red)	— — — — —
Railroad, multiple tracks		Location doubtful (red)	— — — —
Trail	— — — — — —	Protracted (red)	- - - - - - - -
Spring or seep (blue)	●	Found section corner (red)	
Perennial stream (blue)		Found closing corner (red)	
Intermittent stream (blue)		Benchmark with elevation	BM ☐ 9134 BM ✛ 277
Disappearing stream (blue)		Permanent mark	△ Neace ✛ Neace
Falls: small, large (blue)		Spot elevation	✕ 7523
Rapids: small, large (blue)		Checked spot elevation	△ 1012
Marsh or swamp		Benchmark coincident with found section corner (red/black)	BM ✛ 5280
Perennial; intermittent lake (blue)			
Dry lake/pond (brown/blue)	Dry Lake		

Figure B.1 Standard symbols for topographic maps published by the USGS.

Map Scale

The scale of a map indicates the relationship between the distance separating two points on the map and the actual distances between those two points on the Earth's surface. In keeping with the fundamental reason for making maps, map dimensions always represent a greater distance on the Earth's surface than their own actual length.

Three scales are commonly used on topographic maps: (1) fractional, (2) graphic, and (3) verbal. A proper map should have its scale represented in one of these ways and many maps indicate the scale in at least two of them.

The fractional scale or ratio scale is given in the form of a fraction, typically with the numerator equal to 1. The fraction is dimensionless so that one unit on the map is equal to the number of those units on the ground as indicated in the denominator. For example, 1:62,500 or 1/62,500 means 1 in on the map is equivalent to 62,500 in on the ground. This also holds for any other unit of measure: 1 cm on the map is equivalent to 62,500 cm on the ground, and so on.

The graphic scale or bar scale is a line or bar drawn on the map and is divided into convenient units showing distances on the ground in meters (feet), kilometers (miles), etc. To determine the distance between two points on the map, the map distance is laid off along the bar scale and the surface distance read directly from it.

The verbal scale is a convenient means for stating the relationship between map distance and ground distance and is commonly used on blueprints and on architectural and engineering drawings. Examples are ⅛″ = 1′, 1″ = 1 mi, or 1″ = 100′ (0.32 cm = 2.5 cm, 2.5 cm = 1.6 km, 2.5 cm = 30 m).

There are advantages to each of the three scales. The fractional scale can be used for any unit and is particularly useful for conversion of SI to British engineering units of length and vice versa. The bar scale lends itself to quick determinations of distance and it has a second advantage. The drawing can be enlarged or reduced and the bar scale information still holds true. This is not the case for fractional scales or verbal scales. In the case of the verbal scale, it is easy to comprehend the relationship between map and ground distances. It is also useful for scaling distances using an architect's scale or mechanical engineer's scale ("rulers" divided according to different scale relationships).

One concept of importance that commonly leads to confusion is the aspect of large-scale maps versus small-scale maps. A large-scale map shows more detail than a small-scale map and to accomplish this it represents a smaller area on the ground. Using the fractional scale a large-scale map will yield a larger number than will a small-scale map. For example, by comparing the 1/24,000 scale to the 1/62,500, the fraction 1/24,000 is larger than 1/62,500, and hence that first map has a larger scale. Since 1/24,000 yields a verbal scale of 1″ = 2000′ (2.5 cm = 610 m) and 1/62,500 yields 1″ = 5208.3′ (2.5 cm = 1588 m), the first map would represent a smaller area in more detail and again be the larger scale map. Refer to Figure B.2 for details.

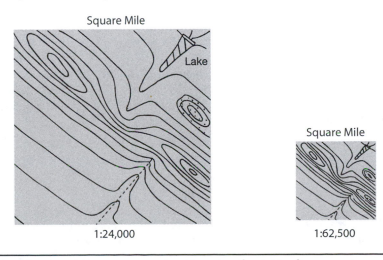

Figure B.2   Topographic detail at different map scales. Note: 1 mi² = 2.6 km².

Converting Scales

We sometimes need to convert from one map scale to another. Conversion from a fractional scale to a verbal scale may be done to provide easier understanding or to a graphic scale for easy distance determination. Only simple arithmetic is involved in any of the scale conversions. The following are examples of these calculations.

Example Problem B.1

Convert the fractional scale 1/250,000 to a verbal scale in terms of in/mi.

Answers

To solve the problem put the fraction in terms of in/in and convert to in/mi.

$$1{:}250{,}000 = 1 \text{ in}{:}250{,}000 \text{ in}$$
$$1 \text{ in} = 250{,}000 \times (1/12 \times 1/5280) \text{ mi}$$
$$1 \text{ in} = 250{,}000/(12 \times 5280) \text{ mi} = 3.95 \text{ mi}$$
$$1 \text{ in on the map} = 3.95 \text{ mi on the ground}$$

Example Problem B.2

Convert the verbal scale 1 in = 150 ft to a fractional scale.

Answers

To solve the problem convert the ground measurement to in and cancel the units to obtain a fractional scale.

$$1 \text{ in} = 150 \text{ ft}$$
$$1 \text{ in} = 150 \times 12 \text{ in}$$
$$1 \text{ in} = 1800 \text{ in}$$
$$1{:}1800 \text{ or } 1/1800$$

Example Problem B.3

Convert the verbal scale 1 in = 6000 ft to a graphic scale.

Answers

To accomplish this you must first choose a convenient, even number to present on the graphic scale or bar graph that would be most useful to the map reader (e.g., 1000 ft, 100 ft, 1 mi, etc.), depending on the actual case. In this example, 1000 ft should be selected, with the following subdivisions determined accordingly: 1 in = 6000 ft or 1/6 in = 1000 ft. Mark off distances 1/6 in apart and label these 1000, 2000, 3000, . . . , 6000. Subdivide in half to obtain 500-ft increments.

It is unacceptable to use ¼ in or some other increment just because it is easy to make it into a bar graph. A ¼-in scale would yield divisions of 1500 ft each, which is not that convenient a distance for the map reader to use.

Example Problem B.4

Convert the fractional scale 1/62,500 to a graphic scale.

Answers

The same procedure is followed as used in Example Problem B.3, but the problem is more challenging:

1:62,500

(1 in:62,500 in)

(1 in:5208.33 ft)

1 in:0.9864 mi

1.014 in:1 mi

Lay off distances of 1.014 in and label them 1 mi; subdivide this distance to obtain ½- and ¼-mi markers.

Topography

Topography refers to the shape or configuration of the Earth's surface. As previously stated topography is most conveniently displayed on a map by means of contour lines, which are imaginary lines of equal elevation. Their elevations are designated with respect to a datum plane, usually mean sea level.

The contour interval, which is indicated on the legend of topographic maps, is the difference in elevation between any two successive contour lines. On standard USGS maps, contour lines are brown and usually every fifth contour line (sometimes every fourth depending on the contour interval) is darker in color and it is labeled with the elevation. Typically only one contour interval is used for any map but when exceptions are made, the map legend will so indicate. Contour intervals may be 1.8 or 3 m (5 or 10 ft) for maps in very flat terrain, 6 to 12 m (20 to 40 ft) in moderately hilly areas, and 15, 24, or 30 m (50, 80, or 100 ft) in mountainous regions. The proper contour interval for a map is one that supplies the maximum detail concerning topography without being so cluttered with closely spaced lines that it becomes unreadable.

Elevations, in addition to those indicated by contour lines, are commonly supplied on topographic maps. Heights of hill summits, road intersections, and lake surfaces may be indicated for selected locations. These are known as *spot elevations* and their accuracy is given to the nearest meter (foot).

Benchmarks, designated BM on the map, are accurate indications of elevations representing permanently fixed points on the ground marked by a brass plate set in concrete. All specific elevations are printed in black on topographic maps.

Relief refers to the difference between the highest and lowest elevations of a feature under consideration. Local relief means the difference between the highest and lowest point in a local area, perhaps within a km^2 (mi^2) or so. Maximum relief for a quadrangle can be determined by subtracting the lowest elevation (e.g., lowest point in the major stream valley) from the highest point of the quadrangle (e.g., a hill, or a mountain) located by careful examination of the elevation.

Contour lines mark the intersection of horizontal, parallel planes passed through the Earth's surface at specific vertical intervals (the contour interval). Because of this manner of construction, they carry certain requirements or restrictions; they also indicate special information because of their behavior in view of these restrictions. A list is provided below.

1. All portions of a contour line have the same elevation.

2. Contour lines never cross or intersect each other. They can merge, as in the case of a vertical or overhanging cliff.

3. Every contour line closes on itself eventually, although this may not be shown on the map portion under study.

4. Contour lines never split or divide.

5. When contour lines cross stream valleys they bend upstream to form a V; that is, the Vs point upstream. They are perpendicular, however, to the water course itself.

6. Uniformly spaced contour lines indicate a uniform slope.

7. Closely spaced contour lines indicate a steep slope.

8. Widely spaced contour lines indicate a gentle slope.

9. Closed contour lines appearing on maps as ellipses represent hills.

10. Closed contour lines with hatchures represent a depression. The hatchures, which are perpendicular to the line, point downslope into the depression.

11. For land surfaces that extend uphill and then into a depression, the first hatchured contour is the same value as the last normal contour.

12. For land surfaces that extend downhill and then into a depression, the first hatchured contour is one interval less than the last normal contour.

13. Maximum contours (ridges) and minimum contours (valleys) always appear in pairs. Therefore, no single lower contour line can lie between two higher ones and no single higher contour line can lie between two lower ones.

Topographic Profiles

A topographic profile is a cross section of the Earth's surface taken along a specific direction. It is the view one would observe of the upper surface of a trench as seen in the side view. This is known as a side, profile, or elevation view as contrasted to the map or plan view, which represents the Earth's surface as seen from vertically above. Topographic profiles can be constructed along any given line on a topographic map. See Figure B.3 as an illustration.

Profiles are easily constructed using common graph paper. For the horizontal scale of the profile, the map scale is selected for convenience. A vertical scale is then arbitrarily chosen that will accentuate the surface features as needed. Because this exaggerates the vertical dimension it is known as vertical exaggeration. Only when the vertical and horizontal scales are the same is the profile a true representation of the cross section. Slope angles of the land surface as depicted on the profile will also be increased if the vertical scale is exaggerated.

Instructions for preparing a topographic profile are as follows:

1. Select the line on the map along which the profile will be made. The student should appreciate that this line can be viewed from either of two sides. Two different profiles, one the mirror image of the other, will be obtained depending on the direction in which the cross section is viewed. Arrows can be placed on the map view to show the direction of sight.

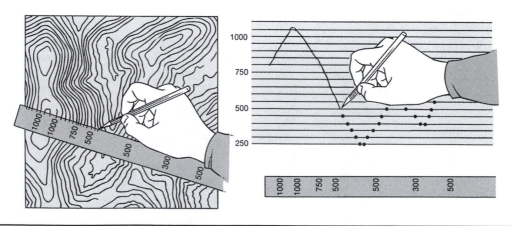

Figure B.3   Construction of a topographic profile.

2. Set up a vertical scale on the graph paper beginning with a minimum elevation that is less than the lowest elevation along the profile. A scale of 1/10 or 1/8 in equal to a contour interval may be convenient depending on the divisions of the graph paper itself. Label the horizontal lines appropriately.

3. Place the graph paper along the line of profile. It may be necessary to fold under the margin of the graph paper so the graph subdivisions extend directly to the folded edge. Opposite each contour line that crosses the line of the profile place a small dash or tick mark and label the elevation. Also mark stream locations, depressions, hilltops, and so on, which may be useful in smoothing the final profile. If the contour lines are too closely spaced, only the dark lines should be considered.

4. Extend the points downward on the graph paper until they cross the appropriate lines marking those elevations.

5. Connect the points by a smooth line taking into account the information about stream location and so on noted earlier. Title the profile and indicate the vertical exaggeration (Figure B.4). The following explanation provides more detail on how it is determined.

Vertical Exaggeration

Vertical exaggeration (VE) is determined by comparing the horizontal scale to the vertical scale. The vertical scale is used as the numerator and the horizontal scale as the denominator to obtain the vertical exaggeration in the following way: Horizontal scale 1″ = 1000′; vertical scale 1″ = 300′. Note: 1 in = 2.54 cm; 1 ft = 0.3048 m.

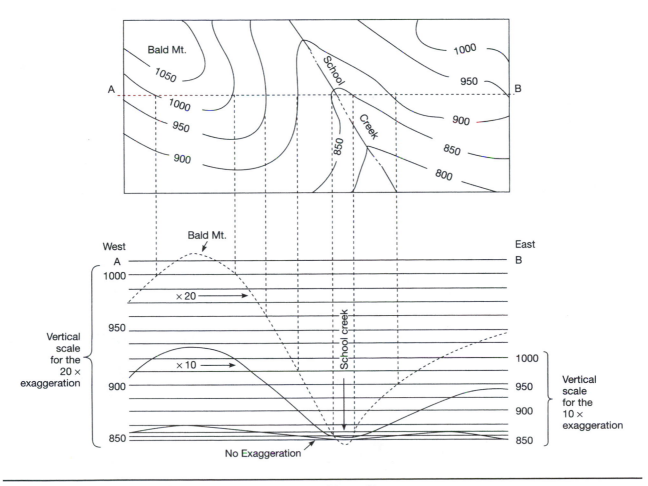

Figure B.4 Effects of vertical exaggeration on a topographic profile.

$$VE = \frac{\text{vertical scale } 1''/300'}{\text{horizontal scale } 1''/1000'}$$

The units cancel and vertical exaggeration = 3.33. As another example, consider a horizontal scale 1/24,000, vertical 1″ = 50′. First put both scales into dimensionless form: vertical 1″ = 50′ or 1″ = (12 × 50″) or 1″ = 6000″ or 1/6000:

$$VE = \frac{1/6000}{1/24,000} = 4$$

Figure B.4 on page 551 illustrates the effect of vertical exaggeration.

Location by the Township and Range System

Under the Public Land Survey System (PLSS), much of the United States has been subdivided by a grid method known as the township and range system or the Congressional system. It was established by Congress in 1785 to provide a system of land survey that could be applied throughout the developing western territories. The United States has been divided to provide a basic unit, which is a square measuring 9.6 km (6 mi) on a side. Not included in the system are the original 13 colonies plus Maine, West Virginia, Kentucky, Tennessee, and Texas (Figure B.5).

A principal meridian and baseline are established as the coordinate axes for the grid. Typically,

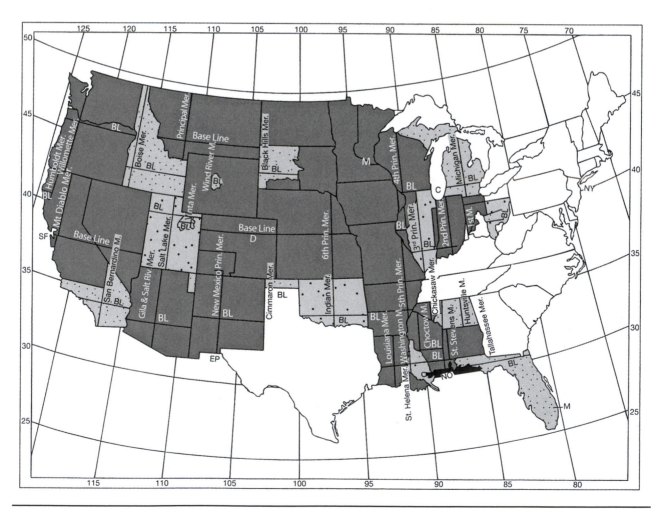

Figure B.5   Principal meridians and baselines for the United States.

each state of the United States has a designated principal meridian and baseline for surveying reference. Strips or tiers 9.6 km (6 mi) wide, termed *townships,* are laid off north and south from the baseline and labeled accordingly: T1N, T2N, . . . , T1S, T2S, . . . , whereas similar strips laid off east and west from the principal meridian, known as *ranges,* are also labeled accordingly: R1W, R2W, . . . , R1E, R2E, ... A designated square 6 mi (9.6 km) on a side, for example, T10N, R5W, is known as a township and consists of 36 mi^2 (92 km^2). The townships are subdivided into 36 pieces, each 1 mi^2 (2.6 km^2) in size, known as *sections,* which are numbered consecutively from 1 to 36 beginning in the northeast corner.

The sections in turn are divided into four quarter sections, these quarter sections into sixteenth sections, and they in like manner into sixty-fourth sections. A possible designation at this level would be NE¼ of NW¼ of SE¼, Section 10, T10N, R5W. Refer to Figure B.6 for further details.

The area contained in a section is 2.6 km^2 (1 mi^2) or 640 acres. Therefore, 1/64 of a section (as indicated in the example above) is 10 acres in size, a relatively small portion of land. An acre is 4041 m^2 (43,500 ft^2) or a square 63.6 m (208.7 ft) on a side. The ability to locate a piece of land contained in a state of the United States to within 10 acres by this simple procedure is a significant accomplishment. Figure B.7 on page 555 shows the township and range system in Indiana.

Gradients or Slopes

The slope of the land surface or of a stream bed is known as the *gradient.* This is equal to the vertical distance divided by the horizontal (or map) distance. For stream gradients, the units of m/km (ft/mi) are used. For most streams this yields convenient numbers between a tenth to several tens of m/km (ft/mi). Mountainous regions may have steep gradients in excess of 19 m/km (100 ft/mi), whereas the largest rivers may yield 1 m (several ft) or fractions of m/km (ft/mi). The profile of a stream along its stream bed is called its *long profile.*

Human-made features extending over long distances typically have their gradients supplied in percent, which equals (vertical distance/horizontal distance) × 100. Some of these features may have a slope of several percent, which would be quite steep compared to stream gradients.

Dips of rock strata are also represented as slopes or gradients. Dipping rock units will not be considered in this section, but for purposes of comparison, the dip of such units is commonly expressed in degrees of the angle measured from the horizontal. One degree of dip is equal to about a 2% slope.

Compass Directions

Meridian lines bound a topographic map on its left and right margins. These extend in the true north–south direction so that true north is toward the top of the map. Other maps used as engineering drawings should, along with an indication of scale, contain a north arrow indicating the true north direction.

For most topographic maps, true north will not coincide with the direction determined by a magnetic compass, that is, with magnetic north. The angle between true north and magnetic north measured in the plane of the map is known as magnetic declination.

On standard topographic maps, the magnetic declination is shown along the bottom margin by means of an arrow with an angle measured from true north. The angle is labeled and the direction can be either east or west of true north in various parts of the United States. Magnetic declination listed on such maps depicts the known value at the time of map preparation. Owing to an annual drift in the location of magnetic north, the latest declination must be applied when using compass measurements in a field study. The general details of the magnetic declination in the contiguous states of the United States are shown in Figure B.8 on page 556.

Geologic Maps of Horizontal Rocks

The extension from topographic maps to geologic maps that depict horizontal rocks is easily accomplished. Geologic maps show the surface distribution of various types of rock. They also indicate the relative age of the rocks and suggest their manner of association in the subsurface. The plan view of contacts between units are shown on these maps as well. Topographic contour lines may or may not be included on a geologic map.

A geologic contact is the surface of intersection between two different rock units or formations.

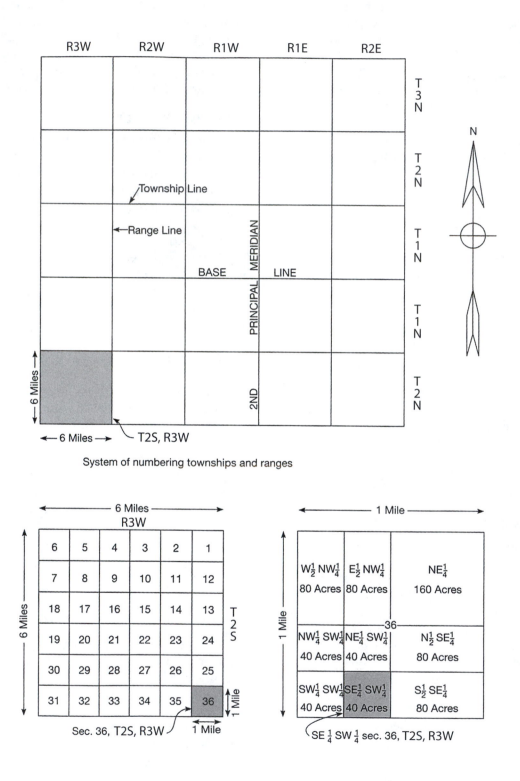

Figure B.6   Land division according to the township and range system.

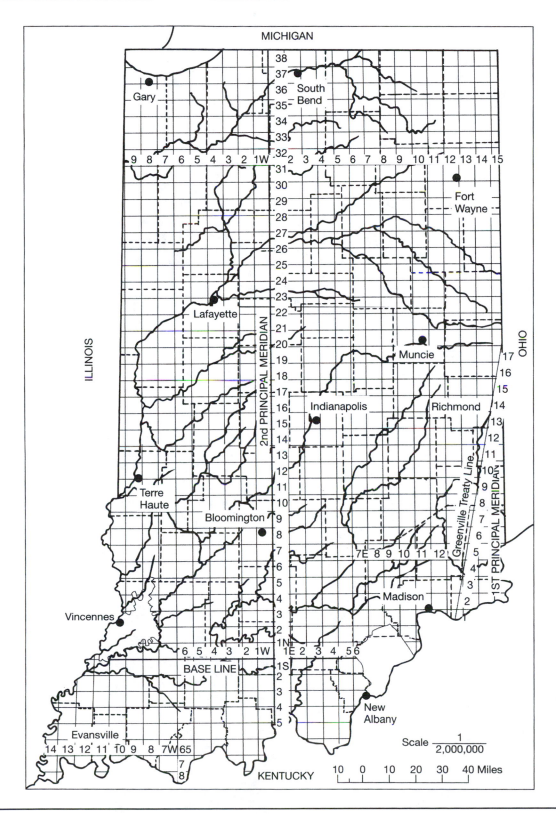

Figure B.7   Township and range system for Indiana.

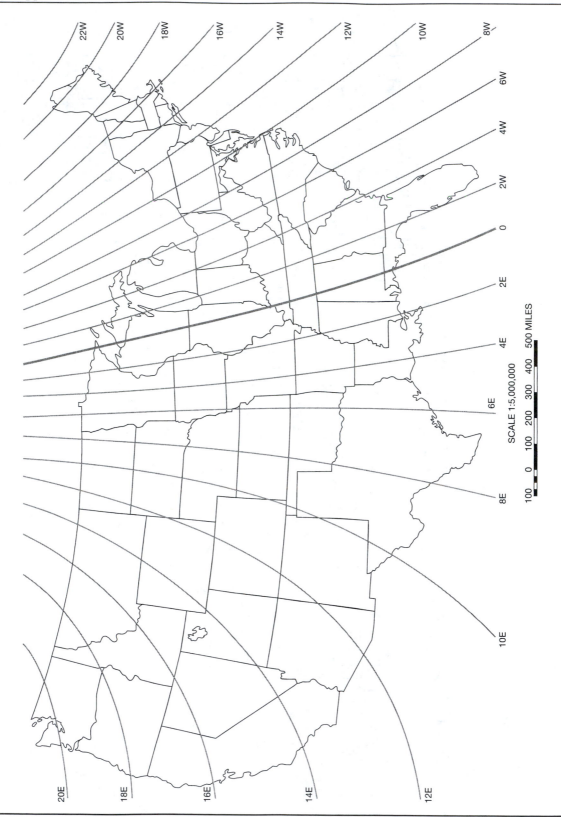

Figure B.8   Magnetic declination in the United States.

The term *formation* pertains to a rock unit that is thick enough and sufficiently widespread to be represented on a geologic map. It typically contains a distinctive lithology or rock type that can be traced in the field. On a geologic map, the contact between two adjacent formations will appear as a line.

When the formations are horizontal, their contacts are horizontal as well. Such contacts will occur parallel to the contour lines on a topographic map because contour lines, of course, depict the intersection of the Earth's surface with horizontal planes passing through the Earth. It is an easy matter to trace these horizontal contacts on a topographic map if the contact elevations are known.

Geologic cross sections of horizontal rocks are also an easy extension from topographic cross sections. A geologic cross section shows the distribution of formations in the vertical section. It is simply a topographic profile with the geology of the section added. For horizontal rocks such addition is an easy matter. The contact elevations are added to the cross section and the formations are illustrated by different colors.

EXERCISES ON TOPOGRAPHIC MAPS

1. Given the graphic scale on a map of 1¼ in represents 1 mi on the ground, compute the fractional scale.

2. Given the fractional scale of 1:84,000, convert this to a convenient verbal scale. Convert it also into two graphic scales, one reading in thousands of ft and the second in mi, ½ mi, and ¼ mi.

MAP READING

For these exercises, the US Geological Survey offers TopoView (https://ngmdb.usgs.gov/topoview/). This software will allow you to view and download topographical maps produced from 1880–2010 for locations throughout the United States. Use a map that has a scale of 1:24,000.

BELLEFONTE, PENNSYLVANIA

3. Why was the township and range system not used for this map? What is the approximate area of the map? Give the scale, contour interval, and magnetic declination including direction. What is the area of each rectangle on this map?

4. Using latitude and longitude, determine as accurately as possible the center of the town of Julian. What is the elevation at this point?

5. Determine the straight line distance between the center of State College to the center of Julian. Give the distance to the nearest tenth of a km (mi).

6. What is the general direction of flow of Bald Eagle Creek? What is the maximum relief in this area? Give the location of the maximum and minimum elevations to the nearest 1/16 of a rectangle.

7. The pass through Eagle Mountain just north of Bellefonte is called a *water gap*. What would be the effects of flooding on this narrow portion of Logan Branch Creek?

8. What is the overall gradient of Brower Hollow from its beginning to the confluence with Bald Eagle Creek? Concerning distance, assume it consists of two straight-line segments, one from the stream's beginning to the bend near the 985 elevation and the second from that point to Bald Eagle Creek. Give the gradient in m/km (ft/mi).

9. Using graph paper, plot the long profile of Bush Hollow from Grindstone Gap to Bald Eagle Creek. Plot accurately the 30-m (100-ft) contours only, approximating 15-m (50-ft) increments by interpolation. For the vertical scale, use 1 cm = 48 m (1′ = 400″); for the horizontal distance use the map distance. (Locate the contour lines of interest on a folded edge of a piece of scrap paper and transfer these points to the cross section.)

Describe the shape of the long profile. Calculate the vertical exaggeration of your drawing.

10. A railroad tunnel is planned that will connect the railroad at Unionville through the mountain to the Bellefonte Central Railroad on the opposite side. The tunnel would run perpendicular to the trend of the mountain. Determine the following information:

a. length of the tunnel in m (ft),

b. the gradient of the tunnel in percent, and

c. the maximum height of rock cover over the tunnel.

To accomplish these three tasks, sketch the topographic cross section showing the two portals (entrances to the tunnel), the point of maximum elevation of the mountain, and the tunnel gradient.

d. What would be the total overburden stress at the maximum depth of rock cover if the unit weight of rock = 2643 kg/m^3 (165 lb/ft^3)?

e. Why would the rock load on the tunnel roof not likely be this great? Explain.

f. If 1.5 m (5 ft) of rock load must be carried by the tunnel supports, what stress in kg/m^2 (lb/ft^2) would this exert on the tunnel roof?

g. If limestone, sandstone, and shale were present at various locations along the tunnel, what kinds of problems might develop during tunnel excavation?

West Lafayette, Indiana

11. What is the contour interval of this map? Indicate in latitude and longitude the area covered by the map. What is the map scale and the magnetic declination?

12. What is the difference in elevation between the top of Happy Hollow Creek at Grant and Salisbury Streets and its confluence with the Wabash River? Using the straight-line distance between these two points, determine the average gradient.

13. Locate the pond in section 13 near the center of the map using the township and range system. What is the approximate area of the lake as shown?

14. Note the change in relief about 1.6 km (1 mi) north of the Wabash River near the southern boundary of the map. The bluff marks the intersection of glacial outwash on the upper terrace of the Wabash River system with the upland surface of the glacial till plain.

a. Trace (mentally, no pencils or pens please) this intersection across the Purdue University campus. What is the general direction of these features?

b. The terrace consists of sand and gravel deposits, whereas the till plain is mostly silt and clay. What effect would this have for buildings (residence halls) that straddle these two landforms?

c. Make a cross-sectional drawing in the area of the football stadium. Show the geologic materials in the subsurface.

d. On which landform is the university water supply wells located? Why? On which landform is the water tower located? Why?

15. Obtain a copy of the Lafayette East Quadrangle as well as the Lafayette West Quadrangle. Based on topographic features and other geographical constraints explain why West Lafayette is growing northward whereas Lafayette is growing southward.

Topographic Map and Horizontal Sedimentary Rocks

16. Refer to Exercise Figure B.1 on the following page. The sedimentary rocks in this area are horizontal and have negligible soil cover. Streams are shown as dotted lines.

a. Given the following information, draw in the contacts between formations and color the formations on the map and on the key to the right of the map. The youngest beds are at the top of the key.

(i) The contact between the Savage Sandstone and the Lydon Conglomerate occurs at elevation 366 m (1200 ft).

(ii) A well is located at point 1. Savage Sandstone crops out at the surface. At a depth of 15 m (50 ft) the contact between the Savage Sandstone and the Morgan Dolomite is reached and at 46 m (150 ft) the contact between the Morgan Dolomite and Churp Limestone is reached.

(iii) Another well is located at point 2. Churp Limestone crops out at the surface. Fairfield Shale is found at a depth of 30 m (100 ft).

b. Draw a geologic cross section (topographic cross section showing geologic units) along line A–A′ looking in the direction indicated by the arrows. Use a vertical scale of 1″ = 400′ (1 cm = 48 m) and a horizontal scale equal to the map scale.

c. A dam is being considered for location at point 3. The local consulting geologist suggested that the reservoir level be maintained below the 183-m (600-ft) contour level. On what basis do you think he or she made this recommendation?

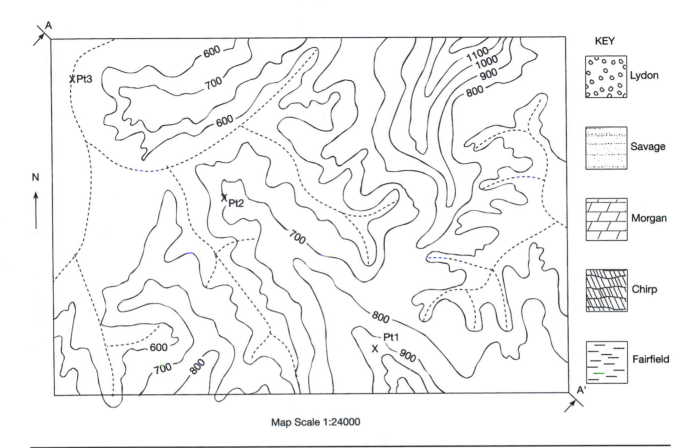

Map Scale 1:24000

Exercise Figure B.1